Introduction to Ordinary Differential Equations with *Mathematica*®

ALFRED GRAY   MICHAEL MEZZINO   MARK A. PINSKY

Introduction to Ordinary Differential Equations with *Mathematica*®

An Integrated Multimedia Approach

 **Web-Enhanced**   **Includes CD-ROM**

Springer ⑤ **TELOS**® THE ELECTRONIC LIBRARY OF SCIENCE

Alfred Gray
Department of
    Mathematics
University of Maryland
College Park, MD 20742
USA

Michael Mezzino
Department of
    Mathematics
University of Houston–
    Clear Lake
Houston, TX 77058
USA

Mark A. Pinsky
Department of
    Mathematics
Northwestern University
Evanston, IL 60208
USA

Publisher: Allan M. Wylde
Publishing Associate: Keisha Sherbecoe
Production Editor: Victoria Evarretta
Manufacturing Supervisor: Joe Quatela
Cover Designer: Nikita Pristouris
Editorial Consultant: Paul Wellin

Printed on acid-free paper.

Library of Congress Cataloging-in-Publication Data
Gray, Alfred, 1939–
    Introduction to ordinary differential equations with *Mathematica* /
    Alfred Gray, Michael Mezzino, Mark A. Pinsky.
        p. cm.
    Includes bibliographical references (p. –) and indexes.
    ISBN 0-387-94481-8 (hc : alk. paper)
    1. Differential equations—Data processing.   2. Mathematica
    (Computer file)   I. Mezzino, Michael.   II. Pinsky, Mark A., 1940–
        III. Title.
    QA371.5.D37G74   1997
    515'.352'078553042—dc21   96-54732

Photocomposed pages prepared from the authors' TeX files.
Printed and bound by Hamilton Printing Co., Rensselaer, NY.
Printed in the United States of America.

9 8 7 6 5 4 3 2 1

ISBN 0-387-94481-8 Springer-Verlag New York Berlin Heidelberg   SPIN 10496122

THE
ELECTRONIC
® LIBRARY
OF
SCIENCE

TELOS, The Electronic Library of Science, is an imprint of Springer-Verlag New York with publishing facilities in Santa Clara, California. Its publishing program encompasses the natural and physical sciences, computer science, mathematics, economics, and engineering. All TELOS publications have a computational orientation to them, as TELOS' primary publishing strategy is to wed the traditional print medium with the emerging new electronic media in order to provide the reader with a truly interactive multimedia information environment. To achieve this, every TELOS publication delivered on paper has an associated electronic component. This can take the form of book/diskette combinations, book/CD-ROM packages, books delivered via networks, electronic journals, newsletters, plus a multitude of other exciting possibilities. Since TELOS is not committed to any one technology, any delivery medium can be considered. We also do not foresee the imminent demise of the paper book, or journal, as we know them. Instead we believe paper and electronic media can coexist side-by-side, since both offer valuable means by which to convey information to consumers.

The range of TELOS publications extends from research level reference works to textbook materials for the higher education audience, practical handbooks for working professionals, and broadly accessible science, computer science, and high technology general interest publications. Many TELOS publications are interdisciplinary in nature, and most are targeted for the individual buyer, which dictates that TELOS publications be affordably priced.

Of the numerous definitions of the Greek word "telos," the one most representative of our publishing philosophy is "to turn," or "turning point." We perceive the establishment of the TELOS publishing program to be a significant step forward towards attaining a new plateau of high quality information packaging and dissemination in the interactive learning environment of the future. TELOS welcomes you to join us in the exploration and development of this exciting frontier as a reader and user, an author, editor, consultant, strategic partner, or in whatever other capacity one might imagine.

TELOS, The Electronic Library of Science
Springer-Verlag Publishers
3600 Pruneridge Avenue, Suite 200
Santa Clara, CA 95051

THE
ELECTRONIC
LIBRARY
OF
SCIENCE

TELOS Diskettes

Unless otherwise designated, computer diskettes packaged with TELOS publications are 3.5" high-density DOS-formatted diskettes. They may be read by any IBM-compatible computer running DOS or Windows. They may also be read by computers running NEXTSTEP, by most UNIX machines, and by Macintosh computers using a file exchange utility.

In those cases where the diskettes require the availability of specific software programs in order to run them, or to take full advantage of their capabilities, then the specific requirements regarding these software packages will be indicated.

TELOS CD-ROM Discs

For buyers of TELOS publications containing CD-ROM discs, or in those cases where the product is a stand-alone CD-ROM, it is always indicated on which specific platform, or platforms, the disc is designed to run. For example, Macintosh only; Windows only; cross-platform, and so forth.

TELOSpub.com (Online)

Interact with TELOS online via the Internet by setting your World-Wide-Web browser to the URL: http://www.telospub.com.

The TELOS Web site features new product information and updates, an online catalog and ordering, samples from our publications, information about TELOS, data-files related to and enhancements of our products, and a broad selection of other unique features. Presented in hypertext format with rich graphics, it's your best way to discover what's new at TELOS.

TELOS also maintains these additional Internet resources:

gopher://gopher.telospub.com
ftp://ftp.telospub.com

For up-to-date information regarding TELOS online services, send the one-line e-mail message:

send info to: *info@telospub.com.*

PREFACE

Our Perspective

The purpose of this book is to provide a traditional treatment of elementary ordinary differential equations while introducing the computer-assisted methods that are now available with *Mathematica*. We have chosen *Mathematica* over other systems of computer algebra because of its combination of easy access and computational power, as evidenced through symbolic, numerical, and graphical output.

In order to make this work totally self-contained, we have developed the fundamentals of differential equations from the very beginning. This includes the solution methods for the traditional classes of solvable equations (first-order linear, second-order constant-coefficients, linear systems, Laplace transforms, power series solutions and so forth) as well as a presentation of the basic theory of existence/uniqueness and the traditional numerical methods for first-order equations. In the process some new mathematical points have been developed, to be noted below. It is our firm belief that a solid mastery of the subject of differential equations can only be achieved through a strong traditional course. This is enhanced by the graphical capabilities of *Mathematica*, which have allowed the incorporation of many more graphs than are normally available in books at this level.

When we move beyond the traditional course and envisage the integration of computers, many potential problems arise in redesigning the syllabus. A well-known first attempt was the development of BASIC and similar programs in the 1970s. The ease with which numerical routines such as the Euler method could be implemented made the new software a very useful pedagogical device. However, computer technology distracted students from learning about differential equations.

The scene changed dramatically with the advent of symbolic manipulation programs—first *Macsyma*, then *Derive*, *Maple*, *Mathematica* and others—which made symbolic as well as numerical solutions of differential equations possible via computers. In principle symbolic manipulation programs can perform routine calculations, permitting students to cover more theory and applications. In practice, it may be necessary for students to learn a fair amount about a symbolic manipulation program for a course in differential equations. Certainly, using *Mathematica* is far simpler than writing programs in C or BASIC. Nevertheless, students and professors frequently become frustrated when *Mathematica* does not behave exactly as mathematics does, and valuable time is wasted. It is for this reason that our text provides a parallel development of classical solution methods as well as a special *Mathematica* package, **ODE.m**. No prior knowledge of *Mathematica* is required either to use this book or the programs contained in **ODE.m**.

Prerequisites

We presume knowledge of calculus, but a full course in linear algebra is not required. Vectors and matrices are not needed until Chapter 15; students can easily learn the necessary material from the linear algebra review provided in Appendix A.

Features for Instructors

- **Power series**. In Chapters 20 and 21 we present a new and efficient method for obtaining the power series solution or Frobenius solution to a differential equation. The method uses an algorithm that goes back to Cauchy. (The standard method is also explained.)

- **Phase portraits**. We have also achieved some coherence in the discussion of two-dimensional linear systems with constant-coefficients. Thirteen separate cases are enumerated and discussed in detail in Chapter 16.

- **Laplace transforms**. Chapter 14 contains a detailed treatment of Laplace transforms because of their importance in engineering. Discontinuous forcing functions and the Dirac delta function are fully treated. Furthermore, ODE automates the use of Laplace transforms, much more so than is done with *Mathematica*'s built-in packages.

- **Seminumerical solutions**. Since it is usually impossible to find the general solution of a constant-coefficient linear differential equation whose order is greater than 4, we show how to find numerical approximations to the roots of the associated characteristic equation and to use them to construct an approximate general solution.

- **Mathematical details**. Proofs of many important theorems are given in optional subsections. A criterion for global existence of first-order nonlinear equations is presented as a complement to the classical Picard local existence proof in Chapter 5.

- **Other solution methods**. For historical completeness we have included the solution methods of Bernoulli, Clairaut, and Lagrange. Since these are rarely used in practice, these methods can be skipped in all but the most extensive differential equations courses. However, since they are available in **ODE.m**, they can be invoked at a moment's notice on those infrequent occasions when they are needed.

Features for Students

- *Mathematica*. With some experience, it will be found that the use of *Mathematica*, and in particular **ODE.m**, will simplify a student's ability to deal with differential equations. The principal command **ODE** is as easy to use as a hand calculator, but much more powerful. While we emphasize the importance of mastering the traditional methods, it is no longer necessary to continually suffer (as students did 50 years ago) doing lengthy hand calculations to solve differential equations.

- **Exercises and worked examples**. Most people learn at least in part by imitation and repetition. Accordingly, we have included more than 300 worked examples in the text and more than 650 exercises. Past experience has shown that this method will guarantee efficient mastery of the traditional methods of solution. There are also many worked examples, exercises, and solutions related to the *Mathematica* material. All of these will be included in the CD-ROM with the book. Students can have practice solving differential equations, with and without *Mathematica*.

- **Procedure boxes**. Important solution techniques are summarized in boxes; these convenient summaries are easy to find.

- **Computer graphics**. All graphics in this book have been generated by *Mathematica*; this is a guarantee of their accuracy. When writing this book, we have found time after time that a graph adds a new dimension to understanding a problem, much more so than either a formula or a table of values. The CD-ROM also includes color graphics.

- **Historical footnotes**. The history of differential equations is both fascinating and important. Historical footnotes, including photo portraits of these individuals, provide details about some of the scientists involved in its development.

- **Extensive indices**. There are three indices: a general index, a name index, and a *Mathematica* index. The *Mathematica* index can be used by those who already know *Mathematica* to find examples of how various *Mathematica* commands are used.

Features for Programmers

Although the package **ODE.m** can be used to mask many details of the solving of differential equations, we provide in optional sections details on how **ODE.m** does its job. In most cases this involves translating certain mathematical formulas to the formalism of *Mathematica*.

Complete descriptions of the miniprograms that make up **ODE.m** are provided on the CD-ROM. These easy-to-understand miniprograms provide valuable insight not only on how *Mathematica* goes about its job, but also about the solution process for many types of ordinary differential equations.

Connections with the Traditional Course

This book can be used as a traditional text, with no reference to *Mathematica* or computers. In every case, the computer enhancement always *follows* the traditional text. The following points are noted for reference:

- The mathematical theory of first-order differential equations is given in Chapter 3, and the implementation using **ODE.m** is given in Chapter 4.

- The mathematical theory of second-order linear differential equations is given in Chapters 8 and 9, and the implementation using **ODE.m** is given in Chapter 10.

- The mathematical theory of higher-order linear differential equations is given in Sections 12.1–12.3, and the implementation using **ODE.m** is given in Sections 12.4–12.5.

- The mathematical theory of numerical solutions to differential equations is given in Sections 13.1–13.3, and the implementation using **ODE.m** is given in Sections 13.4–13.8.

- The mathematical theory of Laplace transforms is given in Sections 14.1–14.9, and the implementation using **ODE.m** is given in Section 14.10.

- The mathematical theory of linear systems is given in Chapter 15, and the implementation using **ODE.m** is given in Chapter 16.

Use of the CD-ROM

Together with this book, we have developed a CD-ROM that allows the user to have unprecedented access to aspects of differential equations, that were previously unavailable with conventional texts. This allows the instructor and student to have easy access to the symbolic, graphical, and dynamic aspects of the subject. In detail, the CD-ROM has a number of directories, which can be accessed as follows.

> **The *Mathematica* solution of worked examples**: In order to access the *Mathematica* notebook containing Example 5.10, one opens Chapter_05.*ma* and then searches for Example 5.10.

The *Mathematica* solution of exercises: In order to find the solution of Exercise 3 in Section 14.5, one selects the subdirectory Chapter 14, then opens Sec1405.ma and searches for Exercise 3.

Portrait gallery: We have included postscript files of the portraits of each of the forty-eight mathematicians who have made significant contributions to the field of differential equations. They require a postscript viewer or postscript printer for access.

Miscellaneous *Mathematica* notebooks: Among the additional features available in this directory we have a comparison of different techniques of numerical integration, various direction fields of two-dimensional systems, two and three-dimensional phase portraits, a tour of highlights of the **ODE.m** package, phase portraits and other aspects of the pendulum, and examples of resonance using *Mathematica*'s **Play** function.

***Mathematica* movies**: The relevant subdirectories are (i) LinearSystems, (ii) NonlinearSystems, (iii) SeriesApproximations, (iv) VectorFields. These cover, respectively, 2D and 3D phase portraits related to Chapter 16; springs, pendula, Lorentz attractors, van der Pol attractors, and predator-prey models; the Picard series approximation to the solution of $y' = t^2 - y$, $y(1) = 2$ on the interval $-2 < t < 6$; a *Mathematica* movie of the flow corresponding to the equation $y' = \sin(t)/y$ on a rectangle centered at $(0, 0)$.

Packages: This directory contains **ODE.m**, which includes the *Mathematica* functions that are frequently used throughout the book. Other *Mathematica* functions are found in the auxiliary package **ODEx.m**.

ODE Reference Manual: This contains the Manual which documents the *Mathematica* code in **ODE.m**. It can be used as a beginners handbook and quick reference manual for ordinary differential equations. Although this manual can be read with any HTML browser, it is also present in encapsulated postscript form in the "psrefman" folder for those with a postscript viewer or postscript printer.

Sample Labs: For the student who wants extra practice, we have organized seven laboratory assignments using *Mathematica*. The entire collection gives a representative sampling of the entire book. Each assignment opens with a review of the relevant theory, followed by a sample problem preceding the new exercise. The topics include first-order equations, applications to cell growth, second-order constant-coefficient equations, numerical solutions, Laplace transforms, series solutions, and linear systems.

Support for Different Versions of Mathematica

As we go to press, Wolfram Research Inc. is announcing the release of *Mathematica* 3.0, as documented in *The Mathematica Book*, by Stephen Wolfram, Third Edition, Cambridge University Press, 1996. All of the programs in our book and in **ODE.m** work both in Version 2.2 and Version 3.0 of *Mathematica*. In addition we are developing some specific new notebooks that are uniquely adapted to *Mathematica* 3.0 and that is available on the Web site version of the CD-ROM, which is accessible at

http://math.cl.uh.edu/ode/ode.html

Supplements and Software

- **Solutions manual**. Short answers to the odd-numbered problems are given at the end of the text. Complete solutions are given in a separate solutions manual.

- **ODE.m**. This package is included with the book and is also available via anonymous ftp.

- **Notebooks**. *Mathematica* notebooks, both for Version 2.2 and Version 3.0 of *Mathematica*, are provided as explained above in "Use of the CD-ROM."

- **Solutions notebooks**. Solutions are also provided in *Mathematica* notebooks as explained above in "Use of the CD-ROM." Students can experiment to produce new problems and graphs.

Addresses

The latest version of **ODE.m** and explanatory *Mathematica* notebooks are available via anonymous ftp from **math.cl.uh.edu**. The e-mail addresses of the authors are:

Alfred Gray	**gray@bianchi.umd.edu**
Michael Mezzino	**mezzino@gauss.cl.uh.edu**
Mark A. Pinsky	**pinsky@math.nwu.edu**

ACKNOWLEDGMENTS

We wish to express our sincere gratitude to the following individuals who made valuable comments on various stages of the evolution of the manuscript:

Douglas Alexander (TRACOR, Incorporated)

James Anderson (University of Maryland)

Luis A. Cordero (University of Santiago de Compostela)

Albert Currier (University of Maryland)

Gautam Dasgupta (Columbia University)

Bill Davis (Ohio State University)

Dave Diller (Northwestern University)

Kit Dodson (Manchester University)

Chris Flannery (Northwestern University)

Mary Gray (American University)

Joe Grohens (Wolfram Research Incorporated)

Denny Gulick (University of Maryland)

Garry Helzer (University of Maryland)

Harry Hughes (Southern Illinois University)

Steve Izen (Case Western Reserve University)

Alfredo Jiménez (Pennsylvania State University, Hazelton)

Joe Kaiping (Wolfram Research Incorporated)

Jerry Keiper (Wolfram Research Incorporated)

Lee Lorch (York University)

William M. MacDonald (University of Maryland)

Kirk Mathews (Air Force Institute of Technology)

Ed Packel (Lake Forest College)

Patrick Reich (University of Houston - Clear Lake)

Norman Richert (University of Houston - Clear Lake)

Clark Robinson (Northwestern University)

Tarek G. Shawki (University of Illinois)

Renming Song (University of Michigan)

Nancy Stanton (Notre Dame University)

Volker Wihstutz (University of North Carolina)

Calvin Wilcox (University of Utah)

David Withoff (Wolfram Research Incorporated)

We also wish to express our heartfelt thanks and appreciation to our editors - Allan Wylde and Paul Wellin, for their encouragement and unwavering support during the months of development of both the text and the software. To the staff of Springer-Verlag, especially Keisha Sherbecoe and Victoria Evarretta, for their careful attention to the details of production.

And finally, to our families, who have always provided the necessary love and emotional support, especially during the challenging periods of this project.

TABLE OF CONTENTS

1

BASIC CONCEPTS

In this chapter we describe the subject of differential equations in general terms and touch upon its history. An introduction to *Mathematica* is given in Chapter 2. The study of ordinary differential equations is begun in earnest in Chapter 3.

1.1   *The Notion of a Differential Equation*

A **differential equation** is an equation involving a function y and its derivatives. Simple examples of differential equations include

$$y''(t) + 4y'(t) + 4y(t) = 0 \qquad \text{and} \qquad y' + \sin(t)y^2 = 3e^{t^2}. \tag{1.1}$$

When the equation involves one or more derivatives with respect to a particular variable, that variable is called an **independent variable**. A variable is called **dependent** if a derivative of that variable occurs in the equation. In both equations of (1.1), t is the independent variable and y is the dependent variable. We exclude from the class of differential equations those that are identities, such as

$$\frac{d}{dt}(t\,y) = t\frac{dy}{dt} + y.$$

In the case of one independent variable t, we write $y = y(t)$ and speak of an **ordinary differential equation**, whereas for a function of several independent variables $u = u(x_1, x_2, \ldots, x_n)$ we speak of a **partial differential equation**. Many books use "ODE" as an abbreviation for "ordinary differential equation" and "PDE" as an abbreviation for "partial differential equation." An example of a partial differential equation is the **Laplace equation**

$$\frac{\partial^2 u}{\partial x^2} + \frac{\partial^2 u}{\partial y^2} = 0. \tag{1.2}$$

1

In this book we shall only solve ordinary differential equations, but occasionally we shall need partial differential equations to study ordinary differential equations. The partial derivatives

$$\frac{\partial u}{\partial x}, \quad \frac{\partial^2 u}{\partial x^2}, \ldots$$

are frequently written as u_x, u_{xx}, and so forth. Thus (1.2) can be rewritten as $u_{xx} + u_{yy} = 0$.

Definition. *The* **order of a differential equation** *is the highest order of differentiation that appears in the equation.*

The most general form of an nth order ordinary differential equation is

$$F\left(t, y, y', \ldots, y^{(n)}\right) = 0, \tag{1.3}$$

where $y^{(n)}$ denotes the nth derivative of y, and F is some function of $n + 2$ variables. For example, the first differential equation of (1.1) is an ordinary differential equation of second-order because a second derivative is the highest derivative that appears in it. Similarly, the second equation is of first-order. Laplace's equation (1.2) is also of second-order.

Definition. *A* **solution** *of an ordinary differential equation* (1.3) *is a function ϕ (possessing derivatives up to order at least n) such that when ϕ is substituted for y in* (1.3), *the left-hand side of* (1.3) *vanishes identically. In other words, a solution of* (1.3) *is a function ϕ for which*

$$t \longmapsto F\left(t, \phi(t), \phi'(t), \ldots, \phi^{(n)}(t)\right)$$

vanishes identically for all t in some interval $a < t < b$.

Some differential equations have many solutions, others none. In general, higher-order differential equations are more difficult to solve than lower-order differential equations. First-order differential equations will be studied in Chapters 3–6.

There is a strong analogy between differential equations and algebraic equations. For example, the first equation of (1.1) looks very much like

$$y^2 + 4y + 4 = 0. \tag{1.4}$$

Notice, however, that the solution of (1.4) is the *number* -2, whereas the solutions of (1.1) are *functions*. In fact, it is easy to check by direct substitution that the functions y_1 and y_2 defined by $y_1(t) = e^{-2t}$ and $y_2(t) = t\,e^{-2t}$ are solutions of the first equation of (1.1).

Here is another important way to classify differential equations:

Definition. *An ordinary differential equation $F\left(t, y, y', \ldots, y^{(n)}\right) = 0$ is said to be* **linear** *provided F is a linear function of the variables $y, y', \ldots, y^{(n)}$. (F may or may not* be linear in t.) Equivalently, a differential equation of order n is linear provided that it can be expressed in the form

$$a_n(t)y^{(n)}(t) + a_{n-1}(t)y^{(n-1)}(t) + \cdots + a_1(t)y'(t) + a_0(t)y(t) = g(t) \qquad (1.5)$$

with $a_n(t)$ not identically 0. Otherwise, an ordinary differential equation is said to be **nonlinear**.

For example,

$$y''(t) + t\, y(t) = 0$$

is a linear second-order differential equation, but

$$y'(t) + y(t)^2 = 0$$

is a nonlinear first-order differential equation because of the $y(t)^2$ term. Of course, these two differential equations can be abbreviated to $y'' + t\, y = 0$ and $y' + y^2 = 0$, or as

$$\frac{d^2 y}{dt^2} + t\, y = 0 \qquad \text{and} \qquad \frac{dy}{dt} + y^2 = 0.$$

Another example is

$$2y\, y' + y'' = 0;$$

this differential equation is nonlinear because of the $y\, y'$ term.

As we shall see in Section 3.2, there is a general method for solving first-order linear differential equations. However, higher-order linear differential equations are so much more complicated that they can be solved only in special cases, the most important of which is the case when the a_i's in (1.5) are constants. Second-order linear differential equations will be studied in Chapters 8–11 and 20–21.

In Chapters 15–17 we shall also consider **systems of differential equations** for several unknown functions. An example is

$$\frac{dx}{dt} = x + y, \qquad \frac{dy}{dt} = y + z, \qquad \frac{dz}{dt} = z + x.$$

Exercises for Section 1.1

For each of the following differential equations, determine whether it is *ordinary* or *partial*, *linear* or *nonlinear*, and find the *order*.

1. $y' + y^2 = \sin(t)$

2. $y'' + 100y = \sin(t)$

3. $y'' + 4y = e^y$

4. $\dfrac{\partial u}{\partial t} = \dfrac{\partial (u^2)}{\partial x}$

5. $y' + t\,y = \sin(t)$ **7.** $u_{tt} = u_{xx} + u_{yy}$

6. $y'' + y^2 = e^t$ **8.** $\dfrac{\partial u}{\partial t} = \dfrac{\partial^2 u}{\partial x^2}$

1.2   Sources of Differential Equations

A profound change in science took place in the seventeenth century. Until that time mathematics had been dominated by the techniques of Euclidean geometry. Calculus was not available as a tool for science. All of that was changed by the discovery of calculus by Newton[1] and Leibniz.[2]

The history of differential equations is intertwined with that of calculus. Physical and geometrical problems gave rise to differential equations, and the solution of differential equations required calculus. The first edition of Newton's **Principia** (see [Newton]) was published in 1687; Leibniz's first papers on calculus appeared a few years earlier. Since then, differential equations have become an indispensable tool for describing natural phenomena.

Newton's Laws of Motion

Newton's laws of motion provide one of the most fruitful sources of differential equations.

1

Sir Isaac Newton (1642–1727). English mathematician, physicist, and astronomer. Newton's contributions to mathematics encompass not only his fundamental work on calculus and his discovery of the binomial theorem, but substantial work in algebra, number theory, classical and analytic geometry, methods of computation and approximation, and probability.

As Lucasian professor at Cambridge, Newton was required to lecture once a week, but his lectures were so abstruse that he frequently had no auditors. As a result of his invention of the refracting telescope, Newton was elected a fellow of the Royal Society in 1672. Twice elected as Member of Parliament for the University, Newton was also appointed warden of the mint; his knighthood was awarded primarily for his services in that role. In **Philosophiæ Naturalis Principia Mathematica**, Newton set forth fundamental mathematical principles and then applied them to the development of a world system. This is the basis of the Newtonian physics that determined how the universe was perceived until the twentieth-century work of Einstein.

2

Baron Gottfried Wilhelm Leibniz (1646–1716). German mathematician. Trained as a lawyer, Leibniz spent most of his life in Hanover, where he made contributions to diplomacy, history, law, logic, mathematics, metaphysics, philology, politics, and theology. He also invented a calculating machine. Although Leibniz discovered calculus a few years later than Newton, it is the notation of Leibniz (such as dt and $\int$) that has gained the widest acceptance. Leibniz never thought of the derivative as a limit.

Law I: Every body continues in a state of rest, or of uniform motion in a straight line, unless it is compelled to change that state by forces impressed upon it.

Law II: The rate of change of motion of a body is proportional to the force impressed and takes place in the direction of a straight line in which the force acts.

Law III: To every action there is an equal and opposite reaction; or, the mutual actions of two bodies are always equal and oppositely directed.

Newton's second law implies that the total force on a body is equal to the time rate of change of the momentum, which in the case of a constant mass is equal to the product of the mass and the acceleration. In symbols,

$$F = m\,a. \tag{1.6}$$

The acceleration is the second derivative of the position y with respect to the time t, that is

$$a = \frac{d^2 y}{dt^2}.$$

The force(s) on the body are then expressed in terms of the time t, the displacement y, the velocity dy/dt, and other parameters, to yield a differential equation. In the absence of any forces we have the differential equation $y'' = 0$, which yields uniform motion, as stated also in **Newton's first law of motion**. The solution of the differential equation $y'' = 0$ is easily found to be $y = A\,t + B$, where A and B are constants.

Another important case is that of a spring. **Hooke's**[3] **law of elastic restoration** can be expressed as the differential equation

$$m\,y'' = -k\,y, \tag{1.7}$$

where k is the elastic constant.

Temperature

A second source of differential equations is found in the laws of temperature equalization in solids, in particular, **Newton's law of cooling.**[4] A common postulate is that the time rate

---

[3] Robert Hooke (1635–1703). English scientist. For thirty years Hooke was professor of geometry at Gresham College, London. He achieved fame in 1665 with his book **Micrographia**, which contains beautiful pictures and a number of fundamental biological discoveries. He became embroiled in a priority dispute over the inverse square law with Newton, and consequently for the rest of his life was one of Newton's chief antagonists.

[4] The details were worked out by G. W. Richmann, J. B. Biot, and J. Stefan.

of change of temperature is proportional to the difference between the temperature of the solid body and the temperature of its surroundings. For example, an apple removed from a refrigerator warms most quickly when the temperature difference is large. The temperature of the apple may be modeled by the differential equation

$$\frac{d\mathbf{T}}{dt} = -k(\mathbf{T} - \mathbf{T}_e),$$

where $\mathbf{T}(t)$ denotes the temperature of the apple and $\mathbf{T}_e$ denotes the room temperature, for example, 68°F.

The above examples are developed in greater detail in Chapters 6, 7 and 11.

The Catenary

In 1691 Jakob Bernoulli[5] gave a solution to the problem of finding the curve assumed by a flexible inextensible cord hung freely from two fixed points; Leibniz has called such a curve a catenary (which stems from the Latin word *catena*, meaning chain). The solution is based on the differential equation

$$\frac{d^2y}{dx^2} = \frac{1}{c}\sqrt{1 + \left(\frac{dy}{dx}\right)^2}. \tag{1.8}$$

To derive (1.8), we consider a portion $\overline{\mathbf{pq}}$ of the cable between the lowest point $\mathbf{p}$ and an arbitrary point $\mathbf{q}$. Three forces act on the cable: the weight $\overline{\mathbf{pq}}$ and the tensions $\mathbf{T}$ and $\mathbf{U}$ at $\mathbf{p}$ and $\mathbf{q}$. If w is the linear density and s is the length of $\mathbf{p}$ and $\mathbf{q}$, then the weight of $\mathbf{p}$ and $\mathbf{q}$ is $w\,s$.

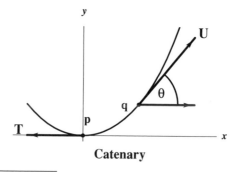

Catenary

5

Jakob Bernoulli (1654–1705). Jakob and his brother Johann were the first of a Swiss mathematical dynasty. The work of the Bernoullis was instrumental in establishing the dominance of Leibniz's methods of exposition. Jakob Bernoulli laid basic foundations for the development of the calculus of variations, as well as working in probability, astronomy, and mathematical physics.

Let |**T**| and |**U**| denote the magnitudes of the forces **T** and **U**, and write

$$\mathbf{U} = \big(|\mathbf{U}|\cos(\theta), |\mathbf{U}|\sin(\theta)\big).$$

Because of equilibrium we have

$$|\mathbf{T}| = |\mathbf{U}|\cos(\theta) \qquad \text{and} \qquad w\,s = |\mathbf{U}|\sin(\theta). \tag{1.9}$$

Let $\mathbf{q} = (x, y)$, where x and y are functions of s. From (1.9) we obtain

$$\frac{dy}{dx} = \tan(\theta) = \frac{w\,s}{|\mathbf{T}|}. \tag{1.10}$$

Since the length of $\overline{\mathbf{pq}}$ is

$$s = \int_0^x \sqrt{1 + \left(\frac{dy}{dx}\right)^2}\, dx,$$

the fundamental theorem of calculus tells us that

$$\frac{ds}{dx} = \sqrt{1 + \left(\frac{dy}{dx}\right)^2}. \tag{1.11}$$

When we differentiate (1.10) with respect to x and use (1.11), we get (1.8).

Although at first glance the catenary looks like a parabola, it is in fact the graph of the hyperbolic cosine.

1.3   Solving Differential Equations

It is very important to clarify what we mean by the **solution** of a differential equation. Different equations admit different forms of solution, so we must be very clear on the different possibilities for **solving** a differential equation.

Elementary Functions as Solutions: Explicit and Implicit Forms

The most basic sense in which we may solve a differential equation is in the form of an explicit **elementary function**, that is, a finite arithmetic combination of powers, radicals, exponentials, logarithmic, trigonometric, and inverse trigonometric functions, as known from calculus courses. For example, the general solution of the differential equation $y'' = 0$ can be obtained in terms of elementary functions as follows. Indeed, y' must be a constant, say $y' = B$, and so $y = y(t)$ must be of the form $y = A + B\,t$ for suitable constants A and B. The linear function $A + B\,t$ is a prime example of an elementary function.

Here is a more complicated example:

Example 1.1. *Check that* $y = (t/2)(\sin(t))^2$ *is a solution of*

$$y'' + 4y = t + \sin(2t). \tag{1.12}$$

Solution. Using various trigonometric identities, we compute

$$y' = \frac{(\sin(t))^2}{2} + t\sin(t)\cos(t) \qquad \text{and} \qquad y'' = \sin(2t) + t\cos(2t). \tag{1.13}$$

Then (1.12) follows directly from (1.13). ∎

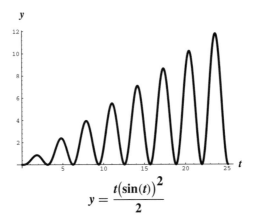

$$y = \frac{t\big(\sin(t)\big)^2}{2}$$

In some cases a differential equation may be solved in terms of elementary functions, although only **implicitly**. For example, as we shall see in Section 3.4 the differential equation

$$3y^2y' + t\,y' + y = 0 \tag{1.14}$$

is solved by the implicit equation

$$y^3 + t\,y = C, \tag{1.15}$$

where C is a constant. (It is easy to check using the chain rule that (1.15) implies (1.14).) The cubic equation (1.15) can be further solved for y, but the formula is quite complicated. For example, one of the three solutions of (1.15) turns out to be

$$y = \frac{-2^{\frac{1}{3}}t}{\left(27\,C + \sqrt{729\,C^2 + 108\,t^3}\right)^{\frac{1}{3}}} + \frac{\left(27\,C + \sqrt{729\,C^2 + 108\,t^3}\right)^{\frac{1}{3}}}{3(2)^{\frac{1}{3}}}. \tag{1.16}$$

In this and more complicated cases it is best to say that we have solved (1.14) when we have found the **implicit solution** (1.15), and not to bother expressing y explicitly in terms of t. Here is the plot of (1.15) when $C = 1$.

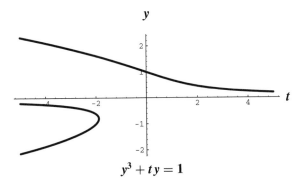

$$y^3 + ty = 1$$

On the other hand, the differential equation

$$y'(t) = \sin(t^2) \tag{1.17}$$

cannot be solved in terms of elementary functions. The solution of (1.17) can be written either as an indefinite integral or as a definite integral:

$$y(t) = \int \sin(t^2)dt = \int_0^t \sin(s^2)ds + C, \tag{1.18}$$

where C is a constant. Neither integral in (1.18) can be reduced to an expression involving elementary functions. Nevertheless, (1.18) can be a very useful form of the solution, since any definite integral can be approximated numerically, for example by expanding the integrand in a power series. Furthermore, we can get a good understanding of the definite integral numerically by graphing it.

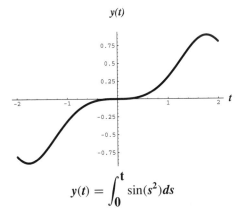

$$y(t) = \int_0^t \sin(s^2)ds$$

Guessing the Solution versus Verifying the Solution

In comparison with other parts of mathematics, the subject of differential equations is known for its dependence on intuitive reasoning and related heuristic arguments. Some of these **ad hoc** methods of reasoning have put off serious students of mathematics and led to the impression in some quarters that this is not "honest mathematics." Before we embark upon further study of the subject, it is important to be clear about the role of intuition versus rigor in dealing with differential equations.

We shall often use our past experience, or make questionable shortcuts, to obtain an explicit solution of a differential equation. But once we have arrived at a proposed formula for the solution, there is no further room for heuristic reasoning. The **verification** of the proposed solution formula follows the rigorous steps of analysis, whether they be the principles for integration and differentiation of elementary functions of calculus or theorems on the convergence and differentiability of infinite series of functions. **There is no substitute for the verification process**: it provides the meeting ground for heuristic methods and rigorous mathematical reasoning.

As a simple illustration of the above remarks, consider the problem of solving the differential equation

$$\frac{dy}{dt} = 2t\, y. \tag{1.19}$$

We may simply rewrite (1.19) as

$$\frac{dy}{y} = 2t\, dt. \tag{1.20}$$

We integrate both sides of (1.20) obtaining[6]

$$\log(y) = t^2 + C, \tag{1.21}$$

where C is a constant. Finally, we exponentiate both sides of (1.21) to get

$$y = e^{\log(y)} = e^{t^2 + C} = e^C e^{t^2}. \tag{1.22}$$

A mathematical critic might object to the differential notation (1.20) or to division in (1.20) by y, which may be zero. Indeed, we cannot say in advance that $y \neq 0$. Nor does the

---

[6]We shall use the notation $\log(x)$ to denote the **natural logarithm** of x. Thus, for $x > 0$ we have

$$\log(x) = \int_1^x \frac{du}{u}.$$

Some authors use the notation $\ln(x)$ to denote the natural logarithm of x.

form of the proposed solution settle this point; we cannot "prove" in advance that $y \neq 0$. However, the *verification* that y given by (1.22) is a solution to (1.19) is unambiguous:

$$\frac{dy}{dt} = \frac{d}{dt}\left(e^C e^{t^2}\right) = e^C e^{t^2}(2t) = 2t\, y.$$

In general, when "solving" differential equations, most of our efforts are spent in the formal process of finding the form of the solution. Nevertheless, the solution process is not complete until we have verified that the proposed formula is indeed a solution. For one thing, we might have made an arithmetic or computational error in obtaining the formula, which will be revealed in the verification process. Even more important, the verification process puts on a common ground the mathematical steps that are used to establish the validity of the solution, diminishing the prominence of the intuitive steps.

Particular Solutions versus General Solutions

Differential equations that model problems in the sciences and engineering are usually expected to have unique solutions. We expect to be able to predict the future from the past and present conditions. However, when we begin solving simple differential equations, we find that there are certain **arbitrary constants** that enter from the process of indefinite integration. Of course, we may ignore such a constant and obtain a **particular solution** of the equation. However, the conditions of the model may further dictate certain **initial conditions**, or **boundary conditions**, which may further specify the "arbitrary constant". For example, we saw on page 10 that the differential equation $y' = 2t\, y$ has the solution $y = e^{C+t^2}$ for any constant C. If we further specify that $y(0) = 6$, then the constant is uniquely determined, and the solution must be

$$y = 6e^{t^2};$$

no further ambiguity is possible.

In order to adjust to the different possible initial conditions of the problem, we always look for the most general solution of the equation containing as many arbitrary constants as arise in the solution process; particular solutions may be obtained by specializing the values of the constants. In this regard, it will be useful to distinguish between the **definite integral** and the **indefinite integral** of a function.

Definite and Indefinite Integrals

The expression

$$\int_a^b f(t)dt$$

produces a **number** when we are given a continuous function $f(t)$ defined for $a \leq t \leq b$. For example, the methods of elementary calculus provide us with such definite integrals as

$$\int_0^1 t^2 dt = \frac{1}{3}, \qquad \int_0^\pi \sin(t) dt = 2 \quad \text{and} \quad \int_0^\infty e^{-t} dt = 1.$$

Of course, the integration variable t plays no role in the final answer, which is the same whatever "dummy variable" is used; equivalently, we could have written

$$\int_0^1 z^2 dz = \frac{1}{3}, \qquad \int_0^\pi \sin(\theta) d\theta = 2, \qquad \text{and} \qquad \int_0^\infty e^{-\psi} d\psi = 1.$$

The letter "d" is a "control character". It tells us that we are to integrate with respect to the variable following the letter d.

The expression

$$\int f(t) dt$$

denotes the **indefinite integral** of f. It stands for a **class** of functions, all of which differ by a constant. Any one of these functions, denoted by F, has the property that

$$F'(t) = f(t). \tag{1.23}$$

Because of (1.23) an indefinite integral is sometimes called an **antiderivative**. If $F_1'(t) = f(t)$ and $F_2'(t) = f(t)$ for all t, then $F_1(t) - F_2(t) = C$ for some constant C. We say that F_1 and F_2 are **indefinite integrals** of f. For example, the indefinite integral

$$\int t^2 dt$$

is the class of functions of the form

$$\frac{t^3}{3} + C,$$

where C is a constant.

From calculus we know that a **definite integral**

$$\int_a^b f(t) dt$$

can be evaluated from the knowledge of any **indefinite integral** F of f by the formula

$$\int_a^b f(t) dt = F(b) - F(a).$$

The value of the constant by which any two antiderivatives differ disappears in this computation.

The definite integral can also be used to manufacture new functions from old ones. In order to do this we fix a lower limit $t = a$ and introduce a "dummy variable of integration" s and define a function Φ by

$$\Phi(t) = \int_a^t f(s)\,ds.$$

A theorem of calculus states that the function $\Phi(t)$ so defined satisfies the "differential equation" $\Phi'(t) = f(t)$, provided that the function $f(t)$ is continuous throughout the interval of definition. The dummy variable of integration is essential to make an unambiguous system of mathematics. An attempt to deal with an expression such as

$$\int_a^t f(t)\,dt$$

may lead to confusing and incorrect results, because t is used both as a dummy *variable* and as a *constant* limit of integration.

We now illustrate, by means of a simple example, how to obtain the "general solution" of a differential equation by using the indefinite integral.

Example 1.2.   *Find the general solution of the differential equation $y' = 2t\,y$ by means of indefinite integration.*

Solution.   We rewrite $y' = 2t\,y$ in the form

$$\frac{y'}{y} = 2t. \tag{1.24}$$

The left-hand side of (1.24) is the derivative of the function $\log(y)$, and the right-hand side is the derivative of the function t^2. Taking the indefinite integral of both sides of (1.24), we get

$$\log\left(y(t)\right) + C_1 = \int \frac{y'}{y}\,dt = \int 2t\,dt = t^2 + C_2. \tag{1.25}$$

The constants C_1 and C_2 in (1.25) can be combined into a single constant C, resulting in

$$\log(y) = t^2 + C. \tag{1.26}$$

Next, we exponentiate both sides of (1.26) to obtain the solution of (1.24):

$$y = e^{\log(y)} = e^{t^2 + C} = e^C\, e^{t^2}. \tag{1.27}$$

It can be verified by direct substitution that the solution y given by (1.27) indeed satisfies the differential equation $y' = 2t\,y$. Furthermore, if $z(t)$ is **any** solution of the equation (possibly obtained by a different procedure), then the function

$$\tilde{z}(t) = z(t)e^{-t^2}$$

satisfies $\tilde{z}'(t) = 0$; hence $\tilde{z}(t)$ is a constant, and so $z(t)$ is a multiple of $y(t)$. This proves that the general solution of $y' = 2t\,y$ is indeed given by (1.27). ∎

The differential equation (1.24) may also be solved using definite integrals, as follows.

Example 1.3. *Solve the differential equation $y' = 2t\,y$ by definite integration.*

Solution. We integrate both sides of $y'(s) = 2s\,y(s)$ from 0 to t, obtaining

$$\log\big(y(t)\big) - \log\big(y(0)\big) = \int_0^t \big(\log(y)\big)'(s)ds = \int_0^t \frac{y'(s)ds}{y(s)} = \int_0^t 2s\,ds = t^2. \quad (1.28)$$

When we exponentiate both sides of (1.28), we obtain

$$\frac{y(t)}{y(0)} = \frac{e^{\log(y(t))}}{e^{\log(y(0))}} = e^{\log(y(t)) - \log(y(0))} = e^{t^2},$$

so that

$$y(t) = y(0)e^{t^2}. \ ∎ \quad\quad\quad\quad (1.29)$$

Thus, (1.29) is the same as (1.27), except that the "arbitrary" constant C has been replaced by $y(0)$. ∎

Numerical Solutions

In scientific and engineering applications we are often less interested in a **formula** for a solution than we are in the **numerical values** of the solution. With easy access to computing facilities and attendant software, it becomes increasingly important to be familiar with numerical methods and their implementation. In Chapter 13 we shall discuss several numerical methods and their implementations in *Mathematica*.

Of course, if we have already obtained the solution in explicit form or by a series or integral, then the numerical computation is direct, depending on the evaluation of the functions obtained. Although it is generally preferable to have a simple formula for the solution of a differential equation, sometimes this is impossible. Nevertheless, in the course of our work we shall see that many first-order differential equations and systems of the form

$$y' = f(t, y)$$

can be solved numerically. We shall develop several numerical methods for solving $y' = f(t, y)$ in Chapter 13. It will be especially useful to compare the numerical results obtained with the true values of the solution in those cases in which the differential equation is solvable in terms of elementary functions. This will give us a greater appreciation of numerical methods and serve as a guide to what we may expect in cases when a solution in terms of elementary functions is unavailable.

Plotting Solutions

Even if the solution of a differential equation can be given symbolically or numerically, the nature of the solution is usually best understood by a picture. For this reason, the reader should attempt to plot all solutions either with *Mathematica* or by hand.

Exercises for Section 1.3

Compute the following indefinite integrals:

1. $\displaystyle\int t^{3/2}dt$

2. $\displaystyle\int t\sin(2t)dt$

3. $\displaystyle\int \frac{dt}{t(t-1)}$

4. $\displaystyle\int t^2 e^{-t^3}dt$

Compute the following definite integrals:

5. $\displaystyle\int_1^2 t^{3/2}dt$

6. $\displaystyle\int_a^b t\sin(2t)dt$

7. $\displaystyle\int_e^{2e} \frac{dt}{t(t-1)}$

8. $\displaystyle\int_0^\infty t^2 e^{-t^3}dt$

9. Show that $y(x) = (1/c)\cosh(c\,x)$ is a solution of (1.8).

2

USING MATHEMATICA

Mathematica is a very powerful tool for studying differential equations. It can be used to find both symbolic and numerical solutions of differential equations and to graph the solutions. Since no previous knowledge of *Mathematica* is presumed, in this chapter we give a brief introduction to its use. Details are given in relevant chapters as they are needed. In particular, the package **ODE.m**, the main package for this book for solving ordinary differential equations, is explained in Chapters 4 and 10.

Mathematica is a *symbolic manipulation program*; this means that not only numbers but also symbols can be used in calculations. A *Mathematica* session consists of a series of questions and answers. Although *Mathematica* notation differs in several crucial ways from ordinary mathematical notation, the translation between the two is straightforward. In addition to symbolic manipulation, *Mathematica* has powerful graphing capabilities. This important feature will allow us to graph solutions to differential equations with ease.

2.1 *Getting Started with* **Mathematica**

There are two types of implementations of *Mathematica*, (1) the textual (or command line) interface and (2) the Notebook (or graphical) interface.

The Notebook Front End

Notebook versions of *Mathematica* are available for PC's equipped with Microsoft Windows or NeXTstep, Macintoshes, NeXT workstations, and X-windows running on Unix workstations. In this section we explain the NeXT, Macintosh, and PC notebook interfaces; notebook interfaces on other computers are similar and can be found by clicking on the *Mathematica* "Help" menu.

17

To bring up *Mathematica*, double click on the *Mathematica* icon, which is a small polyhedron. This brings up a window called "Untitled"; it is actually a notebook. To enter a formula just start typing. For example, the polynomial $x^{15} - 1$ can be factored by typing

```
Factor[x^15 - 1]
```

The result is

$$(-1 + x) (1 + x + x^2)(1 + x + x^2 + x^3 + x^4)(1 - x + x^3 - x^4 + x^5 - x^7 + x^8)$$

Notebooks are divided into cells. Each formula is entered into a "cell", recognizable by a bracket on the right side of the window. After typing a formula into a cell, the formula can be evaluated by hitting the key "ENTER". (Notice that "ENTER" is different from "RETURN"; the latter is used to go to the next line of a cell. However, "SHIFT/RETURN" has the same effect as "ENTER".) Cells can be used to display either text or graphics and can be printed individually.

Help

To obtain help on any command, type **?** in front of the command and hit enter. For example, **?Plot** gives a short description of how to plot a function. Wild cards can also be used. Thus to see all commands that start with **D** type **?D***. To get more extensive help, use **??**.

The Function Browser

The Function Browser in the Help menu explains *Mathematica* functions and commands; it allows functions and commands to be pasted into cells, usually as templates. Although the function browser has been a feature on the NeXT and Macintosh notebook versions of *Mathematica* for some time, this will be a standard feature on all notebook versions of *Mathematica* beginning with version 3.0. On a NeXT or Macintosh it is also possible to access the Function Browser with the key combination "COMMAND-SHIFT-F". Furthermore, if the name of a function or command in a cell is selected with the mouse, the key combination "COMMAND-SHIFT-F" will open the function browser and give a short explanation.

Key Equivalences Between Platforms

On notebook versions of *Mathematica*, many frequently used menu options can also be performed with specific key combinations. For example, on the NeXT computer, all cells in a notebook can be selected with the key combination "COMMAND-SHIFT-A". On the Macintosh it would be "APPLE-SHIFT-A" and on the PC it would be "CONTROL-SHIFT-A". In general, whenever a reference is made to a key sequence on the NeXT using the

"COMMAND" key, you may substitute the "APPLE" key on a Macintosh or the "CONTROL" key when using the Microsoft Windows version on a PC.

Copying Input and Output from Above

A series of calculations frequently requires evaluating similar expressions, one after the other. Two key combinations greatly facilitate this process. The precise keystrokes can be obtained from the *Mathematica* "Help" Menu. In the case of NeXT workstations, the key combination "COMMAND-L" copies the input from the cell above. This new input can be edited and reevaluated. Sometimes the key combination "COMMAND-SHIFT-L", which copies the output from the cell above, is also useful.

Command Completion

A partially written command or function can be completed in the NeXT workstation with the key combination "COMMAND-K". For other computers, consult the *Mathematica* "Help" menu. If there is only one possible completion, *Mathematica* pastes it in at the insertion point; otherwise, a pop-up menu listing the possible completions is activated. Clicking on the selection completes the command or function.

Copying, Cutting, and Pasting

Highlighted text or cells can be copied to a clipboard in the NeXT workstation with with the key combination "COMMAND-C". For other computers, consult the *Mathematica* "Help" Menu.(The clipboard is a part of the computer memory that temporarily holds text and graphics. It is not usually possible to view the clipboard directly.) The contents of the clipboard can be copied to another location in a *Mathematica* notebook with key combination "COMMAND-V". The key combination "COMMAND-X" is similar to "COMMAND-C"; it also transfers text or cells to a clipboard, but "COMMAND-X" deletes the original material.

Printing

To print a whole notebook, click the mouse on the "Print" icon. To print an individual cell, first use the mouse to move the cursor to a cell bracket and click on it. Then move the cursor to the "Print Selection" icon and click the mouse button. This brings up a dialog box that allows the graphics to be sent to the printer.

Saving

Any *Mathematica* notebook (titled or untitled) can be saved to disk by selecting the Menu option labeled "Save", clicking on the "X" in the upper right corner of the window or by means of the key combination "COMMAND-S". To save a titled notebook under a different name, while retaining the current titled notebook, use "COMMAND-SHIFT-S". On all notebook versions of *Mathematica* except the Macintosh, a saved notebook consists of two files, one with an extension **.ma**, the other with the extension **.mb**. A **.ma** file is an ASCII file, which means that it can be accessed independently of *Mathematica* with a text editor (although a large number of formatting commands will be visible). On the other hand, a **.mb** file contains binary information. In most situations the information contained in a **.mb** file duplicates that of the corresponding **.ma** file, but the **.mb** information loads more quickly into a *Mathematica* session. To save disk space, **.mb** files can be deleted. Usually, **.mb** files created by *Mathematica* on different computers are incompatible with one another.

Graphics

A notebook contains several different kinds of information. In addition to *Mathematica* definitions and textual information, a notebook can contain graphics. For example, *Mathematica* has many plotting commands, each of which produces a graphic cell that contains the plot. The key combinations "COMMAND-C", "COMMAND-X" and "COMMAND-V" work with graphics cells exactly the same way that they work with cells containing text.

Textual Interface: Direct Access to **Mathematica** *Through a Terminal Window*

Except for older Macintosh versions, the notebook implementation of *Mathematica* is split into two parts: a kernel and a front end. The user communicates with the front end, which in turn sends messages back and forth to the kernel. It is also possible to access the *Mathematica* kernel directly through a terminal window. Moreover, for some implementations of *Mathematica* this is the only way to use the program. To use *Mathematica* in a terminal window simply type "math". Then all of the symbolic commands of *Mathematica* are available to the user. There is also a primitive form of graphics available called Terminal graphics. On most workstations graphics can also be displayed in a separate window. In the non-Windows version of *Mathematica* for PC's the display of graphics is full-screen.

The use of *Mathematica* in a terminal window is especially useful with modems. Typically, a user works on a small computer but uses a modem to connect via a telephone line to a more powerful computer. When *Mathematica* is used in a terminal window that communicates over a telephone line to a large computer equipped with *Mathematica*, all of *Mathematica*'s symbolic capabilities are available. Although the transfer of data via telephone lines is usually too slow for high quality graphics, primitive graphics display is possible in a terminal window.

2.2 *Mathematica* Notation versus Ordinary Mathematical Notation

Parentheses

Parentheses, as used in ordinary mathematics, have at least four distinct meanings according to the context. One of these is the notation for an open interval on the real number line:

$$(a, b) = \{ t \in \mathbb{R} \mid a < t < b \};$$

This formalism is not needed by *Mathematica*. Different notation is employed by *Mathematica* for each of two of the other three uses of parentheses.

Use of Parenthesis	Ordinary Mathematical Notation	*Mathematica* Notation
Grouping	$a(b + c)$	**a(b+c)**
Point in $\mathbb{R}^2$	(p, q)	**{p,q}**
Function f applied to a variable t	$f(t)$	**f[t]**

Defining Functions

Mathematica usually uses **:=** instead of **=** in function definitions. Thus to define in *Mathematica* a function **y** that assigns a real number or some other expression, we use

y[t_]:= some expression in **t**.

Here the underscore after the **t** is an important part of the syntax. Whereas **t** denotes the symbol t, **t_** means generic t. Note that square brackets are used instead of parentheses. For example, if we define

y[t_]:= t^2 + 3t + 6

then we can use this form numerically. Thus

y[2]

elicits the response

16

On the other hand, if we evaluate **y** on a symbol, the result is a symbol; thus

y[s]

yields

```
        2
6 + 3 s + s
```

The Equal Sign

Mathematica also has different symbols for equality depending on the usage in ordinary mathematics. Here are some examples.

Use of Equality	Ordinary Mathematical Notation	*Mathematica* Notation
Assignment	$a = 5$	**a = 5**
Defining a function	$f(x) = e^x \sin(x)$	**f[x_]:= E^x Sin[x]**
An equation to be solved	$2x + 3y = 5$	**2x + 3y == 5**

On rare occasions *Mathematica* also employs a triple equals "**===**". For an example of its use, see subSection 10.2.1.

Multiplication and Division

Most of the mathematical operations, for example addition (+), subtraction (-), division (/), and exponentiation (^), use the symbols that have become standard in modern programming languages, but multiplication in *Mathematica* is represented by either an asterisk ***** or a space. For aesthetic reasons we usually prefer to denote multiplication by a space. However, it is sometimes necessary to use an asterisk for multiplication at the end of an intermediate line in the middle of a multi-line expression.

Notice that **xy** represents a single expression in *Mathematica*; it is never the same as **x y**, when both **x** and **y** are symbols. But the following expressions are all the same in *Mathematica*:

2y   **2 y**   **2*y**

because they indicate multiplication of a symbol by a number. Note also that **x2** denotes a single expression, whereas **x x2** denotes **x** times **x2**.

   Mathematica uses the same symbol for division as ordinary mathematics, namely **/**. However, one must be careful to write

$$\frac{1}{2x}$$

as **1/(2x)** or **1/2/x**. *Mathematica* interprets **1/2x** as

$$\left(\frac{1}{2}\right)x.$$

Universal Constants and Numerical Values

The numbers $e \approx 2.71828\dots$, $\pi \approx 3.14159\dots$, and $i = \sqrt{-1}$ are represented in *Mathematica* by **E**, **Pi**, and **I**. *Mathematica* distinguishes between symbolic values and numerical values. A numerical approximation to π can be found with the command **N[Pi]**:

```
3.14159
```

Mathematica gives exact results whenever possible. Although *Mathematica* responds to **Sin[1]** by repeating the expression, *Mathematica*'s answer to **N[Sin[1]]** is the six-digit decimal approximation

```
0.841471
```

Nevertheless, some expressions such as **Sin[0]** and **Sin[Pi/4]** are automatically simplified, since their values are well-known. Anytime a numerical value is required, apply the operator **N**.

Lists

A list consisting of three elements   **a1, a2, a3**   is denoted in *Mathematica* by **{a1,a2,a3}**. Note that braces are used instead of parentheses. Thus a list of n numbers can be thought of as a point or vector in Euclidean space $\mathbb{R}^n$. Lists of vectors form a matrix; see Appendix B.

Internal Functions

All internal *Mathematica* functions begin with a capital letter. Thus the *Mathematica* notation for $\sin(x)$ is **Sin[x]**. A user-defined function can begin with either a lowercase or an uppercase letter, but if it coincides with an internal function, *Mathematica* probably will not work properly. For this reason, it is a good idea to use lower case letters to define new functions to avoid collision with *Mathematica*'s internally defined functions.

Differentiation

In ordinary mathematics there are two symbols for differentiation. We write

$$\frac{d}{dt}$$

for the total derivative with respect to t, when t is the only independent variable. On the other hand,

$$\frac{\partial}{\partial t}$$

is used for the partial derivative with respect to t, when t is one of several independent variables. If we assume that $y = y(t)$ is differentiable with respect to t, then the total derivative of y with respect to t can be computed by applying the operator d/dt to y, usually written as

$$y' = \frac{dy}{dt}, \qquad y'' = \frac{d^2 y}{dt^2},$$

and so forth. If y is a function of only one variable, then the partial derivative and the total derivative are the same.

Mathematica also has two symbols for differentiation. **Dt**, read "derivative total", and **D** for partial differentiation. In operator notation,

$$\frac{d}{dt} \quad \longleftrightarrow \quad \mathtt{Dt[\ ,t]}, \qquad\qquad \frac{d^n}{dt^n} \quad \longleftrightarrow \quad \mathtt{Dt[\ ,\{t,n\}]},$$

$$\frac{\partial}{\partial t} \quad \longleftrightarrow \quad \mathtt{D[\ ,t]}, \qquad\qquad \frac{\partial^n}{\partial t^n} \quad \longleftrightarrow \quad \mathtt{D[\ ,\{t,n\}]}.$$

D is used much more frequently in *Mathematica* than **Dt**. To see the difference between **Dt[,t]** and **D[,t]**, we define a function $y = y(t)$ in *Mathematica* by

```
y[t_]:= t^2 + 3t + a
```

Then

```
Dt[y[t],t]
```

yields

```
3 + 2 t + Dt[a, t]
```

because **a** is considered a function of **t**. On the other hand,

```
D[y[t],t]
```

results in

```
3 + 2 t
```

because the use of **D[,t]** tells *Mathematica* to treat **t** as the only variable for the purposes of differentiation.

The Prime Notation

In the case of a function **f** of a single variable **t** , *Mathematica* has another notation for the derivative of **f** with respect to **t** , namely **f' [t]**. This notation is useful since it imitates ordinary mathematical notation closely. For example, if we define

```
f[t_]:= a t^2 + b t + c
```

then *Mathematica*'s response to **f' [t]** is

```
b + 2 a t
```

Integration

Just as in ordinary mathematics there are two kinds of integration in *Mathematica*: indefinite and definite. To compute

$$\int t\, e^t \sin(t) dt$$

we use

```
Integrate[t E^t Sin[t],t]
```

The answer is

```
 t            t            t
E  Cos[t] - E  t Cos[t] + E  t Sin[t]
─────────────────────────────────────
                  2
```

Mathematica suppresses the constant of integration. For definite integration *Mathematica* also uses the command **Integrate**, but with a different syntax. To find

$$\int_0^1 t\sqrt{1-t}\, dt$$

we use

```
Integrate[t Sqrt[1-t],{t,0,1}]
```

to obtain

```
 4
───
 15
```

Many definite integrals are too complicated to compute symbolically. In such cases **NIntegrate** can be used. To find[1]

$$\int_0^1 (\sin(t))^{1/3} dt$$

we enter the command

NIntegrate[Sin[t]^(1/3),{t,0,1}]

The approximate value of the integral is

```
0.733269
```

The command **NIntegrate** can be quite slow. We shall see in Chapter 13 that the command **NDSolve**, which we shall introduce for finding numerical solutions of differential equations, can usually be used as a faster alternative for **NIntegrate**.

Simplifying Expressions

Mathematica's command **Simplify** attempts to simplify expressions. For example

D[(t^4 + 1)/(t^4 - 1),t]

gives the complicated answer

```
    3          3         4
  4 t       4 t  (1 + t )
 ------  -  -----------
       4          4 2
 -1 + t      (-1 + t )
```

We get a much nicer result when we apply **Simplify**:

Simplify[D[(t^4 + 1)/(t^4 - 1),t]]

```
      3
  -8 t
 --------
      4 2
(-1 + t )
```

An equivalent form of **Simplify[expression]** is **expression//Simplify**, which is useful when one forgets to write **Simplify** before writing **expression**. Also, useful is **Simplify[%]**, which simplifies the previous output.

Mathematica contains several other commands that simplify in different ways. For example, **Together** combines fractions, and **Apart** rewrites a rational expression as a sum of terms with minimal denominators.

---

[1]Notice that we use the notation $(\sin(t))^{1/3}$ instead of $\sin^{1/3} t$. In the first place, $(\sin(t))^{1/3}$ more closely corresponds to *Mathematica*'s notation. Secondly, $(\sin(t))^{1/3}$ is more logical than $\sin^{1/3} t$, because it is the quantity $\sin(t)$ that is being raised to the $1/3$ power, not the function sin. We avoid completely the notation $\sin^{-1}$, writing arcsin instead.

Clearing Values

Mathematica retains a value assigned to a variable unless it is specifically cleared.

```
Clear[symbol1, symbol2,...]
```

clears values for the specified symbols. A useful shorthand for **Clear** is "**=.**". On the other hand,

```
Remove[symbol1, symbol2,...]
```

removes symbols completely, so that their names are no longer recognized by *Mathematica*. To clear or remove all variables starting with a lower case letter use **Clear["Global`@*"]** or **Remove["Global`@*"]**.

Solving Equations

Linear and other simple algebraic equations can be solved with **Solve**. Here are two examples:

```
Solve[{x + y == 1, x - y == 2},{x,y}]
```

```
        3          1
{{x -> -, y -> -(-)}}
        2          2
```

```
Solve[x^2 - 5 x + 6 == 0,x]
```

```
{{x -> 2}, {x -> 3}}
```

For more information on **Solve** see Section A.8 of Appendix B.

Functional Notation in Mathematica

Mathematica, like ordinary mathematics, distinguishes between functions and the values of functions. **Exp** is a function whose value on **z** (which can be either a number or a symbol) is **Exp[z]**. (Note that **Exp[z]** is the same as **E^z**.) One way to generate functions is to use *Mathematica*'s command **Function**. For example, the function that assigns z^2 to z is **Function[z,z^2]**.

Mathematica has another way of representing functions which requires less writing. In this shorthand notation the pound sign **#** is a place holder in a function and the ampersand **&** terminates the function. So another way to write **Function[z,z^2]** is **#^2&**. For example, both **Function[z,z^2][3]** and **#^2&[3]** give the result

9

For more details see Wolfram's book **Mathematica** [Wm1, pages 207–209].[2]

Exercises for Section 2.2

Work the following exercises with *Mathematica*, using **Simplify** when it is appropriate.

1. Express the differential equation $y''(t) + t\, y'(t) + (1 + t^2)y(t) = t/(1 + t)$ using the *Mathematica*'s "prime" notation. Do not evaluate the expression.

2. Express the differential equation in Problem 1 using the operator **D**. Do not evaluate the expression.

3. Define the function $f(t) = 1 + t + \sin(t) + e^t - \sqrt{5 - t^2}$ and compute the exact and numerical values of $f(2)$.

4. Define the function $f(t, c) = \cos(t) + t^{3/2} + \dfrac{2}{3}c$.

5. Compute the exact value of
$$\frac{1}{3 \cdot 5 \cdot 7 \cdot 9};$$
    then use **N** to find a numerical approximation.

6. Compute the following derivatives:
$$\frac{d}{dt}\left(\frac{\log(\sin(t))}{\tan(t)}\right), \qquad \frac{d^2}{dt^2}\left(\frac{t^3 + 1}{t\,\sin(t)}\right).$$

7. Compute the following integrals:
$$\int \left(t^3\, e^{5t}\, \cos(t)\right)dt, \qquad \int_0^\pi (\sin(t))^7 dt.$$

---

[2]The bibliography begins on page 869. Since it explains the software *Mathematica*, the clearly written book **Mathematica** is the basic reference for software questions.

8. Verify the fundamental theorem of calculus for $g(t) = t\,e^t + \sec(t)$. [In other words, show that the derivative of the integral of $g(t)$ equals $g(t)$.]

9. Write the function f defined by $f(t) = t^3 + \tan(t)$ as a *Mathematica* function and compute the symbolic and numerical values of $f(\pi/4)$.

2.3   *Plotting in **Mathematica***

Mathematica has several different commands for plotting.

Using `Plot` *to Graph Functions*

Functions of one variable can be plotted with *Mathematica*'s command `Plot`. For example, the *Mathematica*'s command to plot $t\,\sin(t)$ from -2π to 2π is

```
Plot[t Sin[t],{t,-2Pi,2Pi}]
```

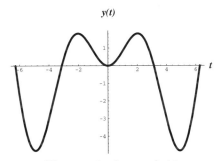

The graph of $y = t\sin(t)$

The two basic parts of `Plot` are the function specification (in this case `t  Sin[t]`) and the domain (in this case `{t,0,-2Pi,2Pi}`). In addition, it is possible to use plotting options to modify the display generated by `Plot`. The most important option is `PlotPoints`. For example, the graph drawn by

```
Plot[t Sin[t^3],{t,1,4.5}]
```

is not good:

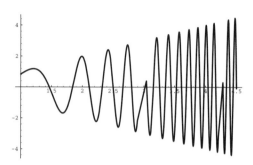

The problem is that *Mathematica*'s plotting algorithm has chosen too few points. We get a much better result when we increase the resolution of the graph by changing **PlotPoints** from its default value of 25 to 100:

```
Plot[t Sin[t^3],{t,1,4.5},PlotPoints->100]
```

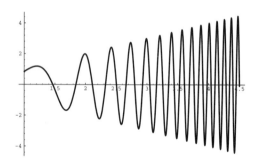

Using **ParametricPlot** *to Plot Plane Curves*

Mathematica's command **ParametricPlot** is used for plotting curves in the plane. For example, the following command plots a cardioid:

```
ParametricPlot[{2Cos[t](1 + Cos[t]),2(1 + Cos[t])Sin[t]},
{t,0,2Pi},AspectRatio->Automatic]
```

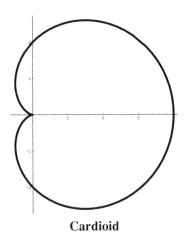

Cardioid

The option **AspectRatio->Automatic**, which makes the vertical and horizontal scales the same, should always be used with **ParametricPlot**.

Using **ContourPlot** *to Plot Plane Curves*

Several contours defined implicitly by an equation of the form $F(x, y) = c$ can be plotted using *Mathematica*'s command **ContourPlot**. For example,

```
ContourPlot[Cos[x]Cos[y],{x,-2Pi,2Pi},{y,-2Pi,2Pi},
PlotPoints->100]
```

gives the plot of $\cos(x)\cos(y) = c$ for various values of c:

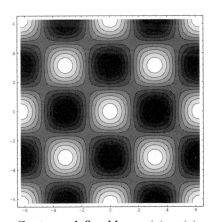

Contours defined by cos(x)cos(y)

For more information on parametrically defined curves and implicitly defined curves, see Section 17.1.

Using **ParametricPlot3D** *to Plot Space Curves*

Mathematica's command for plotting space curves is **ParametricPlot3D**. For example, the following command plots a helix:

```
ParametricPlot3D[{Cos[t],Sin[t],0.3t},{t,0,6Pi}]
```

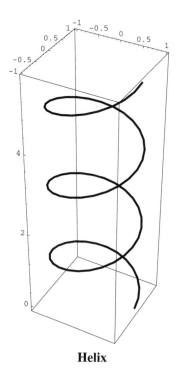

Helix

Exercises for Section 2.3

Do the following plots with *Mathematica*.

   1. Plot $\left(\sin(t^2)\right)^2$ from -2π to 2π.

2. Use **ParametricPlot** to plot the following plane curves.

   a.   The **ellipse** as $t \longrightarrow \big(\sin(t), 2\cos(t) \big)$, with $0 \le t \le 2\pi$.

   b.   The **parabola** as $t \longrightarrow \big(4t, t^2 \big)$, with $-2 \le t \le 2$.

   c.   The **astroid** as $t \longrightarrow \big((\sin(t))^3, (\cos(t))^3 \big)$, with $0 \le t \le 2\pi$.

3. Viviani's curve is the space curve $t \longrightarrow \big(1 + \cos(t), \sin(t), 2\sin(t/2) \big)$.   Use **ParametricPlot3D** to plot it from -2π to 2π.

4. Use **ContourPlot** to plot contours defined by $x^2 y^3$. Use the ranges $-2\pi \le x \le 2\pi$ and $-2\pi \le y \le 2\pi$.

3

FIRST-ORDER DIFFERENTIAL EQUATIONS

Having established the basic concepts, we now proceed to identify a class of first-order differential equations that can be solved by the methods of calculus. This will serve as a foundation for all of the future mathematics as well as eventually permit us to develop some interesting applications based on the solution formulas obtained. As a platform for the entire chapter, we first formulate the initial value problem in Section 3.1.

The most basic class comprises the first-order linear equations that are completely solved explicitly in Section 3.2 by two integrations, where the *integrating factor* plays a central role. In Section 3.3 we focus attention on a class of nonlinear first-order equations, the *separable equations*, which can also be solved (implicitly) by means of two integrations. Both linear and separable first-order equations have many applications, to be studied in Chapter 6. We conclude this chapter with three other classes of first-order equations that are solvable by calculus: the exact equations (Section 3.4), the homogeneous equations (Section 3.5), and the Bernoulli equations (Section 3.6).

3.1 Introduction to First-Order Equations

The most general form of a first-order (ordinary) differential equation is

$$F(t, y, y') = 0, \tag{3.1}$$

where F is a function whose first partial derivatives are continuous. We shall usually assume that (3.1) can be solved for y'; in other words, (3.1) can be rewritten in the **standard form**

$$y' = f(t, y). \tag{3.2}$$

The implicit function theorem from calculus implies that such a reduction is possible under the mild assumption that $\partial F/\partial y'$ does not vanish. There very well may be several differential equations of the form (3.2) that correspond to one equation of the form (3.1). When this is the case, we attempt to find all of them and study them individually.

Example 3.1. *Reduce the differential equation $2(y')^2 - 10y' + 12 = 0$ to the standard form* (3.2).

Solution. Since $2(y')^2 - 10y' + 12$ factors as $2(y' - 2)(y' - 3)$, the study of the differential equation $2(y')^2 - 10y' + 12 = 0$ is equivalent to the study of the differential equations

$$y' = 2 \quad \text{and} \quad y' = 3. \ \blacksquare$$

A special case of the differential equation (3.2) is already familiar from calculus, namely the case when $f(t, y)$ does not depend on y. After renaming $f(t, y)$ as $q(t)$, (3.2) becomes

$$y' = q(t). \tag{3.3}$$

We want to find a function $y(t)$ whose derivative is $q(t)$. For that we take the indefinite integral of both sides of (3.3):

$$y(t) = \int q(t)dt + C, \tag{3.4}$$

where C is a constant of integration. We call (3.4) the **general solution** of (3.3).

Example 3.2. *Find the general solution to $y' = \sin(3t) + \cos(t)$.*

Solution. We have

$$y(t) = \int \left(\sin(3t) + \cos(t) \right) dt = -\frac{1}{3}\cos(3t) + \sin(t) + C, \tag{3.5}$$

where C is a constant of integration. $\blacksquare$

The constant C in (3.4) can be determined, provided that we know the value Y_0 of y at some t_0, that is, if we know that $y(t_0) = Y_0$. Such a condition is called an **initial condition**.

Definition. *A first-order differential equation together with an initial condition is called a (first-order)* **initial value problem**.

Substitution of the initial condition $y(t_0) = Y_0$ into (3.4) yields

$$Y_0 = \int q(t)dt \Big|_{t=t_0} + C. \tag{3.6}$$

Obviously (3.6) can be solved for C.

Example 3.3. *Solve the initial value problem*

$$\begin{cases} y' = (\sin(t))^2, \\ y(\pi) = 1. \end{cases} \tag{3.7}$$

Solution. We compute

$$y(t) = \int (\sin(t))^2 dt = \int \frac{1 - \cos(2t)}{2} \, dt = \frac{t}{2} - \frac{\sin(2t)}{4} + C, \tag{3.8}$$

where C is a constant of integration. We substitute the initial condition $y(\pi) = 1$ into (3.8) to obtain

$$1 = \frac{\pi}{2} + C. \tag{3.9}$$

When we solve (3.9) for C and substitute into (3.8), we get

$$y = \frac{t}{2} - \frac{\sin(2t)}{4} + 1 - \frac{\pi}{2}. \ \blacksquare$$

We can also solve an initial value problem corresponding to (3.3) by means of definite integration. First, we change the variable t to u in (3.3):

$$y'(u) = q(u). \tag{3.10}$$

Integrating both sides of (3.10) from t_0 to t yields

$$y(t) - y(t_0) = \int_{t_0}^{t} q(u) \, du, \quad \text{or} \quad y(t) = y(t_0) + \int_{t_0}^{t} q(u) \, du. \tag{3.11}$$

Example 3.4. *Use definite integration to solve the initial value problem*

$$\begin{cases} y' = \log(t), \\ y(1) = 2. \end{cases} \tag{3.12}$$

Solution. In (3.11) we take $t_0 = 1$ and $q(t) = \log(t)$. The result is

$$y(t) = y(1) + \int_{1}^{t} \log(u)du = 2 + \left(u \log(u) - u \right) \Big|_{1}^{t} = 3 + t \log(t) - t. \ \blacksquare$$

Of course, the general equation (3.2) can be much more complicated than (3.3), so much so that we cannot hope to find an integral formula for the general solution. Nevertheless, there are many special cases when either a formula or at least a procedure can be given to find the solution of (3.2); this is the subject of the present chapter. After the general solution of (3.2) has been found, the constant of integration is easy to determine from any initial condition.

Exercises for Section 3.1

Find the general solution of each of the following first-order differential equations.

1. $y' = (t^3 + 2)^3 t^2$ **3.** $y' = t \sin(t)$

2. $y' = e^{3t}$ **4.** $t\, y' = 2$

Solve the following initial value problems.

5. $\begin{cases} y' = t^3 + 4t, \\ y(0) = 1 \end{cases}$ **7.** $\begin{cases} y' = t\, e^{3t}, \\ y(1) = 0 \end{cases}$

6. $\begin{cases} y' = \dfrac{\sin(t)}{\cos(t)}, \\ y(\pi/4) = 1 \end{cases}$ **8.** $\begin{cases} t\, y' = \log(t), \\ y(1) = 0 \end{cases}$

Mathematica's command **Integrate** can be used to solve differential equations of the form (3.3). To find the general solution of the differential equation in Example 3.2 we use

```
Integrate[Sin[3t] + Cos[t],t]
```

to get

$$\frac{-\text{Cos}[3\ t]\ +\ 3\ \text{Sin}[t]}{3}$$

This expression, except for the constant of integration, which has been omitted, is the same as the right-hand side of (3.5).

9. Use **Integrate** to solve problems 1–4.

3.2   First-Order Linear Equations

First-order linear equations constitute one of the most important types of first-order equations, because they frequently arise in applications. In this section we solve the first-order linear equation

$$a(t)y' + b(t)y = c(t),\tag{3.13}$$

where $a(t)$ is nonzero for t in some interval. First, let us reduce (3.13) to the **leading-coefficient-unity-form**

$$y' + p(t)y = q(t)\tag{3.14}$$

by dividing both sides of (3.13) by $a(t)$. Thus

$$p(t) = \frac{b(t)}{a(t)} \quad \text{and} \quad q(t) = \frac{c(t)}{a(t)}.$$

We reduce (3.13) to (3.14) because (3.13) involves the three functions a, b, and c, while (3.14) involves only the two functions p and q.

The most obvious way to try to solve (3.14) is to integrate both sides of the equation. If q is a simple function such as a polynomial or trigonometric function, we can integrate the right-hand side of (3.14) by standard integration techniques from calculus. Thus if p vanishes identically, we can easily solve (3.14) simply by integrating both sides of the equation. This was the approach that we used in Section 3.1.

However, if p does not vanish identically, standard integration techniques do not help us with the integration of the left-hand side of (3.14). Therefore, we are forced to look for a trick in order to solve (3.14). We proceed as follows. We first multiply both sides of (3.14) by an unknown function μ with the aim of transforming (3.14) into an equation that is simple to integrate:

$$\mu(t)\big(y'(t) + p(t)y(t)\big) = \mu(t)q(t).\tag{3.15}$$

If we are lucky, the left-hand side of (3.15) becomes the derivative of a product, and we get

$$\frac{d}{dt}\big(\mu(t)y(t)\big) = \mu(t)q(t).\tag{3.16}$$

Then we can integrate both sides of (3.16):

$$\mu(t)y(t) = \int \frac{d}{dt}\big(\mu(t)y(t)\big)dt = \int \mu(t)q(t)dt + C_0,$$

so that

$$y(t) = \frac{1}{\mu(t)}\left(\int \mu(t)q(t)dt + C_0\right),\tag{3.17}$$

where C_0 is a constant of integration. Then (3.17) is called the **general solution** of (3.14); it is given in terms of the function μ, which we still need to determine. We call μ an **integrating factor**.

How do we go about finding an integrating factor in (3.15)? Certainly we must have

$$\mu(t)\big(y'(t) + p(t)y(t)\big) = \frac{d}{dt}\big(\mu(t)y(t)\big) = \mu'(t)y(t) + \mu(t)y'(t). \qquad (3.18)$$

It is easy to simplify (3.18) algebraically to

$$\frac{\mu'(t)}{\mu(t)} = p(t). \qquad (3.19)$$

But we can integrate both sides of (3.19):

$$\log\big(\mu(t)\big) = \int p(t)dt + C. \qquad (3.20)$$

We might as well make our integrating factor μ as simple as possible, so we declare the constant of integration C in (3.20) to be zero. (It can be proved that taking the constant of integration to be nonzero does not yield additional solutions to (3.14).) Taking exponentials of both sides of (3.20) yields

$$\mu(t) = e^{\int p(t)dt} = \exp\left(\int p(t)dt\right). \qquad (3.21)$$

In fact, (3.21) is the formula for the integrating factor of (3.14). Thus the solution of (3.14) is given by (3.17) and (3.21).

Here is a summary of the solution process to find the general solution of the first-order differential equation (3.13):

To find the general solution of $a(t)y' + b(t)y = c(t)$ carry out the following steps:

Step 1: Write the differential equation in the leading-coefficient-unity-form $y' + p(t)y = q(t)$.

Step 2: Compute the integrating factor $\mu(t) = \exp\left(\int p(t)dt\right)$.

Step 3: Compute the integral $\int \mu(t)q(t)dt$.

Step 4: The general solution is $y(t) = \dfrac{1}{\mu(t)}\left(\int \mu(t)q(t)dt + C_0\right)$, where C_0 is a constant of integration.

Example 3.5. *Find the general solution of the equation*

$$t\, y' + 5y = 3t. \tag{3.22}$$

Solution. To solve (3.22), we first write it in the leading-coefficient-unity-form

$$y' + \frac{5}{t} y = 3. \tag{3.23}$$

An integrating factor for (3.23) is given by

$$\mu(t) = \exp\left(\int \frac{5}{t} dt\right) = e^{C + 5\log(t)} = e^C\, t^5. \tag{3.24}$$

The constant C is immaterial and may be taken equal to 0; then $\mu(t) = t^5$. Multiplying (3.23) by t^5 gives

$$(t^5 y)' = 3t^5. \tag{3.25}$$

Both sides of (3.25) can be easily integrated to yield

$$t^5 y = \frac{t^6}{2} + C_0, \qquad \text{or} \qquad y(t) = \frac{t}{2} + \frac{C_0}{t^5}, \tag{3.26}$$

where C_0 is a constant of integration. $\blacksquare$

(Although the constant of integration in (3.24) may be assumed to be zero, it would be a mistake to take the constant of integration in (3.26) to be zero.)

Linear differential equations have the property that the solutions are **global**, meaning that they exist in any region in which the coefficient functions $p(t)$ and $q(t)$ in (3.14) are continuous. In particular, if $p(t)$ and $q(t)$ are continuous on the entire line $-\infty < t < \infty$, then a solution $y(t)$ also exists on the entire line $-\infty < t < \infty$. Similarly, a solution to (3.23) is defined on any subset of $-\infty < t < \infty$ that does not contain 0. We shall see in Section 5.2 that solutions of nonlinear differential equations may not be global.

First-Order Linear Equations with Initial Conditions

We shall also need to solve $a(t)y' + b(t)y = c(t)$ when it is given together with an initial condition. As we know from page 36, an initial condition is the specification $y(t_0) = Y_0$ for some t_0. The initial condition serves to determine the constant of integration C_0 that appeared when we obtained the solution (3.17). It is easy to find C_0: we just substitute $t = t_0$ in both sides of (3.17) and use the fact that $y(t_0) = Y_0$. The resulting linear equation can be solved for C_0.

Recall (see page 36) that a first-order differential equation together with an initial condition is called a (first-order) initial value problem.

Example 3.6. *Solve the initial value problem*

$$\begin{cases} t\,y' + 5y = 3t, \\ y(1) = 0. \end{cases} \tag{3.27}$$

Solution. We already found the general solution of (3.27) in Example 3.5, namely $y(t) = t/2 + C_0/t^5$. Now we substitute $t = 1$ into (3.26) and use $y(1) = 0$; we obtain

$$0 = \frac{1}{2} + C_0.$$

Therefore, $C_0 = -1/2$, and so

$$y(t) = \frac{t}{2} - \frac{1}{2t^5}. \quad \blacksquare$$

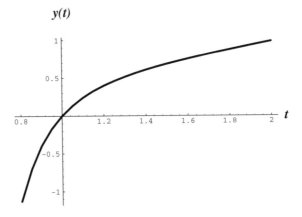

Solution to $t\,y' + 5y = 3t,$ $y(1) = 0$,
plotted over the interval $0.8 < t < 2$

We can also solve the initial value problem corresponding to a first-order linear equation (3.14) by means of definite integration. Let the initial condition be given by $y(t_0) = Y_0$. We make a specific choice of integrating factor that satisfies $\mu(t_0) = 1$, namely

$$\mu(t) = \exp\!\left(\int_{t_0}^{t} p(u)\,du\right).$$

With this choice of integrating factor, we take the definite integral from t_0 to t of both sides of (3.16), obtaining

$$\mu(t)\,y(t) - \mu(t_0)\,y(t_0) = \int_{t_0}^{t} \mu(u)q(u)\,du. \tag{3.28}$$

But $\mu(t_0) = 1$ by construction, and $y(t_0)$ is required to be Y_0. Thus, we can write the solution of (3.14) in the form

$$y(t) = \frac{1}{\mu(t)}\left(Y_0 + \int_{t_0}^t \mu(u)q(u)du\right). \tag{3.29}$$

We can summarize the above work in the form of a theorem.

Theorem 3.1. (Existence and uniqueness theorem) *Let $p(t)$ and $q(t)$ be continuous functions on an interval $a < t < b$, and let t_0 be some point in this interval. Then for any real number Y_0, the first-order linear differential equation*

$$y' + p(t)y = q(t) \tag{3.30}$$

has a unique solution defined for $a < t < b$ that satisfies the condition $y(t_0) = Y_0$.

Proof. The derivation of (3.29) proves the existence of a solution $y(t)$ of (3.30) satisfying $y(t_0) = Y_0$. The uniqueness of the solution is proved as follows. Suppose that $y_1(t)$ and $y_2(t)$ are two functions that satisfy (3.30) and pass through (t_0, Y_0), that is, $y_1(t_0) = y_2(t_0) = Y_0$. The function $y_3(t)$ defined as the difference $y_3(t) = y_1(t) - y_2(t)$ satisfies the differential equation

$$y_3' + p(t)y_3 = 0 \tag{3.31}$$

and the initial condition $y_3(t_0) = 0$. Multiplying (3.31) by the integrating factor

$$\mu(t) = \exp\left(\int_{t_0}^t p(u)\,du\right)$$

and integrating yields

$$\mu(t)y_3(t) = C, \tag{3.32}$$

where C is a constant. Since $y_3(t_0) = 0$, the constant must be zero, and we conclude from (3.32) that $y_3(t)$ is identically zero. Thus $y_1(t) - y_2(t) \equiv 0$, which proves that the two solutions must coincide. ∎

Here is a summary of the steps to follow to solve an initial value problem corresponding to the first-order differential equation (3.13):

Carry out the following steps to solve the initial value problem

$$\begin{cases} a(t)y' + b(t)y = c(t), \\ y(t_0) = Y_0 : \end{cases}$$

Step 1: Write the differential equation in the leading-coefficient-unity-form $y' + p(t)y = q(t)$.

Step 2: Compute the integrating factor $\mu(t) = \exp\left(\int p(t)dt\right)$.

Step 3: Compute the integral $\int \mu(t)q(t)dt$.

Step 4: The solution is then $y(t) = \dfrac{1}{\mu(t)}\left(Y_0\mu(t_0) + \int_{t_0}^{t} \mu(u)q(u)du\right)$.

Summary of Techniques Introduced

A first-order linear equation can always be solved by means of an integrating factor. The general solution is expressed in terms of two integrations—the first to find the integrating factor and the second integration to solve the equation. The solution involves only one arbitrary constant. This constant is uniquely determined by an initial condition.

Exercises for Section 3.2

In each of Problems 1–10 find the general solution of the differential equation.

1. $y' + 4y = 2t + 3e^{3t}$

2. $y' - 2y = t\,e^{3t}$

3. $y' + 2t\,y = 3t\,e^{-t^2}$

4. $y' + \dfrac{1}{t}y = e^{t^2}$

5. $t\,y' + 5y = t^3$

6. $(1 + t^2)y' + 9y = 0$

7. $y' - y\,e^t = 2t\,e^{e^t}$

8. $y' + y = \sin(t)$

9. $y' + y\cot(t) = 5e^{\cos(t)}$

10. $y' + 2y\cos(t) = (\sin(t))^2\cos(t)$

In each of Problems 11–20 solve the initial value problem.

11. $\begin{cases} y' + 4y = 2t + 3e^{3t}, \\ y(0) = 5 \end{cases}$

16. $\begin{cases} (1 + t^2)y' + 9y = 0, \\ y(3) = 1 \end{cases}$

12. $\begin{cases} y' - 2y = t\, e^{3t}, \\ y(3) = 1 \end{cases}$

17. $\begin{cases} y' - y\, e^t = 2t\, e^{e^t}, \\ y(0) = 1 \end{cases}$

13. $\begin{cases} y' + 2t\, y = t^3 e^{-t^2}, \\ y(0) = 4 \end{cases}$

18. $\begin{cases} y' + \dfrac{3}{t} y = \dfrac{\sin(t)}{t^3}, \\ y(\pi) = 0 \end{cases}$

14. $\begin{cases} y' + \dfrac{1}{t} y = e^{t^2}, \\ y(1) = 2 \end{cases}$

19. $\begin{cases} t\, y' + 2y = \cos(t), \\ y(\pi/2) = 0 \end{cases}$

15. $\begin{cases} t\, y' + 5y = t^3, \\ y(1) = 1 \end{cases}$

20. $\begin{cases} \cos(t)y' - \sin(t)y = 1, \\ y(0) = 1 \end{cases}$

21. Find the general solution of the differential equation

$$\frac{dy}{dt} = \frac{1}{e^y - t}. \tag{3.33}$$

[Hint: Consider t as a function of y.]

22. (Continuation of Exercise 21) Solve the differential equation (3.33) with the initial condition $y(t_0) = Y_0$, where t_0 and Y_0 are such that $e^{Y_0} \neq t_0$.

23. (Continuation of Problem 21) Solve the differential equation (3.33) with the initial condition $y(t_0) = Y_0$, where t_0 and Y_0 are such that $e^{Y_0} = t_0$. Show that there are *two* solutions of (3.33) satisfying $y(t_0) = Y_0$, but these solutions are only defined for $t^2 \geq t_0^2$.

24. Prove that the following **superposition principle** holds for first-order linear equations: if y_1 is a solution of the equation

$$y_1' + p(t)y_1 = q_1(t)$$

and y_2 is a solution of the equation

$$y_2' + p(t)y_2 = q_2(t),$$

then $y = y_1 + y_2$ is a solution of the equation

$$y' + p(t)y = q_1(t) + q_2(t).$$

25. Show that the following **subtraction principle** holds for first-order linear equations: if y_1 and y_2 are both solutions of the equation $y' + p(t)y = q(t)$, then the difference $y = y_1 - y_2$ is a solution of the differential equation $y' + p(t)y = 0$.

26. Use Problem 24 to show that the solution of the initial value problem for

$$y' + p(t)y = q(t)$$

with $y(t_0) = Y_0$ can be written in the form $y = y_P + y_H$, where y_P is the particular solution of the equation with $y(t_0) = 0$ and y_H is the solution of the differential equation $y' + p(t)y = 0$ with $y(t_0) = Y_0$.

27. This problem gives a version of the method **variation of parameters** (see Section 9.4) that is suited to a first-order linear equation. Suppose that y_H is a solution of the differential equation $y' + p(t)y = 0$ and that y is a solution of the first-order linear equation $y' + p(t)y = q(t)$; define $z(t) = y/y_H$.

    a. Show that $z(t)$ satisfies the equation $z'(t) = q(t)/y_H(t)$.

    b. Use part **a** to show that $y(t) = y_H(t) \int \left(\dfrac{q(t)}{y_H(t)} \right) dt.$

28. This exercise illustrates the possibility of uniquely solving an equation by means of conditions at infinity.

    a. Find the general solution of the equation

    $$y' - y = \cos(t) - \sin(t). \tag{3.34}$$

    b. Find the unique solution of (3.34) that remains bounded when t tends to ∞.

29. Show that the general solution to the first-order linear equation

$$a(t)y'(t) + b(t)y(t) = c(t)$$

is given by the formula

$$y(t) = \exp\left(-\left(\int \frac{b(t)dt}{a(t)} \right) \right)\left(C + \int \exp\left(\int \frac{b(s)ds}{a(s)} \bigg|_{s=t} \right)\left(\frac{c(t)}{a(t)} \right) dt \right), \tag{3.35}$$

where C is a constant.

3.3 Separable Equations

In this section we solve **separable differential equations**, which are differential equations of the form

$$y' = a(t)b(y). \tag{3.36}$$

Here $a(t)$ and $b(y)$ are assumed to be continuous functions with $b(y)$ different from zero everywhere that it is defined. Included among separable equations is the elementary equation $y' = a(t)$ as well as the first-order linear equation $y' = a(t)y$, which is a special case of equation (3.14) discussed in Section 3.2.

In order to solve (3.36), we first proceed formally and write it in terms of differentials with all expressions involving y on the left-hand side of the equation and all expressions involving t on the right-hand side of the equation. This process is known as **separation of variables**:

$$\frac{dy}{b(y)} = a(t)dt. \tag{3.37}$$

Then the solution of (3.36) is obtained simply by integrating both sides of (3.37):

$$\int \frac{dy}{b(y)} = \int a(t)dt. \tag{3.38}$$

Note that the indefinite integrals in (3.38) involve arbitrary constants. The left-hand side of (3.38) will be some function of y, say $B(y)$. When the inverse function B^{-1} of B is known we can express y in terms of t. Otherwise, (3.38) is considered to be the solution of (3.36).

Example 3.7. *Find the general solution of the equation*

$$y' = t\, y^2. \tag{3.39}$$

Solution. We separate the variables by rewriting (3.39) as

$$\frac{dy}{y^2} = t\, dt. \tag{3.40}$$

By integrating both sides of (3.40) we obtain

$$-\frac{1}{y} = \frac{t^2}{2} + C, \tag{3.41}$$

where C is a constant. We can solve (3.41) for y; thus, the solution of (3.39) obtained by the method of separation of variables is

$$y(t) = \frac{-1}{C + \dfrac{t^2}{2}}. \tag{3.42}$$

Note, however, that (3.40) has another solution that is not of the form (3.42), namely $y \equiv 0$. ∎

Sometimes it is not possible to give an explicit formula for the independent variable y in the solution to a separable equation.

Example 3.8. *Find the general solution of the equation*

$$\left(e^y\, y + 1\right) y' = 2t. \tag{3.43}$$

Solution. This time when we separate the variables we obtain

$$\left(e^y\, y + 1\right) dy = 2t\, dt. \tag{3.44}$$

When we integrate both sides of (3.44) we obtain

$$e^y y - e^y + y = t^2 + C, \tag{3.45}$$

where C is a constant. Since the inverse function of the left-hand side is not expressible in terms of elementary functions, we consider (3.45) to be the solution of (3.43) obtained by the method of separation of variables. ∎

Here is a summary of the method of separation of variables to solve (3.36):

To solve $y' = a(t)b(y)$ carry out the following steps:

Step 1: Find an indefinite integral $A(t)$ of $a(t)$, that is, a function A such that $A'(t) = a(t)$.

Step 2: Find an indefinite integral $B(y)$ of $1/b(y)$, that is, a function B such that $B'(y) = 1/b(y)$.

Step 3: The solution of $y' = a(t)b(y)$ obtained by the method of separation of variables is then

$$
\begin{cases}
y = B^{-1}\left(A(t) + C\right), & \text{if the inverse function } B^{-1} \text{ is known,} \\
B(y) = A(t) + C, & \text{if the inverse function } B^{-1} \text{ is not known,}
\end{cases}
$$

where C is a constant.

Step 4: Look for other solutions of $y' = a(t)b(y)$. (They can usually be found by inspection.)

Next, let us discuss the solution of $y' = a(t)b(y)$ when an initial condition of the form $y(t_0) = Y_0$ is also given. We first follow the four steps described above. Then the constant is determined from the condition $y(t_0) = Y_0$.

The following example shows that the general solution to a separable equation must sometimes be restricted in order to satisfy given initial conditions.

Example 3.9. *Solve the differential equation*

$$y' = \frac{3t^2 + 2t + 4}{2(y-2)} \qquad (3.46)$$

with the initial condition $y(0) = 0$.

Solution. First, we move all expressions involving y to the left-hand side of (3.46):

$$2(y-2)\frac{dy}{dt} = 3t^2 + 2t + 4. \qquad (3.47)$$

Next, we rewrite (3.47) in terms of differentials:

$$2(y-2)dy = (3t^2 + 2t + 4)dt. \qquad (3.48)$$

By integrating both sides of (3.48), we obtain

$$y^2 - 4y = t^3 + t^2 + 4t + C, \qquad (3.49)$$

where C is the constant of integration. To determine the value of C, we use the initial condition $y(0) = 0$. When we substitute $t = 0$ and $y = 0$ into (3.49), we obtain $C = 0$. Thus, (3.49) becomes

$$y^2 - 4y = t^3 + t^2 + 4t. \qquad (3.50)$$

The quadratic formula can be used to solve (3.50) for y in terms of t:

$$y = 2 \pm \sqrt{t^3 + t^2 + 4t + 4}. \qquad (3.51)$$

To decide on the correct sign in (3.51), we again use the initial condition $y(0) = 0$. Thus, we must choose the minus sign, and the solution of (3.46) is finally obtained as

$$y = 2 - \sqrt{t^3 + t^2 + 4t + 4}. \; \blacksquare \qquad (3.52)$$

It is interesting to determine the interval of validity of the solution obtained in Example 3.9. For $t > 0$ the solution (3.52) is well-defined, since all of the terms under the square root are positive. For $t < 0$ we note that the cubic polynomial under the square root on the right-hand side of (3.52) can be written in the form

$$t^3 + t^2 + 4t + 4 = (t^2 + 4)(t + 1). \qquad (3.53)$$

The first factor $(t^2 + 4)$ is always positive, but the second factor $(t + 1)$ becomes zero when $t = -1$, and then becomes negative for $t < -1$. Therefore, we cannot expect to define the square root for $t < -1$. At the point $t = -1$ the solution is well-defined by $y(-1) = 2$, but the derivative is infinite at this point. In summary, *the solution exists in the region where $t > -1$*. The graph is given below.

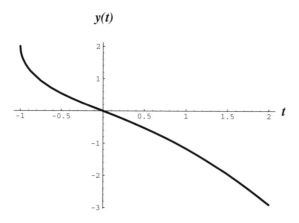

y(t)

The graph of $y = 2 - \sqrt{(t^2 + 4)(t + 1)}$

In Chapter 6 we shall see several examples of separable differential equations that arise in applications.

The following theorem[1] gives a necessary and sufficient condition for a differential equation to be separable.

Theorem 3.2.   *Let f be a differentiable function of two variables.*

(i)   *If there are functions $\phi(t)$ and $\psi(y)$ such that*

$$f(t, y) = \phi(t)\psi(y), \tag{3.54}$$

then

$$f\frac{\partial^2 f}{\partial t \partial y} = \left(\frac{\partial f}{\partial t}\right)\left(\frac{\partial f}{\partial y}\right). \tag{3.55}$$

(ii)   *Conversely, if (3.55) holds and f is nonzero, then there exist functions ϕ and ψ for which f can be written in the form (3.54).*

---

[1] This theorem is based on the article "When is an ordinary differential equation separable" by David Scott in the **American Mathematical Monthly 92** (1985), 422-423.

Proof. If (3.54) holds, then

$$
\begin{cases}
\dfrac{\partial f}{\partial t} = \phi'(t)\psi(y), \\[2mm]
\dfrac{\partial f}{\partial y} = \phi(t)\psi'(y), \\[2mm]
\dfrac{\partial^2 f}{\partial t\,\partial y} = \phi'(t)\psi'(y).
\end{cases}
\tag{3.56}
$$

Then (3.56) implies (3.55) by an easy calculation.

Conversely, if (3.55) holds and f is nonzero, then

$$
\frac{\partial^2 \log\big(f(t,y)\big)}{\partial y\,\partial t} = \frac{\partial}{\partial y}\left(\frac{\dfrac{\partial f}{\partial t}}{f(t,y)}\right) = \frac{f(t,y)\dfrac{\partial^2 f}{\partial t\,\partial y} - \dfrac{\partial f}{\partial t}\dfrac{\partial f}{\partial y}}{f(t,y)^2} = 0.
\tag{3.57}
$$

We conclude from (3.57) that $\partial \log\big(f(t,y)\big)/\partial t$ is a function of t alone, so we may write

$$
\frac{\partial \log\big(f(t,y)\big)}{\partial t} = \beta(t).
\tag{3.58}
$$

When we integrate both sides of (3.58), we get

$$
\log\big(f(t,y)\big) = \int \beta(t)dt + \gamma(y),
\tag{3.59}
$$

where $\gamma(y)$ denotes a constant of integration. Putting

$$
\phi(t) = \exp\left(\int \beta(t)dt\right)
$$

and $\psi(y) = e^{\gamma(y)}$, we obtain (3.54). ∎

Example 3.10. *Show that the differential equation*

$$
\frac{dy}{dt} = 1 + t^2 + y^3 + t^2 y^3
\tag{3.60}
$$

is separable.

Solution. Letting $f = 1 + t^2 + y^3 + t^2 y^3$, we compute

$$
\frac{\partial f}{\partial t}\frac{\partial f}{\partial y} = 6t\,y^2 + 6t^3 y^2 + 6t\,y^5 + 6t^3 y^5 = \frac{\partial^2 f}{\partial t\,\partial y} f.
$$

In fact, (3.60) can be rewritten as

$$
\frac{dy}{1+y^3} = (1+t^2)dt,
$$

both sides of which can be integrated. ∎

Summary of Techniques Introduced

The general solution to the **separable equation** $dy/dt = a(t)b(y)$ can always be found by rewriting the equation as

$$\frac{dy}{b(y)} = a(t)dt$$

and integrating both sides of the equation. If several solutions are found, those not satisfying given initial conditions must be excluded when solving a separable initial value problem.

Exercises for Section 3.3

Find the general solution to each of the following separable equations.

1. $y' = \dfrac{5y}{t(y-3)}$

2. $y't^2 = y(y-3)$

3. $t^3 y' = \sqrt{1 - y^2}$

4. $e^{y^3} = \csc(t)y^2 y'$

5. $(1 + t + y + t\,y)y' = 1$

6. $\log(y\,y') = t$

7. $\big(y\cos(y) + \sin(y)\big)y' = t + \log(t)$

8. $\dfrac{y'}{\sin(3t)} - t^3 e^{2y} = 0$

9. $\dfrac{\sin(y)y'}{t} = \dfrac{t\,e^t}{y}$

10. $y' = \big(\sin(t)\big)^2\big(\sin(2y)\big)^2$

Solve the following initial value problems.

11. $\begin{cases} y' = t(1 + y^2), \\ y(0) = 1 \end{cases}$

12. $\begin{cases} y\,y' = t^2, \\ y(1) = 3 \end{cases}$

13. $\begin{cases} y^2 y' = e^t, \\ y(1) = 4 \end{cases}$

14. $\begin{cases} t\,y\,y' = 1, \\ y(1) = 5 \end{cases}$

15. $\begin{cases} y' = -\dfrac{t}{y}, \\ y(1) = 1 \end{cases}$

16. $\begin{cases} e^t y' = y, \\ y(0) = 4 \end{cases}$

17. $\begin{cases} t\,y' + \sin(y) = 0, \\ y(1) = \pi \end{cases}$

18. $\begin{cases} y' = -t\,y, \\ y(0) = \dfrac{1}{\sqrt{2\pi}} \end{cases}$

19. $\begin{cases} y\,y'\sqrt{1-t^2} = 1, \\ y(0) = 1 \end{cases}$

20. $\begin{cases} y' = t^2 y, \\ y(0) = 4 \end{cases}$

Use Theorem 3.2 to determine whether the following equations are separable.

21. $y' = \dfrac{t\,y + t + y + 1}{t\,y}$

23. $y' = \cos(y)\big(\cos(t+y) + \cos(t-y)\big)$

22. $t(y^2 + 5y + t)\,dt + (t^2 - 1)\,dy = 0$

24. $e^{t\,y}(y^2 + 5y + 4)\,dt + (t^2 - 1)\,dy = 0$

3.4   Exact Equations and Integrating Factors

The function $f(t, y)$ in the differential equation (3.2) can be written (in many different ways) as the negative of the quotient of two functions $M = M(t, y)$ and $N = N(t, y)$; then (3.2) becomes

$$\frac{dy}{dt} = -\frac{M(t, y)}{N(t, y)}. \tag{3.61}$$

We can formally rewrite (3.61) as

$$M(t, y)dt + N(t, y)dy = 0. \tag{3.62}$$

In (3.62) the expressions dt and dy are called **differentials**. More generally, the **differential** of a function $\Phi(t, y)$ is defined by

$$d\Phi = \frac{\partial \Phi}{\partial t}dt + \frac{\partial \Phi}{\partial y}dy, \tag{3.63}$$

as is well known from calculus. Although a precise mathematical definition of the notion of differential can be given, there is no need to do so here. We call (3.62) the **symmetric form using differentials** of a first-order differential equation.

Definition.   *The differential equation* (3.62) *is called* **exact** *provided that*

$$\frac{\partial M(t, y)}{\partial y} = \frac{\partial N(t, y)}{\partial t}. \tag{3.64}$$

Exact differential equations arise in a natural manner. If we are given a function Φ of two variables, we can consider an equation of the form

$$\Phi(t, y) = C, \qquad (3.65)$$

where C is a constant. When we take differentials of both sides of (3.65) we get

$$d\Phi = \frac{\partial \Phi}{\partial t} dt + \frac{\partial \Phi}{\partial y} dy = 0, \qquad (3.66)$$

because the differential of a constant is zero. Let $M = \partial\Phi/\partial t$ and $N = \partial\Phi/\partial y$. Since

$$\frac{\partial^2 \Phi}{\partial y \partial t} = \frac{\partial^2 \Phi}{\partial t \partial y},$$

it follows that $M\, dt + N\, dy = 0$ is an exact differential equation in the special case that $M = \partial\Phi/\partial t$ and $N = \partial\Phi/\partial y$.

Next, we consider the converse question. We are given the differential equation (3.62) and we seek a function Φ (called a **potential function**) for which $M = \partial\Phi/\partial t$ and $N = \partial\Phi/\partial y$. If we can find such a function, then (3.62) is transformed into

$$d\Phi = \frac{\partial \Phi}{\partial t} dt + \frac{\partial \Phi}{\partial y} dy = 0.$$

We can integrate $d\Phi = 0$; the result is (3.65). In other words, (3.65) is the solution of $M\, dt + N\, dy = 0$.

Example 3.11. *Find the general solution of the exact differential equation*

$$2t\, y\, dt + (t^2 - y)dy = 0. \qquad (3.67)$$

Solution. Let $M = 2t\, y$ and $N = t^2 - y$. Since

$$\frac{\partial M}{\partial y} = 2t = \frac{\partial N}{\partial t},$$

the equation (3.67) is indeed exact. By inspection we see that a potential function Φ is given by

$$\Phi(t, y) = t^2 y - \frac{y^2}{2}.$$

Hence the solution of (3.67) is

$$t^2 y - \frac{y^2}{2} = C. \qquad (3.68)$$

A more informal way to solve (3.67) is as follows. First, we regroup the terms:

$$(2t\, y\, dt + t^2 dy) - y\, dy = 0. \qquad (3.69)$$

Then we rewrite $2t\,y\,dt + t^2 dy$ as $d(t^2 y)$; thus (3.69) becomes

$$d(t^2 y) - y\,dy = 0. \tag{3.70}$$

It is now easy to integrate (3.69); the result is (3.68). ∎

The following theorem establishes the existence of a potential function for an exact differential equation.

Theorem 3.3. *Consider the differential equation $M\,dt + N\,dy = 0$. There exists a potential function Φ with continuous second partial derivatives such that*

$$M = \frac{\partial \Phi}{\partial t} \qquad and \qquad N = \frac{\partial \Phi}{\partial y} \tag{3.71}$$

if and only if M and N satisfy condition (3.64); that is, if and only if

$$M\,dt + N\,dy = 0$$

is an exact differential equation.

Proof. Suppose there exists Φ with continuous second partial derivatives such that (3.71) holds. Then

$$\frac{\partial^2 \Phi}{\partial y \partial t} = \frac{\partial^2 \Phi}{\partial t \partial y}. \tag{3.72}$$

But (3.71) and (3.72) imply that (3.64) holds. Hence $M\,dt + N\,dy = 0$ is exact.

Conversely, suppose that $M\,dt + N\,dy = 0$ is an exact differential equation. Consider the expression

$$h = N - \int \frac{\partial M}{\partial y} dt, \tag{3.73}$$

where the indefinite integral on the right-hand side of (3.73) denotes the integral of $\partial M/\partial y$ with respect to t, treating y as a constant. Then (3.64) and the fundamental theorem of calculus imply that

$$\frac{\partial h}{\partial t} = \frac{\partial N}{\partial t} - \frac{\partial}{\partial t} \int \frac{\partial M}{\partial y} dt = \frac{\partial N}{\partial t} - \frac{\partial M}{\partial y} = 0.$$

Therefore, h is a function of y alone, and so we may write $h = h(y)$. Now define

$$\Phi = \Phi(t, y) = \int M\,dt + \int h(y)dy, \tag{3.74}$$

where the first integral on the right-hand side of (3.74) denotes the integral of M with respect to t, again treating y as a constant. From (3.74) and (3.73) we get

$$\frac{\partial \Phi}{\partial t} = \frac{\partial}{\partial t} \int M\,dt + 0 = M$$

and

$$\frac{\partial \Phi}{\partial y} = \frac{\partial}{\partial y}\left(\int M\,dt\right) + h(y) = \int \frac{\partial M}{\partial y}\,dt + h(y) = N.$$

Hence (3.71) is satisfied. ∎

The method of proof of Theorem 3.3 yields a formula for the solution of an exact differential equation.

Corollary 3.4. *The solution of the exact differential equation $M\,dt + N\,dy = 0$ is given by $\Phi(t, y) = C$, where C is a constant of integration and*

$$\Phi(t, y) = \int M\,dt + \int\left(N - \int \frac{\partial M}{\partial y}\,dt\right)dy. \tag{3.75}$$

Here is a summary of how to find the general solution of an exact differential equation.

To find the general solution of the exact differential equation

$$M\,dt + N\,dy = 0$$

carry out the following steps:

Step 1: Check that $M\,dt + N\,dy = 0$ is an exact equation.

Step 2: Try to guess a function $\Phi(t, y)$ such that

$$\frac{\partial \Phi}{\partial t} = M \qquad \text{and} \qquad \frac{\partial \Phi}{\partial y} = N.$$

Step 3: If Step 2 is not possible, use formula (3.75) to determine $\Phi(t, y)$.

Step 4: The general solution of $M\,dt + N\,dy = 0$ is given by

$$\Phi(t, y) = C, \qquad \text{where } C \text{ is a constant.}$$

Example 3.12. *Find the general solution of the differential equation*

$$3t^2 y^4 + 4t^3 y^3\, y' = 0. \tag{3.76}$$

Solution. The symmetric form of (3.76) using differentials is

$$3t^2 y^4\, dt + 4t^3 y^3\, dy = 0.$$

We have $M = 3t^2 y^4$ and $N = 4t^3 y^3$, so that

$$\frac{\partial M}{\partial y} = 12t^2 y^3 \qquad \text{and} \qquad \frac{\partial N}{\partial t} = 12t^2 y^3, \tag{3.77}$$

which are equal; hence equation (3.77) is exact. The potential function Φ is obtained by means of formula (3.75):

$$\Phi = \int M\,dt + \int \left(N - \int \frac{\partial M}{\partial y}dt\right)dy$$

$$= \int 3t^2 y^4\,dt + \int \left(4t^3 y^3 - \int 12t^2 y^3 dt\right)dy$$

$$= t^3 y^4 + \int (4t^3 y^3 - 4t^3 y^3)dy = t^3 y^4.$$

The general solution of the differential equation (3.76) is thus

$$t^3 y^4 = C. \tag{3.78}$$

If the integration constant C in (3.78) is positive, we can find two solutions y of (3.78) when $t > 0$ by the explicit formulas $y = \pm C^{1/4}t^{-3/4}$. For $t < 0$ these solutions do not exist, since y^4 is positive for all values of y.

If the integration constant C in (3.78) is negative, let $C = -C_1$; then we can find two solutions of (3.78) when $t < 0$ by the explicit formula $y = \pm C_1^{1/4}(-t)^{-3/4}$, which is well-defined for $t < 0$; these do not exist for $t > 0$ in this case. The graph below depicts the integral curves of the equation $t^3 y^4 = C$ for various values of C. $\blacksquare$

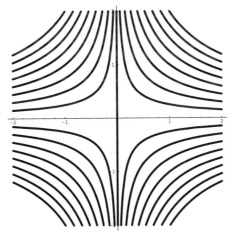

Several solutions to the differential equation
$$3t^2 y^4 + 4t^3 y^3 y' = 0$$

Integrating Factors That Make a First-Order Equation Exact

We next consider the problem of writing a given first-order differential equation in the form

$$M \, dt + N \, dy = 0. \tag{3.79}$$

It is apparent that we have a great deal of freedom in the choice of M and N; we can multiply each of these by the same function $\mu(t, y)$ and leave the equation untouched. This may transform an exact equation into a nonexact equation or transform a nonexact equation into an exact equation. The second possibility is especially useful; we formalize it as follows:

Given a (possibly nonexact) differential equation $M \, dt + N \, dy = 0$, can we find a function $\mu(t, y)$ such that the equation

$$\mu M \, dt + \mu N \, dy = 0$$

is exact? We call μ an **integrating factor**; it is a generalization of the integrating factors that we used in Section 3.2. The criterion for exactness (3.65) becomes

$$\frac{\partial(\mu M)}{\partial y} = \frac{\partial(\mu N)}{\partial t}. \tag{3.80}$$

Writing out (3.80) using the product rule and simplifying, we have

$$\frac{\partial \mu}{\partial y} M - \frac{\partial \mu}{\partial t} N + \mu \left(\frac{\partial M}{\partial y} - \frac{\partial N}{\partial t} \right) = 0. \tag{3.81}$$

Equation (3.81) is a **partial differential equation** for the function $\mu(t, y)$; unfortunately, it might be as difficult to solve as the original equation (3.79). Frequently, however, progress can be made if we assume that the integrating factor $\mu(t, y)$ is a function of one variable, either t or y. In what follows we abbreviate $\partial M/\partial y$ to M_y and $\partial N/\partial t$ to N_t.

Case 1. μ is a function of t alone: $\mu = \mu(t)$

In this case the exactness condition (3.81) reduces to

$$-N \frac{\partial \mu}{\partial t} + \mu \left(\frac{\partial M}{\partial y} - \frac{\partial N}{\partial t} \right) = 0,$$

resulting in the ordinary differential equation

$$\frac{1}{\mu} \frac{d\mu}{dt} = \frac{M_y - N_t}{N}. \tag{3.82}$$

Since both μ and $d\mu/dt$ are functions of t alone, the quotient $(M_y - N_t)/N$ must also be a function of t. Therefore, (3.82) is a separable differential equation for the function μ, which can be solved by integrating both sides of the equation. We get

$$\log(\mu(t)) = \int \left(\frac{M_y - N_t}{N}\right) dt,$$

or

$$\mu(t) = \exp\left(\int \left(\frac{M_y - N_t}{N}\right) dt\right). \tag{3.83}$$

Thus, we have proved:

Theorem 3.5. *The differential equation $M\, dt + N\, dy = 0$ has an integrating factor μ which depends on t alone if and only if*

$$\frac{M_y - N_t}{N}$$

is a function of t alone. In this case μ is given by (3.83).

Example 3.13. *Find the general solution of $12t\, y + 6y^2 + (3t^2 + 4t\, y)y' = 0$.*

Solution. In this case we have

$$M = 12t\, y + 6\, y^2 \qquad \text{and} \qquad N = 3\, t^2 + 4t\, y.$$

Thus, $M_y = 12\, t + 12\, y$ and $N_t = 6\, t + 4y$, so that

$$\frac{M_y - N_t}{N} = \frac{6t + 8\, y}{3t^2 + 4t\, y} = \frac{2}{t}.$$

Hence an integrating factor is $\mu(t) = t^2$. The original equation is transformed into

$$0 = (12t^3 y + 6t^2 y^2)dt + (3t^4 + 4t^3 y)dy = d\left(3t^4 y + 2t^3 y^2\right). \tag{3.84}$$

The potential function is found by integration; thus, the solution is

$$\Phi(t, y) = 3t^4 y + 2t^3 y^2 = C,$$

where C is a constant. This is a quadratic equation in y that can be solved by the quadratic formula to obtain an explicit solution. ∎

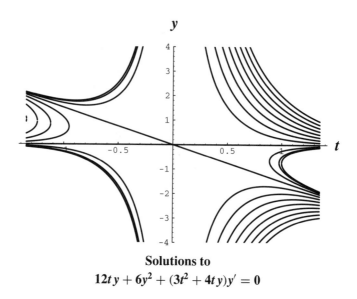

Solutions to
$$12t\,y + 6y^2 + (3t^2 + 4t\,y)y' = 0$$

Case 2. μ is a function of y alone: $\mu = \mu(y)$

This case is entirely parallel to the case just discussed. The equation for an integrating factor is $(\mu M)_y = (\mu N)_t$, or

$$\frac{\mu'}{\mu} = \frac{N_t - M_y}{M}.$$

If the quantity $(N_t - M_y)/M$ depends only on y, then an integrating factor can be found in the form

$$\mu(y) = \exp\left(\int \frac{N_t - M_y}{M}dy\right).$$

Example 3.14. *Find an integrating factor for and solve the initial value problem*

$$\begin{cases} y - (t + y^3)y' = 0, \\ \quad\quad y(1) = 1. \end{cases} \tag{3.85}$$

Solution. We take $M = y$ and $N = -(t + y^3)$ and compute

$$\frac{N_t - M_y}{M} = -\frac{2}{y},$$

so that $\mu(y) = y^{-2}$ is an integrating factor. Multiplying by this integrating factor converts the differential equation of (3.85) to

$$\frac{y\,dt - t\,dy}{y^2} = y\,dy. \tag{3.86}$$

When we integrate both sides of (3.86) we obtain $t/y = y^2/2 + C$, where C is a constant of integration. The initial condition of (3.85) implies that $C = 1/2$, and so we obtain

$$\frac{t}{y} - \frac{y^2}{2} = \frac{1}{2}. \ \blacksquare$$

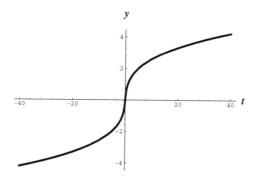

Solution to $y - (t + y^3)y' = 0,$ $y(1) = 1$

Summary of Techniques Introduced

Implicit solutions can be obtained for an **exact equation**, written

$$\frac{dy}{dt} = -\frac{M(t, y)}{N(t, y)} \qquad \text{with} \qquad \frac{\partial M}{\partial y} = \frac{\partial N}{\partial t}.$$

If an equation of this type does not satisfy the condition of exactness, it is possible to find an integrating factor of the form $\mu = \mu(t)$ in the case that

$$\frac{1}{N}\left(\frac{\partial M}{\partial y} - \frac{\partial N}{\partial t}\right)$$

is a function of t. Similarly, it is possible to find an integrating factor $\mu = \mu(y)$ in the case that

$$\frac{1}{M}\left(-\frac{\partial M}{\partial y} + \frac{\partial N}{\partial t}\right)$$

is a function of y.

Exercises for Section 3.4

Find the general solution of each of the following exact equations.

1. $(3y + t\,y^2) + (3t + t^2 y)y' = 0$ 3. $(t + y)y' + y = 0$

2. $(2y - e^y) + t(2 - e^y)y' = 0$ 4. $e^y\left(2t + (t^2 - y^2 - 2y)y'\right) = 0$

Find the general solution of each of the following equations by finding a suitable integrating factor.

5. $(y - t)y' + y = 0$ 8. $y - t\,y' = 0$

6. $(2y + y^2) - t\,y' = 0$ 9. $(1 + t)e^t + (y\,e^y - t\,e^t)y' = 0$

7. $(2t^2 + 3t\,y - y^2) + (t^2 - t\,y)y' = 0$ 10. $2t\,y + (y^2 - t^2)y' = 0$

Solve the following initial value problems.

11. $\begin{cases} 2t - y + (2y - t)y' = 0, \\ y(2) = 4 \end{cases}$

13. $\begin{cases} \dfrac{y\cos(t)}{t} + \dfrac{2y\sin(t)}{t^2} + \dfrac{\sin(t)y'}{t} = 0, \\ y(\pi/2) = 1 \end{cases}$

12. $\begin{cases} e^t + y + (2y + t + y\,e^y)y' = 0, \\ y(0) = 1 \end{cases}$

14. $\begin{cases} y\cos(t)\sin(y) + \cos(y)\sin(t)y\,y' = 0, \\ y(\pi/4) = \pi/4 \end{cases}$

15. Show that the differential equation $y\,f(t\,y)dt + t\,g(t\,y)dy = 0$ has an integrating factor of the form
$$\mu(t, y) = \frac{1}{t\,y(f - g)}.$$

16. Use Problem 15 to show that the differential equation
$$y(1 - t\,y)dt - t(1 + t\,y)dy = 0$$
has $\mu = 1/(2t\,y)$ as an integrating factor. This leads to the solution $y\,e^{t\,y} = C\,t$.

17. Show that if a differential equation has the form $t^a y^b(p\,y\,dt + q\,t\,dy) + t^d y^e(r\,y\,dt + s\,t\,dy) = 0$, where a, b, d, e, p, q, r, s are constants, then an integrating factor of the form $t^k y^n$ for specific values of k and n can be found.

18. For which values of the constants a, b, c, and f does the first-order differential equation
$$(a\,t\,y + b\,y^2) + (c\,t^2 + f\,t\,y)y' = 0 \tag{3.87}$$

have an integrating factor that is a function $\mu(t)$ of t alone? Find a formula for μ.

19. Suppose that μ is an integrating factor for $M\,dt + N\,dy = 0$ leading to the solution $\Phi(t, y) = C$ and that F is any function of Φ. Show that $\mu\,F(\Phi)$ is also an integrating factor for $M\,dt + N\,dy = 0$.

20. Show that if

$$\frac{M_y - N_t}{Ny - Mt} = f(z)$$

where $z = t\,y$, then

$$\mu = e^{\int f(z)\,dz}$$

is an integrating factor for $M\,dt + N\,dy = 0$.

Use the method of Exercise 20 to solve the following differential equations.

21. $(t\,y^2 + y)\,dt + t\,dy = 0$ **23.** $(y^3 - 2y\,t^2)\,dt + (2t\,y^2 - t^3)\,dy = 0$

22. $(3y + 8t\,y^2)\,dt + (2t + 6y\,t^2)\,dy = 0$ **24.** $(y\,t^2 + y^2)\,dt - t^3\,dy = 0$

3.5   *Homogeneous First-Order Equations*

A first-order differential equation is called **homogeneous** if it has the form

$$y' = F\left(\frac{y}{t}\right), \tag{3.88}$$

where F is a differentiable function of one variable. In order to transform (3.88) into a separable equation, we define a new function z by the formula

$$z = \frac{y}{t}. \tag{3.89}$$

From (3.89) we get

$$y = t\,z \quad \text{and} \quad y' = t\,z' + z. \tag{3.90}$$

The substitutions (3.89) and (3.90) transform the differential equation (3.88) into

$$t\,z' + z = F(z). \tag{3.91}$$

We separate the variables in (3.91), obtaining

$$\frac{dz}{F(z) - z} = \frac{dt}{t}. \tag{3.92}$$

Just as in Section 3.3, we solve (3.92) by integrating both sides of the equation. After solving (3.92), we use the substitution (3.89) to get an equation in y and t.

Of course, we can make the substitution (3.89) in a differential equation whether or not the differential equation is known to be homogeneous. So if we think a first-order differential equation may be homogeneous, we can go ahead and try the substitution (3.89); if it converts the differential equation into another, simpler differential equation, we try to solve the simpler equation.

Example 3.15. *Find the general solution of the differential equation*

$$y' = \frac{2y^2 + t^2}{2t\,y}.\tag{3.93}$$

Solution. The substitution $z = y/t$ transforms (3.93) into

$$t\,z' + z = \frac{2(t\,z)^2 + t^2}{2t\,(t\,z)} = \frac{2z^2 + 1}{2z}.$$

Hence

$$t\,z' = \frac{2z^2 + 1}{2z} - z = \frac{1}{2z},$$

so that we obtain the separable equation

$$2z\,dz = \frac{dt}{t}.\tag{3.94}$$

We integrate both sides of (3.94) to get

$$z^2 = \log(t) + C,\tag{3.95}$$

where C is a constant of integration. Thus, the general solution of (3.93) is

$$y = t\,z = \pm t\sqrt{C + \log(t)}.\ \blacksquare\tag{3.96}$$

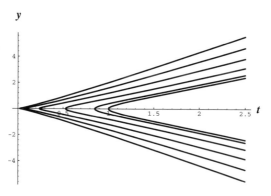

Several solutions to $\ \ y' = (2y^2 + t^2)/(2t\,y)$

We summarize the steps to find the general solution of a homogeneous differential equation:

If the differential equation $y' = f(t, y)$ appears to be homogeneous of the form $y' = F(y/t)$, carry out the following steps:

Step 1: Substitute $y = t\,z$ and $y' = t\,z' + z$ and obtain a differential equation of the form $t\,z' + z = F(z)$.

Step 2: Separate the variables in $t\,z' + z = F(z)$ and integrate both sides of the resulting differential equation.

Step 3: Substitute $z = y/t$ to obtain an equation involving y and t.

Step 4: If possible, solve for y.

Exercises for Section 3.5

Find the general solution of each of the following homogeneous differential equations.

1. $y' = \dfrac{t - y}{t}$

2. $y' = \dfrac{2t + y}{t}$

3. $y' = \dfrac{t^2 + 5y^2}{t\,y}$

4. $y' = \dfrac{t^2 - 3y^2}{t\,y}$

5. $y' = \dfrac{4y^4 + t^4}{4t\,y^3}$

6. $y' = \dfrac{6y^4 + t^4}{4t\,y^3}$

7. $y' = \dfrac{6y\,t}{2t^2 - y^2}$

8. $y' = \dfrac{y + \sqrt{y^2 + t^2}}{t}$

9. $y'(t^2y^2 - t^4) = t^2y^2 - y^4$

10. $y' = \dfrac{4t\,y + 8y^2}{4t^2 + y^2}$

Solve the following initial value problems.

11. $\begin{cases} y' = \dfrac{4t\,y + 8y^2}{4t^2 + y^2}, \\ y(1) = 2 \end{cases}$

12. $\begin{cases} y' = \dfrac{y}{t} + \dfrac{t}{y}, \\ y(4) = 0 \end{cases}$

13. $\begin{cases} y' = \dfrac{y - \sqrt{t^2 + y^2}}{t}, \\[2mm] y(1) = 0 \end{cases}$

16. $\begin{cases} y' = \dfrac{y + 2t}{2y - t}, \\[2mm] y(1) = 1 \end{cases}$

14. $\begin{cases} y' = \dfrac{t\,y - y^2}{t^2}, \\[2mm] y(1) = 1 \end{cases}$

17. $\begin{cases} y' = \dfrac{t^2 + 2t\,y - y^2}{t^2 - 2t\,y - y^2}, \\[2mm] y(1) = 1 \end{cases}$

15. $\begin{cases} y' = \dfrac{t + y}{t}, \\[2mm] y(1) = 1 \end{cases}$

18. $\begin{cases} y' = \dfrac{y(y^2 + 3t^2)}{2t^3}, \\[2mm] y(1) = 1 \end{cases}$

19. Consider a first-order differential equation of the form

$$\frac{dy}{dt} = f\left(\frac{a_1 t + b_1 y + c_1}{a_2 t + b_2 y + c_2}\right), \tag{3.97}$$

where $a_1, b_1, c_1, a_2, b_2, c_2$ are constants with a_2, b_2, c_2 not all zero.

a. Show that the transformation

$$\tilde{t} = t - h, \qquad \tilde{y} = y - k$$

puts (3.97) in the form

$$\frac{d\tilde{y}}{d\tilde{t}} = f\left(\frac{a_1\tilde{t} + b_1\tilde{y} + a_1 h + b_1 k + c_1}{a_2\tilde{t} + b_2\tilde{y} + a_2 h + b_2 k + c_2}\right).$$

b. Show that if $a_1 b_2 - a_2 b_1 \neq 0$, then h and k can be determined such that

$$a_1 h + b_1 k + c_1 = 0 \qquad \text{and} \qquad a_2 h + b_2 k + c_2 = 0,$$

yielding a differential equation that is homogeneous in $\tilde{t}$ and $\tilde{y}$.

Use Exercise 19 to solve the following equations.

20. $y' = \dfrac{t - y}{t + y + 2}$

21. $y' = \dfrac{t + y + 1}{t - y}$

22. $y' = (y - t)^2$ [Hint: Let $z = y - t$.]   **23.** $y' = \dfrac{1}{t + y}$ [Hint: Let $z = t + y$.]

24. A function $f(t, y)$ is said to be **generalized homogeneous of degree k** if

$$f(t, y) = t^{k-1} f(y/t^k).$$

Consider the differential equation $y' = f(t, y)$.

    a. Prove that if $f(t, y)$ is generalized homogeneous of degree k, then

$$t \frac{\partial (tf/y)}{\partial t} = -k \left(y \frac{\partial (tf/y)}{\partial y} \right).$$

    b. Prove that if $f(t, y)$ is generalized homogeneous of degree k, then the transformation $y = z t^k$ applied to $y' = f(t, y)$ leads to a separable equation in t and z.

Show that the following differential equations are generalized homogeneous and solve them by the method of Exercise 24.

25. $y' = (t^4 y^2 + 1)/t^3$ **27.** $y' = (y t^2 + 1)/t^3$

26. $y' = (t y + 1)/t^2$ **28.** $y' = (t^2 y^2 + 1)/t^2$

3.6   Bernoulli Equations

A **Bernoulli equation** is a first-order differential equation of the form

$$a(t)y' + b(t)y = c(t)y^n. \tag{3.98}$$

We assume that n is different from 0 or 1, so that (3.98) is nonlinear. To solve (3.98) we first reduce it to the leading-coefficient-unity-form by dividing both sides of the equation by $a(t)$. Letting $p(t) = b(t)/a(t)$ and $q(t) = c(t)/a(t)$, we get

$$y' + p(t)y = q(t)y^n. \tag{3.99}$$

Next, we make the change of variables

$$u = y^{1-n}. \tag{3.100}$$

The chain rule tells us that $u' = (1-n)y^{-n}y'$, or equivalently,

$$y' = \frac{y^n u'}{1-n}. \tag{3.101}$$

Substitution of (3.101) into (3.99) yields

$$\frac{y^n u'}{1-n} + p(t)y = q(t)y^n. \tag{3.102}$$

We divide both sides of (3.102) by $y^n/(1-n)$ and use (3.100) to obtain

$$u' + (1-n)p(t)u = (1-n)q(t). \tag{3.103}$$

Clearly, (3.103) is a first-order linear equation in u. To solve it, we follow the steps given on page 40. Thus, an integrating factor for (3.103) is

$$\mu(t) = \exp\left((1-n)\int p(t)dt\right). \tag{3.104}$$

After completing the steps on page 40, we find the general solution y to (3.98) by using the reverse change of variables

$$y = u^{1/(1-n)}. \tag{3.105}$$

Example 3.16. *Find the general solution of the Bernoulli equation*

$$y' - t\,y = t\,y^2. \tag{3.106}$$

Solution. We have $n = 2$, so we transform (3.106) by means of the substitution $y = u^{-1}$; the result is

$$u' + t\,u = -t. \tag{3.107}$$

An integrating factor for this first-order linear differential equation is easily found to be $\exp(t^2/2)$, so that the solution of (3.107) is

$$u = -1 + Ce^{-t^2/2},$$

where C is a constant. Thus, the general solution of (3.106) is

$$y = \frac{1}{-1 + Ce^{-t^2/2}}. \quad\blacksquare$$

Example 3.17. *Solve the initial value problem*

$$\begin{cases} t^2 y' + 2t\,y - y^3 = 0, \\ y(1) = 1. \end{cases} \tag{3.108}$$

Solution. The substitution $u = y^{-2}$ converts the differential equation of (3.108) to the first-order linear differential equation

$$u' - 4t^{-1}u + 2t^{-2} = 0. \qquad (3.109)$$

Using the integrating factor t^{-4}, we find the general solution to (3.109) to be

$$y^{-2} = u = \frac{2}{5}t^{-1} + Ct^4, \qquad (3.110)$$

where C is a constant. Then the initial condition $y(1) = 1$ implies that $C = 3/5$. When we substitute this into (3.110) and solve for y we get

$$y = \sqrt{\frac{5t}{2 + 3t^5}}. \quad \blacksquare$$

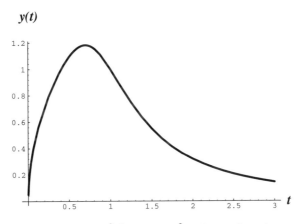

Solution to $\quad t^2y' + 2ty - y^3 = 0, \quad y(1) = 1$, plotted over the interval $\quad 0 < t < 3$

Here is a summary of the steps to follow to solve a Bernoulli equation.

To find the general solution of $a(t)y' + b(t)y = c(t)y^n$ carry out the following steps:

Step 1: Write the differential equation in the leading-coefficient-unity-form $y' + p(t)y = q(t)y^n$.

Step 2: Use the substitution $u = y^{1-n}$ to convert $y' + p(t)y = q(t)y^n$ to the form $u' + (1-n)p(t)u = (1-n)q(t)$.

Step 3: Compute the integrating factor

$$\mu(t) = \exp\left((1-n)\int p(t)dt\right).$$

Step 4: Compute the integral $\int (1-n)\mu(t)q(t)dt$.

Step 5: The solution is then given by

$$y(t) = \left(\frac{1}{\mu(t)}\left(C + \int \mu(t)(1-n)q(t)dt\right)\right)^{\frac{1}{1-n}},$$

where C is a constant of integration.

Exercises for Section 3.6

Find the general solution of each of the following Bernoulli equations.

1. $y' + y = y^4 \sin(t)$

2. $t^3 y' - 2t\, y = y^3$

3. $t\, y' + 2y = \dfrac{e^t}{y}$

4. $t\, y' + 2y = \dfrac{t}{y^3}$

Solve the following initial value problems.

5. $\begin{cases} y' + \dfrac{y}{4t} = -e^{\sqrt{t}}y^3, \\ y(1) = 1 \end{cases}$

6. $\begin{cases} y' - 2y\tan(t) + y^2\big(\tan(t)\big)^4 = 0, \\ y(0) = 2 \end{cases}$

7. $\begin{cases} y' - \dfrac{t\,y}{2} = t\,y^5, \\ y(1) = 1 \end{cases}$

9. $\begin{cases} 2y' - \dfrac{y}{t} + \cos(t)y^3 = 0, \\ y(2) = 1 \end{cases}$

8. $\begin{cases} y' + \dfrac{y}{t} = t^3 y^3, \\ y(2) = 1 \end{cases}$

10. $\begin{cases} t\,y' + y = \dfrac{1}{y^2}, \\ y(1) = 2 \end{cases}$

4

THE PACKAGE **ODE.m**

Mathematica's built-in command to find a symbolic solution of an ordinary differential equation or an initial value problem is **DSolve**. Although this command solves a fair number of differential equations, the user has little control over the method used for the solution. Furthermore, the internal procedures employed by **DSolve** are hidden from the user.

Included on the diskette that accompanies this book is the *Mathematica* package **ODE.m**, including documentation. (See the preface for information about getting the package on the Internet.) The principal command in the package is called **ODE**; it does everything that **DSolve** does, but options can be used to specify the method of solution. With no options, **ODE** calls **DSolve**.

Some of the possible methods for first-order differential equations that can be specified with **ODE** are:

```
Method->Separable                Method->FirstOrderLinear
Method->FirstOrderHomogeneous    Method->Bernoulli
Method->FirstOrderExact          Method->IntegratingFactor
Method->Clairaut                 Method->Riccati
Method->Lagrange                 Method->AlmostLinear
```

Furthermore, the user has control of the form of the answer with the options

```
Form->Rule         Form->Explicit
Form->Equation     Form->Function
Form->LogEquation  Form->ExpEquation
```

A solution to an initial value problem can be plotted with the option **PlotSolution** as explained below.

In Sections 4.1–4.3 we give the basics of **ODE**. Then in Sections 4.4–4.8 we demonstrate the usage of the options **Method->FirstOrderLinear**, **Method->Separable**,

73

Method->FirstOrderHomogeneous and **Method->Bernoulli**. The *Mathematica* implementation of these options is discussed in discretionary subsections for those interested in programming.

Clairaut and Lagrange equations are treated in Section 4.9, and *Mathematica*'s implementation of a few nonelementary integrals is discussed in Section 4.10. Then in Section 4.11 we show how to use **ODE** to define new functions by means of a differential equation. This powerful feature will be used frequently throughout the rest of the book. Finally, in Section 4.12 we give a brief introduction to Riccati equations and show how to solve Riccati equations using **ODE**.

4.1   Getting Started with ODE

For the installation details of **ODE.m** see the README file on the enclosed CD-ROM. The package **ODE.m** can be read into a *Mathematica* session on any computer with the command

Needs["ODE`"]

On most computers the command

<<ODE.m

can also be used. This makes the command **ODE** available and displays the help message

To get more information type ?ODE

The information given in response to **?ODE** provides a short introduction to **ODE**. The methods available as options for **ODE** can be seen using **?Method** in a *Mathematica* session. Similarly, **?Form** describes the various output forms.

In this chapter we explain how to use **ODE**, concentrating on first-order equations. Of course, **ODE** cannot find symbolic solutions to all first-order differential equations; however, almost all of the differential equations discussed in Chapter 3 can be solved using **ODE**, although the form of the solution is a little different from what the user would find using hand calculation.

The use of **ODE** to solve higher-order differential equations will be discussed in Chapter 10.

If initial conditions are not specified, **ODE** tries to find the general solution to the differential equation. More precisely,

ODE[diffeq,y[t],t]   or more generally,   **ODE[diffeq,y[t],t,options]**

attempts to solve the differential equation **diffeq** for the dependent variable **y[t]** in terms of the independent variable **t**. The output is written as

```
{{y[t] -> some expression}}
```

Example 4.1. *Find the general solution of the separable equation* $y' = \sec(y)$ *using* **ODE**.

Solution. The command

ODE[y'[t] == Sec[y[t]],y[t],t,Method->Separable]

generates the output

```
{{y[t] -> ArcSin[t + C[1]]}}
```

The constant of integration is written in *Mathematica* as **C[1]**. Thus the general solution to $y' = \sec(y)$ is

$$y = \arcsin(t + C),$$

where C is a constant of integration. ∎

An initial value problem can be solved with **ODE** using the command

ODE[{diffeq,initialcond},y[t],t]

or more generally,

ODE[{diffeq,initialcond},y[t],t,options].

Example 4.2. *Use* **ODE** *to solve the initial value problem*

$$\begin{cases} y' = \sec(y), \\ y(0) = \pi/2. \end{cases} \tag{4.1}$$

Solution. We use

ODE[{y'[t] == Sec[y[t]],y[0] == Pi/2},y[t],t,
Method->Separable]

to get

```
{{y[t] -> ArcSin[1 + t]}}
```

Thus, the solution to (4.1) is $\sin(y) = 1 + t$. ∎

Forms of output

The default form of the solution for **ODE** has the form **y[t]->expression** and is called a **Rule**; it is useful for references to the solution in subsequent calculations. However, **ODE** (but not **DSolve**) admits several other forms of output. To get the general solution to $y' + 5y = t$ as an equation we use

```
ODE[y'[t] + 5y[t] == t,y[t],t,
Method->FirstOrderLinear,Form->Equation]
```

obtaining

```
          1     t   C[1]
y[t] == -(--) + - + ----
          25    5    5 t
                    E
```

When it is possible to do so, the option **Form->Explicit** gives the solution with no reference to **y[t]**:

```
ODE[y'[t] + 5y[t] == t,y[t],t,
Method->FirstOrderLinear,Form->Explicit]
```

```
   1    t   C[1]
-(--) + - + ----
   25   5    5 t
            E
```

Finally, **Form->Function** obtains the solution as a *Mathematica* function:

```
ODE[y'[t] + 5y[t] == t,y[t],t,
Method->FirstOrderLinear,Form->Function]
```

```
                    1     t   C[1]
{{y -> Function[t, -(--) + - + ----]}}
                    25    5    5 t
                                E
```

(See page 27 for a description of **Function**.)

Changing the name of the constant of integration

The name of the constant of integration can be changed from its default **C** to some other letter, say **q**, with the option **Constants->q**:

```
ODE[y'[t] + 5y[t] == t,y[t],t,
Method->FirstOrderLinear,Form->Equation,Constants->q]
```

```
          1     t   q[1]
y[t] == -(--) + - + ----
          25    5    5 t
                    E
```

The constant of integration should not be a *Mathematica* reserved word, such as **D** or **N**. ∎

4.2   Features of ODE

ODE *May Succeed When* DSolve *Fails*

The syntax of **ODE** in its simplest form is the same as **DSolve**, but a method of solution can be specified. For example, if we suspect that a first-order differential equation is separable, we can use the option **Method->Separable**.

Example 4.3.   *Use Mathematica to find the general solution to the differential equation*

$$y' = \frac{1}{1 + y^{3/2}}. \tag{4.2}$$

Solution. **DSolve** (using version 2.2 of *Mathematica*) fails to solve this simple separable equation. To the command

```
DSolve[y'[t] == 1/(1 + y[t]^(3/2)),y[t],t]
```

Mathematica signals that **DSolve** cannot find a solution by answering

```
                        1
DSolve[y'[t] ==  ───────────── ,  y[t],  t]
                          3/2
                 1 + y[t]
```

However, we can use the command **ODE** with the option **Method->Separable** to find the general solution to (4.2). Thus

```
ODE[y'[t] == 1/(1 + y[t]^(3/2)),y[t],t,Method->Separable]
```

yields

```
                5/2
       2 y[t]
y[t] + ─────────  == t + C[1]
          5
```

Thus, the implicit solution to (4.2) is

$$y + \frac{2y^{5/2}}{5} = t + C,$$

where C is a constant of integration. ∎

ODE *Can Call* DSolve

ODE with the option **Method->DSolve** calls **DSolve** to find a symbolic solution of a differential equation or an initial value problem. (So in Example 4.3 the command

```
ODE[y'[t] == 1/(1 + y[t]^(3/2)),y[t],t,Method->DSolve]
```

would fail.) When **DSolve** works, so does ODE with the option **Method->DSolve**, as in the following example:

Example 4.4. *Solve the initial value problem* $y' = -y$, $y(0) = 2$.

Solution. The commands

```
ODE[{y'[t] == -y[t],y[0] == 2},y[t],t,Method->DSolve]
```

and

```
DSolve[{y'[t] == -y[t],y[0] == 2},y[t],t]
```

give the same output:

```
            2
{{y[t] -> ---}}
           t
          E
```

So the solution is $y = 2e^{-t}$. ∎

The default method of ODE is **DSolve**, which means that when no method is specified, ODE calls **DSolve**. So in Example 4.4 the command

```
ODE[{y'[t] == -y[t],y[0] == 2},y[t],t]
```

could have been used.

Notation for ODE *Input*

In ordinary mathematics it is customary to write a differential equation in a form such as $y' + 3y = 2$. In order to input this equation to *Mathematica*'s internal command **DSolve** this notation must be changed to

$$\texttt{y'[t] + 3y[t] == 2}$$

Thus, y and y' must be changed to **y[t]** and **y'[t]**, and = must be changed to **==**. However, ODE is more flexible; although = still must be changed to **==**, the dependent variable can be entered as just **y** (instead of **y[t]**). Similarly, **y'** can be used in place of **y'[t]**.

Example 4.5. *Find the general solution to* $y' + 3y = 2$ *using* **ODE** *with the option* **Method->FirstOrderLinear**.

Solution. We use either

ODE[y' + 3y == 2,y,t,Method->FirstOrderLinear]

or

ODE[y'[t] + 3y[t] == 2,y,t,Method->FirstOrderLinear]

to get

```
          2    C[1]
{{y ->    - + ----}}
          3    3 t
              E
```

On the other hand, if we want the output in terms of **y[t]**, we can use either of the commands

ODE[y' + 3y == 2,y[t],t,Method->FirstOrderLinear]

or

ODE[y'[t] + 3y[t] == 2,y[t],t,Method->FirstOrderLinear]

to get the output

```
            2    C[1]
{{y[t]  -> - + ----}}
            3    3 t
                E
```

where **y** has been replaced by **y[t]**. ∎

Automatic Solution of Differential Equations

ODE can also solve differential equations automatically with **Method->Automatic**. This option forces **ODE** to try one solution method after another until it either finds a solution (other than the trivial solution $y = 0$) or exhausts its repertoire of methods. Here is how **ODE** goes about solving Exercise 21 of Section 3.2.

Example 4.6. *Find the general solution of the equation*

$$\frac{dy}{dt} = \frac{1}{e^y - t}.$$

(4.3)

Solution. We use

ODE[y'[t] == 1/(E^y[t] - t),y[t],t,Method->Automatic]   (4.4)

to get

```
First-order nonlinear differential equation.

Checking for Trivial solutions:
No Trivial solutions
Checking FirstOrderLinear:
Not FirstOrderLinear
Checking Bernoulli:
Not Bernoulli
Checking Separable:
Not Separable
Checking FirstOrderExact:
Not FirstOrderExact
Checking IntegratingFactor:
IntegratingFactor - failed
Checking FirstOrderHomogeneous:
Not FirstOrderHomogeneous
Checking GeneralizedHomogeneous:
Not GeneralizedHomogeneous
Checking AlmostLinear:
Not AlmostLinear
Checking LinearFractional:
Not LinearFractional
Checking Clairaut:
Not Clairaut
Checking Lagrange:
Not Lagrange
Checking Riccati:
Not Riccati
```

So far **ODE** has been unable to solve the differential equation. But next it has success by interchanging the variables **y** and **t**. Here is the continuation of the output of (4.4):

```
Checking ExchangeVariables:
            y
t'[y] == E  - t[y]

First-order linear differential equation.

Checking FirstOrderLinear:
        y
       E    C[1]
t[y] == - + ----
       2      y
            E
```
$$
\{\{y[t] \to \text{Log}[t - \text{Sqrt}[t^2 - 2\,C[1]]]\}, \{y[t] \to \text{Log}[t + \text{Sqrt}[t^2 - 2\,C[1]]]\}\}
$$

Thus, the general solution to (4.3) is

$$
t = \frac{e^y}{2} + \frac{C_1}{e^y},
$$

or

$$y = \log\left(t \pm \sqrt{t^2 - C_1}\right),$$

where C_1 is a constant. (We have changed *Mathematica*'s output **C[1]** to C_1.)▌

    ODE has another option **Method->All**; with this option **ODE** attempts to solve a differential equation by successively trying all appropriate symbolic methods known to **ODE**. It does not terminate until it has tried all methods.

4.3   Plotting with ODE

There are several plotting options included with **ODE**. The most common of these options is **PlotSolution**, which is used to plot the solution to an initial value problem. In its simplest form **PlotSolution** calls *Mathematica*'s command **Plot**. The following example shows how the option **PlotSolution** works.

Example 4.7. *Solve the initial value problem*

$$\begin{cases} y' + y = e^{-t}\cos(5t), \\ y(0) = 0, \end{cases}$$

and plot the solution over the interval $-1 < t < 1$.

Solution. We use

```
ODE[{y'[t] + y[t] == E^-t Cos[5t],y[0] == 0},y,t,
Method->Automatic,PlotSolution->{{t,-1,1}}]
```

to get

```
          Sin[5 t]
{{y ->   --------- }}
             t
           5 E
```

and the plot

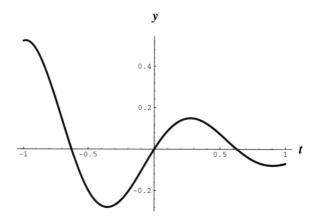

Solution to the initial value problem

$$y' + y = e^{-t}\cos(5t), \quad y(0) = 0$$

When the solution to an initial value problem is given implicitly, it can be plotted with **ODE**, but the option **PlotSolution** requires a slightly different syntax. The ranges of both the dependent and independent variables must be specified. The following example illustrates the procedure.

Example 4.8. *Solve the initial value problem*

$$\begin{cases} 3t^2 - 3y - 3t\, y' + 3y^2 y' = 0, \\ y(0) = 0, \end{cases}$$

and plot the solution over the rectangle $R = \left\{ (t, y) \,\middle|\, -2 < t < 2, \ -2 < y < 2 \right\}.$

Solution. We use

```
ODE[{3t^2 - 3y - 3t y' +  3y^2 y' == 0,y[0] == 0},y,t,
Method->IntegratingFactor,Form->Equation,
PlotSolution->{{{t,-2,2},{y,-2,2}}}]
```

to get

```
 3            3
t  - 3 t y + y  == 0
```

and the plot

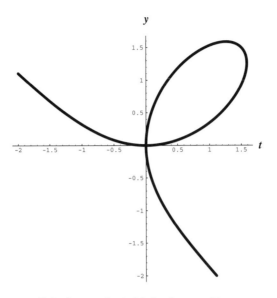

Solution to the initial value problem

$$3t^2 - 3y - 3t\,y' + 3y^2y' = 0, \quad y(0) = 0$$

The curve just obtained is called the **folium of Descartes**.[1]

ODE has an option **Parameters** that can be used to find several solutions to the same differential equation, provided that the differential equation has a free parameter. For example, if **a** and **b** are non-negative integers, the command **Parameters->{p,a,b}** can be used to create solutions for each integer **p** with **a** $\leq$ **p** $\leq$ **b**. **Parameters** is especially useful when used with **PlotSolution**, as the following example shows.

Example 4.9. *Solve the initial value problem*

$$\begin{cases} y' + a\,y = t, \\ y(0) = 0, \end{cases}$$

where a takes on integer values from 1 *to* 6. *Plot the solutions over the interval* $-1 < t < 1$.

---

[1]

René du Perron Descartes (1596–1650). French mathematician and philosopher. Descartes' great work **Discours de la méthode pour bien conduire sa raison, et chercher la vérité dans les sciences** (Discourse on the Method for Reasoning Well and Seeking Truth in the Sciences), contains an appendix, **La géométrie**, which includes what developed into Cartesian geometry. He used coordinate systems (sometimes nonorthogonal) to prove classical geometry theorems.

Solution. We use

```
ODE[{y'[t] + a y[t] == t,y[0] == 0},y[t],t,
Method->FirstOrderLinear,Parameters->{{a,1,6}},
PlotSolution->{{t,-1,1}}]
```

to get

```
              -2        1        t
{{y -> -a        + ─────────── + -}}
                    2   a t      a
                   a   E
```

and the plot

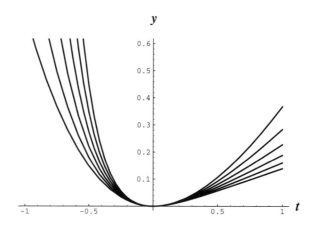

y

Solution to the initial value problem

$$y' + a\,y = t, \quad y(0) = 0, \quad \text{for} \quad a = 1, 2, 3, 4, 5, 6$$

A three-dimensional display of solutions to a differential equation can be generated with the option **StackPlotSolution**.

Example 4.10. *Use* **StackPlotSolution** *to show how the solutions to the initial value problem*

$$\begin{cases} y' - y = e^{-t}, \\ y(0) = a, \end{cases}$$

change with the initial condition.

Solution. The command

```
ODE[{y' - y == E^-t,y[0] == a},y,t,
Method->FirstOrderLinear,Parameters->{{a,-2,2,0.1}},
StackPlotSolution->{{t,-2,2}}]
```

gives the output

```
                   t
           -1      E       t
{{y ->    ---- +  ---  + a E }}
            t       2
          2 E
```

and the plot

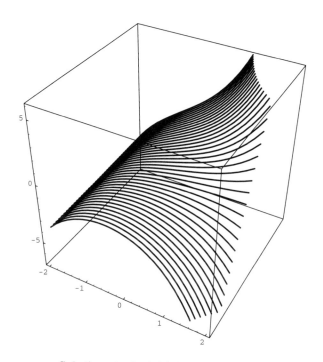

Solutions to the initial value problem

$$y' - y = e^{-t}, \quad y(0) = a, \quad \text{for} \quad -2 \le a \le 2$$

4.4   *First-Order Linear Equations via* ODE

For first-order *linear* differential equations the commands **DSolve** and **ODE** give the same output. Let us begin with a simple example.

Example 4.11. *Find the general solution of the first-order linear differential equation*

$$y' + 5y = t. \tag{4.5}$$

Solution. We use any of the following three commands:

```
ODE[y'[t] + 5y[t] == t,y[t],t,Method->FirstOrderLinear]
```

```
ODE[y'[t] + 5y[t] == t,y[t],t,Method->DSolve]
```

```
DSolve[y'[t] + 5y[t] == t,y[t],t]
```

The answer is

```
                 1      t    C[1]
{{y[t] -> -(---) + - + ----}}
                25     5     5 t
                             E
```

Note that both **ODE** and **DSolve** require three arguments: an equation, the dependent variable, and finally the independent variable. Furthermore, **ODE** allows **Method** options.

   ODE can also be used to solve a first-order initial value problem. This time the syntax is **ODE[{diffeq,initcond},y[t],t]**.

Example 4.12. *Solve the initial value problem*

$$\begin{cases} y' + 2y = t\,e^{-2t}, \\ y(1) = 0. \end{cases} \tag{4.6}$$

Solution. This initial value problem actually consists of two equations, $y' + 2y = t\,e^{-2t}$ and $y(1) = 0$. In *Mathematica* they constitute the list

```
{y'[t] + 2y[t] == t E^(-2t),y[1] == 0}
```

To solve (4.6) we use

```
ODE[{y'[t] + 2y[t] == t E^(-2t),y[1] == 0},y[t],t,
Method->FirstOrderLinear]
```

Mathematica gives the answer to (4.6) as a rule:

```
                              2
                  -1         t
{{y[t] -> ----- + -----}}
                  2 t        2 t
               2 E         2 E
```

The initial value Problem (4.6) can be solved and the solution plotted over the interval $-1 < t < 2$ with the command

```
ODE[{y'[t] + 2y[t] == t E^(-2t),y[1] == 0},y[t],t,
Method->FirstOrderLinear,
PlotSolution->{{t,-1,2}}]
```

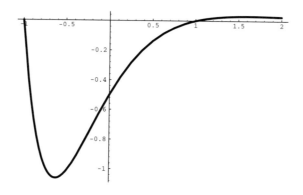

The **PlotSolution** option can be modified to include the names of the horizontal and vertical axes:

```
ODE[{y'[t] + 2y[t] == t E^(-2t),y[1] == 0},y[t],t,
Method->FirstOrderLinear,
PlotSolution->{{t,-1,2},AxesLabel->{"t","y(t)"}}]
```

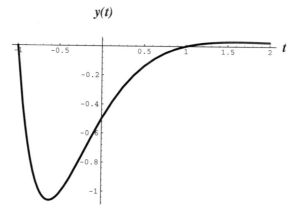

The solution to $y' + 2y = te^{-2t}$, $y(1) = 0$

Imitating the Mathematical Steps to Solve a First-Order Linear Equation with **Mathematica**

In this subsection[2] we explain the command **FirstOrderLinear**; it is the engine behind the option **Method->FirstOrderLinear** of the command **ODE**. The main idea is to utilize the following formula to find the solution to any first-order linear differential equation $a(t)y' + b(t)y = c(t)$:

$$y(t) = \exp\left(-\left(\int \frac{b(t)dt}{a(t)}\right)\right)\left(C + \int \exp\left(\int \left.\frac{b(s)ds}{a(s)}\right|_{s=t}\right)\left(\frac{c(t)}{a(t)}\right) dt\right). \qquad (4.7)$$

Because of its complexity, this formula is of little use in hand calculation, where it is better to follow the steps given on page 40. On the other hand, *Mathematica* has no difficulty in executing (4.7).

In order to define the *Mathematica* version of (4.7) we need **Module**. This important command can be used to create a miniprogram inside another *Mathematica* command. The syntax of **Module** is:

```
Module[list of local variables,
    statement 1;
        .
        .
        .
    statement k]
```

We use **Module** to define a miniprogram **FirstOrderLinear** to solve the general first-order linear equation $a(t)y'(t) + b(t)y(t) = c(t)$. First, we give the miniprogram, and then we explain it.

```
FirstOrderLinear[a_,b_,c_,const_][t_]:=
    Module[{p,q,intfac},
        p = b[t]/a[t];
        q = c[t]/a[t];
        intfac = Exp[Integrate[p,t]];
        Simplify[(Integrate[intfac q,t] +
            const)/intfac]]
```

The command **FirstOrderLinear** is a function of four parameters and one variable; the parameters consist of the coefficient functions **a**, **b**, and **c**, together with the constant of integration, denoted by **const**. The variable is the independent variable **t**. Let us

---

[2]This subsection contains technical *Mathematica* details on how ODE works. It is not needed in order to use ODE, nor is it referred to later.

examine the definition of **FirstOrderLinear** in terms of **Module**. The local variables are **p**, **q**, and **intfac**. The first two statements in the miniprogram assign to **p** the value **b[t]/a[t]** and to **q** the value **c[t]/a[t]**. The next statement,

$$\texttt{intfac = Exp[Integrate[p,t]];}$$

is the translation into *Mathematica* of the defining equation of the integrating factor, namely

$$\mu(t) = \exp\left(\int p(t)dt\right).$$

Similarly,

$$\texttt{(Integrate[intfac q,t] + const)/intfac} \qquad (4.8)$$

is the translation of

$$\frac{1}{\mu(t)}\left(\int \mu(t)q(t)dt + C\right).$$

The last statement of the miniprogram applies **Simplify** to (4.8).

Let us see how to use the command **FirstOrderLinear** to find the general solution of a first-order linear equation.

Example 4.13. *Find the general solution of* $2t\,y' + 6y = t\sin(t)$ *using the command* **FirstOrderLinear***.*

Solution. We have $a(t) = 2t$, $b(t) = 6$, and $c(t) = t\sin(t)$; these functions can be defined in *Mathematica* by

```
a[t_]:= 2t
b[t_]:= 6
c[t_]:= t Sin[t]
```

Therefore, we use

$$\texttt{FirstOrderLinear[a,b,c,C][t]//Together} \qquad (4.9)$$

The result is

```
              3                   2
2 C + 6 t Cos[t] - t  Cos[t] - 6 Sin[t] + 3 t  Sin[t]
-----------------------------------------------------
                         3
                      2 t
```

where **C** denotes the constant of integration. We have used *Mathematica*'s command **Together** to combine fractions. The general solution of $2t\,y' + 6y = t\sin(t)$ is therefore

$$y = \frac{2C + 6t\cos(t) - t^3\cos(t) - 6\sin(t) + 3t^2\sin(t)}{2t^3}$$

We note that we could have used

FirstOrderLinear[2#&,6&,# Sin[#]&,C][t]//Together

instead of defining the functions **a**, **b**, and **c** and using (4.9). ∎

Since the syntax of the command **FirstOrderLinear** is a little hard to remember, it is usually preferable to use the equivalent command **ODE** with the option **Method->FirstOrderLinear**, as explained on page 85. In fact, the option **Method->FirstOrderLinear** tells **ODE** to call the command **FirstOrderLinear**. Thus in Example 4.13 we could have used

**ODE[2t y'[t] + 6y[t] == t Sin[t],y[t],t,
Method->FirstOrderLinear]**

Exercises for Section 4.4

Use **ODE** with the option **Method->FirstOrderLinear** to find the general solution of each of the following differential equations:

1. $y' - t\,y = t$

2. $t\,y' + y = t\log(t)$

3. $t\,y' + y = t\sin(t^2)$

4. $t^2 y' + y - 2t\,y = t^2$

5. $y' + \dfrac{y}{t\log(t)} = \dfrac{1}{t}$

6. $y' + y = t^5\cos(t)^2$

7. $y' - \dfrac{3y}{t^2} = \dfrac{1}{t^2}$

8. $y' - \dfrac{2y}{t} = t\log(t)$

9. $t\,y' + 2y = \cos(t)^2$

10. $(1 + t^2)^3 y' + 4t\,y = \dfrac{t}{(1 + t^2)^2}$

Use **ODE** (or **DSolve**) to solve the following initial value problems:

11. $\begin{cases} t\,y' + y = t^4 + t^3, \\ y(1) = 1/2 \end{cases}$

16. $\begin{cases} 2(1+t)y + t(2+t)y' = 1 + 3t^2, \\ y(-1) = 1 \end{cases}$

12. $\begin{cases} 2y + t\,y' = e^t, \\ y(1) = 1 \end{cases}$

17. $\begin{cases} y'\cos(t) - y\sin(t) = t^3 e^{t^2}, \\ y(0) = 1 \end{cases}$

13. $\begin{cases} y\cot(t) + y' = 2\csc(t), \\ y(\pi/2) = 1 \end{cases}$

18. $\begin{cases} 3.1y' + \log(t)y = 0, \\ y(1.2) = 2.3 \end{cases}$

14. $\begin{cases} 2y + t\,y'(t) = 2\sin(t), \\ y(\pi) = 1/\pi \end{cases}$

19. $\begin{cases} 0.05y' + t\,y = t^2, \\ y(0.1) = 0 \end{cases}$

15. $\begin{cases} y'\cot(t) + y = 4\sin(t), \\ y(-\pi) = 0 \end{cases}$

20. $\begin{cases} y' - 0.5t\,y = e^{0.25t^2}, \\ y(0) = 1 \end{cases}$

Plot the solutions to the following initial value problems:

21. $y' + y = t\sin(t), \quad y(0) = 0.$ Use the range $-2\pi < t < 2\pi$.

22. $y'\sin(t) + y\cos(t) = \sin(t)^2 e^t, \quad y(\pi/2) = 0.$ Use the range $-\pi < t < \pi$.

23. $y' - \dfrac{2y}{t} = t, \qquad y(1) = 1.$ Use the range $0.01 < t < 1$.

24. Use **FirstOrderLinear** to find the general solution of each of the differential equations 1–10.

25. Use **FirstOrderLinear** to find the general solution of each of the differential equations 1–10 of Section 3.2.

4.5 *Separable Equations via* ODE

Frequently, **DSolve** (or **ODE** with the option **Method->DSolve**) can be used to solve a separable equation.

Example 4.14. *Find the general solution to* $t^{-2}y' = y^2\sin(t)$.

Solution. We use either

ODE[t^-2 y'[t] == y[t]^2 Sin[t],y[t],t,Method->DSolve]

or

DSolve[t^-2 y'[t] == y[t]^2 Sin[t],y[t],t]

to get

```
                                1
{{y[t]  ->  -----------------------------------------------},
                                2
            C[1]  - 2 Cos[t] + t  Cos[t] - 2 t Sin[t]

   {y[t] -> 0}}
```

The constant of integration is denoted by $C[1]$. We have found solutions

$$y = \frac{1}{C[1] - 2\cos(t) + t^2 \cos(t) - 2t\sin(t)} \quad \text{and} \quad y \equiv 0. \blacksquare$$

In contrast with the situation for first-order linear differential equations, *Mathematica*'s built-in command **DSolve** does not always solve a separable first-order equation.

Example 4.15. *Find the general solution to*

$$\frac{dy}{dt} = \frac{1 + t^3 \sin(t)}{1 + y^{3/2}}. \tag{4.10}$$

Solution. **DSolve** does not handle (4.10) well because it tries to find a formula for y in terms of t. A better scheme is to separate the variables and then use *Mathematica* to integrate each side of the equation. **ODE** with the option **Method->Separable** performs this task automatically:

**ODE[y'[t] == (1 + t^3 Sin[t])/(1 + y[t]^(3/2)),y[t],t,
Method->Separable]**

```
              5/2
        2 y[t]
y[t] +  --------  == t + C[1] + 6 t Cos[t] -
           5

    3                         2
   t  Cos[t] - 6 Sin[t] + 3 t  Sin[t]
```

Hence the solution of (4.10) is

$$y + \frac{2}{5} y^{5/2} = t + 6t \, \cos(t) - t^3 \, \cos(t) - 6 \, \sin(t) + 3 \, t^2 \, \sin(t) + C,$$

where C is a constant of integration. ∎

A separable equation with an initial condition $y(t_0) = Y_0$ can be solved by definite integration. One integrates the left-hand side of the equation from Y_0 to y and the right-hand side of the equation from t_0 to t. The following example illustrates the technique.

Example 4.16. *Solve the initial value problem*

$$\begin{cases} y^2 e^y t^{-3} e^t y' = 1, \\ y(0) = 1. \end{cases} \tag{4.11}$$

Solution. The simplest way to solve (4.11) is to use **ODE** with the option **Method->Separable**:

```
ODE[{y[t]^2 E^y[t] t^-3 E^t y'[t] == 1,y[0] == 1},y[t],t,
Method->Separable]
```

$$E^{y[t]} \; (2 - 2 \, y[t] + y[t]^2) \; == \; \frac{-6 + E^t \, (6 + E) - 6 \, t - 3 \, t^2 - t^3}{E^t}$$

However, we can also solve (4.11) making use only of *Mathematica*'s built-in command **Integrate**. First, we separate the variables in (4.11) (by hand), obtaining

$$e^y y^2 dy = t^3 e^{-t} dt. \tag{4.12}$$

Next, we use *Mathematica* to integrate both sides of (4.12); the initial condition requires that the left-hand side be integrated from 1 to y and the right-hand side from 0 to t.

```
Integrate[v^2 E^v,{v,1,y}] == Integrate[u^3 E^-u,{u,0,t}]
```

The result is

$$-E + E^y \; (2 - 2 \, y + y^2) \; == \; 6 + \frac{-6 - 6 \, t - 3 \, t^2 - t^3}{E^t}$$

Therefore, the solution of (4.11) is

$$e^y (2 - 2y + y^2) = e + 6 - \frac{6 + 6t + 3t^2 + t^3}{e^t}. \; ∎$$

Exercises for Section 4.5

Use **ODE** with the option **Method->Separable** to find the general solution of each of the following differential equations.

1. $\sin(t)(\sin(y))^2 - (\cos(t))^2 y' = 0$

5. $y^2 + 2y + 5 - y' = 0$

2. $y^3 + t^6 y' = 0$

6. $y^2 \cos(\sqrt{t}) - 2\sqrt{t}\, e^{1/t} y' = 0$

3. $1 + \sqrt{a^2 - t^2}\, y' = 0$

7. $y + (1 + t^2) \arctan(t) y' = 0$

4. $2y(t^2 - t - 1) + (t^3 - t) y' = 0$

8. $y' = (y^3 + 8) t^3 \sin(t)$

4.6   First-Order Equations with Integrating Factors via ODE

The option **Method->IntegratingFactor** tells **ODE** to try to solve a first-order equation by finding an integrating factor.

Example 4.17. *Find the general solution to*

$$2t^4 e^y + (t^5 - t^3 y^2 - 2t^3 y) e^y \frac{dy}{dt} = 0. \tag{4.13}$$

Solution. We use

```
ODE[2t^4 E^y[t] + (t^5 - t^3y[t]^2 -
2t^3y[t])E^y[t]y'[t] == 0,y[t],t,
Method->IntegratingFactor]
```

to get

```
                                       -3
Integrating factor u = u[t] = t

  y[t]    2        2
 E      (t   - y[t] ) == C[1]
```

Thus, the solution to (4.13) is

$$e^y(t^2 - y^2) = C_1,$$

where C_1 is a constant. ∎

ODE also has an option **Method->FirstOrderExact** to solve an exact first-order differential equation.

Example 4.18. *Find the general solution to*

$$3t^2 y^2 + \frac{1}{t} + \left(\frac{2t^3 y^2 - 1}{y}\right)\frac{dy}{dt} = 0. \qquad (4.14)$$

Solution. We use

```
ODE[3t^2y^2 + 1/t + ((2t^3y^2 - 1)/y)y' == 0,y,t,
Method->FirstOrderExact]
```

to get

```
  3  2
 t  y
    t
E
--------- == C[1]
   y
```

Thus, the general solution of (4.14) is

$$\frac{t\, e^{t^3 y^2}}{y} = C_1,$$

where C_1 is a constant. ∎

Of course, **Method->IntegratingFactor** will solve any exact first-order differential equation, so it can be used instead of **Method->FirstOrderExact**. For that matter, an easy argument shows that any separable first-order differential equation has an integrating factor, so **Method->IntegratingFactor** can also be used in place of **Method->Separable**. Similarly, a first-order linear differential equation can also be solved with **Method->IntegratingFactor**. The options

```
Method->FirstOrderExact        Method->Separable
Method->FirstOrderLinear       Method->Bernoulli
```

are included in the package **ODE.m**, because their implementations consist of *Mathematica* commands that mimic the mathematical solution techniques. Furthermore, they may be used to test whether a differential equation is first-order linear, first-order exact or separable. Note also that the outputs of the different forms may be slightly different even though they are mathematically equivalent.

Imitating the Mathematical Steps to Solve a First-Order Exact Equation with **Mathematica**

The command **FirstOrderExact** is the engine behind **ODE**'s option **Method->FirstOrderExact**.[3]

---

[3] This subsection contains technical *Mathematica* details on how **ODE** works. It is not needed in order to use **ODE**, nor is it referred to later.

```
FirstOrderExact[m_,n_,const_][t_,y_]:=
    Module[{},
        If[Simplify[D[m[t,y],y]/D[n[t,y],t]] =!= 1,
        Print["(",m[t,y],") d",t," + (",n[t,y],") d",y,
            " = 0 is not exact."],
    Simplify[Integrate[m[t,y],t] + Integrate[n[t,y] -
            D[Integrate[m[t,y],t],y],y] == const]]]
```

The following example shows how to use **FirstOrderExact** as a command instead of an option to **ODE**.

Example 4.19. *Use Mathematica to solve*

$$2t \, y \, dt + (t^2 - 1)dy = 0. \tag{4.15}$$

Solution. We use

FirstOrderExact[2 #1 #2&,#1^2-1&,C][t,y]

to get

```
        2
(-1 + t ) y == C
```

This is the solution to (4.15). ∎

Exercises for Section 4.6

Use **ODE** with the option **Method->IntegratingFactor** to find the general solution of each of the following differential equations. Also, determine which equations are solvable with the option **Method->FirstOrderExact**.

1. $t^3 + t \, y^4 + 2y^3 \dfrac{dy}{dt} = 0$

2. $t^2 + y^2 + 2t \, y \dfrac{dy}{dt} = 0$

3. $2t \, y + (y^2 - 3t^2) \dfrac{dy}{dt} = 0$

4. $1 - 2t \, y^2 + 2t \, y(1 - t - t \, y^2) \dfrac{dy}{dt} = 0$

5. $y(1 + t \, y) + t(1 - t \, y) \dfrac{dy}{dt} = 0$

6. $t\left(y \dfrac{dy}{dt} - 1\right) - (t \, y^2 - 2t) = 0$

7. $y^2 e^{ty^2} + 1 + (2t\, y\, e^{ty^2} - 3y^2)\dfrac{dy}{dt} = 0$

8. $4t^3 y^3 - 2y + \left(3t^4 y^2 - 2t + 2y + \dfrac{2}{y}\right)\dfrac{dy}{dt} = 0$

9. $2t \sin(t^2) + 3\cos(y) - 3t \sin(y)\dfrac{dy}{dt} = 0$

10. $y^2 - 3 - \dfrac{y}{t^2 + t\, y} + \left(\dfrac{1}{t + y} + 2\, y + 2y\right)\dfrac{dy}{dt} = 0$

4.7 *First-Order Homogeneous Equations via* **ODE**

The option **Method->FirstOrderHomogeneous** tells **ODE** to attempt to solve a first-order differential equation as a homogeneous equation.

Example 4.20. *Find the general solution to the homogeneous first-order differential equation*

$$t + y\, y' = 2y. \tag{4.16}$$

Solution. **ODE** has the option **Form->LogEquation**, which is similar to the option **Form->Equation**, the difference being that **Form->LogEquation** takes logarithms of both sides of the equation. To solve (4.16) we use

```
ODE[t + y[t]y'[t] == 2y[t],y[t],t,
Method->FirstOrderHomogeneous,Form->Equation]
```

to get

```
  1/(-1 + y[t]/t)
E                   t
----------------- == t C[1]
     t - y[t]
```

So the solution to (4.16) is

$$\frac{e^{t/(-t+y)}}{t - y} = t\, C_1,$$

where C_1 is a constant. ∎

Similarly, **Method->FirstOrderHomogeneous** can be used to solve an initial value problem.

Example 4.21. *Solve the homogeneous first-order initial value problem*

$$\begin{cases} 3y + 7t = (3t + 7y)y', \\ y(1) = -2. \end{cases} \qquad (4.17)$$

Solution. The command

```
ODE[{3y[t] + 7t ==(3t + 7 y[t])y'[t],y[1] == -2},y[t],t,
Method->FirstOrderHomogeneous]
```

gives the output

$$\frac{1}{(-1 + \frac{y[t]}{t})^{5/7}\, (1 + \frac{y[t]}{t})^{2/7}} == -(\frac{t}{3^{5/7}})$$

Hence the implicit solution to (4.17) is

$$\frac{t}{(-t + y)^{5/7}(t + y)^{2/7}} = -\frac{t}{3^{5/7}}. \ \blacksquare$$

Imitating the Mathematical Steps to Solve a Homogeneous First-Order Equation with **Mathematica**

The engine behind the option **Method->FirstOrderHomogeneous** is the command **FirstOrderHomogeneous**.[4] It is a miniprogram that imitates the mathematical steps that were used on page 65.

```
FirstOrderHomogeneous[f_,const_][t_,y_]:=
    Module[{s,z,tmp},
        tmp = Together[f[t,y] /. y -> z t];
        Simplify[Integrate[1/(tmp - z),z] ==
            (Log[t] + const) /. z -> y/t]]
```

Example 4.22. *Use* **FirstOrderHomogeneous** *to find the general solution of*

$$y' = \frac{2y^4 + t^4}{3t\, y^3}. \qquad (4.18)$$

---

[4]This subsection contains technical *Mathematica* details on how ODE works. It is not needed in order to use ODE, nor is it referred to later.

Solution. We first define the right-hand side of (4.18) as a pure function:

g:= (2#2^4 +#1^4)/(3#1#2^3)&

Here *t* is denoted by **#1** and *y* is denoted by **#2**. We get the solution of (4.18) with the command

FirstOrderHomogeneous[g,C][t,y]

```
          4
         y  -(3/4)        t
(1 -  ---)          ==   -
          4               C
         t
```

Thus, the solution of (4.18) is

$$-\frac{3}{4} \log\left(1 - \frac{y^4}{t^4}\right) = A + \log(t),$$

where $A = -\log(\mathbf{C})$ is a constant. ∎

Exercises for Section 4.7

Use **ODE** with the option **Method->FirstOrderHomogeneous** to find the general solution of each of the following differential equations.

1. $y' = \dfrac{y^2}{t^2 - t\,y}$

2. $y' = \dfrac{2t\,y}{3t^2 + y^2}$

3. $y' = -\dfrac{15t + 11y}{9t + 5y}$

4. $y' = \dfrac{y - t}{y + t}$

5. $t\,y' - y = \sqrt{t^2 + y^2}$

6. $y' = -\dfrac{y(2t^3 - y^3)}{t(2y^3 - t^3)}$

7. $y' = \dfrac{y - t - t\,\cos(y/t)}{t}$

8. $y' = \dfrac{y + t\,e^{y/t}}{t}$

9. $y' = \dfrac{y - 2t\,\tanh(y/t)}{t}$

10. $y' = -\dfrac{t^2 + t\,y + y^2}{t^2}$

4.8   Bernoulli Equations via ODE

The following example shows how to use **Method->Bernoulli** to find the general solution to a Bernoulli equation. (See Section 3.6.)

Example 4.23. *Use* ODE *with option* **Method->Bernoulli** *to find the general solution of $y' + 2t\,y = 2a\,t^3y^3$. Solve and plot the initial value problem*

$$\begin{cases} y' + 2t\,y = 2a\,t^3y^3, \\ y(0) = 1. \end{cases} \tag{4.19}$$

Solution. The command

```
ODE[y'[t] + 2t y[t] == 2a t^3 y[t]^3,y[t],t,
Method->Bernoulli]
```

yields

```
                          Sqrt[2]
{{y[t] -> ─────────────────────────────────}, {y[t] -> 0}}
                                      2
                  2        2 t
          Sqrt[a + 2 a t + 2 E    C[1]]
```

Thus, the general solution is

$$y = \left(\frac{2}{a + 2a\,t^2 + 2e^{2t^2}C} \right)^{1/2},$$

where C is a constant. To solve and plot the initial value Problem (4.19) we use

```
ODE[{y'[t] + 2t y[t] == 2a t^3 y[t]^3,y[0] == 1},y[t],t,
Method->Bernoulli,Parameters->{{a,-1,4}},
PlotSolution->{{t,-1,1}}]
```

to get the output

```
                          Sqrt[2]
{{y[t] -> ─────────────────────────────────────}}
                       2            2
                   2 t          2 t         2
          Sqrt[a + 2 E    - a E    + 2 a t ]
```

and the plot

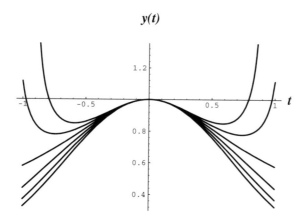

Solution to the initial value problem

$$y' + 2ty = 2a\,t^3 y^3, \quad y(0) = 1, \quad -1 \le a \le 4$$

Exercises for Section 4.8

Use **ODE** with the option **Method->Bernoulli** to find the general solution of each of the following differential equations.

1. $y' - y = t\,e^{5t} y^3$

2. $y' + 5y = t\sin(t)e^t y^5$

3. $y' + \dfrac{y}{t-3} = 5(t-3)y^{3/2}$

4. $t^3 y' - 2t\,y = \dfrac{t}{y^{10}}$

Use **ODE** with the option **Method->Bernoulli** to solve each of the following initial value problems.

5. $\begin{cases} t\,y' = t\,y^2 + (1+t)y, \\ y(1) = 1 \end{cases}$

6. $\begin{cases} 3(1+t^2)y' = 2t\,y(y^3 - 1), \\ y(0) = 2 \end{cases}$

7. $\begin{cases} y' + \tan(2t)y = y^2 \cos(2t)\sin(2t), \\ y(0) = 1 \end{cases}$

8. $\begin{cases} y' = \dfrac{y^2 + 2t\,y}{t^2}, \\ y(1) = 2 \end{cases}$

4.9   *Clairaut and Lagrange Equations via* ODE

A **Clairaut**[5] **differential equation** is a first-order differential equation of the form

$$y = p t + f(p),\tag{4.20}$$

where p stands for dy/dt. If we differentiate (4.20) with respect to t we obtain

$$p = p + (t + f'(p))\frac{dp}{dt},$$

which reduces to

$$0 = (t + f'(p))\frac{dp}{dt}.\tag{4.21}$$

Thus, either

$$\frac{dp}{dt} = 0,$$

or

$$t + f'(p) = 0.\tag{4.22}$$

If $dp/dt = 0$, then $p = C$, where C is a constant; this yields the **general solution** to (4.20) of the form

$$y_{\text{gen}}(t) = C t + f(C).\tag{4.23}$$

Each value of C in (4.23) gives rise to a straight line.

On the other hand, if (4.22) holds, we obtain a plane curve parametrized by p, namely

$$\begin{cases} t &= -f'(p), \\ y &= -f'(p)p + f(p). \end{cases}\tag{4.24}$$

Then (4.24) represents the **singular solution** to (4.20). Since it is nonlinear, (4.24) is not a special case of (4.23). In fact, it turns out that the envelope of all straight lines of the form (4.23) is (4.24).

Example 4.24. *Find the general solution of*

$$y = t y' + (y')^2\tag{4.25}$$

and also the singular solution.

Solution. We let $p = y'$ and rewrite (4.25) as

$$y = t\,p + p^2. \tag{4.26}$$

Taking derivatives of both sides of (4.26) and canceling two terms, we obtain

$$0 = t\,p' + 2p\,p' = p'(t + 2p). \tag{4.27}$$

Then (4.27) implies that either $p' = 0$ or $t + 2p = 0$. If $p' = 0$, then

$$p = C_1, \tag{4.28}$$

where C_1 is a constant. Substituting (4.28) into (4.26), we obtain the general solution of (4.25), namely

$$y_{\text{gen}}(t) = t\,C_1 + C_1^2.$$

If, on the other hand, $t = -2p$, we substitute $p = -t/2$ into (4.26) to obtain the singular solution of (4.25):

$$y_{\text{sing}}(t) = -\frac{t^2}{4}. \quad \blacksquare$$

y(t)

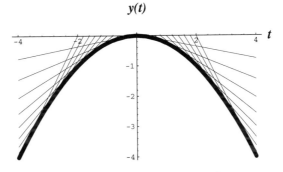

Solutions to $y = t\,y' + (y')^2$

ODE's option **Method->Clairaut** can be used to solve a Clairaut equation.

Example 4.25. *Use* **ODE** *to find the general solution of* $y = t\,y' + (y')^{1/5}$, *and also the singular solution. Plot the singular solution.*

Solution. The command

```
ODE[y == t y' + y'^(1/5),y,t,
Method->Clairaut,Form->Equation,
PlotSingularSolution->{{{t,-3,1},{y,-2,2}}}]
```

results in the following output:

```
                                    4
Singular solution(s) detected: 3125 t y   == -256

          1/5                     4
{y == C[1]    + t C[1], 3125 t y   == -256}
```

and gives the plot

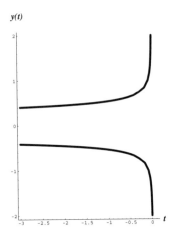

Singular solution to $y = t y' + (y')^{1/5}$

Thus, we have found the general solution $y_{\text{gen}}(t) = C_1 + t\,C_1$, where C_1 is a constant, and the singular solution

$$y_{\text{sing}}(t) = \left(\frac{256}{-3125t}\right)^{1/4} . \blacksquare$$

For any single **ODE** command, all plots generated can be superimposed onto a single plot with the option **SuperimposePlots->True**. For Clairaut equations, this can be very illuminating, since the envelope of the family of general solutions is defined by the singular solution as seen in Example (4.25).

Example 4.26. *Use* **ODE** *to plot a family of general solutions together with the singular solution to the Clairaut equation* $y = t\,y + y'^3$.

Solution. The command

```
ODE[y == t y' + y'^3,y,t,
Method->Clairaut,Form->Equation,
Parameters->{{C[1],-1,1,0.1}},
PlotSingularSolution->{{t,-2,0},
  PlotStyle->{AbsoluteThickness[3]}},
PlotSolution->{{t,-2,0},
  PlotStyle->{GrayLevel[0.5]}},
SuperimposePlots->True]
```

results in the following output:

```
                          3        2
Singular solution detected: {4 t   == -27 y }
```

```
               3    3        2
{y == t C[1] + C[1] , 4 t   == -27 y }
```

and gives the plot

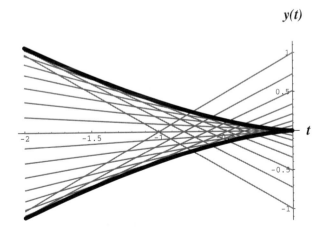

**A family of general solutions and
the singular solution to** $y = t y' + (y')^3$

Lagrange Equations

A **Lagrange**[6] **equation** is a first-order differential equation of the form

$$y = f(p)t + g(p), \qquad\qquad (4.29)$$

where p stands for dy/dt. Clearly, a Clairaut equation is a special case of a Lagrange equation. To solve (4.29) the technique used to solve the Clairaut equation (4.20) must be generalized. We illustrate the use of **ODE**'s option **Method->Lagrange** to solve a Lagrange equation.

Example 4.27. *Use* **ODE** *to find the general solution of* $y = 2t\, y'^2 + \cos(y')^2$ *and also the singular solution.*

Solution. We use

```
ODE[y == 2t y'^2 + Cos[y']^2,y,t,Method->Lagrange] /. ODEs->s
```

to get

```
       C - Cos[2 s] - SinIntegral[2 s]
{t == ---------------------------------- ,
                         2
                 (1 - 2 s)

                 2
             2  2 s  (C - Cos[2 s] - SinIntegral[2 s])
  y == Cos[s]  + ------------------------------------- ,
                                     2
                             (1 - 2 s)

                  t        1 2
  y == 1,  y == - + Cos[-]  }
                  2        2
```

SinIntegral is explained in Section 4.10. ∎

Exercises for Section 4.9

Use **ODE** with the option **Method->Clairaut** to find both the general solution and the singular solution of each of the following differential equations.

---

1. $y = t\,y' + (1 + y'^2)^{1/2}$

2. $y = t\,y' + (y')^3$

3. $y = t\,y' + \sin(y')$

4. $y = t\,y' + 3 - 1/y'$

5. $y = t\,y' + \exp(y')$

6. $y = t\,y' + \log(y')$

7. $y = t\,y' + (y')^4/4$

8. $y = t\,y' + 2\exp(y')$

Use **ODE** with the option **Method->Lagrange** to find both the general solution and the singular solution of each of the following differential equations.

9. $y = 2t\,y' + \log(y')$

10. $y = t(1 + y') + y'^2$

11. $y = (3/2)t\,y' + \exp(y')$

12. $y = 2t\,y' + \cos(y')^2$

4.10   Nonelementary Integrals

Frequently, the solution of a differential equation requires an integration that cannot be carried out by the standard methods of calculus. Here is a short list of such nonelementary integrals:

Definition	*Mathematica* **Name**
$\mathbf{erf}(t) = \dfrac{2}{\sqrt{\pi}} \displaystyle\int_0^t e^{-u^2} du$	`Erf[t]`
$\mathbf{erfi}(t) = \dfrac{2}{\sqrt{\pi}} \displaystyle\int_0^t e^{u^2} du$	`Erfi[t]`
$\mathbf{li}(t) = \displaystyle\int_0^t \dfrac{du}{\log(u)}$	`LogIntegral[t]`
$\mathbf{S}(t) = \displaystyle\int_0^t \sin\left(\dfrac{u^2 \pi}{2}\right) du$	`FresnelS[t]`
$\mathbf{C}(t) = \displaystyle\int_0^t \cos\left(\dfrac{u^2 \pi}{2}\right) du$	`FresnelC[t]`
$\mathbf{Si}(t) = \displaystyle\int_0^t \dfrac{\sin(u)}{u} du$	`SinIntegral[t]`

The table above contains only a few of the special functions defined in *Mathematica*. Others are encountered in subsequent chapters. For example, the Gamma function is defined in Section 14.4 for use with Laplace transforms, and Jacobi elliptic functions are briefly discussed in Section 19.4 in connection with the pendulum equation. Airy functions are treated in Section 20.4, Bessel functions are studied in Sections 21.4–21.7, and hypergeometric functions are dealt with in Section 21.9.

Concerning the functions in the above table, the Fresnel integrals $\mathbf{S}(t)$ and $\mathbf{C}(t)$ are used to describe the plane curve called the clothoid in Section 19.5. The function **erf** is the so-called **error function**; it is related to the **probability integral** of mathematical statistics, defined by $\Phi(t) = \dfrac{1}{\sqrt{2\pi}} \displaystyle\int_{-\infty}^{t} e^{-w^2/2} dw$. The two functions are connected by the formula

$$\mathbf{erf}(t) = 2\Phi\left(t\sqrt{2}\right) - 1. \tag{4.30}$$

The solution of many simple differential equations involves **erf**, as in the following example.

Example 4.28. *Find the general solution to the differential equation*

$$y' - 2t\,y = 1. \tag{4.31}$$

Plot several solutions.

Solution. To find the general solution, it suffices to solve the initial value problem

$$\begin{cases} y' - 2t\,y = 1, \\ y(0) = a, \end{cases}$$

where a is an arbitrary constant. For this we use

```
ODE[{y' - 2 t y == 1,y[0] == a},y[t],t,
Method->FirstOrderLinear,Parameters->{{a,-1,1,0.5}},
PlotSolution->{{t,-2,2}}]
```

to get

```
                2
             2  t
            t   E   Sqrt[Pi] Erf[t]
{{y[t] -> a E   + ------------------}}
                         2
```

and the plots

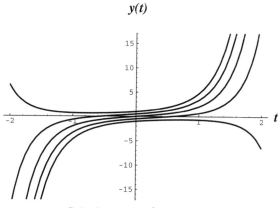

y(t)

Solutions to $y' - 2t\,y = 1$

Hence the general solution to (4.31) is

$$y = Ce^{t^2} + \frac{e^{t^2}\sqrt{\pi}\,\mathbf{erf(t)}}{2},$$

where C is a constant. ∎

The **log integral** (denoted by **li**) is a function that arises in number theory.

Example 4.29. *Solve the initial value problem*

$$\begin{cases} y' = 1/\log(t), \\ y(0) = 1. \end{cases}$$

Plot the solution over the interval $0 \le t < 1$.

Solution. We use

```
ODE[{y' == 1/Log[t],y[0] == 1},y[t],t,
Method->FirstOrderLinear,
PlotSolution->{{t,0,1}}]
```

to get

```
{{y[t] -> 1 + LogIntegral[t]}}
```

and the plot

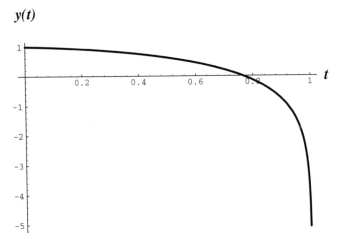

Solution to $y' = 1/\log(t),$ $y(0) = 1$

Using the notation introduced on page 108, we can write $y(t) = 1 + \mathbf{li}(t)$. ∎

Exercises for Section 4.10

Use **ODE** to solve each of the following initial value problems. Plot the solutions over appropriate intervals. [These may take a while - be patient.]

1. $\begin{cases} y' = \sin(t^2), \\ y(0) = 0 \end{cases}$

3. $\begin{cases} y' = t\,y + y^5, \\ y(0) = 4 \end{cases}$

2. $\begin{cases} y' = t^2 \cos(t^2), \\ y(0) = 0 \end{cases}$

4. $\begin{cases} y' = 1 - t^2 + y^2, \\ y(0) = 1 \end{cases}$

5. Prove Equation (4.30).

6. Show that $\mathbf{erfi}(t) = -i\,\mathbf{erf}(i\,t)$.

4.11   Using ODE *to Define New Functions*

In Section 4.10 we made use of *Mathematica*'s extensive collection of functions to solve certain differential equations in terms of nonelementary functions. Unfortunately, there are many other differential equations that cannot be solved using this collection. In this section we reverse the usual proceedure of solving a differential equation in terms of known functions: we show how to use **ODE** to define a new function from an initial value problem. This new function can be algebraically manipulated, differentiated, and integrated with *Mathematica*. Furthermore, all of *Mathematica*'s plotting routines are available to aid in the visual understanding of the function. Before we describe the proceedure, we need to discuss the difference between **=** and **:=**.

On page 21 we discussed function definitions using **:=**. This is usually the best way to define a function, but sometimes it is useful to use **=** instead. The difference between **:=** and **=** is as follows.

- **lhs = rhs** evaluates **rhs** and assigns the result to be the value of **lhs**. From then on **lhs** is replaced by **rhs** whenever it appears.

- In contrast, **lhs:= rhs** assigns **rhs** to be the delayed value of **lhs**, and **rhs** remains in unevaluated form. When **lhs** appears, it is replaced by **rhs**, evaluated anew each time.

Although **a = 2** and **a:= 2** give the same output, there is a difference for functions. For example, let us first put **b = 5**; then we define two functions by

```
f[x_] = b + x
```

and

```
g[x_]:= b + x
```

Then `f[t]` and `g[t]` produce the same result, namely

```
5 + t
```

However, let us redefine **b = 3**; then `f[t]` gives the same result as before, but `g[t]` gives instead

```
3 + t
```

A rule of thumb is that `lhs:= rhs` should ordinarily be used, but `lhs = rhs` should be employed for complicated functions, for which repeated evaluation would be too time-consuming.

To define a new function with **ODE**, it is essential to use `=` and not `:=`, because we want the computation done immediately. (The use of `:=` is possible, but then **ODE** would attempt to solve the differential equation each time the function is evaluated.) A functional definition also requires the option **Form->Explicit**. The following example illustrates the procedure.

Example 4.30. *Use* **ODE** *to define the function* $y(t) = \int_{\pi/4}^{t} \left(\frac{\tan s}{s}\right) ds.$

Solution. Since *Mathematica* cannot automatically integrate $\tan(s)/s$, we use

```
y[t_] = ODE[{yy' == Tan[t]/t,yy[Pi/4] == 1},yy,t,
        Method->FirstOrderLinear,Form->Explicit] /. ODEs->s
```

to obtain

```
     4 t                Tan[s]      Pi
2 - --- + Integrate[------, {s, --, t}]
     Pi                  s          4
```

(We have used **yy** inside the **ODE** command for convenience. It reminds us that in future *Mathematica* commands **yy** should not be used as a dummy variable unless its value is first cleared. Also the replacement **ODEs->s** just simplifies the notation.) Now the function **y** is available for further processing. Thus, we can plot **y** with

```
Plot[Evaluate[y[t]],{t,-Pi/2,Pi/2}]
```

obtaining

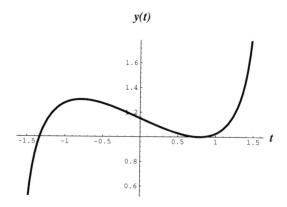

Exercises for Section 4.11

Use **ODE** to define solutions to the following initial value problems. Then plot the solutions over an appropriate interval.

1. $\begin{cases} y' - t\,y = \sin(t), \\ y(0) = 0 \end{cases}$

5. $\begin{cases} y' + t\,y = \dfrac{\sin(t)}{t}, \\ y(1) = 0 \end{cases}$

2. $\begin{cases} y' = \sin(t^3), \\ y(0) = 0 \end{cases}$

6. $\begin{cases} y' + y = e^{-t^2}, \\ y(0) = 1 \end{cases}$

3. $\begin{cases} y' = \sin(e^t), \\ y(0) = 0 \end{cases}$

7. $\begin{cases} y' = e^{-t^2}\cos(t), \\ y(0) = 1 \end{cases}$

4. $\begin{cases} y' = e^{\cos(t)}, \\ y(0) = 1 \end{cases}$

8. $\begin{cases} y' + t\,y = \sin(e^{-t})y^2, \\ y(0) = 1 \end{cases}$

4.12   Riccati Equations

A first-order differential equation of the form

$$a(t)y'(t) = b(t) + c(t)y(t) + d(t)y(t)^2$$

is called a (generalized) **Riccati**[7] **equation**. Assuming that $a(t)$ is nonzero, it can be reduced to

$$y'(t) = p(t) + q(t)y(t) + r(t)y(t)^2. \qquad (4.32)$$

In the case that p, q, and r are constants, (4.32) is a separable equation, whose solution is given implicitly by

$$\int \frac{dy}{p + q\,y + r\,y^2} = t + C,$$

where C is a constant. When p, q, and r are nonconstant functions, the situation is much more complicated. Since (4.32) is nonlinear, we cannot expect the sum of solutions to be a solution. Nevertheless, if we know one solution to (4.32), we can find the general solution.

Theorem 4.1. *Suppose that $y(t)$ and y_κ are solutions of (4.32). Define a new function $z(t)$ by the equation*

$$y(t) = y_\kappa(t) + \frac{1}{z(t)}. \qquad (4.33)$$

Then $z(t)$ satisfies the linear equation $z'(t) + \big(q(t) + 2r(t)y_\kappa(t)\big)z(t) = -r(t)$, for which an integrating factor is

$$\mu(t) = \exp\left(\int \big(q(t) + 2y_\kappa(t)r(t)\big)\, dt\right). \qquad (4.34)$$

Hence the general solution to (4.32) is given by

$$y(t) = y_\kappa(t) - \frac{\mu(t)}{C + \displaystyle\int r(t)\mu(t)\, dt}, \qquad (4.35)$$

where C is a constant and y_κ is a known solution.

Proof. We define z by (4.33) and use (4.32) to calculate

$$y_\kappa' - \frac{z'}{z^2} = y' = p + q\left(y_\kappa + \frac{1}{z}\right) + r\left(y_\kappa + \frac{1}{z}\right)^2. \qquad (4.36)$$

The fact that y_κ is a solution to (4.32) reduces (4.36) to

$$-\frac{z'}{z^2} = \frac{q}{z} + \frac{2r\,y_\kappa}{z} + \frac{r}{z^2}, \qquad \text{or} \qquad z' + (q + 2r\,y_\kappa)z = -r.$$

7

Count Jacopo Francesco Riccati (1676–1754). Venetian mathematician and founder of a family of mathematicians. He solved special cases of (4.32) in 1724. Many books use the misspelling "Ricatti."

Clearly, this first-order differential equation has μ defined by (4.34) as an integrating factor. Thus

$$\mu z = -C - \int \mu r \, dt, \tag{4.37}$$

where C is a constant. Then (4.35) follows from (4.37). ∎

Example 4.31. *Find the general solution to*

$$t \, y' - y^2 + (2t + 1)y = t^2 + 2t, \tag{4.38}$$

given the solution $y_K = t$.

Solution. In hand calculations it is usually better to use a *method* of solution instead of a *formula* for a solution; therefore we proceed as follows. We first reduce (4.38) to

$$y' = t + 2 - \left(2 + \frac{1}{t}\right)y + \frac{y^2}{t}. \tag{4.39}$$

Let $y = t + 1/z$; from (4.39) we get

$$1 - \frac{z'}{z^2} = y' = t + 2 - \left(2 + \frac{1}{t}\right)\left(t + \frac{1}{z}\right) + \frac{1}{t}\left(t + \frac{1}{z}\right)^2 = 1 - \frac{1}{t\,z} + \frac{1}{t\,z^2},$$

or

$$z' - \frac{z}{t} = -\frac{1}{t}.$$

The general solution of this first-order linear differential equation is $z = 1 + C\,t$, where C is a constant. Hence the general solution of (4.38) is

$$y = t + \frac{1}{1 + C\,t}. \quad ∎$$

The *Mathematica* command to find the general solution of a Riccati equation, given one solution **yk**, closely resembles (4.35):

```
Riccati[a_,b_,c_,d_,yk_,const_Symbol][t_]:=
    Module[{intfac,p,q,r},
        p = b[t]/a[t];
        q = c[t]/a[t];
        r = d[t]/a[t];
        intfac = Exp[Integrate[q + 2yk[t] r,t]];
        Simplify[yk[t] -
            intfac/(Integrate[r intfac,t] + const)]]
```

Easier to use is **ODE**'s option **Method->Riccati**; it is illustrated in the following example:

Example 4.32. *Solve and plot the initial value problem*

$$\begin{cases} y' + y^2 - 2y\sin(t) + \big(\sin(t)\big)^2 - \cos(t) = 0, \\ y(\pi/2) = 0, \end{cases}$$

given the known solution $y_K(t) = \sin(t)$ *to* $y' + y^2 - 2y\sin(t) + \big(\sin(t)\big)^2 - \cos(t) = 0.$

Solution. We use

```
ODE[{y' + y^2 - 2y Sin[t] + Sin[t]^2 - Cos[t] == 0,
y[Pi/2] == 0},y,t,Method->Riccati,KnownSolution->Sin[t],
PlotSolution->{{t,-2Pi,2Pi}}]
```

to obtain

```
         -2 + 2 Sin[t] + Pi Sin[t] - 2 t Sin[t]
{{y -> ─────────────────────────────────────────}}
                    2 + Pi - 2 t
```

and the plot

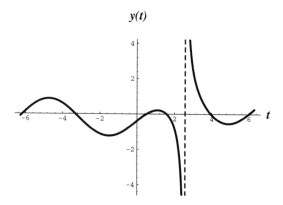

y(t)

Solution to $y' + y^2 - 2y\sin(t) + \big(\sin(t)\big)^2 - \cos(t) = 0,$
$y(\pi/2) = 0,$ **plotted over the interval** $-2\pi < t < 2\pi$

Exercises for Section 4.12

In each of Exercises 1–6 find the general solution to the Riccati equation given a known solution y_K. Work each problem by hand and with *Mathematica*. (Exercise 6 involves the function **Erfi**.)

1. $\begin{cases} y' = 6 + 5y + y^2, \\ y_K = -2 \end{cases}$

4. $\begin{cases} y' = e^{2t} + (1 + 2e^t)y + y^2, \\ y_K = -e^t \end{cases}$

2. $\begin{cases} y' = 9 + 6y + y^2, \\ y_K = -3 \end{cases}$

5. $\begin{cases} y' = \sec(t)^2 - \tan(t)y + y^2, \\ y_K = \tan(t) \end{cases}$

3. $\begin{cases} y' = -2 - y + y^2, \\ y_K = 2 \end{cases}$

6. $\begin{cases} y' = 1 - t - y + t\,y^2, \\ y_K = 1 \end{cases}$

In each of the Exercises 7–10 solve the initial value problem, both by hand and with *Mathematica*.

7. $\begin{cases} y'e^{-t} + y^2 - 2y\,e^t = 1 - e^{2t}, \\ y(0) = 2, \quad y_K = e^t \end{cases}$

9. $\begin{cases} y' = e^{2t} + (1 + 2e^t)y + y^2, \\ y(0) = 1, \quad y_K = -e^t \end{cases}$

8. $\begin{cases} t^2 y' = t^2 y^2 + ty + 1, \\ y(1) = 1, \quad y_K = -1/t \end{cases}$

10. $\begin{cases} y' = -4/t^2 - y/t + y^2, \\ y(1) = 1, \quad y_K = 2/t \end{cases}$

11. Suppose that $y(t)$ is a solution of the Riccati equation $y' = p(t) + q(t)y + r(t)y^2$ with $r(t) \neq 0$. Define

$$w(t) = \exp\left(-\int r(t)y(t)dt\right).$$

Show that $y = -w'/(r(t)w)$ and that $w(t)$ satisfies the second-order linear equation

$$w'' - \left(q(t) + \frac{r'(t)}{r(t)}\right)w' + p(t)r(t)w = 0.$$

12. Suppose that $w(t)$ is a solution of the linear second-order equation $w'' + P(t)w' + Q(t)w = 0$, and that $w(t) \neq 0$. Define $y(t) = w'(t)/w(t)$ and show that $y(t)$ is a solution of the Riccati equation $y' = -Q(t) - P(t)y - y^2$.

5

EXISTENCE AND UNIQUENESS OF SOLUTIONS OF FIRST-ORDER DIFFERENTIAL EQUATIONS

In Chapter 3 we discussed solution methods for a variety of first-order differential equations. It is clear, however, that some differential equations are too complicated to find formulas for their solutions. For example, there is no way to find a formula for the solution of a complicated differential equation such as

$$y' + \sin(t + 3\sin(y)) = 0.$$

In this chapter we address the problem of the **existence** of a solution to a first-order differential equation. It turns out that we can show that many first-order differential equations possess solutions, even though we have no clue on how to find a formula for the solution. On the other hand, in Chapter 13 we shall show how to find numerical approximations to many differential equations. For example, the initial value problem

$$\begin{cases} y' + \sin(t + 3\sin(y)) = 0, \\ y(0) = -10 \end{cases} \tag{5.1}$$

has an approximate numerical solution whose plot is

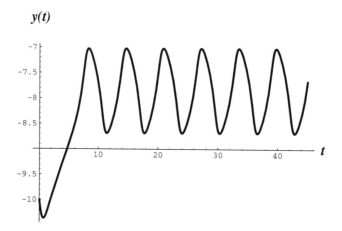

A related problem is the **uniqueness** of the solution of an initial value problem. In Section 5.1 we give a basic existence and uniqueness theorem (Theorem 5.1) that states that under mild differentiability assumptions a first-order initial value problem has a unique solution. Nevertheless, we give an example of a first-order initial value problem that has more than one solution. Explosions of differential equations are discussed in Section 5.2. Picard iterations are treated in Section 5.3; they are used in Section 5.4 to give proofs of the theorems of Section 5.1. Direction fields are introduced in Section 5.5.

Stability of autonomous nonlinear equations is discussed in Section 5.6. This section contains the critical point method, which because of its simplicity is especially useful.

5.1   The Existence and
Uniqueness Theorem

Recall that the most general form of a first-order differential equation is

$$F(t, y, y') = 0. \tag{5.2}$$

Of course, (5.2) might have no solutions for algebraic reasons. For example,

$$y'^2 + t^2 = -1$$

has no (real) solutions because there are no real numbers a and b for which $a^2 + b^2 = -1$. Since we are interested in criteria for the solutions of differential equations and not of algebraic equations, we shall assume that (5.2) can be solved algebraically for y'; then we rewrite (5.2) as

$$y' = f(t, y). \tag{5.3}$$

Here f is a continuous function of two variables t and y in some region of the (t, y) plane. Let us define precisely what we mean by a solution to (5.3).

Definition. A **solution** *on the interval $a < t < b$ of the differential equation $y' = f(t, y)$ (if it exists) is a function $y = y(t)$ with a continuous derivative y' that identically satisfies the equation*

$$y'(t) = f\big(t, y(t)\big) \tag{5.4}$$

for $a < t < b$.

Frequently, the solution to (5.3) involves a constant of integration C_1, as we saw in Section 3.2. In order to pin down the solution it is convenient to consider the corresponding **initial value problem**.

Definition. *Let $a < t_0 < b$. A **solution** on the interval $a < t < b$ of the initial value problem*

$$\begin{cases} y'(t) = f\big(t, y(t)\big), \\ y(t_0) = Y_0 \end{cases} \tag{5.5}$$

is a solution to $y' = f(t, y)$ for which $y(t_0) = Y_0$.

This means that we look for a solution whose graph passes through the point (t_0, Y_0). A word on the terminology is in order. The term **initial value problem** is suggested by the interpretation of the independent variable t as time. In this interpretation we are given the state of the differential equation at an initial moment of time, and we are required to determine the state of the differential equation for all future time. Strictly speaking, this leads to an equation that is to be solved for $t \geq t_0$, although usually the solution exists for $t < t_0$ as well.

It is not necessary that f in (5.3) be defined everywhere; it is sufficient that the domain of definition contain an open rectangle about some point (t_0, Y_0) in whose neighborhood we search for a solution.

Example 5.1. *In what region can we expect a solution to the differential equation*

$$y' = \frac{1 + t\, y}{4 - y^2}? \tag{5.6}$$

Solution. The function f given by

$$f(t, y) = \frac{1 + t\,y}{4 - y^2}$$

is well-defined in any region for which $y \neq \pm 2$. For example, f is defined in the open rectangle $\{\,(t, y) \mid |t| < 2.67, |y| < 1.999\,\}$ containing the point $(0, 0)$. ∎

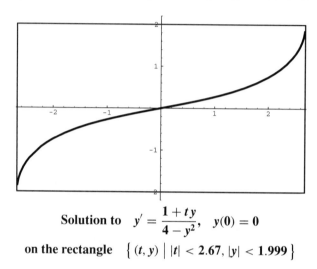

Solution to $y' = \dfrac{1 + t y}{4 - y^2}, \quad y(0) = 0$

on the rectangle $\{\,(t, y) \mid |t| < \mathbf{2.67}, |y| < \mathbf{1.999}\,\}$

We can now state the existence and uniqueness theorem for first-order differential equations.

Theorem 5.1.   (**Local existence and uniqueness**) *Suppose that* $f(t, y)$ *is a function defined on the rectangle*

$$\{\,(t, y) \mid a < t < b, c < y < d\,\},$$

which contains the point (t_0, Y_0). *Assume that both*

$$f(t, y) \qquad and \qquad \frac{\partial f}{\partial y}(t, y)$$

are continuous in this rectangle.

(i)   *There is an interval* $t_0 - h < t < t_0 + h$ *inside the interval* $a < t < b$ *on which is defined a solution* $y = \phi(t)$ *of the initial value problem*

$$\begin{cases} y' = f(t, y), \\ y(t_0) = Y_0. \end{cases}$$

(ii) *If $\psi(t)$ is another solution defined on the interval $t_0 - h < t < t_0 + h$, then $\phi(t) = \psi(t)$ on that interval.*

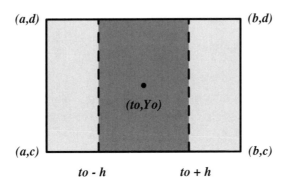

A proof of Theorem 5.1 is presented in Section 5.4.

Example 5.2. *Use Theorem 5.1 to show that the initial value problem* (5.1) *has a solution near the point* $(0, -10)$.

Solution. In this case $(t_0, Y_0) = (0, -10)$ and $f(t, y) = -\sin(t + 3\sin(y))$. Clearly, $f(t, y)$ is continuous for all t and y. We compute

$$\frac{\partial f}{\partial y} = -\cos(t + 3\sin(y))3\sin(y),$$

so that $\partial f / \partial y$ is continuous for all t and y. Theorem 5.1 implies that $y' = \sin(t + 3\sin(y))$ has a solution near any (t_0, Y_0), in particular near $(0, -10)$. ∎

  In order to understand the meaning of Theorem 5.1, it is helpful to study an example in which the hypotheses are *not* satisfied. If the hypothesis of continuity of the partial derivative $\partial f / \partial y$ is dropped, then we may have several solutions passing through the same point.

Example 5.3. *Consider the initial value problem*

$$\begin{cases} y' = y^{1/3}, \\ y(0) = 0. \end{cases} \tag{5.7}$$

Show that there are three solutions of the form $y = A t^a$ for suitable choices of the constants A and $a > 0$.

Solution. We make the educated guess that the solution of (5.7) has the form $y = A\,t^a$. We compute

$$y' = A\,a\,t^{a-1} \quad \text{and} \quad y^{1/3} = (A\,t^a)^{1/3}.$$

In order that these two expressions be equal for all t, we must have equality of the coefficients:

$$A\,a = A^{1/3}, \tag{5.8}$$

and if $A \ne 0$ equality of the exponents:

$$a - 1 = a/3. \tag{5.9}$$

One possible choice is $A = 0$, leading to the solution $y \equiv 0$ of (5.7). To obtain a nonzero solution of (5.7), we solve (5.9), obtaining $a = 3/2$; substitution of this value into (5.8) leads to $A^{2/3} = 2/3$, that is, $A = \pm\sqrt{8/27}$. Three solutions of (5.7) are thus

$$y_1(t) \equiv 0, \qquad y_2(t) = +t^{3/2}\sqrt{\frac{8}{27}}, \qquad y_3(t) = -t^{3/2}\sqrt{\frac{8}{27}}. \; \blacksquare \tag{5.10}$$

It is interesting that although **DSolve** fails, **ODE** finds all of the solutions (5.10) to (5.7) using the option **Method->AllSymbolic** (see the exercises).

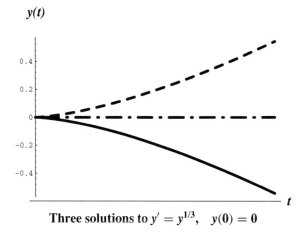

Three solutions to $y' = y^{1/3}, \quad y(0) = 0$

Example 5.3 can be "explained" in terms of Theorem 5.1. The function $f(t, y) = y^{1/3}$ is continuous, but the partial derivative

$$\frac{\partial f}{\partial y} = \frac{1}{3y^{2/3}}$$

is discontinuous at $y = 0$, violating the hypothesis of Theorem 5.1. Therefore, we may expect several solutions passing through $(0, 0)$.

Theorem 5.1 can be used to show that the graphs of two solutions of the same differential equation cannot cross each other—or even touch each other in the (t, y) plane. We formulate this as a corollary.

Corollary 5.2. *Suppose that $y_1(t)$ and $y_2(t)$ are solutions of the same differential equation $y' = f(t, y)$ and satisfy $y_1(t_0) = y_2(t_0)$ for some t_0. Assume that the hypotheses of Theorem 5.1 are satisfied. Then $y_1(t) = y_2(t)$ for all values of t in the interval of definition.*

In practical situations Corollary 5.2 can be used to obtain important information about the solutions of an equation that may be difficult to solve explicitly.

Example 5.4. *Consider the initial value problem*

$$\begin{cases} y' = \sin(y), \\ y(0) = Y_0, \end{cases} \tag{5.11}$$

where $0 < Y_0 < \pi$. Show that $0 < y(t) < \pi$ for all t.

Solution. The differential equation $y' = \sin(y)$ has the constant solutions $y_1(t) \equiv 0$ and $y_2(t) \equiv \pi$, since $\sin(0) = \sin(\pi) = 0$. Suppose $y(t)$ were another solution of (5.11) for which $y(t_1) \geq \pi$ for some t_1. By continuity there would be a smallest value of t_1 for which this happens; call it t_0. Then $y(t_0) = \pi$. Consider the initial value problem

$$\begin{cases} y' = \sin(y), \\ y(t_0) = \pi. \end{cases} \tag{5.12}$$

But we already know the solution to (5.12), namely $y_2(t) \equiv \pi$. Corollary 5.2 implies the uniqueness of the solution of (5.12). Hence $y(t) \equiv y_2(t) \equiv \pi$. Thus we get a contradiction to the assumption that $0 < y(0) < \pi$.

A similar argument shows that any solution $y(t)$ to (5.11) is always greater than zero. Hence every solution $y(t)$ must remain between 0 and π. ∎

Let us visually check the argument of Example 5.4 by plotting the solution of the initial value Problem (5.11).

Example 5.5. *Use* **ODE** *to find and plot the solution of the initial value problem*

$$\begin{cases} y' = \sin(y), \\ y(0) = a, \end{cases} \tag{5.13}$$

for $a = 0.1\pi,\ldots,a = 0.9\pi$.

Solution. We use the options **Parameters** and **PlotSolution** as follows:

```
ODE[{y' == Sin[y],y[0] == a},y,t,
Method->Separable,Parameters->{{a,0.1Pi,0.9Pi,0.1Pi}},
PlotSolution->{{t,-2Pi,2Pi}}]
```

Mathematica answers

```
                  t      a
{{y -> 2 ArcTan[E   Tan[-]]}}
                         2
```

and gives the plot

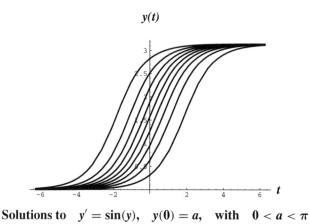

$$y(t)$$

Solutions to $y' = \sin(y),$ $y(0) = a,$ with $0 < a < \pi$

Thus, the general solution to (5.13) is $y = 2 \arctan\left(e^t \tan\left(\dfrac{a}{2}\right)\right).$ ∎

Exercises for Section 5.1

For each of the following equations, find the region in the (t, y) plane where the existence and uniqueness theorem (Theorem 5.1) guarantees the existence of a unique solution through any specified initial point.

1. $y' = t/y$

3. $y' = (t^2 + y^2 - 4)^{1/3}$

2. $y' = \dfrac{3t + 2y}{2t - 5y}$

4. $y' = \dfrac{\log|(t + 1)y|}{y - t^3}$

5. For which values of α can one expect to have a unique solution to the equation $y' = |y|^\alpha$ for any specified initial point in the (t, y) plane?

6. Use **ODE** to find and plot three solutions to the initial value Problem (5.7).

7. Show that the solution $y(t)$ to the initial value problem

$$
\begin{cases}
y' = \dfrac{\sin(y)}{\cosh(t)}, \\[2mm]
y(0) = \pi/2
\end{cases}
$$

satisfies $0 < y(t) < \pi$ for all t. Use **ODE** to find and plot the solution.

5.2 Explosions and a Criterion for Global Existence

In general we can expect a nonlinear differential equation $y' = f(t, y)$ to have only **local solutions**. In other words, a solution may exist only for some proper subset of the interval $a < t < b$ even though $f(t, y)$ is defined on all of the interval $a < t < b$. This is in marked contrast to the linear equations studied in Section 3.2, where the solutions are **global**: any solution of a linear equation of the form $y' + p(t)y = q(t)$ is defined on the entire interval on which the functions $p(t)$ and $q(t)$ are continuous. In order to describe this property of nonlinear equations, one often uses the term **explosion**. Other authors use the term **blowup**.

Definition. *The* **explosion time** T *of a solution* $y(t)$ *of an initial value problem with* $y(t_0) = Y_0$ *is the largest time* $T > t_0$ *for which* $y(t)$ *is defined for* $t < T$, *but not for* $t = T$.

The solution of a nonlinear differential equation may explode in a finite time, which cannot be predicted from the form of the equation. This is best illustrated with an example.

Example 5.6. *Consider the initial value problem*

$$
\begin{cases}
y'(t) = y(t)^2, \\[2mm]
y(0) = 25
\end{cases}
\tag{5.14}
$$

defined for $-\infty < t < \infty$. *Show that the hypotheses of Theorem 5.1 are satisfied and that the solution to (5.14) is*

$$
y(t) = \frac{1}{1/25 - t} = \frac{25}{1 - 25t}.
\tag{5.15}
$$

Find the interval on which this solution exists.

Solution.   Both the function $f(t, y) = y^2$ and its partial derivative $\partial f/\partial y = 2y$ are continuous for $-\infty < t < \infty$ and $-\infty < y < \infty$. Thus the hypotheses of Theorem 5.1 are satisfied. Since (5.14) is separable, its solutions can be easily found by hand using the methods of Section 3.3 or with *Mathematica*. For example, the command

```
ODE[{y'[t]  ==  y[t]^2,y[0]  ==  25},y[t],t,Method->Separable]
```

yields

```
                 25
{{y[t]  ->  ──────────}}
            1 - 25 t
```

telling us that the solution to (5.14) is indeed given by (5.15). The solution $y(t)$ is defined only on the interval $-\infty < t < 0.04$, even though the differential equation is defined for $-\infty < t < \infty$. ∎

The following graph of the solution to (5.14) shows the explosion at 0.04.

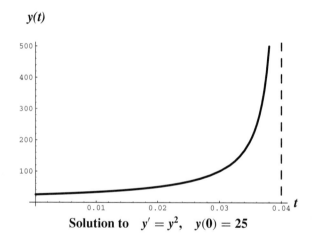

Solution to $y' = y^2$, $y(0) = 25$

Next, we give an example of a differential equation with two explosions.

Example 5.7.   *Show that the initial value problem*

$$\begin{cases} y' = 1 + y^2, \\ y(0) = 0 \end{cases} \tag{5.16}$$

suffers explosions at $t = \pm\pi/2$. Plot the solution.

Solution.   We use

```
ODE[{y' == 1 + y^2,y[0] == 0},y,t,
Method->Separable,
PlotSolution->{{t,-0.475Pi,0.475Pi}}]
```

to get

```
{{y -> Tan[t]}}
```

and the plot

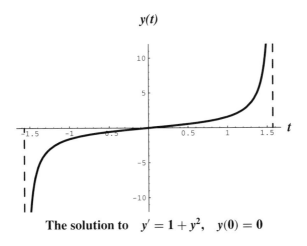

The solution to $y' = 1 + y^2,$ $y(0) = 0$

Thus, the solution to (5.16) is $y = \tan(t)$, which clearly has singularities at $\pm\pi/2$. ∎

The behavior of Examples 5.6 and 5.7 cannot occur for first-order linear differential equations. For a first-order linear equation $y' + p(t)y = q(t)$, the solution is defined on the entire interval of definition of the coefficient functions $p(t)$ and $q(t)$. The following theorem (proved in Section 5.4) gives a condition on a nonlinear equation sufficient to exclude explosions.

Theorem 5.3. *Suppose that we have the differential equation $y' = f(t, y)$ defined on the interval $a < t < b$, where $f(t, y)$ is a continuous function with a continuous partial derivative $\partial f/\partial y$. Suppose in addition that the function f satisfies the following growth condition:*

$$|f(t, y)| \le p(t)|y| + q(t), \tag{5.17}$$

for $a < t < b$ and $-\infty < y < \infty$, where $p(t)$ and $q(t)$ are continuous functions on the interval $a < t < b$. Then the solution $y(t)$ can be defined on the entire interval $a < t < b$.

Theorem 5.3 can be paraphrased as saying that *if a nonlinear differential equation can be dominated by a linear equation, then explosions cannot occur.* We shall prove Theorem 5.3 in Section 5.4.

Corollary 5.4. *Suppose that $p(t)$ and $q(t)$ are continuous functions on the interval $a <$ $t < b$. Then any solution $y(t)$ to the first-order linear differential equation $y' + p(t)y = q(t)$ can be defined on the whole interval $a < t < b$.*

Proof. We take $f(t, y) = -p(t)y + q(t)$ and define $\tilde{p}(t) = |p(t)|$ and $\tilde{q}(t) = |q(t)|$. Clearly

$$|f(t, y)| \le \tilde{p}(t)|y| + \tilde{q}(t),$$

so that (5.17) is satisfied (using $\tilde{p}$ and $\tilde{q}$). Hence the corollary follows from Theorem 5.3. ∎

Theorem 5.3 can also be used to prove global existence of solutions to certain nonlinear equations.

Example 5.8. *Show that the equation $y' = (t + y)\sin(1 + y^2)$ has solutions that are defined on the entire real line $-\infty < t < \infty$.*

Solution. We know that $|\sin(\theta)| \le 1$ for any real number θ. So we can take $p(t) = 1$ and $q(t) = |t|$ in Theorem 5.3 and apply the theorem in the form $|f(t, y)| \le |y| + |t|$ to conclude the global existence of the solution. ∎

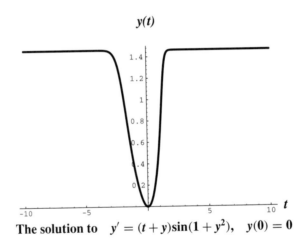

$$y(t)$$

The solution to $y' = (t + y)\sin(1 + y^2), \quad y(0) = 0$

It is interesting to note that the example $y' = y^2$ considered in Example 5.6 does not satisfy the hypotheses of Theorem 5.3, since there is no choice of the functions $p(t)$ and $q(t)$ that will render valid an inequality of the form $y^2 \le p(t)|y| + q(t)$. Indeed, we know that this equation has explosions—global existence is not satisfied.

Exercises for Section 5.2

For each of the following differential equations, determine whether or not the hypotheses of the theorem on global existence (Theorem 5.3) are satisfied, by finding appropriate functions $p(t)$ and $q(t)$.

1. $y' = t + y^2$

2. $y' = \log(1 + y^2)$

3. $y' = \cos(t^2 + y)$

4. $y' = t \sin(t)$

5. $y' = \sinh(y)$

6. $y' = \tanh(y)$

7. $y' = \sinh(t)\sin(y) + e^{t^2}$

8. $y' = t \log(t^2 + y^2)$

9. $y' = e^{-y^2}$

10. $y' = y\,e^{-y^2}$

For each of the following initial value problems, show explicitly that an explosion occurs by solving the equation by separation of variables and finding the explosion time T.

11. $\begin{cases} y' = y^4, \\ y(0) = 1 \end{cases}$

13. $\begin{cases} y' = y^2 + 3y + 2, \\ y(0) = 0 \end{cases}$

12. $\begin{cases} y' = 4 + y^2, \\ y(0) = 0 \end{cases}$

14. $\begin{cases} y' = e^y, \\ y(0) = 0 \end{cases}$

5.3   *Picard Iteration*

In Chapter 3 we gave several techniques that can be used to solve first-order differential equations or initial value problems. In this section we take a different approach by discussing an iterative technique called the Picard[1] method. This method is more important theoretically than practically; we use it in Section 5.4 to prove Theorem 5.1.

First, we observe that we can rewrite an initial value problem

$$\begin{cases} y'(t) = f\big(t, y(t)\big), \\ y(t_0) = Y_0 \end{cases} \tag{5.18}$$

as a single equation. This is done by taking the definite integral of both sides of the differential equation of (5.18) and using the initial condition. The fundamental theorem

---

[1]

Charles Émile Picard (1856–1941). French mathematician who made important contributions to real and complex analysis. He proved that if $f(z)$ is a complex function of a complex variable that has an essential singularity at a, then in a neighborhod of a the equation $f(z) = A$ has infinitely many solutions, except possibly for two values of A.

of calculus states that for any differentiable function y with a continuous derivative y', the function can be retrieved as

$$y(t) = y(t_0) + \int_{t_0}^t y'(s)ds. \tag{5.19}$$

If y satisfies (5.18), then (5.19) can be rewritten as

$$y(t) = Y_0 + \int_{t_0}^t f(s, y(s))ds. \tag{5.20}$$

Conversely, it is easy to see that (5.20) implies (5.18). Equation (5.20) is an **integral equation**, which incorporates both the differential equation and the initial condition. For many purposes (5.20) is a more efficient statement of the initial value problem, since it consists of one equation instead of two.

Equation (5.20) gives rise to a sequence $\{y_n\}$ of **Picard approximating solutions** to (5.18) as follows. We begin with a constant function as a trial solution

$$y_0(t) = Y_0. \tag{5.21}$$

The graph of this function passes through the point (t_0, Y_0), but the function y_0 does not satisfy the differential equation, unless $f(t, y_0) = 0$ for all t. Excluding this trivial case, we define the next approximating function y_1 by

$$y_1(t) = Y_0 + \int_{t_0}^t f(s, y_0(s))ds. \tag{5.22}$$

The new function y_1 satisfies $y_1'(t) = f(t, y_0(t))$ by the fundamental theorem of calculus, and its graph also passes through (t_0, Y_0). To obtain higher approximations, we continue this procedure to define functions $y_2(t)$, $y_3(t)$, and so on, so that in general

$$y_n(t) = Y_0 + \int_{t_0}^t f(s, y_{n-1}(s))ds, \tag{5.23}$$

for $n = 1, 2, \ldots$. The y_n's are called the **Picard approximating functions**. We shall use them in Section 5.4 to prove the existence and uniqueness theorem (Theorem 5.1). This is the method of **Picard iteration**.

Example 5.9. *Find the first three Picard approximating functions y_1, y_2, y_3 for the initial value problem*

$$\begin{cases} y' = 1 + y^2, \\ y(0) = 0. \end{cases} \tag{5.24}$$

Solution. In this case we have

$$y_0(t) = 0, \qquad y_1(t) = 0 + \int_0^t (1 + 0^2)\,ds = t,$$

$$y_2(t) = 0 + \int_0^t (1 + s^2)\,ds = t + \frac{t^3}{3},$$

$$y_3(t) = 0 + \int_0^t \left(1 + \left(s + \frac{s^3}{3}\right)^2\right) ds$$

$$= \int_0^t \left(1 + s^2 + \frac{2s^4}{3} + \frac{s^6}{9}\right) ds = t + \frac{t^3}{3} + \frac{2t^5}{15} + \frac{t^7}{63}. \ \blacksquare$$

The alert reader will note that the first three terms of the Taylor series for the tangent function appear in these computations.

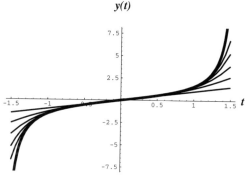

The first five Picard approximations to $y' = 1 + y^2,\ \ y(0) = 0,$
together with the exact solution $y(t) = \tan(t)$

Using *Mathematica* for *Picard Iteration*

Recursive definitions are possible in *Mathematica*. To define a function **f** recursively, we first define **f[0]**; then **f[n]** is defined in terms of the values **f[m]**, where **m** is an integer greater than or equal to **0** but less than **n**. The simplest example of such a function is the factorial function. Although *Mathematica* has this function defined internally (it is simply **n!**), we could also define an equivalent function by

fac[0]:= 1   and   **fac[n_]:= n fac[n - 1]**

For example, **fac[6]** gives the expected result 720. We can use a modification of **fac** to define a sequence of Picard approximating functions in *Mathematica*. The following example illustrates the procedure.

Example 5.10. *Find the first three Picard approximating functions* y_1, y_2, y_3 *for the initial value problem*

$$\begin{cases} y' = t^2 - y, \\ y(1) = 2. \end{cases} \tag{5.25}$$

Solution. To define the sequence of approximating functions we use

```
y[0,t_]:= 2
y[n_,t_]:= y[0,t] + Integrate[u^2 - y[n-1,u],{u,1,t}]
```

The definitions of the approximating functions `y[0,t]`, `y[1,t]`, `y[2,t]` and `y[3,t]` can be computed and conveniently displayed with the command

```
Table[y[n,t],{n,0,3}]//TableForm
```

This gives

$$2$$

$$\frac{11}{3} - 2t + \frac{t^3}{3}$$

$$\frac{53}{12} - \frac{11t}{3} + t^2 + \frac{t^3}{3} - \frac{t^4}{12}$$

$$\frac{93}{20} - \frac{53t}{12} + \frac{11t^2}{6} - \frac{t^4}{12} + \frac{t^5}{60}$$

as the result. ∎

Let us automate the procedure of Example 5.10. The following command **Picard** first creates a sequence of approximating functions `y[i,t]`, and then uses **Table** to create a list of them.

```
Picard[f_,{t0_,Y0_},iterations_][t_]:=
  Module[{s,y,nn,tt},
    y[0,s_]:= Y0;
    y[nn_,tt_]:= Y0 + Integrate[f[s,y[nn-1,s]],{s,t0,tt}];
    Table[y[i,t],{i,0,iterations}]]
```

Furthermore, **ODE** with the option **Method->Picard** has the same effect as the command **Picard**. Thus the command

```
Picard[(#1^2 - #2)&,{1,2},3][t]//TableForm
```

gives the same output as that of Example 5.10. To obtain the simultaneous plot of the Picard approximating solutions $y_0(t), \ldots, y_5(t)$ of (5.25) we use

```
ODE[{y'== t^2 - y,y[1] == 2},y,t,
Method->Picard,Iterations->a,
GraphLabel->approxsols,
Parameters->{{a,0,5,1}},PlotSolution->{{t,-3,6}}]
```

to obtain

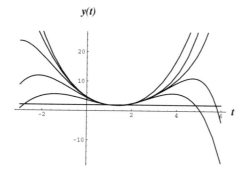

Since (5.25) is a first-order linear equation, its exact solution can be obtained and plotted as a thick line with the command

```
ODE[{y' == t^2 - y,y[1] == 2},y,t,
Method->FirstOrderLinear,GraphLabel->plotexact,
PlotSolution->{{t,-3,6},PlotStyle->{AbsoluteThickness[2]}}]
```

The result is

```
                 1 - t        2
{{y[t] -> 2 + E       - 2 t + t }}
```

and the graph is

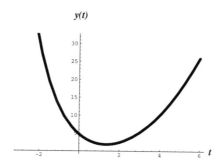

The plots of the approximating functions and the exact solution can be combined with

`Show[{Graph[plotexact],Graph[approxsols]}]`

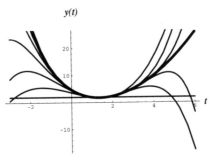

The exact solution and Picard approximations to

$$y' = t^2 - y, \quad y(1) = 2$$

Exercises for Section 5.3

Use the method of Picard iteration to find approximations to the solutions of the following equations.

1. $\begin{cases} y' = 2y, \\ y(0) = 1 \end{cases}$
5. $\begin{cases} y' = -y^3/2, \\ y(0) = 1 \end{cases}$

2. $\begin{cases} y' = y^2, \\ y(0) = 1 \end{cases}$
6. $\begin{cases} y' = 1 + y, \\ y(1) = 0 \end{cases}$

3. $\begin{cases} y' = 1 + 3y, \\ y(0) = 2 \end{cases}$
7. $\begin{cases} y' = 1 + 4y + 2y^2, \\ y(0) = 0 \end{cases}$

4. $\begin{cases} y' = -2t\,y, \\ y(0) = 1 \end{cases}$
8. $\begin{cases} y' = 1 - t\,y^2, \\ y(0) = 0 \end{cases}$

5.4   *Proofs of Existence Theorems*

We give the proofs[2] of the existence theorems of Sections 5.1 and 5.2.

5.4.1   *Proof of the Convergence of Picard's Method*

In this subsection we use the method of Picard iteration to prove the existence and uniqueness theorem (Theorem 5.1). The idea of the proof is to show that the Picard iterations (defined by (5.23)) converge uniformly and in the limit satisfy the integral equation (5.20).

We are given a continuous function $f(t, y)$ with a continuous first partial derivative $\partial f / \partial y$ in the rectangle

$$\tilde{R} = \big\{ (t, y) \mid a < t < b, c < y < d \big\},$$

which contains the point (t_0, Y_0). These continuous functions are also bounded, that is,

$$|f(t, y)| \le K_0 \qquad \text{and} \qquad \left| \frac{\partial f}{\partial y} \right| \le K_1 \tag{5.26}$$

for some constants K_0 and K_1. By the mean value theorem from calculus, (5.26) implies that

$$|f(t, y_1) - f(t, y_2)| \le K_1 |y_1 - y_2|, \tag{5.27}$$

whenever (t, y_1) and (t, y_2) are points in $\tilde{R}$. The inequality (5.27) is called a **local Lipschitz condition**.

In order to simplify the notation, we work with a rectangle

$$R = \big\{ (t, y) \mid |t - t_0| \le \alpha \text{ and } |y - Y_0| \le \beta \big\}.$$

Clearly, this is no loss of generality, since the given rectangle $\tilde{R}$ contains a rectangle of the form R, centered on (t_0, Y_0).

From (5.26) it follows that each Picard approximating function $y_n(t)$ (given by (5.23)) satisfies

$$|y_n(t) - Y_0| = \left| \int_{t_0}^{t} f\big(s, y_{n-1}(s)\big) ds \right| \le K_0 |t - t_0|. \tag{5.28}$$

Now we restrict t so that $|t - t_0| \le \beta / K_0$, leading to the choice of $h = \min(\alpha, \beta / K_0)$. Clearly, $y_0(t) = Y_0$ lies in the rectangle R. If $y_{n-1}(t)$ lies in the rectangle R, then (5.28) shows that $y_n(t)$ also lies in the rectangle R. Therefore, by mathematical induction it follows that all of the Picard approximating functions remain in the rectangle R.

Having done this, we define

$$M_n(t) = |y_n(t) - y_{n-1}(t)|,$$

---

[2]This section is theoretical and may be omitted by those not interested in the details of the proof.

for $n = 1, 2, \ldots$. Clearly, $M_n(t_0) = 0$ for all n. From (5.27) it follows that

$$
\begin{aligned}
M_{n+1}(t) &= \left| y_{n+1}(t) - y_n(t) \right| = \left| \int_{t_0}^{t} \Big(f\big(s, y_n(s)\big) - f\big(s, y_{n-1}(s)\big) \Big) ds \right| \\
&\le K_1 \left| \int_{t_0}^{t} \left| y_n(s) - y_{n-1}(s) \right| ds \right| = K_1 \left| \int_{t_0}^{t} M_n(s) ds \right|
\end{aligned}
\tag{5.29}
$$

for $n = 1, 2, \ldots$. (In (5.29) we have been careful to take care of the case $t \le t_0$ as well as the case $t \ge t_0$.)

We now use mathematical induction to show that (5.29) implies the inequalities

$$
M_n(t) \le K_0 K_1^{n-1} \frac{|t - t_0|^n}{n!}
\tag{5.30}
$$

for all n. Clearly, (5.30) holds for $n = 1$, since

$$
M_1(t) \le \left| \int_{t_0}^{t} |f(s, y_0)| ds \right| \le K_0 |t - t_0|.
$$

Assuming that (5.30) has been proved for the value n, we compute

$$
M_{n+1}(t) \le K_1 \left| \int_{t_0}^{t} M_n(s) ds \right| \le K_1 \left| \int_{t_0}^{t} K_0 K_1^{n-1} \frac{|s - t_0|^n}{n!} ds \right| = K_0 K_1^{n} \frac{|t - t_0|^{n+1}}{(n+1)!},
$$

completing the inductive step.

Furthermore, the assumption that $|t - t_0| \le h$ and (5.30) imply that

$$
|y_{n+1}(t) - y_n(t)| = M_{n+1}(t) \le \left(\frac{K_0}{K_1} \right) \frac{(K_1 h)^{n+1}}{(n+1)!}.
\tag{5.31}
$$

Since the infinite series

$$
\sum_{n=0}^{\infty} \frac{(K_1 h)^{n+1}}{(n+1)!}
$$

converges, (5.31) implies that the infinite series

$$
\sum_{n=0}^{\infty} \big(y_{n+1}(t) - y_n(t) \big)
$$

converges uniformly. But

$$
y_m(t) - Y_0 = y_m(t) - y_0(t) = \sum_{n=0}^{m-1} \big(y_{n+1}(t) - y_n(t) \big),
\tag{5.32}
$$

since the sum in (5.32) collapses. Therefore, the limit

$$y(t) = \lim_{m \to \infty} y_m(t)$$

exists. We now take the limit of both sides of equation (5.23). The left-hand side tends to $y(t)$. By uniform convergence, the right-hand side tends to

$$Y_0 + \int_{t_0}^{t} f(s, y(s)) ds.$$

Therefore, $y(t)$ is a solution of the integral equation (5.20). In particular, $y(t)$ satisfies the differential equation and initial condition (5.18). ∎

5.4.2   *Proof of the Criterion for Global Existence*

In this subsection we prove the global existence theorem (Theorem 5.3). We are given that the function $f(t, y)$ satisfies the inequality $|f(t, y)| \leq p(t)|y| + q(t)$ where $p(t)$ and $q(t)$ are continuous functions on an interval $a \leq t \leq b$. In particular, on any subinterval $a_1 \leq t \leq b_1$ there are constants P and Q such that $|p(t)| \leq P$ and $|q(t)| \leq Q$ on this subinterval. Let $a < t_0 < b$. We propose to show that for $t_0 \leq t \leq b$ each Picard iteration $y_n(t)$ satisfies

$$|y_n(t)| \leq |y_0| \left(1 + P(t - t_0) + \frac{P^2}{2}(t - t_0)^2 + \cdots + \frac{P^n}{n!}(t - t_0)^n \right)$$

$$+ \left(Q(t - t_0) + \frac{Q P(t - t_0)^2}{2} + \cdots + \frac{Q P^{n-1}}{n!}(t - t_0)^n \right) \qquad (5.33)$$

for $n = 1, 2, \ldots$ This will be accomplished by the method of mathematical induction. For $n = 1$ we have by definition

$$|y_1(t)| \leq \left| y_0 + \int_{t_0}^{t} f(s, y_0(s)) ds \right| \leq |y_0| + \int_{t_0}^{t} \left(|y_0|P + Q \right) ds$$

$$= |y_0|\left(1 + P(t - t_0) \right) + Q(t - t_0). \qquad (5.34)$$

Thus, (5.33) holds for $n = 1$. Assuming the truth of (5.33) for the value $n - 1$, we compute

$$|y_n(t)| = \left| y_0 + \int_{t_0}^{t} f(s, y_{n-1}(s)) ds \right| \leq |y_0| + \int_{t_0}^{t} \left(Q + P|y_{n-1}(s)| \right) ds$$

$$\leq |y_0| + \int_{t_0}^{t} \left(Q + P|y_0| \left(1 + P(s - t_0) + \cdots + \frac{P^{n-1}(s - t_0)^{n-1}}{(n-1)!} \right) \right) ds$$

$$+ P \int_{t_0}^{t} \left(Q(s - t_0) + \frac{Q P(s - t_0)^2}{2} + \cdots + \frac{Q P^{n-2}(s - t_0)^{n-1}}{(n-1)!} \right) ds. \qquad (5.35)$$

The integrals on the right-hand side of (5.35) are easily computed, so we obtain

$$|y_n(t)| \le |y_0| \left(1 + P(t - t_0) + \frac{P^2(t - t_0)^2}{2} + \cdots + \frac{P^n(t - t_0)^n}{n!} \right)$$

$$+ Q \left((t - t_0) + \frac{P(t - t_0)^2}{2} + \cdots + \frac{P^{n-1}(t - t_0)^n}{(n - 1)!} \right). \tag{5.36}$$

Then (5.34)–(5.36) give the inequality (5.33) for all values of n by mathematical induction in the case that $t_0 \le t \le b$. From (5.33) we get

$$|y_n(t)| \le |y_0| \left(1 + P|t - t_0| + \frac{P^2}{2}|t - t_0|^2 + \cdots + \frac{P^n}{n!}|t - t_0|^n \right)$$

$$+ \left(Q|t - t_0| + \frac{Q\,P|t - t_0|^2}{2} + \cdots + \frac{Q\,P^{n-1}}{n!}|t - t_0|^n \right). \tag{5.37}$$

Moreover, a similar proof shows that (5.37) also holds in the case that $a \le t \le t_0$. Thus, (5.37) holds for $a \le t \le b$.

The first term on the right-hand side of (5.37) is a partial sum of the power series for the exponential function

$$e^{P|t-t_0|} = \sum_{k=0}^{\infty} \frac{P^k}{k!}|t - t_0|^k;$$

hence

$$\sum_{k=0}^{n} \frac{P^k}{k!}|t - t_0|^k \le e^{P|t-t_0|}. \tag{5.38}$$

Similarly, the second term on the right-hand side of (5.37) is the partial sum of the power series for

$$\frac{Q}{P}\left(e^{P|t-t_0|} - 1\right) = \sum_{k=1}^{\infty} Q\frac{P^k}{k!}|t - t_0|^k,$$

so that

$$\sum_{k=1}^{\infty} Q\frac{P^k}{k!} \le \frac{Q}{P}\left(e^{P|t-t_0|} - 1\right). \tag{5.39}$$

Hence (5.37)–(5.39) imply that for $a \le t \le b$ we have

$$|y_n(t)| \le |y_0|e^{P|t-t_0|} + \frac{Q}{P}\left(e^{P|t-t_0|} - 1\right). \tag{5.40}$$

Therefore, the Picard iterations remain in a common rectangular region for all values of t with $a \le t \le b$. Letting K_0 be the maximum of $|f(t, y)|$ and K_1 the maximum of $|\partial f/\partial y|$

over this rectangle, we can refer to the proof of Section 5.4.1, to see that the argument presented there can be made here, leading to the estimates

$$M_n(t) \le K_0 K_1^{n-1} \frac{|t - t_0|^n}{n!}$$

for $n = 1, 2, \ldots$ and to the convergence of the Picard iterations $y_n(t)$ for $a \le t \le b$. ∎

5.5   *Direction Fields and Differential Equations*

Whether or not we can obtain an explicit elementary solution to $y' = f(t, y)$, we can always obtain some geometric idea of the solution by plotting a short line segment or an arrow through each point (t, y) with slope $f(t, y)$. The collection of all such segments is called a **direction field** for the differential equation. Computing a direction field requires only the evaluation of f on some collection of points; this is much easier than solving and plotting the differential equation itself. By constructing short segments through sufficiently many points, we may form an idea of the nature of the solution.

Here is the direction field for the differential equation

$$y' = -y \sin(t) + \cos(t),$$

together with the solution passing through the point $(0, 1)$.

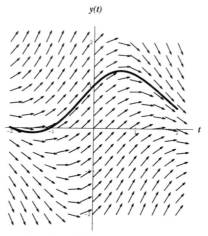

Direction field for the equation
$$y' = -y \sin(t) + \cos(t)$$

Direction fields provide an excellent method of getting some idea of what the solution of a differential equation looks like without the aid of a computer or hand calculator. Frequently it is possible to use the function f to draw the short segments by hand. This information can then be used to approximate graphically the solutions of the differential equation by fitting the solutions between the arrows.

It is also possible to use **ODE** with any method and the option **PlotField** to draw a direction field.

Example 5.11. *Find the direction field for $y' = 1 - 3t\,y$ using the* **PlotField** *option of* **ODE**.

Solution. We use

```
ODE[y' == 1 - 3t y,y,{t,-2,2},
Method->FirstOrderLinear,
PlotField->{{{t,-2,2},{y,-2,2}}}]
```

to get

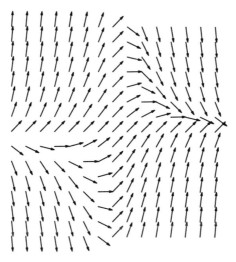

Direction field for the equation
$$y' = 1 - 3t\,y$$

The option **PlotField** is a plotting function like **PlotSolution**, but there are two syntactical differences. Firstly, both a **t** and a **y** range must be specified for **PlotField**. Secondly, since **PlotField** requires no solution to the differential equation it plots, the value of **Method** is irrelevant. With no method specified, **ODE** uses **DSolve**. To prevent

ODE from attempting to solve a differential equation, the option **Method->None** should be used.

A direction field plot and a solution plot can be combined using the option **SuperimposePlots->True**. For example,

```
ODE[{y' == 1 - 3t y,y[0]==0},y,t,
Method->FirstOrderLinear,
PlotField->{{{t,-2,2},{y,-2,2}}},
PlotSolution->{{t,-2,2},
  PlotStyle->{{AbsoluteThickness[2]}}},
SuperimposePlots->True]
```

yields the output

```
           Pi              3
      Sqrt[—] Erfi[Sqrt[-] t]
           6              2
{{y ->  —————————————————————}}
                  2
              (3 t )/2
            E
```

and the plot

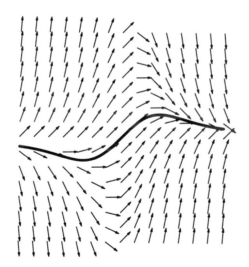

Direction field with solution of the equation
$$y' = 1 - 3t\,y$$

(The definition of **Erfi** is given on page 108.)

Finally, we give an example of the use of **PlotField** to give an idea of the solution curves of a differential equation whose solution cannot be found exactly. We use **Method->None** to prevent **ODE** from looking for a solution.

Example 5.12. *Find the direction field for* $y' = \cos(t) + \sin(y(y - 1))$ *using the* **PlotField** *option of* **ODE** *without solving the differential equation.*

Solution. We use

```
ODE[y' == Cos[t] + Sin[y(y - 1)],y,{t,-2,2},
Method->None,
PlotField->{{{t,-2,2},{y,-2,2}}}]
```

to get the plot

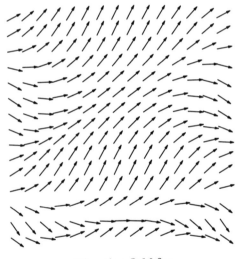

Direction field for

$$y' = \cos(t) + \sin(y(y-1))$$

Exercises for Section 5.5

Use **ODE** with the options **PlotField** and **Method->None** to plot the direction fields for the following differential equations.

1. $y' = \dfrac{y+t}{y-t}$ **4.** $y' = e^{2t} + (1 + 2e^t)y + y^2$

2. $y' = \sin\left(\dfrac{y+t}{y-t}\right)$ **5.** $y' = \sin(t)/t$

3. $y' + y = t\,y^3$ **6.** $y' = t + y^2$

Use **ODE** with the options `PlotField`, `PlotSolution` and `SuperimposePlots->`
`True` to plot the direction fields together with one solution curve for the following differential equations.

7. $y' = y + y^3$ **11.** $y' = \cos(t)/\sin(y)$

8. $y' = 1 - y/t$ **12.** $y' = e^{\sin(t)}\cos(t)$

9. $y' = \sin(t)/y$ **13.** $y' = y - \cos\big((\pi/2)t\big)$

10. $y' = t/y$ **14.** $y' = t\,e^y$

5.6   *Stability Analysis of Nonlinear First-Order Equations*

In Sections 3.1–3.6 we described several techniques for solving first-order ordinary differential equations. Each involves some amount of ingenuity. In this section we describe a
completely different approach to the study of solutions of differential equations of the form

$$y' = f(y). \tag{5.41}$$

A differential equation of the form (5.41), whose right-hand side has no explicit dependence
on t, is called **autonomous**. Of course, (5.41) is a separable equation; from Section 3.3 we
know that an implicit solution is given by

$$\int \frac{dy}{f(y)} = t + C, \tag{5.42}$$

where C is a constant. However, if f is even mildly complicated, the integral on the
left-hand side of (5.42) may be tough to compute. Instead of actually finding formulas
for solutions of (5.41), we study their *behavior*. This method for studying the solutions of
(5.41) is called the **critical point method**.

It is remarkable that the critical point method is generally simpler than the techniques discussed in Chapters 3 and 4. In addition to being important in its own right, the critical point method will be extended to autonomous systems of differential equations, which will be studied in Chapter 17.

Critical Points and Critical Solutions

In order to obtain information about the solutions of (5.41), we first find those values of y for which the right-hand side of (5.41) vanishes.

Definition. *The values* $Y_1, Y_2, \ldots$ *for which* $f(Y_i) = 0$ *are called the* **critical points** *of the differential equation* $y' = f(y)$. *The solution* $y_i(t) \equiv Y_i$ *of the initial value problem*

$$\frac{dy}{dt} = f(y) \quad \text{with} \quad y(0) = Y_i \tag{5.43}$$

is called a **critical solution**, *or an* **equilibrium solution**, *of* $y' = f(y)$.

The critical point method gives qualitative information about a differential equation without actually solving the equation. It can be summarized as follows:

> To study the behavior of solutions of $y' = f(y)$ using the critical point method, carry out the following steps:
>
> **Step 1:**   Determine certain easily found solutions called **critical solutions**.
>
> **Step 2:**   Understand the other solutions in terms of the **critical solutions**.

Example 5.13.   *Find the critical points and critical solutions of the differential equation*

$$\frac{dy}{dt} = 5y - y^2. \tag{5.44}$$

Solution. In this case we have

$$f(y) = 5y - y^2 = y(5 - y),$$

which is zero when either $y = 0$ or $y = 5$. Therefore, the differential equation (5.44) has two critical points. We have found quite easily two solutions to (5.44), namely the critical solutions $y(t) \equiv 0$ and $y(t) \equiv 5$. ∎

Once we have found the critical points and critical solutions of (5.41), we must determine the behavior of the other solutions. Since f may be complicated, we want to determine the behavior of each noncritical solution without finding a formula for it. In order to do this, it is essential to know when the function $f(y)$ that appears in the differential equation $y' = f(y)$ is positive and when it is negative. In those regions where $f(y) > 0$ we have $y' > 0$, signifying that the solution will be *increasing*. In those regions in which $f(y) < 0$ we have $y' < 0$, signifying that the solution will be *decreasing*. We expect that the solution will increase or decrease toward a critical solution as time becomes large, without ever attaining it for a finite time.

An easy way to determine the sign of $f(y)$ for various values of y is to draw the graph of the function f. In other words, we draw an **auxiliary graph** of dy/dt versus y.

Example 5.14.   *Determine the behavior of the solutions of the equation* $y' = 5y - y^2$ *and graph them.*

Solution. We first graph $f(y) = y(5 - y)$:

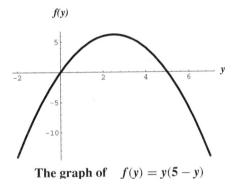

The graph of $f(y) = y(5 - y)$

From this auxiliary graph it is clear that the product $y(5 - y)$ is negative when $y < 0$ or $y > 5$, but that $y(5 - y)$ is positive for $0 < y < 5$. Thus

$$y < 0 \text{ or } y > 5 \quad \Longrightarrow \quad y' = y(5 - y) < 0 \quad \Longrightarrow \quad y \text{ is decreasing,}$$

and

$$0 < y < 5 \quad \Longrightarrow \quad y' = y(5 - y) > 0 \quad \Longrightarrow \quad y \text{ is increasing.}$$

Therefore, any solution $y(t)$ with $y(0) > 0$ must approach the critical solution $y(t) \equiv 5$ as $t \longrightarrow \infty$. On the other hand, any solution $y(t)$ with $y(0) < 0$ diverges from the critical solution $y(t) \equiv 0$. The graphs of some of the solutions to $y' = y(5 - y)$ are given below. ∎

$y(t)$

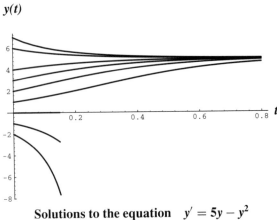

Solutions to the equation $y' = 5y - y^2$

No solution can cross any other solution. This is a consequence of the existence and uniqueness theorem (Theorem 5.1), or more precisely, of Corollary 5.2.

Note that the critical solutions $y \equiv 0$ and $y \equiv 5$ of $y' = y(5 - y)$ are of a different nature, as one can see from the graphs of the solutions. In the first case the nearby solutions tend to move away from the critical solution as time increases, whereas in the second case nearby solutions tend to approach the critical solution as time increases. This is the dichotomy between **instability** and **stability**.

Definition. *A critical solution of a differential equation is said to be* **unstable** *if all solutions that start nearby eventually wander away from the critical solution; the critical solution is said to be* **stable** *if all solutions that start nearby remain nearby.*

In Example 5.16 the critical solution $y(t) \equiv 0$ is unstable, while the critical solution $y(t) \equiv 5$ is stable. This is evident from the plot of the solutions of the equation $y' = 5y - y^2$.

It is interesting to compare the information obtained by the above method with the analysis that results when one solves the equation $y' = f(y)$ explicitly. In general, it is too complicated to use hand calculations to find a formula for the solution of $y' = f(y)$. But we can do it in Example 5.13, since the equation $y' = y(5 - y)$ is separable and the integrals can be straightforwardly evaluated.

Example 5.15. *Solve* $y' = y(5 - y)$ *using the method of separation of variables, and examine the behavior of a solution* $y(t)$ *for large* t.

Solution. To solve $y' = y(5 - y)$ explicitly, we first observe that $y \equiv 0$ is a solution corresponding to the initial condition $y(0) = 0$; similarly, $y \equiv 5$ is a solution corresponding to the initial condition $y(0) = 5$. Now assume that $y(0) \neq 0$ and $y(0) \neq 5$; then any solution of $y' = y(5 - y)$ will be nonzero for small t, and we can rewrite $y' = y(5 - y)$ as

$$\frac{dy}{y(5 - y)} = dt. \tag{5.45}$$

The left-hand side of (5.45) can be rewritten using partial fractions:

$$\left(\frac{1}{y} - \frac{1}{y - 5} \right) dy = 5dt. \tag{5.46}$$

Integrating both sides of (5.46), we obtain

$$\log(y) - \log(y - 5) = 5t + \log(C_1),$$

or

$$\frac{y}{y - 5} = C_1 e^{5t}, \tag{5.47}$$

where C_1 is a constant. Substitution of $t = 0$ in (5.47) yields

$$\frac{y(0)}{y(0) - 5} = C_1. \tag{5.48}$$

From (5.47) and (5.48) it follows that

$$\frac{y}{y - 5} = \frac{y(0)e^{5t}}{y(0) - 5}. \tag{5.49}$$

When we solve (5.49) for y we obtain

$$y(t) = y = \frac{5y(0)}{y(0)(1 - e^{-5t}) + 5e^{-5t}}. \tag{5.50}$$

Moreover, it is easy to check that $y(t)$ given by (5.50) is indeed a solution of (5.47). If $y(0) \geq 0$, then the denominator on the right-hand side of (5.50) cannot vanish for $t > 0$. Furthermore,

$$\lim_{t \to \infty} y(t) = 5,$$

which demonstrates the stability of the critical solution $y \equiv 5$. On the other hand, if $y(0) < 0$, then the denominator on the right-hand side of (5.50) vanishes for $t = t_1$, where t_1 is given by

$$e^{-5t_1} = \frac{-y(0)}{5 - y(0)}.$$

It follows that $y(t_1) = -\infty$; in particular, a solution y for which $y(0) < 0$ diverges from the critical solution $y(t) \equiv 0$. Thus $y \equiv 0$ is an unstable critical solution. ∎

The detailed analysis of Example 5.15 illustrates both the power and the limitations of the **qualitative stability analysis**. We can determine the stable and unstable points of the equation, and we can determine some, but not all of the behavior of the noncritical solutions.

The tortuous hand calculations necessary to solve the initial value problem

$$\begin{cases} y' = y(5 - y), \\ y(0) = Y_0 \end{cases}$$

can be carried out with ease using **ODE**. For example, the command

```
ODE[{y' == y(5 - y),y[0] == Y0},y,t,Method->Separable]
```

gives the output

```
            5 t
       5 E      Y0
{{y -> ---------------}}
            5 t
     5 - Y0 + E    Y0
```

confirming (5.50).

For more general equations, there may be critical solutions that are neither stable nor unstable.

Definition. *Let y_i be a critical point of the differential equation $y' = f(y)$, and let $y(t) \equiv y_i$ be the corresponding critical solution. We say that y_i is a **semistable critical point** of the differential equation $y' = f(y)$ if some solutions that start nearby remain nearby to $y(t) \equiv y_i$, while other solutions that start nearby diverge from $y(t) \equiv y_i$ with increasing time. This typically happens because $f(y)$ maintains the same sign on both sides of the critical point y_i.*

Example 5.16. *For the differential equation*

$$y' = y^2(5 - y)(5 + y),$$

find the stable critical points, unstable critical points, and semistable critical points.

Solution. We have $f(y) = y^2(5 - y)(5 + y)$, so that the critical points are at $y = 0$, $y = 5$, and $y = -5$. Here is the graph of f:

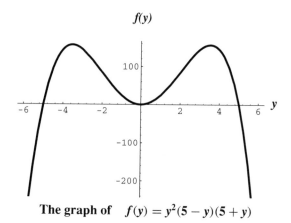

The graph of $f(y) = y^2(5 - y)(5 + y)$

From this auxiliary graph we see that $f(y) < 0$ for $y < -5$ or $y > 5$, while $f(y) \geq 0$ for $-5 \leq y \leq 5$. Therefore, the critical point -5 is unstable, the critical point 5 is stable, while the critical point 0 is semistable. The graphs of some of the solutions are indicated below. ▮

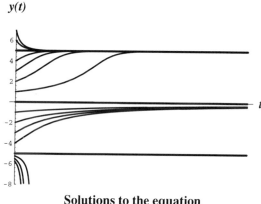

Solutions to the equation
$$y' = y^2(5 - y)(5 + y)$$

Hand calculations to solve $y' = y^2(5 - y)(5 + y)$ are too complicated to be worthwhile; however, the following command finds the implicit symbolic solution to the differential equation:

```
ODE[y' == y^2(5 - y)(5 + y),y,t,Method->Separable]
```

The result is

```
              1/250
     (5 + y)
{ ───────────────────── ==  E^t  C[1],  y == 0}
   1/(25 y)        1/250
 E          (5 - y)
```

or

$$\frac{(5+y)^{1/250}}{e^{1/(25y)}(5-y)^{1/250}} = e^{t+C_1},$$

where C_1 is a constant. ∎

We now state a general theorem to explain Examples 5.13–5.16.

Theorem 5.5. *Suppose that the differential equation $y' = f(y)$ has critical points $Y_1, Y_2, \ldots, Y_n$ and that the function f has a continuous derivative at each Y_i.*

(i) *If $f'(Y_i) < 0$, then the critical point Y_i is stable.*

(ii) *If $f'(Y_i) > 0$, then the critical point Y_i is unstable.*

(iii) *If $f'(Y_i) = 0$ and $f''(Y_i) \neq 0$, then the critical point Y_i is semistable.*

(iv) *If the critical point is semistable, then $f'(Y_i) = 0$.*

Proof. The proof of Theorem 5.5 depends on examining the sign of $f(y)$ in a neighborhood of $y = Y_i$. For convenience, we take $Y_i = 0$. For example, to prove (i), we note that $f(y) < 0$ in an interval of the form $0 < y \leq a$ for some $a > 0$. This implies that the solution of $y' = f(y)$ is a decreasing function of t if $0 < y(0) \leq a$. But the existence and uniqueness theorem (Theorem 5.1) requires that such a solution cannot cross $y = 0$, so it must satisfy $y(t) > 0$ for all $t > 0$. By a fundamental principle of real analysis, the limit of $y(t)$ exists when $t \longrightarrow \infty$; the limit must be a critical solution satisfying $0 \leq y \leq a$. But the only such critical solution is $y = 0$, which shows that $y(t) \longrightarrow 0$ when $t \longrightarrow \infty$. Similarly, $f(y) > 0$ in some interval of the form $-a \leq y < 0$, and we may argue as above that $y(t) \longrightarrow 0$ when $t \longrightarrow \infty$ in the case that $-a \leq y(0) < 0$. The proofs of the other statements are carried out similarly. ∎

A more comprehensive treatment of stability, which applies to an arbitrary system of n nonlinear differential equations, is given in Chapter 17. For example, part (i) of Theorem 5.5 is a special case of Theorem 17.9 when $n = 1$.

Conditions (i)–(iii) are *sufficient conditions* for stability, instability, or semistability, while (iv) is a *necessary condition* for semistability. As an illustration, consider the previous example, where $f(y) = y^2(5 - y)(5 + y)$. From the definition of the derivative or by direct computation, we see that

$$f'(0) = 0, \qquad f'(5) = -250, \qquad f'(-5) = +250, \qquad \text{and} \qquad f''(0) = 50.$$

Referring to parts (i)–(iii) of Theorem 5.5, we again conclude that $y = -5$ is unstable, $y = 0$ is semistable, and $y = 5$ is stable.

Summary of Techniques Introduced

A nonlinear equation of the form $y' = f(y)$ can be analyzed by first finding the **critical points** Y_i, which are the solutions of the equation $f(Y_i) = 0$. A critical point is **stable** if $f'(Y_i) < 0$ and **unstable** if $f'(Y_i) > 0$. If $f'(Y_i) = 0$, the critical point *can* be **semistable**, depending on whether or not $f(y)$ maintains a constant sign. This information is used to obtain accurate graphs of the solution curves.

To study the behavior of solutions of $y' = f(y)$ carry out the following steps:

Step 1: Determine the critical points of $y' = f(y)$, that is, those values $Y_1, Y_2, \ldots$ for which $f(Y_i) = 0$.

Step 2: The critical solutions of $y' = f(y)$ are then the solutions of the form $y(t) \equiv Y_i$.

Step 3: Determine the stability type of each critical solution.

Step 4: Solutions that start nearby to a stable critical solution $y(t) \equiv y_{\text{stab}}$ approach that critical solution as $t \longrightarrow \infty$.

Step 5: Solutions that start nearby to an unstable critical solution $y(t) \equiv y_{\text{unstab}}$ diverge from that critical solution as $t \longrightarrow \infty$.

Step 6: Solutions that start nearby to a semistable critical solution $y(t) \equiv y_{\text{semistab}}$ diverge from that critical solution on one side as $t \longrightarrow \infty$ but converge on the other.

Exercises for Section 5.6

Use **ODE** with the options **PlotField** and **Method->None** to plot the direction fields of the following differential equations and estimate the critical values. If possible, find the exact locus of critical values for each problem.

1. $y' = 2y - e^{-y}$ 3. $y' = y^2 - y - \cos(y)$

2. $y' = y^4 + 3y^2 + y - 7$ 4. $y' = e^{-y^2} - y^4$

Find the critical points of each of the following equations and classify each of the critical points as stable, unstable, or semistable. Plot the auxiliary graph and some of the solutions. If possible, use **ODE** to solve the equation.

5. $y' = y - 3y^2$ 12. $y' = y^{2n+1}, \quad n = 2, 3, 4, \ldots$

6. $y' = y - y^3$ 13. $y' = y^{2n}, \quad n = 2, 3, 4, \ldots$

7. $y' = 1 - \cos(y)$ 14. $y' = \sinh(y)$

8. $y' = \sin(y^2)$ 15. $y' = y(e^y - 1)$.

9. $y' = y^2(6 - y)(4 + y)$

10. $y' = y^2$ 16. $y' = -y^{2n+1}, \quad n = 2, 3, 4, \ldots$

11. $y' = y^3$ 17. $y' = -y^{2n}, \quad n = 2, 3, 4, \ldots$

6

APPLICATIONS OF FIRST-ORDER EQUATIONS I

In this chapter we describe several simple applications of first-order differential equations. The process of representing real world problems by the language of mathematics is known as **mathematical modeling**. The first step in the modeling process is the translation into mathematics of the nontechnical language used to describe the problem. Usually, many variables must be ignored for there to be hope of solving the problem in a reasonable length of time. The end result of the modeling process will ideally be a logical structure, simple but effective, to describe the problem. For us this structure will consist of one or more differential equations. We must then attempt to solve the differential equations either symbolically or numerically. Finally, the solution must be interpreted to see whether it gives a reasonable solution to a real-world problem. This interpretation is often best done with a plot. The model may need to be modified several times to achieve a good result.

Much of mathematical modeling involves the use of functional dependence with *parameters*. This means that we assume a functional relation $y = y(t)$ that contains one or more unknown constants, whose values are estimated from experimental data. Then we can use these resulting functions to make predictions. For example, the equation $y = g\,t^2/2$ contains the single parameter g, while the equation $y = C\,e^{kt}$ contains the two parameters C and k. The parameters may appear both in the equations and in the initial conditions associated with the differential equations.

We begin by applying the solutions found in Section 3.2 to problems of exponential growth. In Section 6.1 we show that a population with a constant growth rate satisfies a simple first-order equation that is both linear and separable. When the growth rate is nonconstant, it still may be the case that the population obeys a first-order linear equation;

this is the situation discussed in Section 6.2.

The logistic population model is discussed in Section 6.3; it is based on a nonlinear separable equation that can be understood by the methods of Section 5.6 or solved explicitly. In Section 6.4 we discuss population growth with harvesting, which also gives rise to a nonlinear separable equation.

Details of the exponential growth and logistic models are worked out in Section 6.5 for the population of the United States; the predicted values are compared with the actual values.

Section 6.6 is devoted to temperature equalization models.

6.1   Population Models with Constant Growth Rate

Consider a population of individuals whose number is so large that individual changes are relatively small and therefore the state of the system can be modeled by a continuous variable. For example, a colony of ants or the number of molecules in a mold of bacteria are both well described by this assumption. The simplest law of time evolution of such a system is reflected in the statement,

> *The time rate of change of population is proportional to the present population size.*

In order to translate this assumption into mathematics, we let $\Pi(t)$ be the population size at time t. In absolute terms the values of $\Pi(t)$ are integers and change by integer amounts. We can measure $\Pi(t)$ in any convenient system of units, such as hundreds or millions. But for a large population an increase by one or two over a short time span is so small as to be infinitesimal relative to the total, so that $\Pi(t)$ is extremely well approximated by a differentiable function, denoted by $P(t)$.

The time rate of change of $P(t)$ is given by the derivative $P' = dP/dt$. Let us assume that P' is proportional to P. (Predictions based on this assumption will be compared with real population data in Section 6.5.) The constant of proportionality k is called the **growth rate**; it is measured in units of time^{-1}.

Putting all of this together, we have the differential equation

$$\frac{dP}{dt} = k\,P. \tag{6.1}$$

In particular, if the population size is a nonzero constant, the growth rate is 0. We also allow k to be negative in (6.1), in which case the population declines.

Equation (6.1) is a simple separable equation of the type that we solved in Section 3.3. (It is also a first-order linear equation.) One way to solve (6.1) is to rewrite it as

$$\frac{dP}{P} = k\, dt \tag{6.2}$$

and integrate both sides of the equation. It is also possible to solve (6.1) as a first-order linear equation using the methods of Section 3.2; then e^{-kt} is an integrating factor, and (6.1) becomes

$$\frac{d}{dt}\left(e^{-kt}P\right) = 0. \tag{6.3}$$

The solution of both (6.2) and (6.3) is

$$P(t) = C\, e^{kt}, \tag{6.4}$$

where C is a constant.

Suppose we are given an initial condition of the form $P(t_0) = P_0$, where t_0 is some moment of time and P_0 is the population size at t_0. We can determine the constant C in (6.4) by solving the equation $P_0 = C\, e^{kt_0}$; then (6.4) becomes

$$P(t) = P_0 e^{-kt_0} e^{kt} = P_0 e^{k(t-t_0)}. \tag{6.5}$$

We call (6.5) the **exponential law of growth**, or **Malthusian law of growth**. It was first enunciated by Malthus[1] in his influential **An Essay on the Principles of Population as It Affects the Future Improvement of Society** published in 1798. He wrote,

> *Population, when unchecked, increases in a geometrical ratio.*
> *Subsistence only increases in an arithmetical ratio.*

The exponential growth law is the simplest of several growth laws that we shall examine. It has the following important feature.

Lemma 6.1. *Suppose the population P is given by the exponential growth law (6.5), where the growth rate k is positive. For some t_0 let T_d be the amount of time required for the starting population P_0 to double. Then*

$$T_d = \frac{\log(2)}{k} \approx \frac{0.693}{k}. \tag{6.6}$$

In particular, the doubling time T_d does not depend on the starting population P_0.

---

[1]

The Reverend Thomas Robert Malthus (1766–1834). English professor of history and political economy.

Proof. By definition $P(t_0 + T_d) = 2P(t_0)$. But (6.5) implies that

$$2 = \frac{P(t_0 + T_d)}{P(t_0)} = \frac{P_0 e^{kT_d}}{P_0} = e^{kT_d}.$$

Hence we get (6.6). ∎

Example 6.1. *There are* 100 *rabbits in a colony, which multiplies according to the rate constant* $k = 3 \times 10^{-6}$ *second*$^{-1}$. *How long does it take for the rabbit population to double to* 200?

Solution. The size of the colony is irrelevant. According to (6.6) the doubling time T_d is given by

$$T_d \approx \frac{0.693}{0.000003} = 231049 \text{ seconds} \approx 2.67 \text{ days.} \quad ∎$$

To plot the rabbit population size, we first observe that since 1 second is the same as $1/86400$ of a day, we can express k as

$$k = 3 \times 10^{-6} \text{ second}^{-1} = 3 \times 10^{-6} \times 86400 \approx 0.259 \text{ day}^{-1}.$$

Therefore, to plot the number of rabbits over a 3 day period assuming an initial population size of $P(0) = P_0$, where P_0 ranges from 50 to 150 in increments of 25, we use

```
ODE[{P' == 0.259 P,P[0] == P0},P,t,
Method->FirstOrderLinear,Parameters->{{P0,50,150,25}},
PlotSolution->{{t,0,3}}]
```

with the output

```
         0.259 t
{{P -> E        P0}}
```

and the plot

Rabbits

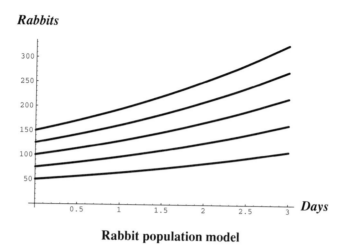

Rabbit population model

It is possible to use experimental data to determine the growth rate.

Lemma 6.2.   *Assume that P is the size of a population obeying an exponential growth law. Let P_1 and P_2 be values of P at distinct times $t_1 < t_2$. Then the growth rate k is given by*

$$k = \frac{1}{t_2 - t_1} \log\left(\frac{P_2}{P_1}\right). \tag{6.7}$$

Proof. From (6.5) we have

$$P_1 = P_0 e^{k(t_1 - t_0)} \qquad \text{and} \qquad P_2 = P_0 e^{k(t_2 - t_0)}.$$

Hence

$$\frac{P_2}{P_1} = e^{k(t_2 - t_1)}. \tag{6.8}$$

When we take logarithms of both sides of (6.8) we get (6.7). ∎

Example 6.2.   *The first two columns of the following table contain the population sizes of selected countries for 1978 and 1994. Assuming an exponential growth law, use this data to compute the growth rates of the countries.*

Country	1978 Population	1994 Population	Yearly Growth Rate 1978—1994
Argentina	26.4	33.5	$0.015 = 1.5\%$
Brazil	115.4	156.6	$0.019 = 1.9\%$
China	933.0	1205.2	$0.016 = 1.6\%$
Denmark	5.1	5.2	$0.001 = 0.1\%$
India	638.4	896.6	$0.021 = 2.1\%$
Indonesia	145.1	194.6	$0.018 = 1.8\%$
Japan	114.9	125.0	$0.005 = 0.5\%$
United States	218.7	257.8	$0.010 = 1.0\%$

Solution. For example, for Argentina, the yearly growth rate is given by

$$k = \frac{1}{1994 - 1978} \log\left(\frac{33.5}{26.4}\right) = \frac{1}{16}(0.238) \approx 0.015 = 1.5\%.$$

For the other entries in the table see Exercise 2. ∎

Population with Migration

We can also use the theory of first-order linear equations to describe a population that obeys an **exponential law of growth with migration**. This means that in addition to the growth due to reproduction of members of the population, there is also a constant flow of new members into or out of the population. In symbols this means that the population size $P(t)$ is a solution of the differential equation

$$\frac{dP}{dt} = kP + r, \tag{6.9}$$

where the **migration rate** r is measured in units per second for whatever units pertain to P. As before, we call k the **growth rate**; in this subsection we assume that k is positive. Equation (6.9) is a first-order linear differential equation that can be solved with the methods of Section 3.2. An integrating factor for (6.9) is e^{-kt}, and (6.9) becomes

$$\frac{d}{dt}\left(P(t)e^{-kt}\right) = r\,e^{-kt}. \tag{6.10}$$

Integrating both sides of (6.10), we obtain

$$P(t)e^{-kt} = C - \left(\frac{r}{k}\right)e^{-kt},$$

where C is a constant. Thus the general solution of (6.9) is

$$P(t) = -\frac{r}{k} + C\,e^{kt}. \tag{6.11}$$

If we are given an initial condition in the form $P(t_0) = P_0$, then we can obtain the constant C in (6.11) by solving the equation

$$P_0 = -\frac{r}{k} + C\,e^{kt_0}$$

to yield

$$C = \left(P_0 + \frac{r}{k}\right)e^{-kt_0}.$$

Thus, (6.11) becomes

$$P(t) + \frac{r}{k} = \left(P_0 + \frac{r}{k}\right)e^{k(t-t_0)}. \tag{6.12}$$

This equation can be also written in the form

$$P(t) = P_0 e^{k(t-t_0)} + \frac{r}{k}\left(e^{k(t-t_0)} - 1\right). \tag{6.13}$$

In this form we have separated the effects of the migration and the initial condition of the population. We also note that we have made no assumption on the sign of r; when r is positive we have **immigration**, and when r is negative we have **emigration**.

When migration is present, it is no longer true that the doubling time is independent of the starting population; nevertheless, we can find a formula for it.

Lemma 6.3.   *Let P be the population size of a population that obeys an exponential growth with migration law (6.9). Let $M_d(P_0)$ be the amount of time required for P_0 to double, and let k and r denote the growth and migration rates. Assume that $k > 0$.*

(i) *$M_d(P_0)$ is given by the formula*

$$M_d(P_0) = \frac{1}{k}\log\left(2 - \frac{r}{k\,P_0 + r}\right). \tag{6.14}$$

(ii) *The doubling time in the case that $r > 0$ is strictly less than the doubling time in the case $r = 0$, that is, $M_d(P_0) < T_d$.*

(iii) *If $r > 0$, then $M_d(P_0)$ is an increasing function of the initial population size P_0.*

You are asked to prove Lemma 6.3 in Exercise 14.

Example 6.3.   *Assume that a bacteria population obeys an exponential law with migration, resulting in the differential equation*

$$\frac{dP}{dt} = (0.0015)P + 0.0125, \tag{6.15}$$

where time is measured in seconds. How long does it take for the population to increase from P = 100 to P = 200? Plot several solutions of (6.15).

Solution.   The growth rate is $k = 0.0015$ second^{-1} and the migration rate is $r = 0.0125$ second^{-1}. From (6.14) it follows that the doubling time is given by

$$M_d(100) = \frac{1}{0.0015} \log\left(2 - \frac{0.0125}{0.0015 \times 100 + 0.0125}\right)$$

$$\approx 435.951 \text{ seconds} \approx 0.121098 \text{ hours}.$$

Note that in Example 6.3, we cannot unambiguously speak of the time necessary for the population to *double*, as we could for the models with $r = 0$. In Example 6.3 to increase from $P = 10$ to $P = 20$ takes *less time* than to increase from $P = 100$ to $P = 200$, because

$$M_d(10) = \frac{1}{0.0015} \log\left(2 - \frac{0.0125}{0.0015 \times 10 + 0.0125}\right)$$

$$\approx 290.212 \text{ seconds} \approx 0.0806 \text{ hours}.$$

Thus, $M_d(100) > M_d(10)$, as predicted by part (iii) of Lemma 6.3.

To obtain a plot of solutions to (6.15), we first convert the growth and migration rates from seconds to hours:

$$k = 0.0015 \text{ second}^{-1} = 3600 \times 0.0015 = 5.4 \text{ hour}^{-1}$$

and

$$r = 0.0125 \text{ bacteria/second} = 3600 \times 0.0125 = 45 \text{ bacteria/hour}.$$

Therefore, we use

```
ODE[{P' == 5.4 P + 45,P[0] == P0},P,t,
Method->FirstOrderLinear,
Parameters->{{P0,50,150,25}},
PlotSolution->{{t,0,1}}]
```

to obtain the output

$$\{\{P \rightarrow -8.33 + 8.33 \; E^{5.4 \; t} + E^{5.4 \; t} \; P0\}\}$$

and the plot

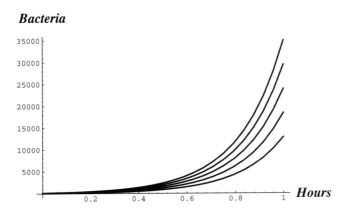

Immigration model

In practical situations, one does not generally know in advance the values of the parameters k and r that define the growth rate and the migration rate. We can use observed data consisting of the population at three different times to *compute k and r*. Let us find the exact formulas.

Lemma 6.4. *Assume that $t_0 < t_1 < t_2$ are equally spaced time values, that is, $t_2 - t_1 = t_1 - t_0$. We let the corresponding population values be denoted by P_0, P_1, and P_2. Then the growth rate k and migration rate r of a population that obeys an exponential law of population growth with migration are given by*

$$k = \frac{1}{t_1 - t_0} \log\left(\frac{P_2 - P_1}{P_1 - P_0}\right) \tag{6.16}$$

and

$$r = \left(\frac{P_1^2 - P_0 P_2}{P_2 - 2P_1 + P_0}\right) k. \tag{6.17}$$

Proof. From (6.12) we have

$$P_1 = \left(P_0 + \frac{r}{k}\right) e^{k(t_1 - t_0)} - \frac{r}{k} \tag{6.18}$$

and

$$P_2 = \left(P_1 + \frac{r}{k}\right) e^{k(t_2 - t_1)} - \frac{r}{k}. \tag{6.19}$$

When we subtract (6.18) from (6.19) and use the assumption $t_2 - t_1 = t_1 - t_0$, we find that

$$P_2 - P_1 = (P_1 - P_0)e^{k(t_1 - t_0)}, \tag{6.20}$$

so that we can compute the growth rate k according to (6.16).

In order to determine the migration rate, we use (6.18) to compute

$$\frac{r}{k}\left(e^{k(t_1 - t_0)} - 1\right) = P_1 - P_0 e^{k(t_1 - t_0)}.$$

Using (6.20), we express $e^{k(t_1 - t_0)}$ in terms of P_0, P_1, and P_2:

$$e^{k(t_1 - t_0)} = \frac{P_2 - P_1}{P_1 - P_0}. \tag{6.21}$$

After some calculation we obtain the following formula from (6.21) and (6.16):

$$\frac{r}{k} = \frac{P_1^2 - P_0 P_2}{P_2 - 2P_1 + P_0}. \tag{6.22}$$

Then (6.17) follows from (6.22). ∎

Notice that (6.20) implies that $P_2 - P_1 > P_1 - P_0$ when $k > 0$. Furthermore, the sign of the quantity $P_1^2 - P_0 P_2$ determines the sign of the migration rate r; also, $r = 0$ if and only if $P_1^2 = P_0 P_2$. The migration rate and growth rate will both be positive if and only if the observed values $P_0 < P_1 < P_2$ satisfy the inequality

$$\sqrt{P_0 P_2} < P_1 < \frac{1}{2}(P_0 + P_2). \tag{6.23}$$

Example 6.4. *A population with yearly growth rate k and migration rate r is found to have the values $P_0 = 1.40$, $P_1 = 1.70$, $P_2 = 2.05$ at the times $t_0 = 1.0$, $t_1 = 1.2$, $t_2 = 1.4$, respectively. Find the yearly growth rate k and the migration rate r.*

Solution. From (6.16) we get

$$k = \frac{1}{t_1 - t_0} \log\left(\frac{P_2 - P_1}{P_1 - P_0}\right) = \frac{1}{0.2} \log\left(\frac{0.35}{0.30}\right) \approx 0.154.$$

Furthermore, (6.17) implies that

$$\frac{r}{k} = \frac{P_1^2 - P_0 P_2}{P_2 - 2P_1 + P_0} = \frac{0.02}{0.05} = 0.4.$$

Hence $r = 0.154 \times 0.4 = 0.0616$. ∎

Exercises for Section 6.1

1. A colony of bacteria doubles in 4 hours and 20 minutes. Assume an exponential growth law. Find the growth rate k.

2. Assuming an exponential growth law, compute the yearly growth rates of the countries listed in Example 6.2.

3. For each of the following population models with constant growth, determine and solve the initial value problem that describes it.

   a. The yearly growth rate is $k = 5$ and $P(2) = 6$.

   b. The population doubles in one unit of time and $P(0) = 3$.

   c. The yearly growth rate is $k = 0.0135/\text{year}$ and $P(1993) = 3 \times 10^9$.

4. A population with constant growth rate k is found to have the values $P_0 = 151$ million when $t_0 = 1950$ and $P_1 = 203$ million when $t_1 = 1970$. Find the growth rate.

5. A report by the 1994 UN Conference on World Population estimates that the world population will increase from 5.7 billion in 1994 to 8.3 billion in 2025. Assuming zero migration, estimate the yearly growth rate k.

6. According to the 1994 UN Conference on World Population, it is estimated that the U.S. population will increase from 260.75 million in 1994 to 322 million in 2025. Assuming zero migration, estimate the yearly growth rate k.

7. According to the 1994 UN Conference on World Population, it is estimated that the population of China will increase from 1.192 billion in 1994 to 1.504 billion in 2025. Assuming zero migration, estimate the yearly growth rate k.

8. An environmental organization states that in the next 6 seconds, 24 people will be added to the population of the earth. Assuming a constant yearly growth rate, no migration and a current population of 3 billion, compute:

   a. The increase of the earth's population in one hour;

   b. The increase of the earth's population in one day;

   c. The increase of the earth's population in one year;

   d. The time necessary for the earth's population to double in size.

9. Solve Exercise 8 if we assume a *linear* rate of growth, that is, $P(t) = P_0 + r\,t$ with 24 people added every 6 seconds. Find **a.** The increase in one hour, **b.** The increase in one day, **c.** The increase in one year and **d.** The time necessary to double the population.

10. Find the population size $P(t)$ for all $t > 0$, given a growth rate of $k = 0.03$, a migration rate of $r = 0.15$, and an initial population size of $P(0) = 2.4 \times 10^8$.

11. Find the population $P(t)$ for all $t > 0$, given that growth rate is $k = 2$, the migration rate is $r = 3$ and the population at $t = 1$ is $P(1) = 10^{23}$.

12. Solve the following initial value problem involving growth with migration:

$$\begin{cases} 4P' = P + 1, \\ \\ P(0) = 16. \end{cases}$$

13. A population with constant growth rate and migration rate is found to have the values $P(1800) = 5.31$ million, $P(1810) = 7.24$ million and $P(1820) = 9.64$ million. Find the growth and migration rates.

14. Prove Lemma 6.3.

15. Derive the inequalities (6.23).

6.2   *Population Models with Variable Growth Rate*

In many realistic situations the growth rate or migration may be nonconstant. Consider, for example, a population with **seasonal migration**, which is larger in the summer than in the winter. Similarly, the rate of growth may also be considered to vary with time, corresponding to a "mating season" of increased fertility. The general form of the corresponding mathematical model is the first-order linear equation

$$\frac{dP}{dt} = k(t)P + r(t),   \tag{6.24}$$

where $k(t)$ is the time-dependent growth rate and $r(t)$ is the time-dependent migration rate. For seasonal variations it is natural to assume that these are periodic functions of time. We consider several separate cases.

Case 1. *Periodic growth rate with no migration*

This is modeled by the differential equation

$$\frac{dP}{dt} = \big(k + \varepsilon \cos(a t)\big)P.   \tag{6.25}$$

The constant $a > 0$ gives the **frequency** of the time-dependent variation, while the constant $\varepsilon > 0$ gives the **amplitude** of the time-dependent variation. We suppose that $\varepsilon < k$, so that the growth rate is always positive.

Equation (6.25) is a first-order linear equation and can be solved by the methods of Section 3.2. We can also use **ODE** as follows:

```
ODE[{P' == (k + epsilon Cos[a t])P,P[0] == P0},P,t,
Method->FirstOrderLinear]
```

to get

```
            k t + (epsilon Sin[a t])/a
{{P -> E                                P0}}}
```

Thus, the solution of (6.25) with $P(0) = P_0$ is

$$P(t) = P_0 e^{kt + \varepsilon \sin(at)/a}.$$

It is interesting to note that there is a **phase lag** in the solution $P(t)$ in comparison with the growth rate $k + \varepsilon \cos(a t)$. Indeed, the growth rate is largest at the times $t = 0, 2\pi/a$, $4\pi/a, \ldots$, while the solution $P(t)$ is largest at the times $t = \pi/2a, 5\pi/2a, 9\pi/2a, \ldots$. At first this may seem surprising, but in fact it is to be expected, since the increasing rate of reproduction is not fully felt until the new individuals begin reproduction, which this model predicts to be one-quarter of a cycle later. For example, if $a = 2\pi/365$ days, then we may expect to wait 90 days for the increased population due to time variations. Similarly, a decline in the birth rate is not felt until 90 days later for analogous reasons.

Case 2. Periodic migration rate with constant growth rate

In this model we assume that the population is governed according to the differential equation

$$\frac{dP}{dt} = kP + r + \varepsilon \cos(a t). \tag{6.26}$$

As before, r, ε, and k are positive constants with $\varepsilon < r$. Equation (6.26) is solved by the integrating factor $\mu(t) = e^{-kt}$. We obtain the solution by writing

$$\frac{d}{dt}\left(P(t)e^{-kt}\right) = e^{-kt}\left(r + \varepsilon \cos(a t)\right). \tag{6.27}$$

From calculus we have the indefinite integral

$$\int e^{-kt} \cos(a t)\, dt = \frac{e^{-kt}\left(-k \cos(a t) + a \sin(a t)\right)}{a^2 + k^2}.$$

The solution of the initial value problem with $P(0) = P_0$ can therefore be obtained by integrating both sides of (6.27) from 0 to t:

$$e^{-kt}P(t) - P_0 = \left(\frac{r}{k}\right)(1 - e^{-kt}) + \left(\frac{\varepsilon}{a^2 + k^2}\right)\left(e^{-kt}\left(- k\cos(a t) + a \sin(a t)\right) + k\right),$$

or

$$P(t) = P_0 e^{kt} + \left(\frac{r}{k}\right)(e^{kt} - 1) + \frac{\varepsilon k e^{kt}}{a^2 + k^2} + \frac{\varepsilon\left(- k\cos(a t) + a \sin(a t)\right)}{a^2 + k^2}. \tag{6.28}$$

It is interesting to observe the effects of the periodic variation of the migration rate, in comparison with the solution (6.13) for constant migration rate. The first two terms on the right-hand side of (6.28) coincide with the right-hand side of (6.13) (when $t_0 = 0$). The last term includes the effect of the periodic variation of migration and is bounded by a constant, independent of t. The third term includes the effect of the variation and also increases at an exponential rate with t. The numerical effect of this term depends on the relative values of ε, k, and P_0. We illustrate with an example.

Example 6.5. *Suppose that the growth rate is* $k = 0.01$, *the migration rate is* $r(t) = 5 \times 10^4 + 10^4 \cos(t)$, *and the initial population size is* $P_0 = 3 \times 10^5$. *Compare with the corresponding model with constant migration in which* $r(t) = 5 \times 10^4 + 10^4 \cos(t)$ *is replaced by* $r = 5 \times 10^4$.

Solution. The initial value problem in the case of nonconstant migration is

$$\begin{cases} P' = 0.01P + 5 \times 10^4 + 10^4 \cos(t), \\ P(0) = 3 \times 10^5, \end{cases} \tag{6.29}$$

and the initial value problem in the case of constant migration is

$$\begin{cases} P' = 0.01P + 5 \times 10^4, \\ P(0) = 3 \times 10^5. \end{cases} \tag{6.30}$$

The solutions to these two differential equations can be found by substitution into formulas (6.28) and (6.13). We choose, however, to solve the initial value problems directly using **ODE**. For (6.29) we use

```
PeriodicImigration[t_]=
ODE[{P' == 0.01 P + 5 10^4 + 10^4 Cos[t],P[0] == 3 10^5},
P[t],t,Method->FirstOrderLinear,Form->Explicit]
```

resulting in

$$-5.\ 10^6 + 5.3001\ 10^6\ E^{0.01\ t} - 99.99\ \text{Cos[t]} + 9999.\ \text{Sin[t]}$$

Similarly, for (6.29) we use

```
ConstantImigration[t_]=
ODE[{P' == 0.01 P + 5 10^4,P[0] == 3 10^5},P[t],t,
Method->FirstOrderLinear,Form->Explicit]
```

resulting in

$$-5.\ 10^6 + 5.3\ 10^6\ E^{0.01\ t}$$

The simultaneous plot of the two functions **PeriodicImigration[t]** and **ConstantImigration[t]** is obtained with

```
Plot[Evaluate[{PeriodicImigration[t],
ConstantImigration[t]}],{t,0,2Pi}]
```

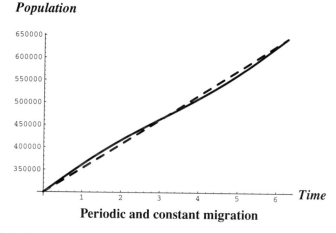

Population

Periodic and constant migration

We see that initially, when $0 \le t \le 3$, the periodic model gives a slightly larger value for the population size, but that when $t = 4$ the model with constant rate gives a slightly larger value. ∎

Exercises for Section 6.2

Solve the following population models with time-dependent growth and/or migration:

1. $\begin{cases} P' = 4 + \cos(t), \\ P(0) = 3 \end{cases}$

3. $\begin{cases} P' = 4\cos(t) + \cos(t)P, \\ P(0) = 3 \end{cases}$

2. $\begin{cases} P' = 4P + \cos(t), \\ P(0) = 3 \end{cases}$

4. $\begin{cases} P' = 5 - \sin(t) + 3\cos(t), \\ P(1) = 4 \end{cases}$

5. Suppose that the growth rate and migration rate are given by

$$k = 0.02 \quad \text{and} \quad r(t) = 3 \times 10^4 + 2 \times 10^4 \cos(t),$$

and that the initial population is $P_0 = 7 \times 10^5$. Find the population at $t = 1, 2, 3, 4$ and compare with the corresponding model of constant migration with $k = 0.02$ and $r = 3 \times 10^4$.

6. Suppose that the growth rate and migration rate are given by $k(t) = k$ and $r(t) = r_0 + r_1 t$, where k, r_0, and r_1 are constants.

    a. Solve the resulting differential equation (6.24) with $P(0) = P_0$.

    b. Assume the following census data for the 1960 US population:

$$k = 0.014, \quad r_0 = 2.5 \times 10^5, \quad r_1 = 5.2 \times 10^3, \quad P_0 = 1.8 \times 10^8.$$

Use the solution of part **a** to compute the population in the years 1970, 1980, 1990, 2000 (corresponding to $t = 10, 20, 30, 40$).

6.3   Logistic Model of Population Growth

Population growth has been modeled by first-order linear equations in Sections 6.1 and 6.2. In the models in those sections there is a single critical point, which is either stable or unstable, leading either to unlimited growth or to convergence to a unique steady-state solution, independent of the initial conditions.

In many applied problems we do not expect such simple behavior; rather, it may happen that the ultimate behavior of the system is radically different depending on the details of the initial conditions. In population models, the conclusion of unlimited population growth may be unrealistic. When what is observed is substantially different from what is predicted by a differential equation, then the differential equation provides the wrong model; it must be modified.

As Malthus noted, unbridled growth of a population that keeps doubling every so many years cannot continue forever. Therefore, let us consider the following differential equation, which is a modification of $P' = k P$, namely,

$$\frac{dP}{dt} = P(k - bP), \tag{6.31}$$

where k and b are positive constants. We call (6.31) the **logistic equation**; it is a separable equation of the type we studied qualitatively in Section 5.6. The graph of a solution to (6.31) is called a **logistic curve**. A population satisfying (6.31) is said to obey a **logistic law of population growth**. The logistic model was first considered by Verhulst[2] in the 1840s. Notice that if we were to take $b = 0$, the logistic equation (6.31) would reduce to (6.1), whose solution is the exponential law of population growth (6.5).

It is substantially easier to obtain qualitative information about (6.31) using the methods of Section 5.6 than it is to find a formula for the solution. Clearly, the critical points are 0 and k/b. Let $f(P) = P(k - bP)$. Then

$$f'(0) = k > 0 \qquad \text{and} \qquad f'(k/b) = -k < 0.$$

Since k and b are positive, Theorem 5.5 implies that the critical point 0 is unstable and the critical point k/b is stable. In particular, any solution for which $P(0) > 0$ must approach the value k/b, called the **carrying capacity** of the model. Also, the auxiliary graph given below for (6.31) is a parabola bounded from above. The auxiliary graph can then be used to determine the general shape of a solution to (6.31) with the initial condition $P(0) = P_0$, where $0 < P_0 < k/b$.

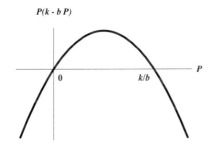

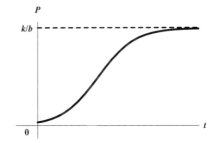

Auxiliary graph for the logistic equation   **Typical logistic curve**

To solve the differential equation (6.31) explicitly, we treat it as a separable equation, and solve it by the methods of Section 3.3. First, we assume that neither P_0 nor $k - b P_0$ vanishes. We rewrite (6.31) as

$$\frac{dP}{P(k - bP)} = dt. \tag{6.32}$$

[2]Pierre François Verhulst (1804–1849). Belgian mathematician. His theory of population increase, first proposed in 1835, was supported by Quetelet but was later attacked by Doubleday in 1841.

The left-hand side of (6.32) can be expanded by the method of partial fractions; thus, (6.32) becomes

$$\frac{1}{k}\left(\frac{1}{P} + \frac{b}{k - bP}\right)dP = dt. \tag{6.33}$$

Integrating both sides of (6.33) leads to the general solution of (6.31):

$$\log(P) - \log(k - bP) = kt + C. \tag{6.34}$$

Let us assume the initial condition $P(t_0) = P_0$. The constant C in (6.34) is found by setting $t = t_0$ in (6.34); thus

$$\log(P_0) - \log(k - bP_0) = kt_0 + C. \tag{6.35}$$

From (6.34) and (6.35) we get

$$\log(P) - \log(k - bP) = k(t - t_0) + \log(P_0) - \log(k - bP_0). \tag{6.36}$$

Taking the exponential of both sides of (6.36) results in

$$\frac{P}{k - bP} = \frac{P_0}{k - bP_0}e^{k(t-t_0)}. \tag{6.37}$$

Equation (6.37) can be solved for P:

$$P(t) = P = \frac{kP_0}{(k - bP_0)e^{-k(t-t_0)} + bP_0}. \tag{6.38}$$

To check these computations, we use **ODE** to solve (6.31) as a Bernoulli equation. Thus

ODE[{P′ == P(k - b P),P[t0] == P0},P,t,Method->Bernoulli]

yields

```
                    k t
                 E      k P0
{{P ->  ──────────────────────────────}}
          k t0         k t        k t0
         E      k + b E    P0 - b E    P0
```

which is equivalent to (6.38).

    Equation (6.38) has been derived assuming that both $P_0 \neq 0$ and $k - bP_0 \neq 0$. But it also correctly represents the solution of (6.31) if $k - bP_0 = 0$, since in that case we have $P(t) \equiv k/b$. Similarly, (6.38) is correct when $P_0 = 0$. Notice however that when $P(t)$ is interpreted as population size (and not just a solution of $P' = P(k - bP)$), we must have $P_0 \geq 0$.

We can use (6.38) to analyze the solution of (6.31) for large t. Since $e^{-kt} \longrightarrow 0$ as $t \longrightarrow \infty$, we conclude that

$$\lim_{t \to \infty} P(t) = \begin{cases} 0, & \text{for } P_0 = 0, \\ \dfrac{k}{b} = \text{the carrying capacity,} & \text{for } P_0 > 0. \end{cases}$$

One advantage of finding the solution of the logistic equation (6.31) is that we can solve (6.38) for t. Thus, from (6.37) we get

$$t = t_0 + \frac{1}{k} \log\left(\frac{k\, P_0^{-1} - b}{k\left(P(t)\right)^{-1} - b} \right). \tag{6.39}$$

Example 6.6.   *Find the critical solutions and carrying capacity of the logistic equation $P' = P(4 - P)$; then solve and plot the solution. Finally, if $P(0) = 2$, find the time t for which $P(t) = 3$.*

Solution. We have $k = 4$ and $b = 1$. The critical solutions are clearly $P \equiv 0$ and $P \equiv 4$, and the carrying capacity is 4. Also, $t_0 = 0$ and $P_0 = 2$. As a special case of (6.38), we obtain the solution in the form

$$P(t) = \frac{4P_0}{(4 - P_0)e^{-4t} + P_0} = \frac{4}{e^{-4t} + 1}.$$

If we are given that $t_0 = 0$, $P_0 = 2$, and $P(t) = 3$, then we can use (6.39) to compute

$$t = \frac{1}{4} \log\left(\frac{4(1/2) - 1}{4(1/3) - 1} \right) = \frac{\log(3)}{4} \approx 0.274. \ \blacksquare$$

We can solve and plot several solutions of $P' = P(4 - P)$ using

```
ODE[{P' == 4P - P^2,P[0] == P0},P,t,
Method->Bernoulli,Parameters->{{P0,0,4,0.4}},
PlotSolution->{{t,0,2}}]
```

A more appealing plot is obtained with

```
ODE[{P' == 4P - P^2,P[0] == Min[P0,P0^2]},P,t,
Method->Bernoulli,Parameters->{{P0,0,4,0.4}},
PlotSolution->{{t,0,2}}]
```

P(t)

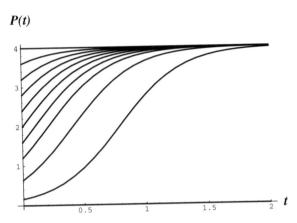

Stability analysis of the logistic equation $P' = P(4 - P)$

Formula (6.38) can be used to find the parameters k and b from the observed data of the population at three different equally spaced times.

Lemma 6.5. *Assume that $t_0 < t_1 < t_2$ are equally spaced time values, that is, $t_2 - t_1 = t_1 - t_0$. We denote the corresponding population values by P_0, P_1 and P_2. The growth rate k and carrying capacity k/b of a population that obeys a logistic law of population growth are given by*

$$k = -\frac{1}{t_1 - t_0} \log \left(\frac{P_2^{-1} - P_1^{-1}}{P_1^{-1} - P_0^{-1}} \right) \tag{6.40}$$

and

$$\frac{k}{b} = \frac{2P_1^{-1} - P_0^{-1} - P_2^{-1}}{P_1^{-2} - P_0^{-1} P_2^{-1}}. \tag{6.41}$$

Proof. We write the solution of (6.31) in the form

$$\frac{1}{P(t)} = \frac{b}{k} \left(1 - e^{-k(t - t_0)} \right) + \frac{1}{P_0} e^{-k(t - t_0)}. \tag{6.42}$$

Then from (6.42) we have

$$\frac{1}{P_1} = \frac{b}{k} \left(1 - e^{-k(t_1 - t_0)} \right) + \frac{1}{P_0} e^{-k(t_1 - t_0)} \tag{6.43}$$

and

$$\frac{1}{P_2} = \frac{b}{k} \left(1 - e^{-k(t_2 - t_1)} \right) + \frac{1}{P_1} e^{-k(t_2 - t_1)}. \tag{6.44}$$

Subtracting these two equations yields

$$\frac{1}{P_2} - \frac{1}{P_1} = \left(\frac{1}{P_1} - \frac{1}{P_0}\right) e^{-k(t_1-t_0)}. \tag{6.45}$$

We can solve (6.45) for the parameter k, obtaining (6.40). Similarly, we can solve (6.43) and (6.44) for b/k, obtaining

$$\frac{b}{k} = \frac{P_1^{-2} - P_0^{-1} P_2^{-1}}{2P_1^{-1} - P_0^{-1} - P_2^{-1}}. \tag{6.46}$$

Then (6.41) follows from (6.46). ∎

Example 6.7. *A population size is observed to have the values $P_0 = 0.25$ million at time $t_0 = 0$, $P_1 = 0.40$ million at time $t_1 = 1$, and $P_2 = 0.50$ million at time $t_2 = 2$. Find the growth rate k and the carrying capacity k/b.*

Solution. We substitute the observed data into (6.40) to obtain

$$k = -\log\left(\frac{0.50^{-1} - 0.40^{-1}}{0.40^{-1} - 0.25^{-1}}\right) = 1.10.$$

To obtain b we use (6.46) to write

$$\frac{b}{k} = \frac{(0.40)^{-2} - (0.25)^{-1}(0.50)^{-1}}{2(0.40)^{-1} - (0.25)^{-1} - (0.50)^{-1}} = 1.75/\text{million},$$

and the value $b = 1.75 \times 1.09 \approx 1.92$. The carrying capacity is

$$\frac{k}{b} = \frac{1}{1.75} \approx 0.57 \text{ million}. ∎$$

Finally, we indicate the appropriate formulation of doubling time for a logistic model. Since $P(t) < k/b$ for all t, the doubling time is only well-defined if the initial population satisfies $0 < P_0 < k/(2b)$, because $P_0 = 0$ implies that $P(t) = 0$ for all t.

Lemma 6.6. *The doubling time for the logistic model is defined for $0 < P_0 < k/(2b)$ and given by the formula*

$$L_d(P_0) = L_d = \frac{1}{k}\log\left(2 + \frac{2b\,P_0}{k - 2b\,P_0}\right). \tag{6.47}$$

In particular, the doubling time for the logistic growth model is larger than the corresponding doubling time for the exponential growth model given by (6.6). Furthermore, in the limiting case when $P_0 \longrightarrow 0$, formula (6.47) coincides with (6.6).

Proof. We have

$$2P_0 = P(L_d) = \frac{k\,P_0}{(k - b\,P_0)e^{-kL_d} + b\,P_0}. \tag{6.48}$$

When we solve (6.48) for L_d, we get (6.47). The statement concerning the limiting case follows immediately from (6.47). ∎

Exercises for Section 6.3

1. Consider the logistic equation $dP/dt = P(6 - P)$ with the initial condition $P(0) = P_0$.

    a. Find the critical points and classify their stability.

    b. Solve the equation.

    c. If $P_0 = 1$, find the time t for which $P(t) = 5$.

    d. Plot representative solutions of $dP/dt = P(6 - P)$.

    e. Use **ODE** with the option **Method->AllSymbolic** to determine which methods solve $dP/dt = P(6 - P)$.

2. The Gompertz[3] model of population growth is described by the equation

$$\frac{dP}{dt} = P\big(k - b\log(P)\big), \tag{6.49}$$

    where k and b are positive constants.

    a. Find the critical points of (6.49) and discuss their stability. [Hint: Use an auxiliary graph and not Theorem 5.5 to determine the stability of the critical points.]

    b. Show that if $0 < P(0) < e^{k/b}$, then $P(t)$ tends to $e^{k/b}$ when $t \longrightarrow \infty$.

    c. Solve the differential equation (6.49) as a separable equation.

    d. Plot representative solutions of (6.49) when $k = 1$ and $b = 4$.

3. Suppose that the logistic equation (6.31) is used to model the natural growth of halibut in the Pacific Ocean. Let $P(t)$ be the total mass at time t of halibut, measured in kilograms. Assume that $k = 0.71/\text{year}$, $b = 8.82 \times 10^{-9}$, and $P(0) = 0.25 \, k/b$.

    a. Find the value of P after two years.

    b. Find the time t for which $P(t) = 0.75 \, k/b$.

    c. Plot the predicted total mass from 0 years to 10 years.

4. Consider the solution (6.38) of the logistic equation with $k > 0$ and $P(t_0) = P_0$, where $0 < P_0 < k/b$.

    a. Show that $P(t) < P_0 e^{k(t-t_0)}$ for any fixed $t > t_0$.

---

[3] Benjamin Gompertz (1779–1865). English pioneer of actuarial science. Denied admission to universities because he was Jewish, Gompertz was self-educated, reading Newton and Maclaurin. He applied the calculus to actuarial questions; in 1825 he showed that the mortality rate increases in a geometric progression. His rigid adherence to Newton's fluxional notation prevented wide recognition of his work.

b. Show that if $k > 0$ is fixed and $b \longrightarrow 0$, then $P(t) \longrightarrow P_0 e^{k(t-t_0)}$ for any fixed $t > t_0$.

5. Suppose that the logistic equation (6.31) is modified to

$$\frac{dP}{dt} = -P(k - bP),$$

where $k > 0$. (This models exponential decay in the presence of a *threshold*.) Assume that $P(t_0) = P_0$.

    a. Find the critical points and discuss their stability.

    b. Show that if $0 \le P(0) < k/b$, then $P(t)$ tends to zero when $t \longrightarrow \infty$.

    c. Show that if $P(0) > k/b$, then $P(t)$ becomes infinite at some positive time.

6. Suppose that the logistic equation is further modified to

$$\frac{dP}{dt} = -P(k - bP)(k - cP),$$

where k, b and c are positive constants with $b > c$. (This models exponential decay in the presence of a *double threshold*.) Assume that $P(t_0) = P_0$.

    a. Find the critical points and discuss their stability.

    b. Show that if $0 \le P(t_0) < k/b$, then $P(t)$ tends to zero when $t \longrightarrow \infty$.

    c. Show that if $P(t_0) > k/b$, then $P(t)$ tends to b/c when $t \longrightarrow \infty$.

7. Suppose that the logistic model is modified to

$$\frac{dP}{dt} = P(k - bP)^2, \tag{6.50}$$

where k and b are positive constants. Assume that $P(t_0) = P_0$.

    a. Find the critical points of (6.50) and classify their stability.

    b. Solve the differential equation (6.50) implicitly.

    c. Find the time t for which $P(t) = 0.9k/b$, if $P_0 = 0.1k/b$.

8. Show that (6.48) implies (6.47).

9. Find the doubling time $L_d(P_0)$ for the logistic model in the case that $k = 3.06$, $b = 0.02$, and $P_0 = 25.4$.

10. Let $t_0 < t_1 < t_2$ be three equally spaced times, that is, $t_2 - t_1 = t_1 - t_0$. Denote the corresponding population values by P_0, P_1 and P_2. Prove that

$$\frac{1}{P_1} < \frac{1}{2}\left(\frac{1}{P_0} + \frac{1}{P_2}\right) \quad \text{and} \quad \frac{1}{P_1} < \sqrt{\frac{1}{P_0 P_2}}.$$

11. A certain contagious virus is known to die out if it is present in sufficiently small quantities but will multiply without bound if enough is present. Assume the model of Exercise 5 with the values $k = 0.14$ and $b = 0.42$. Determine the critical level below which the virus will die out.

6.4   *Population Growth with Harvesting*

In our next model we modify the logistic equation (6.31) of Section 6.3 to account for the possibility of **harvesting**, that is, removal of a part of the population over time. This removal could be due to hunting, fishing or the spread of disease.

The harvesting rate is denoted by H. Maintaining the notation from Section 6.3 for the population size $P(t)$, we postulate the differential equation

$$\frac{dP}{dt} = P(k - bP) - H, \tag{6.51}$$

where k, b, and H are nonnegative constants. Clearly, (6.51) reduces to the logistic equation (6.31) when $H = 0$; it also reduces to the growth with migration equation (6.9) when $b = 0$. We want to solve (6.51) with the initial condition $P(0) = P_0 > 0$. Because of the population interpretation, we are only interested in solutions when $P(t) > 0$.

In order to analyze (6.51), we first find the critical points of (6.51). They are obtained as solutions of the equation

$$P(k - bP) - H = 0. \tag{6.52}$$

This quadratic equation can be solved by the quadratic formula to yield two possible solutions P_+ and P_-, where

$$P_\pm = \frac{k \pm \sqrt{k^2 - 4bH}}{2b}. \tag{6.53}$$

If the expression within the radical in (6.53) is positive, there will be two real roots, whereas there will be no real roots if the expression within the radical is negative. In case the radical is zero, there will be exactly one real root. This leads us to consider three cases:

Case 1 (**subthreshold case**): $0 < H < \dfrac{k^2}{4b}$;

Case 2 (**threshold case**): $H = \dfrac{k^2}{4b}$;

Case 3 (**superthreshold case**): $H > \dfrac{k^2}{4b}$.

The solution will have different behavior depending on the case. The number $H_c = k^2/(4b)$ is called the **harvesting threshold**. We shall see that if the harvesting level H exceeds the harvesting threshold H_c, then the population becomes extinct in a finite time.

Definition. *Let $P(t)$ be a population size that satisfies the differential equation* (6.51) *with the initial condition $P(0) = P_0 > 0$. The* **extinction time** *T_{extinct} (if it exists) is the first value of t for which $P(t) = 0$.*

Next, we determine the stability of the critical points of the differential equation (6.51) in each of the three cases.

Subthreshold case: $0 < H < k^2/(4b)$

In Case 1 the quadratic equation (6.52) has two distinct real roots $P_{\pm}$; they satisfy

$$0 < P_- < \frac{k}{2b} < P_+ < \frac{k}{b}. \tag{6.54}$$

In order to determine their stability, we compute the derivative with respect to P of the right-hand side of (6.51):

$$\frac{d}{dP}\left(P(k - bP) - H\right) = k - 2bP. \tag{6.55}$$

From (6.54) it follows that the right-hand side of (6.55) is positive for $P = P_-$ and negative for $P = P_+$. We conclude from Theorem 5.5 that the critical point P_- is unstable, while the critical point P_+ is stable. This has the following consequences:

- If the initial population satisfies $P(0) < P_-$, then the population $P(t)$ becomes extinct in a finite time, since the solution moves away from the unstable critical point.

- If the initial population $P(0)$ satisfies $P_- < P(0) < P_+$, then the population $P(t)$ tends to the stable critical point $P = P_+$. This is the **steady-state** value of the population.

- If the initial population $P(0)$ satisfies $P(0) > P_+$, then it is also the case that the population $P(t)$ tends to the stable critical point $P = P_+$.

Example 6.8. *Find the critical points of the harvesting equation*

$$P' = P(4 - P) - 3;$$ (6.56)

then solve the equation.

Solution. Since $P(4 - P) - 3 = (P - 3)(1 - P)$, the critical points of (6.56) are $P_- = 1$ and $P_+ = 3$. To find the formula for the solution of (6.56) we use

```
ODE[{P' == 4P - P^2 - 3,P[0] == P0},P,t,Method->Separable]
```

to get

```
                2 t             2 t
       3 - 3 E     - P0 + 3 E     P0
{{P -> ---------------------------------}}
                2 t             2 t
       3 -   E     - P0 +   E     P0
```

Thus, the formula for the solution of (6.56) is

$$P(t) = \frac{(3 - P_0)e^{-2t} + 3(P_0 - 1)}{(3 - P_0)e^{-2t} + (P_0 - 1)} \cdot \ \blacksquare$$

P(t)

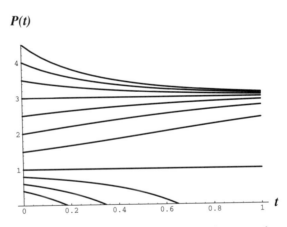

Solutions to the subthreshold harvesting equation
$$P' = P(4 - P) - 3$$

Threshold case: $H = k^2/(4b)$

In Case 2 (the threshold case) the quadratic equation (6.52) has only one solution, so that we have exactly one critical point $P = k/(2b)$. At this point the derivative of $P(k - bP) - H$

is zero, so we cannot infer the stability from this information. But the second derivative is nonzero; indeed,

$$\frac{d^2}{dP^2}\big(P(k - bP) - H\big) = -2b \neq 0.$$

Then Theorem 5.5 implies that $k/(2b)$ is a semistable critical point. The right-hand side of (6.51) is negative for all P except $P = k/(2b)$, where it is zero. This means that all solutions are *decreasing* functions of time. In particular, if $P(0) > k/(2b)$, the solution decreases and approaches the critical point $P = k/(2b)$ when t approaches ∞. If $P(0) < k/(2b)$, then the solution also decreases and eventually reaches the value zero in a finite time; this means that the population becomes extinct.

Example 6.9. *Find the critical points of the harvesting equation*

$$P' = P(4 - P) - 4 = -(P - 2)^2; \tag{6.57}$$

then solve the equation.

Solution. The unique critical point of (6.57) is $P = 2$. To find the formula for the solution of (6.57) we use

ODE[{P' == -(P - 2)^2,P[0] == P0},P,t,
Method->Separable]

to get

```
         P0 - 4 t + 2 P0 t
{{P ->  ───────────────────}}
          1 - 2 t + P0 t
```

Thus, the formula for the solution of (6.57) is

$$P(t) = \frac{P(0) - 4t + 2P(0)t}{1 - 2t + P(0)t}. \quad \blacksquare$$

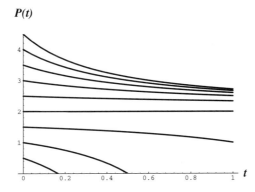

Solutions to the threshold harvesting equation
$$P' = -(P - 2)^2$$

Superthreshold case: $H > k^2/(4b)$

Finally, we analyze Case 3, the superthreshold case. In this case the quadratic equation (6.52) has no real roots. The right-hand side of (6.51) is negative for all values of P, so that $dP/dt < 0$ for all t. The solution decreases until it reaches the value zero, when the population becomes extinct. This is depicted in the figure below.

Example 6.10. *Find the critical points of the harvesting equation*

$$P' = P(4 - P) - 6; \tag{6.58}$$

then solve the equation.

Solution. There are no real critical points of (6.58). To find the formula for the solution of (6.58) we use

```
ODE[{P' == -P^2 + 4P - 6,P[0] == P0},P,t,
Method->Separable]
```

to get

```
                                   -2 + P0
{{P -> 2 - Sqrt[2] Tan[Sqrt[2] t - ArcTan[-------]]}}
                                   Sqrt[2]
```

Thus, the formula for the solution of (6.56) is

$$P(t) = 2 - \sqrt{2}\tan\left(\sqrt{2}t + \arctan\left(\frac{2 - P(0)}{\sqrt{2}}\right)\right). \quad \blacksquare$$

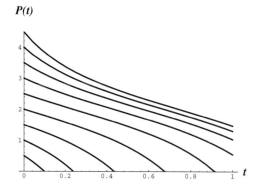

Solutions to the superthreshold harvesting equation
$$P' = P(4 - P) - 6$$

The main conclusions to be drawn from the above analysis are as follows: In order to sustain population growth, the harvesting rate H must be less than the threshold value $H_c = k^2/(4b)$, which is the maximum growth rate in the absence of harvesting. If the initial population is sufficiently large, the steady-state value with subthreshold harvesting is strictly less than the steady-state value in the absence of harvesting ($P_+ < k/b$).

Exercises for Section 6.4

1. Consider the (subthreshold) harvesting model governed by

$$\frac{dP}{dt} = P(4 - P) - 3.$$

   Find the extinction time, given that $0 < P(0) < 1$.

2. Consider the (threshold) harvesting model governed by

$$\frac{dP}{dt} = P(4 - P) - 4.$$

   Find the extinction time, given that $0 < P(0) < 2$.

3. Consider the (subthreshold) harvesting model governed by

$$\frac{dP}{dt} = P(6 - P) - 5.$$

   a. Find the critical points and classify their stability.

   b. Solve the equation.

4. Consider the (threshold) harvesting model governed by

$$\frac{dP}{dt} = P(6 - P) - 9.$$

   a. Find the critical points and classify their stability.

   b. Solve the equation.

   c. Find the time t such that $P(t) = 0$ if $P(0) < 3$.

   d. For which values of $P(0) = P_0$ do we have $P(t) = 0$ for some t? What happens to $P(t)$ for other values of $P(0)$?

5. Consider the (superthreshold) harvesting model with

$$\frac{dP}{dt} = P(6 - P) - 13.$$

    a. Show that there are no critical points.

    b. Solve the equation.

    c. Find the time t for which $P(t) = 0$, given that $P(0) = P_0$.

6. Suppose that the harvesting model is modified to

$$\frac{dP}{dt} = P(k - bP) - hP,$$

where $0 < h < k$ are constants.

    a. Find the critical points P_- and P_+ and classify their stability.

    b. Show that $P(t) > 0$ for all t if $P(0) > 0$.

    c. Find the value of h that maximizes the harvesting yield per unit time when we reach the stable critical value P_+; that is, maximize hP_+.

6.5 Population Models for the United States

The founding fathers of the United States showed great foresight in legislating that a census should be taken every ten years, starting in 1790. In Article I of the constitution it is written:

> *The actual enumeration shall be made within three years after the first meeting of the Congress of the United States, and within every subsequent term of ten years, in such manner as they shall by law direct.*

In this section we apply the theory of Sections 6.1 and 6.3 to analyze the population growth of the United States. The basic problem is to find some natural law that can be used to predict future population. For simplicity we do not consider the effects of major wars, immigration, and technological changes. We follow the pioneering work [PeRe] of Pearl and Reed published in 1920.

The second column in the table below consists of the population sizes of the United States given in millions.

Year	US Census	Exponential Growth	Logistic Growth
1790	3.93	3.93	3.93
1800	5.31	5.11	5.33
1810	7.24	6.65	7.23
1820	9.64	8.64	9.76
1830	12.87	11.24	13.11
1840	17.07	14.62	17.51
1850	23.19	19.01	23.19
1860	31.44	24.72	30.41
1870	39.82	32.15	39.37
1880	50.16	41.81	50.17
1890	62.95	54.38	62.76
1900	75.99	70.72	76.86
1910	91.97	91.97	91.97
1920	105.71	119.61	107.34
1930	122.78	155.54	122.41
1940	131.67	202.28	136.35
1950	151.33	263.06	148.72
1960	179.32	342.11	159.29
1970	203.21	444.91	168.01
1980	226.50	578.60	175.02
1990	249.63	752.45	180.53

Estimating the US Population by an Exponential Law of Growth

The growth rate k of a population obeying an exponential growth law is given in terms of the population size P by (6.7). This formula requires population sizes at two times. A translation of (6.7) into *Mathematica* is given by

```
MalthusGrowth[{t0_,P0_},{t1_,P1_}]:= Log[P1/P0]/(t1 - t0)
```

We need two base years for t_0 and t_1; we choose 1790 and 1910. (A different choice would yield different formulas in the following analysis.) The second column of the table on page 185 tells us that the population of the United States in 1790 was approximately 3.93 million and the population in 1910 was approximately 91.97 million. Using these figures we calculate the growth rate with

```
MalthusGrowth[{1790,3.93},{1910,91.97}]
```

obtaining

```
0.0262735
```

Thus, the formula for the size of the US population using the exponential growth model is

$$P(t) = 3.93e^{0.0262735(t-1790)}. \tag{6.59}$$

The general formula for the size of a population that obeys an exponential growth law is given by (6.5); a translation of this formula into *Mathematica* is

```
Malthus[k_,{t0_,P0_}][t_]:= P0 E^(k t -  k t0)
```

The *Mathematica* version of (6.59) is found with

```
Malthus[0.0262735,{1790,3.93}][t]
```

which gives the output

```
        -47.0296 + 0.0262735 t
3.93 E
```

Furthermore, the third column in the table on page 185 can be calculated using

```
Table[Malthus[0.0262735,{1790,3.93}][t],{t,1790,1990,10}]
```

with the result

```
{3.93, 5.11089, 6.64663, 8.64382, 11.2411, 14.6189,
  19.0116, 24.7242, 32.1534, 41.815, 54.3796, 70.7197,
  91.9697, 119.605, 155.544, 202.282, 263.065, 342.111,
  444.909, 578.596, 752.454}
```

We can plot the US population predicted by the exponential law (derived from the points $(t_0, P_0) = (1790, 3.93)$ and $(t_1, P_1) = (1910, 91.97)$) with

```
Plot[Evaluate[Malthus[0.0262735,{1790,3.93}][t]],
{t,1790,1990}]
```

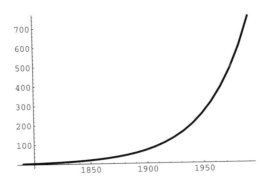

Exponential model of US population

Estimating the US Population by a
Logistic Law of Growth

The growth rate k and carrying capacity k/b of a population obeying a logistic growth law are given in terms of the population size P by (6.40) and (6.41); *Mathematica* versions of these two quantities are

```
LogisticGrowth[{t0_,P0_},{t1_,P1_},{t2_,P2_}]:=
    -Log[(P2^-1 - P1^-1)/(P1^-1 - P0^-1)]/(t1 - t0)
```

and

```
LogisticCarryingCapacity[{t0_,P0_},{t1_,P1_},{t2_,P2_}]:=
    (2P1^-1 - P0^-1 - P2^-1)/(P1^-2 - P0^-1 P2^-1)
```

(As in (6.40) and (6.41) the times **t0**, **t1**, and **t2** must be equally spaced.) The *Mathematica* version of the parameter b is given as the quotient of the growth and the carrying capacity, that is,

```
LogisticRate[{t0_,P0_},{t1_,P1_},{t2_,P2_}]:=
    LogisticGrowth[{t0,P0},{t1,P1},{t2,P2}]/
    LogisticCarryingCapacity[{t0,P0},{t1,P1},{t2,P2}]
```

We need three base years for a logistic model; we chose 1790, 1850, and 1910 in order to compare our results with the exponential model discussed on page 185. (Note that 1850 is the average of 1790 and 1910.) The following command finds the growth rate of the US population, assuming a logistic model from the population figures for 1790, 1850, and 1910:

```
LogisticGrowth[{1790,3.93},{1850,23.19},{1910,91.97}]
```

giving

```
0.0313323
```

Similarly, the parameter b is found using

```
LogisticRate[{1790,3.93},{1850,23.19},{1910,91.97}]
```

with the result

```
0.000158722
```

The formula for the size of a population that obeys a logistic law of growth is given by (6.38); a translation of this formula into *Mathematica* is given by

```
Logistic[{k_,b_},{t0_,P0_}][t_]:=
    k P0 /((k - b P0)E^(-k t + k t0) + b P0)
```

Then

```
Logistic[{0.0313323,0.000158722},{1790,3.93}][t]
```

results in

$$\frac{0.123136}{0.000623777 + 0.0307085\ E^{56.0848\ -\ 0.0313323\ t}}$$

Thus, the formula for the size of the US population using the logistic growth model is approximately

$$P(t) \approx \frac{0.03 \times 3.93}{(0.03 - 0.000062)e^{0.03(t-1790)} + 0.000062 \times 3.93}$$

$$\approx \frac{0.123}{0.00062 + 0.031e^{56.085-0.031t}}. \tag{6.60}$$

Furthermore, the fourth column in the table on page 185 is found using

```
Table[Logistic[{0.0313323,0.000158722},{1790,3.93}][t],
{t,1790,1990,10}]
```

which gives the result

```
{3.93, 5.337, 7.2289, 9.7574, 13.1093, 17.5053, 23.1899, 30.4085,
 39.3664, 50.1703, 62.7618, 76.8636, 91.9697, 107.399, 122.412,
 136.345, 148.718, 159.285, 168.012, 175.021, 180.527}
```

We can plot the US population predicted by the logistic law (derived from the points $(t_0, P_0) = (1790, 3.93)$, $(t_1, P_1) = (1850, 23.19)$, and $(t_2, P_2) = (1910, 91.97)$) with

```
Plot[Evaluate[Logistic[
{0.0313323,0.000158722},{1790,3.93}][t]],{t,1790,1990}]
```

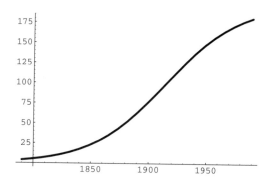

Logistic model of the US population

Comparing Models of US Population Growth

Mathematica has a very useful command **Interpolation** that constructs an **interpolating function** from data. Internally, *Mathematica* represents such an interpolating function by piecewise polynomials, but it is not useful to manipulate these piecewise polynomials directly.

The advantage of an interpolating function over the raw data that it represents is that an interpolating function has many of the properties of an ordinary function. In particular, an interpolating function can be differentiated. Thus an interpolating function permits not only the graphical representation of a function defined by data, but also the graphical representation of the derivatives of the function.

Example 6.11. *Plot the actual US population from 1790 to 1990 by means of an interpolating function and compare the graph with that of the exponential and logistic models.*

Solution. We need to make the population data for the US in the second column of the table on page 185 into a list of points, so we use

```
uspopulation=
{{1790,3.93},{1800,5.31},{1810,7.24},{1820,9.64},{1830,12.87},
{1840,17.07},{1850,23.19},{1860,31.44},{1870,39.82},
{1880,50.16},{1890,62.95},{1900,75.99},{1910,91.97},
{1920,105.71},{1930,122.78},{1940,131.67},{1950,151.33},
{1960,179.32},{1970,203.21},{1980,226.50},{1990,249.63}}
```

Then the interpolating function is defined by

```
uspopulationfunction=Interpolation[uspopulation]
```

```
InterpolatingFunction[{1790, 1990}, <>]
```

This terse output indicates that *Mathematica* has created an
InterpolatingFunction. For example, we can estimate the population of the US in
1843 with

uspopulationfunction[1843]

yielding

```
18.6675
```

So the population of the US in 1843 was approximately 18.67 million. The simultaneous plot
is of **uspopulationfunction[t]**, and its exponential and logistic approximations are
drawn with

```
Plot[Evaluate[{uspopulationfunction[t],
Malthus[0.0262735,{1790,3.93}][t],
Logistic[{0.0313323,0.000158722},{1790,3.93}][t]}],
{t,1790,1990}]
```

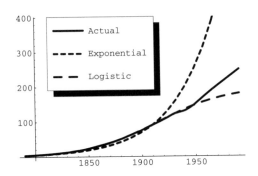

US Population with exponential and logistic models

From this graph we see that the actual US population size is well approximated by an
exponential law curve up to about 1910; after that an exponential law increasingly errs on
the high side. On the other hand, a logistic curve well approximates the US population size
until about 1930; after that the logistic curve predicts too small a population.

On the other hand, the graphs of the first and second derivatives of the three functions
plotted in Example 6.12 show considerably more variation.

Example 6.12. *Compare the derivative of the Mathematica constructed approximation
to the US population from 1790 to 1990 with the derivatives of the exponential and logistic
approximations.*

Solution. The first derivatives are given in *Mathematica* by

```
uspopulationfunction'[t]
Malthus[0.0262735,{1790,3.93}]'[t]
```

and

```
Logistic[{0.0313323,0.000158722},{1790,3.93}]'[t]
```

So we can obtain the simultaneous plot of the first derivatives of each approximation with

```
Plot[Evaluate[{uspopulationfunction'[t],
Malthus[0.0262735,{1790,3.93}]'[t],
Logistic[{0.0313323,0.000158722},{1790,3.93}]'[t]}],
{t,1790,1990}]
```

The resulting graph shows that the derivatives of the exponential and logistic approximations resemble the rate of change of the US population very poorly.

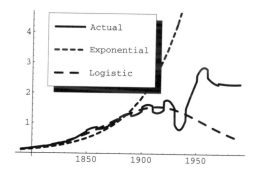

**First derivatives of the US population function and its
exponential and logistic approximations**

The second derivatives are plotted in a similar fashion.

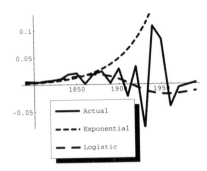

**Second derivatives of the US population function and its
exponential and logistic approximations**

Thus, even though the exponential and logistic approximations explain well the US population over certain intervals, their first and second derivatives do not.

Exercises for Section 6.5

Plot the actual population (expressed in millions) together with the exponential and logistic approximations for the populations of each of the following states.

1. Maryland. Use the data

   ```
   marylandpopulation=
   {{1700,0.03},{1725,0.07},{1750,0.14},{1775,0.23},
   {1800,0.34},{1825,0.44},{1850,0.58},{1875,0.83},
   {1900,1.19},{1925,1.50},{1950,2.34},{1975,4.10}}
   ```

2. Texas. Use the data

   ```
   texaspopulation=
   {{1850,0.21},{1875,1.71},{1900,3.05},
   {1925,4.85},{1950,7.71},{1975,12.24}}
   ```

3. Illinois. Use the data

   ```
   illinoispopulation=
   {{1850,0.85},{1875,2.95},{1900,4.82},
   {1925,6.67},{1950,8.71},{1975,11.15}}
   ```

6.6   Temperature Equalization Models

The theory of first-order linear equations will now be applied to study the approach to **thermal equilibrium** in solid bodies. Consider, for example, an apple that is removed from a refrigerator that is kept at 38°F. The room temperature is maintained at 68°F. In the first few minutes after removal, the temperature of the apple will increase quickly until it approaches room temperature. Of course, the temperature of the apple never *exactly* equals room temperature, but it comes closer and closer with the passage of time; hence the rate of change of temperature becomes smaller and smaller with the passage of time. Thus we make the following postulate (known as **Newton's law of cooling**) for the temperature of a solid:

> ***The time rate of change of the temperature is proportional to the difference between the temperature of the solid and the temperature of the surroundings.***

To translate this into mathematics, we let $T(t)$ be the temperature of the solid at time t and T_e be the temperature of the external surroundings. The constant $k > 0$ gives the postulated proportionality; thus, we have the differential equation

$$\frac{dT}{dt} = -k(T - T_e). \tag{6.61}$$

The minus sign is inserted because we expect $T(t)$ to *decrease* if $T(t) > T_e$, whereas we expect $T(t)$ to *increase* if $T(t) < T_e$. Equation (6.61) is a first-order linear equation that has $\mu(t) = e^{kt}$ as an integrating factor. Thus

$$\frac{d}{dt}\left(e^{kt}T(t)\right) = e^{kt}k\,T_e. \tag{6.62}$$

We integrate both sides of (6.62) from 0 to T to obtain

$$e^{kt}T(t) - T(0) = \int_0^t e^{ks}k\,T_e\,ds = T_e(e^{kt} - 1),$$

which we rewrite as

$$T(t) = T_e + \left(T(0) - T_e\right)e^{-kt}. \tag{6.63}$$

When $t \longrightarrow \infty$ the exponential terms in (6.63) tend to zero, so that

$$\lim_{t \to \infty} T(t) = T_e.$$

We use the term **steady-state** for the ultimate temperature T_e. The mathematical model predicts that as time increases the temperature of the solid approaches ever closer to the steady-state temperature. However, the exponential function is never zero, so that we always have $T(t) \neq T_e$ for any finite value of t; the steady-state is never *exactly* attained.

Example 6.13. *Find and plot the solution to* (6.61) *in the case that* $T_e = 68°F$, $T(0) = 38°F$ *for* $k = 0.1, 0.3, 0.5, 0.7, 0.9$.

Solution. We use

```
ODE[{T' == -k(T - 68),T[0] == 38},T,t,
Method->FirstOrderLinear,Parameters->{{k,0.1,0.9,0.2}},
PlotSolution->{{t,0,15}}]
```

to get

```
          30
{{T -> 68 - ---}}
          k t
         E
```

and the plot

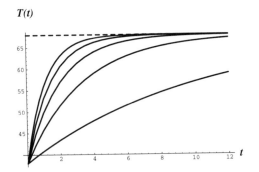

Solution to $T' = -k(T - 68)$, $T(0) = 38$
for $k = 0.1, 0.3, 0.5, 0.7, 0.9$

The solution (6.63) can also be rewritten in the form

$$T(t) - T_e = e^{-kt}\big(T(0) - T_e\big), \tag{6.64}$$

which we now use to define the concept of **relaxation time**. This is the number $\tau > 0$ such that

$$\frac{T(\tau) - T_e}{T(0) - T_e} = e^{-1} \approx 0.368. \tag{6.65}$$

(The reference level $e^{-1} \approx 0.368$ is chosen for convenience in order to simplify the subsequent calculations. Some authors use the reference level $1/2$, which introduces the factor $\log(2)$ in the subsequent formulas.) The physical interpretation is,

> *The relaxation time is the time necessary for the temperature to reach a fixed fraction of the steady-state temperature.*

From (6.64) and (6.65) it follows that

$$e^{-k\tau} = e^{-1},$$

so that the relaxation time is given by

$$\tau = \frac{1}{k}.$$

Example 6.14. *An apple at* 38°F *is removed from a refrigerator, placed in a room at* 68°F, *and allowed to warm according to the above theory (equation (6.61)). We observe that after* 2 *hours the apple has reached a temperature of* 60°F. *Find the relaxation time* τ. *How much longer must we wait for the temperature to reach* 65°F?

Solution. We must find the parameter k. We have $T(0) = 38°$F, $T(2) = 60°$F, and $T_e = 68°$F. When we substitute these values into (6.64) and measure time in hours, we get

$$e^{-2k} = \frac{60 - 68}{38 - 68} = \frac{8}{30}.$$

Thus

$$-2k = \log\left(\frac{8}{30}\right) \approx -1.322, \qquad \text{or} \qquad k \approx \frac{1.322}{2} \approx 0.661,$$

so that

$$\tau = \frac{1}{0.661} \approx 1.513 \text{ hours.}$$

The time t_{65} required to reach 65°F is found by solving

$$e^{-kt_{65}} \frac{65 - 68}{38 - 68} = \frac{3}{30},$$

which yields $k\, t_{65} = \log(30/3) = 2.303$, or

$$t_{65} = \frac{2.303}{0.661} = 3.483,$$

about 3.5 hours. Thus, we must wait an additional 1.5 hours. ∎

In many situations the temperature of the surroundings is time-dependent. For example, the daily temperature may vary by as much as 40°F, which can cause periodic changes in the temperatures of unheated buildings such as garages, barns, or warehouses. It is natural to study the **steady-state temperature** as $t \longrightarrow \infty$, which will also be time-dependent.

The basic hypothesis of this section translates into the differential equation

$$\frac{dT}{dt} = -k(T - T_e(t)),\tag{6.66}$$

where $k > 0$ is a constant, $T_e(t)$ is the temperature of the surroundings and $T(t)$ is the sought after temperature. We illustrate with an example.

Example 6.15. *Suppose that the external temperature $T_e(t)$ varies according to*

$$T_e(t) = A_0 + A_1 \cos(\omega t),\tag{6.67}$$

where A_0, A_1, and ω are positive constants. Assume that the temperature is governed by (6.66). Solve for $T(t)$ and determine the steady-state temperature when $t \longrightarrow \infty$. Find the maximum and minimum steady-state temperatures and their time of occurrence.

Solution. From (6.66) and (6.67) we have the equation

$$\frac{dT}{dt} + kT = k\big(A_0 + A_1 \cos(\omega t)\big),$$

which is solved using the integrating factor e^{kt}; thus

$$\frac{d}{dt}\big(e^{kt}T(t)\big) = k\,e^{kt}\big(A_0 + A_1 \cos(\omega t)\big),$$

which we integrate to obtain

$$\begin{aligned}
e^{kt}T(t) &= T(0) + \int_0^t k\,e^{ks}\big(A_0 + A_1 \cos(\omega s)\big)\,ds\\
&= T(0) + A_0\big(e^{kt} - 1\big) + \frac{A_1\,k}{k^2 + \omega^2}\big(k\,e^{kt}\cos(\omega t) - k + \omega\,e^{kt}\sin(\omega t)\big)
\end{aligned}$$

and the solution

$$T(t) = T(0)e^{-kt} + A_0(1 - e^{-kt}) + \frac{A_1\,k\big(k\,\cos(\omega t) + \omega \sin(\omega t) - k\,e^{-kt}\big)}{k^2 + \omega^2}.\tag{6.68}$$

The steady-state temperature $T_{\text{steady}}(t)$ is obtained by neglecting the exponential terms in (6.68), which are very small when t is large; thus

$$T_{\text{steady}}(t) = A_0 + \frac{A_1\,k}{k^2 + \omega^2}\big(k\,\cos(\omega t) + \omega \sin(\omega t)\big).\tag{6.69}$$

The maximum and minimum values of $T_{\text{steady}}(t)$ are obtained by taking the derivative of (6.69) and setting it to zero. We compute

$$T'_{\text{steady}}(t) = \frac{A_1\,k}{k^2 + \omega^2}\big(-k\omega\,\sin(\omega t) + \omega^2 \cos(\omega t)\big).\tag{6.70}$$

From (6.70) it is clear that $T'_{\text{steady}}(t)$ vanishes precisely when $\tan(\omega t) = \omega/k$. There will be infinitely many such values of t; they are of the form ωT_0, $\omega T_0 + \pi$, $\omega T_0 + 2\pi$, ..., where $0 < \omega T_0 < \pi/2$.

The maximum and minimum are obtained by noting that

$$\sin(\omega T_0) = \frac{\omega}{\sqrt{\omega^2 + k^2}} \qquad \text{and} \qquad \cos(\omega T_0) = \frac{k}{\sqrt{\omega^2 + k^2}},$$

whereas

$$\sin(\omega T_0 + \pi) = -\frac{\omega}{\sqrt{\omega^2 + k^2}} \qquad \text{and} \qquad \cos(\omega T_0 + \pi) = -\frac{k}{\sqrt{\omega^2 + k^2}},$$

from which we conclude that for all t,

$$-\sqrt{\omega^2 + k^2} \le k\cos(\omega t) + \omega\sin(\omega t) \le \sqrt{\omega^2 + k^2}.$$

It follows from (6.69) that the maximum and minimum steady-state temperatures are

$$T_{\max} = A_0 + \frac{A_1 k}{\sqrt{\omega^2 + k^2}} \qquad \text{and} \qquad T_{\min} = A_0 - \frac{A_1 k}{\sqrt{\omega^2 + k^2}}. \qquad (6.71)$$

Example 6.16. *Find and plot the solution to the initial value problem*

$$\begin{cases} \dfrac{dT}{dt} = -0.5\left(T - 65 - 15\cos\left(\dfrac{\pi t}{12}\right)\right), \\ T(0) = 20. \end{cases} \qquad (6.72)$$

Find the maximum and minimum steady-state temperatures and their times of occurrence.

Solution. The solution to (6.72) can be found and plotted using

```
ODE[{T' == -(1/2)(T - (65 + 15Cos[t Pi/12])),T[0] == 20},
T,t,Method->FirstOrderLinear,Form->Explicit,
PlotSolution->{{t,0,24Pi}},PostSolution->{Cancel}]
```

resulting in

$$\frac{2340}{36 + \text{Pi}^2} + \frac{65\,\text{Pi}^2}{36 + \text{Pi}^2} - \frac{2160}{36\,E^{t/2} + E^{t/2}\,\text{Pi}^2} - \frac{45\,\text{Pi}^2}{36\,E^{t/2} + E^{t/2}\,\text{Pi}^2} +$$

$$\frac{540\,\text{Cos}[\dfrac{\text{Pi}\,t}{12}]}{36 + \text{Pi}^2} + \frac{90\,\text{Pi}\,\text{Sin}[\dfrac{\text{Pi}\,t}{12}]}{36 + \text{Pi}^2}$$

and the plot

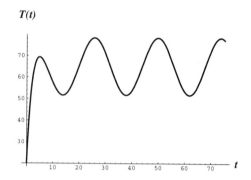

Solution to
$$T' = -0.5\big(T - 65 - 15\cos(\pi t/12)\big), \quad T(0) = 20$$

Thus

$$T(t) = 65 + \frac{540\cos(\pi t/12) + 90\pi\,\sin(\pi t/12) - e^{-t/2}(2160 + 45\pi^2)}{36 + \pi^2}$$

$$\approx 65 + 11.773\cos(0.262t) + 6.164\sin(0.262t) - 56.773e^{-t/2}.$$

Since (6.72) is a special case of the differential equation considered in Example 6.15, we can use the results of that example to find T_{max} and T_{min}. In the case when $k = 1/2$, $A_0 = 65$, $A_1 = 15$, and $\omega = \pi/12 \approx 0.2618$, we use

```
N[A0 + A1 k/Sqrt[omega^2 + k^2] /.
{A0->65,A1->15,k->1/2,omega->Pi/12}]
```

and

```
N[A0 - A1 k/Sqrt[omega^2 + k^2] /.
{A0->65,A1->15,k->1/2,omega->Pi/12}]
```

to compute
$$T_{max} = 78.2886°F \quad \text{and} \quad T_{min} = 51.7114°F.$$

The times of occurrence are given by

$$T = \frac{1}{\omega}\left(\arctan\left(\frac{\omega}{k}\right) - n\pi\right)$$

for $n = 0, 1, \ldots$. In the case at hand the first five of these values are computed via *Mathematica* with

```
N[Table[(ArcTan[omega/k] + n Pi)/omega,{n,0,4}] /.
{k->1/2,omega->Pi/12}]
```

{1.84243, 13.8424, 25.8424, 37.8424, 49.8424}

Thus, the first value is $T \approx 1.8424$ hours. Therefore, the maximum steady-state temperature occurs approximately 1 hour and 50 minutes later than the maximum external temperature.

To compare the external temperature (6.67) with the steady-state temperature (6.69) we put

```
TSteadyState[t_]:=
   65 + (540Cos[Pi t/12]+90Pi Sin[Pi t/12])/(36 + Pi^2)
```

Then a simultaneous plot of the steady-state and external temperatures is found with

```
Plot[Evaluate[{TSteadyState[t],65 + 15Cos[t Pi/12]}],
{t,0,12Pi}]
```

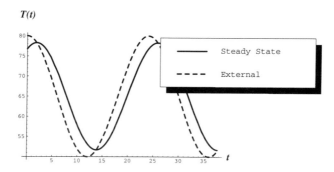

Steady-state solution to
$$T' = -0.5\bigl(T - 65 - 15\cos(\pi t/12)\bigr), \quad T(0) = 20$$

Exercises for Section 6.6

Solve the following temperature equalization models and find the relaxation time:

1. $\begin{cases} T'(t) = -3(T - 68), \\ T(0) = 40 \end{cases}$

2. $\begin{cases} T'(t) = -k(T - T_e), \\ T(t_0) = 0, \qquad (k, T_e \text{ constant}) \end{cases}$

3. $\begin{cases} T'(t) = -4(T - 68), \\ \lim\limits_{t\to\infty} e^{4t} T'(t) = -3 \end{cases}$

4. $\begin{cases} T'(t) + 4T = 100, \\ T(0) = -5 \end{cases}$

5. A glass of cold water at $35°$F reaches a temperature of $42°$F after 5 minutes in a room at temperature $68°$F. Determine the relaxation time. Find the temperature of the water 15 minutes after its temperature was $35°$F.

6. A cup of coffee initially at $200°$F cools to $175°$F in 3 minutes in a room at temperature $70°$F. Find the relaxation time. When will the temperature of the coffee reach $112°$F?

7. A building is cooled to a temperature of $75°$F on a $90°$F day, at which time the air conditioner is turned off. Assuming a relaxation time of 4 hours and that the air conditioner is not turned back on, find the temperature 2 hours after the air conditioner is turned off. When will the temperature reach $85°$F?

8. An oven, whose maximum temperature is $400°$F, can reach $380°$F in 15 minutes. A potato must be cooked for 1 hour between $350°$F and $400°$F. How long does it take to cook a potato if the oven and the potato are initially at room temperature of $75°$F? What can be said if the oven is preheated to $350°$F?

9. Experimental data suggests that a baked potato will drop to a temperature of $160°$F 10 minutes after it is removed from an oven at $400°$F and placed in a room whose temperature is $75°$F. If the human palate cannot tolerate food temperatures exceeding $125°$F, how long must a person wait before eating the potato?

10. A barn has no heating or cooling and is subject to the daily variation of external temperature, which varies from $40°$F to $80°$F according to the formula

$$T_e(t) = 60 + 20\cos\left(\frac{\pi t}{12}\right),$$

where t is measured in hours. Assuming a relaxation time of 3 hours, find the maximum and minimum steady-state temperatures. If the maximum external temperature occurs at 2 P.M., when does the maximum temperature of the barn occur?

11. Two friends sit down for a cup of coffee. The first person pours the cream into her coffee immediately. The second person waits t seconds before pouring cream into her coffee. Which cup of coffee is hotter at this time, the pre-mixed cup or the just-mixed cup? [Hint: Assume that the coffee and cream both have the same relaxation time and that when mixed the new temperature instantaneously becomes the weighted average of the two temperatures.]

12. It is required to find the time of a person's death, given measurements of the temperature of the corpse at two later times.

   a. Assuming that $t_1 < t_2$ and we are given $T(t_1) = T_1$ and $T(t_2) = T_2 < T_1$, find the time $t_0 < t_1$ for which $T(t_0) = 98.6°$F.

   b. Illustrate the calculations of part **a** in the case that $t_1 = 1$ hour, $t_2 = 2$ hours, $T_1 = 95°$F, $T_2 = 92°$F, and room temperature is assumed to be $T_e = 68°$F.

7

APPLICATIONS OF
FIRST-ORDER
EQUATIONS II

Several applications of first-order equations to elementary mechanics are given in Section 7.1. Rocket propulsion is described in 7.2. Electric circuits are treated in Section 7.3. First-order equations also arise in mixing problems (Section 7.4).

Mathematica offers a unique contribution to modeling, because frequently it can be used to construct realistic models. Pursuit curves are treated in Section 7.5; using *Mathematica*, we show how to draw the curve taken by a ship pursuing another ship traveling in a straight line.

7.1   Application of First-Order
Equations to Elementary Mechanics

First-order linear equations arise in certain special cases of Newtonian mechanics. In general, Newton's second law of motion (see page 5) applied to a particle of mass m moving in a straight line gives rise to a second-order differential equation, which may be written in the form

$$m\frac{d^2y}{dt^2} = F\left(t, y, \frac{dy}{dt}\right), \tag{7.1}$$

where the force F depends on time t, the position $y(t)$ and the velocity $v(t) = y'(t)$. In this section and the next we consider two important special cases in which (7.1) can be solved by the methods of first-order equations.

In the first case the force does *not* depend on the position; then we have a first-order equation for the function $v = dy/dt$. Indeed, in this case (7.1) has the form

$$m\frac{dv}{dt} = F(t, v). \tag{7.2}$$

In the second case the force is assumed to depend only on the position $y(t)$, so that (7.1) reduces to the second-order equation

$$m\frac{dv}{dt} = F(y). \tag{7.3}$$

We shall see below that (7.3) can be reduced to a first-order equation, leading to a suitable statement of conservation of energy.

First, we consider in detail two special cases of (7.2), which is nonlinear in general; here we restrict attention to two linear models. In Exercises 5, 6, and 7 nonlinear models are considered.

Constant Force with Linear Damping

Suppose that a particle of mass m moves in a constant force field F_0 with an added damping force F_1 that is negatively proportional to the velocity of the particle, that is, $F_1 = -k\,v$, where $k > 0$. Hence (7.2) becomes the first-order linear differential equation

$$m\frac{dv}{dt} = F_0 - k\,v. \tag{7.4}$$

For example, F_0 might be the force due to gravity and F_1 might be air resistance. Clearly, $e^{kt/m}$ is an integrating factor for (7.4), so the solution of (7.4) is

$$v(t) = V_0 e^{-kt/m} + \frac{F_0}{k}\left(1 - e^{-kt/m}\right), \tag{7.5}$$

where $v(0) = V_0$ denotes the initial velocity. From (7.5) we see that the particle has a **limiting velocity** given by

$$\lim_{t \to \infty} v(t) = \frac{F_0}{k}.$$

The position $y(t)$ is obtained through a further integration of (7.5) using the relation $y' = v$, the result is

$$y(t) = Y_0 + \frac{V_0 m}{k}\left(1 - e^{-kt/m}\right) + \frac{F_0}{k}\left(t - \frac{m}{k}\left(1 - e^{-kt/m}\right)\right), \tag{7.6}$$

where $Y_0 = y(0)$ is the initial height of the particle. (The downward direction is positive.) Thus we see that for large t,

$$y(t) \approx \frac{t\, F_0}{k} + Y_0 + \frac{V_0 m}{k} - \frac{F_0 m}{k^2},$$

indicating that the motion resembles that of a free particle with no external forces ($y'' = 0$).

Example 7.1. *Use* **ODE** *to solve (7.4) for the velocity $v(t)$ in the case that $m = 5$, $F_0 = 10$, and $k = 30$. Plot the velocity $v(t)$ and the height $y(t)$. Assume that the initial velocity and position are both* 0.

Solution. To find the velocity we use

```
v[t_] = ODE[{5 vv' == 10 - 30 vv,vv[0] == 0},vv,t,
        Method->FirstOrderLinear,Form->Explicit]
```

```
1     1
- - ---------
3     6 t
    3 E
```

From this output, or from the command **Limit[v[t],t->Infinity]** we see that the limiting velocity is $1/3$. Next, we compute the position vector with

```
y[t_] = Integrate[v[s],{s,0,t}]
```

```
   1        1        t
-(---)  + --------- + -
  18        6 t      3
          18 E
```

This output tells us that $y(t)$ approaches the line $t \longrightarrow t/3 - 1/18$ asymptotically. Therefore, to plot the velocity $v(t)$ from 0 to 1 together with its asymptote we use

```
Plot[v[t]//Evaluate,{t,0,1},
Epilog->{AbsoluteDashing[{6,6}],Line[{{0,1/3},{1,1/3}}]}]
```

and to plot the height $y(t)$ and its asymptote we use

```
Plot[{y[t],t/3 - 1/18}//Evaluate,{t,0,1},
PlotStyle->{{},{AbsoluteDashing[{6,6}]}}]
```

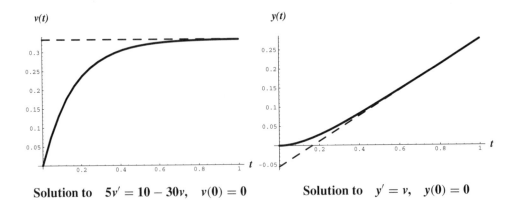

Solution to $5v' = 10 - 30v,$ $v(0) = 0$ Solution to $y' = v,$ $y(0) = 0$

Example 7.2. *A parachutist weighing* 120 *pounds jumps from an airplane on a calm day. She experiences a frictional force that equals* $-7.5v$, *where* v *denotes her downward velocity. Find her limiting velocity when* $t \longrightarrow \infty$.

Solution. We choose coordinates such that the downward direction corresponds to *positive* values of v. The constant force of gravity is equal to her weight: $F_0 = m\,g = 120$ pounds. The problem specifies that $k = 7.5$ pounds/(feet/second); hence

$$\lim_{t \to \infty} v(t) = \frac{F_0}{k} = \frac{m\,g}{k} = \frac{120}{7.5} = 16.0 \text{ feet/second.} \ \blacksquare$$

Example 7.3. *An object weighing* 5 *pounds is released from rest* 200 *feet above the ground and is allowed to fall under the influence of gravity. Assume that the force due to air resistance is proportional to the velocity of the object, with proportionality constant* $k = 2$ *pounds/(feet/second). Determine when the object will strike the ground.*

Solution. The differential equation for the velocity is

$$\frac{5v'}{32} = 5 - 2v,$$

where the mass is $5/32$. We solve it with

```
v[t_] = ODE[{5/32 vv' == 5 - 2 vv,vv[0] == 0},vv,t,
        Method->FirstOrderLinear,Form->Explicit];
```

(The semicolon at the end of the command suppresses the output.) The height function $y(t)$ is then found with

```
y[t_] = ODE[{yy' == v[t],yy[0] == 0},yy,t,
        Method->FirstOrderLinear,Form->Explicit]
```

$$-\left(\frac{25}{128}\right) + \frac{25}{128\ E^{(64\ t)/5}} + \frac{5\ t}{2}$$

To find the approximate value of t for which $y(t) = 200$, we use **FindRoot**. This useful command can be frequently used to solve complicated equations. It requires an estimate of the solution. In the case at hand we guess $t = 25$. The command

FindRoot[y[t] == 200,{t,25}]

gives the output

{t -> 80.0781}

So the object hits the ground after approximately 80 seconds. ∎

Periodic Force with Linear Damping

In this subsection we modify the previous model by replacing the constant force field by a sinusoidally varying function. It turns out that although the velocity no longer tends to a limit, we still have a limiting **average velocity**. We illustrate this with the following example.

Example 7.4. *A particle moves in a force field $F(t)$ that varies sinusoidally with time according to*

$$F(t) = F_0 + F_1 \cos(\omega t),$$

where F_0 and F_1 are constants. The particle also has a damping force that is negatively proportional to the velocity. Find the velocity at all future times and discuss the asymptotic behavior of the particle.

Solution. The appropriate differential equation for this example is

$$m\frac{dv}{dt} + k\,v = F_0 + F_1 \cos(\omega t), \tag{7.7}$$

where k denotes the constant of proportionality of the damping force. An integrating factor for (7.7) is $e^{kt/m}$, leading to

$$(e^{kt/m}v)' = \left(\frac{F_0}{m} + \frac{F_1}{m}\cos(\omega t)\right)e^{kt/m}.$$

The integral of the constant term is done as before. When combined with the integral of the trigonometric term, we get the solution to (7.7) in the form

$$
v(t) \;=\; V_0 e^{-kt/m} + \frac{F_0}{k}\left(1 - e^{-kt/m}\right) + \frac{F_1\left(k\cos(\omega t) - k\,e^{-kt/m} + m\,\omega\sin(\omega t)\right)}{k^2 + m^2\omega^2}
$$

$$
\;=\; \frac{F_0}{k} + \left(V_0 - \frac{F_0}{k} - \frac{k}{k^2 + m^2\omega^2}\right) e^{-kt/m} + \frac{F_1\left(k\cos(\omega t) + m\,\omega\sin(\omega t)\right)}{k^2 + m^2\omega^2},
$$

where V_0 is the initial velocity. In this example we do not have a limiting velocity when $t \longrightarrow \infty$. Nevertheless, it is possible to define a **limiting average velocity** $\tilde{v}$ as follows. First, we compute

$$
\int_0^t v(s)\,ds \;=\; \frac{F_0}{k}s - \frac{m}{k}\left(V_0 - \frac{F_0}{k} - \frac{k}{k^2 + m^2\omega^2}\right)e^{-ks/m}
$$

$$
+ \frac{F_1}{k^2 + m^2\omega^2}\left(\frac{k}{\omega}\sin(\omega s) - m\cos(\omega s)\right)\Bigg|_0^t
$$

$$
= \frac{F_0}{k}t - \frac{m}{k}\left(V_0 - \frac{F_0}{k} - \frac{k}{k^2 + m^2\omega^2}\right)(e^{-kt/m} - 1)
$$

$$
+ \frac{F_1}{k^2 + m^2\omega^2}\left(\frac{k}{\omega}\sin(\omega t) - m\cos(\omega t) + m\right). \tag{7.8}
$$

From (7.8) it follows that the limit as $t \longrightarrow \infty$ of

$$
\frac{1}{t}\int_0^t v(s)\,ds \tag{7.9}
$$

exists. Furthermore, the integral (7.9) represents the average of $v(s)$ over the interval $0 \le s \le t$. Therefore, it is reasonable to define the limiting average velocity to be

$$
\tilde{v} = \lim_{t\to\infty}\frac{1}{t}\int_0^t v(s)\,ds = \frac{F_0}{k}. \quad\blacksquare
$$

Example 7.5. *Use* **ODE** *to solve and plot the solution $v(t)$ of (7.7) if the initial velocity is $v(0) = 0$, the mass is $m = 6$, the damping force is given by $-2v$, and the force field $F(t)$ is given by*

$$
F(t) = 10 - 2\cos(20t).
$$

Solution. To find the velocity $v(t)$ we use

```
v[t_] = ODE[{5 vv' ==   10 - 20 vv - 2Cos[20t],vv[0] == 0},
          vv,t,Method->FirstOrderLinear,Form->Explicit]
```

obtaining

$$\frac{1}{2} - \frac{129}{260\ E^{4\ t}} - \frac{Cos[20\ t]}{260} - \frac{Sin[20\ t]}{52}$$

Then the average velocity is computed with

```
av[t_] = (1/t)Integrate[v[s],{s,0,t}]//Expand
```

giving

$$\frac{1}{2} - \frac{1}{8\ t} + \frac{129}{1040\ E^{4\ t}\ t} + \frac{Cos[20\ t]}{1040\ t} - \frac{Sin[20\ t]}{5200\ t}$$

Then the plot of $v(t)$ together with the line $v = 0.5$ is obtained with the command

```
Plot[Evaluate[{v[t],av[t]}],{t,0.001,2},
Epilog->{AbsoluteDashing[{6,6}],Line[{{0,1/2},{2,1/2}}]},
PlotRange->All]
```

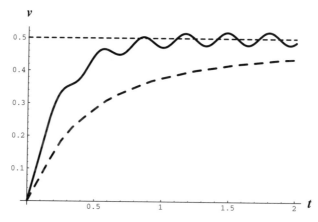

Solution to $5v' = 10 - 20v - 2\cos(20t)$, $v(0) = 0$ and its average

(We have started the plot from 0.001 instead of 0 because *Mathematica* thinks wrongly that **av** has a singularity at 0.) From this plot we see that the graph of $v(t)$ repeatedly crosses the line $v = 0.5$, and that this line is an asymptote of the average velocity. ∎

Example 7.6. *A parachutist weighing* 120 *pounds jumps from an airplane on a windy day. She experiences a frictional force that equals* $-7.0v - 0.25\cos(4t)$, *where* v *denotes her downward velocity. Find her limiting average velocity when* $t \longrightarrow \infty$.

Solution.   As in Example 7.2, the constant force of gravity is equal to her weight: $F_0 = m\,g = 120$ pounds. The problem specifies that $k = 7.0$ pounds/(feet/second); hence

$$\lim_{t \to \infty} \frac{1}{t} \int_0^t v(s)\,ds = \frac{F_0}{k} = \frac{m\,g}{k} = \frac{120}{7.0} = 17.1429 \text{ feet/second.} \quad \blacksquare$$

Conservation of Energy

It is also possible to solve (7.1) in case that the force F depends only on the position y. In this case (7.1) reduces to the differential equation

$$m \frac{d^2 y}{dt^2} = m \frac{dv}{dt} = F(y). \tag{7.10}$$

We can solve (7.10) by writing

$$\frac{dy}{dt} = v \quad \text{and} \quad \frac{dv}{dt} = \frac{dv}{dy} \frac{dy}{dt}, \tag{7.11}$$

to obtain

$$m\,v \frac{dv}{dy} = F(y). \tag{7.12}$$

Then (7.12) is a separable equation that can be solved to obtain v as a function of y, as follows. Let $V(y)$ be an indefinite integral of $-F(y)$; thus

$$V'(y) = -F(y). \tag{7.13}$$

We call $V(y)$ the **potential energy**; it can be interpreted as the work necessary to move the particle from y to a reference point y_0. Then (7.12) can be integrated with respect to y, yielding

$$\frac{1}{2} m\,v^2 + V(y) = E, \tag{7.14}$$

where E is a constant called the **total energy** of motion. The term $(1/2)m\,v^2$ is the **kinetic energy** of motion, interpreted as the energy that the particle would acquire purely from its motion, apart from any external forces. The resultant equation (7.14) is the statement of **conservation of energy**. It provides a mathematical relation between the variables v and y that does not involve time. Although conservation of energy does not give all of the details of the motion, certain useful consequences can often be derived from it.

We illustrate conservation of energy with Hooke's law of elasticity from page 5:

$$m\,y'' = -k\,y. \tag{7.15}$$

We use (7.10)–(7.15) to reduce (7.15) to a first-order equation:

$$m\,v \frac{dv}{dy} = -k\,y.$$

For convenience we take $y_0 = 0$. Since $F(y) = -k\,y$, the potential energy is given by $V(y) = k\,y^2/2$. Thus, we obtain the conservation of energy equation

$$\frac{1}{2}m\,v^2 + \frac{1}{2}k\,y^2 = E. \tag{7.16}$$

Equation (7.16) shows that the motion takes place along a family of ellipses in the y, v plane. The ratio of the axes depends only on the ratio of the constants m and k, but the size of the axes depends on the initial conditions through the energy. We illustrate with an example. A more detailed study of equation (7.15) will be conducted in Chapter 11.

Example 7.7. *A spring governed by Hooke's law has the values $m = 1$ and $k = 4$ in a suitable system of units. If the system is started with $V_0 = 4$ and $Y_0 = 0$, find the energy E and the maximum displacement.*

Solution. In this case (7.16) is the equation $v^2/2 + 2y^2 = E$. The initial conditions lead to the value $E = 8$. The resultant curve is an ellipse in the (y, v)-plane.

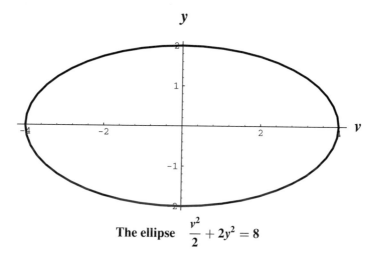

The ellipse $\dfrac{v^2}{2} + 2y^2 = 8$

The maximum value of y is obtained when $v = 0$; thus $y_{\max} = 2$. ∎

Summary of Techniques Introduced

Certain problems in Newtonian mechanics may be described in terms of first-order equations. The two main classes are (i) the force does not depend on the position, (ii) the force depends *only* on the position. Case (i) is a first-order equation that can be solved explicitly in certain cases. Case (ii) can be partially solved to obtain a suitable statement of conservation of energy, which is illustrated for Hooke's law of elasticity.

Exercises for Section 7.1

1. Suppose that a particle moves in a constant force field F_0 with no frictional forces. Find the velocity $v(t)$ and the position $y(t)$ for all later times t.

2. Suppose that a particle moves in a constant force field with a damping force proportional to the negative of the velocity, and suppose that the initial velocity V_0 is positive.

   a. Find formulas for the velocity $v(t)$ and the position $y(t)$.

   b. Show that $v(t) < \bar{v}(t)$ and $y(t) < \bar{y}(t)$, where $\bar{v}(t)$ and $\bar{y}(t)$ are obtained from Exercise 1.

3. Suppose that a fully equipped parachutist weighing 175 pounds falls from a height of 6000 feet and opens the parachute after 12 seconds of free fall. Assume an air resistance of $0.8v$ with the parachute closed and an air resistance of $10v$ with the parachute open, where v denotes the velocity.

   a. Find the speed of fall and the distance fallen when the parachute opens.

   b. What is the limiting velocity after the parachute opens?

   c. What is the total time of descent?

4. A particle moves in a force field that varies sinusoidally with time according to

$$F(t) = F_0 + F_1 \cos(\omega t - \alpha),$$

   where F_0, F_1, and α are constants. Assume further that there is a damping force that is negatively proportional to the velocity. Find the velocity $v(t)$ and the position $y(t)$. Show that $\lim_{t \to \infty} v(t)$ does not exist, but the limit of $y(t)/t$ when $t \longrightarrow \infty$ does exist, and find it.

5. The upward motion of a body of mass m traveling in the presence of nonlinear air resistance is governed by the differential equation

$$m \frac{dv}{dt} = -m g - k v^2, \qquad\qquad v > 0, \qquad\qquad (7.17)$$

   where m, g, and k are positive constants.

   a. Solve the given differential equation with $v(0) = V_0 > 0$.

   b. If $v(0) = V_0 > 0$, find the time t for which $v(t) = 0$.

   c. If $v(0) = V_0 > 0$, find the total height to which the body travels.

6. The downward motion of a body of mass m traveling in the presence of nonlinear air resistance is governed by the differential equation

$$m\frac{dv}{dt} = -m\,g + k\,v^2, \qquad\qquad v < 0, \qquad\qquad (7.18)$$

where m, g, and k are positive constants.

    a. Solve the given differential equation with $v(0) = V_0 < 0$.

    b. Find the limit of $v(t)$ when $t \longrightarrow \infty$.

7. The models described in Exercises 5 and 6 can be subsumed in the form

$$m\frac{dv}{dt} + k\,v|v| = -m\,g. \qquad\qquad (7.19)$$

    a. Write (7.19) in the form $dv/dt = F(v)$.

    b. Show that $F(v) < 0$ for $v > -\sqrt{m\,g/k}$ and $F(v) > 0$ for $v < -\sqrt{m\,g/k}$.

    c. Using the theory developed in Section 5.6, show that (7.19) has exactly one critical point, which is a stable critical point.

    d. Use this information to sketch the graph of the solution of (7.19) for $t \geq 0$ in the following cases: (i) $V_0 > 0$; (ii) $V_0 = 0$; (iii) $-\sqrt{m\,g/k} < V_0 < 0$; (iv) $V_0 < -\sqrt{m\,g/k}$.

8. It is observed that a raindrop measuring one sixteenth of an inch in diameter falls with a limiting velocity of 15 miles per hour. Assuming a model of the form (7.4), find the frictional constant k in this model. (Recall that the density of water is 62.4 pounds/foot^3.)

9. A ball of mass m is thrown upward with initial velocity V_0 and experiences a frictional force proportional to the velocity, with proportionality constant k.

    a. Find the formula for the height y.

    b. Find the maximum height y_{max} attained by the ball.

    c. Illustrate the result of part **b** in case $m = 0.25$ kilograms, $V_0 = 20$ meters/second, $k = 1/30$ kilograms/second.

    d. Using the values from part **c**, find the speed at which the ball hits the ground. [Hint: The solution will involve a transcendental equation, which can be solved using *Mathematica*'s command **FindRoot**.]

    e. Using the values from part **c**, plot the height as a function of the velocity.

7.2   *Rocket Propulsion*

In this section we consider the case of gravitational force. According to **Newton's**[1] **law of gravitation**, the attractive force $\mathbf{F}_g$ between two objects varies inversely as the square of the distance between them. More precisely, the magnitude of $\mathbf{F}_g$ is given by

$$\|\mathbf{F}_g\| = \frac{G\,m_1 m_2}{r^2}, \tag{7.20}$$

where m_1 and m_2 are the masses of the two objects, r is the distance between them, and G is the universal gravitational constant, that is,

$$G \approx \frac{6.6726 \times 10^{-11}\ \text{meters}^3}{\text{kilogram second}^2}.$$

Let $\mathbf{M}_{\text{Earth}}$ and $\mathbf{R}$ denote the mass and radius of the earth; the approximate values of these constants are given by

$$\mathbf{M}_{\text{Earth}} \approx 5.976 \times 10^{24}\ \text{kilograms} \qquad \text{and} \qquad \mathbf{R} \approx 6378.164\ \text{kilometers}.$$

For an object of mass m on the surface of the earth, (7.20) reduces to

$$F = m\,g, \tag{7.21}$$

where $F = \|\mathbf{F}_g\|$ and

$$g = \frac{G\mathbf{M}_{\text{Earth}}}{\mathbf{R}^2} \approx \frac{9.80\ \text{meters}}{\text{second}^2} \approx \frac{32\ \text{feet}}{\text{second}^2}.$$

Equation (7.21) is the formula for the weight of an object on the earth's surface. The constant g is called the **acceleration due to gravity**. Since the earth is an imperfect rotating sphere, formula (7.21) will vary slightly over the surface of the earth.

Suppose that a particle (no longer assumed to lie on the earth's surface) moves under the influence of the earth's gravitational field. We let y denote the distance from the surface of the earth, with the positive direction upward. We can write the equation of motion of the particle as

$$m\frac{dv}{dt} = -\frac{m\,g\,\mathbf{R}^2}{(\mathbf{R}+y)^2}. \tag{7.22}$$

---

[1]Newton discovered his law of gravitation in 1666 but held up publication for nearly twenty years because of technical difficulties. One concerned the correct values of the constants involved; the other was the assumption that the earth could attract as though all of its mass were concentrated at the center. Finally, Edmond Halley ((1656–1742), the discoverer of Halley's comet) convinced him to publish. In fact, Halley paid for the publication of Newton's **Principia**.

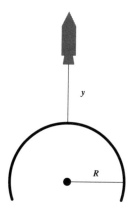

Rocket launch

Since (7.22) is a special case of (7.10), we can rewrite (7.22) as

$$m\, v\frac{dv}{dy} = -\frac{m\, g\, \mathbf{R}^2}{(\mathbf{R}+y)^2}.$$ (7.23)

We cancel the common factor m and separate the variables in (7.23) to obtain

$$v\, dv = -\frac{g\, \mathbf{R}^2}{(\mathbf{R}+y)^2}dy.$$ (7.24)

When we integrate both sides of (7.24), we get

$$\frac{1}{2}v^2 = \frac{g\, \mathbf{R}^2}{\mathbf{R}+y} + C,$$ (7.25)

which is the statement of conservation of energy for this model. If we are given the initial velocity $v = V_0$ when $y = 0$, then the constant C in (7.25) is computed as $V_0^2/2 - g\,\mathbf{R}$. Thus (7.25) becomes

$$v^2 = V_0^2 - 2g\,\mathbf{R} + \frac{2g\,\mathbf{R}^2}{\mathbf{R}+y}.$$ (7.26)

From (7.26) we can compute the **escape velocity** V_{Earth}. Here V_{Earth} is defined as the smallest V_0 such that $v(t)$ is well-defined for all y. Thus we need the right-hand side of (7.26) to be nonnegative for all $y > 0$, in particular when $y \longrightarrow \infty$. Hence (7.26) implies

$$V_0^2 - 2g\,\mathbf{R} \geq 0.$$ (7.27)

The smallest V_0 satisfying (7.27) is

$$\mathbf{V}_{\text{Earth}} = \sqrt{2g\,\mathbf{R}}.$$

The value of $\mathbf{V}_{\text{Earth}}$ is approximately

$$\mathbf{V}_{\text{Earth}} \approx \sqrt{2 \left(\frac{0.0098 \text{ kilometers}}{\text{second}^2} \right) 6378.164 \text{ kilometers}} \approx \frac{11.2 \text{ kilometers}}{\text{second}}$$

or approximately 6.9 miles per second. We would like to solve equation (7.26) for y as a function of t. For general values of V_0 the solution of equation (7.26) involves transcendental functions, so that it is difficult to solve (7.26) explicitly for the position as a function of time. However, this can be done explicitly in case V_0 is the escape velocity, as shown in the following example.

Example 7.8.   *Solve (7.22) for the position $y(t)$ in the case that the initial velocity is the escape velocity, that is, $v(0) = \mathbf{V}_{\text{Earth}}$.*

Solution.   In this case (7.26) becomes

$$v = \frac{dy}{dt} = \sqrt{\frac{2g\,\mathbf{R}^2}{\mathbf{R} + y}}, \tag{7.28}$$

which is also a separable equation. The solution of (7.28) is

$$\frac{2}{3}(\mathbf{R} + y)^{3/2} = \mathbf{R}\,t\sqrt{2g} + C_2, \tag{7.29}$$

where C_2 is a constant. We have $y(0) = 0$, and so $C_2 = (2/3)\mathbf{R}^{3/2}$. Thus (7.29) becomes

$$y(t) = \left(\frac{3\sqrt{2g}}{2} \mathbf{R}\,t + \mathbf{R}^{3/2} \right)^{2/3} - \mathbf{R} \tag{7.30}$$

$$\approx \left((1339.41\,t + 509382)^{2/3} - 6378 \right) \text{ kilometers.} \quad\blacksquare$$

Example 7.9.   *Use the result of Example 7.8 to find the time T_M necessary for a rocket to reach the moon, if it is launched at the escape velocity. Assume that the distance to the moon is $238,855$ miles.*

Solution.   We first solve (7.30) for t; thus

$$t = \frac{2}{3(2g)^{1/2}\mathbf{R}} \left((y(t) + \mathbf{R})^{3/2} - \mathbf{R}^{3/2} \right). \tag{7.31}$$

Since

$$\mathbf{R} = 3963 \text{ miles}, \qquad y(T_M) = 238,855 \text{ miles},$$

and

$$g = 32/5280 = 0.0061 \text{ miles/second}^2,$$

(7.31) implies that

$$T_M = \frac{2}{3(2g)^{1/2}\mathbf{R}} \left((y(T_M) + \mathbf{R})^{3/2} - \mathbf{R}^{3/2} \right)$$

$$= \frac{2}{3(2 \times 0.0061)^{1/2}3963} \left((238,855 + 3963)^{3/2} - (3963)^{3/2} \right)$$

$$= 181853 \text{ seconds} = 50.52 \text{ hours}. \ \blacksquare$$

Exercises for Section 7.2

1. The following table lists the radius and mass of each of the nine planets and Earth's moon. Use it to compute the corresponding acceleration due to gravity and escape velocity for each planet and the moon.

Planet	Radius		Mass	
Mercury	$\mathbf{R}_{\text{Mercury}}$	$= 2439$ kilometers	$\mathbf{M}_{\text{Mercury}}$	$= 3.303 \times 10^{23}$ kilograms
Venus	$\mathbf{R}_{\text{Venus}}$	$= 6051$ kilometers	$\mathbf{M}_{\text{Venus}}$	$= 4.870 \times 10^{24}$ kilograms
Earth	$\mathbf{R}_{\text{Earth}}$	$= 6378$ kilometers	$\mathbf{M}_{\text{Earth}}$	$= 5.976 \times 10^{24}$ kilograms
Mars	$\mathbf{R}_{\text{Mars}}$	$= 3393$ kilometers	$\mathbf{M}_{\text{Mars}}$	$= 6.421 \times 10^{23}$ kilograms
Jupiter	$\mathbf{R}_{\text{Jupiter}}$	$= 71492$ kilometers	$\mathbf{M}_{\text{Jupiter}}$	$= 1.900 \times 10^{27}$ kilograms
Saturn	$\mathbf{R}_{\text{Saturn}}$	$= 60286$ kilometers	$\mathbf{M}_{\text{Saturn}}$	$= 5.688 \times 10^{26}$ kilograms
Uranus	$\mathbf{R}_{\text{Uranus}}$	$= 25559$ kilometers	$\mathbf{M}_{\text{Uranus}}$	$= 8.684 \times 10^{25}$ kilograms
Neptune	$\mathbf{R}_{\text{Neptune}}$	$= 24764$ kilometers	$\mathbf{M}_{\text{Neptune}}$	$= 1.024 \times 10^{26}$ kilograms
Pluto	$\mathbf{R}_{\text{Pluto}}$	$= 1150$ kilometers	$\mathbf{M}_{\text{Pluto}}$	$= 1.29 \times 10^{22}$ kilograms
Moon	$\mathbf{R}_{\text{Moon}}$	$= 1738$ kilometers	$\mathbf{M}_{\text{Moon}}$	$= 7.349 \times 10^{22}$ kilograms

2. Suppose that a projectile is sent into space from Earth at the escape velocity $V_{Earth} = \sqrt{2g\,R}$. We would like to discuss the average velocity for large time or find a suitable numerical substitute.

     a. Show that both $v(t)$ and $y(t)/t$ tend to zero when $t \longrightarrow \infty$.

     b. Show that the limit $\lim\limits_{t\to\infty} \dfrac{y(t)}{t^{2/3}}$ exists and find its value.

3. Suppose that a projectile is sent into space from Earth at an initial velocity less than the escape velocity: $V_0 < V_{Earth}$. Find a formula for the maximum distance reached by the projectile. [Hint: At the maximum we must have $v = 0$.]

4. Suppose that a projectile is sent against a *constant* gravitational field $y'' = -g$ with initial velocity V_0. Find the maximum distance reached by the projectile and show that it is strictly greater than the maximum distance found in Exercise 3.

5. Suppose that a projectile is sent into space from a platform at height $z > 0$ above the surface of the earth. Find the escape velocity.

6. **a.** Suppose that a projectile is sent into space from Earth at an initial velocity greater than the escape velocity: $V_0 > V_{Earth}$. Find the limiting velocity when $t \longrightarrow \infty$.

     b. Illustrate with $R_{Earth} = 4000$ miles and $V_0 = 10$ miles/second.

7. Assume that the distance from the earth to the sun is $93,000,000$ miles. Use the result of Example 7.8 to find the time necessary for a rocket to reach the sun, if it is launched at the escape velocity.

8. In 1989 the unmanned NASA spacecraft *Magellan* took 100 days to reach the planet Venus. Assuming that it was launched at the escape velocity, use the result of Example 7.8 to estimate the distance from Earth to Venus.

7.3   Electrical Circuits

The theory of electrical networks deals with the flow of electrical currents through assemblies of basic elements under the influence of given impressed voltages. The basic elements are

- Resistor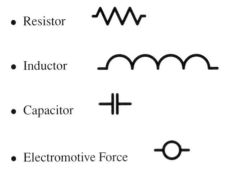

- Inductor

- Capacitor

- Electromotive Force

The **current** I depends on time t and flows through each of these elements; the unit for I is the **ampere**.[2] The current through an element is related to the **voltage** E (also time-dependent); the unit for E is the **volt**.[3]

Certain devices such as a battery or an electric generator provide electromotive force, measured in volts. A resistor produces **resistance** R, a capacitor **capacitance** C, and an inductor **inductance** L. In this section we consider only resistors and inductors; more general circuits that include capacitors are considered in Sections 11.3 and 18.2. The unit for R is the **ohm**[4], and the unit for L is the **henry**.[5]

---

[2] André Marie Ampère (1775–1836). French physicist who made important contributions to electrodynamics and partial differential equations.

[3] Alessandro Giuseppe Volta (1745–1827). Italian physicist who constructed the first batteries that would produce electricity continuously.

[4] Georg Simon Ohm (1789–1854). German physicist noted for his contributions to mathematics, acoustics, and the measurement of electrical resistance.

[5] Joseph Henry (1797–1878). American physicist, famous for experiments on electromagnetic induction. First director of the Smithsonian Institution and first president of the National Academy of Sciences.

The analysis of electric circuits is governed by the two fundamental laws of Kirchhoff:[6]

- **Kirchhoff voltage law**: In a closed circuit the impressed voltage equals the sum of the voltage drops in the rest of the circuit.

- **Kirchhoff current law**: At a junction of two or more elements of a circuit the amount of current into the junction point equals the amount of current away from the junction point.

Voltage drops are given as follows:

$$\textbf{Voltage drop across the resistor} \; = \; R \, I, \qquad (7.32)$$

$$\textbf{Voltage drop across the inductor} \; = \; L \, \frac{dI}{dt}. \qquad (7.33)$$

Equation (7.32) is called **Ohm's law**; the proportionality constant R is called the **resistance**. Equation (7.33) is called Faraday's[7] law; the proportionality constant L is called the **inductance**.

Single Loop Circuit

We consider a simple circuit (called an **LR circuit**) consisting of a single loop containing a resistor R, an inductor L, and an electromotive force $E(t)$ introducing voltage into the circuit. The Kirchhoff current law indicates merely that the current is the same throughout. From the Kirchhoff voltage law we get

$$L \frac{dI(t)}{dt} + R \, I(t) = E(t). \qquad (7.34)$$

6

Gustav Robert Kirchhoff (1824–1887), professor at Breslau, Heidelberg, and Berlin, was one of the leading physicists of the nineteenth century. While a student in 1845, he discovered the basic laws of circuits. He became one of the founders of spectroscopy.

7

Michael Faraday (1791–1867). English physicist. Discover of benzene, inventor of the dynamo, and a main architect of classical field theory.

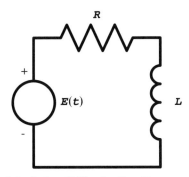

Electrical circuit illustrating the equation

$$L\frac{dI(t)}{dt} + R\,I(t) = E(t)$$

We also assume that the initial current $I(0)$ equals a known value I_0. To solve (7.34) with this initial condition we use the integrating factor $\mu(t) = e^{Rt/L}$ to convert (7.34) to

$$\frac{d}{dt}\big(I(t)e^{Rt/L}\big) = \frac{E(t)e^{Rt/L}}{L}. \tag{7.35}$$

When we integrate both sides of (7.35) from 0 to t, we obtain

$$I(t)e^{Rt/L} - I_0 = \int_0^t \frac{E(s)e^{Rs/L}\,ds}{L}. \tag{7.36}$$

Then (7.36) can be rewritten as

$$I(t) = I_0 e^{-Rt/L} + \int_0^t \frac{E(s)e^{R(s-t)/L}\,ds}{L}. \tag{7.37}$$

The function $E(t)$ is typically a constant E_0 (the case of a battery) or alternating (the case of household current).

Example 7.10. *Solve (7.34) explicitly in case of constant voltage $E(t) = E_0$ and find the limiting behavior when $t \longrightarrow \infty$.*

Solution. In this case we can explicitly calculate the integral on the right-hand side of (7.37) as

$$\int_0^t \frac{E_0 e^{R(s-t)/L}\,ds}{L} = \left(\frac{E_0}{R}\right)\big(1 - e^{-Rt/L}\big).$$

Thus, the solution to (7.34) in the case of constant voltage is

$$I(t) = I_0 e^{-Rt/L} + \left(\frac{E_0}{R}\right)(1 - e^{-Rt/L})$$

$$= \left(I_0 - \frac{E_0}{R}\right) e^{-Rt/L} + \left(\frac{E_0}{R}\right). \tag{7.38}$$

When $t \longrightarrow \infty$, the exponential terms in (7.38) tend to zero, and we have

$$\lim_{t \to \infty} I(t) = \frac{E_0}{R}. \ \blacksquare$$

The quantity $(I_0 - E_0/R)e^{-Rt/L}$ is called the **transient current**, since this part of the current dies out as t tends to ∞. The term E_0/R is called the **steady-state current**.

Example 7.11. *An LR circuit has an electromotive force of* 50 *volts, a resistance of* 500 *ohms, an inductance of* 1 henry, *and an initial current of* 1 ampere. *Find the current, the transient current, and the steady-state current.*

Solution. We have $E = 50$, $R = 500$, $L = 1$, and $I_0 = 1$. From (7.38) we find that

$$I(t) = e^{-500t} + \frac{50}{500}(1 - e^{-500t}) = 0.1 + 0.9 e^{-500t} \text{ amperes.}$$

The transient current is $0.9 e^{-500t}$ amperes, and the steady-state current is 0.1 amperes. $\blacksquare$

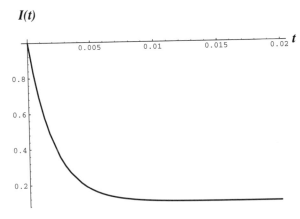

**Solution to $I'(t) + 500\,I(t) = 50$
with the initial condition $I(0) = 1$**

Next, we consider a typical case in which $E(t)$ is nonconstant.

Example 7.12. *Plot the solutions to the initial value problem*

$$\begin{cases} L\dfrac{dI(t)}{dt} + R\,I(t) = E(t), \\ I(0) = 0, \end{cases} \tag{7.39}$$

in the case that $L = \ell$ *henrys* $(\ell = 0.01, 0.015, 0.02, 0.25, 0.03)$, $R = 20$ *ohms, and* $E(t) = 100\cos(240\pi\, t)$ *volts.*

Solution. To avoid collision with *Mathematica*'s symbol **I** (which denotes $\sqrt{-1}$), we use **ii** to denote the current. The simultaneous plot of the solutions is found with

```
ODE[{L ii' + 20ii == 100Cos[240Pi t],ii[0] == 0},ii,t,
Method->FirstOrderLinear,Form->Explicit,
Parameters->{{L,0.01,0.03,0.005}},
PlotSolution->{{t,0,0.01},PlotPoints->100,PlotRange->All}]
```

to obtain the plot

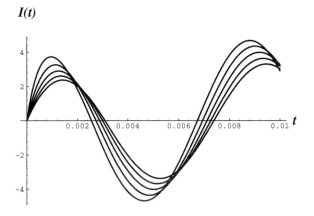

Solutions to $L\,I'(t) + R\,I(t) = E(t)$
plotted over the interval $0 < t < 0.01$

and the solution

```
      -5                      5 (Cos[240 Pi t] + 12 L Pi Sin[240 Pi t])
 ──────────────────────  +  ──────────────────────────────────────────
  (20 t)/L        2   2                         2    2
 E          (1 + 144 L  Pi )              1 + 144 L  Pi
```

Hence

$$I(t) = \frac{5\cos(240\pi\, t) + 60L\,\pi\sin(240\pi\, t)}{144L^2\pi^2 + 1} - \frac{5}{e^{20t/L}(144L^2\pi^2 + 1)}. \; \blacksquare$$

Exercises for Section 7.3

Solve the following problems involving LR circuits, assuming that L, R, E_0, and ω are constants.

1. $\begin{cases} L\dfrac{dI}{dt} + RI = E_0, \\ I(0) = 0 \end{cases}$

3. $\begin{cases} L\dfrac{dI}{dt} + RI = E_0\cos(\omega t), \\ I'(0) = 0 \end{cases}$

2. $\begin{cases} L\dfrac{dI}{dt} + RI = E_0\cos(\omega t), \\ I(0) = 0 \end{cases}$

4. $\begin{cases} L\dfrac{dI}{dt} + RI = E_0\big(\cos(\omega t)\big)^2, \\ I(0) = 0 \end{cases}$

5. Suppose that an LR circuit contains an inductor with $L = 0.02$ henry and a resistor with $R = 300$ ohms in series with an impressed voltage $E_0 = 5000$ volts.

    a. Find the formula for the current if initially $I(0) = 0$.

    b. Find the limit of $I(t)$ as $t \longrightarrow \infty$.

    c. Find the time t_1 for which $|I(t_1) - I(\infty)| = I(\infty)/2$.

    d. Plot $I(t)$ for $0 \le t \le 0.0005$.

6. Let $I(t)$ be the solution of the differential equation of Exercise 2

    a. Show that the limit of $I(t)$ as $t \longrightarrow \infty$ does not exist.

    b. Compute $\displaystyle\lim_{T\to\infty} \frac{1}{T} \int_0^T I(t)\,dt$.

    c. Compute $\displaystyle\lim_{T\to\infty} \frac{1}{T} \int_0^T I(t)^2\,dt$.

7. A resistor of 150 ohms and an inductor of 3 henrys are connected in an LR circuit with a 50 cycle voltage source having amplitude 220 volts. Assume the initial current is 1 ampere. Find the formula for the current and plot it from 0 to $\pi/10$.

8. Suppose that an LR circuit containing an inductor with $L = 0.5$ henry and a resistor with $R = 500$ ohms is subjected to an external source with $E(t) = 15$ volts. Use **ODE** to plot simultaneously the currents when the initial current assumes the values $I_0 = 0, 0.1, 0.2, 0.3, 0.4, 0.5$.

7.4 *Mixing Problems*

As our next application of first-order equations, we consider flow problems involving *brine*, which is simply a solution of salt in water. The analysis that we give here can be used for other dissolved substances, for example pollutants. We imagine a tank of brine that is equipped with an entrance pipe and an exit pipe. We assume that the brine with a density of ρ grams of salt per cubic centimeter enters the tank at a rate r_{in} cubic centimeters per second. The solution inside the tank is thoroughly mixed; then it leaves through the exit pipe at a rate of r_{out} cubic centimeters per second.

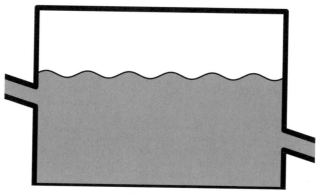

Brine tank

Let

$$x(t) = \text{the mass of salt in the tank at time } t.$$

In order to derive a differential equation for x, we first observe that in a time interval $(t, t + \delta t)$ the net change in salt can be written as:

$$x(t + \delta t) - x(t) = \text{change in salt mass in } (t, t + \delta t)$$

$$= \text{mass of salt that enters in } (t, t + \delta t)$$

$$-\text{mass of salt that leaves in } (t, t + \delta t).$$

Let us assume that the derivative of x exists. If we divide $x(t + \delta t) - x(t)$ by δt and let δt tend to zero, we obtain

$$x'(t) = \lim_{\delta t \to 0} \frac{x(t + \delta t) - x(t)}{\delta t}$$

$$= \text{rate at which salt enters} - \text{rate at which salt exits.} \qquad (7.40)$$

The rate at which salt enters is the product of the entrance rate r_{in} of the brine and the salt density ρ; thus

$$\text{rate at which salt enters} = r_{in}\rho. \tag{7.41}$$

The rate at which salt exits is the product of the exit rate r_{out} of the solution with the instantaneous density of salt. Hence

$$\text{rate at which salt exits} = r_{out}\frac{x(t)}{V(t)}, \tag{7.42}$$

where $V(t)$ denotes the volume of the brine at time t. Let V_0 be the total volume of brine in the tank at time $t = 0$. The rate of change of $V(t)$ is

$$\frac{dV}{dt} = (r_{in} - r_{out}). \tag{7.43}$$

Integrating this differential equation and incorporating the initial condition $V(0) = V_0$, we find that the total volume of brine in the tank at time t is given by

$$V(t) = V_0 + (r_{in} - r_{out})t. \tag{7.44}$$

From (7.40)–(7.44) we obtain a differential equation for x:

$$x'(t) = r_{in}\rho - r_{out}\frac{x(t)}{V_0 + bt}, \tag{7.45}$$

where $b = r_{in} - r_{out}$. If $b \neq 0$, this first-order linear equation is solved by means of the integrating factor

$$\exp\left(\int \frac{r_{out}}{V_0 + bt}dt\right) = (V_0 + bt)^{r_{out}/b}. \tag{7.46}$$

Thus, (7.45) becomes

$$\left((V_0 + bt)^{r_{out}/b}x\right)' = r_{in}\rho(V_0 + bt)^{r_{out}/b},$$

and the solution to (7.45) is

$$x(t)(V_0 + bt)^{r_{out}/b} = \frac{r_{in}\rho}{b\left(1 + \dfrac{r_{out}}{b}\right)}(V_0 + bt)^{1 + r_{out}/b} + C,$$

where C is a constant. Noting that $1 + r_{out}/b = r_{in}/b$, we conclude that

$$x(t)(V_0 + bt)^{r_{out}/b} = \rho(V_0 + bt)^{1 + r_{out}/b} + C,$$

or

$$x(t) = \rho(V_0 + bt) + C(V_0 + bt)^{-r_{out}/b}. \tag{7.47}$$

The constant C is determined from the initial salt level by

$$C = x(0)V_0^{r_{out}/b} - \rho V_0^{r_{in}/b}.$$

Thus

$$x(t) = \rho(V_0 + bt) + \left(x(0)V_0^{r_{out}/b} - \rho V_0^{r_{in}/b}\right)(V_0 + bt)^{-r_{out}/b}. \tag{7.48}$$

We consider separate cases:

Case I: $r_{in} > r_{out}$

Since $b > 0$, the second term on the right-hand side of (7.48) tends to zero as $t \longrightarrow \infty$, and thus

$$\lim_{t \to \infty} \frac{x(t)}{t} = \rho b = \rho(r_{in} - r_{out}).$$

We illustrate with a numerical example.

Example 7.13. *Suppose that a tank initially contains* 5 *grams of pure salt in a solution of* 1000 *cubic centimeters. A brine solution with a salt concentration of* 0.03 *grams per cubic centimeter is added at the rate of* 25 *cubic centimeters per second to the tank. The resulting brine solution is removed at the rate of* 20 *cubic centimeters per second. Let* $x(t)$ *denote the mass of salt in the tank at time* t. *Find the differential equation for* $x(t)$, *solve it, and find the limiting behavior as* $t \longrightarrow \infty$.

Solution. In this case we have $r_{in} = 25$, $r_{out} = 20$, $\rho = 0.03$, and $V_0 = 1000$; thus, $V(t) = 1000 + (25 - 20)t = 1000 + 5t$, and we obtain the initial value problem

$$\begin{cases} x'(t) = (25)(0.3) - \dfrac{20x(t)}{1000 + 5t} = 0.75 - \dfrac{20x(t)}{1000 + 5t}, \\ x(0) = 5, \end{cases} \tag{7.49}$$

which falls into Case I. First, we rewrite the differential equation of (7.49) as

$$x' + \frac{4x}{t + 200} = \frac{3}{4}. \tag{7.50}$$

This first-order differential equation has the integrating factor $(t + 200)^4$; thus, the general solution to (7.50) is

$$x(t) = \frac{3}{20}(t + 200) + C(t + 200)^4. \tag{7.51}$$

Since $x(0) = 5$, the constant of integration in (7.51) is given by $C = -4(10)^{-10}$. Hence the solution to (7.49) is

$$x(t) = 30 + \frac{3t}{20} - \frac{4 \times 10^{10}}{(200 + t)^4}.$$

The limiting behavior is

$$\lim_{t\to\infty} \frac{x(t)}{t} = 5 \times 0.03 = 0.15 \text{ grams/second.} \quad \blacksquare$$

Example 7.14. *Use* **ODE** *to solve and plot the initial value Problem* (7.49) *and to determine the limit of* $x(t)/t$ *as* $t \longrightarrow \infty$.

Solution. To solve (7.49), we employ the method of Section 4.11, expressing the solution as a function:

```
x[t_] = ODE[{xx'[t] == 3/4 - 20 xx[t]/(1000 + 5 t),
        xx[0] == 5},xx,t,Method->FirstOrderLinear,
        Form->Explicit,PostSolution->{Apart}]
```

yielding

$$30 + \frac{3\,t}{20} - \frac{40000000000}{(200 + t)^4}$$

Then

```
Limit[x[t]/t,t->Infinity]
```

yields

$$\frac{3}{20}$$

We have made use of *Mathematica*'s command **Limit**. The command

```
Plot[x[t]/t,{t,1,100},Epilog->{AbsoluteDashing[{5,5}],
Line[{{0,3/20},{100,3/20}}]}]
```

yields the plot of $x(t)/t$ from 1 to 100 together with the asymptote drawn as a dashed line. $\blacksquare$

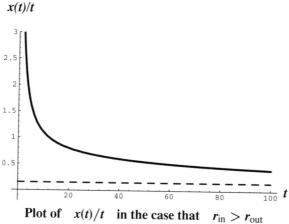

Plot of $x(t)/t$ in the case that $r_{\text{in}} > r_{\text{out}}$

Case II: $r_{\text{in}} < r_{\text{out}}$

In this case the same solution (7.48) is valid, but with a different interpretation. Since $b < 0$, the exponent $-r_{\text{out}}/b$ on the second term on the right-hand side of (7.47) is positive. Furthermore, $V_0 + b\,t$ is zero at the time $t = V_0/|b|$; thus

$$\lim_{t \to V_0/|b|} x(t) = 0.$$

The solution (7.48) is undefined for $t > V_0/|b|$ in this case.

Case III: $r_{\text{in}} = r_{\text{out}}$

In this case the integrating factor of (7.45) is not given by (7.46) but by

$$\exp\left(\frac{r\,t}{V_0}\right),$$

where $r = r_{\text{in}} = r_{\text{out}}$. Thus in this case the solution of (7.45) is

$$x(t) = x(0)e^{-rt/V_0} + \rho\, V_0(1 - e^{-rt/V_0}),$$

with the steady-state behavior

$$\lim_{t \to \infty} x(t) = \rho\, V_0.$$

Exercises for Section 7.4

Solve the following problems, of which 1,2,4 involve salt solutions.

1. Suppose that a tank initially contains 5 grams of pure salt in a solution of 1000 cubic centimeters. A brine solution with a salt concentration of 0.03 grams per cubic centimeter is added at the rate of 20 cubic centimeters per second to the tank. The resulting brine solution is removed at the rate of 25 cubic centimeters per second. Find the differential equation for the mass of salt in the tank at time t, solve it, and determine the limiting behavior.

2. Suppose that a tank initially contains 5 grams of pure salt in a solution of 1000 cubic centimeters. A brine solution with a salt concentration of 0.03 grams per cubic centimeter is added at the rate of 25 cubic centimeters per second to the tank. The resulting brine solution is removed at the same rate. Find the differential equation for the mass of salt in the tank at time t, solve it, and determine the limiting behavior.

3. The pollution of a lake of constant volume is governed by the differential equation

$$\frac{dP}{dt} = \frac{P_I - P}{\tau},\qquad\qquad (7.52)$$

where $P(t)$ is the density of pollution, $P_I(t)$ is the input pollution density, and the constant τ is the so-called *average retention time of water*, related to the time necessary to drain the entire lake.

   a. Solve (7.52) if the input pollution density is constant, that is $P_I(t) = K$.

   b. Graph the solutions obtained in part **a** in the case of Lake Superior ($\tau = 189$ years), Lake Michigan ($\tau = 30.8$ years), Lake Ontario ($\tau = 7.8$ years), and Lake Erie ($\tau = 2.6$ years). Consider separately the values $K/P(0) = 0, 0.25, 0.50, 0.75$.

   c. Solve (7.52) if the input pollution density is given by $P_I(t) = K_0 e^{-\alpha t}$ with $0 < \alpha \neq \tau$.

4. A tank contains 300 gallons of water. By mistake 400 pounds of salt are poured into the tank instead of 200 pounds. To correct the mistake, the stopper is removed from the bottom of the tank, allowing 4 gallons of the brine to flow out each minute. Simultaneously, 4 gallons of fresh water per minute are pumped into the tank. The mixture is kept uniform by constant stirring. Write down the initial value problem for the mass of salt, solve it, and determine how long it will take for the brine to contain 200 pounds of salt.

7.5 *Pursuit Curves*

The problem of pursuit probably originated with Leonardo da Vinci. It is to find the curve by which a vessel moves while pursuing another vessel, supposing that the speeds of the two vessels are always in the same ratio. Let us formulate this problem mathematically.

Definition. *Let α and β be plane curves parametrized on an interval $a < t < b$. We say that α is a **pursuit curve of** β provided that*

(i) *the velocity vector $\alpha'(t)$ points towards the point $\beta(t)$ for $a < t < b$, that is, $\alpha'(t)$ is a multiple of $\alpha(t) - \beta(t)$;*

(ii) *the speeds of α and β are related by $\|\alpha'\| = k\|\beta'\|$, where k is a positive constant. We call k the **speed ratio**.*

*A **capture point** is a point $\mathbf{p}$ for which $\mathbf{p} = \alpha(t_1) = \beta(t_1)$ for some t_1.*

In the diagram below α is the curve of the pursuer, and β is the curve of the pursued.

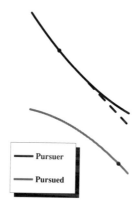

Pursuit curve

When the speed ratio k is larger than 1, the pursuer travels faster than the pursued. Although this would usually be the case in a physical situation, it is not a necessary assumption for the mathematical analysis of the problem.

We derive differential equations for pursuit curves in terms of coordinates.

Lemma 7.1. *Write $\alpha = (x, y)$ and $\beta = (f, g)$, and assume that α is a pursuit curve of β. Then*

$$x'^2 + y'^2 = k^2(f'^2 + g'^2) \tag{7.53}$$

and

$$x'(y - g) - y'(x - f) = 0. \tag{7.54}$$

Proof. Equation (7.53) is the same as $\|\alpha\| = k\|\beta\|$. To prove (7.54), we observe that $\alpha(t) - \beta(t) = \big(x(t) - f(t), y(t) - g(t)\big)$ and $\alpha'(t) = \big(x'(t), y'(t)\big)$. Note that the vector $\big(-y(t) + g(t), x(t) - f(t)\big)$ is perpendicular to $\alpha(t) - \beta(t)$. The condition that $\alpha'(t)$ is a multiple of $\alpha(t) - \beta(t)$ is conveniently expressed by saying that $\alpha'(t)$ is perpendicular to $\big(-y(t) + g(t), x(t) - f(t)\big)$, that is,

$$\begin{aligned} 0 &= \big(x'(t), y'(t)\big) \cdot \big(-y(t) + g(t), x(t) - f(t)\big) \\ &= x'(t)\big(-y(t) + g(t)\big) + y'(t)\big(x(t) - f(t)\big). \end{aligned} \tag{7.55}$$

Then (7.55) is equivalent to (7.54). ∎

Next, we specialize to the case when the curve of the pursued is a straight line.

Example 7.15. *Assume that the curve β of the pursued is a vertical straight line passing through the point $(a, 0)$, and that the speed ratio k is larger than 1. Find the curve α of the pursuer, assuming the initial conditions $\alpha(0) = (0, 0)$ and $\alpha'(0) = (1, 0)$.*

Solution. We can parameterize β as

$$\beta(t) = \big(a, g(t)\big).$$

Furthermore, the curve α of the pursuer can be parametrized as

$$\alpha(t) = \big(t, y(t)\big).$$

The condition (7.53) becomes

$$1 + y'^2 = k^2 g'^2, \tag{7.56}$$

and (7.54) reduces to

$$(y - g) - y'(t - a) = 0. \tag{7.57}$$

Differentiation of (7.57) with respect to t yields

$$-y''(t - a) = g'. \tag{7.58}$$

From (7.56) and (7.58) we get

$$1 + y'^2 = k^2(a - t)^2 y''^2. \tag{7.59}$$

Let $p = y'$; then (7.59) can be rewritten as

$$\frac{k \, dp}{\sqrt{1 + p^2}} = \frac{dt}{a - t}. \tag{7.60}$$

This separable first-order equation has the solution

$$\operatorname{arcsinh}(p) = -\frac{1}{k}\log\!\left(\frac{a-t}{a}\right) = \log\!\left(\left(\frac{a-t}{a}\right)^{-1/k}\right), \tag{7.61}$$

when we make use of the initial condition $y'(0) = 0$. Then (7.61) can be rewritten as

$$y' = p = \sinh\big(\operatorname{arcsinh}(p)\big) = \frac{1}{2}\left(e^{\operatorname{arcsinh}(p)} - e^{-\operatorname{arcsinh}(p)}\right)$$

$$= \frac{1}{2}\left(\left(\frac{a-t}{a}\right)^{-1/k} - \left(\frac{a-t}{a}\right)^{1/k}\right). \tag{7.62}$$

Integration of (7.62) (making use of the initial condition $y(0) = 0$) yields

$$y = \frac{ak}{k^2 - 1} + \frac{1}{2}\left(\frac{ak}{k+1}\left(\frac{a-t}{a}\right)^{1+1/k} - \frac{ak}{k-1}\left(\frac{a-t}{a}\right)^{1-1/k}\right). \tag{7.63}$$

The curve of the pursuer is then $\alpha(t) = \big(t, y(t)\big)$, where y is given by (7.63). Since $\alpha(t_1) = \beta(t_1)$ if and only if $t_1 = a$, the capture point is

$$\mathbf{p} = \left(a, \frac{ak}{k^2 - 1}\right). \; \blacksquare \tag{7.64}$$

The graph below depicts the case when $a = 1$ and k has the values 1.5, 2.0, 2.5, 3.0, and 3.5. As the speed ratio k becomes smaller and smaller, the capture point goes higher and higher.

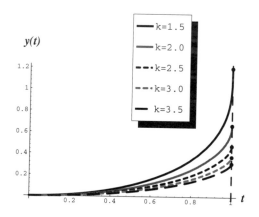

**Pursuit curves for the case when the
pursued moves in a straight line**

Exercises for Section 7.5

1. Suppose that a truck starts at a point $(a, 0)$ on a desert and moves north. Let a police car start at the point $(0, 0)$ with speed 0 and pursue the truck across the desert. Assume that the police car travels twice as fast as the truck. Show that the police car captures the truck at the point $(a, 2a/3)$.

2. Consider the equation for the pursuit curve of Example 7.15. In the case that the speed ratio is 1, show that the equation reduces to

$$y = \frac{a}{4}\left(\left(\frac{a-t}{a}\right)^2 - 1 - 2\log\left(\frac{a-t}{a}\right)\right),$$

and that the pursuer never catches the pursued.

3. Equation (7.63) can be used to define a function in *Mathematica* by

```
linearpursuit[a_,k_][t_]:= (a k)/(k^2 - 1) +
    (k(a - t)(a - t)^k^(-1))/(2a^(1/k)(1 + k)) -
    (a^(1/k)k(a - t)^(1 - 1/k))/(2(k - 1))
```

Use this function to plot a pursuit curve with $a = 1$ and $k = 1.2$.

4. Generalize Example 7.15 to the case when the initial velocity of the curve α is nonzero.

5. Solve the differential equation (7.60) using **ODE**. Compare the results using

Method->Separable,

Method->IntegratingFactor,

Method->DSolve.

8

SECOND-ORDER LINEAR DIFFERENTIAL EQUATIONS

In Chapter 3 we solved several types of first-order differential equations. Many of them are naturally adapted to problems of growth as we saw in Chapter 6. However, when we want to model physical situations that involve *oscillatory* phenomena, first-order equations are no longer sufficient. Many fundamental laws of physics are naturally expressed mathematically in terms of second-order differential equations. Among these, by far the most important are second-order *linear* differential equations; we begin to study them in this important chapter.

In the next few sections we develop the general theory of second-order linear equations. The most important kind of second-order linear equations are those with constant-coefficients; they are discussed in Chapter 9.

In Section 8.1 we solve some especially simple second-order linear equations by techniques that we already know from Chapter 3. Existence and uniqueness theory for second-order equations is given in Section 8.2. Fundamental sets of solutions of the second-order equations are treated in Section 8.3, and the Wronskian in Section 8.4. The relation between linear independence and the Wronskian is discussed in Section 8.5.

Reduction of order is an important method for reducing a second-order linear equation to a first-order linear equation when one solution to the second-order linear equation is known; it is discussed in Section 8.6. Finally, in Section 8.7 we describe how to construct a second-order linear differential equation that has two given functions as solutions.

8.1 General Forms and Examples

Before beginning our systematic study, let us solve a few second-order linear differential equations using techniques that are available to us from Chapter 3. At the same time this will allow us to fix our ideas on what new solution methods are needed.

The most general form of a second-order linear differential equation is

$$a(t)y'' + b(t)y' + c(t)y = g(t), \tag{8.1}$$

where $a(t)$ is nowhere zero on some interval. However, (8.1) is always reducible to the **leading-coefficient-unity-form**

$$y'' + p(t)y' + q(t)y = r(t); \tag{8.2}$$

it can be obtained by dividing both sides of (8.1) by $a(t)$ and writing

$$p(t) = \frac{b(t)}{a(t)}, \qquad q(t) = \frac{c(t)}{a(t)}, \qquad \text{and} \qquad r(t) = \frac{g(t)}{a(t)}.$$

Unless otherwise stated, we shall assume that this reduction has been performed. Here $p(t), q(t)$, and $r(t)$ are given continuous functions defined on some interval. The unknown function $y(t)$ is to be found among the class of functions with continuous first and second derivatives.

If both functions $p(t)$ and $q(t)$ in (8.2) are zero, then the solution can be found by elementary calculus, as indicated in the following example.

Example 8.1. *Find the general solution to the second-order equation* $y'' = 4e^{2t}$.

Solution. The general solution can be found by two integrations. We have

$$y'(t) = 2e^{2t} + C_1 \qquad \text{and} \qquad y(t) = e^{2t} + C_1 t + C_2.$$

This solution involves two arbitrary constants, which have been denoted by C_1 and C_2. ∎

It is also possible to solve the second-order equation in the case where only the function $q(t)$ is zero, by rewriting the equation as a *first-order linear differential equation* for the function y'. This equation can then be solved by the method of integrating factors from Section 3.2, as indicated in the next example.

Example 8.2. *Find the general solution to the second-order linear equation*

$$t\, y'' + y' = 9t^2. \tag{8.3}$$

Solution. We first rewrite (8.2) in the leading-coefficient-unity-form

$$(y')' + \frac{y'}{t} = 9t. \tag{8.4}$$

We consider (8.4) to be a first-order linear differential equation for the function y'. By the techniques of Section 3.2 we can solve (8.4) by multiplying both sides of (8.4) by the integrating factor

$$\mu(t) = \exp\left(\int \frac{dt}{t}\right) = t.$$

Thus, we can write (8.4) as

$$(t\,y')' = 9t^2,$$

which is integrated to yield

$$t\,y' = 3t^3 + C_1. \tag{8.5}$$

Then we rewrite (8.5) in the leading-coefficient-unity-form

$$y' = 3t^2 + \frac{C_1}{t}. \tag{8.6}$$

We get the general solution y to (8.3) by integrating both sides of (8.6):

$$y = t^3 + C_1 \log(t) + C_2. \ \blacksquare$$

In both examples (8.1) and (8.2), the solution involved *two* arbitrary constants. We shall see that this is a characteristic property of the solutions of second-order linear differential equations. The next example introduces an important equation that will occur repeatedly in our work.

Example 8.3.   *Find the general solution to the second-order linear equation*

$$y'' + k^2 y = 0, \tag{8.7}$$

where k is a real constant.

Solution. We can solve (8.7) by the substitution

$$v = y' = \frac{dy}{dt}.$$

From the chain rule for composite functions we have

$$y'' = \frac{dv}{dt} = \left(\frac{dv}{dy}\right)\left(\frac{dy}{dt}\right) = v\left(\frac{dv}{dy}\right).$$

This substitution converts the second-order equation (8.7) into the first-order *separable* equation

$$v\left(\frac{dv}{dy}\right) + k^2 y = 0,$$

or

$$v\,dv = -k^2 y\,dy. \tag{8.8}$$

We integrate both sides of (8.8), obtaining

$$\frac{v^2}{2} + \frac{k^2 y^2}{2} = \tilde{C}_1. \tag{8.9}$$

The constant $\tilde{C}_1$ is positive, since it is a sum of two squares; therefore we can write $2\tilde{C}_1 = k^2 C_1^2$, where C_1 is another constant. Now, (8.9) is also a separable equation, since it can be written as

$$\frac{dy}{dt} = v = \pm k\sqrt{C_1^2 - y^2},$$

or in terms of differentials as

$$\frac{dy}{\sqrt{C_1^2 - y^2}} = \pm k\,dt. \tag{8.10}$$

Integration of both sides of (8.10) leads to

$$\arcsin\left(\frac{y}{C_1}\right) = \pm(kt + C_2),$$

where C_2 is another constant of integration. Thus the solution to (8.7) is

$$y = C_1 \sin\big(\pm(kt + C_2)\big) = \pm C_1 \sin(kt + C_2). \tag{8.11}$$

Using the addition law for the sine function, we can also write (8.11) in the form

$$y = A\sin(kt) + B\cos(kt),$$

where $A = \pm C_1 \cos(C_2)$ and $B = \pm C_1 \sin(C_2)$. $\blacksquare$

With hindsight one might have simply *guessed* the solutions $y = \sin(kt)$ and $y = \cos(kt)$ in Example 8.3. This is much simpler than the cumbersome solution procedure indicated above; furthermore, it is mathematically justified by direct verification: both of these functions have the property that they are negatively proportional to their own second derivatives, with the same constant of proportionality. It is arguably easier to recall these simple facts when needed than to repeat the lengthy solution process described above.

Exercises for Section 8.1

Reduce the following second-order linear differential equations to the leading-coefficient-unity-form (8.2).

1. $\sin(t)y'' + t\,y' + 3y = \sin(t)$

2. $t\,y'' + 4y' + t\,y = 0$

3. $\dfrac{y''}{t} + y' + t\,y = t^3$

4. $y'' + 4y' + 3y = 0$

Use the method of Example 8.1 to solve the following differential equations by elementary calculus.

5. $y'' = 3t^2 + 2t + \sin(t)$

6. $y'' = t\,e^t$

Use the method of Example 8.2 to solve the following differential equations, which involve only y'', y', and t.

7. $y'' + y'/t = 0$

8. $y'' + 4y' = 17\sin(2t)$

Use the method of Example 8.3 to solve the following differential equations, which involve only y'' and y.

9. $y'' + 9y = 0$

10. $5y'' + y = 0$

Solution Strategy for Second-Order Linear Differential Equations

In each of the three examples above, we have obtained the general solution of a second-order linear equation in terms of a particular solution and the general solution of the associated homogeneous equation, containing two arbitrary constants. Anticipating the general theory of Section 8.2, let us formulate some solution techniques that will apply in the general case.

The differential equation

$$a(t)y'' + b(t)y' + c(t)y = g(t) \tag{8.12}$$

is similar to the differential equation

$$a(t)y' + b(t)y = c(t) \tag{8.13}$$

that we considered in Section 3.2. But the presence of y'' together with y' and y in (8.12) makes it much more complicated than (8.13). In fact, (8.12) is so complicated that there is no hope of finding a formula for the solution that works in all cases.

Nevertheless, we can formulate a strategy that will allow us to solve (8.12) in many important cases. For this it is essential to divide second-order linear differential equations into two classes.

Definition.   *The equation (8.12) is said to be* **homogeneous**[1] *if $g(t) \equiv 0$; otherwise the equation is said to be* **inhomogeneous**.

Even the homogeneous equation

$$a(t)y'' + b(t)y' + c(t)y = 0 \tag{8.14}$$

is too complicated to solve in all cases. When possible, however, we find what is called the **general solution** y_H of (8.14); it has the form

$$y_H = C_1 y_1 + C_2 y_2,$$

where y_1 and y_2 are **fundamental solutions** of (8.14), and C_1 and C_2 are constants of integration. Note that the general solution to (8.14) is the same as the general solution of the leading-coefficient-unity-form

$$y'' + p(t)y' + q(t)y = 0, \tag{8.15}$$

where $p(t) = b(t)/a(t)$ and $q(t) = c(t)/a(t)$, so we can solve whichever is more convenient.

The next step is to find a **particular solution** y_P of the inhomogeneous equation (8.15). Then the general solution to (8.15) is

$$y_H + y_P.$$

Here is a summary of all the steps involved.

---

[1] This use of the word "homogeneous" for second-order differential equations is completely different from its use in the study of first-order differential equations in Section 3.5.

Solution strategy for second-order linear differential equations

1. Convert $a(t)y'' + b(t)y' + c(t)y = g(t)$ to $y'' + p(t)y' + q(t)y = r(t)$.

2. Find the general solution y_H to the homogeneous equation

$$y'' + p(t)y' + q(t)y = 0.$$

3. Find a particular solution y_P to the inhomogeneous equation

$$y'' + p(t)y' + q(t)y = r(t).$$

4. The general solution of the inhomogeneous equation is

$$y = y_H + y_P.$$

(Step 1 is unnecessary for *homogeneous* differential equations, but safe to use. Step 1 should always be used for *inhomogeneous* differential equations.)

8.2   Existence and Uniqueness Theory

As it is not possible to solve explicitly the general second-order linear equation (8.12), we need to ask whether or not a solution exists. In fact, it can be proved that under very general conditions a solution to (8.12) does indeed exist. We state the fundamental existence and uniqueness theorem in terms of an **initial value problem**. This consists in finding a solution to (8.12) that satisfies, in addition, the initial conditions

$$y(t_0) = Y_0, \qquad y'(t_0) = Y_1. \tag{8.16}$$

Here t_0 is a given point, and Y_0 and Y_1 are given real numbers.

Theorem 8.1. (**Existence and uniqueness theorem**) *Suppose that $p(t)$, $q(t)$, and $r(t)$ are given continuous functions defined on an interval $a < t < b$. Then there exists one and only one function $y(t)$ that satisfies the second-order linear equation*

$$y'' + p(t)y' + q(t)y = r(t)$$

for $a < t < b$, together with the initial conditions (8.16).

Theorem 8.1 (the proof of which is similar to that of Theorem 5.1) asserts only the existence and uniqueness of a solution. It does not tell us how to find the formula for the solution.

Example 8.4. *For what values of* t_0 *can we expect to find a solution to the initial value problem*

$$\begin{cases} t^2 y'' + t\, y' - y = 0, \\ y(t_0) = Y_0,\ y'(t_0) = Y_1? \end{cases} \tag{8.17}$$

Solution. In order to determine the interval, we write the differential equation of (8.17) in the leading-coefficient-unity-form (8.12), that is,

$$y'' + \frac{y'}{t} - \frac{y}{t^2} = 0.$$

The coefficients $1/t$ and $1/t^2$ are continuous on any interval not containing $t = 0$. Hence Theorem 8.1 implies that we may solve the initial value problem (8.17) as long as $t_0 \neq 0$. The solution will exist on any interval not containing 0. ∎

Notice that Theorem 8.1 gives no clue how to *find* the solution to (8.17). Nevertheless, we shall develop techniques in Chapter 20 that permit us to solve (8.17) explicitly. It turns out that the function y defined by

$$y(t) = \frac{t\, Y_0}{2t_0} + \frac{t_0 Y_0}{2t} + \frac{t\, Y_1}{2} - \frac{t_0^2 Y_1}{2t} \tag{8.18}$$

solves (8.17); this can be checked by direct substitution.

y(t)

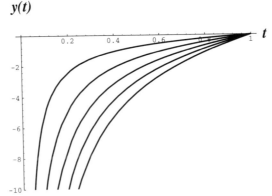

Solutions to $t^2 y'' + t y' - y = 0$ **with the initial conditions**
$y(1) = 0$ **and** $y'(1) = a$ **for** $1 \le a \le 5$

Important methods for combining solutions of the second-order equation are contained in the following two statements.

Superposition principle

If y_1 and y_2 are both solutions of the **homogeneous equation**

$$y'' + p(t)y' + q(t)y = 0,$$

then so is $C_1 y_1 + C_2 y_2$ for any choice of the constants C_1 and C_2.

Subtraction principle

If y_1 and y_2 are both solutions of the second-order linear **inhomogeneous equation**

$$y'' + p(t)y' + q(t)y = r(t),$$

then $y = y_1 - y_2$ is a solution to the homogeneous equation

$$y'' + p(t)y' + q(t)y = 0.$$

The proofs of both of these principles depend on the *linearity* property of the differential equation, expressed through the **operator** L defined by

$$L[y] = y'' + p(t)y' + q(t)y.$$

The operator L is well-defined for any twice differentiable function $y(t)$. If y_1 and y_2 are any two such functions, we have

$$L[y_1 + y_2] = L[y_1] + L[y_2] \quad \text{and} \quad L[y_1 - y_2] = L[y_1] - L[y_2], \tag{8.19}$$

and if C is any real number we have

$$L[C\, y] = C\, L[y]. \tag{8.20}$$

Properties (8.19) and (8.20) follow from the fact that the operation of differentiation is a linear operation. In other words, the derivative of a sum or difference of two functions is the sum or difference of the derivatives of the two functions. Similarly, the derivative of a constant multiple of a function is that multiple of the derivative of the function. The details of the proofs of (8.19) and (8.20) are as follows:

$$
\begin{aligned}
L[y_1 \pm y_2] &= (y_1 \pm y_2)'' + p(t)(y_1 \pm y_2)' + q(t)(y_1 \pm y_2) \\
&= \left(y_1'' + p(t)y_1' + q(t)y_1 \right) \pm \left(y_2'' + p(t)y_2' + q(t)y_2 \right) \\
&= L[y_1] \pm L[y_2],
\end{aligned}
$$

and

$$L[C\,y] = (C\,y)'' + p(t)(C\,y)' + q(t)(C\,y) = C\bigl(y'' + p(t)y' + q(t)y\bigr) = C\,L[y].\;\blacksquare$$

The subtraction principle can be used to write the general solution of a second-order linear equation in terms of a particular solution and a solution of the corresponding homogeneous equation. We illustrate with an example.

Example 8.5. *What is the form of the general solution to the differential equation*

$$y'' + 4y = 4? \tag{8.21}$$

Solution. It is easily seen that the function $y_P \equiv 1$ is a solution to (8.21). If y is another solution of (8.21), then by the subtraction principle, the function $y_H = y - y_P$ must be a solution to the homogeneous equation $y_H'' + 4y_H = 0$. But in Example 8.3 we found the general solution of the homogeneous equation to be $A\cos(2t) + B\sin(2t)$. Therefore, the general solution to (8.21) is

$$y = 1 + A\cos(2t) + B\sin(2t).$$

Exercises for Section 8.2

In which intervals can we expect to find a solution to the following initial value problems?

1. $\begin{cases} y'' + y' + y = 0, \\ y(t_0) = Y_0, \;\; y(t_0) = Y_1 \end{cases}$

4. $\begin{cases} t\,y'' + t^2 y = \cos(t), \\ y(t_0) = Y_0, \;\; y(t_0) = Y_1 \end{cases}$

2. $\begin{cases} y'' + t y' + y = \sin(t), \\ y(t_0) = Y_0, \;\; y(t_0) = Y_1 \end{cases}$

5. $\begin{cases} \sin(t)y'' + 3y = 0, \\ y(t_0) = Y_0, \;\; y'(t_0) = Y_1 \end{cases}$

3. $\begin{cases} t\,y'' + t\,y' + y = \sin(t), \\ y(t_0) = Y_0, \;\; y(t_0) = Y_1 \end{cases}$

6. $\begin{cases} \cot(t)y'' + y\,t = 0, \\ y(t_0) = Y_0, \;\; y'(t_0) = Y_1 \end{cases}$

7. Verify that (8.18) is a solution of (8.17).

8.3   Fundamental Sets of Solutions of the Homogeneous Equation

In this section we discuss the structure of the set of solutions of a second-order linear homogeneous equation

$$y'' + p(t)y' + q(t)y = 0. \tag{8.22}$$

Definition.   *A pair of functions* $\{y_1, y_2\}$ *is called a* **fundamental set of solutions** *of* (8.22) *on an interval* $a < t < b$ *if both functions satisfy the equation and every solution* $y(t)$ *of* (8.22) *can be written in the form*

$$y(t) = C_1 y_1(t) + C_2 y_2(t) \tag{8.23}$$

for $a < t < b$ *for suitable constants* C_1 *and* C_2. *We call a function* y *given by* (8.23) *with arbitrary* C_1 *and* C_2 *the* **general solution** *of* (8.22).

Let us show by example that fundamental sets of solutions exist for certain equations.

Example 8.6.   *Find a fundamental set of solutions for the equation* $y'' = 0$ *and also the general solution for* $-\infty < t < \infty$.

Solution.   If $y(t)$ is any solution to $y'' = 0$, then $y'(t)$ must be a constant, say $y'(t) \equiv C_1$. A second integration results in $y(t) = C_1 t + C_2$. We have written the general solution in the form (8.23) with $y_1 = 1$ and $y_2 = t$. Therefore, the functions 1 and t constitute a fundamental set of solutions for the differential equation $y'' = 0$. Furthermore, the general solution is

$$y(t) = C_1 t + C_2. \ \blacksquare$$

Example 8.7.   *Find a fundamental set of solutions for the equation*

$$y'' + 4y' = 0, \tag{8.24}$$

and also the general solution.

Solution.   If y satisfies (8.24), then the function y' satisfies a first-order linear equation; thus, $y' = C_1 e^{-4t}$. A second integration yields

$$y = -\frac{C_1}{4} e^{-4t} + C_2, \tag{8.25}$$

where C_1 and C_2 are constants. The general solution of (8.24) has been written in the form (8.23) with $y_1 = e^{-4t}$ and $y_2 = 1$. We conclude that e^{-4t} and 1 constitute a fundamental set of solutions for the differential equation (8.24) and that the general solution is (8.25). $\blacksquare$

The *existence* of a fundamental set of solutions follows from the existence and uniqueness theorem (Theorem 8.1). We record the statement and proof as follows.

Theorem 8.2. *Any second-order linear homogeneous equation has a fundamental set of solutions consisting of y_1 and y_2, where $y_1(t_0) = y_2'(t_0) = 1$ and $y_1'(t_0) = y_2(t_0) = 0$. The formula that expresses any solution $y(t)$ in terms of $y_1(t)$ and $y_2(t)$ is*

$$y(t) = y(t_0)y_1(t) + y'(t_0)y_2(t). \tag{8.26}$$

Proof. Let the differential equation be written in the leading-coefficient-unity-form

$$y'' + p(t)y' + q(t)y = 0 \tag{8.27}$$

on the interval $a < t < b$. Let t_0 be any point of this interval. By the existence part of Theorem 8.1 there exists a solution y_1 of (8.27) satisfying the initial conditions $y_1(t_0) = 1$ and $y_1'(t_0) = 0$. By the same token there exists a solution y_2 satisfying the initial conditions $y_2(t_0) = 0$ and $y_2'(t_0) = 1$. Now let $y(t)$ be *any* solution to (8.27). We claim that (8.26) holds. To show this, we note that both sides of (8.26) are solutions of the same second-order linear equation and have the same initial conditions at the point $t = t_0$. Therefore, by the uniqueness part of Theorem 8.1 they must be the same function, which proves the asserted equality, establishing that the pair $\{y_1, y_2\}$ constitutes a fundamental set of solutions. ∎

In the language of linear algebra, we have shown that the set of solutions of a second-order homogeneous linear equation forms a *two-dimensional vector space*. (See Section A.6 of Appendix A for more information on abstract vector spaces.) From this abstract viewpoint, the set of solutions of a second-order homogeneous linear equation has a similar structure to the set of points in the Euclidean plane $\mathbb{R}^2$, which can be written in the form $C_1\mathbf{i} + C_2\mathbf{j}$, where $\mathbf{i}$ and $\mathbf{j}$ are the usual basis vectors of $\mathbb{R}^2$ and correspond to the fundamental set $\{y_1, y_2\}$.

We emphasize that a fundamental set of solutions is *not* uniquely determined. For example, the equation $y'' - y = 0$ has the fundamental set $\{y_1, y_2\} = \{e^t, e^{-t}\}$, in terms of exponential functions. But it also has the fundamental set $\{y_1, y_2\} = \{\cosh(t), \sinh(t)\}$ in terms of hyperbolic functions. Of course, the apparent discrepancy is resolved when one notes that the hyperbolic functions are *defined* as linear combinations of exponential functions through the formulas

$$\cosh(t) = \frac{1}{2}(e^t + e^{-t}), \qquad \sinh(t) = \frac{1}{2}(e^t - e^{-t}).$$

For a general equation, the *class* of all solutions is uniquely defined; this class can be represented through several different fundamental sets.

Exercises for Section 8.3

Find a fundamental set of solutions for the following second-order differential equations.

1. $y'' - 3y' = 0$ **3.** $y'' + 16y = 0$

2. $y'' + 5y' = 0$ **4.** $y'' - 9y = 0$

Find the general solution to each of the following second-order equations.

5. $y'' - 3y' = 3$ **7.** $y'' - 9y = 18$

6. $y'' + 16y = 32$ **8.** $4y'' + y = 0$

8.4   The Wronskian

Working directly from the definition, it may be difficult in practice to verify that a given pair of functions $\{y_1, y_2\}$ forms a fundamental set for the differential equation $y'' + p(t)y' + q(t)y = 0$. In this section we formulate and prove a theorem that allows one to check directly that a given pair indeed forms a fundamental set. This theorem utilizes the following important notion:

Definition. *The* **Wronskian**[2] *of two differentiable functions* y_1 *and* y_2 *is the function* $W(y_1, y_2)$ *given by the formula*

$$W(y_1, y_2)(t) = \det \begin{pmatrix} y_1(t) & y_2(t) \\ y_1'(t) & y_2'(t) \end{pmatrix} = y_1(t)y_2'(t) - y_1'(t)y_2(t).$$

For example, if $y_1 = 1$ and $y_2 = t$, then

$$W(y_1, y_2)(t) = 1 \times 1 - 0 \times t = 1 - 0 = 1.$$

Before proving theorems about the Wronskian, we need to recall the details of a basic fact from linear algebra.

2

Józef Maria Hoëné Wronski (1776–1853). Polish mathematician and mystic, who became a French citizen and later moved to England. Although much of Wronski's work was dismissed as nonsense, a closer examination in recent years reveals important mathematical insights.

Review of Cramer's Rule

Frequently, in our study of second-order linear differential equations we shall need to solve systems of two linear *algebraic* equations in two unknowns. The basic fact that we shall need about such systems is the following:

Theorem 8.3. (**Cramer's**[3] **rule**) *Suppose that*

$$\det \begin{pmatrix} a_{11} & a_{12} \\ a_{21} & a_{22} \end{pmatrix} \neq 0. \tag{8.28}$$

Then the solution to the system of linear equations

$$\begin{cases} a_{11}x + a_{12}y = b_1, \\ a_{21}x + a_{22}y = b_2 \end{cases} \tag{8.29}$$

is given by

$$x = \frac{\det \begin{pmatrix} b_1 & a_{12} \\ b_2 & a_{22} \end{pmatrix}}{\det \begin{pmatrix} a_{11} & a_{12} \\ a_{21} & a_{22} \end{pmatrix}} \quad and \quad y = \frac{\det \begin{pmatrix} a_{11} & b_1 \\ a_{21} & b_2 \end{pmatrix}}{\det \begin{pmatrix} a_{11} & a_{12} \\ a_{21} & a_{22} \end{pmatrix}}. \tag{8.30}$$

Proof. We multiply the first equation of (8.29) by a_{22} and the second equation of (8.29) by a_{12} to obtain

$$\begin{cases} a_{22}a_{11}x + a_{22}a_{12}y = a_{22}b_1, \\ a_{12}a_{21}x + a_{12}a_{22}y = a_{12}b_2. \end{cases} \tag{8.31}$$

When we subtract the second equation of (8.31) from the first we get

$$\det \begin{pmatrix} a_{11} & a_{12} \\ a_{21} & a_{22} \end{pmatrix} x = \det \begin{pmatrix} b_1 & a_{12} \\ b_2 & a_{22} \end{pmatrix}; \tag{8.32}$$

similarly,

$$\det \begin{pmatrix} a_{11} & a_{12} \\ a_{21} & a_{22} \end{pmatrix} y = \det \begin{pmatrix} a_{11} & b_1 \\ a_{21} & b_2 \end{pmatrix}. \tag{8.33}$$

Since (8.28) holds, equations (8.32) and (8.33) are equivalent to (8.29). ∎

---

[3] Gabriel Cramer (1704–1752). Swiss mathematician who worked mainly in Geneva. Although other mathematicians of the period, as well as the Chinese centuries earlier, had used similar patterns for solving linear equations, Cramer has been given credit because of his superior notation.

Example 8.8. *Solve the system of equations*

$$\begin{cases} 2x + 3y = 1, \\ 3x - 5y = -3. \end{cases} \qquad (8.34)$$

Solution. We compute

$$x = \frac{\begin{pmatrix} 1 & 3 \\ -3 & -5 \end{pmatrix}}{\begin{pmatrix} 2 & 3 \\ 3 & -5 \end{pmatrix}} = -\frac{4}{19}$$

and

$$y = \frac{\begin{pmatrix} 2 & 1 \\ 3 & -3 \end{pmatrix}}{\begin{pmatrix} 2 & 3 \\ 3 & -5 \end{pmatrix}} = \frac{9}{19}. \ \blacksquare$$

Armed with Cramer's rule, we return to the theory of the Wronskian.

Theorem 8.4. *Suppose that y_1 and y_2 both satisfy the same linear homogeneous second-order equation $y'' + p(t)y' + q(t)y = 0$ on the interval $a < t < b$. The pair $\{y_1, y_2\}$ forms a fundamental set of solutions if and only if the Wronskian $W(y_1, y_2)(t_0) \neq 0$ for some t_0 with $a < t_0 < b$.*

Proof. Let $y(t)$ be a solution to the equation $y'' + p(t)y' + q(t)y = 0$. It is required to find constants C_1 and C_2 such that $y(t) = C_1 y_1(t) + C_2 y_2(t)$. Let t_0 be a point where the Wronskian is nonzero. Consider the system of linear *algebraic* equations in the unknowns C_1 and C_2.

$$\begin{cases} C_1 y_1(t_0) + C_2 y_2(t_0) = y(t_0), \\ C_1 y_1'(t_0) + C_2 y_2'(t_0) = y'(t_0). \end{cases} \qquad (8.35)$$

We can solve (8.35) for C_1 and C_2 because the determinant of (8.35) is

$$\det \begin{pmatrix} y_1(t_0) & y_2(t_0) \\ y_1'(t_0) & y_2'(t_0) \end{pmatrix} = y_1(t_0)y_2'(t_0) - y_1'(t_0)y_2(t_0),$$

which is nothing other than Wronskian of y_1 and y_2 evaluated at t_0. In fact, using Cramer's rule (Theorem 8.3) we have explicit formulas for C_1 and C_2, namely

$$C_1 = \frac{\det\begin{pmatrix} y(t_0) & y_2(t_0) \\ y'(t_0) & y_2'(t_0) \end{pmatrix}}{\det\begin{pmatrix} y_1(t_0) & y_2(t_0) \\ y_1'(t_0) & y_2'(t_0) \end{pmatrix}} \quad \text{and} \quad C_2 = \frac{\det\begin{pmatrix} y_1(t_0) & y(t_0) \\ y_1'(t_0) & y'(t_0) \end{pmatrix}}{\det\begin{pmatrix} y_1(t_0) & y_2(t_0) \\ y_1'(t_0) & y_2'(t_0) \end{pmatrix}}.$$

Having determined the constants C_1 and C_2, we define a new function $\tilde{y}$ by

$$\tilde{y}(t) = C_1 y_1(t) + C_2 y_2(t).$$

The superposition principle on page 243 implies that $\tilde{y}$ is also a solution to $y'' + p(t)y' + q(t)y = 0$. At t_0 we have, by construction, $\tilde{y}(t_0) = y(t_0)$ and $\tilde{y}'(t_0) = y'(t_0)$. Now the existence and uniqueness theorem (Theorem 8.1) implies that we must have $\tilde{y}(t) = y(t)$ for *all* t with $a < t < b$.

Conversely, if the pair $\{y_1, y_2\}$ forms a fundamental set, we must prove that the Wronskian is nonzero. For any pair $\{Y_0, Y_1\}$ of real numbers and for any t_0, there exists a solution to $y'' + p(t)y' + q(t)y = 0$ for which $y(t_0) = Y_0$ and $y'(t_0) = Y_1$. By hypothesis, we can find constants C_1 and C_2 such that

$$y(t) \equiv C_1 y_1(t) + C_2 y_2(t)$$

for all t with $a < t < b$. In particular, we can solve the linear equations

$$\begin{cases} C_1 y_1(t_0) + C_2 y_2(t_0) = Y_0, \\ C_1 y_1'(t_0) + C_2 y_2'(t_0) = Y_1 \end{cases} \tag{8.36}$$

for every choice of Y_0 and Y_1. This can only happen if the determinant of the system (8.36) is nonzero, which is the statement that $W(y_1, y_2)(t_0) \neq 0$. ∎

Example 8.9. *Show that the set $\{\cos(2t), \sin(2t)\}$ forms a fundamental set of solutions for the equation*

$$y'' + 4y = 0 \tag{8.37}$$

on the interval $-\infty < t < \infty$. Also, find the general solution.

Solution. It is easy to check that both $\cos(2t)$ and $\sin(2t)$ satisfy (8.37). The Wronskian of $\cos(2t)$ and $\sin(2t)$ is computed as

$$W(\cos, \sin)(t) = \det\begin{pmatrix} \cos(2t) & \sin(2t) \\ -2\sin(2t) & 2\cos(2t) \end{pmatrix}$$

$$= \cos(2t)\big(2\cos(2t)\big) - \big(-2\sin(2t)\big)\sin(2t)$$

$$= 2\big((\cos(2t))^2 + (\sin(2t))^2\big) = 2,$$

which is nonzero for all t. Therefore, this pair $\{\cos(2t), \sin(2t)\}$ forms a fundamental set of solutions for (8.37). Hence the general solution to (8.37) is

$$y = C_1 \cos(2t) + C_2 \sin(2t). \; \blacksquare$$

The following theorem shows that the Wronskian of two solutions *to the same differential equation* is either identically zero or it is never zero.

Theorem 8.5.   *Suppose that $\{y_1, y_2\}$ are two solutions of the same linear homogeneous second-order equation*

$$y'' + p(t)y' + q(t)y = 0 \tag{8.38}$$

on the interval $a < t < b$. Then the Wronskian $W(y_1, y_2)$ can be explicitly computed by **Abel's[4] formula**

$$W(y_1, y_2)(t) = W(y_1, y_2)(t_0) \exp\left(-\int_{t_0}^{t} p(s)ds \right). \tag{8.39}$$

If the Wronskian $W(y_1, y_2)(t_0) \neq 0$, then $W(y_1, y_2)(t) \neq 0$ for all t with $a < t < b$.

Proof. We first show that the Wronskian satisfies a first-order differential equation. Write

$$W = W(y_1, y_2)(t) = y_1 y_2' - y_1' y_2;$$

then we compute

$$W' = y_1' y_2' + y_1 y_2'' - y_1'' y_2 - y_1' y_2' = y_1 y_2'' - y_1'' y_2. \tag{8.40}$$

Since y_1 and y_2 satisfy (8.38), we have

$$y_1'' = -p(t)y_1' - q(t)y_1 \quad \text{and} \quad y_2'' = -p(t)y_2' - q(t)y_2. \tag{8.41}$$

We substitute (8.41) into (8.38) to obtain

$$
\begin{aligned}
W' &= y_1 y_2'' - y_1'' y_2 \\
   &= y_1\big(-p(t)y_2' - q(t)y_2 \big) - y_2\big(-p(t)y_1' - q(t)y_1 \big) \\
   &= -p(t)\big(y_1 y_2' - y_1' y_2 \big) = -p(t)W.
\end{aligned}
$$

---

4   Niels Henrik Abel (1802–1829). Norwegian mathematician. He proved the impossibility of solving algebraically the general equation of the fifth degree and also revolutionized the understanding of elliptic functions by studying the inverse of functions of elliptic integrals.

This shows that the Wronskian is a solution to the first-order linear differential equation

$$W'(t) + p(t)W(t) = 0. \tag{8.42}$$

An integrating factor for (8.42) is

$$\exp\left(\int_{t_0}^{t} p(s)ds\right),$$

so we obtain

$$\frac{d}{dt}\left(\exp\left(\int_{t_0}^{t} p(s)ds\right)W(t)\right) = 0. \tag{8.43}$$

Then (8.39) results when (8.43) is integrated from 0 to t_0. The exponential function is never zero. Therefore, $W(t_0) \neq 0$ implies that $W(t) \neq 0$ for all t. ∎

It is important to realize that the Wronskian of two nonzero functions, if they are not solutions of the *same* second-order linear differential equation, might sometimes be zero and sometimes nonzero.

Theorems 8.4 and 8.5 show that the Wronskian is a powerful theoretical tool in dealing with second-order linear homogeneous equations. How do we obtain a set of solutions $\{y_1, y_2\}$ whose Wronskian is nonzero?

The existence of such a set is furnished, in principle, by the existence and uniqueness theorem (Theorem 8.1). To construct the solution, we pick a point t_0 in the interval $a < t < b$ and solve the differential equation $y'' + p(t)y' + q(t)y = 0$ with the initial conditions $y(t_0) = 1$ and $y'(t_0) = 0$, calling the resultant solution y_1. Next, we solve the same differential equation with the initial conditions $y(t_0) = 0$ and $y'(t_0) = 1$, calling the resultant solution y_2. At the point t_0 we compute the Wronskian as

$$W(y_1, y_2)(t_0) = \det\begin{pmatrix} 1 & 0 \\ 0 & 1 \end{pmatrix} = 1.$$

Therefore, by Theorem 8.4 we have obtained a fundamental set of solutions.

Example 8.10. *Use the Wronskian to find a fundamental set of solutions for the equation*

$$y'' + 4y = 0. \tag{8.44}$$

Solution. The general solution to (8.44) has been found in Example 8.9 to be $y(t) = C_1 \cos(2t) + C_2 \sin(2t)$. For the solution y_1 we take $C_2 = 0$ to obtain $y_1(t) = \cos(2t)$. For the solution y_2, we must take $C_1 = 0$, but the value of C_2 must be chosen such that $y_2'(0) = 1$. This requires that $C_2 = 1/2$ and the solution $y_2(t) = \sin(2t)/2$. Then $\{y_1, y_2\}$ is the required fundamental set of solutions. ∎

The Wronskian via **Mathematica**

The *Mathematica* definition of the Wronskian is essentially the same as the definition that we gave on page 247:

```
Wronskian[y1_,y2_][t_]:= y1[t]y2'[t] - y2[t]y1'[t]
```

Example 8.11. *Use Mathematica to compute the Wronskian of* sin *and* log, *and plot the result.*

Solution. We use

```
Wronskian[Sin,Log][t]
```

to get

$$-(\text{Cos}[t] \ \text{Log}[t]) + \frac{\text{Sin}[t]}{t}$$

To plot the Wronskian of sin and log we use

```
Plot[Wronskian[Sin,Log][t]//Evaluate,{t,0.01,Pi - 0.1}]
```

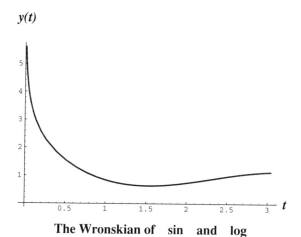

The Wronskian of   sin   and   log

Exercises for Section 8.4

Compute the Wronskian of the following pairs of functions. First, do the computations by hand; then use *Mathematica*.

1. $y_1 = t\,e^t$ and $y_2 = e^t$

3. $y_1 = e^t \cos(2t)$ and $y_2 = e^t \sin(2t)$

2. $y_1 = \cos(2t)$ and $y_2 = \sin(2t)$

4. $y_1 = e^{2t}$ and $y_2 = e^{3t}$

5. Use *Mathematica*'s command **Solve** to solve (8.34).

8.5   *Linear Independence and the Wronskian*

The Wronskian of any two functions y_1 and y_2 can be computed quite apart from any connection with the theory of differential equations. If the functions are proportional, that is, if $y_1(t) = C\,y_2(t)$, then $W(y_1, y_2)(t) = 0$ for all t. This leads to the general notions of **linear dependence** and **linear independence**, which we now discuss. See Section A.6 of Appendix A for more details, where these notions are discussed for abstract vector spaces.

Definition. *Two functions $y_1(t)$ and $y_2(t)$ defined on an interval $a < t < b$ are said to be **linearly dependent** if there exist constants C_1 and C_2, not both zero, such that*

$$C_1 y_1(t) + C_2 y_2(t) \equiv 0 \tag{8.45}$$

*for $a < t < b$. If no such constants exist, then the functions are said to be **linearly independent**.*

Elementary logic shows that the definition of linear independence can be expressed in another way:

Lemma 8.6. *Two functions $y_1(t)$ and $y_2(t)$ on an interval $a < t < b$ are linearly independent if and only if*

$$C_1 y_1(t) + C_2 y_2(t) \equiv 0$$

implies that $C_1 = C_2 = 0$.

In practice, it is usually best to use Lemma 8.6 to establish linear independence.

Example 8.12. *Show that the functions t and e^t are linearly independent on the interval $-2 < t < 2$.*

Solution. Suppose that there are constants C_1 and C_2 such that

$$C_1 t + C_2 e^t \equiv 0 \tag{8.46}$$

for $-2 < t < 2$. Setting $t = 0$ in (8.46) yields $C_2 = 0$. Further, setting $t = 1$ in (8.46) yields $C_1 = 0$. Therefore, (8.6) implies that t and e^t are linearly independent on the interval $-2 < t < 2$. ∎

Example 8.13. *Show that the functions* $\left(\cos(t)\right)^2$ *and* $\left(\sin(t)\right)^2 - 1$ *are linearly dependent on any interval.*

Solution. Since $\sin(t)$ and $\cos(t)$ satisfy the identity

$$\left(\cos(t)\right)^2 + \left(\sin(t)\right)^2 \equiv 1,$$

we have

$$A\left(\cos(t)\right)^2 + B\left(\left(\sin(t)\right)^2 - 1\right) \equiv 0$$

for $A = B = 1$. By definition $\left(\cos(t)\right)^2$ and $\left(\sin(t)\right)^2 - 1$ are linearly dependent. ∎

It is possible that a pair of functions is linearly independent on one interval but linearly dependent on another.

Example 8.14. *Show that the functions t and $|t|$ are linearly independent on the interval* $-2 < t < 2$, *but linearly dependent on the interval* $0 < t < 1$.

Solution. Let C_1 and C_2 be constants such that

$$C_1 t + C_2 |t| \equiv 0 \tag{8.47}$$

on the interval $-2 < t < 2$. We take first $t = -1$ and then $t = 1$ in (8.47), obtaining the system of equations

$$\begin{cases} -C_1 + C_2 = 0, \\ C_1 + C_2 = 0. \end{cases} \tag{8.48}$$

The solution to (8.48) is $C_1 = C_2 = 0$, so that t and $|t|$ are linearly independent on the interval $-2 < t < 2$.

On the other hand, on the interval $0 < t < 1$ the two functions t and $|t|$ coincide, and therefore we may take $C_1 = 1$ and $C_2 = -1$ in (8.45). This shows the linear dependence of t and $|t|$ on the interval $0 < t < 1$. ∎

Here are some useful facts; the proofs are easy.

Lemma 8.7. *If y_1 and y_2 are linearly independent, then neither function can be identically zero.*

Proof. If y_1 were identically zero, we could take $C_1 = 1$ and $C_2 = 0$ in the definition of linear independence. ∎

Lemma 8.8. *If y_1 and y_2 are linearly dependent, then there is a constant C such that either $y_1(t) = C\, y_2(t)$ or $y_2(t) = C\, y_1(t)$ for all t.*

Proof. We can divide the equation $C_1 y_1(t) + C_2 y_2(t) = 0$ by whichever of C_1 or C_2 is nonzero. ∎

In the case that y_1 and y_2 satisfy the *same* second-order linear homogeneous equation on an interval $a < t < b$, their linear independence is equivalent to the Wronskian $W(y_1, y_2)(t)$ being nonzero, according to the following theorem.

Theorem 8.9. *Suppose that y_1 and y_2 are twice differentiable functions on the interval $a < t < b$, and that both satisfy the differential equation $y'' + p(t)y' + q(t)y = 0$. Then the following three properties are equivalent:*

(i) $W(y_1, y_2)(t) \neq 0$ *for all t with a $< t < b$;*

(ii) $W(y_1, y_2)(t_0) \neq 0$ *for some t_0 with a $< t_0 < b$;*

(iii) *the functions y_1 and y_2 are linearly independent for a $< t < b$.*

(iv) *the functions y_1 and y_2 constitute a fundamental set of solutions for $y'' + p(t)y' + q(t)y = 0$.*

Proof. Clearly, (i) implies (ii). Next, suppose (ii) holds, and that C_1 and C_2 are constants such that

$$C_1 y_1(t) + C_2 y_2(t) \equiv 0 \tag{8.49}$$

on the interval $a < t < b$. Differentiating (8.49) yields

$$C_1 y_1'(t) + C_2 y_2'(t) \equiv 0 \tag{8.50}$$

on the interval $a < t < b$. When we take $t = t_0$ in (8.49) and (8.50), we obtain the following system of linear algebraic equations in C_1 and C_2:

$$\begin{cases} C_1 y_1(t_0) + C_2 y_2(t_0) = 0, \\ C_1 y_1'(t_0) + C_2 y_2'(t_0) = 0. \end{cases} \tag{8.51}$$

By assumption $W(y_1, y_2)(t_0) \neq 0$, so we can solve the system (8.51) by Cramer's rule, obtaining $C_1 = C_2 = 0$. Hence y_1 and y_2 are linearly independent.

To prove that (iii) implies (i), let y_1 and y_2 be linearly independent solutions of $y'' + p(t)y' + q(t)y = 0$. Suppose that $W(y_1, y_2)(t_0) = 0$ for some t_0 with $a < t_0 < b$. This means that the system of linear equations

$$\begin{cases} C_1 y_1(t_0) + C_2 y_2(t_0) = 0, \\ C_1 y_1'(t_0) + C_2 y_2'(t_0) = 0 \end{cases}$$

has a solution $(C_1, C_2) \neq (0, 0)$, since the determinant of the system is precisely the Wronskian $W(y_1, y_2)(t_0)$. We form a new function $y(t) = C_1 y_1(t) + C_2 y_2(t)$. By the superposition principle on page 243, $y(t)$ is a solution to the homogeneous differential equation $y'' + p(t)y' + q(t)y = 0$. Also, by construction we have $y(t_0) = y'(t_0) = 0$. Therefore, from the uniqueness part of Theorem 8.1 we must have $y(t) \equiv 0$. This contradicts the linear independence of the pair $\{y_1, y_2\}$, since $(C_1, C_2) \neq (0, 0)$

Finally, Theorem 8.4 implies the equivalence of (ii) and (iv). $\blacksquare$

For a general pair of functions y_1 and y_2 that do *not* satisfy the same second-order differential equation, parts (i), (ii), (iii) of Theorem 8.9 are no longer equivalent. In fact, (i) implies (ii) and (ii) implies (iii), but (iii) no longer implies (i), and the implication is *strict* according to the following theorem.

Theorem 8.10.   *Let $a < t_0 < b$.*

(i) *There exists a pair of functions y_1 and y_2 such that $W(y_1, y_2)(t_0) = 0$ for some t_0 with $a < t_0 < b$, but $W(y_1, y_2)(t) \not\equiv 0$ on the interval $a < t < b$.*

(ii) *There exists a pair of functions y_1 and y_2 that are linearly independent, but*

$$W(y_1, y_2)(t) = 0$$

for all t with $a < t < b$.

Proof.   First, consider the pair of functions $y_1(t) = 1$ and $y_2(t) = t^2$ on the interval $-2 < t < 2$. The Wronskian is

$$W(y_1, y_2)(t) = \det \begin{pmatrix} 1 & t^2 \\ 0 & 2t \end{pmatrix} = 2t,$$

which is nonzero for $t \neq 0$, and we have proved (i).

For (ii) consider the pair of functions $y_1(t) = t^3$ and $y_2(t) = |t|^3$ on the interval $-2 < t < 2$. It can be checked that y_2 is differentiable (even at 0) and that $y_2'(t) = 3|t|t$. To prove that the functions y_1 and y_2 are linearly independent, suppose that we had a linear relation of the form $C_1 y_1(t) + C_2 y_2(t) \equiv 0$. Then by setting $t = 1$ and $t = -1$, we would conclude that $C_1 + C_2 = 0$ and $-C_1 + C_2 = 0$, from which it would follow that $C_1 = C_2 = 0$. In spite of the fact that y_1 and y_2 are linearly independent, the Wronskian is computed as

$$W(y_1, y_2)(t) = \det \begin{pmatrix} t^3 & |t|^3 \\ 3t^2 & 3t|t| \end{pmatrix} = 3|t|^5 - 3|t|^5 = 0$$

for *all t*. $\blacksquare$

The above discussion can be summarized as follows. The properties of linear independence and nonzero Wronskian are entirely equivalent to one another when applied to a pair of solutions of the same second-order linear differential equation. However, if we try to extend this equivalence to functions that do not satisfy the same differential equation, we find that these properties are *not* equivalent to one another. There exist simple examples of linearly independent functions with zero Wronskian.

Exercises for Section 8.5

In each of the following problems, determine whether the pair of functions $\{y_1, y_2\}$ are linearly independent or dependent.

1. $y_1 = t^3 - 3t, \quad y_2 = t^3 + 3t$ **3.** $y_1 = \log(t), \quad y_2 = \log(2t)$

2. $y_1 = e^{at}\cos(bt), \quad y_2 = e^{at}\sin(bt)$ **4.** $y_1 = e^{at}, \quad y_2 = e^{at+b}$

8.6   Reduction of Order

A useful application of the Wronskian occurs in the method of **reduction of order**. This procedure, due to D'Alembert[5], goes as follows. We are given one solution y_1 of a linear second-order homogeneous equation

$$y'' + p(t)y' + q(t)y = 0, \tag{8.52}$$

and we are required to find a second linearly independent solution y_2. By Theorem 8.9 we know that this is equivalent to finding a solution y_2 for which the Wronskian $W(y_1, y_2)(t) \neq 0$. Theorem 8.5 tells us that $W(y_1, y_2)(t)$ can be expressed directly in terms of $p(t)$ as

$$W(t) = W(y_1, y_2)(t) = C \exp\left(-\int p(t)dt\right),$$

where C is a constant. The definition of the Wronskian, namely

$$y_1 y_2' - y_1' y_2 = W, \tag{8.53}$$

5

Jean le Rond D'Alembert (1717–1783). French mathematician who worked in mechanics and hydrodynamics. He pioneered the use of differential equations in physics. In 1754 he wrote an article suggesting that the theory of limits be put on a firm foundation.

can be viewed as a first-order differential equation in y_2; the leading-coefficient-unity-form of (8.53) is

$$y_2' - \left(\frac{y_1'}{y_1}\right) y_2 = \frac{W}{y_1}. \tag{8.54}$$

From Section 3.2 we know that an integrating factor for (8.54) is given by

$$\exp\left(-\int \frac{y_1'}{y_1} dt\right) = \frac{1}{y_1}. \tag{8.55}$$

When we multiply both sides of (8.54) by (8.55), we get

$$\left(\frac{y_2}{y_1}\right)' = \frac{y_2'}{y_1} - \frac{y_1' y_2}{y_1^2} = \frac{W}{y_1^2}. \tag{8.56}$$

Then we integrate both sides of (8.56), obtaining

$$\frac{y_2(t)}{y_1(t)} = \int \frac{W(y_1, y_2)(t) dt}{y_1(t)^2}.$$

We summarize these calculations as a theorem.

Theorem 8.11.   (**Method of reduction of order**) *If y_1 is a nonzero solution to the second-order equation $y'' + p(t)y' + q(t)y = 0$ on the interval $a < t < b$, then a second linearly independent solution is given by*

$$y_2(t) = y_1(t) \int \frac{W(y_1, y_2)(t)}{y_1(t)^2} dt, \tag{8.57}$$

where the Wronskian $W(y_1, y_2)(t)$ is given by

$$W(y_1, y_2)(t) = C \exp\left(-\int p(s) ds\right) \tag{8.58}$$

for any nonzero constant C.

Proof.   It remains to show that y_1 and y_2 are linearly independent. If y_1 and y_2 were linearly dependent, then (8.57) would imply that

$$\int \frac{W(y_1, y_2)(t)}{y_1(t)^2} dt \tag{8.59}$$

was a constant. But then the integrand in (8.59) would have to vanish, contradicting (8.58). Hence y_1 and y_2 are linearly independent. ▌

Note that the method of reduction of order involves *two* integrations, the first to find the Wronskian from the function $p(t)$ and the second to integrate

$$\frac{W(y_1, y_2)(t)}{y_1(t)^2}.$$

Example 8.15. *Find a second linearly independent solution to the differential equation* $y'' + 2y' + y = 0$ *if we are given the solution* $y_1(t) = e^{-t}$.

Solution. The Wronskian is given by

$$W = C \exp\left(-2\int dt\right) = Ce^{-2t}.$$

Theorem 8.11 gives

$$y_2 = y_1 \int \frac{W}{y_1^2} dt = e^{-t} \int \frac{Ce^{-2t}}{(e^{-t})^2} dt = e^{-t} \int C\,dt = \frac{Ct}{e^t}. \quad \blacksquare$$

Exercises for Section 8.6

Use the method of reduction of order to find a second linearly independent solution to each of the following second-order equations when we are given the indicated first solution.

1. $\begin{cases} y'' - 3y' + 2y = 0, \\ y_1(t) = e^{2t} \end{cases}$

5. $\begin{cases} t^2 y'' + t\,y' + (t^2 - 1/4)y = 0, \\ y_1(t) = \sin(t)/\sqrt{t} \end{cases}$

2. $\begin{cases} y'' + 4y' + 4y = 0, \\ y_1(t) = e^{-2t} \end{cases}$

6. $\begin{cases} t\,y'' - (t+1)y' + y = 0, \\ y_1(t) = t + 1 \end{cases}$

3. $\begin{cases} t^2 y'' + 4t\,y' = 0, \\ y_1(t) = 6 \end{cases}$

7. $\begin{cases} t\,y'' - (t+2)y' + 2y = 0, \\ y_1(t) = e^t \end{cases}$

4. $\begin{cases} t^2 y'' + 5t\,y' - 5y = 0, \\ y_1(t) = t \end{cases}$

8. $\begin{cases} t\,y'' - y' + 4t^3 y = 0, \\ y_1(t) = \sin(t^2) \end{cases}$

8.7   *Equations with Given Solutions*

In this section[6] we show that *any two* functions with a nonzero Wronskian may be considered as solutions of some linear second-order homogeneous equation.

---

[6]This section contains optional material. It is used only in Section 10.4.

Theorem 8.12. *Suppose that* y_1 *and* y_2 *are twice continuously differentiable functions on the interval* $a < t < b$, *for which* $W(y_1, y_2)(t) \neq 0$ *for* $a < t < b$. *Then there exist continuous functions* $p(t)$ *and* $q(t)$ *such that both* y_1 *and* y_2 *satisfy the differential equation* $y'' + p(t)y' + q(t) = 0$.

Proof. The functions $p(t)$ and $q(t)$ must satisfy the system of linear equations

$$\begin{cases} p(t)y_1' + q(t)y_1 = -y_1'', \\ p(t)y_2' + q(t)y_2 = -y_2''. \end{cases} \tag{8.60}$$

The determinant of the system (8.60) is

$$\det\begin{pmatrix} y_1' & y_1 \\ y_2' & y_2 \end{pmatrix} = -W(y_1, y_2)(t) \neq 0,$$

by hypothesis. Therefore, we may solve (8.60) for $p(t)$ and $q(t)$ using Cramer's rule (Theorem 8.3). Explicitly,

$$p(t) = \frac{\det\begin{pmatrix} -y_1'' & y_1 \\ -y_2'' & y_2 \end{pmatrix}}{\det\begin{pmatrix} y_1' & y_1 \\ y_2' & y_2 \end{pmatrix}} = \frac{y_1'' y_2 - y_1 y_2''}{W(y_1, y_2)(t)} \tag{8.61}$$

and

$$q(t) = \frac{\det\begin{pmatrix} y_1' & -y_1'' \\ y_2' & -y_2'' \end{pmatrix}}{\det\begin{pmatrix} y_1' & y_1 \\ y_2' & y_2 \end{pmatrix}} = \frac{-y_1'' y_2' + y_1' y_2''}{W(y_1, y_2)(t)}. \ \blacksquare \tag{8.62}$$

Example 8.16. *The functions* $y_1 = t$ *and* $y_2 = e^t$ *are required to be solutions of a differential equation* $y'' + p(t)y' + q(t)y = 0$. *Find the equation.*

Solution. The Wronskian is

$$W(y_1, y_2)(t) = \det\begin{pmatrix} t & e^t \\ 1 & e^t \end{pmatrix} = (t-1)e^t,$$

which is nonzero whenever $t \neq 1$. Therefore, we expect to find the functions $p(t)$ and $q(t)$ on any interval on which $t \neq 1$. To do this we must solve the equations

$$\begin{cases} p(t) + t\, q(t) = 0, \\ e^t p(t) + e^t q(t) = -e^t. \end{cases} \tag{8.63}$$

for $p(t)$ and $q(t)$. We simplify (8.63) to

$$\begin{cases} p(t) + t\,q(t) = 0, \\ p(t) + q(t) = -1, \end{cases}$$

which is solved to yield

$$p(t) = \frac{t}{1-t} \quad \text{and} \quad q(t) = \frac{1}{t-1}.$$

The required equation is thus

$$y'' + \frac{t\,y'}{1-t} - \frac{y}{1-t} = 0.$$

It is clear that the restriction $t \neq 1$ is essential here. ∎

Exercises for Section 8.7

The following pairs of functions form a fundamental set of solutions of a linear second-order differential equation $y'' + p(t)y' + q(t)y = 0$. Find the functions $p(t)$ and $q(t)$.

1. $t\,e^t$ and $t^2 e^t$

2. $\cos(2t)$ and $\sin(2t)$

3. $e^t \cos(2t)$ and $e^t \sin(2t)$

4. t and t^4

5. 2 and $\log(t)$

6. t^3 and t^{-3}

Summary of Techniques Introduced in Chapter 8

The general solution to a second-order homogeneous linear differential equation

$$y'' + p(t)y' + q(t)y = 0$$

can always be expressed in terms of a **fundamental set of solutions** $\{y_1, y_2\}$:

$$y = C_1 y_1 + C_2 y_2.$$

The functions y_1 and y_2 can be explicitly found in the case of the equation $y'' + p(t)y' = 0$ (by reduction to a first-order linear equation) and for the equation $y'' + k^2 y = 0$ (by trigonometric functions). For the general equation the superposition principle can be used to write the general solution to the homogeneous equation in terms of a fundamental set.

The subtraction principle can be used to write the general solution to the **inhomogeneous equation** $y'' + p(t)y' + q(t)y = r(t)$ in terms of a **particular solution** and a fundamental set of solutions for the homogeneous equation. The **Wronskian** of two solutions of the homogeneous equation is defined by

$$W(y_1, y_2)(t) = \det\begin{pmatrix} y_1(t) & y_2(t) \\ y_1'(t) & y_2'(t) \end{pmatrix};$$

it can be used to test whether or not a given set of solutions forms a fundamental set (depending on whether $W(t) \neq 0$). The Wronskian can be computed directly as

$$W(y_1, y_2)(t) = W(y_1, y_2)(t_0) \exp\left(-\int_{t_0}^{t} p(s)ds\right).$$

This can be used to effect the method of **reduction of order**, which allows one to find the second solution y_2 of a fundamental set when one knows the first solution y_1.

9

SECOND-ORDER LINEAR DIFFERENTIAL EQUATIONS WITH CONSTANT COEFFICIENTS

In the previous sections we have established the general structure and properties of solutions of linear second-order equations. We now proceed to find explicit solution formulas for the general equation with constant-coefficients, written

$$a\,y'' + b\,y' + c\,y = g(t), \tag{9.1}$$

where a, b, and c are constants and we assume that $a \neq 0$, so that (9.1) is second-order and not first-order.

The paramount importance of this class of equations rests on two basic facts: (1) these equations arise in a variety of applications in mechanics and electricity, and (2) one can explicitly find all solutions of the homogeneous equation ($g(t) \equiv 0$) and use this information to find many solution formulas for the general (inhomogeneous) case.

In detail, the homogeneous equation, written

$$a\,y'' + b\,y' + c\,y = 0, \tag{9.2}$$

is solved in Section 9.1 and Section 9.2. Then in Section 9.3 we develop an algebraic method to solve the inhomogenous equation in the case that the right side is of a special

265

form. Finally, in Section 9.4 we deal with the case of a general right-hand side, where the solution of the inhomogeneous equation is obtained in terms of an integral.

9.1 *Constant-Coefficient Second-Order Homogeneous Equations*

In this section we find the general solution of a second-order linear homogeneous differential equation with constant-coefficients. The equation has the form (9.2), and is relatively simple to solve. In this section we assume that a, b, and c are real and that $a \neq 0$. Let us consider a **trial solution** of (9.2) of the form

$$y = e^{rt}, \tag{9.3}$$

where r is to be determined. We compute

$$y' = r\, e^{rt} \quad \text{and} \quad y'' = r^2 e^{rt}. \tag{9.4}$$

Substituting (9.4) and (9.3) into (9.2), we find that

$$0 = a\, y'' + b\, y' + c\, y = (a\, r^2 + b\, r + c)e^{rt}. \tag{9.5}$$

In order for the right-hand side of (9.5) to be zero, the factor $a\, r^2 + b\, r + c$ must be zero, since the exponential function e^{rt} is never zero. Thus we get the quadratic equation

$$a\, r^2 + b\, r + c = 0. \tag{9.6}$$

Hence (9.6) is a necessary condition that the trial solution (9.3) solve (9.2). But it is easy to see that it is also a sufficient condition. Therefore, we have

Proposition 9.1. *The trial solution $y = e^{rt}$ is a solution of $a\, y'' + b\, y' + c\, y = 0$ if and only if $a\, r^2 + b\, r + c = 0$.*

Definition. *We call* (9.6) *the* **associated characteristic equation** *of the differential equation* (9.2).

The roots of (9.6) are given by the **quadratic formula**:

$$r = \frac{-b \pm \sqrt{b^2 - 4a\, c}}{2a}. \tag{9.7}$$

The detailed nature of the solutions of (9.2) depends on whether or not (9.6) has distinct real roots, repeated real roots, or complex conjugate roots. This, in turn, depends on whether the **discriminant** $D = b^2 - 4a\, c$ is positive, zero, or negative. In the following subsections we consider each of these cases separately.

9.1.1   Case of Real and Distinct Roots

In this subsection we find two linearly independent solutions of the differential equation $a\,y'' + b\,y' + c\,y = 0$ in the case that the discriminant is positive, that is, when $D = b^2 - 4a\,c > 0$. In this case, the associated characteristic equation $a\,r^2 + b\,r + c = 0$ has the two real roots

$$r_1 = \frac{-b + \sqrt{b^2 - 4a\,c}}{2a} \qquad \text{and} \qquad r_2 = \frac{-b - \sqrt{b^2 - 4a\,c}}{2a}. \tag{9.8}$$

Since $a \neq 0$ and $D > 0$, we have

$$r_1 - r_2 = \frac{\sqrt{b^2 - 4a\,c}}{a} \neq 0$$

so that $r_1 \neq r_2$.

Proposition 9.2. *If the discriminant $D = b^2 - 4a\,c$ is positive, then a fundamental set of solutions of the differential equation $a\,y'' + b\,y' + c\,y = 0$ is provided by the functions $y_1 = e^{r_1 t}$ and $y_2 = e^{r_2 t}$, where r_1 and r_2 are given by (9.8). Furthermore, the general solution of $a\,y'' + b\,y' + c\,y = 0$ is*

$$y = C_1 e^{r_1 t} + C_2 e^{r_2 t}. \tag{9.9}$$

Proof. It is immediate that both of the functions y_1 and y_2 are solutions of $a\,y'' + b\,y' + c\,y = 0$, as both r_1 and r_2 satisfy the associated characteristic equation $a\,r^2 + b\,r + c = 0$. To show that $\{y_1, y_2\}$ is a fundamental set of solutions, we compute the Wronskian, using the definition given in Section 8.4:

$$W(y_1, y_2)(t) = \det\begin{pmatrix} y_1(t) & y_2(t) \\ y_1'(t) & y_2'(t) \end{pmatrix}$$

$$= \det\begin{pmatrix} e^{r_1 t} & e^{r_2 t} \\ r_1 e^{r_1 t} & r_2 e^{r_2 t} \end{pmatrix}$$

$$= (r_2 - r_1)e^{r_1 t} e^{r_2 t}. \tag{9.10}$$

Neither of the exponential factors on the right-hand side of (9.10) can vanish; furthermore, $r_2 - r_1 \neq 0$. Therefore, the Wronskian is everywhere nonzero. Now Theorem 8.4 implies that $\{y_1, y_2\}$ is a fundamental set of solutions for the differential equation $a\,y'' + b\,y' + c\,y = 0$. ∎

In practice, the associated characteristic equation $a\,r^2 + b\,r + c = 0$, in the case that its roots are real, can be solved in either of two ways. It is best to try first to factor $a\,r^2 + b\,r + c$. If $a\,r^2 + b\,r + c$ has no obvious factorization, the quadratic formula can be used.

Example 9.1. *Find a fundamental set of solutions for the differential equation*

$$y'' - 4y' + 3y = 0, \tag{9.11}$$

and also determine the general solution.

Solution. The associated characteristic equation of (9.11) is

$$0 = r^2 - 4r + 3 = (r - 3)(r - 1),$$

so that roots are $r_1 = 3$ and $r_2 = 1$. A fundamental set of solutions of (9.11) is provided by the functions $y_1 = e^{3t}$ and $y_2 = e^t$. Furthermore, the general solution of (9.11) is

$$y = C_1 e^{3t} + C_2 e^t. \ \blacksquare$$

Example 9.2. *Find a fundamental set of solutions for the differential equation*

$$y'' - 5y' + 3y = 0, \tag{9.12}$$

and also determine the general solution.

Solution. The associated characteristic equation of (9.12) is

$$0 = r^2 - 5r + 3, \tag{9.13}$$

which is inconvenient to factor. However, we can find the roots of (9.13) by means of the quadratic formula:

$$r = \frac{5 \pm \sqrt{5^2 - 4 \times 3}}{2} = \frac{5 \pm \sqrt{13}}{2}.$$

Thus, a fundamental set of solutions of (9.12) is provided by the functions

$$y_1 = e^{(5+\sqrt{13})t/2} \qquad \text{and} \qquad y_2 = e^{(5-\sqrt{13})t/2}.$$

The general solution of (9.12) is then

$$y = C_1 \exp\left(\left(\frac{5 + \sqrt{13}}{2}\right)t\right) + C_2 \exp\left(\left(\frac{5 - \sqrt{13}}{2}\right)t\right). \ \blacksquare$$

Once we have found a fundamental set of solutions $\{y_1, y_2\}$ of a homogeneous second-order linear equation, we can then solve the corresponding initial value problem by taking a suitable linear combination of y_1 and y_2.

Example 9.3. *Find the solution of the differential equation* $y'' - 4y' + 3y = 0$ *that satisfies the initial conditions* $y(0) = 1$ *and* $y'(0) = 5$.

Solution. In Example 9.1 we found the general solution of $y'' - 4y' + 3y = 0$ to be $y = C_1 e^{3t} + C_2 e^t$. The initial conditions require that we solve the system of algebraic equations

$$\begin{cases} C_1 + C_2 = 1, \\ 3C_1 + C_2 = 5. \end{cases} \tag{9.14}$$

Subtracting the first equation of (9.14) from the second yields $2C_1 = 4$, or $C_1 = 2$. Back substitution into the first equation of (9.14) yields $C_2 = -1$, and we have the solution $y(t) = 2e^{3t} - e^t$. ∎

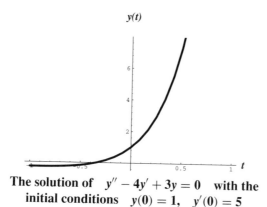

y(t)

The solution of $y'' - 4y' + 3y = 0$ with the
initial conditions $y(0) = 1, \quad y'(0) = 5$

9.1.2   Case of Real Repeated Roots

Now we return to $a\,y'' + b\,y' + c\,y = 0$ and suppose that the discriminant $D = b^2 - 4a\,c = 0$. In this case the associated characteristic equation $a\,r^2 + b\,r + c = 0$ has the solution

$$r = \frac{-b \pm 0}{2a},$$

which yields two *equal* roots. It is clear that one solution of $a\,y'' + b\,y' + c\,y = 0$ is

$$y_1(t) = e^{rt},$$

but how do we find a second linearly independent solution? Let us try $y = t\,e^{rt}$. The best reason for this apparently wild guess is that it works. But it can also be explained by the method of reduction of order, from Section 8.6. We have the solution $y_1(t) = e^{rt}$. We know from Section 8.4 that the Wronskian W of any two solutions satisfies the differential equation $W' + p(t)W = 0$, which in this case reads

$$W' + \frac{b}{a}W = 0;$$

the solution of this separable first-order differential equation is

$$W(t) = C\, e^{-bt/a} = C\, e^{2rt},$$

where C is a constant. The method of reduction of order now gives the second solution y_2 from the integral

$$\frac{y_2}{y_1} = \int \frac{W(t)}{y_1(t)^2} dt = \int \frac{C\, e^{2rt}}{e^{2rt}} dt = C\, t.$$

Hence we have $y_2(t) = C\, t\, e^{rt}$.

Now let us find the general solution of $a\, y'' + b\, y' + c\, y = 0$ in the case of repeated roots in a simpler way.

Proposition 9.3. *If the discriminant $D = b^2 - 4ac$ is zero, then a fundamental set of solutions of the differential equation $a\, y'' + b\, y' + c\, y = 0$ with $a \neq 0$ is provided by the functions $y_1(t) = e^{rt}$ and $y_2(t) = t\, e^{rt}$, where $r = -b/2a$. The general solution is*

$$y = C_1 e^{rt} + C_2 t\, e^{rt}. \tag{9.15}$$

Proof. We have

$$y_2 = t\, e^{rt}, \qquad y_2' = r\, t\, e^{rt} + e^{rt}, \qquad \text{and} \qquad y_2'' = r^2 t\, e^{rt} + 2r\, e^{rt},$$

so that

$$
\begin{aligned}
a\, y_2'' + b\, y_2' + c\, y_2 &= a(r^2 t + 2r)e^{rt} + b(r\, t + 1)e^{rt} + c\, t\, e^{rt} \\
&= (a\, r^2 + b\, r + c)t\, e^{rt} + (2a\, r + b)e^{rt} = 0.
\end{aligned}
$$

Hence y_2 is a solution of $a\, y'' + b\, y' + c\, y = 0$. The Wronskian of y_1 and y_2 is given by

$$W(y_1, y_2)(t) = \det\begin{pmatrix} y_1 & y_2 \\ y_1' & y_2' \end{pmatrix} = \det\begin{pmatrix} e^{rt} & t\, e^{rt} \\ r\, e^{rt} & (1 + r\, t)e^{rt} \end{pmatrix} = e^{2rt}.$$

Thus, $W(y_1, y_2)(t)$ is never zero, since the exponential is never zero. Again Theorem 8.4 implies that $\{y_1, y_2\}$ is a fundamental set of solutions for the differential equation $a\, y'' + b\, y' + c\, y = 0$ and that the general solution is given by

$$y = C_1 e^{rt} + C_2 t\, e^{rt}. \ \blacksquare$$

Example 9.4. *Find a fundamental set of solutions of the differential equation*

$$y'' - 4y' + 4y = 0,$$

and also determine the general solution.

Solution. The discriminant is $D = 4^2 - 4 \times 4 = 16 - 16 = 0$, so we have repeated roots with $r = 2$. Thus, the functions $y_1(t) = e^{2t}$ and $y_2(t) = t\,e^{2t}$ constitute a fundamental set of solutions of $y'' - 4y' + 4y = 0$, and the general solution is

$$y = C_1 e^{2t} + C_2 t\, e^{2t}. \;\blacksquare \tag{9.16}$$

Example 9.5.   *Find the solution of the equation $y'' - 4y' + 4y = 0$ satisfying the initial conditions $y(0) = 1$ and $y'(0) = 5$.*

Solution. From Example 9.4 we know that the general solution of $y'' - 4y' + 4y = 0$ is given by (9.16). The constants are determined from solving the equations

$$\begin{cases} C_1 + 0 = 1, \\ 2C_1 + C_2 = 5. \end{cases}$$

We find that $C_1 = 1$ and $C_2 = 3$; hence

$$y(t) = e^{2t} + 3t\, e^{2t}. \;\blacksquare$$

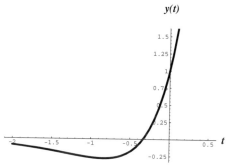

The solution of $y'' - 4y' + 4y = 0$ with the
initial conditions $y(0) = 1, \quad y'(0) = 5$

9.1.3   Case of Complex Conjugate Roots

The third and final case occurs when the associated characteristic equation $a\,r^2 + br + c = 0$ has negative discriminant $D = b^2 - 4a\,c$. This implies that $a\,r^2 + br + c = 0$ has two distinct roots that are complex conjugates of one another:

$$r = \frac{-b \pm \sqrt{b^2 - 4a\,c}}{2a} = \lambda \pm i\,\mu.$$

We have written the complex roots in terms of their real and imaginary parts

$$\lambda = -\frac{b}{2a} \quad \text{and} \quad \mu = \frac{\sqrt{4a\,c - b^2}}{2a}, \tag{9.17}$$

where we have taken the positive square root of the positive number $4a\,c - b^2 = -D$.

The case of complex roots can be analyzed by making a substitution. We define a new function z through the equation

$$y = e^{\lambda t} z. \tag{9.18}$$

The derivatives of y can be expressed in terms of those of z:

$$\begin{cases} y' = e^{\lambda t} z' + \lambda\, e^{\lambda t} z, \\ y'' = e^{\lambda t} z'' + 2\lambda\, e^{\lambda t} z' + \lambda^2 e^{\lambda t} z. \end{cases} \tag{9.19}$$

From (9.18) and (9.19) it follows that

$$a\, y'' + b\, y' + c\, y = \left(a\left(z'' + 2\lambda\, z' + \lambda^2 z\right) + b\left(z' + \lambda\, z\right) + c\, z \right) e^{\lambda t}$$

$$= \left(a\, z'' + (2a\,\lambda + b)z' + (a\,\lambda^2 + b\,\lambda + c)z \right) e^{\lambda t}. \tag{9.20}$$

The term $(2a\,\lambda + b)$ on the right-hand side of (9.20) is zero, since we have chosen $\lambda = -b/2a$. The third term on the right-hand side of (9.20) can also be simplified as follows:

$$a\,\lambda^2 + b\,\lambda + c = a\left(\frac{-b}{2a}\right)^2 + b\left(\frac{-b}{2a}\right) + c = c - \frac{b^2}{4a} = \frac{1}{4a}(4a\,c - b^2) = a\,\mu^2, \tag{9.21}$$

where μ is given by (9.17). From (9.20) and (9.21) it follows that the function $z(t)$ must be a solution of the second-order equation

$$z'' + \mu^2 z = 0. \tag{9.22}$$

The general solution of (9.22) was found in Example 8.3 to be

$$z = A\,\cos(\mu\,t) + B\,\sin(\mu\,t). \tag{9.23}$$

Combining (9.23) and (9.18), we have the following result.

Proposition 9.4. *If the discriminant $D = b^2 - 4a\,c$ is negative, then a fundamental set of solutions of the homogeneous differential equation*

$$a\, y'' + b\, y' + c\, y = 0 \tag{9.24}$$

is $\{y_1, y_2\}$, where

$$y_1 = e^{\lambda t}\,\cos(\mu\,t) \qquad and \qquad y_2 = e^{\lambda t}\,\sin(\mu\,t). \tag{9.25}$$

In (9.25), $\lambda = -b/2a$ and $\mu = \sqrt{4a\,c - b^2}/2a$ are the real and imaginary parts of either of the complex roots $r = \left(-b \pm \sqrt{b^2 - 4a\,c}\right)/2a$. Furthermore, the general solution of (9.24) is

$$y = A\,e^{\lambda t}\,\cos(\mu\,t) + B\,e^{\lambda t}\,\sin(\mu\,t). \tag{9.26}$$

Proof. The above analysis shows that both y_1 and y_2 are solutions of (9.24). It remains to compute the Wronskian of y_1 and y_2. We have

$$\begin{cases} y_1' = \lambda e^{\lambda t} \cos(\mu t) - \mu e^{\lambda t} \sin(\mu t), \\ y_2' = \lambda e^{\lambda t} \sin(\mu t) + \mu e^{\lambda t} \cos(\mu t). \end{cases} \tag{9.27}$$

Using (9.25) and (9.27), we compute

$$W(y_1, y_2)(t) = \det \begin{pmatrix} y_1 & y_2 \\ y_1' & y_2' \end{pmatrix}$$

$$= \det \begin{pmatrix} e^{\lambda t} \cos(\mu t) & e^{\lambda t} \sin(\mu t) \\ \lambda e^{\lambda t} \cos(\mu t) - \mu e^{\lambda t} \sin(\mu t) & \lambda e^{\lambda t} \sin(\mu t) + \mu e^{\lambda t} \cos(\mu t) \end{pmatrix}$$

$$= e^{2\lambda t} \big(\cos(\mu t)\big(\lambda \sin(\mu t) + \mu \cos(\mu t)\big) - \sin(\mu t)\big(\lambda \cos(\mu t) - \mu \sin(\mu t)\big) \big)$$

$$= e^{2\lambda t} \mu \big(\cos(\mu t)^2 + \sin(\mu t)^2 \big) = \mu e^{2\lambda t}.$$

Thus, $W(y_1, y_2)(t) \neq 0$, since $\mu \neq 0$ and the exponential function is never zero. We have the desired fundamental set of solutions. ∎

Example 9.6.   *Find a fundamental set of solutions of the differential equation*

$$y'' + y' + 1.25 y = 0, \tag{9.28}$$

and also determine the general solution.

Solution.   The discriminant is $D = 1^2 - 4 \times 1.25 = -4 < 0$, so the associated characteristic equation of (9.28) has complex roots of the form $\lambda + i\mu$. We have

$$\lambda = -\frac{b}{2a} = -\frac{1}{2} = -0.5 \quad \text{and} \quad \mu = \frac{\sqrt{4ac - b^2}}{2a} = \frac{\sqrt{5} - 1}{2} = 1,$$

so that the required fundamental set of solutions consists of $y_1 = e^{-0.5t} \cos(t)$ and $y_2 = e^{-0.5t} \sin(t)$. The general solution of (9.28) is

$$y = e^{-0.5t} \big(A \cos(t) + B \sin(t) \big). \quad ∎ \tag{9.29}$$

Example 9.7.   *Find the solution of the differential equation $y'' + y' + 1.25 y = 0$ that satisfies the initial conditions $y(0) = 1$ and $y'(0) = 0$.*

Solution. From Example 9.6 we know that the general solution is given by (9.29). We differentiate (9.29), obtaining

$$y' = -0.5e^{-0.5t}\big(A\cos(t) + B\sin(t)\big) + e^{-0.5t}\big(-A\sin(t) + B\cos(t)\big). \qquad (9.30)$$

Substitution of the initial conditions into (9.29) and (9.30) yields the system of linear equations

$$\begin{cases} 1 = y(0) = A + 0, \\ 0 = y'(0) = -0.5A + B. \end{cases} \qquad (9.31)$$

The solution of (9.31) is obviously $A = 1$ and $B = 0.5$. Thus

$$y(t) = e^{-0.5t}\big(\cos(t) + 0.5\sin(t)\big). \quad \blacksquare$$

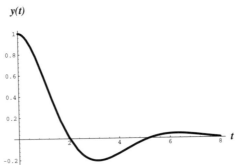

y(t)

The solution of $y'' + y' + 1.25y = 0$
with the initial conditions $y(0) = 1, \quad y'(0) = 0$

Summary of Techniques Introduced

The constant-coefficient homogeneous linear equation $a\,y'' + b\,y' + c\,y = 0$ with $a \neq 0$ can be solved according to three cases, depending on the sign of the discriminant $D = b^2 - 4a\,c$.

- In the case that D is positive, a fundamental set of solutions consists of

$$y_1(t) = e^{r_1 t} \qquad \text{and} \qquad y_2(t) = e^{r_2 t},$$

where the real numbers $r_1 \neq r_2$ are the roots of the associated characteristic equation $a\,r^2 + b\,r + c = 0$. The general solution is

$$y(t) = C_1 e^{r_1 t} + C_2 e^{r_2 t}.$$

- In the case that $D = 0$, a fundamental set of solutions consists of

$$y_1(t) = e^{rt} \quad \text{and} \quad y_2(t) = t \, e^{rt},$$

  where $r = -b/2a$ is the unique root of the associated characteristic equation $a \, r^2 + b \, r + c = 0$. The general solution is

$$y(t) = C_1 e^{rt} + C_2 t \, e^{rt}.$$

- In the case that D is negative, a fundamental set of solutions consists of

$$y_1(t) = e^{\lambda t} \cos(\mu \, t) \quad \text{and} \quad y_2(t) = e^{\lambda t} \sin(\mu \, t),$$

  where $r = \lambda \pm i \, \mu$ are the complex conjugate roots of the associated characteristic equation $a \, r^2 + b \, r + c = 0$. The general solution is

$$y(t) = e^{\lambda t} \big(A \cos(\mu \, t) + B \sin(\mu \, t) \big).$$

Exercises for Section 9.1

Find a fundamental set of solutions and the general solution for the following constant-coefficient equations:

1. $y'' + 4y' + 3y = 0$

2. $2y'' + 4y' - 6y = 0$

3. $y'' - 6y' + 9y = 0$

4. $5y'' + 50y' + 250y = 0$

5. $y'' + 4y' + 5y = 0$

6. $5y'' - 20y' + 30y = 0$

7. $y'' + k \, y' = 0$

8. $y'' + k^2 y = 0$

9. $y'' - k^2 y = 0$

10. $y'' - 2k \, y' + k^2 y = 0$

Solve the following initial value problems:

11. $\begin{cases} y'' - 4y' + 3y = 0, \\ y(0) = 7, \quad y'(0) = 16 \end{cases}$

12. $\begin{cases} 2y'' + 4y' - 6y = 0, \\ y(0) = 4, \quad y'(0) = 0 \end{cases}$

13. $\begin{cases} y'' - 6y' + 9y = 0, \\ y(0) = 4, \ y'(0) = 17 \end{cases}$

17. $\begin{cases} y'' + k\,y' = 0, \\ y(0) = a, \ y'(0) = b \end{cases}$

14. $\begin{cases} 5y'' + 50y' + 250y = 0, \\ y(0) = 0, \ y'(0) = -5 \end{cases}$

18. $\begin{cases} y'' + k^2 y = 0, \\ y(0) = a, \ y'(0) = b \end{cases}$

15. $\begin{cases} y'' + 4y' + 5y = 0, \\ y(0) = 3, \ y'(0) = -2 \end{cases}$

19. $\begin{cases} y'' - k^2 y = 0, \\ y(0) = a, \ y'(0) = b \end{cases}$

16. $\begin{cases} 5y'' - 20y' + 30y = 0, \\ y(0) = 2, \ y'(0) = 4 \end{cases}$

20. $\begin{cases} y'' - 2k\,y' + k^2 y = 0, \\ y(0) = a, \ y'(0) = b \end{cases}$

9.2   Complex Constant-Coefficient Second-Order Homogeneous Equations

In our study of *real* second-order differential equations we have been forced to use complex numbers when the associated characteristic equation of $a\,y'' + b\,y' + c\,y = 0$ has complex roots. In the present section we study the differential equation $a\,y'' + b\,y' + c\,y = 0$ assuming that the coefficients a, b, and c are complex constants. This theory is important not only as a generalization, but also for gaining perspective and simplifying formulas in the real case. But first we need to recall some important facts about complex numbers.

9.2.1   Review of Complex Numbers

A **complex number** is, by definition, an expression of the form

$$a = \alpha + i\beta,$$

where α and β are real numbers. We call α the **real part** and β the **imaginary part** of a, and we write

$$\alpha = \mathfrak{Re}(a), \qquad\qquad \beta = \mathfrak{Im}(a).$$

We can think of a complex number as a point in the plane if we identify $\alpha + i\beta$ with (α, β).

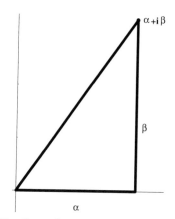

**The Cartesian representation of a
complex number**

The sum and product of complex numbers $a_1 = \alpha_1 + i\beta_1$ and $a_2 = \alpha_2 + i\beta_2$ are defined in terms of addition and multiplication of real numbers:

$$\left\{ \begin{array}{l} a_1 + a_2 = (\alpha_1 + \alpha_2) + i(\beta_1 + \beta_2), \\[2mm] a_1 a_2 = (\alpha_1\beta_1 - \alpha_2\beta_2) + i(\alpha_1\beta_2 + \alpha_2\beta_1). \end{array} \right. \tag{9.32}$$

Note in particular that the complex number i satisfies $i^2 = -1$. The complex number 0 is represented as $0 + i0$, and the complex number 1 by $1 + i0$. These function as the additive and multiplicative identity elements respectively, just as in the real number system. Addition and multiplication of complex numbers satisfy the commutative, associative, and distributive laws:

$$\left\{ \begin{array}{ll} a_1 + a_2 = a_2 + a_1, & a_1 a_2 = a_2 a_1, \\[2mm] a_1 + (a_2 + a_3) = (a_1 + a_2) + a_3, & a_1(a_2 a_3) = (a_1 a_2)a_3, \\[2mm] \multicolumn{2}{c}{a_1(a_2 + a_3) = a_1 a_2 + a_1 a_3 = (a_2 + a_3)a_1.} \end{array} \right. \tag{9.33}$$

Indeed, (9.33) follows from (9.32) and the fact that the real numbers satisfy the commutative, associative, and distributive laws.

Definition.   *The* **modulus** $|a|$ *of a complex number* $a = \alpha + i\beta$ *is defined by*

$$|a| = \sqrt{\alpha^2 + \beta^2}.$$

The **argument** $\mathrm{Arg}(a)$ *of a nonzero complex number* $a = \alpha + i\beta$ *is defined by*

$$\mathrm{Arg}(a) = \theta,$$

where θ is the unique real number satisfying $-\pi < \theta \le \pi$ and

$$\cos(\theta) = \frac{\alpha}{|a|}, \qquad \sin(\theta) = \frac{\beta}{|a|}.$$

$\mathrm{Arg}(0)$ *is undefined.*

It is easy to prove that

$$|a\,b| = |a||b| \qquad \text{and (when } \alpha \ne 0) \qquad \tan\big(\mathrm{Arg}(a)\big) = \frac{\beta}{\alpha}.$$

In *Mathematica*, $i = \sqrt{-1}$ is denoted by **I**. Operations such as addition and multiplication can be performed on complex numbers just as for real numbers. The modulus and argument are denoted in *Mathematica* by **Abs** and **Arg**. To check that *Mathematica* interprets $\mathrm{Arg}(-1)$ as π and not $-\pi$ we use

Arg[-1]

to get

```
Pi
```

Similarly, *Mathematica* uses **Re[a]** and **Im[a]** for the real and imaginary parts of a complex number **a**.

The **polar representation** of a complex number $a = \alpha + i\beta$ is obtained by writing $\alpha = r\cos(\theta)$ and $\beta = r\sin(\theta)$, leading to

$$a = r\big(\cos(\theta) + i\,\sin(\theta)\big).$$

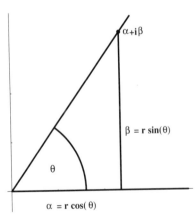

**The polar representation of a
complex number**

The polar representation gives a convenient method for multiplying two complex numbers. By using the trigonometric addition formulas

$$\cos(\theta_1 + \theta_2) = \cos(\theta_1) \cos(\theta_2) - \sin(\theta_1) \sin(\theta_2),$$

$$\sin(\theta_1 + \theta_2) = \sin(\theta_1) \cos(\theta_2) + \cos(\theta_1) \sin(\theta_2),$$

it follows that

$$\big(r_1(\cos(\theta_1) + i \sin(\theta_1))\big)\big(r_2(\cos(\theta_2) + i \sin(\theta_2))\big)$$

$$= r_1 r_2 \big(\cos(\theta_1) \cos(\theta_2) - \sin(\theta_1) \sin(\theta_2) + i (\cos(\theta_1) \sin(\theta_2) + \sin(\theta_1) \cos(\theta_2))\big)$$

$$= r_1 r_2 \big(\cos(\theta_1 + \theta_2) + i \sin(\theta_1 + \theta_2)\big).$$

Thus, multiplication of the complex numbers is effected by multiplying moduli and adding arguments modulo 2π.

Convergence of a series of complex numbers is defined as in the real-valued case. The series

$$\sum_{k=1}^{\infty} a_k \tag{9.34}$$

has the sum S if and only if the sequence of complex numbers $\{S_k\}$, where $S_k = a_1 + \cdots + a_k$, satisfies

$$\lim_{k \to \infty} S_k = S.$$

The test for absolute convergence states that if the terms of the series (9.34) satisfy a system of inequalities of the form $|a_k| \le b_k$ for a convergent series of positive *real* numbers $\{b_k\}$, then the series (9.34) is also convergent. Furthermore, we have the inequality

$$\left| \sum_{k=1}^{\infty} a_k \right| \le \sum_{k=1}^{\infty} b_k.$$

We need to define the exponential function $z \longmapsto e^z$, where z is a complex number. We want the definition to preserve four important properties of the real exponential function; these properties are:

$$e^{x+y} = e^x e^y, \tag{9.35}$$

$$e^x \text{ is never zero,} \tag{9.36}$$

$$e^x = \sum_{k=0}^{\infty} \frac{x^k}{k!} = 1 + x + \frac{x^2}{2} + \cdots, \tag{9.37}$$

$$\frac{d \, e^{ax}}{dx} = a \, e^{ax}. \tag{9.38}$$

Since the right-hand side of (9.37) makes sense whether or not x is a real number, we use it to define the **complex exponential function**:

$$e^z = \sum_{k=0}^{\infty} \frac{z^k}{k!}. \tag{9.39}$$

The terms of the series satisfy $|z^k| = |z|^k$, so we can appeal to the convergence of the real exponential series to deduce the convergence of the right-hand side of (9.39).

Lemma 9.5. *Let z and w be complex numbers. Then*

$$e^z e^w = e^{z+w}.$$

Proof. Multiplication of the series for e^z and e^w and the binomial theorem yield

$$e^z e^w = \left(\sum_{k=0}^{\infty} \frac{z^k}{k!}\right)\left(\sum_{\ell=0}^{\infty} \frac{w^\ell}{\ell!}\right) = \sum_{m=0}^{\infty} \left(\sum_{k+\ell=m} \left(\frac{z^k}{k!}\right)\left(\frac{w^\ell}{\ell!}\right)\right)$$

$$= \sum_{m=0}^{\infty} \left(\sum_{k+\ell=m} \left(\frac{(k+\ell)!}{k!\,\ell!}\right)\left(\frac{z^k w^\ell}{(k+\ell)!}\right)\right)$$

$$= \sum_{m=0}^{\infty} \frac{1}{m!}\left(\sum_{k+\ell=m} \left(\frac{(k+\ell)!}{k!\,\ell!}\right) z^k w^\ell\right) = \sum_{m=0}^{\infty} \frac{(z+w)^m}{m!} = e^{z+w}. \ \blacksquare$$

In the case that the complex number is purely imaginary, $z = i\,y$, we have

$$e^{iy} = \sum_{k=0}^{\infty} \frac{(i\,y)^k}{k!} = \sum_{m=0}^{\infty} \frac{(-1)^m y^{2m}}{(2m)!} + i \sum_{m=0}^{\infty} \frac{(-1)^m y^{2m+1}}{(2m+1)!}. \tag{9.40}$$

We recognize the first series on the right-hand side of (9.40) as the series for cos, and the second as the series for sin. Therefore,

$$e^{iy} = \cos(y) + i\sin(y). \tag{9.41}$$

Frequently, (9.41) is called **Euler's**[1] **formula.** From (9.35) and (9.41) we deduce

$$e^{x+iy} = e^x\big(\cos(y) + i\sin(y)\big) \tag{9.42}$$

---

[1]

Leonhard Euler (1707–1783). Born in Basel, Switzerland, Euler worked most of his life in Berlin and Saint Petersburg. Euler was the most prolific mathematician of all time. He contributed greatly to the evolution and systematization of analysis, in particular to the founding of the calculus of variations and the theories of differential equations, functions of complex variables, and special functions, and also he laid the foundations of number theory as a rigorous discipline. Moreover, he concerned himself with applications of mathematics to fields as diverse as lotteries, hydraulic systems, shipbuilding and navigation, actuarial science, demography, fluid mechanics, astronomy, and ballistics.

for the exponential of the complex number $z = x + iy$. Formula (9.42) can be used to reduce the proof of any result about complex exponentials to a computation involving only real exponentials and trigonometric functions. In particular, (9.36) and (9.38) follow from (9.42).

9.2.2   Using Complex Numbers to Solve Differential Equations

We can now return to the problem of solving $a y'' + b y' + c y = 0$ in the case that the associated characteristic equation has complex roots; we also allow the coefficients a, b and c to be complex. By the quadratic formula (which holds in the complex case as well as the real case) the roots of the associated characteristic equation $a r^2 + b r + c = 0$ are given by

$$r_1 = \frac{-b + \sqrt{b^2 - 4a c}}{2a} \qquad \text{and} \qquad r_2 = \frac{-b - \sqrt{b^2 - 4a c}}{2a}. \tag{9.43}$$

Furthermore, the Wronskian in the complex case is given by the same formula as in the real case on page 247.

In the case of distinct complex roots we have

Proposition 9.6.   *Let a, b, and c be complex constants. Suppose that the discriminant $D = b^2 - 4a c$ is nonzero. A fundamental set of solutions of the homogeneous differential equation $a y'' + b y' + c y = 0$ is provided by the functions $y_1 = e^{r_1 t}$ and $y_2 = e^{r_2 t}$, where r_1 and r_2 are given by (9.43). Furthermore, the general solution of $a y'' + b y' + c y = 0$ is*

$$y = C_1 e^{r_1 t} + C_2 e^{r_2 t}.$$

Similarly, in the case of repeated complex roots we have

Proposition 9.7.   *Let a, b and c be complex constants. If the discriminant $D = b^2 - 4a c$ is zero, then a fundamental set of solutions of the differential equation $a y'' + b y' + c y = 0$ with $a \neq 0$ is provided by the functions $y_1(t) = e^{r t}$ and $y_2(t) = t e^{r t}$, where $r = -b/2a$. The general solution is*

$$y = C_1 e^{r t} + C_2 t e^{r t}.$$

The proof of Proposition 9.6 is the same as that of Proposition 9.2, and the proof of Proposition 9.7 is the same as that of Proposition 9.3.

Example 9.8.   *Find a fundamental set of solutions for the differential equation*

$$y'' - (2 + i)y' + 2i y = 0, \tag{9.44}$$

and also determine the general solution.

Solution. The associated characteristic equation of (9.44) is

$$0 = r^2 - (2+i)r + 2i = (r-i)(r-2),$$

so that roots are $r_1 = i$ and $r_2 = 2$. A fundamental set of solutions of (9.44) is provided by the functions $y_1 = e^{it}$ and $y_2 = e^{2t}$. Furthermore, the general solution of (9.44) is

$$y = C_1 e^{it} + C_2 e^{2t},$$

where C_1 and C_2 are complex constants. ∎

An important reason for the introduction of complex numbers is that they can be used to solve real differential equations.

Example 9.9.   *Use complex numbers to find the general solution of the differential equation*

$$y'' + y' + 1.25y = 0. \tag{9.45}$$

Solution. The roots of the associated characteristic equation $r^2 + r + 1.25 = 0$ are

$$r_1 = -0.5 + i \qquad \text{and} \qquad r_2 = -0.5 - i,$$

and so the general solution of (9.45) can be written as

$$y = C_1 e^{(-0.5+i)t} + C_2 e^{(-0.5-i)t}. \tag{9.46}$$

It is important to realize that the constants in (9.46) are in general complex. In most applications we need to convert the right-hand side of (9.46) to a form that uses only real constants and real functions. To accomplish the conversion, we use Euler's formula (9.42). Thus

$$y = C_1 e^{(-0.5+i)t} + C_2 e^{(-0.5-i)t} = C_1 e^{-0.5t}\big(\cos(t) + i\sin(t)\big) \tag{9.47}$$

$$+ C_2 e^{-0.5t}\big(\cos(t) - i\sin(t)\big)$$

$$= e^{-0.5t}\big((C_1 + C_2)\cos(t) + i(C_1 - C_2)\sin(t)\big).$$

When we put $A = C_1 + C_2$ and $B = i(C_1 - C_2)$, we can rewrite (9.47) as

$$y = e^{-0.5t}\big(A\cos(t) + B\sin(t)\big), \tag{9.48}$$

which is the same result that we achieved in Example 9.6. ∎

We shall see in Chapter 12 that the use of complex numbers is essential for solving higher-order differential equations.

Exercises for Section 9.2

In Problems 1–6 find the general solution of the stated differential equation:

1. $y'' - 3y' + 3y = 0$

4. $y'' - (2 + 4i)y' - (3 - 4i)y = 0$

2. $y'' + 4y = 0$

5. $y'' - (3 + 2i)y' + 6i\ y = 0$

3. $y'' - 4y' + 8y = 0$

6. $y'' - (1 - 4i)y' - (3 + 3i)y = 0$

7. Use equation (9.42) to establish equations (9.36) and (9.38).

8. The five most important numbers in mathematics are perhaps 0, 1, i, e, and π, and the three most important operations are probably addition, multiplication, and exponentiation. Show that are they are all related by the formula

$$e^{i\pi} + 1 = 0.$$

9. Suppose that a, b, and c are real numbers and that z is a complex solution of $a\,y'' + b\,y' + c\,y = 0$. Show that the real and imaginary parts of z are also solutions of $a\,y'' + b\,y' + c\,y = 0$.

9.3   The Method of Undetermined Coefficients

In this section and in Section 9.4 we find the general solution of the inhomogeneous equation with constant-coefficients

$$L[y] = a\,y'' + b\,y' + c\,y = g(t), \tag{9.49}$$

where a, b, c are real constants with $a \neq 0$, and $g(t)$ is a function of t, called a **forcing function**. In this generality it is always possible to find a particular solution of (9.49) by the method of **variation of parameters**, to be treated in Section 9.4. However, if $g(t)$ is a product of polynomials, exponential functions, and trigonometric functions, or a linear combination of such functions, we can find a particular solution of (9.49) that is an elementary function of the same type by the method of **undetermined coefficients**, which we discuss in the present section. Here is a comparison of the merits and demerits of each method.

- **Undetermined coefficients**

    Advantage: relatively easy to use in hand calculation.

    Disadvantage: does not always work.

- **Variation of parameters**

    Advantage: always works.

    Disadvantage: relatively difficult to use in hand calculation.

Some Basic Examples

The method of undetermined coefficients consists in making an initial assumption about the form of a particular solution y_P of a differential equation, but with coefficients left unspecified—hence the name *undetermined coefficients*. The proposed particular solution y_P is then substituted into the differential equation; we must attempt to determine the coefficients such that y_P satisfies the equation. Eventually, we shall describe the method of undetermined coefficients in full generality, but first we begin with some simple examples.

Example 9.10. *Find a particular solution of*

$$y'' - 5y' + 6y = 2e^{4t}, \tag{9.50}$$

and also determine the general solution.

Solution. Since the right-hand side of (9.50) contains the factor e^{4t}, we guess that a particular solution has the form $y_P \equiv A\,e^{4t}$, where A is a constant. Since

$$y_P' = 4A\,e^{4t} \qquad \text{and} \qquad y_P'' = 16A\,e^{4t},$$

it follows from (9.50) that

$$2e^{4t} = y_P'' - 5y_P' + 6y_P = (16A - 20A + 6A)e^{4t} = 2Ae^{4t}.$$

Thus, $A = 1$, so that a particular solution is given by

$$y_P = e^{4t};$$

furthermore, the general solution to (9.50) is $y = C_1 e^{2t} + C_2 e^{3t} + e^{4t}$. ∎
   The next example introduces a substantial complication.

Example 9.11. *Find a particular solution of*

$$y'' - 5y' + 6y = 7e^{2t}, \tag{9.51}$$

and also determine the general solution.

Solution. Our first guess $y_P = A\,e^{2t}$ does not work because substitution into (9.51) results in

$$7e^{2t} = (4Ae^{2t} - 10Ae^{2t} + 6Ae^{2t}) = 0,$$

which is impossible. In fact, we should have known that $A\,e^{2t}$ could not be a solution of the inhomogeneous equation, because it is already a solution of the homogeneous equation.
   Our experience with the repeated-root case in Section 9.1.2 suggests that we try a particular solution of the form $y_P = A\,t\,e^{2t}$. We compute

$$y_P' = A(e^{2t} + 2t\,e^{2t}) \qquad \text{and} \qquad y_P'' = A(4\,e^{2t} + 4t\,e^{2t}). \tag{9.52}$$

We substitute (9.52) into (9.51) and get

$$7e^{2t} = A(4e^{2t} + 4t\,e^{2t}) - 5A(e^{2t} + 2t\,e^{2t}) + 6A\,t\,e^{2t}$$

$$= (4A - 5A)e^{2t} + (4A - 10A + 6A)t\,e^{2t} = -A\,e^{2t}.$$

Thus, $A = -7$, yielding the particular solution $y_\mathrm{P} = -7t\,e^{2t}$ and the general solution $y = C_1 e^{2t} + C_2 e^{3t} - 7t\,e^{2t}$. ∎

Next, we consider the case of a trigonometric forcing function.

Example 9.12. *Find a particular solution of*

$$y'' - 5y' + 6y = \cos(2t), \tag{9.53}$$

and also determine the general solution.

Solution. For a particular solution we try $y_\mathrm{P} = A\cos(2t) + B\sin(2t)$. We compute

$$y_\mathrm{P}' = -2A\sin(2t) + 2B\cos(2t) \qquad \text{and} \qquad y_\mathrm{P}'' = -4A\cos(2t) - 4B\sin(2t).$$

Hence

$$\cos(2t) = y_\mathrm{P}'' - 5y_\mathrm{P}' + 6y_\mathrm{P} \tag{9.54}$$

$$= -4A\cos(2t) - 4B\sin(2t) - 5\big(-2A\sin(2t) + 2B\cos(2t)\big)$$

$$+ 6\big(A\cos(2t) + B\sin(2t)\big)$$

$$= (2A - 10B)\cos(2t) + (10A + 2B)\sin(2t).$$

Since (9.54) holds for all t, we must have

$$\begin{cases} 2A - 10B = 1, \\ 10A + 2B = 0. \end{cases} \tag{9.55}$$

The solution of (9.55) is

$$A = \frac{1}{52} \qquad \text{and} \qquad B = -\frac{5}{52}.$$

Hence a particular solution of (9.53) is

$$y_\mathrm{P} = \frac{\cos(2t)}{52} - \frac{5\sin(2t)}{52},$$

and the general solution is

$$y = C_1 e^{2t} + C_2 e^{3t} + \frac{\cos(2t)}{52} - \frac{5\sin(2t)}{52}. \quad \blacksquare$$

If the right-hand side of an inhomogeneous differential equation has several terms, then particular solutions corresponding to the individual terms can be found and added together to form the particular solution for the differential equation. This follows from the fact that a linear operator L has the properties (8.19); thus, if y_1 satisfies $L[y_1] = g_1(t)$ and y_2 satisfies $L[y_2] = g_2(t)$, then the equation $L[y] = g_1(t) + g_2(t)$ is solved by taking $y(t) = y_1(t) + y_2(t)$. This allows us to consider the terms of the forcing function one by one and then to add the particular solutions corresponding to the individual terms. The following example illustrates the procedure.

Example 9.13.   *Find a particular solution of*

$$y'' - 5y' + 6y = 7e^{2t} + \cos(2t). \tag{9.56}$$

Solution. From Example 9.11 we know that a particular solution to $y'' - 5y' + 6y = 7e^{2t}$ is $y_{P1} = -7t\,e^{2t}$, and from Example 9.12 we know that a particular solution of $y'' - 5y' + 6y = \cos(2t)$ is

$$y_{P2} = \frac{\cos(2t)}{52} - \frac{5\sin(2t)}{52}.$$

It is easily checked that the sum

$$y_{P1} + y_{P2} = -7t\,e^{2t} + \frac{\cos(2t)}{52} - \frac{5\sin(2t)}{52}.$$

is a particular solution of (9.56). $\blacksquare$

When solving an intial value problem involving an inhomogeneous differential equation, it is important to evaluate the constants of integration only after the general solution to the inhomogeneous equation has been found.

Example 9.14.   *Solve the initial value problem*

$$\begin{cases} y'' - 5y' + 6y = 7e^{2t} + \cos(2t), \\ y(0) = y'(0) = -2. \end{cases} \tag{9.57}$$

Solution. From Example 9.13 we know that the general solution to $y'' - 5y' + 6y = 7e^{2t} + \cos(2t)$ is

$$y = C_1 e^{2t} + C_2 e^{3t} - 7t\,e^{2t} + \frac{\cos(2t)}{52} - \frac{5\sin(2t)}{52}. \tag{9.58}$$

Hence
$$y' = 2C_1e^{2t} + 3C_2e^{3t} - 7e^{2t} - 14t\,e^{2t} - \frac{5\cos(2t)}{26} - \frac{\sin(2t)}{26}. \qquad (9.59)$$

From (9.58) and (9.59) we get
$$\begin{cases} -2 = C_1 + C_2 + 1/52, \\ -2 = 2C_1 + 3C_2 - 7 - 5/26. \end{cases}$$

The solution of this system is $C_1 = -45/4$ and $C_2 = 120/13$. Hence the solution to (9.57) is

$$y = \frac{-45e^{2t}}{4} + \frac{120e^{3t}}{13} - 7t\,e^{2t} + \frac{\cos(2t)}{52} - \frac{5\sin(2t)}{52}. \quad\blacksquare$$

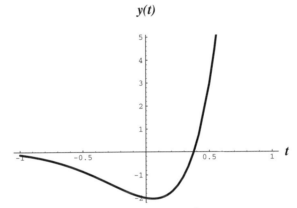

y(t)

**The solution of $y'' - 5y' + 6y = 7e^{2t} + \cos(2t)$ with the
initial conditions $y(0) = y'(0) = -2$**

Having introduced these examples, we now pass to a general description of the method of undetermined coefficients to solve equations of the form (9.49).

Purely Exponential Right-Hand Side

To find a particular solution to $L[y] = a\,y'' + b\,y' + c\,y = e^{\alpha t}$, we first try $y_P = A\,e^{\alpha t}$. Computing the necessary derivatives, we find that

$$e^{\alpha t} = L[y_P] = A\big(a(\alpha^2 e^{\alpha t}) + b(\alpha e^{\alpha t}) + c(e^{\alpha t})\big) = A(a\,\alpha^2 + b\,\alpha + c)e^{\alpha t},$$

or
$$1 = A(a\,\alpha^2 + b\,\alpha + c). \qquad (9.60)$$

This leads us to consider three subcases.

Nonresonant case: $a\alpha^2 + b\alpha + c \neq 0$

When $a\alpha^2 + b\alpha + c \neq 0$, we can solve (9.60) for A and obtain a particular solution as

$$y_P = \frac{e^{\alpha t}}{a\alpha^2 + b\alpha + c}.$$

This is illustrated by Example 9.10.

In order to extend the method of Example 9.11 we make the following general computation

$$
\begin{aligned}
L[t\,e^{\alpha t}] &= a(\alpha^2 t + 2\alpha)e^{\alpha t} + b(1 + \alpha\,t)e^{\alpha t} + c\,t\,e^{\alpha t} \\
&= (a\alpha^2 + b\alpha + c)t\,e^{\alpha t} + (2a\alpha + b)e^{\alpha t}.
\end{aligned}
\tag{9.61}
$$

First resonant case: $a\alpha^2 + b\alpha + c = 0$ and $2a\alpha + b \neq 0$

In this case it follows from (9.61) that a particular solution to $a\,y'' + b\,y' + c\,y = e^{\alpha t}$ is obtained as

$$y_P = \frac{t\,e^{\alpha t}}{2a\alpha + b}.$$

For the final case we first note that

$$L[t^2 e^{\alpha t}] = (a\alpha^2 + b\alpha + c)t^2 e^{\alpha t} + 2(2a\alpha + b)t\,e^{\alpha t} + 2a\,e^{\alpha t}.
\tag{9.62}$$

Second resonant case: $a\alpha^2 + b\alpha + c = 0$ and $2a\alpha + b = 0$

In this case we still have $a \neq 0$, since we are dealing with a second-order equation. Therefore, using (9.61) we obtain a particular solution by solving $2a\,A = 1$, leading to the particular solution

$$y_P(t) = \frac{t^2 e^{\alpha t}}{2a}.$$

Purely Trigonometric Right-Hand Side

We next consider the case when the function $g(t)$ is a combination of sine and cosine functions. To solve either of the equations

$$L[y] = a\,y'' + b\,y' + c\,y = \cos(\beta t) \quad \text{or} \quad L[y] = a\,y'' + b\,y' + c\,y = \sin(\beta t) \tag{9.63}$$

with $\beta \neq 0$, we first look for a particular solution in the form

$$y_P = A\cos(\beta t) + B\sin(\beta t).$$

The derivatives of y_P are given by

$$y_P' = -A\beta \sin(\beta t) + B\beta \cos(\beta t) \qquad \text{and} \qquad y_P'' = -A\beta^2 \cos(\beta t) - B\beta^2 \sin(\beta t).$$

Hence

$$L[y_P] = \left(A(-a\beta^2 + c) + Bb\beta\right)\cos(\beta t) + \left(B(-a\beta^2 + c) - Ab\beta\right)\sin(\beta t).$$

In order for $L[y_P] = \cos(\beta t)$ to hold, the following system of two simultaneous linear equations must be solved for A and B.

$$\begin{cases} A(c - a\beta^2) + B(\beta b) = 1, \\ A(-\beta b) + B(c - a\beta^2) = 0. \end{cases} \qquad (9.64)$$

The determinant of the system (9.64) is $D = (c - a\beta^2)^2 + (\beta b)^2$. Therefore, we must consider two cases, depending on whether or not D is zero. In the first case $(D \neq 0)$ the complex number $i\beta$ is *not* a root of the associated quadratic equation $ar^2 + br + c = 0$. This is called the

Nonresonant case: either $c - a\beta^2 \neq 0$ or $b \neq 0$

In this case at least one of the terms in D is nonzero, so that the determinant D is non-zero, and we can solve the system of equations (9.64) by Cramer's rule (see page 248). Thus,

$$A = \frac{(c - a\beta^2)}{D} \qquad \text{and} \qquad B = \frac{\beta b}{D}.$$

For this choice of constants we have $L[A\cos(\beta t) + B\sin(\beta t)] = \cos(\beta t)$.

Example 9.15. *Find a particular solution of the equation* $y'' + 4y' + 3y = \cos(4t)$.

Solution. We have

$$D = (c - a\beta^2)^2 + (\beta b)^2 = (3 - 4^2)^2 + (4 \times 4)^2 = 169 + 256 = 425 \neq 0,$$

so that we are in the nonresonant case, and the particular solution is of the form

$$y_P(t) = A\cos(4t) + B\sin(4t),$$

where the coefficients A and B are determined as above by

$$A = \frac{(3 - 4^2)}{425} = -\frac{13}{425} \qquad \text{and} \qquad B = \frac{4 \times 4}{425} = \frac{16}{425}.$$

Thus $y_P(t) = -\dfrac{13\cos(4t)}{425} + \dfrac{16\sin(4t)}{425}.$ ∎

The analysis is entirely similar in the case that the right-hand side of the equation is the function $\sin(\beta t)$; the only change necessary is that the numbers 1 and 0 are interchanged in the right-hand side of the equations for A and B. The formulas for the coefficients in this case are

$$A = -\frac{\beta b}{D} \quad \text{and} \quad B = \frac{(c - a\beta^2)}{D}.$$

For this choice of constants we have $L[A \cos(\beta t) + B \sin(\beta t)] = \sin(\beta t)$.

Finally, we consider the case in which the determinant $D = 0$; equivalently, the complex number $r = i\beta$ is a root of the associated characteristic equation $a r^2 + br + c = 0$. This is the

Resonant case: both $c - a\beta^2 = 0$ and $b = 0$

In this case the differential equation (9.63) has a simpler form, since

$$L[y] = a y'' + b y' + c y = a y'' + 0 \times y' + a\beta^2 y = a(y'' + \beta^2 y).$$

By analogy with the case of purely exponential right-hand side, we look for a particular solution of the form

$$y_P = t\big(A \cos(\beta t) + B \sin(\beta t)\big).$$

Making use of the formula $(f\, g)'' = f\, g'' + 2f'g' + f''g$, we have

$$y_P'' = t\big(-\beta^2 A \cos(\beta t) - \beta^2 B \sin(\beta t)\big) + 2\big(-A\beta \sin(\beta t) + B\beta \cos(\beta t)\big),$$

so that

$$L[y] = a\big(y_P'' + \beta^2 y_P\big) = 2a\left(-A\beta \sin(\beta t) + B\beta \cos(\beta t)\right).$$

To solve $L[y] = \cos(\beta t)$, we must take $A = 0$ and $2a\, B\, \beta = 1$, obtaining the solution

$$y_P = \frac{t \sin(\beta t)}{2a\, \beta}.$$

Similarly, to solve $L[y] = \sin(\beta t)$, we must take $B = 0$ and $-2a\, A\, \beta = 1$ leading to the particular solution

$$y_P = -\frac{t \cos(\beta t)}{2a\, \beta}.$$

This completes the analysis of the purely trigonometric right-hand side.

Example 9.16. *Find a particular solution of the equation* $y'' + 4y = \cos(2t)$.

Solution. In this case the right-hand side is already a solution of the homogeneous equation $y'' + 4y = 0$, and we look for a particular solution in the form $y_P = t\big(A \cos(2t) + B \sin(2t)\big)$. The above method produces the particular solution

$$y_P = \frac{t \sin(2t)}{4}. \quad \blacksquare$$

The General Case

It is helpful to summarize the method of undetermined coefficients as a set of procedures that can be followed to produce a particular solution of the differential equation

$$a\,y'' + b\,y' + c\,y = \begin{cases} t^n e^{\alpha t} \cos(\beta\,t), \\[2mm] t^n e^{\alpha t} \sin(\beta\,t). \end{cases} \tag{9.65}$$

1. First, decide whether the complex number $\alpha + i\beta$ is a solution of the associated characteristic equation $a\,r^2 + b\,r + c = 0$.

2. If not, then a particular solution y_P can be found in the form

$$y_P = \big(A_n(t)\cos(\beta\,t) + B_n(t)\sin(\beta\,t)\big)e^{\alpha t},$$

   where $A_n(t)$ and $B_n(t)$ are polynomials of degree n. This is the **nonresonant case**.

3. If $r = \alpha + i\beta$ is a solution of $a\,r^2 + b\,r + c = 0$ and $2a\,r + b \neq 0$, then a particular solution y_P can be found in the form

$$y_P = t\big(A_n(t)\cos(\beta\,t) + B_n(t)\sin(\beta\,t)\big)e^{\alpha t},$$

   where $A_n(t)$ and $B_n(t)$ are polynomials of degree n. This is the **first resonant case**.

4. If $r = \alpha + i\beta$ is a solution of $a\,r^2 + b\,r + c = 0$ and $2a\,r + b = 0$, then r is real and a particular solution y_P can be found in the form

$$y_P = t^2 A_n(t)\,e^{\alpha t},$$

   where $A_n(t)$ is a polynomial of degree n. This is the **second resonant case**.

5. In each case the coefficients of the polynomials $A_n(t)$ and $B_n(t)$ can be obtained by solving suitable systems of simultaneous linear equations.

6. If the right-hand side of (9.65) consists of several terms of the form $t^n e^{\alpha t}\cos(\beta\,t)$ and $t^n e^{\alpha t}\sin(\beta\,t)$, then the above procedures can be applied to each term to obtain a particular solution of the required form. The required solution is then obtained by adding the particular solutions thus, obtained.

In the analysis summarized above, the concept of resonance is applied more generally than in the above examples. The right-hand side is said to be in **resonance** precisely when the complex number $r = \alpha + i\beta$ is a root of the associated characteristic equation $ar^2 + br + c = 0$. In Section 11.2 we shall show that this mathematical definition of resonance corresponds to the physical definition of resonance.

The method of undetermined coefficients may require a fair amount of experimentation. If one assumes too little for the particular solution $y_P(t)$, then a contradiction is reached. But the contradiction usually indicates how $y_P(t)$ should be modified. On the other hand, if one assumes that $y_P(t)$ is too general, unnecessary work must be carried out, but at least the correct answer is obtained.

Summary of Techniques Introduced

The method of undetermined coefficients is a convenient method for finding a particular solution to $a\,y'' + b\,y' + c\,y = g(t)$ in the case that $g(t)$ is a sum of terms each of which contains only polynomials, exponentials, sin, and cos.

Exercises for Section 9.3

In Exercises 1–10 use the method of undetermined coefficients to find a particular solution of the indicated equation.

1. $y'' + 3y' + 2y = e^{-5t}$ **6.** $y'' + 3y' + 2y = t^2$

2. $y'' + 3y' - 10y = e^{-5t}$ **7.** $y'' + 3y' + 2y = t^2 e^t$

3. $y'' + 4y' + 4y = e^{-2t}$ **8.** $y'' + 3y' + 2y = t^2 e^{-t}$

4. $y'' + 4y' + 4y = \cos(2t)$ **9.** $y'' + 4y' + 2y = (\sin(t))^2$

5. $y'' + 4y = \cos(2t)$ **10.** $y'' + 6y' + 3y = t\,e^t \cos(2t)$

In Exercises 11–20 solve the stated initial value problem

11. $\begin{cases} y'' + 3y' + 2y = e^{-5t}, \\ y(0) = 0,\, y'(0) = 0 \end{cases}$ **13.** $\begin{cases} y'' + 4y' + 4y = e^{-2t}, \\ y(0) = 2,\, y'(0) = 0 \end{cases}$

12. $\begin{cases} y'' + 3y' - 10y = e^{-5t}, \\ y(0) = 1,\, y'(0) = 2 \end{cases}$ **14.** $\begin{cases} y'' + 4y' + 4y = \cos(2t), \\ y(0) = 3,\, y'(0) = 0 \end{cases}$

15. $\begin{cases} y'' + 4y = \cos(2t), \\ y(0) = 2,\ y'(0) = 5 \end{cases}$

18. $\begin{cases} y'' + 3y' + 2y = t^2 e^{-t}, \\ y(0) = 5,\ y'(0) = 2 \end{cases}$

16. $\begin{cases} y'' + 3y' + 2y = t^2, \\ y(0) = 4,\ y'(0) = 2 \end{cases}$

19. $\begin{cases} y'' + 4y' + 2y = \sin(t)^2, \\ y(0) = 3,\ y'(0) = 1 \end{cases}$

17. $\begin{cases} y'' + 3y' + 2y = t^2 e^t, \\ y(0) = 8,\ y'(0) = 2 \end{cases}$

20. $\begin{cases} y'' + 6y' + 3y = t\,e^t \cos(2t), \\ y(0) = 2,\ y'(0) = 1 \end{cases}$

In Exercises 21–30 determine a suitable form for a particular solution of the indicated equation. Do not evaluate the constants.

21. $y'' + 4y' = 3t^2 + t^4 e^{-4t} + \sin(2t)$ **22.** $y'' + 2y = 2t + 3t \sin(4t)$

23. $y'' + 5y' + 6y = e^{-t} \cos(2t) + 6e^{2t} t^2 \sin(t)$

24. $y'' + 2y' + 2y = 3e^{-t} + 2e^{-t} \cos(t) + 4e^{-t} t^2 \sin(t)$

25. $y'' + 2y' + 5y = 4t\,e^{-t} \cos(2t) + 2t\,e^{-2t} \cos(t)$

26. $y'' + 9y = t^2 \sin(3t) + (3t + 4) \cos(3t)$

27. $y'' + 6y' + 5y = (t + 1)e^t \sin(2t) + 3t^2 e^{5t} + 2t^3 e^{-5t}$

28. $y'' + 4y' + 5y = 3t^2 e^{-2t} \cos(t) + 4t\,e^{2t} \sin(t)$

29. $y'' + 6y' + 8y = 4e^{-2t} + 3t\,e^{-3t} + 2t^2 e^{-4t} + t^3 e^{-5t}$

30. $y'' + 4y' + 4y = t^2 e^{-2t} + 4 + t^2 e^{4t}$

31. Suppose that $c \neq 0$. Show that there is a solution of the differential equation $a\,y'' + b\,y' + c\,y = t^n$ of the form

$$y_P(t) = P_n(t) = A_0 + A_1 t + A_2 t^2 + \cdots + A_n t^n.$$

32. Suppose that $b \neq 0$. Show that there is a solution of the differential equation $a\,y'' + b\,y' = t^n$ of the form

$$y_P(t) = t\,P_n(t) = t\left(A_0 + A_1 t + A_2 t^2 + \cdots + A_n t^n\right).$$

33. Suppose that $a \neq 0$. Show that there is a solution of the differential equation $a\, y'' = t^n$ of the form

$$y_{\mathrm{P}} = A_{n+2} t^{n+2}.$$

34. Suppose that $r = \alpha + i\beta$ is a complex number for which $a\, r^2 + b\, r + c \neq 0$. Show that there is a solution of the differential equation

$$a\, y'' + b\, y' + c\, y = t^n\, e^{\alpha t}\, \cos(\beta\, t) \qquad (9.66)$$

of the form

$$y(t) = e^{\alpha t}\big(P_n(t)\cos(\beta t) + Q_n(t)\sin(\beta t)\big),$$

where $P_n(t)$ and $Q_n(t)$ are polynomials of degree n.

35. Suppose that $r = \alpha + i\beta$ is a complex number for which $a\, r^2 + b\, r + c = 0$ and $2a\, r + b \neq 0$. Show that there is a solution of (9.66) of the form

$$y(t) = t\, e^{\alpha t}\big(P_n(t)\cos(\beta t) + Q_n(t)\sin(\beta t)\big),$$

where $P_n(t)$ and $Q_n(t)$ are polynomials of degree n.

36. Suppose that $r = \alpha + i\beta$ is a complex number for which $a\, r^2 + b\, r + c = 0$ and $2a\, r + b = 0$. Show that there is a solution of (9.66) of the form

$$y(t) = t^2\, P_n(t) e^{\alpha t},$$

where $P_n(t)$ is a polynomial of degree n.

9.4   The Method of Variation of Parameters

In Section 9.3 we found an explicit particular solution of the linear second-order inhomogeneous equation $a\, y'' + b\, y' + c\, y = g(t)$ in the case that $g(t)$ is of a special form. In the present section we develop a method that treats the case of a general functional form of $g(t)$ by representing a particular solution as an *integral* that involves $g(t)$ and other functions that depend only on the solutions of the associated homogeneous differential equation $a\, y'' + b\, y' + c\, y = 0$. For the purposes of mathematical generality and for use in various applications, we develop the method in the case of a general second-order linear equation, written in the leading-coefficient-unity-form

$$y'' + p(t)y' + q(t)y = g(t). \qquad (9.67)$$

9.4.1   Formulation of the Method

We solve (9.67) in terms of a fundamental set of solutions of the homogeneous equation

$$y'' + p(t)y' + q(t)y = 0. \tag{9.68}$$

Let the fundamental set of solutions of (9.68) be denoted by $\{y_1, y_2\}$. Since the general solution of (9.68) is

$$y = C_1 y_1 + C_2 y_2, \tag{9.69}$$

a good guess for a particular solution of (9.67) is

$$y_P(t) = u_1(t)y_1(t) + u_2(t)y_2(t), \tag{9.70}$$

where the constants C_1 and C_2 have been replaced by possibly nonconstant functions $u_1(t)$ and $u_2(t)$. This method is called the method of **variation of parameters** because the constants C_1 and C_2 in (9.69) are allowed to vary. The first step in finding u_1 and u_2 is to compute the derivative of y_P:

$$y_P' = u_1 y_1' + u_1' y_1 + u_2 y_2' + u_2' y_2, \tag{9.71}$$

a sum of four terms. Clearly, the second derivative y_P'' involves eight terms in general. In order to simplify this expression, we require as a first guess that the terms involving the derivatives $u_1'(t)$ and $u_2'(t)$ vanish. Thus we set

$$u_1' y_1 + u_2' y_2 = 0. \tag{9.72}$$

Then (9.72) simplifies (9.71) to

$$y_P' = u_1 y_1' + u_2 y_2'. \tag{9.73}$$

Using (9.73) we compute

$$y_P'' = u_1 y_1'' + u_1' y_1' + u_2 y_2'' + u_2' y_2'. \tag{9.74}$$

From (9.70), (9.73), and (9.74) it follows that

$$
\begin{aligned}
y_P'' + p(t)y_P' + q(t)y_P &= u_1 y_1'' + u_1' y_1' + u_2 y_2'' + u_2' y_2' \\
&\quad + p(t)\bigl(u_1 y_1' + u_2 y_2'\bigr) + q(t)\bigl(u_1 y_1 + u_2 y_2\bigr).
\end{aligned} \tag{9.75}
$$

We rearrange the terms on the right-hand side of (9.75) to obtain

$$
\begin{aligned}
y_P'' + p(t)y_P' + q(t)y_P &= u_1\bigl(y_1'' + p(t)y_1' + q(t)y_1\bigr) + u_1' y_1' \\
&\quad + u_2\bigl(y_2'' + p(t)y_2' + q(t)y_2\bigr) + u_2' y_2'.
\end{aligned} \tag{9.76}
$$

Since y_1 and y_2 both satisfy the homogeneous equation (9.68), we simplify (9.76) to

$$y_P'' + p(t)y_P' + q(t)y_P = u_1'y_1' + u_2'y_2'. \tag{9.77}$$

Now let us suppose that y_P satisfies the inhomogeneous equation (9.67). From (9.72) and (9.77) we obtain a system of two linear algebraic equations in the unknowns u_1' and u_2', namely

$$\begin{cases} u_1'y_1 + u_2'y_2 = 0, \\ u_1'y_1' + u_2'y_2' = g(t). \end{cases} \tag{9.78}$$

Notice that the determinant of the coefficients on the right-hand side of (9.78) is precisely the Wronskian $W(y_1, y_2)(t)$. We can solve (9.78) by Cramer's rule (see page 248). Thus

$$u_1' = \frac{\det\begin{pmatrix} 0 & y_2 \\ g(t) & y_2' \end{pmatrix}}{\det\begin{pmatrix} y_1 & y_2 \\ y_1' & y_2' \end{pmatrix}} = \frac{-y_2 g(t)}{W(y_1, y_2)(t)}, \tag{9.79}$$

and

$$u_2' = \frac{\det\begin{pmatrix} y_1 & 0 \\ y_1' & g(t) \end{pmatrix}}{\det\begin{pmatrix} y_1 & y_2 \\ y_1' & y_2' \end{pmatrix}} = \frac{y_1 g(t)}{W(y_1, y_2)(t)}. \tag{9.80}$$

We integrate (9.79) and (9.80) to get

$$\begin{cases} u_1(t) = - \int \dfrac{y_2(s)g(s)}{W(y_1, y_2)(s)} ds \Big|_{s\to t}, \\[2mm] u_2(t) = \int \dfrac{y_1(s)g(s)}{W(y_1, y_2)(s)} ds \Big|_{s\to t}. \end{cases} \tag{9.81}$$

From (9.70) and (9.81) it follows that

$$y_P(t) = y_1(t)u_1(t) + y_2(t)u_2(t)$$

$$= -y_1(t) \int \frac{y_2(s)g(s)}{W(y_1, y_2)(s)} ds \Big|_{s\to t} + y_2(t) \int \frac{y_1(s)g(s)}{W(y_1, y_2)(s)} ds \Big|_{s\to t}. \tag{9.82}$$

We combine the two integrals on the right-hand side of (9.82) to write

$$y_P(t) = \int \frac{y_1(s)y_2(t) - y_1(t)y_2(s)}{W(y_1, y_2)(s)} g(s) ds \Big|_{s\to t}. \tag{9.83}$$

Let us summarize these results as a theorem.

Theorem 9.8. **(Variation of parameters)** *If $\{y_1, y_2\}$ is a fundamental set of solutions of the homogeneous equation $y'' + p(t)y' + q(t)y = 0$, then a particular solution y_P to the inhomogeneous equation $y'' + p(t)y' + q(t)y = g(t)$ is given by (9.82) or (9.83).*

Example 9.17. *Find a particular solution of the differential equation*

$$y'' - y = e^t \tag{9.84}$$

both by the method of variation of parameters and by the method of undetermined coefficients, and compare the results.

Solution. $\{e^t, e^{-t}\}$ is a fundamental set of solutions of the homogeneous equation $y'' - y = 0$; the Wronskian is

$$W = \det\begin{pmatrix} e^t & e^{-t} \\ e^t & -e^{-t} \end{pmatrix} = -2.$$

Therefore a particular solution of (9.84) is given by

$$y_P(t) = \int \left(\frac{e^s e^{-t} - e^{-s} e^t}{-2} \right) e^s \, ds \bigg|_{s \to t} = \int \left(\frac{e^{2s} e^{-t} - e^t}{-2} \right) ds \bigg|_{s \to t}$$

$$= \left(-\frac{e^{2s} e^{-t}}{4} + \frac{e^t s}{2} \right) \bigg|_{s \to t} = -\frac{e^t}{4} + \frac{t\, e^t}{2}.$$

Hence the general solution of (9.84) found by the method of undetermined coefficients is

$$y_P = C_1 e^t + C_2 e^{-t} - \frac{e^t}{4} + \frac{t\, e^t}{2}, \tag{9.85}$$

where C_1 and C_2 are constants.

Now let us solve (9.84) by the method of undetermined coefficients. Since e^t is already a solution of the homogeneous equation $y'' - y = 0$, we assume a particular solution of (9.84) is of the form $y_P(t) = A\,t\,e^t$. Then $y_P''(t) = A(t + 2)e^t$, so when we substitute $y_P(t)$ into (9.84) we get

$$e^t = A(t + 2)e^t - A\,t\,e^t = 2A\,e^t;$$

thus, $A = 1/2$. Therefore, $y_P(t) = t\,e^t/2$, and the general solution of (9.84) found by the method of variation of parameters is

$$y_P = A_1 e^t + A_2 e^{-t} + \frac{t\, e^t}{2}, \tag{9.86}$$

where A_1 and A_2 are constants.

At first glance, (9.85) and (9.86) seem to be different. But the first and third terms on the right-hand side of (9.85) can be combined, yielding

$$y_P = \left(C_1 - \frac{1}{4}\right)e^t + C_2 e^{-t} + \frac{t\,e^t}{2}. \tag{9.87}$$

By taking $A_1 = C_1 - 1/4$ and $A_2 = C_2$, we see that the general solutions (9.85) produced by the two methods are the same. ∎

Thus, whenever the method of variation of parameters is used, one should check whether parts of the homogeneous and particular solutions can be combined.

It was easier to solve (9.84) using undetermined coefficients than it was using variation of parameters. But the method of variation of parameters can yield a solution in cases when the method of undetermined coefficients does not work.

Example 9.18.   *Find the general solution of*

$$y'' + y = \cot(t). \tag{9.88}$$

Solution. We cannot solve (9.88) by the method of undetermined coefficients because it is not of the form (9.65). To solve (9.88) by the method of variation of parameters, we first note that $\{\cos(t), \sin(t)\}$ is a fundamental set of solutions of the homogeneous equation $y'' + y = 0$; the Wronskian is

$$W = \det\begin{pmatrix} \cos(t) & \sin(t) \\ -\sin(t) & \cos(t) \end{pmatrix} = 1.$$

We use (9.83) to obtain

$$
\begin{aligned}
y_P(t) &= \int \Big(\cos(s)\sin(t) - \sin(s)\cos(t)\Big)\cot(s)ds\Big|_{s\to t} \\[2mm]
&= \int \left(\frac{(\cos(s))^2}{\sin(s)}\right)\sin(t) - \cos(s)\cos(t)\Big)ds\Big|_{s\to t} \\[2mm]
&= \int \left(\left(\frac{1}{\sin(s)} - \sin(s)\right)\sin(t) - \cos(s)\cos(t)\right)ds\Big|_{s\to t} \\[2mm]
&= \log(\tan(t/2))\sin(t) + \cos(t)\sin(t) - \sin(t)\cos(t) = \sin(t)\log(\tan(t/2)).
\end{aligned}
$$

Hence the general solution of (9.88) is

$$y = C_1\cos(t) + C_2\sin(t) + \sin(t)\log(\tan(t/2)). \ ∎$$

Sometimes it is useful to solve an inhomogeneous second-order differential equation when the forcing function is a general function. The differential equation can be solved by the method of variation of parameters, but the answer will involve an integral. The following example illustrates the technique.

Example 9.19.   *Find a particular solution of the equation $y'' + \omega^2 y = g(t)$, where ω is a nonzero constant and $g(t)$ is an unspecified function.*

Solution. A fundamental set of solutions of the homogeneous equation $y'' + \omega^2 y = 0$ is provided by $y_1(t) = \cos(\omega t)$ and $y_2(t) = \sin(\omega t)$. The Wronskian is

$$W(y_1, y_2)(t) = \det\begin{pmatrix} y_1 & y_2 \\ y_1' & y_2' \end{pmatrix} = \det\begin{pmatrix} \cos(\omega t) & \sin(\omega t) \\ -\omega\sin(\omega t) & \omega\cos(\omega t) \end{pmatrix} = \omega.$$

From (9.82) it follows that a particular solution is given by

$$y_P(t) = \int \left(\frac{\cos(\omega s)\sin(\omega t) - \cos(\omega t)\sin(\omega s)}{\omega} \right) g(s)ds \bigg|_{s \to t} \qquad (9.89)$$

$$= \int \left(\frac{\sin(\omega(t-s))}{\omega} \right) g(s)ds \bigg|_{s \to t}. \quad \blacksquare$$

If in Example 9.19 the function $g(s)$ were a combination of exponentials and polynomials, we could also solve the problem by the method of undetermined coefficients developed in Section 9.3. We would have to consider, for each term of the right-hand side, whether or not that term is resonant or nonresonant. By comparison, the method of variation of parameters makes no distinction between the resonant case and the nonresonant case. As long as we know the fundamental set of solutions of the homogeneous equation, all solutions of the inhomogeneous equation are simultaneously represented by the same integral formula.

Example 9.20.   *Find the general solution of $y'' + \omega^2 y = \cos(\omega t)$.*

Solution. From (9.89) it follows that

$$y_P(t) = \left(\frac{\sin(\omega t)}{\omega} \right) \int (\cos(\omega s))^2 ds \bigg|_{s \to t} - \left(\frac{\cos(\omega t)}{\omega} \right) \int \sin(\omega s)\cos(\omega s)ds \bigg|_{s \to t}$$

$$= \left(\frac{\sin(\omega t)}{2\omega} \right) \int (1 + \cos(2\omega s))ds \bigg|_{s \to t} - \left(\frac{\cos(\omega t)}{2\omega} \right) \int \sin(2\omega s)ds \bigg|_{s \to t}$$

$$= \left(\frac{\sin(\omega t)}{2\omega} \right) \left(s + \frac{\sin(2\omega s)}{2\omega} \right) \bigg|_{s \to t} + \left(\frac{\cos(\omega t)}{2\omega} \right) \frac{\cos(2\omega s)}{2\omega} \bigg|_{s \to t}$$

$$= \frac{t\sin(\omega t)}{2\omega} + \frac{(\sin(\omega t))^2\cos(\omega t)}{2\omega^2} + \frac{\cos(\omega t)(1 - 2(\sin(\omega t))^2)}{4\omega^2}$$

$$= \frac{t\sin(\omega t)}{2\omega} + \frac{\cos(\omega t)}{4\omega^2}.$$

We have obtained

$$\frac{t\,\sin(\omega t)}{2\omega} + \frac{\cos(\omega t)}{4\omega^2}$$

as a particular solution. But since $\cos(\omega t)/(4\omega^2)$ is already a solution of the homogeneous equation, we see that $t\,\sin(\omega t)/(2\omega)$ is also a particular solution. Thus the general solution of $y'' + \omega^2 y = \cos(\omega t)$ is

$$y = A\cos(\omega t) + B\sin(\omega t) + \frac{t\,\sin(\omega t)}{2\omega}. \quad\blacksquare$$

Summary of Techniques Introduced

The method of variation of parameters can always be used to construct a particular solution y_P to

$$a(t)y'' + b(t)y' + c(t)y = g(t) \tag{9.90}$$

in terms of a fundamental set $\{y_1, y_2\}$ of solutions to the homogeneous equation $a(t)y'' + b(t)y' + c(t)y = 0$. The expression for y_P consists of an integral involving y_1, y_2 and their Wronskian.

> ### Warning
>
> Before using the method of variation of parameters, it is essential to convert (9.90) to the form
>
> $$y'' + p(t)y' + q(t)y = r(t).$$

Exercises for Section 9.4

Use the method of variation of parameters to find a particular solution of each of the following equations.

1. $y'' - 4y' + 4y = (t+1)e^{2t}$

2. $4y'' + 36y = \csc(3t)$

3. $y'' + y = e^{-t}$

4. $y'' + 9y = 9(\sec(3t))^2$

5. $y'' + y = e^t \sin(t)$

6. $y'' + y = \sec(t)\tan(t)$

7. $y'' - 2y' + y = \dfrac{e^t}{t}$

8. $y'' - 2y' + y = \dfrac{e^t}{t^2}$

9. $y'' + y = \tan(t)$

10. $y'' - 3y' + 2y = \dfrac{1}{1 + e^{-t}}$ **11.** $y'' + 2y' + y = e^{-t} \log(t)$

12. $t^2 y'' - 2t\, y' + 2y = t^3 \sin(t)$. [Hint: $y_1 = t,\ y_2 = t^2$.]

13. $y'' - 2y' + y = t^n e^t$, where $n \neq -1, -2$. (Notice that n need not be an integer.)

14.   **a.** Use undetermined coefficients to find a particular solution of

$$y'' + 2y' + y = 4t^2 - 3.$$

   b. Use variation of parameters to find a particular solution of

$$y'' + 2y' + y = \frac{e^{-t}}{t}.$$

   c. Use the results of parts **a** and **b** together with the superposition principle to find a particular solution of

$$y'' + 2y'' + y = 4t^2 - 3 + \frac{e^{-t}}{t}.$$

Use variation of parameters to solve the following initial value problems.

15. $\begin{cases} y'' - 4y' + 3y = 9t^2 + 4 \\ y(0) = 6,\ y'(0) = 8 \end{cases}$

20. $\begin{cases} y'' + 2y' - 8y = 2e^{-2t} - e^{-t} \\ y(0) = 1,\ y'(0) = 0 \end{cases}$

16. $\begin{cases} y'' + 5y' + 4y = t + e^t \\ y(0) = 0,\ y'(0) = 3 \end{cases}$

21. $\begin{cases} y'' - 4y' + 4y = (12t^2 - 6t)e^{2t} \\ y(0) = 1,\ y'(0) = 0 \end{cases}$

17. $\begin{cases} y'' - 8y' + 15y = 9t\, e^{2t} \\ y(0) = 5,\ y'(0) = 10 \end{cases}$

22. $\begin{cases} y'' - 2y' + y = e^t \arctan(t) \\ y(0) = 0,\ y'(0) = 1 \end{cases}$

18. $\begin{cases} y'' + 7y' + 10y = 4t\, e^{-3t} \sin(t) \\ y(0) = 0,\ y'(0) = -1 \end{cases}$

23. $\begin{cases} y'' + 3y' + 2y = \sin(e^t) \\ y(0) = 1,\ y'(0) = 1 \end{cases}$

19. $\begin{cases} 4y'' - y = t\, e^{t/2} \\ y(0) = 1,\ y'(0) = 0 \end{cases}$

24. $\begin{cases} y'' + 3y' + 2y = (1 - e^t)^{3/2} \\ y(0) = 1,\ y'(0) = 0 \end{cases}$

The method of variation of parameters can be used to find a particular solution to $y'' + p(t)y' + q(t)y = r(t)$ whether or not $p(t)$ and $q(t)$ are constant, provided that a fundamental set of solutions $\{y_1, y_2\}$ is known. Methods for finding such a fundamental set for a wide variety of equations will be given in Chapters 20 and 21. In the meantime, find the general solutions to the following differential equations, for which the fundamental set is provided.

25. $\begin{cases} t^2 y'' - 2t\, y' + 2y = t \\ y_1(t) = t,\ y_2(t) = t^2 \end{cases}$

28. $\begin{cases} (t-1)y'' - t\, y' + y = 1 \\ y_1(t) = t,\ y_2(t) = e^t \end{cases}$

26. $\begin{cases} t^2 y'' - 4t\, y' + 6y = \log(t) \\ y_1(t) = t^2,\ y_2(t) = t^3 \end{cases}$

29. $\begin{cases} (1+t)y'' - (2+t)y' = e^t \\ y_1(t) = 1,\ y_2(t) = t\, e^t \end{cases}$

27. $\begin{cases} t^2 y'' - 2t\, y' + 2y = \big(\log(t)\big)^2 \\ y_1(t) = t,\ y_2(t) = t^2 \end{cases}$

30. $\begin{cases} \big(\cos(t) - \sin(t)\big)y'' + 2\sin(t)y' \\ -\big(\cos(t) + \sin(t)\big)y = 1 \\ y_1(t) = \sin(t),\ y_2(t) = e^t \end{cases}$

Solve the following initial value problems.

31. $\begin{cases} t^2 y'' - 2t\, y' + 2y = t^2 \\ y(1) = 0,\ y'(1) = 0 \\ y_1(t) = t,\ y_2(t) = t^2 \end{cases}$

34. $\begin{cases} (t-1)y'' - t\, y' + y = 1 \\ y(0) = 0,\ y'(0) = 0 \\ y_1(t) = t,\ y_2(t) = e^t \end{cases}$

32. $\begin{cases} t^2 y'' - 3t\, y' + 3y = \log(t) \\ y(1) = 0,\ y'(1) = 0 \\ y_1(t) = t,\ y_2(t) = t^3 \end{cases}$

35. $\begin{cases} (1+t)y'' - (2+t)y' = e^t \\ y(0) = 0,\ y'(0) = 0 \\ y_1(t) = 1,\ y_2(t) = t\, e^t \end{cases}$

33. $\begin{cases} t^2 y'' - 2y = \big(\log(t)\big)^2 \\ y_1(t) = 1/t,\ y_2(t) = t^2 \\ y(1) = 0,\ y'(1) = 0 \end{cases}$

36. $\begin{cases} \big(\cos(t) - \sin(t)\big)y'' + 2\sin(t)y' \\ -\big(\cos(t) + \sin(t)\big)y = 1 \\ y(0) = 0,\ y'(0) = 0 \\ y_1(t) = \sin(t),\ y_2(t) = e^t \end{cases}$

10

USING **ODE** *TO SOLVE*
SECOND-ORDER LINEAR
DIFFERENTIAL
EQUATIONS

In Chapter 9 we learned how to solve many second-order differential equations of the form

$$a\,y'' + b\,y' + c\,y = f(t), \tag{10.1}$$

where a, b, and c are constants and $f(t)$ is a continuous function. Finding the solution of $a\,y'' + b\,y' + c\,y = 0$ is not too difficult, and it can be carried out in a few minutes. However, if $f(t)$ is nonzero, it may be very tedious to solve (10.1). For example, to find the solution of

$$y'' - 2y' + 2y = t^3 e^t \cos(t)$$

by hand is difficult because integration by parts is required several times. The solution is actually

$$y(t) = e^t \left(C_1 \cos(t) + C_2 \sin(t) + \frac{(-3t + 2t^3)\cos(t) + (-3t^2 + t^4)\sin(t)}{8} \right).$$

Mathematica is an effective tool for solving symbolically second-order differential equations such as (10.1). In this chapter we explain how to use **ODE** to solve second-order linear equations with constant-coefficients, both homogeneous and inhomogeneous. Any of the following options can be used:

```
Method->SecondOrderLinear          Method->NthOrderLinear
Method->NthOrderLinearComplex      Method->DSolve
Method->ApproximateNthOrderLinear  Method->Laplace
```

The options `Method->SecondOrderLinear` and `Method->NthOrderLinear` are discussed in Section 10.1; they are equivalent for second-order equations. Both `Method->NthOrderLinear` and `Method->NthOrderLinearComplex` can be used to solve nth-order equations; they will be described in Chapter 12. As was the case with first-order equations, ODE with the option `Method->DSolve` is equivalent to *Mathematica*'s internal command `DSolve`. The option `Method->Laplace` uses Laplace transforms and will be discussed in Chapter 14. A seminumerical solution to an nth-order equation can be found using `Method->ApproximateNthOrderLinear`; it will be discussed in Section 12.5.

Each of the six options mentioned above works equally well with homogeneous or inhomogeneous equations. Furthermore, each can be used either to find the general solution to a second-order differential equation with constant-coefficients or to solve an initial value problem. In addition, the option `Form->FundamentalSet` produces a fundamental set of solutions.

In Section 10.3 we describe `Method->ReductionOfOrder`. This option together with the option `KnownSolution` can be used to solve certain linear second-order differential equations with nonconstant-coefficients. A similar option `ProposedSolution` can be used to test solutions to a differential equation or initial value problem. The command `EquationFromSolutions`, which constructs a homogeneous second-order linear differential equation from two of its solutions, is described in Section 10.4.

10.1   *Using* ODE *to Solve Second-Order Constant-Coefficient Equations*

In this section we show how to use ODE to solve second-order differential equations with constant-coefficients. The differential equation may be either homogeneous or inhomogeneous. Both `Method->SecondOrderLinear` and `Method->NthOrderLinear` produce the same output. The only difference is that `SecondOrderLinear` uses simple routines (described below) for second-order equations, but `NthOrderLinear` uses more complicated routines that work for higher-order equations (see Chapter 12). In those cases *not* involving complex numbers, `DSolve` can be used interchangeably with `SecondOrderLinear` and `NthOrderLinear`. However, in the case of complex roots of the associated characteristic equation, the output of `DSolve` in version 2.2 of *Mathematica* frequently involves complex exponents, and `NthOrderLinearComplex` always involves complex exponents. On the other hand, when `SecondOrderLinear` or **NthOrderLinear** solves a differential equation with real coefficients, the output is always real. Since one usually needs to solve differential equations with real coefficients, **SecondOrderLinear** and **NthOrderLinear** are the preferred commands.

Homogeneous Constant-Coefficient Equations

The syntax for finding the general solution to a second-order differential equation is identical to that for a first-order differential equation. Let us begin with a homogeneous second-order constant-coefficient equation whose associated characteristic equation has real distinct roots.

Example 10.1. *Use* **ODE** *to find the general solution of*

$$y'' + 13y' + 12y = 0.$$

Solution. We use either of the following commands:

```
ODE[y'' + 13y' + 12y == 0,y,t,Method->SecondOrderLinear]

ODE[y'' + 13y' + 12y == 0,y,t,Method->DSolve]
```

to obtain

```
         C[1]    C[2]
{{y ->  ------ + ----}}
         12 t     t
        E        E
```

Thus, the general solution is $y = C_1 e^{-12t} + C_2 e^{-t}$. $\blacksquare$

In Example 10.1 we can obtain a fundamental set of solutions using the command

```
ODE[y'' + 13y' + 12y == 0,y,t,
Method->SecondOrderLinear,Form->FundamentalSet]

  -12 t    -t
{E      , E  }
```

In the next example we solve a second-order homogeneous equation whose associated characteristic equation has repeated roots.

Example 10.2. *Find the general solution of* $y'' - 10y' + 25y = 0$ *with* **ODE***, expressing the answer as an equation.*

Solution. We use

```
ODE[y'' - 10y' + 25y == 0,y,t,
Method->SecondOrderLinear,Form->Equation]
```

to get

$$y == E^{5\,t} \, C[1] + E^{5\,t} \, t\, C[2]$$

Thus, the general solution is $y = C_1 e^{5t} + C_2 t\, e^{5t}$. ∎

We turn now to second-order homogeneous equations whose associated characteristic equations have complex roots.

Example 10.3. *Find the general solution of* $y'' + 3y = 0$ *with* **ODE** (*or* **DSolve**).

Solution. When we use

DSolve[y''[t] + 3 y[t] == 0,y[t],t]

or

ODE[y''[t] + 3 y[t] == 0,y[t],t,Method->DSolve]

we get

$$\{\{y[t] \to E^{-I\,\mathrm{Sqrt}[3]\,t}\, C[1] + E^{I\,\mathrm{Sqrt}[3]\,t}\, C[2]\}\}$$

We do not want complex exponents here. **Method->SecondOrderLinear**, however, gives an answer that is free of complex exponents:

ODE[y''[t] + 3 y[t] == 0,y[t],t,Method->SecondOrderLinear]

`{{y[t] -> C[2] Cos[Sqrt[3] t] + C[1] Sin[Sqrt[3] t]}}`

Thus, the general solution is $y = C_1 \cos\left(t\sqrt{3}\right) + C_2 \sin\left(t\sqrt{3}\right)$. ∎

On the other hand, if one of a, b, or c is complex, the solution of $a\,y'' + b\,y' + c\,y = 0$ is given in terms of complex exponents.

Example 10.4. *Find the general solution of* $y'' - 2i\,y' - 2y = 0$ *using* **ODE** (*or* **DSolve**).

Solution. Either

ODE[y'' - 2I y' - 2y == 0,y,t,Method->SecondOrderLinear]

or

ODE[y'' - 2I y' - 2y == 0,y,t,Method->DSolve]

yields

```
                (-1 + I) t              (1 + I) t
{{y -> E                    C[1] + E                C[2]}}
```

Thus, the general solution of $y'' - 2i\, y' - 2y = 0$ is $y = C_1 e^{(-1+i)t} + C_2 e^{(1+i)t}$. ∎

To solve a second-order linear equation with initial conditions, we must provide **ODE** with a list of three equations: the differential equation and two initial conditions. The syntax is

$$\textbf{ODE[\{diffeq,initcond1,initcond2\},y[t],t,Method]}$$

Example 10.5. *Use **ODE** to solve the initial value problem*

$$\begin{cases} y'' + 13y' + 12y = 0, \\ y(0) = 2,\ y'(0) = 3. \end{cases}$$

Solution. We use

ODE[{y'' + 13y' + 12y == 0,y[0] == 2,y'[0] == 3},y,t, Method->SecondOrderLinear]

to obtain

```
                 -5          27
{{y ->  ————————  +  ———— }}
               12 t          t
          11 E          11 E
```

Thus, the solution is $-\dfrac{5e^{-12t}}{11} + \dfrac{27e^{-t}}{11}$. ∎

The solution to the initial value problem of Example 10.5 can be found and plotted with the command

ODE[{y'' + 13y' + 12y == 0,y[0] == 2,y'[0] == 3},y,t, Method->SecondOrderLinear,PlotSolution->{{t,-0.2,1}}]

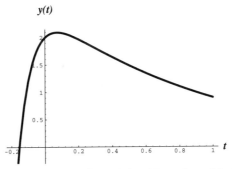

The solution of $y'' + 13y' + 12y = 0$ **with the initial conditions** $y(0) = 2$ **and** $y'(0) = 3$

The next initial value problem involves square roots.

Example 10.6. *Use* ODE *to find and plot the solution of* $2y'' - 7y = 0$ *with the initial conditions* $y(0) = 5$ *and* $y'(0) = -2$.

Solution. We use

```
ODE[{2y''[t] - 7 y[t] == 0,y[0] == 5,y'[0] == -2},y[t],t,
Method->SecondOrderLinear,PlotSolution->{{t,-1,1}}]
```

obtaining

```
           5     2
           - + Sqrt[-]
           2     7               5     2     Sqrt[7/2] t
{{y[t] ->  ─────────────── + (- - Sqrt[-]) E            }}
             Sqrt[7/2] t        2     7
           E
```

Thus, the solution is

$$y = \left(\frac{5}{2} + \sqrt{\frac{2}{7}}\right) e^{-\sqrt{7/2}\,t} + \left(\frac{5}{2} - \sqrt{\frac{2}{7}}\right) e^{\sqrt{7/2}\,t}$$

$$= 5\cosh\left(\sqrt{\frac{7}{2}}\,t\right) - \sqrt{\frac{8}{7}}\sinh\left(\sqrt{\frac{7}{2}}\,t\right). \ \blacksquare$$

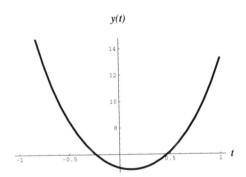

y(t)

The solution of $2y'' - 7y = 0$ with the initial conditions $y(0) = 5$ and $y'(0) = -2$

Here is an example involving trigonometric functions.

Example 10.7. *Use* ODE *to solve* $3y'' - 2y' + 15y = 0$ *with the initial conditions* $y(0) = 0$ *and* $y'(0) = -1$.

Solution. We use

```
ODE[{3y'' - 2y' + 15y == 0,y[0] == 0,y'[0] == -1},y,t,
Method->SecondOrderLinear]
```

to get

```
             t/3       2 Sqrt[11] t
        -3 E    Sin[---------------]
                           3
{{y -> -----------------------------}}
             2 Sqrt[11]
```

Thus, the solution is $y(t) = \left(\dfrac{-3}{2\sqrt{11}}\right) e^{t/3} \sin\left(\dfrac{(2\sqrt{11})t}{3}\right)$. The solution can be plotted with the command

```
ODE[{3y'' - 2y' + 15y == 0,y[0] == 0,y'[0] == -1},y,t,
Method->SecondOrderLinear,PlotSolution->{{t,-2Pi,2Pi}}]
```

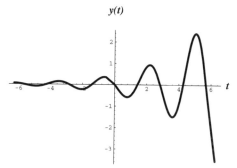

The solution of $3y'' - 2y' + 15y = 0$ **with the
initial conditions** $y(0) = 0$ **and** $y'(0) = -1$

Inhomogeneous Constant-Coefficient Equations

We can use **ODE** to find the general solution to inhomogeneous second-order linear differential equations; the syntax is the same as we used in the homogeneous case. The following example would be complicated to do by hand.

Example 10.8. *Use* **ODE** *to find the general solution of* $y'' + 13y' + 12y = t^5 e^{-t}$.

Solution. We use

```
ODE[y'' + 13y' + 12y == t^5 E^-t,y,t,
Method->SecondOrderLinear]
```

to obtain

```
          120 t        60 t²        20 t³        5 t⁴
{{y -> - --------- + ---------- - ---------- + ---------- -
              t             t             t             t
         1771561 E   161051 E    14641 E     1331 E

     5         6
    t         t        C[1]       C[2]
  ------- + ------- + ------ + -------}}
        t         t     12 t        t
   121 E     66 E      E          E
```

Thus, the general solution is

$$y(t) = \left(-\frac{120t}{1771561} + \frac{60t^2}{161051} - \frac{20t^3}{14641} + \frac{5t^4}{1331} - \frac{t^5}{121} + \frac{t^6}{66} + C_2\right)e^{-t}$$

$$+ C_1 e^{-12t},$$

where C_1 and C_2 are constants. ∎

Next, we use **ODE** to solve an inhomogeneous initial value problem.

Example 10.9. *Use **ODE** to find the solution of* $y'' - 2y' + 2y = t^3 e^t \cos(t)$ *with the initial conditions* $y(0) = 0$ *and* $y'(0) = 1$.

Solution. We use

```
ODE[{y'' - 2y' + 2y == t^3 E^t Cos[t],y[0] == 0,y'[0] == 1},
y,t,Method->SecondOrderLinear]
```

to obtain

```
            t                 t                  t
      -3 E  t Cos[t]     E  t³ Cos[t]      11 E  Sin[t]
{{y -> --------------- + -------------- + -------------- -
             8                 4                 8

        t                 t
     3 E  t² Sin[t]     E  t⁴ Sin[t]
     -------------- + --------------}}
            8                8
```

Thus, the solution is

$$y(t) = \frac{e^t\left(-3t\,\cos(t) + 2t^3\,\cos(t) + 11\,\sin(t) - 3t^2\,\sin(t) + t^4\,\sin(t)\right)}{8}. ∎$$

The solution can be plotted with the command

```
ODE[{y'' - 2y' + 2y == t^3 E^t Cos[t],y[0] == 0,y'[0] == 1},
y,t,Method->SecondOrderLinear,PlotSolution->{{t,-Pi/2,2Pi}}]
```

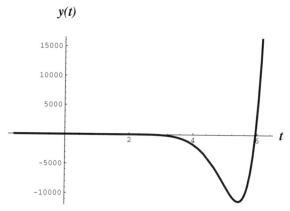

The solution of $y'' - 2y' + 2y = t^3 e^t \cos(t)$ with the
initial conditions $y(0) = 0$ and $y'(0) = -1$

Plotting Multiple Solutions

As in Chapter 4, **ODE**'s option **Parameters** can be used to plot multiple solutions. In the
following example we vary the differential equation, keeping the initial conditions fixed.

Example 10.10. *Use* **ODE** *to solve*

$$
\begin{cases}
y'' + y = e^{nt/20}, \\
y(0) = 0, \ y'(0) = 1;
\end{cases}
$$

plot the solutions for n = 0, 1, 2, 3, 4, 5 *on one graph.*

Solution. We use

```
ODE[{y'' + y == E^(n t/20),y[0] == 0,y'[0] == 1},y,t,
Method->SecondOrderLinear,Form->Explicit,
Parameters->{{n,0,5,1}},
PlotSolution->{{t,-2Pi,4Pi},PlotPoints->100},
PostSolution->{Apart}]
```

to get

$$\frac{400\, E^{(n\, t)/20}}{400 + n^2} + \frac{-400\, \text{Cos}[t] + 400\, \text{Sin}[t] - 20\, n\, \text{Sin}[t] + n^2\, \text{Sin}[t]}{400 + n^2}$$

and the graph

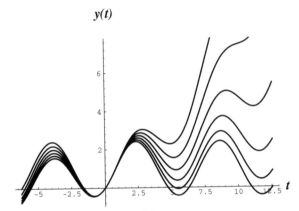

y(t)

Solutions of $y'' + y = e^{nt/20}$ **for** $0 \le n \le 5$ **with the initial conditions** $y(0) = 0$ **and** $y'(0) = 1$

Similarly, we may plot solutions of the same differential equation with different initial conditions.

Example 10.11. *Use* **ODE** *to solve*

$$\begin{cases} y'' - 2y' + 2y = 0, \\ y(0) = 0, \, y'(0) = a; \end{cases}$$

plot the solutions for $a = 0, 0.1, 0.2, 0.3, 0.4$ *on one graph.*

Solution. We use

```
ODE[{y'' - 2y' + 2y == 0,y[0] == 0,y'[0] == a},y,t,
Method->SecondOrderLinear,Parameters->{{a,0,0.4,0.1}},
PlotSolution->{{t,0,Pi},PlotPoints->100}]
```

to get

```
{{y -> a E  Sin[t]}}
         t
```

and the graph

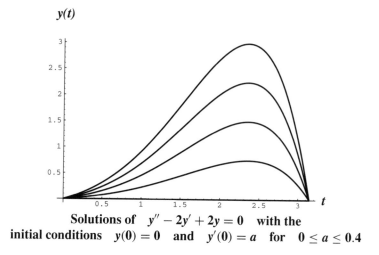

Solutions of $\quad y'' - 2y' + 2y = 0 \quad$ with the
initial conditions $\quad y(0) = 0 \quad$ and $\quad y'(0) = a \quad$ for $\quad 0 \le a \le 0.4$

Here is an example of the use of **Parameters** and **StackPlotSolution**.

Example 10.12. *Use* **ODE** *to solve*

$$\begin{cases} y'' + 6y' + 9y = 10\sin(t), \\ y(0) = a,\, y'(0) = 0. \end{cases}$$

Use **ODE**'s *option* **StackPlotSolution** *to plot the solutions for* $0 \le a \le 4$ *at intervals of* 0.02.

Solution. We use

```
ODE[{y'' + 6y' + 9y == 10Sin[t],y[0] == a,y'[0] == 0},y,t,
Method->SecondOrderLinear,Parameters->{{a,0,4,0.2}},
StackPlotSolution->{{t,0,2Pi},PlotPoints->100}]
```

to get

$$\{\{y \to \frac{3}{5\,E^{3\,t}} + \frac{a}{E^{3\,t}} + \frac{t}{E^{3\,t}} + \frac{3\,a\,t}{E^{3\,t}} - \frac{3\,\text{Cos}[t]}{5} + \frac{4\,\text{Sin}[t]}{5}\}\}$$

and the graph

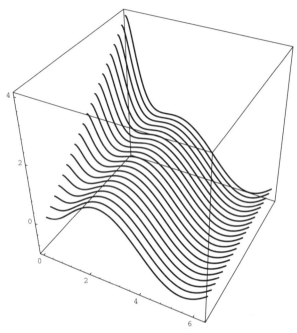

Solutions of $y'' + 6y' + 9y = 10\sin(t)$ **with the**
initial conditions $y(0) = a$ **and** $y'(0) = 1$ **for $0 \le a \le 4$**

Exercises for Section 10.1

Use **ODE** to find the general solution of each of the following differential equations.

1. $3y'' + 4y' - 2y = 0$ **6.** $y'' + 2i\,y' - 3y = 0$

2. $y'' + 24y' - 2y = 0$ **7.** $i\,y'' + y' - 3y = \sin(i\,t)$

3. $32y'' + 25y' - 4y = 0$ **8.** $y'' + 6y' + 9 = t^2 e^{-3t}$

4. $2y'' - y' + 5y = 0$ **9.** $y'' - 4y' + 12y = 15e^t - 4\cos(t)$

5. $y'' + 16y = e^t - 9t^2$ **10.** $y'' + 12y = \sin(12t)$

Use **ODE** to solve and plot each of the following initial value problems. This requires some experimentation to find a good interval over which to plot the solution. In each problem begin by using the interval $-1 < t < 1$, and then try to determine a larger or smaller interval for which the graph of the solution is more interesting.

11. $\begin{cases} 2y'' - 10y' - 3y = 0, \\ y(0) = 5, \quad y'(0) = -3 \end{cases}$

16. $\begin{cases} y'' + 5y = t\sin(t), \\ y(0) = 1, \quad y'(0) = \pi \end{cases}$

12. $\begin{cases} y'' + 10y = 0, \\ y(0) = 1, \quad y'(0) = -1 \end{cases}$

17. $\begin{cases} y'' - 4y = (1 + t)\cos(10t), \\ y(0) = 0, \quad y'(0) = 0 \end{cases}$

13. $\begin{cases} y'' - 3y' + 10y = 0, \\ y(0) = 1, \quad y'(0) = -1 \end{cases}$

18. $\begin{cases} y'' - 8y = (1 + t - t^2)e^t, \\ y(0) = 0, \quad y'(0) = 1 \end{cases}$

14. $\begin{cases} 8y'' + 3y' - 5y = 0, \\ y(0) = 1, \quad y'(0) = 2 \end{cases}$

19. $\begin{cases} y'' - 2y = \sin(t) + e^t, \\ y(0) = 1, \quad y'(0) = 0 \end{cases}$

15. $\begin{cases} 8y'' + 5y = t^2, \\ y(0) = 1, \quad y'(0) = \pi/\sqrt{5/2} \end{cases}$

20. $\begin{cases} y'' - 2y' + y = t\sin(4t) + e^t\cos(t), \\ y(0) = 1, \quad y'(0) = 0 \end{cases}$

Use **ODE** to solve and stack plot families of solutions to each of the following initial value problems. This requires some experimentation to find a good interval over which to plot the solution. Also, it may be necessary to adjust the viewpoint using the option **ViewPoint**. In each problem begin by using the interval $-1 < t < 1$, and then try to determine a larger or smaller interval for which the graph of the solution is more interesting.

21. $\begin{cases} y'' + y = \sin(2t), \\ y(0) = 0, \quad 0 \le y'(0) \le 4 \end{cases}$

24. $\begin{cases} y'' + 5y = t\sin(t), \\ 0 \le y(0) \le 4, \quad y'(0) = \pi \end{cases}$

22. $\begin{cases} y'' - 2y' + y = t, \\ -2 \le y(0) \le 2, \quad y'(0) = 0 \end{cases}$

25. $\begin{cases} y'' + y = \cos(k\,t), 5 \le k \le 10 \\ y(0) = 1, \quad y'(0) = 0 \end{cases}$

23. $\begin{cases} 3y'' - 2y' + 15y = 0, \\ 0 \le y(0) \le 4, \quad y'(0) = -1 \end{cases}$

26. $\begin{cases} y'' + k\,y = \exp(t), 1 \le k \le 5 \\ y(0) = 0, \quad y'(0) = 0 \end{cases}$

27. $\begin{cases} y'' + y = e^{kt}, 0 \le k \le 5 \\ y(0) = 0, \;\; y'(0) = 0 \end{cases}$ **28.** $\begin{cases} a\,y'' + y' + a\,y = 0, -1 \le a \le 1 \\ y(0) = 0, \;\; y'(0) = 1 \end{cases}$

Find and plot the solutions to the following initial value problems, whose solutions involve nonelementary integrals.

29. $\begin{cases} y'' - 4y' + 4y = \sqrt{t}, \\ y(0) = y'(0) = 0 \end{cases}$ **31.** $\begin{cases} y'' - y = e^{t^2}, \\ y(0) = y'(0) = 0 \end{cases}$

30. $\begin{cases} y'' - y = e^{-t^2}, \\ y(0) = y'(0) = 0 \end{cases}$ **32.** $\begin{cases} y'' + y' - 2y = \log t, \\ y(1) = y'(1) = 0 \end{cases}$

10.2   *Details of* ODE *for Second-Order Constant-Coefficient Equations*

When ODE[1] using the option **Method->SecondOrderLinear** encounters a second-order linear constant-coefficient equation, it mimics the mathematical procedures that we described in Chapter 9. First, it calls **HomogeneousSecondOrderLinear**, which finds the general solution y_H of the associated homogeneous equation; this routine is described in Section 10.2.1. Next, a call to **ParticularSolution** is made to determine whether **UnderterminedCoefficients** can be used or whether the more robust but slower **VariationOfParameters** must be used to find a particular solution y_P to the inhomogeneous equation; the routine **VariationOfParameters** is described inSection 10.2.2. Finally, ODE adds y_H and y_P and outputs the result.

Note that **HomogeneousSecondOrderLinear**, **ParticularSolution**, **UnderterminedCoefficients** and **VariationOfParameters** are all *Mathematica* commands in the package **ODE.m**; they can be used independently of the command **ODE**.

---

[1]This section contains technical *Mathematica* details on how ODE works. It is not needed in order to use ODE, nor is it referred to later.

10.2.1   How ODE *Solves a Second-Order Homogeneous Constant-Coefficient Equation*

As we observed in Chapter 9, solving a second-order homogeneous constant-coefficient equation is complicated by the fact that there are several cases to consider. The command **HomogeneousSecondOrderLinear** uses *Mathematica*'s command **If** to determine which case is appropriate. We first give the command, then explain it.

```
HomogeneousSecondOrderLinear[a_,b_,c_,const_Symbol][t_]:=
   Module[{m1,m2,sol},
      If[a === 0,
         Print["Equation is not second-order."],
         m1 = PowerExpand[(-b - Sqrt[b^2 - 4a c])/(2a)];
         m2 = PowerExpand[(-b + Sqrt[b^2 - 4a c])/(2a)];
      If[m1 === m2,
         sol = const[1] Exp[m1 t] + const[2] t Exp[m1 t],
      If[FreeQ[m1,Complex] && FreeQ[m2,Complex],
         sol = const[1] Exp[m1 t] +  const[2] Exp[m2 t],
      If[!FreeQ[{a,b,c},Complex],
         sol = const[1] Exp[m1 t] + const[2] Exp[m2 t],
         sol = Exp[Re[m2] t] (const[1] Sin[Im[m2] t] +
            const[2] Cos[Im[m2] t])]]]];
      sol]]
```

First, **HomogeneousSecondOrderLinear** verifies that $a\,y'' + b\,y' + c\,y = 0$ is actually second-order by checking whether a is nonzero; to do this it uses the triple equal **===**, which is the *Mathematica* command for "identical with". If $a = 0$, then **HomogeneousSecondOrderLinear** outputs (using **Print**) the phrase "Equation is not second-order"; otherwise it defines

$$\texttt{m1,m2} = \frac{-b \pm \sqrt{b^2 - 4a\,c}}{2a}.$$

The output of **HomogeneousSecondOrderLinear** will vary, depending on the nature of the roots **m1** and **m2** of the associated characteristic equation $a\,r^2 + b\,r + c = 0$. First, **HomogeneousSecondOrderLinear** does the repeated root case. If **m1 === m2**, then **HomogeneousSecondOrderLinear** assigns the solution to **sol** using

$$\texttt{sol = const[1] Exp[m1 t] + const[2] t Exp[m1 t]}. \qquad (10.2)$$

Up to this point it makes no difference whether the coefficients a, b, and c are real or complex. But if **m1** and **m2** are different, then **HomogeneousSecondOrderLinear** needs to check whether **m1** and **m2** are complex or real. This is accomplished with *Mathematica*'s command

FreeQ[m1,Complex] && FreeQ[m2,Complex].

In the case that both **m1** and **m2** are real, **HomogeneousSecondOrderLinear** assigns the solution to **sol** using

$$\text{sol = const[1] Exp[m1 t] + const[2] Exp[m2 t].} \qquad (10.3)$$

Finally, **HomogeneousSecondOrderLinear** must confront the complex root case; there are two subcases. First, **HomogeneousSecondOrderLinear** determines using **!FreeQ[a,b,c,Complex]** whether a, b, or c is complex. If one of a, b, or c is complex, then **HomogeneousSecondOrderLinear** will use the complex exponents and assign the solution to **sol** using (10.2). On the other hand, if a, b, and c are all real, then trigonometric functions are called for; in this case **HomogeneousSecondOrderLinear** assigns the solution to **sol** using

$$\begin{aligned}\text{sol = Exp[Re[m2] t]*} \\ \text{(const[1] Sin[Im[m2] t] + const[2] Cos[Im[m2] t]).}\end{aligned} \qquad (10.4)$$

Finally, **sol** is outputted in the appropriate form, depending on whether (10.2), (10.3), or (10.4) has been chosen.

Example 10.13. *Use* **HomogeneousSecondOrderLinear** *to find the general solution to* $y'' - 6y' + 9y = 0$.

Solution. The command

HomogeneousSecondOrderLinear[1,-6,9,C][t]

yields

```
 3 t          3 t
E    C[1] + E    t C[2]
```

Hence the general solution is

$$y(t) = C_1 e^{3t} + C_2 t\, e^{3t},$$

where C_1 and C_2 are constants. ∎

10.2.2   Variation of Parameters via ODE

Here is a *Mathematica* miniprogram that computes the particular solution of an inhomogeneous second-order linear differential equation using variation of parameters. The *Mathematica* code is directly derived from the corresponding mathematical formula (formula (9.83) on page 296).

```
VariationOfParameters[y1_,y2_,g_][t_]:=
    Module[{w,u1,u2},
        w = Simplify[y1[t]y2'[t] - y1'[t]y2[t]];
        u1 = Simplify[Integrate[-y2[t]g[t]/w,t]];
        u2 = Simplify[Integrate[y1[t]g[t]/w,t]];
        Simplify[y1[t]u1 + y2[t]u2]]
```

To explain how the miniprogram **VariationOfParameters** works, let us use it to find a particular solution of a second-order linear differential equation.

Example 10.14. *Use* **VariationOfParameters** *to find a particular solution of*

$$y'' + y = \sec(t). \tag{10.5}$$

Solution.   A fundamental set of solutions of the homogeneous differential equation $y'' + y = 0$ is given by

$$y_1(t) = \cos(t) \qquad \text{and} \qquad y_2(t) = \sin(t).$$

If we type in

VariationOfParameters[Cos,Sin,Sec][t]

Mathematica answers with

```
Cos[t] Log[Cos[t]] + t Sin[t]
```

Thus, a particular solution of (10.5) is given by

$$y_p(t) = \cos(t) \log\big(\cos(t)\big) + t \sin(t).$$

The general solution of (10.5) can be obtained with the command

ODE[y'' + y == Sec[t],y,t,Method->SecondOrderLinear]

```
{{y -> C[2] Cos[t] + Cos[t] Log[Cos[t]] + t Sin[t] + C[1] Sin[t]}}
```

Hence the general solution to (10.5) is

$$y(t) = C_1 \cos(t) + C_2 \sin(t) + \cos(t) \log\big(\cos(t)\big) + t \sin(t)$$

where C_1 and C_2 are constants. ∎

Exercises for Section 10.2

Use **HomogeneousSecondOrderLinear** to find the general solution to the following homogeneous second-order linear differential equations.

1. $y'' + y = 0$

2. $y'' - y = 0$

3. $y'' + 3y' + 2y = 0$

4. $y'' + y' = 0$

5. $y'' - 3y' + 2y = 0$

6. $y'' + k^2 y = 0$

7. $y'' + i\,y = 0$

8. $y'' + 3i\,y' - 2y = 0$

Use the command **VariationOfParameters** to find a particular solution to the following second-order linear differential equations which have the functions y_1 and y_2 as solutions of the corresponding homogeneous equation.

9. $\begin{cases} y'' + y = \sec(t) \\ y_1 = \sin(t), \quad y_2 = \cos(t) \end{cases}$

10. $\begin{cases} y'' - 4y' + 3y = \dfrac{e^t}{1 + e^t} \\ y_1 = e^t, \quad y_2 = e^{3t} \end{cases}$

11. $\begin{cases} y'' - y = \cosh(t) \\ y_1 = e^t, \quad y_2 = e^{-t} \end{cases}$

12. $\begin{cases} y'' + 3y' + 2y = \sin(e^t) \\ y_1 = e^{-t}, \quad y_2 = e^{-2t} \end{cases}$

13. $\begin{cases} y'' - y = 2^t \\ y_1 = e^t, \quad y_2 = e^{-t} \end{cases}$

14. $\begin{cases} y'' + y = \sec(t)\tan(t) \\ y_1 = \sin(t), \quad y_2 = \cos(t) \end{cases}$

15. $\begin{cases} y'' - 2y' + 2y = e^t \sec(t) \\ y_1 = e^t \sin(t), \quad y_2 = e^t \cos(t) \end{cases}$

16. $\begin{cases} y'' - 2y' + y = 4\,e^t \log(t) \\ y_1 = e^t, \quad y_2 = t\,e^t \end{cases}$

10.3   Reduction of Order and Trial Solutions via ODE

To use *Mathematica* to find a second solution to a second-order homogeneous linear differential equation

$$a(t)y'' + b(t)y' + c(t)y = 0,$$

we mimic the corresponding mathematical steps (see Theorem 8.11 on page 259) with the miniprogram **ReductionOfOrder**

```
ReductionOfOrder[a_,b_,y1_,const_][t_]:=
    Module[{intfac},
        p = b[t]/a[t];
        intfac = Exp[-Integrate[b[t]/a[t],t]];
        const[1]*y1[t] + const[2]*
            Together[y1[t] Integrate[intfac/y1[t]^2,t]]]
```

Example 10.15. *Use Mathematica to find a second linearly independent solution to the differential equation* $2t^2 y'' + 3t\, y' - y = 0$ *given the solution* $y_1(t) = 1/t$.

Solution. We use

ReductionOfOrder[2#^2&,3#&,1/#&,C][t]

to which *Mathematica* responds

```
C[1]    2 Sqrt[t] C[2]
----  + --------------
 t            3
```

It is easier to use **ODE** with the options **Method->ReductionOfOrder** and **KnownSolution** than it is to use the command **ReductionOfOrder**. Thus

**ODE[2t^2 y'' + 3t y' - y == 0,y,t,
Method->ReductionOfOrder,KnownSolution->1/t]**

yields

```
         C[1]    2 Sqrt[t] C[2]
{{y ->   ----  + --------------}}
          t            3
```

Hence $y_2(t) = 2\sqrt{t}/3$ is a second solution to $2t^2 y'' + 3t\, y' - y = 0$. Since any multiple of a solution to a homogeneous linear differential equation is also a solution, we could have taken $y_2(t) = C\sqrt{t}$ instead. ∎

How can one find the "known solution" to use with **Method->ReductionOfOrder**? **ProposedSolution** is the **ODE** option that has been specifically designed to search for solutions. Thus a reasonable guess can be made for the first solution to a second-order differential equation. Then **ODE** with the option **ProposedSolution** checks whether the guess is in fact a solution. In the case of failure it may be possible to modify the guess (perhaps several times) to find a solution.

Example 10.16. *Use* **ODE** *with the option* **ProposedSolution** *to find a solution to* $y'' + y' + e^{-2t} y = 0$. *Then use* **ReductionOfOrder** *to find the general solution.*

Solution. The factor e^{-2t} suggests that there may be a solution of the form $f(e^{-2t})$. In fact

```
ODE[y'' + y' + E^(-2t)y == 0,y,t,
ProposedSolution->Cos[E^-t]]
```

yields

```
Solution verified
               -t
{{y -> Cos[E   ]}}
```

This tells us that $y_1(t) = \cos(e^{-t})$ is a solution. Then we use

```
ODE[y'' + y' + E^(-2t)y == 0,y,t,
Method->ReductionOfOrder,KnownSolution->Cos[E^-t]]
```

to get

```
            -t              -t
{{y -> C[1] Cos[E  ] - C[2] Sin[E  ]}}
```

Hence we find the general solution in the form

$$y(t) = C_1 \cos(e^{-t}) + C_2 \sin(e^{-t}),$$

where C_1 and C_2 are constants. ∎

Exercises for Section 10.3

Use the command **ReductionOfOrder** to find a second linearly independent solution to each of the following second-order equations when given the indicated first solution.

1. $\begin{cases} t^2 y'' - 6t\, y' + 12y = 0, \\ y_1(t) = t^3 \end{cases}$

2. $\begin{cases} 2t^2 y'' + 4t\, y' - 2y = 0, \\ y_1(t) = t^{(-1+\sqrt{5})/2} \end{cases}$

3. $\begin{cases} y'' + \left(-\dfrac{1}{t} + \dfrac{3}{16t^2}\right)y = 0, \\ y_1(t) = t^{1/4} \exp\left(2t^{1/2}\right) \end{cases}$

4. $\begin{cases} t\, y'' - \left(2 - t\tan(t)\right)y' \\ \quad + \left(\dfrac{2}{t} - \tan(t)\right)y = 0, \\ y_1(t) = t \end{cases}$

5. $\begin{cases} t^2 y'' - 2y = 0, \\ y_1(t) = t^2 \end{cases}$

7. $\begin{cases} (t^2 - t)y'' + (3t - 1)y' + y = 0, \\ y_1(t) = 1/(1 - t) \end{cases}$

6. $\begin{cases} y'' - t\,y' - y = 0, \\ y_1(t) = \exp(t^2/2) \end{cases}$

8. $\begin{cases} t\,y'' + (1 - t)y' + 2y = 0, \\ y_1(t) = t^2 - 4t + 2 \end{cases}$

10.4   Equations with Given Solutions via ODE

The following *Mathematica* miniprogram **EquationFromSolutions** mimics the mathematical procedure described in Section 8.7 for finding a second-order linear differential equation with two given linearly independent functions as solutions.[2]

```
EquationFromSolutions[y1_,y2_][t_,y_]:=
    Module[{w,p,q},
    w = Simplify[y1[t]y2'[t] - y1'[t]y2[t]];
    p = Simplify[(y1''[t]y2[t] - y1[t]y2''[t])/w];
    q = Simplify[(-y1''[t]y2'[t] + y1'[t]y2''[t])/w];
    y''[t] + p y'[t] + q y[t] == 0]
```

Example 10.17. *Use Mathematica to find a differential equation whose solutions are* $y_1(t) = t\,e^t$ *and* $y_2(t) = e^{2t}$.

Solution. We use

EquationFromSolutions[# E^#&,E^(2#)&][t,y]

to get

```
2 t y[t]     (2 - 3 t) y'[t]
--------  +  ---------------  + y''[t]  == 0
 -1 + t           -1 + t
```

Thus, the differential equation is

$$y'' + \frac{2 - 3t}{t - 1}y' + \frac{2t}{t - 1}y = 0. \tag{10.6}$$

To check that $t\,e^t$ is indeed a solution to (10.6) we use

---

[2]This section contains optional material. It uses material from Section 8.7.

```
Simplify[y''[t] + (2 - 3 t)/( t - 1) y'[t]
    + 2 t y[t]/( t - 1) /.
{y[t]->t E^t,y'[t]->D[t E^t,t],
y''[t]->D[t E^t,{t,2}]}]
```

Mathematica answers

```
0
```

Alternatively, we could use

```
ODE[y''[t] + (2 - 3 t)/(t - 1) y'[t] +
    2 t y[t]/(t - 1) == 0,y[t],t,
Method->DSolve,DSolvePackage->True]
```

to which *Mathematica* answers

```
              1/2 + t              -(1/2) + 2 t              1/2 + t
{{y[t] -> E          t C[1] + 2 E               C[2] - 2 E          t C[2]}}
```

We change the names of the constants and write the general solution to (10.6) as

$$y = C_1 t\, e^t + C_2 e^{2t}. \ \blacksquare$$

Exercises for Section 10.4

In each of the following problems use **EquationFromSolutions** to find a second-order linear differential equation that has the functions y_1 and y_2 as solutions.

1. $\begin{cases} y_1 = \sin(2t) \\ y_2 = \cos(2t) \end{cases}$

5. $\begin{cases} y_1 = t\, e^t \\ y_2 = \log(t) \end{cases}$

2. $\begin{cases} y_1 = t\sin(t) \\ y_2 = \cos(t) \end{cases}$

6. $\begin{cases} y_1 = t\, e^t \\ y_2 = \sin(t) \end{cases}$

3. $\begin{cases} y_1 = t^2 \\ y_2 = \log(t) \end{cases}$

7. $\begin{cases} y_1 = t^2 + t \\ y_2 = \cos(t) \end{cases}$

4. $\begin{cases} y_1 = e^t \\ y_2 = \left(\log(t)\right)^2 \end{cases}$

8. $\begin{cases} y_1 = \sin(t) \\ y_2 = \exp(t)\cos(t) \end{cases}$

11

APPLICATIONS OF LINEAR SECOND-ORDER EQUATIONS

In this chapter we study physical systems that change with time t and whose mathematical model is the second-order linear constant-coefficient equation

$$a\frac{d^2 y}{dt^2} + b\frac{dy}{dt} + c\,y = g(t).$$

The function $g(t)$ is called the **forcing function**.

Remarkably, this differential equation can be used to model several different physical processes. In Section 11.1 we model mass-spring systems with zero forcing functions and in Section 11.2 we model mass-spring systems with sinusoidal forcing functions. Analogous models for electrical circuits are given in Section 11.3.

Not only does the theoretical material that we developed in Chapter 9 provide an accurate description of this mechanical system, but the converse is also true. Understanding mass-spring systems provides us with a method for visualizing the solution to a constant-coefficient second-order linear equation. In fact, in certain circumstances we can *hear* the solution; the details are given in Section 11.4.

11.1 Mass-Spring Systems

In this section we apply the theory from Chapter 9 of *homogeneous* constant-coefficient second-order linear differential equations to mass-spring systems with free vibrations. We derive the mass-spring equation in Section 11.1.1. The derivation that we give applies to the general case of *forced vibrations*, which will be solved completely in Section 11.2. Free vibrations constitute the special case in which the forcing function is identically zero. Section 11.1.2 contains preparatory material from trigonometry. Undamped free vibrations are discussed in Section 11.1.3, and damped free vibrations are discussed in Section 11.1.4.

11.1.1   Derivation of the Mass-Spring Equation

The basic model to be studied here consists of a mass suspended from a fixed post by means of a spring, under the influence of gravity. We call this model a **mass-spring system**. Newton's second law of motion (see page 5) can be used to derive a differential equation for the displacement of the mass from its rest position. For this purpose we need to analyze the forces that operate on the mass-spring system.

The mass is denoted by m, measured in suitable units. The gravitational force exerted is written $w = m\,g$, where g is the acceleration due to gravity, approximately 32 feet/second^2 in British units, or 981 centimeters/second^2 in c.g.s. units. The reader should recall the distinction between the concepts of **mass** m and **weight** w; the former refers to the intrinsic resistance of a body to motion, while the latter refers to the gravitational force exerted on the body—thus, the formula $w = m\,g$.

More details about systems of units and physical constants are given in Appendix B. There are also *Mathematica* packages that can be used to facilitate the conversion from one system to another. The following command reads these packages, as well as a package containing the values of many physical constants, into a *Mathematica* session:

```
Needs["Miscellaneous`Master`"]
```

Then the acceleration due to gravity g can be found using

```
AccelerationDueToGravity
```

```
9.80665 Meter
─────────────
      2
  Second
```

To find the value of g in terms of feet/second^2 we use

```
Convert[AccelerationDueToGravity,Feet/Second^2]
```

to get

```
32.174 Feet
───────────
      2
  Second
```

In addition to the gravitational force, there is the elastic force due to the elongation of the spring. The spring is supposed to have length l when no mass is present. When a mass is attached at rest, the spring becomes elongated by an amount Δl. The basic assumption of linear elasticity (**Hooke's law**) is that the elastic force is proportional to the displacement, where the constant of proportionality (called **Hooke's constant**, or the **spring constant**) is denoted by the letter k. For the mass-spring system at rest, this translates into the equation

$$F_{\text{gravitational}} = k\,\Delta l = m\,g. \tag{11.1}$$

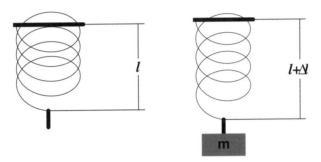

Spring with and without mass

If we know the mass and the elongation, equation (11.1) allows us to compute Hooke's constant for the spring.

Example 11.1. *It is known that a* 10 pound *mass stretches a massless spring* 2 *inches. Find Hooke's constant k for the spring.*

Solution. From the data given, we have

$$w = m\,g = 10 \text{ pounds} \qquad \text{and} \qquad \Delta l = \frac{1}{6} \text{ foot.}$$

Therefore

$$k = \frac{m\,g}{\Delta l} = \frac{10}{1/6} = 60 \text{ pounds/foot.} \quad \blacksquare$$

We introduce a vertical coordinate axis such that its origin is located precisely at the position where the mass is at rest and the positive direction is down. When the mass-spring is set into motion, there is an additional elongation from its equilibrium position. This is denoted by $u(t)$ and is measured so that positive values of u correspond to displacement in the *downward* vertical direction.

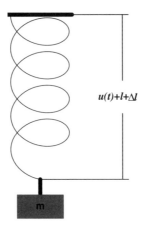

Spring at time *t*

The elastic force acts to restore the spring to its equilibrium position, where the spring has length $l + \Delta l$, and therefore the sign is taken to be negative. We have

$$F_{\text{elastic}} = -k\big(\Delta l + u(t)\big). \tag{11.2}$$

In addition to the forces due to gravity and elasticity, we must consider forces due to friction and forces from the outside. The former might be due to air resistance or other sources of viscosity; they act to retard the motion. For small velocities we assume that the magnitude of this frictional force is proportional to the speed $|du/dt|$. To obtain the correct sign, we note that when the spring moves down, we have $du/dt > 0$, whereas we expect the frictional force to act in the upward direction; similarly, when $du/dt < 0$, the spring is moving up, and we expect that the friction acts in the downward direction. In either case we see that it is correct to write this force in the form

$$F_{\text{frictional}} = -c\frac{du}{dt} = -c\,u'(t), \tag{11.3}$$

where $c \geq 0$ is a constant called the **frictional constant**.

Finally, we may have an external force F_{external}, which can be constant or depend on time. For example, F_{external} might have the form

$$F_{\text{external}}(t) = F_0 \cos(\omega t - \alpha).$$

Later we shall call ω the angular frequency and F_0 the amplitude. However, we can consider a more general external forcing function, which will simply be written $F(t)$ without further specification.

Newton's Second Law of Motion states that the sum of all the forces is equal to the product of the mass and the acceleration. The latter is the second derivative $d^2u/dt^2 = u''(t)$. This leads to the equation

$$F_{\text{gravitational}} + F_{\text{elastic}} + F_{\text{frictional}} + F_{\text{external}} = m\frac{d^2u}{dt^2}. \tag{11.4}$$

Substitution of (11.1)–(11.3) into (11.4) yields

$$m\,g - k\big(\Delta l + u(t)\big) - c\frac{du}{dt} + F(t) = m\frac{d^2u}{dt^2}. \tag{11.5}$$

By noting that $m\,g = k\Delta l$, we simplify the first two terms of (11.5) to get the differential equation

$$m\,u''(t) + c\,u'(t) + k\,u(t) = F(t). \tag{11.6}$$

In this section we study (11.6) in the case that $F(t) \equiv 0$, corresponding to free vibrations. The general case of forced vibrations will be studied in Section 11.2. But first we need to utilize some notions from trigonometry.

11.1.2   The Amplitude-Phase Representation

From Section 9.2 we know that in the case of complex roots the general solution of a constant-coefficient second-order homogeneous equation can be written as

$$y(t) = e^{\lambda t}\big(A\cos(\mu\,t) + B\sin(\mu\,t)\big), \tag{11.7}$$

where λ, μ, A, and B are real constants. Equation (11.7) is well adapted to describe an initial value problem when we are given the initial position and velocity. However, it will sometimes be necessary to write the solution of the initial value problem in terms of an **amplitude** and a **phase angle**, in the form

$$y(t) = R\,e^{\lambda t}\cos(\mu\,t - \alpha). \tag{11.8}$$

Here $R \geq 0$ is the amplitude of the solution and $-\pi < \alpha \leq \pi$ is the phase angle. We call (11.8) the **amplitude-phase representation** of $y(t)$.

To make the required connection between (11.7) and (11.8), we expand the cosine function using trigonometry and equate the two forms of y. In detail, we rewrite (11.8) as

$$y(t) = R\,e^{\lambda t}\cos(\mu\,t)\cos(\alpha) + R\,e^{\lambda t}\sin(\mu\,t)\sin(\alpha). \tag{11.9}$$

Comparing (11.9) with (11.7) and equating the respective coefficients of $\cos(\mu\,t)$ and $\sin(\mu\,t)$ leads to

$$R\cos(\alpha) = A \quad\text{and}\quad R\sin(\alpha) = B. \tag{11.10}$$

In other words, R and α are simply the **polar coordinates** of the point (A, B). If $(A, B) = (0, 0)$, then $R = 0$, and α is undefined. Otherwise, the equations (11.10) can be solved for R and α to yield

$$R = \sqrt{A^2 + B^2} \qquad \text{and} \qquad \tan(\alpha) = \frac{B}{A}.$$

Notice that $B/A = \tan(\alpha) = \tan(\alpha + \pi) = \tan(\alpha - \pi)$, so we need more than the ordinary arctangent function to choose among α, $\alpha - \pi$, and $\alpha + \pi$ when we try to solve the equation $\tan(\alpha) = B/A$ for α. If the point (A, B) is not in the first quadrant, then the phase angle is not in the interval $0 < \alpha < \pi/2$. In order to find the required phase angle in all possible cases, we proceed according to the following rules:

$$A \geq 0 \text{ and } B > 0 \quad \Longrightarrow \quad 0 < \alpha \leq \pi/2,$$

$$A < 0 \text{ and } B \geq 0 \quad \Longrightarrow \quad \pi/2 < \alpha \leq \pi,$$

$$A > 0 \text{ and } B \leq 0 \quad \Longrightarrow \quad -\pi/2 < \alpha \leq 0,$$

$$A \leq 0 \text{ and } B < 0 \quad \Longrightarrow \quad -\pi < \alpha \leq -\pi/2.$$

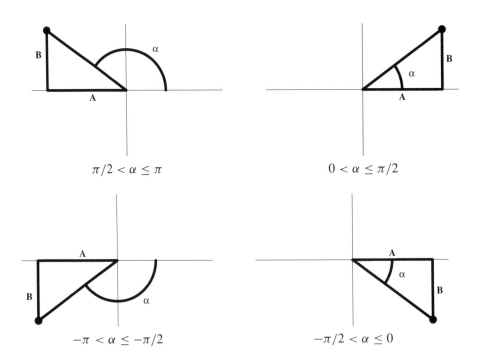

For this reason we define a 2-argument version of the arctangent function as follows:

Definition. *The function of two variables* $\arctan(A, B)$ *is given by*

$$\arctan(A, B) = \begin{cases} \arctan(B/A) & \text{if } A \geq 0, \\ \arctan(B/A) + \pi & \text{if } A < 0 \text{ and } B \geq 0, \\ \arctan(B/A) - \pi & \text{if } A < 0 \text{ and } B < 0. \end{cases}$$

The *Mathematica* commands for $\arctan(B/A)$ and $\arctan(A, B)$ are **ArcTan[B/A]** and **ArcTan[A,B]**.

Example 11.2. *Write each of the functions*

$$y_1(t) = 3\cos(2t) + 4\sin(2t), \qquad y_2(t) = -3\cos(2t) + 4\sin(2t),$$

$$y_3(t) = -3\cos(2t) - 4\sin(2t), \qquad y_4(t) = 3\cos(2t) - 4\sin(2t)$$

in the amplitude-phase representation.

Solution. We have $R = \sqrt{3^2 + 4^2} = 5$. To find the numerical values of $\arctan(3, 4)$, $\arctan(-3, 4)$, $\arctan(3, -4)$, and $\arctan(-3, -4)$, we use the *Mathematica* command

N[{ArcTan[3,4],ArcTan[-3,4],ArcTan[3,-4],ArcTan[-3,-4]}]

obtaining

```
{0.927295, 2.2143, -0.927295, -2.2143}
```

Thus, we have the approximations

$$y_1(t) = 5\cos(2t - 0.9273), \qquad y_2(t) = 5\cos(2t - 2.2143),$$

$$y_3(t) = 5\cos(2t + 2.2143), \qquad y_4(t) = 5\cos(2t + 0.9273). \ \blacksquare$$

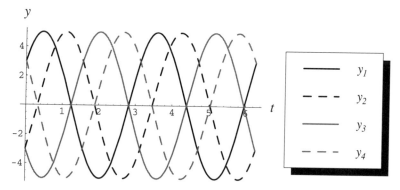

Plots of $y_1(t), y_2(t), y_3(t), y_4(t)$

We define a *Mathematica* function **AmplitudePhaseAngle** that converts a sum of the form $C_1 \cos(\omega) + C_2 \sin(\omega)$ to the amplitude-phase representation as follows:

```
AmplitudePhaseAngle[c1_. Cos[w_] + c2_. Sin[w_]]:=
        N[Sqrt[c1^2 + c2^2]Cos[N[w - ArcTan[c1,c2]]]]


AmplitudePhaseAngle[c1_. Cos[w_]]:= N[c1]Cos[N[w]]


AmplitudePhaseAngle[c2_. Sin[w_]]:= N[c2]Cos[N[w - Pi/2]]
```

(Notice that in the definition of **AmplitudePhaseAngle** we have written **c1_.** and not **c1_** in order to be able to carry out the conversion when **c1** does not explicitly occur.) The calculations of Example 11.2 can be confirmed with

```
{AmplitudePhaseAngle[ 3Cos[2t] + 4Sin[2t]],
 AmplitudePhaseAngle[-3Cos[2t] + 4Sin[2t]],
 AmplitudePhaseAngle[-3Cos[2t] - 4Sin[2t]],
 AmplitudePhaseAngle[ 3Cos[2t] - 4Sin[2t]]}
```

to which we get the answer

```
{5 Cos[0.927295 - 2. t], 5 Cos[2.2143 - 2. t],
  5 Cos[2.2143 + 2. t], 5 Cos[0.927295 + 2. t]}
```

11.1.3   *Undamped Free Vibrations (Simple Harmonic Motion)*

First, let us clarify some of the terms used to describe mass-spring systems.

Definition. *A mass-spring system is said to have* **free vibrations** *if the forcing function is identically zero. Otherwise a mass-spring system is said to have* **forced vibrations**.

We study mass-spring systems with free vibrations in the remaining subsections of the present section. Forced vibrations are postponed to Section 11.2.

Definition. *A mass-spring system is said to be* **undamped** *provided that there are no frictional forces. Otherwise a mass-spring system is said to be* **damped**.

We first determine the motion of an undamped mass-spring system with free vibrations. In this important special case the mass-spring system is said to have **simple harmonic motion**. The differential equation (11.6) reduces to

$$m\,u''(t) + k\,u(t) = 0, \qquad\qquad (11.11)$$

where m and k are positive constants interpreted as the mass and spring constants. According to the classification of Section 9.1, the simple harmonic motion equation (11.11) falls into the case in which the associated characteristic equation has complex conjugate roots. Specifically, the roots are $r_1, r_2 = \pm i\sqrt{k/m}$, and the general solution of (11.11) is

$$u(t) = A\cos(\omega_0 t) + B\sin(\omega_0 t), \qquad (11.12)$$

where $\omega_0 = \sqrt{k/m} > 0$ and A and B are constants.

In practice, it is natural to specify initial conditions for (11.11) in the form

$$u(0) = U_0 \qquad \text{and} \qquad u'(0) = U_1. \qquad (11.13)$$

Equation (11.13) determines the constants A and B in (11.12) by setting

$$A = u(0) = U_0 \qquad \text{and} \qquad B\,\omega_0 = u'(0) = U_1. \qquad (11.14)$$

We use (11.14) to rewrite (11.12) as

$$u(t) = U_0\cos(\omega_0 t) + \frac{U_1}{\omega_0}\sin(\omega_0 t). \qquad (11.15)$$

The solution (11.15) can also be written in the amplitude-phase representation (see page 329)

$$u(t) = R\cos(\omega_0 t - \alpha), \qquad (11.16)$$

by solving the equations

$$\begin{cases} U_0 = R\cos(\alpha), \\[2mm] \dfrac{U_1}{\omega_0} = R\sin(\alpha) \end{cases}$$

for the quantities R and α. Thus

$$R = \sqrt{U_0^2 + \frac{U_1^2}{\omega_0^2}}, \qquad \tan(\alpha) = \frac{U_1}{\omega_0 U_0}, \qquad \text{and} \qquad \alpha = \arctan\!\left(U_0, \frac{U_1}{\omega_0}\right).$$

Definition. *We call $\omega_0 = \sqrt{k/m}$ the* **angular frequency** *and R the* **the amplitude** *of the simple harmonic motion given by (11.16). The* **period** *T is given by*

$$T = \frac{2\pi}{\omega_0} = 2\pi\sqrt{\frac{m}{k}}.$$

Finally, α is called the **phase lag**, *or* **phase angle**.

Note that the angular frequency $\omega_0 = \sqrt{k/m}$ and the period T depend only on the differential equation, while the amplitude R and the phase lag α depend also on the initial conditions.

Example 11.3. *A mass weighing* 5 pounds *stretches a massless spring by* 2 inches. *Find Hooke's constant for the spring and determine the subsequent motion if initially the displacement and velocity are given by* $u(0) = 4$ inches *and* $u'(0) = 0$. *Determine the angular frequency* ω_0, *the amplitude* R, *the period* T, *and the formula for* $u(t)$.

Solution. The mass corresponding to the 5 pound weight is

$$m = \frac{w}{g} = \frac{5}{32} \text{ slug,}$$

and the spring constant is $k = \dfrac{m\,g}{\Delta l} = \dfrac{5}{1/6} = 30$ pounds/foot. The angular frequency is

$$\omega_0 = \sqrt{\frac{k}{m}} = \sqrt{\frac{30}{5/32}} = \sqrt{192} = 8\sqrt{3} \approx 13.86 \text{ radians/second.} \tag{11.17}$$

(See Appendix B for information on these units.) Thus (11.11) becomes

$$\frac{5u''(t)}{32} + 30u(t) = 0, \tag{11.18}$$

with the initial conditions $u(0) = 4$ inches $= (1/3)$foot and $u'(0) = 0$. The solution to (11.18) expressed in feet instead of inches is

$$u(t) = \frac{1}{3}\cos(\omega_0 t), \tag{11.19}$$

where ω_0 is given by (11.17). From (11.19) we see that the amplitude is 4 inches $= (1/3)$ft. The period is

$$T = \frac{2\pi}{\omega_0} \approx 0.45 \text{ seconds.} \ \blacksquare$$

The solution to (11.18) can be found and plotted over $0 \le t \le 2$ with **ODE**. We use

```
ODE[{5u''/32 + 30u == 0,u[0] == 1/3,u'[0] == 0},u,t,
Method->SecondOrderLinear,PlotSolution->{{t,0,2}}]
```

to obtain

```
           Cos[8 Sqrt[3] t]
{{u ->  ---------------------}}                    and the plot
                 3
```

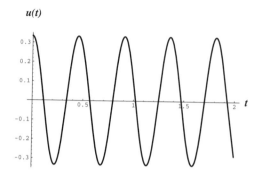

$u(t)$

Solution to $5u''(t)/32 + 30u(t) = 0$
with the initial conditions $u(0) = 1/3$ and $u'(0) = 0$

Let us examine the differential equation $5u''(t)/32 + 30u(t) = 0$ with the more general initial conditions $u(0) = 1/3$ and $u'(0) = a$, where $a = -60, -30, 0, 30, 60$. We use **ODE**'s option **Parameters** to plot the solutions simultaneously. This time we plot the solutions over one period. The command

```
ODE[{5u''/32 + 30u == 0,u[0] == 1/3,u'[0] == a},u,t,
Method->SecondOrderLinear,Parameters->{{a,-60,60,30}},
PlotSolution->{{t,0,2Pi/(8Sqrt[3])}}]
```

yields the graph

$u(t)$

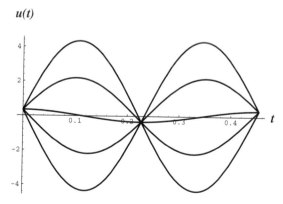

Solutions to $5u''(t)/32 + 30u(t) = 0$ with the initial conditions
$u(0) = 1/3$ and $u'(0) = a$ for $a = -60, -30, 0, 30, 60$

11.1.4   Damped Free Vibrations

We next solve the general case of the mass-spring system with no external forces. The differential equation is

$$m\,u''(t) + c\,u'(t) + k\,u(t) = 0, \tag{11.20}$$

where m, c, and k are all positive. The frictional constant c depends on the viscosity of the medium through which the mass-spring moves; we expect c to be small for air, larger for water, and even larger for oil. We shall see that the assumption $c > 0$ has the consequence that any solution $u(t)$ of (11.20) tends to zero as $t \longmapsto \infty$; this is an essential feature of damped motion that is not shared by undamped motion.

We use the theory of Section 9.1, examining the cases when (11.20) has distinct real roots, repeated real roots, and complex conjugate roots. Because of the physical interpretation, these cases are called **overdamped**, **critically damped**, and **underdamped**. In order to distinguish each case, we need to consider the associated characteristic equation

$$m\,r^2 + c\,r + k = 0 \tag{11.21}$$

of (11.20) and its discriminant $D = c^2 - 4m\,k$. Note that the solutions r_1 and r_2 of (11.21) are given by

$$r_1,\, r_2 = \frac{-c \pm \sqrt{c^2 - 4m\,k}}{2m} = \frac{-c \pm \sqrt{D}}{2m}. \tag{11.22}$$

Case I. The overdamped case $D > 0$

In this case, the frictional constant c is large in relation to the mass m and the spring constant k. The general solution to $m\,u''(t) + c\,u'(t) + k\,u(t) = 0$ has the form

$$u(t) = C_1 e^{r_1 t} + C_2 e^{r_2 t}, \tag{11.23}$$

where C_1 and C_2 are constants and r_1 and r_2 are given by (11.22). Since m and k are positive, $c^2 - 4m\,k < c^2$. Thus, the square root on the right-hand side of (11.22) is less than c. This implies that both r_1 and r_2 are strictly negative, so we can assume that

$$r_2 < r_1 < 0. \tag{11.24}$$

It is clear from (11.23) and (11.24) that

$$\lim_{t \to \infty} u(t) = 0.$$

Furthermore, (11.23) implies that $u(t) = 0$ for at most one value of t.

If $C_1 \neq 0$, we can rewrite (11.23) as

$$u(t) = e^{r_1 t}\left(C_1 + C_2 e^{(r_2 - r_1)t}\right). \tag{11.25}$$

By taking logarithms of the absolute values of both sides of (11.25), we find

$$\log\big(|u(t)|\big) = r_1 t + \log\big|C_1 + C_2 e^{(r_2-r_1)t}\big|.$$

Hence

$$\lim_{t\to\infty} \frac{\log\big(|u(t)|\big)}{t} = \lim_{t\to\infty} \left(r_1 + \frac{1}{t}\log\big|C_1 + C_2 e^{(r_2-r_1)t}\big|\right) = r_1 < 0,$$

since $|C_1 + C_2 e^{(r_2-r_1)t}|$ is bounded as $t \longrightarrow \infty$. (In the special case that $C_1 = 0$ but $C_2 \neq 0$ the limit is r_2.)

Example 11.4. *Find the solution of the mass-spring system governed by*

$$u'' + 8u' + 4u = 0. \tag{11.26}$$

Solve (11.26) *for the initial conditions* $u(0) = 0.1$ *and* $u'(0) = 3.0$.

Solution. In this case, the roots are determined from the associated characteristic equation $r^2 + 8r + 4 = 0$, which is solved to yield

$$r_1, r_2 = \frac{-8 \pm \sqrt{64 - 16}}{2} = \frac{-8 \pm 4\sqrt{3}}{2} = -4 \pm 2\sqrt{3}.$$

Hence the general solution is

$$u_{\mathrm{H}}(t) = C_1 e^{(-4+2\sqrt{3})t} + C_2 e^{(-4-2\sqrt{3})t}.$$

To find and plot the solution to (11.26) with **ODE** we use

```
ODE[{u'' + 8u' + 4u == 0,u[0] == 0.1,u'[0] == 3.0},u,t,
Method->SecondOrderLinear,PlotSolution->{{t,0,3}}]
```

to obtain
$$\left\{\left\{u \to \frac{-0.441}{E^{7.46\,t}} + \frac{0.541}{E^{0.536\,t}}\right\}\right\}$$

Thus, $u(t) \approx 0.541\, e^{-0.536t} - 0.441\, e^{-7.46t}$. ∎

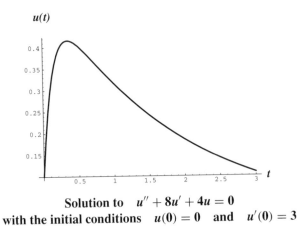

Solution to $u'' + 8u' + 4u = 0$
with the initial conditions $u(0) = 0$ **and** $u'(0) = 3$

We can plot several solutions to $u'' + 8u' + 4u = 0$ (with the same initial position $u(0) = 0.1$ but with varying initial velocity) with the command

```
ODE[{u'' + 8u' + 4u == 0,u[0] == 0.1,u'[0] == a},u,t,
Method->SecondOrderLinear,Parameters->{{a,-4,4,1}},
PlotSolution->{{t,0,3}}]
```

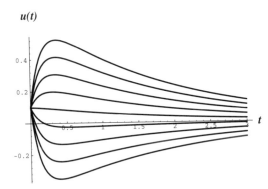

Solutions to $u'' + 8u' + 4u = 0$ **with the initial conditions**
$u(0) = 0.1$ **and** $u'(0) = a$ **for** $a = \pm 4, \pm 3, \pm 2, \pm 1, 0$

This plot confirms our conclusion that $u(t)$ equals zero for at most one value of t.

Case II. The critically damped case $D = 0$

In this case, the frictional constant c exactly equals $2\sqrt{m\,k}$. The general solution to $m\,u''(t) + c\,u'(t) + k\,u(t) = 0$ has the form

$$u(t) = C_1 e^{rt} + C_2 t\, e^{rt}, \tag{11.27}$$

where the root r of the associated characteristic equation is given by $r = -c/(2m)$. We rewrite (11.27) as

$$u(t) = e^{rt}(C_1 + C_2 t); \tag{11.28}$$

by taking logarithms of the absolute values of both sides of (11.28) we find that

$$\log\big(|u(t)|\big) = r\,t + \log|C_1 + C_2 t|.$$

Hence if either C_1 or C_2 is nonzero, we have

$$\lim_{t\to\infty} \frac{\log\big(|u(t)|\big)}{t} = \lim_{t\to\infty}\left(r + \frac{1}{t}\log|C_1 + C_2 t|\right) = r < 0.$$

The qualitative behavior in the critically damped case is exactly the same as in the overdamped case; eventually the solution tends to zero monotonically. It can pass through zero at most once before assuming a definite sign for all sufficiently large values of t.

Example 11.5. *Find the solution of the mass-spring system governed by*

$$u'' + 6u' + 9u = 0 \tag{11.29}$$

with $u(0) = 1$ and $u'(0) = -7$. Find the time t_0 for which $u(t_0) = 0$.

Solution. The general solution to (11.29) is

$$u(t) = C_1 e^{-3t} + C_2 t\, e^{-3t}. \tag{11.30}$$

The constants on the right-hand side of (11.30) are determined by

$$1 = u(0) = C_1 \qquad \text{and} \qquad -7 = u'(0) = -3C_1 + C_2.$$

Thus, $C_1 = 1, C_2 = -7 + 3 = -4$, and the solution is

$$u(t) = e^{-3t}(1 - 4t).$$

Furthermore, $u(t_0) = 0$ when $t_0 = 1/4$. ∎

ODE can be used to find and plot the solution to (11.30) with the command

```
ODE[{u'' + 6u' + 9u == 0,u[0] == 1,u'[0] == -7},u,t,
Method->SecondOrderLinear,PlotSolution->{{t,0,2.5}}]
```

resulting in

$$\{\{u \to E^{-3t} - \frac{4t}{E^{3t}}\}\}$$ and the plot

u(t)

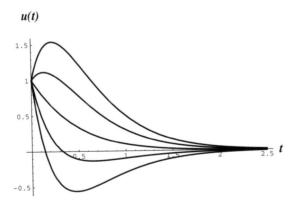

$$u(t) = e^{-3t}(1 - 4t)$$

Several solutions to $u'' + 6u' + 9u = 0$ can be plotted with

```
ODE[{u'' + 6u' + 9u == 0,u[0] == 1,u'[0] == a},u,t,
Method->SecondOrderLinear,Parameters->{{a,-10,6,4}},
PlotSolution->{{t,0,2.5}}]
```

u(t)

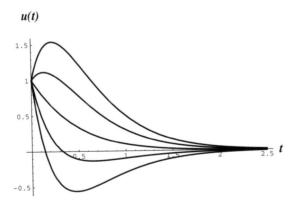

Solutions to $u'' + 6u' + 9u = 0$ with the initial conditions
$u(0) = 1$ and $u'(0) = a$ for $a = -10, -6, -2, 2, 6$

Case III. The underdamped case $D < 0$

In this case, the frictional constant c is small in relation to the mass m and the spring constant k. The general solution to $m\,u''(t) + c\,u'(t) + k\,u(t) = 0$ has the form

$$u(t) = e^{\lambda t}\big(A\cos(\mu\,t) + B\sin(\mu\,t)\big), \tag{11.31}$$

where A and B are real constants and λ and μ are the real numbers determined by

$$\lambda + i\,\mu = -\frac{c}{2m} + i\,\frac{\sqrt{4m\,k - c^2}}{2m}. \tag{11.32}$$

The amplitude-phase representation of $u(t)$ is

$$u(t) = R\,e^{\lambda t}\cos(\mu\,t - \alpha), \tag{11.33}$$

where

$$R = \sqrt{A^2 + B^2}, \qquad \tan(\alpha) = \frac{B}{A}, \qquad \text{and} \qquad \alpha = \arctan(A, B).$$

We call $R\,e^{\lambda t}$ the **damping factor**. Since $\lambda = -c/(2m) < 0$, the solution (11.31) tends to zero when $t \longrightarrow \infty$, as in the previous cases. But the difference here is the presence of **oscillatory behavior**. Instead of at most one zero, the function $u(t)$ has infinitely many zeros, represented by a damped periodic motion. To make this precise, we introduce the notion of **quasiperiod** T_c. This is the time between successive maxima of the function

$$A\cos(\mu\,t) + B\sin(\mu\,t) = R\cos(\mu\,t - \alpha).$$

These successive maxima occur when the argument of the trigonometric function increases by 2π, leading to

$$\mu\,T_c = 2\pi, \qquad \text{or} \qquad T_c = \frac{2\pi}{\mu} = \frac{4\pi\,m}{\sqrt{4m\,k - c^2}}.$$

Equivalently, the quasiperiod can be characterized as twice the time between consecutive zeros of the solution $u(t)$.

   In the limiting case of a frictionless system we have $c = 0$ and a periodic motion with period $T_0 = 2\pi\sqrt{m/k}$. Therefore, the quasiperiod T_c in the general case can be expressed in terms of T_0 by

$$T_c = \frac{2\pi\sqrt{\dfrac{m}{k}}}{\sqrt{1 - \dfrac{c^2}{4m\,k}}} = \frac{T_0}{\sqrt{1 - \dfrac{c^2}{4m\,k}}}. \tag{11.34}$$

The denominator on the right-hand side of (11.34) is strictly less than one, so that we have the inequality

$$T_c > T_0 = 2\pi \sqrt{\frac{m}{k}}.$$

Friction induces a retardation, so that the quasiperiod of the damped motion is *strictly greater than* the period of the corresponding undamped motion.

When the damping is small, we can use the Taylor expansion of the function $1/\sqrt{1-x} = 1 + x/2 + \cdots$ to expand (11.34) as

$$T_c = T_0 \left(1 + \frac{c^2}{8m\,k} + \cdots \right).$$

This shows that the quasiperiod is very close to the period of the associated undamped motion when the damping is small.

Example 11.6. *A mass-spring system is governed by the equation*

$$u'' + 2u' + 17u = 0, \tag{11.35}$$

with the initial conditions $u(0) = 5$ and $u'(0) = 0$. Find the solution and the quasiperiod.

Solution. The associated characteristic equation is $r^2 + 2r + 17 = 0$, whose solutions are

$$r_1, r_2 = \frac{-2 \pm \sqrt{4 - 68}}{2} = -1 \pm 4i,$$

leading to $\lambda = -1$ and $\mu = 4$. The general solution of (11.35) is thus

$$u_H(t) = e^{-t} \big(A\cos(4t) + B\sin(4t) \big).$$

To satisfy the initial conditions, we must have

$$5 = A \qquad \text{and} \qquad 0 = -A + 4B,$$

which are solved to yield $B = A/4 = 5/4$. Thus

$$u(t) = e^{-t} \left(5\cos(4t) + \frac{5\sin(4t)}{4} \right)$$

$$= \frac{5\sqrt{17}\,e^{-t}}{4} \cos\left(4t - \arctan\left(\frac{1}{4}\right) \right) \approx 5.15 e^{-t} \cos(t - 0.24),$$

and the damping factor is approximately $5.15 e^{-t}$. The quasiperiod is

$$T_c = \frac{2\pi}{\mu} = \frac{\pi}{2} \text{ seconds. } \blacksquare$$

Using **ODE**, we can find and plot the solution to (11.35) as follows. First, we define the solution in the phase angle representation by

```
u[t_]=ODE{uu'' + 2uu' + 17uu == 0,uu[0] == 5,uu'[0] == 0},
     uu,t,Method->SecondOrderLinear,Form->Explicit,
     PostSolution->{AmplitudePhaseAngle}]
```

The result is

$$\frac{5 \, \mathrm{Sqrt}[17] \; \mathrm{Cos}[4 \, t \, - \, \mathrm{ArcTan}[5, \, \frac{5}{4}]]}{4 \, E^{t}}$$

The damping factor $5.15 e^{-t} \cos(t - 0.24)$, its negative, and the function $u(t)$ can be simultaneously plotted with

```
Plot[Evaluate[{u[t],u[t]/u[t][[4]],-u[t]/u[t][[4]]}],
{t,0,Pi},PlotStyle->{{AbsoluteThickness[1.5]},
{AbsoluteDashing[{6,6}]},{AbsoluteDashing[{6,6}]}}]
```

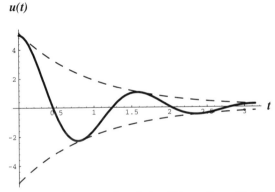

**The solution to $u'' + 2u' + 17u = 0$ with the
initial conditions $u(0) = 5$ and $u'(0) = 0$**

11.1.5   Relaxation Time

We have observed that whenever the frictional constant c is nonzero, the solution $u(t)$ of a mass-spring system with free vibrations tends to zero. In fact

$$u(t) = \begin{cases} e^{r_1 t}\left(C_1 + C_2 e^{(r_2 - r_1)t}\right) & \text{in the overdamped case,} \\ e^{rt}(C_1 + C_2 t) & \text{in the critically damped case,} \qquad (11.36) \\ e^{\lambda t}\left(A\cos(\mu\,t) + B\sin(\mu\,t)\right) & \text{in the underdamped case.} \end{cases}$$

How do we measure the rapidity of the approach of $u(t)$ to zero? Notice that in all three cases of (11.36) the function $u(t)$ has an exponential factor of the form e^{-ht}, where h is positive. It is natural to choose $1/h$ to measure how fast $u(t)$ tends to zero. Thus, we define the **relaxation time**, or **time constant** τ, to be $1/h$. The larger the relaxation time, the more slowly $u(t)$ approaches zero.

The following lemma shows how to compute the relaxation in terms of the mass m, the spring constant k, and the frictional constant c.

Lemma 11.1. *The* **relaxation time** *or* **time constant** τ *is defined by the equation*

$$\tau = \begin{cases} \dfrac{2m}{c - \sqrt{c^2 - 4m\,k}} & \text{if } c^2 > 4m\,k, \\[2ex] \dfrac{2m}{c} & \text{if } c^2 \le 4m\,k. \end{cases} \tag{11.37}$$

Proof. When $c^2 > 4m\,k$, the mass-spring system corresponding to $m\,u''(t) + c\,u'(t) + k\,u(t) = 0$ is overdamped. By definition, $\tau = -1/r_1$, where r_1 is the larger root of the associated characteristic equation. Then the first equation of (11.37) is a consequence of (11.22).

When $c^2 = 4m\,k$, the mass-spring system is critically damped and $\tau = -1/r$, where r is the double root of the associated characteristic equation. Then the second equation of (11.37) is a consequence of (11.22)

Finally, when $c^2 < 4m\,k$, the mass-spring system is underdamped and $\tau = -1/\lambda$, where λ is the real part of either root of the associated characteristic equation. Thus, the third equation of (11.37) follows from (11.33). ∎

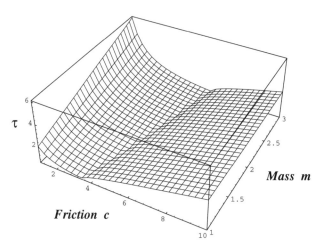

Relaxation time τ **in the case**
$k = 3, \quad 1 \le m \le 3, \quad \text{and} \quad 1 \le c \le 10$

Note that if m and k are fixed, τ as a function of c has a minimum at the critical damping value $c = 2\sqrt{m\,k}$.

Example 11.7. *Find the relaxation time of the following differential equations:*

$$u'' + 8u' + 4u = 0, \tag{11.38}$$

$$u'' + 6u' + 9u = 0, \tag{11.39}$$

$$u'' + 2u' + 17u = 0. \tag{11.40}$$

Solution. For (11.38) the roots of the associated characteristic equation $r^2 + 8r + 4 = 0$ are

$$r_1, r_2 = \frac{-8 \pm \sqrt{64 - 16}}{2} = \frac{-8 \pm 4\sqrt{3}}{2} = -4 \pm 2\sqrt{3}.$$

Thus, (11.38) is overdamped. The relaxation time is

$$\tau = \frac{1}{4 - 2\sqrt{3}} \approx 1.86603 \text{ seconds.}$$

It is easy to see that (11.39) is critically damped, so that in this case

$$\tau = \frac{2}{6} = \frac{1}{3} \text{ second.}$$

Finally, (11.39) has the associated characteristic equation $r^2 + 2r + 17 = 0$, whose roots are

$$r_1, r_2 = \frac{-2 \pm \sqrt{4 - 68}}{2} = -1 \pm 4i.$$

Thus, (11.39) is underdamped, and so in this case $\tau = 1$. ∎

Summary of Techniques Introduced

A mass-spring system in the absence of external forces is governed by the second-order linear homogeneous equation $m\,u'' + c\,u' + k\,u = 0$, where the constant $m > 0$ represents the mass, $c \geq 0$ represents the frictional coefficient, and $k > 0$ represents the spring constant.

    If $c = 0$, the solution is obtained in the form

$$u(t) = A\cos(\omega_0 t) + B\sin(\omega_0 t) = R\cos(\omega_0 t - \alpha).$$

The **angular frequency** $\omega_0 = \sqrt{k/m}$ depends on the equation, while the constants A, B, R, and α depend on the initial conditions. R is the **amplitude** and α is the **phase lag** of the solution

If $c > 0$, then the nature of the solution depends on whether $c^2 > 4m\,k$ (**overdamped case**), $c^2 = 4m\,k$ (**critically damped case**), or $c^2 < 4m\,k$ (**underdamped case**). In the first and second cases all solutions tend to zero when $t \longrightarrow \infty$, passing through zero at most once. In the underdamped case all solutions tend to zero when $t \longrightarrow \infty$ and pass through zero infinitely often. The time between successive passages is given by half of the **quasiperiod** T_c, which is obtained by solving

$$\mu\,T_c = 2\pi,$$

where $\mu = \sqrt{4m\,k - c^2}/(2m)$ is the imaginary part of the root of the associated characteristic equation $m\,r^2 + c\,r + k = 0$.

Exercises for Section 11.1

Problems 1–2 are concerned with Hooke's constant.

1. A weight of 3 pounds stretches a spring by the amount 4 inches. Find Hooke's constant of the spring.

2. Determine Hooke's constant for a spring of natural length 15 inches that is stretched to a distance of 18 inches by an object weighing 5 pounds.

In Exercises 3–8 put $u(t)$ into the amplitude-phase form $u(t) = R\cos(\omega_0 t - \alpha)$.

3. $u(t) = 3\cos(2t) - 7\sin(2t)$ 6. $u(t) = -6\cos(7t) + 6\sin(7t)$

4. $u(t) = \sqrt{3}\cos(5t) + \sin(5t)$ 7. $u(t) = -4\cos(t) - 4\sqrt{3}\sin(t)$

5. $u(t) = 1.2\cos(2t)$ 8. $u(t) = 7\sin(11t)$

Problems 9–19 deal with undamped mass-spring systems $(c = 0)$.

9. A mass of 50 grams is attached to a spring of natural length 100 centimeters. Suppose that the spring is stretched an additional 10 centimeters by the addition of this mass. If the mass is started in motion with an initial velocity of 20 centimeters per second in the downward vertical direction, find the formula for the position of the mass.

10. Suppose that the weight on the spring of Exercise 9 is released from a height of 10 centimeters *below* the equilibrium position with an initial velocity of 2 centimeters per second in the downward direction. Find the formula for the position of the weight.

11. Suppose that the weight on the spring of Exercise 9 is released from a height of 10 centimeters *above* the equilibrium position with an initial velocity of 2 centimeters per second in the downward direction. Find the formula for the position of the weight.

12. Solve the undamped mass-spring equation $m\,u'' + k\,u = 0$ for the initial conditions $u(0) = U_0$ and $u'(0) = 0$ and determine the amplitude and phase lag. Graph the solution in the case that $m = 20$, $k = 5$, and $U_1 = 0$, where U_0 takes on the values $\pm 6, \pm 4, \pm 2$, and 0.

13. Solve the undamped mass-spring equation $m\,u'' + k\,u = 0$ for the initial conditions $u(0) = 0$ and $u'(0) = U_1$, and determine the amplitude and phase lag. Graph the solution in the case that $m = 18$, $k = 2$, and $U_0 = 0$, where U_1 takes on the values $\pm 3, \pm 2, \pm 1$, and 0.

14. Solve the undamped mass-spring equation $m\,u'' + k\,u = 0$ for the initial conditions $u(0) = U_0$ and $u'(0) = U_1$, and determine the amplitude and phase lag. Graph the solution in the case that $m = 25$ and $k = 1$, where U_0 takes on the values ± 50 and 0, and U_1 takes on the values ± 25 and 0.

15. Write the solution of Exercise 14 in the form $u(t) = R\cos(\omega_0 t - \alpha)$, where $R \geq 0$, $-\pi < \alpha \leq \pi$, and $\omega_0 > 0$.

16. Use **ODE** to find the general solution to solve the initial value problem

$$\begin{cases} m\,u''(t) + k\,u(t) = 0, \\ u(0) = U_0, u'(0) = U_1 \end{cases}$$

17. The **total energy** $E(t)$ of a mass-spring system is defined by

$$E(t) = \frac{1}{2}m\,u'(t)^2 + \frac{1}{2}k\,u(t)^2.$$

Show that for an undamped mass-spring system the total energy is a constant independent of t.

18. With reference to Exercise 17, show that in the amplitude-phase representation the amplitude R can be computed in terms of the total energy by the formula

$$R = \sqrt{\frac{2E}{k}}.$$

19. The **phase portrait** of a second-order differential equation is constructed from a Cartesian coordinate system (x_1, x_2), where the horizontal x_1 axis represents the displacement $u(t)$ and the vertical x_2 axis represents the velocity $u'(t)$. Show that a mass-spring system in the absence of friction moves along an ellipse, and find the lengths of the axes of this ellipse.

20. Suppose that the particle of a mass-spring system of mass m and spring constant k is set into motion from a displacement of U_0 with initial velocity U_1.

    a. Find the maximum displacement that the particle attains and the velocity at that point.

    b. Find the maximum velocity the particle attains and the position at which this occurs.

Exercises 21–32 deal with damped mass-spring systems.

21. Find the solution of a mass-spring system for which $m = 2$ grams, $c = 3.5$ gram seconds/centimeter, and $k = 4.7$ grams/centimeter. The initial conditions are $u(0) = 3$ centimeters and $u'(0) = -5$ centimeters/second. Find the quasiperiod and the relaxation time and graph the solution.

22. Find the solution of a mass-spring system for which $m = 1$ gram, $c = 4$ gram seconds/centimeter, and $k = 4$ grams/centimeter. The initial conditions are $u(0) = -1$ centimeter and $u'(0) = 6$ centimeters/second. Find the relaxation time and graph the solution.

23. Find the solution of a mass-spring system for which $m = 3$ grams, $c = 12$ gram seconds/centimeter, and $k = 4$ grams/centimeter. The initial conditions are $u(0) = 1$ centimeter and $u'(0) = -2$ centimeters/second. Find the relaxation time and graph the solution.

24. Assume that a mass-spring system is governed by

$$\begin{cases} u'' + c\,u' + 36u = 0, \\ u(0) = 1, u'(0) = 0, \end{cases}$$

    where $c = 3, 13, 23$. Find the equation of motion and graph the three solutions simultaneously.

25. Suppose that a mass-spring system is overdamped: $c^2 > 4m\,k$. Show that the point $\big(u(t), u'(t)\big)$ moves along a curve defined implicitly by

$$|x_2 - r_1 x_1|^{r_1} = C|x_2 - r_2 x_1|^{r_2},$$

where C is a constant and r_1 and r_2 are the roots of the associated characteristic equation $m r^2 + c r + k = 0$. (See Exercise 18.) Compute the constant C in terms of the initial conditions.

26. Suppose that a mass-spring system is critically damped: $c^2 = 4m k$. Find the implicit equation of the curve along which the point $(u(t), u'(t))$ moves.

27. Suppose that a mass-spring system is underdamped: $c^2 < 4m k$. Show that the quasiperiod T_c and the relaxation time τ are related by the equation

$$\frac{1}{4\pi^2\tau^2} + \frac{1}{T_c^2} = \frac{1}{T_0^2},$$

where $T_0 = 2\pi \sqrt{m/k}$ is the period of the mass-spring system in the absence of friction.

28. Suppose that a mass of 1 gram has a quasiperiod of 24 seconds and a relaxation time of 56 seconds when attached to a spring under the influence of a frictional force. Use the result of Exercise 26 to find the spring constant.

29. Compute the total energy $E(t)$ (see Exercise 17) of a mass-spring system with damping, and show that it is a decreasing function of time.

30. The **Lyapunov function** of an underdamped mass-spring system is defined as $L(t) = m u'(t)^2 + c u(t)u'(t) + k u(t)^2$. Show that $L(t) = L(0)e^{-ct/m}$.

31. Suppose that we have an underdamped mass-spring system, with

$$0 < c^2 < 4m k, \qquad \mu = \frac{\sqrt{4m k - c^2}}{2c}, \qquad \text{and} \qquad \lambda = -\frac{c}{2m}.$$

Define a system of polar coordinates (r, θ) in the phase plane (see Exercise 19) by the formulas

$$\begin{cases} \mu u = r \cos(\theta), \\ u' - \lambda u = r \sin(\theta). \end{cases}$$

Show that in the phase plane the motion takes place along a logarithmic spiral curve whose equation is of the form

$$r = C \, e^{\lambda\theta/\mu},$$

where C is a constant. Find the constant C in terms of the initial conditions and in terms of the amplitude-phase representation of the solution.

32. Consider the initial value problem

$$\begin{cases} u'' + 5u' + 4u = 0, \\ u(0) = 1, u'(0) = a. \end{cases}$$

Find the values of a for which $u(t) \neq 0$ for all $t \geq 0$, and find the values of a for which $u(t_0) = 0$ for some $t_0 \geq 0$.

11.2   *Forced Vibrations of Mass-Spring Systems*

In this section we apply our solution methods to the inhomogeneous equations that arise from the mass-spring system of Section 11.1. We showed that the motion of a mass-spring system is governed by the second-order differential equation

$$m\,u''(t) + c\,u'(t) + k\,u(t) = F(t), \tag{11.41}$$

where m, c, and k are positive constants. We suppose that the forcing function $F(t)$ is given and the unknown function $u(t)$ is to be determined. We confine our attention to the simplest harmonic time dependent forcing function

$$F(t) = F_0 \cos(\omega t - \alpha), \tag{11.42}$$

where F_0 is a constant. Any function of the form $A \cos(\omega t) + B \sin(\omega t)$ can be written in the form (11.42) by making a suitable choice of the constants α and F_0 (see Section 11.1.2). In particular, $\sin(\omega t) = \cos(\omega t - \pi/2)$. We can find a particular solution of (11.42) either by the method of undetermined coefficients (Section 9.3) or by the method of variation of parameters (Section 9.4). The general solution of (11.42) is then the sum of the particular solution and a general solution of the associated homogeneous equation (11.20). By analogy with the steps used to solve the homogeneous equation in Section 11.1, we first consider the case without damping. For simplicity of exposition, we restrict attention below to the case $\alpha = 0$.

Case I. No friction: $c = 0$

In this case, we must solve the differential equation

$$m\,u''(t) + k\,u(t) = F_0 \cos(\omega t). \tag{11.43}$$

According to the method of undetermined coefficients, the form of the solution depends on whether or not the forcing function $F_0 \cos(\omega t)$ is a solution of the homogeneous equation $m\,u''(t) + k\,u(t) = 0$, whose general solution is

$$u_{\mathrm{H}}(t) = C_1 \cos\left(t\sqrt{\frac{k}{m}}\right) + C_2 \sin\left(t\sqrt{\frac{k}{m}}\right). \tag{11.44}$$

There are two subcases.

Case IA. The nonresonant case: $c = 0$ and $\omega \neq \sqrt{k/m}$

When the forcing function $F_0 \cos(\omega t)$ is *not* a solution of the homogeneous equation $m\,u''(t) + k\,u(t) = 0$, we can find a particular solution of (11.43) in the form

$$u_{\mathrm{P}}(t) = A \cos(\omega t). \tag{11.45}$$

Substitution of (11.45) into (11.43) leads to

$$-\omega^2 m\, A + k\, A = F_0, \qquad \text{so that} \qquad A = \frac{F_0}{k - m\,\omega^2}.$$

From (11.44) we see that the general solution of (11.43) is

$$u(t) = C_1 \cos\left(\sqrt{\frac{k}{m}}\,t\right) + C_2 \sin\left(\sqrt{\frac{k}{m}}\,t\right) + \frac{F_0 \cos(\omega t)}{k - m\,\omega^2}. \tag{11.46}$$

We put $\omega_0 = \sqrt{k/m}$ and rewrite (11.46) as

$$u(t) = C_1 \cos(\omega_0 t) + C_2 \sin(\omega_0 t) + \frac{F_0 \cos(\omega t)}{m(\omega_0^2 - \omega^2)}. \tag{11.47}$$

In the following example we give details of the hand calculations needed to solve a mass-spring initial value problem with a periodic forcing function.

Example 11.8. *Find the solution of the differential equation*

$$u'' + 4u = 5\cos(3t), \tag{11.48}$$

with the initial conditions $u(0) = 0$ and $u'(0) = 1$.

Solution. The function $5\cos(3t)$ is not a solution of the homogeneous equation $u'' + 4u = 0$; therefore a particular solution to (11.48) found by the method of undetermined coefficients is

$$u_{\mathrm{P}}(t) = \frac{5\cos(3t)}{4 - 3^2} = -\cos(3t).$$

Thus, the general solution to (11.48) is

$$u(t) = C_1 \cos(2t) + C_2 \sin(2t) - \cos(3t).$$

To satisfy the initial conditions, we must take $0 = u(0) = C_1 - 1$ and $1 = u'(0) = 2C_2$, leading to $C_1 = 1$ and $C_2 = 1/2$. Thus, the solution to (11.48) with the stated initial conditions is

$$u(t) = \cos(2t) + \frac{1}{2}\sin(2t) - \cos(3t).$$

The command

```
ODE[{u'' + 4u == 5 Cos[3t],u[0] == 0,u'[0] == 1},u,t,
Method->SecondOrderLinear,PlotSolution->{{t,0,4Pi}}]
```

with the output

```
                              Sin[2 t]
{{u -> Cos[2 t] - Cos[3 t] + --------}}
                                 2
```

checks the hand calculation and plots the solution. ∎

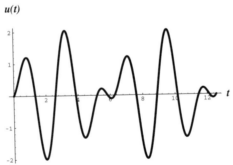

u(t)

The solution to $u'' + 4u = 5\cos(3t)$
with the initial conditions $u(0) = 0$ and $u'(0) = 1$

Beats

When ω is different from but nearly equal to ω_0, it is useful to consider the motion described by (11.47) as a generalization of the harmonic motion $u(t) = R\cos(\omega t - \alpha)$, in which the amplitude R is no longer constant but can vary with time. Suppose the mass is initially at rest and $u(0) = 0$. This leads to a simple initial value problem corresponding to (11.43), namely

$$\begin{cases} m\,u''(t) + k\,u(t) = F_0\cos(\omega t), \\ u(0) = u'(0) = 0. \end{cases} \qquad (11.49)$$

Clearly, $u(0) = u'(0) = 0$ implies that

$$C_1 = -\frac{F_0}{m(\omega_0^2 - \omega^2)} \qquad \text{and} \qquad C_2 = 0$$

in (11.47). Thus, the solution to (11.49) is

$$u(t) = \frac{F_0\big(\cos(\omega t) - \cos(\omega_0 t)\big)}{m(\omega_0^2 - \omega^2)}. \tag{11.50}$$

The right-hand side of (11.50) is the difference of two trigonometric functions with the same amplitude. Using the trigonometric identity

$$\cos(A) - \cos(B) = -2\sin\left(\frac{A - B}{2}\right)\sin\left(\frac{A + B}{2}\right),$$

(see Exercise 5) we rewrite (11.50) as

$$u(t) = \frac{-2F_0}{m(\omega_0^2 - \omega^2)}\sin\left(\frac{(\omega - \omega_0)t}{2}\right)\sin\left(\frac{(\omega + \omega_0)t}{2}\right). \tag{11.51}$$

Suppose now that $|\omega - \omega_0|$ is very small in comparison to $|\omega + \omega_0|$; then the function $t \longmapsto \sin\big((\omega+\omega_0)t/2\big)$ will oscillate much more rapidly than the function $t \longmapsto \sin\big((\omega-\omega_0)t/2\big)$. We can describe (11.51) by saying that $u(t)$ is rapidly varying with angular frequency $(\omega + \omega_0)/2$ but with a slowly varying sinusoidal amplitude

$$\frac{-2F_0}{m(\omega_0^2 - \omega^2)}\sin\left(\frac{(\omega - \omega_0)t}{2}\right).$$

This is the phenomenon of **beats**. It occurs in acoustics when sounds of nearly equal frequency are made by two musical instruments. We illustrate with an example.

Example 11.9. *Find and plot the solution to the initial value problem*

$$\begin{cases} u'' + 16u = \cos(3.5t), \\ u(0) = u'(0) = 0. \end{cases} \tag{11.52}$$

Solution. We need *Mathematica*'s command **TrigFactor** in order to write sums of trigonometric functions as products. It is contained in the package **Trigonometry.m**, which is automatically read in when the package **ODE.m** is read in. To solve and plot (11.52) we use

```
ODE[{u'' + 16u == Cos[3.5t],u[0] == 0,u'[0] == 0},u,t,
Method->SecondOrderLinear,PostSolution->TrigFactor,
PlotSolution->{{t,0,8Pi},PlotPoints->200}]
```

to get the output

```
{{u -> 0.533 Sin[0.25 t] Sin[3.75 t]}}
```

and the plot

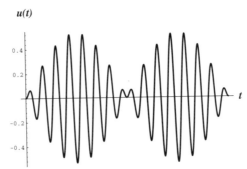

u(t)

A typical graph of beats in forced oscillations:
$$u'' + 16u = \cos(3.5t), \quad u(0) = 0, \quad u'(0) = 0$$

The solution is thus, $u(t) = 0.533 \sin(0.25t) \sin(3.75t)$. ∎

In electronics the variation of amplitude with time is called **amplitude modulation**. The following plot shows how the amplitude of Example 11.9 (that is, $t \longmapsto \pm\sin(0.25t)$) varies in relation to $t \longmapsto \sin(3.75t)$. We use

```
Plot[{0.533 Sin[0.25 t],-0.533 Sin[0.25 t],
0.5333 Sin[0.25 t] Sin[3.75 t]},{t,0,8Pi},PlotPoints->200,
PlotStyle->{{AbsoluteThickness[1],Dashing[{0.03,0.03}]},
{AbsoluteThickness[1],Dashing[{0.03,0.03}]},
{AbsoluteThickness[1.5]}}]
```

to get

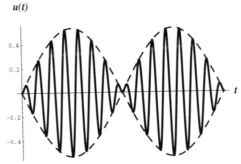

u(t)

Amplitude modulation

Case IB. The resonant case: $c = 0$ and $\omega = \sqrt{k/m}$

In this case, the function $F_0 \cos(\omega t)$ is already a solution of the homogeneous equation $m u''(t) + k u(t) = 0$. From Section 9.3 we know that a particular solution must be sought in the form

$$u_{\mathrm{P}}(t) = t\left(A \cos(\omega t) + B \sin(\omega t)\right). \tag{11.53}$$

We insert (11.53) into (11.43) and after some calculation obtain

$$m u_{\mathrm{P}}'' + k u_{\mathrm{P}} = 2m\, \omega\left(- A \sin(\omega t) + B \cos(\omega t)\right),$$

which will equal $F_0 \cos(\omega t)$ if and only if $A = 0$ and $2m\,\omega B = F_0$. Thus, the general solution to (11.43) is

$$u(t) = C_1 \cos(\omega t) + C_2 \sin(\omega t) + \frac{F_0}{2m\,\omega} t \sin(\omega t). \tag{11.54}$$

It is clear that no matter what the values of the constants C_1 and C_2 are in (11.54), the motion $u(t)$ will become unbounded as $t \longrightarrow \infty$; this is due to the presence of the $t \sin(\omega t)$ term. This is the phenomenon of **resonance**. More generally, the term resonance is applied to a physical system if a bounded forcing function $f(t)$ gives rise to an unbounded response $u(t)$. Resonance occurs in certain electrical systems that will be discussed in Section 11.3.

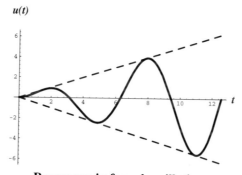

Resonance in forced oscillations

Example 11.10.   *Find the solution of the differential equation*

$$u'' + 16u = \sin(4t), \tag{11.55}$$

with the initial conditions $u(0) = 1$ and $u'(0) = 0$.

Solution.   The function $\sin(4t)$ is already a solution of the homogeneous equation $u'' + 16u = 0$; therefore a particular solution to (11.55) by the method of undetermined coefficients is

$$u_P(t) = -\frac{t\cos(4t)}{8}.$$

Thus, the general solution to (11.55) is

$$u(t) = C_1\cos(4t) + C_2\sin(4t) - \frac{t\cos(4t)}{8},$$

and the solution to (11.55) with the stated initial conditions is

$$u(t) = \cos(4t) + \frac{\sin(4t)}{32} - \frac{t\cos(4t)}{8}.$$

The command

```
ODE[{u'' + 16u == Sin[4t],u[0] == 1,u'[0] == 0},u,t,
Method->SecondOrderLinear,PlotSolution->{{t,0,4Pi}}]
```

with the output

$$\{\{u \; \rightarrow \; \text{Cos}[4\; t] \; - \; \frac{t\; \text{Cos}[4\; t]}{8} \; + \; \frac{\text{Sin}[4\; t]}{32}\}\}$$

checks the hand calculation and plots the solution. ∎

Here is the plot of the solution to (11.55).

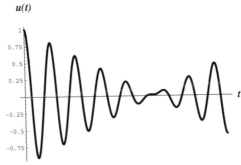

Solution to the initial value problem
$$u'' + 16u = \sin(4t), \quad u(0) = 1, \quad u'(0) = 0$$

Next, we compare resonant and nonresonant solutions.

Example 11.11. *Solve the initial value problem*

$$\begin{cases} u'' + u = 8\cos(a\,t), \\ u(0) = 0,\, u'(0) = 1, \end{cases} \tag{11.56}$$

and use a surface plot to compare the solutions as $a \longrightarrow 1$.

Solution. First, we solve the initial value problem with

```
u[a_,t_] = ODE[{uu'' + uu == 8Cos[a t],
          uu[0] == 0,uu'[0] == 0},uu,t,
          Method->SecondOrderLinear,Form->Explicit]
```

$$\frac{-8\ \text{Cos}[t]}{1 - a^2} + \frac{8\ \text{Cos}[a\ t]}{1 - a^2}$$

The solution has been named **sol** so that we can refer to it. We now have a function of two variables that we can plot using

```
Plot3D[Evaluate[u[a,t]],{t,0,8Pi},{a,0.5,0.9999}]
```

The result is

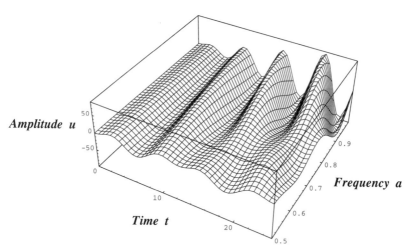

Nonresonant solutions approaching a resonant solution

Next, we add damping to our study of forced vibrations.

Case II. Forced vibrations with damping: c > 0

In this case, we can again obtain a particular solution by the method of undetermined coefficients. The right-hand side of the equation

$$m u''(t) + c u'(t) + k u(t) = F_0 \cos(\omega t) \tag{11.57}$$

cannot be a solution of the homogeneous equation for any value of ω. This is because each of the solutions of the homogeneous equation $m u''(t) + c u'(t) + k u(t) = 0$ contains a term with a factor of the form $e^{\lambda t}$ with $\lambda < 0$. Therefore, the method of undetermined coefficients dictates a particular solution of the form

$$u_P(t) = A \cos(\omega t) + B \sin(\omega t); \tag{11.58}$$

the constants A and B are determined by direct substitution of (11.58) into (11.57); we carried out a similar calculation in Section 9.3, where we solved (9.64). Thus a particular solution of (11.57) is given by

$$u_P(t) = \frac{m(\omega_0^2 - \omega^2) F_0 \cos(\omega t) + c \omega F_0 \sin(\omega t)}{D}, \tag{11.59}$$

where $\omega_0 = \sqrt{k/m}$ and $D = m^2(\omega_0^2 - \omega^2)^2 + c^2 \omega^2$. We can convert (11.59) to the amplitude-phase representation by letting

$$\sin(\delta) = \frac{c \omega}{\sqrt{D}} \qquad \text{and} \qquad \cos(\delta) = \frac{m(\omega_0^2 - \omega^2)}{\sqrt{D}};$$

then (11.59) becomes

$$u_P(t) = \frac{F_0\sqrt{D} \cos(\delta) \cos(\omega t) + F_0\sqrt{D} \sin(\delta) \sin(\omega t)}{D} = \frac{F_0 \cos(\omega t - \delta)}{\sqrt{D}}. \tag{11.60}$$

The particular solution of (11.57) given by (11.60) is also called the **steady-state solution**. It only depends on the coefficients m, c, k, F_0, and ω of the differential equation and not on the initial conditions.

   We can combine the steady-state solution (11.60) with the solution u_H of the homogeneous equation $m u''(t) + c u'(t) + k u(t) = 0$ to yield the general solution of (11.57). We found in Section 9.1 that the form of u_H can be underdamped, overdamped, or critically damped, and that in each case u_H tends to zero as $t \longrightarrow \infty$. For this reason u_H is called the **transient solution**. The detailed form of the transient solution depends on whether the equation (11.57) is overdamped, critically damped, or underdamped. We examine each of the three cases.

Case IIA. The overdamped case: $c^2 > 4mk$

In this case, u_H is a linear combination of two decreasing exponential functions, and the general solution of (11.57) is

$$u(t) = C_1 e^{r_1 t} + C_2 e^{r_2 t} + \frac{F_0 \cos(\omega t - \delta)}{\sqrt{D}},$$

where $r_2 < r_1 < 0$.

Example 11.12. *Solve the initial value problem*

$$\begin{cases} 6u'' + 7u' + u = \cos(t), \\ u(0) = 0, \, u'(0) = 1. \end{cases} \tag{11.61}$$

Plot simultaneously the solution, the steady-state solution, and the transient solution.

Solution. The general solution of the homogeneous equation $6u'' + 7u' + u = 0$ is easily found to be

$$u_H(t) = C_1 e^{-t} + C_2 e^{-t/6}.$$

The method of undetermined coefficients gives the particular solution

$$u_P(t) = \frac{-5\cos(t) + 7\sin(t)}{74} \tag{11.62}$$

for $6u'' + 7u' + u = \cos(t)$. This is the steady-state solution of (11.61); the general solution of (11.61) is

$$u_G(t) = C_1 e^{-t} + C_2 e^{-t/6} + \frac{-5\cos(t) + 7\sin(t)}{74}.$$

To determine C_1 and C_2, we first compute

$$u_G'(t) = -C_1 e^{-t} - \frac{C_2 e^{-t/6}}{6} + \frac{7\cos(t) + 5\sin(t)}{74}.$$

Hence

$$\begin{cases} 0 = u_G(0) = C_1 + C_2 - \dfrac{5}{74}, \\ 1 = u_G'(0) = -C_1 - \dfrac{C_2}{6} + \dfrac{7}{74}. \end{cases} \tag{11.63}$$

The solution of (11.63) is $C_1 = -11/10$ and $C_2 = 216/185$. Hence we find

$$u(t) = -\frac{11}{10}e^{-t} + \frac{216}{185}e^{-t/6} + \frac{-5\cos(t) + 7\sin(t)}{74}$$

$$\approx -\frac{11}{10}e^{-t} + \frac{216}{185}e^{-t/6} + 0.116\cos(2.191 - t) \tag{11.64}$$

for the complete solution to (11.62). From (11.62) and (11.64) it follows that the transient solution to (11.61) is

$$u_{\mathrm{T}}(t) = -\frac{11}{10}e^{-t} + \frac{216}{185}e^{-t/6}.$$

To carry out these calculations using **ODE**, we first define the solution of (11.61) with

```
uall[t_] = ODE[{6u'' + 7u' + u == Cos[t],
          u[0] == 0,u'[0] == 1},u,t,
          Method->SecondOrderLinear,Form->Explicit]
```

obtaining

$$\frac{-11}{10\ E^t} + \frac{216}{185\ E^{t/6}} - \frac{5\ \mathrm{Cos}[t]}{74} + \frac{7\ \mathrm{Sin}[t]}{74}$$

The steady state solution is defined by

```
uss[t_] = AmplitudePhaseAngle[ODE[6u'' + 7u' + u == Cos[t],
          u,t,Method->SecondOrderLinear,Form->Explicit] /.
          {C[1]->0,C[2]->0}]
```

obtaining

```
0.116248 Cos[2.19105 - t]
```

Then a simultaneous plot of the complete solution **uall[t]**, the steady-state solution **uss[t]**, and the transient solution **uall[t] - uss[t]** is obtained using

```
Plot[Evaluate[{uall[t],uss[t],uall[t] - uss[t]}],{t,0,35}]
```

resulting in

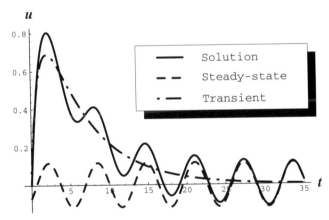

Forced vibration with damping, overdamped case
$$6u'' + 7u' + u = \cos(t), \quad u(0) = 0, \quad u'(0) = 1$$

Case IIB. The critically damped case: $c^2 = 4mk$

In this case, the homogeneous solution is expressed in terms of a single decreasing exponential function, and the general solution takes the form

$$u(t) = e^{rt}(C_1 + C_2 t) + \frac{F_0 \cos(\omega t - \delta)}{\sqrt{D}},$$

where $r < 0$.

Example 11.13. *Solve the initial value problem*

$$\begin{cases} 4u'' + 4u' + u = \cos(5t), \\ u(0) = 1, u'(0) = -1. \end{cases}$$

Plot simultaneously the solution, the steady-state solution and the transient solution.

Solution. We define the solution using

```
uall[t_] = ODE[{4u'' + 4u' + u == Cos[5t],
           u[0] == 1,u'[0] == -1},u,t,
           Method->SecondOrderLinear,Form->Explicit]
```

obtaining

```
  10300         51 t      99 Cos[5 t]    20 Sin[5 t]
 --------- - ---------- - ----------- + -----------
        t/2         t/2      10201          10201
10201 E     101 E
```

Similarly, the steady-state solution is defined by

```
uss[t_] = AmplitudePhaseAngle[ODE[4u'' + 4u' + u == Cos[5t],
           u,t,Method->SecondOrderLinear,Form->Explicit] /.
           {C[1]->0,C[2]->0}]
```

obtaining

```
0.00990099 Cos[2.94226 - 5. t]
```

Then a simultaneous plot of the three solutions is obtained using

```
Plot[Evaluate[{uall[t],uss[t],uall[t] - uss[t]}],{t,0,6Pi}]
```

resulting in

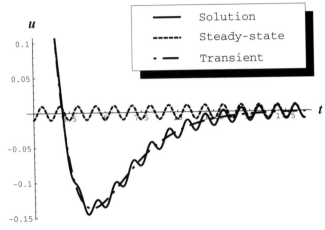

Forced vibration with damping, critically damped case
$$4u'' + 4u' + u = \cos(5t), \quad u(0) = 1, \quad u'(0) = -1$$

Case IIC. The underdamped case: $0 < c^2 < 4mk$

In this case, the homogeneous solution is expressed in terms of decreasing exponentials and trigonometric functions, so that the general solution takes the form

$$u(t) = e^{\lambda t}\left(C_1 \cos(\mu t) + C_2 \sin(\mu t)\right) + \frac{F_0 \cos(\omega t - \delta)}{\sqrt{D}},$$

where $\lambda < 0$.

Example 11.14. *Solve the initial value problem*

$$\begin{cases} 10u'' + 2u' + 5u = \sin(t), \\ u(0) = 1, u'(0) = -1. \end{cases}$$

Plot simultaneously the solution, the steady-state solution, and the transient solution.

Solution. We define the solution using

```
uall[t_] = ODE[{10u'' + 2u' + 5u == Sin[t],
          u[0] == 1,u'[0] == -1},u,t,
          Method->SecondOrderLinear,Form->Explicit]
```

obtaining

$$\frac{31\ \text{Cos}[\frac{7\ t}{10}]}{29\ E^{t/10}} - \frac{2\ \text{Cos}[t]}{29} - \frac{209\ \text{Sin}[\frac{7\ t}{10}]}{203\ E^{t/10}} - \frac{5\ \text{Sin}[t]}{29}$$

Similarly, the steady-state solution is defined by

```
uss[t_] = AmplitudePhaseAngle[ODE[10u'' + 2u' + 5u == Sin[t],
          u,t,Method->SecondOrderLinear,Form->Explicit] /.
          {C[1]->0,C[2]->0}]
```

obtaining

```
0.185695 Cos[1.9513 + t]
```

Then a simultaneous plot of the three solutions is obtained using

```
Plot[Evaluate[{uall[t],uss[t],uall[t] - uss[t]}],{t,0,12Pi}]
```

resulting in

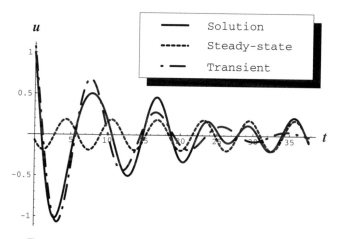

Forced vibration with damping, underdamped case
$$10u'' + 2u' + 5u = \sin(t), \quad u(0) = 0, \quad u'(0) = 1$$

Thus, the solution is

$$u(t) = \frac{e^{-t/10}}{203}\left(217\cos\left(\frac{7t}{10}\right) - 209\sin\left(\frac{7t}{10}\right)\right) - \frac{2\cos(t) + 5\sin(t)}{29}. \ \blacksquare$$

Summary of Techniques Introduced

A mass-spring system in the presence of a sinusoidal forcing term can be solved explicitly by the method of undetermined coefficients. In the frictionless case the form of the solution depends on whether or not the forcing term is in resonance with the homogeneous equation; in the nonresonant case the solution is another sinusoidal oscillation; if the frequency of the forcing term is close to the resonant frequency of the homogeneous equation, then the graph exhibits *beats*. In the resonant case the solution is an unbounded function of time and is 90 degrees out of phase with the forcing function. In the presence of friction, there is no resonant case; the solution is a combination of damped exponentials plus sinusoidal terms, corresponding to the steady-state response of the system to the external force.

Exercises for Section 11.2

1. A mass weighing 12 pounds is attached to a spring with spring constant equal to 2 pounds/inch and is set into motion at $t = 0$ from a height of 1 inch with no velocity by an external force of $8\cos(9t)$ pounds. Assume there is no friction. Find and plot the formula for the position of the mass.

2. A spring is stretched 3 inches by a mass that weighs 4 pounds. The resulting system experiences a frictional force with damping constant equal to 0.10 pound seconds/foot. Further, an external force of $2\cos(2t)$ pounds is applied. Find the transient and steady-state solutions of the resulting initial value problem.

3. Write each of the following functions in the form $A\cos(\omega t - \alpha)\cos(\nu t - \beta)$ for suitable values of $\omega, \nu, \alpha, \beta$:

   a. $\cos(8t) - \cos(6t)$ **c.** $\sin(9t) - \sin(7t)$

   b. $\cos(8t) + \cos(6t)$ **d.** $\sin(9t) + \sin(7t)$

4. Show that a function of the form

$$u(t) = C_1\cos(ct) + C_2\sin(ct) + D_1\cos(ft) + D_2\sin(ft)$$

   with $c < f$ can be written in the form

$$u(t) = A\cos(\omega t - \alpha)\cos(\nu t - \beta),$$

   where $\omega = (c + f)/2$, $\nu = (f - c)/2$ provided that the constants C_1, C_2, D_1, D_2 satisfy the relation $C_1^2 + C_2^2 = D_1^2 + D_2^2$. [Hint: Apply the amplitude-phase representation twice and then use the trigonometric identities for $\sin(x + y)$ and $\cos(x + y)$.]

5. (Converse to Exercise 4)

   a. Show that any function of the form

   $$u(t) = A \cos(\omega t - \alpha) \cos(v t - \beta)$$

   can be written in the form

   $$u(t) = C_1 \cos(c\,t) + C_2 \sin(c\,t) + D_1 \cos(f\,t) + D_2 \sin(f\,t)$$

   for a suitable choice of the constants C_1, C_2, D_1, D_2, c, f, where $C_1^2 + C_2^2 = D_1^2 + D_2^2$. [Hint: Use the trigonometric identities

   $$2\cos(x)\cos(y) = \cos(x - y) + \cos(x + y),$$

   $$2\sin(x)\sin(y) = \cos(x - y) - \cos(x + y),$$

   $$2\sin(x)\cos(y) = \sin(x + y) + \sin(x - y).]$$

   b. Conclude that the set of functions of the form

   $$u(t) = A \cos(\omega t - \alpha) \cos(v t - \beta)$$

   is identical to the set of functions of the form

   $$u(t) = C_1 \cos(c\,t) + C_2 \sin(c\,t) + D_1 \cos(f\,t) + D_2 \sin(f\,t)$$

   where $C_1^2 + C_2^2 = D_1^2 + D_2^2$.

6. Use **ODE** to solve the initial value problem

$$\begin{cases} m\,u''(t) + k\,u(t) = F_0 \cos(\omega t), \\ u(0) = U_0,\ u'(0) = U_1, \end{cases}$$

where $\omega \neq \sqrt{k/m}$.

7. Use **ODE** to solve the initial value problem

$$\begin{cases} m\,u''(t) + k\,u(t) = F_1 \sin(\omega t), \\ u(0) = U_0,\ u'(0) = U_1, \end{cases}$$

where $\omega \neq \sqrt{k/m}$.

8. Use **ODE** to solve the initial value problem

$$\begin{cases} m\,u''(t) + k\,u(t) = F_0 + F_1 t + F_2 t^2, \\ u(0) = U_0,\, u'(0) = U_1, \end{cases}$$

where F_0, F_1, F_2, k, and m are constants.

9. A mass-spring system with constants m and $k > 0$ is acted upon by the forcing function $F(t) = F_0\big(\cos(\omega t)\big)^2$, where $\omega \neq \sqrt{k/m}$. Find the formula for the motion, given the initial conditions $u(0) = 4$ and $u'(0) = 0$.

10. Let $c^2 < 4m\,k$. Find the solution of the differential equation $m\,u''(t) + c\,u'(t) + k\,u(t) = F_0 \cos(\omega t)$ with the following initial conditions:

    a. $u(0) = U_0$ and $u'(0) = 0$;

    b. $u(0) = 0$ and $u'(0) = U_1$;

    c. $u(0) = U_0$ and $u'(0) = U_1$.

11. Consider a mass-spring system with sinusoidal forcing described by the differential equation

$$m\,y'' + c\,y' + k\,y = F_0 \cos(\omega t) \qquad\qquad (11.65)$$

with steady-state solution

$$y(t) = \frac{F_0 \cos(\omega t - \delta)}{\sqrt{D}}. \qquad\qquad (11.66)$$

The **amplification factor** is defined to be

$$\rho = \frac{k}{\sqrt{D}} = \frac{k}{\sqrt{(k - m\omega^2)^2 + c^2\omega^2}} = \frac{1}{\sqrt{\left(1 - \dfrac{\omega^2}{\omega_0^2}\right)^2 + \dfrac{c^2\omega^2}{\omega_0^2}}};$$

it is the factor by which F_0/k must be multiplied to get the amplitude of the steady-state solution of (11.65). Consider ρ as a function of ω/ω_0 for given values of c.

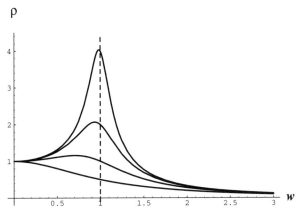

The amplification factor as a function of $w = \omega/\omega_0$

    a. Show that in the limiting case $c = 0$ the amplification factor is infinite when $\omega = \omega_0 = \sqrt{k/m}$.

    b. Show that if $0 < c^2 < 2m\,k$, then the amplification factor is maximum when $\omega^2 = \omega_c^2 = \omega_0^2 - c^2/(2m^2)$.

    c. Show that if $c^2 \geq 2m\,k$, then the amplification factor is maximum when $\omega = 0$.

12. Consider a mass-spring system with sinusoidal forcing described by the differential equation (11.65) with the steady-state solution (11.66). The **steady-state phase lag** δ is defined for $c > 0$ by

$$\tan(\delta) = \frac{c\,\omega}{k - m\omega^2} = \frac{(\omega/\omega_0)(c/\sqrt{m\,k})}{1 - (\omega/\omega_0)^2},$$

with the requirement that $0 \leq \delta < \pi$. This is considered as a function of ω for given values of c, m, k, and F_0.

    a. Show that in the limiting case $c = 0$, the steady-state phase lag has the value $\delta(\omega) = 0$ for $\omega < \omega_0$ and has the value $\delta(\omega) = \pi$ for $\omega > \omega_0$.

    b. Show that for $c > 0$ the steady-state phase lag is a strictly increasing function of ω for $\omega > 0$ with $\delta(0) = 0$, $\delta(\omega_0) = \pi/2$ and $\delta(\omega) \longrightarrow \pi$ as $\omega \longrightarrow \infty$.

13. An undamped mass-spring system with impulsive forcing is described by the differential equation

$$m\,y'' + k\,y = F(t),$$

where $F(t) = F_0$ for $0 < t \le \varepsilon$ and $F(t) = 0$ for $t > \varepsilon$. Assume the initial conditions $y(0) = 0$ and $y'(0) = 0$.

a.   Solve the initial value problem for all t, taking care to maintain the continuity of $y(t)$ and $y'(t)$ at the point $t = \varepsilon$.

b.   Suppose that $\varepsilon \longrightarrow 0$ and $F_0 \longrightarrow \infty$, so that $\varepsilon F_0 \longrightarrow I_0 > 0$. Find the limiting value of the solution $y(t)$ for any t. [Hint: First, find the limiting values of $y(\varepsilon)$ and $y'(\varepsilon)$.]

14.   An underdamped $(c^2 < 4m\,k)$ mass-spring system with impulsive forcing is described by the differential equation

$$m\,y'' + c\,y' + k\,y = F(t),$$

where $F(t) = F_0$ for $0 < t \le \varepsilon$ and $F(t) = 0$ for $t > \varepsilon$. Assume the initial conditions $y(0) = 0$ and $y'(0) = 0$.

a.   Solve the initial value problem for all t, taking care to maintain the continuity of $y(t)$ and $y'(t)$ at the point $t = \varepsilon$.

b.   Suppose that $\varepsilon \longrightarrow 0$ and $F_0 \longrightarrow \infty$, so that $\varepsilon F_0 \longrightarrow I_0 > 0$. Find the limiting value of the solution $y(t)$ for any t. [Hint: First, find the limiting values of $y(\varepsilon)$ and $y'(\varepsilon)$.]

11.3   Electrical Circuits

The methods developed in Chapter 9 can also be applied to certain electrical networks. In Section 7.3 it was shown that the system consisting of a resistor and an inductor gives rise to a first order linear differential equation. Now we add a capacitor and show that the resulting system, called an **LRC series circuit**, gives rise to a second-order linear differential equation.

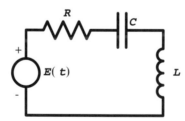

LRC series circuit illustrating the equation

$$L\frac{\mathrm{d}^2 Q}{\mathrm{d}t^2} + R\frac{\mathrm{d}Q}{\mathrm{d}t} + \frac{Q}{C} = E(t)$$

The following table lists commonly used symbols and units for electrical circuits.

Quantity	Symbol	Letter	Unit	Device
Resistance		R	ohms	Resistor
Inductance		L	henry	Inductor
Capacitance		C	farad	Capacitor
Electromotive Force		$E(t)$	volt	Generator or Battery
Charge		Q	coulomb	
Current		I	ampere	

To derive a second-order differential equation, we need the **Kirchhoff voltage law**, given on page 220. In order to apply this law we need to know the voltage drop across each element of the circuit. These voltage drops are as follows:

- According to Ohm's law the voltage drop E_R across a resistor is proportional to the current I passing through the resistor; thus

$$E_R = R\,I;$$

the constant of proportionality R is called the **resistance**.

- According to Faraday's law; the voltage drop E_L across an inductor is proportional to the instantaneous rate of change of the current I passing through the inductor; thus

$$E_L = L\frac{dI}{dt};$$

the constant of proportionality L is called the **inductance**.

- The voltage drop E_C across a capacitor is proportional to the charge Q on the capacitor; thus

$$E_C = \frac{1}{C}Q,$$

where C is called the **capacitance**.

An electromotive force is assumed to add voltage to the circuit. This voltage, denoted by $E(t)$, is assumed to be a given function of time. Application of the Kirchhoff voltage law to our LRC series circuit yields

$$E_L + E_R + E_C = E(t).$$

By combining all of the above, we obtain the equation

$$L \frac{dI}{dt} + R I + \frac{Q}{C} = E(t). \tag{11.67}$$

In the case that the capacitor is absent, (11.67) reduces to equation (7.34) studied in Section 7.3. In the present case, (11.67) is not yet a differential equation, since it contains both the function $I(t)$ and the function $Q(t)$. In order to eliminate one of these, we use the following relation between current (measured in amperes) and charge (measured in coulombs[1]):

$$I(t) = Q'(t). \tag{11.68}$$

Substitution of this relation into (11.67) converts that equation into

$$L Q''(t) + R Q'(t) + \frac{Q(t)}{C} = E(t). \tag{11.69}$$

Then (11.69) is a second-order differential equation whose solution yields the total charge on the capacitor. The current can then be found by differentiation. Alternatively, the current can be found directly as a solution of the differential equation

$$L I''(t) + R I'(t) + \frac{I(t)}{C} = E'(t), \tag{11.70}$$

which results when one differentiates (11.69) and uses the relation (11.68).

We note that (11.69) and (11.70) have the same form as that of the mass-spring system (11.6) studied above. The coefficients in (11.6) correspond to the coefficients in (11.69) according to the following table.

---

[1] Charles Augustin de Coulomb (1736–1806). French physicist.

Mass-Spring System	Electric Circuit
$m\dfrac{d^2u}{dt^2} + c\dfrac{du}{dt} + kt = F(t)$	$L\dfrac{d^2Q}{dt^2} + R\dfrac{dQ}{dt} + \dfrac{Q}{C} = E(t)$
Displacement $\qquad u$	Charge $\qquad Q$
Velocity $\qquad du/dt$	Current $\qquad I = dQ/dt$
Mass $\qquad m$	Inductance $\qquad L$
Friction $\qquad c$	Resistance $\qquad R$
Spring constant $\qquad k$	1/Capacitance $\qquad 1/C$
Forcing function $\qquad F(t)$	Impressed voltage $\qquad E(t)$
Angular frequency $\quad \omega_0 = \sqrt{\dfrac{k}{m}}$	Angular frequency $\quad \omega_0 = \dfrac{1}{\sqrt{LC}}$

We can therefore go through the same mathematical analysis for the electrical system as we did for the mass-spring system.

In the case that the impressed voltage $E(t)$ is identically zero, the charge $Q(t)$ is obtained as a solution of the homogeneous equation

$$L Q'' + R Q' + \frac{Q}{C} = 0. \tag{11.71}$$

The roots of the associated characteristic equation of (11.71) are

$$r_1 = \frac{-R + \sqrt{R^2 - 4L/C}}{2L} \qquad \text{and} \qquad r_2 = \frac{-R - \sqrt{R^2 - 4L/C}}{2L}.$$

Thus, just as with mass-spring systems there are three cases:

(i) overdamped case: $R^2 > 4L/C$.

(ii) critically damped case: $R^2 = 4L/C$.

(iii) underdamped case: $R^2 < 4L/C$.

Example 11.15. *Solve and plot $Q(t)$ and $I(t)$ in the case that $L = 1$ henry, $R = 100$ ohms, $C = 1/(5 \times 10^4)$ farad and $E(t) = 0$. Assume the initial conditions $Q(0) = 0$ and $I(0) = 10$ amperes.*

Solution. In this case, the general solution of (11.71) has the form

$$Q(t) = e^{-Rt/(2L)}\left(C_1 \cos(\omega_0 t) + C_2 \sin(\omega_0 t)\right),$$

where

$$-\frac{R}{2L} = -50 \quad \text{and} \quad \omega_0 = \frac{1}{2L}\sqrt{\frac{4L}{C} - R^2} = \frac{1}{2}\sqrt{4 \times 5 \times 10^4 - 10^4} \approx 217.945.$$

The initial conditions imply that $C_1 = 0$ and $10 = C_2\omega_0$, or $C_2 = 0.0459$. Hence the solution is

$$Q(t) = 0.0459 \sin(217.945t)e^{-50t}.$$

We want to plot both $Q(t)$ and its derivative $I(t)$, so we first use **ODE** to define the function $Q(t)$, using

```
Q[t_] = ODE[{QQ'' + 100.0 QQ' + 5 10^4 QQ == 0,
          QQ[0] == 0,QQ'[0] == 10},QQ,t,
          Method->SecondOrderLinear,Form->Explicit]
```

Then separate plots of $Q(t)$ and $Q'(t)$ are created with

```
Plot[Evaluate[Q[t]],{t,0,0.1}]
```

and

```
Plot[Evaluate[Q'[t]],{t,0,0.1},PlotRange->All]
```

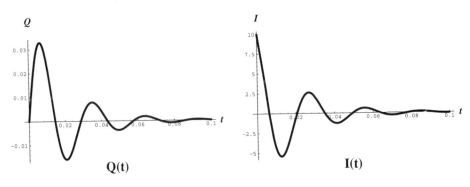

Q(t)   I(t)

Solution to $Q'' + 100.0Q' + 5 \times 10^4 Q = 0$
with the initial conditions $Q(0) = 0$ and $Q'(0) = 10$

LRC circuit with an electromotive force

Next, we consider the case when the impressed voltage of (11.69) has the form

$$E(t) = E_0 \cos(\omega t).$$

In this case, it turns out that the form of the solution of (11.69) depends on the value of the resistance R.

Case IA. The nonresonant case: $R = 0$ and $\omega \neq 1/\sqrt{LC}$

In this case, the forcing function $E_0 \cos(\omega t)$ is *not* a solution of the homogeneous equation $L Q'' + Q/C = 0$. The particular solution obtained by the method of undetermined coefficients is

$$Q_P(t) = \frac{E_0 \cos(\omega t)}{1/C - L\omega^2}.$$

The general solution of (11.69) is then

$$Q(t) = C_1 \cos\left(\frac{t}{\sqrt{LC}}\right) + C_2 \sin\left(\frac{t}{\sqrt{LC}}\right) + \frac{E_0 \cos(\omega t)}{1/C - L\omega^2},$$

where C_1 and C_2 are constants. This solution can be rewritten to show the beat phenomenon, as we did for the mass-spring system.

Case IB. The resonant case: $R = 0$ and $\omega = 1/\sqrt{LC}$

In this case, the function $E_0 \cos(\omega t)$ *is* already a solution of the homogeneous equation. The particular solution obtained by the method of undetermined coefficients is

$$Q_P(t) = \frac{E_0 t \sin(\omega t)}{2L\omega}.$$

Hence the general solution of (11.69) is

$$Q(t) = C_1 \cos(\omega t) + C_2 \sin(\omega t) + \frac{E_0 t \sin(\omega t)}{2L\omega}.$$

Just as in the mass-spring case, $Q(t)$ becomes unbounded no matter what the values of the constants C_1 and C_2 are.

Case II. $R > 0$

In the case of nonzero resistance, resonance is not possible, and a particular solution of (11.69) has the form

$$Q_P(t) = \frac{E_0}{\sqrt{D}} \cos(\omega t - \delta).$$

The general solution then has the form

$$Q(t) = C_1 y_1(t) + C_2 y_2(t) + Q_P(t),$$

where the constants are determined from the initial conditions, and the pair of functions $y_1(t)$ and $y_2(t)$ form a fundamental set of solutions of the homogeneous equation, the details depending on the case (overdamped, critically damped, underdamped).

Summary of Techniques Introduced

Electrical oscillations in LRC circuits can be studied in a similar fashion to the mass-spring system of Sections 11.1 and 11.2. If the impressed voltage is zero, then the solution can be obtained as a combination of exponential functions and trigonometric functions. In the case that the impressed voltage is a sinusoidal function of time, the solution can be obtained by the method of undetermined coefficients. If the resistance is zero, we must distinguish between the resonant and nonresonant cases. In the case of nonzero resistance, the circuit is always nonresonant.

Exercises for Section 11.3

1. Use **ODE** to solve the initial value problem

$$\begin{cases} L\, Q''(t) + Q(t)/C = 0, \\ Q(0) = Q_0,\ Q'(0) = Q_1. \end{cases}$$

2. Use **ODE** to solve the initial value problem

$$\begin{cases} L\, Q''(t) + Q(t)/C = E_0 \cos(\omega t), \\ Q(0) = Q_0,\ Q'(0) = Q_1, \end{cases}$$

where $\omega \neq 1/\sqrt{LC}$.

3. Use **ODE** to solve the initial value problem

$$\begin{cases} L\,Q''(t) + Q(t)/C = E_1 \sin(\omega t), \\ Q(0) = Q_0,\ Q'(0) = Q_1, \end{cases}$$

where $\omega \neq 1/\sqrt{LC}$.

4. Find the form of the solution of the differential equation

$$L\,I''(t) + R\,I'(t) + \frac{I(t)}{C} = 0$$

in the following three cases:

    a. $R > \sqrt{4L/C}$ (overdamped case)

    b. $R = \sqrt{4L/C}$ (critically damped case)

    c. $0 < R < \sqrt{4L/C}$ (underdamped case)

5. Suppose that a circuit contains an inductance, a capacitor, and a resistor in series, with no impressed voltage. Find the formula for the charge if the constants have the following values:

    a. $L = 0.2$ henry, $R = 300$ ohms, $C = 0.0001$ farad.

    b. $L = 0.2$ henry, $R = 100$ ohms, $C = 0.005$ farad.

    c. $L = 10$ henrys, $R = 100$ ohms, $C = 0.002$ farad.

6. Suppose that a circuit contains an inductance, a capacitor, and a resistor in series, with the given impressed voltage. Find the formula for the charge if the constants have the following values:

    a. $L = 10$ henrys, $R = 30$ ohms, $C = 0.02$ farad,
$E(t) = 50 \sin(2t)$ volts.

    b. $L = 3$ henrys, $R = 13$ ohms, $C = 0.25$ farad,
$E(t) = 10 \cos(2t)$ volts.

    c. $L = 1$ henry, $R = 4$ ohms, $C = 0.5$ farad,
$E(t) = 7 \cos(t) + 8 \sin(t)$ volts.

7. Suppose that a circuit contains an inductance, a capacitor, and a resistor in series, with the given impressed voltage. Find the current $I(t)$ given the initial current (in amperes) and charge on the capacitor (in coulombs).

    a. $L = 2$ henrys, $R = 16$ ohms, $C = 0.02$ farad,
$E(t) = 100$ volts, $I(0) = 0$, $Q(0) = 5$.

   b. $L = 2$ henrys, $R = 60$ ohms, $C = 0.0025$ farad,
$E(t) = 100 \exp(-t)$ volts, $I(0) = Q(0) = 0$.

   c. $L = 2$ henrys, $R = 60$ ohms, $C = 0.0025$ farad,
$E(t) = 200 \exp(-10t)$ volts, $I(0) = 0$, $Q(0) = 1$.

8. If $L = 1/5$ henry and $C = 8 \times 10^{-5}$ farad, determine the resistance for which the circuit is critically damped.

9. Determine the steady-state current in a series circuit if $L = 1$ henry, $R = 5000$ ohms, $C = 2.5 \times 10^{-7}$ farad, and the impressed voltage is
$E(t) = 110 \cos(120\pi t)$ volts.

10. Suppose that we have a series circuit with given values of L, R, C, and an impressed voltage of the form $E_0 \cos(\omega t)$. For what value of ω will the amplitude of the steady-state current be a maximum?

11. A series LRC circuit has $L = 0.5$ henry, $R = 20$ ohms, $C = 10^{-2}$ farad, and an impressed voltage $E = 12$ volts. Assume no initial current and no initial charge. Find and plot the subsequent charge and current.

12. A series LRC circuit has $L = 1$ henry, $R = 20$ ohms, $C = 0.005$ farad, and an impressed voltage $E = 6$ volts. Assume no initial current and no initial charge. Find and plot the subsequent charge and current.

13. A series LRC circuit has $L = 1/2$ henry, $R = 30$ ohms, $C = 0.004$ farad, and an impressed voltage $E = 6$ volts. Assume an initial charge of $1/10$ coulomb and an initial current of 2 amperes Find and plot the subsequent charge and current.

14. A series LRC circuit has $L = 0.05$ henry, $R = 10$ ohms, $C = 0.004$ farad, and an impressed voltage $E = 110$ volts. Assume no initial current and no initial charge. Find and plot the subsequent charge and current.

15. A series LRC circuit has $L = 5/100$ henrys, $R = 10$ ohms, $C = 0.004$ farad and an impressed voltage $E = 110 \cos(100t)$volts. Assume no initial current and no initial charge. Find and plot the subsequent charge and current.

16. A series LRC circuit has $L = 20$ henry, $R = 180$ ohms, $C = 1/280$ farad and an impressed voltage $E = 2 \sin(t)$ volts. Assume an initial current of one ampere and no initial charge. Find and plot the subsequent charge and current.

11.4 *Sound*

The notion of resonance can be understood by listening to the solution of differential equations using *Mathematica*'s command **Play**. This command has the syntax

Play[f,{t,tmin,tmax}]

where **f** is the amplitude of a sound that is audible between the times **tmin** and **tmax**. For example

Play[Sin[2Pi440t],{t,0,2}];

plays a sound of 440 cycles per second (called an "A-440").

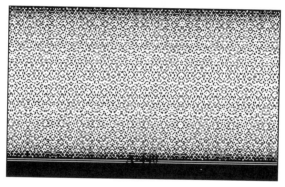

The solution of the differential equation

$$u'' + \omega_0^2 u = \cos(\omega t) \tag{11.72}$$

is given by

$$u(t) = \frac{-2}{m(\omega_0^2 - \omega^2)} \sin\left(\frac{(\omega - \omega_0)t}{2}\right) \sin\left(\frac{(\omega + \omega_0)t}{2}\right). \tag{11.73}$$

If ω is different from but very close to ω_0, then beats occur. If in addition ω_0 is so large that the sound is in the audible range, then we can hear the solution. We can solve (11.72) in the case that $\omega_0 = 2\pi \times 440.0$ and $\omega = 2\pi \times 440.2$ with the command

```
u[t_]=N[ODE[{uu'' + (2Pi 440)^2 uu == Cos[omega t],
     uu[0] == 0, uu'[0] == 0},uu,t,
     Method->SecondOrderLinear,Form->Explicit] /.
     omega-> 2Pi 440.2]
```

which results in

```
0.000143889 Cos[2764.6 t] - 0.000143889 Cos[2765.86 t]
```

The solution $u(t)$ oscillates so rapidly that its graph appears to be a solid black patch, as we see from the command

```
Plot[u[t]//Evaluate,{t,0,16 Pi/5},PlotPoints->1000]
```

which results in

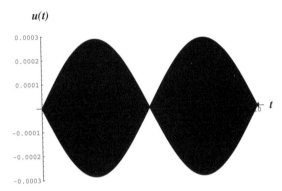

Plot of the solution to

$$u'' + (2\pi \times 440.0)^2 u = \cos(2\pi \times 440.2t)$$

The beats are clearly visible.

The solution to (11.72) can be heard using

```
Play[u[t]//Evaluate,{t,0,16 Pi/5}]
```

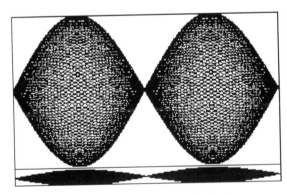

Sound diagram of the solution to

$$u'' + (2\pi \times 440.0)^2 u = \cos(2\pi \times 440.2t)$$

In words, the sound A-440 gradually increases in intensity to its maximum amplitude, then decreases to zero amplitude. Afterwards, the process is repeated.

More information about sound, including many interesting examples, can be found in the book [GrGl1] and the CD-ROM that accompanies it.

12

HIGHER-ORDER LINEAR DIFFERENTIAL EQUATIONS

The purpose of this chapter is to extend to higher-order differential equations the theory of second-order linear differential equations that we developed in Chapters 8 and 9. The techniques needed to solve an nth-order equation are direct extensions of those we have already studied; however, they are frequently more difficult to apply.

Higher-order linear differential equations are intrinsically more difficult than first-order linear equations. Formula (3.35) shows that any first-order linear equation can be solved by two integrations. On the other hand, in 1892 Vessiot[1] proved a proposition that is accurately quoted as, "For $n > 1$ the general linear differential equation of order n cannot be solved by successive integrations". (See [Vess].) The situation is analogous to that of general polynomial equations of degree five and higher, which cannot be solved by successive extraction of radicals and rational operations.

Section 12.1 contains general material on higher-order linear equations. In Section 12.2 we discuss how the theory of homogeneous second-order constant-coefficient equations generalizes to homogeneous higher-order constant-coefficient equations. Then in Section 12.3 we give the general method for finding a particular solution to a higher-order linear equation using the method of variation of parameters. We show in Section 12.4

---

[1]

Ernest Vessiot (1865–1952). French mathematician, a student of Picard. During World War I Vessoit made important discoveries in ballistics. Later, he extended results of the famous French differential geometer E. Cartan and also extended Fredholm integrals to partial differential equations.

how to use **ODE** to solve n^{th}-order linear equations. Seminumerical solutions of nth-order constant-coefficient equations are discussed in Section 12.5.

12.1   General Forms

We shall consider n^{th}-order *linear* differential equations, that is, differential equations that have the form

$$a_n(t)\frac{d^n y}{dt^n} + a_{n-1}(t)\frac{d^{n-1}y}{dt^{n-1}} + \cdots + a_1(t)\frac{dy}{dt} + a_0(t)y = g(t), \qquad (12.1)$$

where the function $a_n(t)$ is nowhere zero on some interval $a < t < b$. We say that (12.1) is **homogeneous** provided that $g(t)$ is identically zero; otherwise the equation is said to be **inhomogeneous**. Other notions that we have defined for second-order linear differential equations have obvious generalizations to n^{th}-order linear differential equations. If necessary, we can divide both sides of (12.1) by $a_n(t)$ to convert (12.1) to the **leading-coefficient-unity-form**

$$y^{(n)} + p_{n-1}(t)y^{(n-1)} + \cdots + p_0(t)y = r(t). \qquad (12.2)$$

Since (12.1) (or (12.2)) involves the nth derivative of y, we can expect that its solution involves n constants of integration. Thus n initial conditions are required to obtain a unique solution to (12.1). The **initial value problem** corresponding to (12.2) is:

$$\begin{cases} y^{(n)} + p_{n-1}(t)y^{(n-1)} + \cdots + p_0(t)y = r(t), \\ y(t_0) = Y_0, \ldots, y^{(n-1)}(t_0) = Y_{n-1}. \end{cases} \qquad (12.3)$$

There is an existence and uniqueness theorem for higher-order equations that generalizes Theorem 8.1:

Theorem 12.1. **(Existence and uniqueness theorem)** *Suppose that the functions* $p_{n-1}(t), \ldots, p_0(t), r(t)$ *are continuous on an interval* $a < t < b$. *Then there exists one and only one function* $y(t)$ *that satisfies the initial value problem* (12.3) *for* $a < t < b$. *Here* t_0 *is a given point with* $a < t_0 < b$, *and* $Y_0, \ldots, Y_{n-1}$ *are given real numbers.*

The strategy for solving (12.1) is quite similar to the strategy we described in Chapter 8. First, we try to find a fundamental set of solutions to the homogeneous equation obtained by replacing $g(t)$ by 0.

Definition. *Functions* $\{y_1, \ldots, y_n\}$ *are said to form a* **fundamental set of solutions** *of the homogeneous differential equation*

$$a_n(t)\frac{d^n y}{dt^n} + a_{n-1}(t)\frac{d^{n-1}y}{dt^{n-1}} + \cdots + a_1(t)\frac{dy}{dt} + a_0(t)y = 0 \qquad (12.4)$$

on an interval $a < t < b$, provided that each y_j $(1 \leq j \leq n)$ satisfies (12.4) and every solution $y(t)$ of (12.4) can be written in the form

$$y(t) = C_1 y_1(t) + C_2 y_2(t) + \cdots + C_n y_n(t) \tag{12.5}$$

*for suitable constants $C_1, \ldots, C_n$. We call (12.5) the **general solution** of (12.4).*

It is possible to generalize the definition of the Wronskian from two functions to n functions.

Definition.   *The **Wronskian** of differentiable functions $y_1, \ldots, y_n$ is the function $W(y_1, y_2, \ldots, y_n)$ given by the formula*

$$W(y_1, y_2, \ldots, y_n)(t) = \det \begin{pmatrix} y_1(t) & y_2(t) & \cdots & y_n(t) \\ y_1'(t) & y_2'(t) & \cdots & y_n'(t) \\ \vdots & \vdots & & \vdots \\ y_1^{(n-1)}(t) & y_2^{(n-1)}(t) & \cdots & y_n^{(n-1)}(t) \end{pmatrix}.$$

Just as in the case of second-order equations, the Wronskian can be used to determine when several solutions constitute a fundamental set of solutions. Theorem 8.4 generalizes to n^{th}-order equations as

Theorem 12.2.   *Suppose that $y_1, \ldots, y_n$ satisfy the same n^{th}-order homogeneous linear differential equation*

$$a_n(t) \frac{d^n y}{dt^n} + a_{n-1}(t) \frac{d^{n-1} y}{dt^{n-1}} + \cdots + a_1(t) \frac{dy}{dt} + a_0(t) y = 0, \tag{12.6}$$

on the interval $a < t < b$. Then $\{y_1, \ldots, y_n\}$ forms a fundamental set of solutions to (12.6) if and only if the Wronskian $W(y_1, \ldots, y_n)(t_0) \neq 0$ for some t_0 with $a < t_0 < b$.

Furthermore, there is a generalization to higher-order linear differential equations of Abel's formula (8.39):

Theorem 12.3.   *Suppose that $y_1, \ldots, y_n$ are solutions to the same n^{th}-order homogeneous linear differential equation*

$$y^{(n)} + p_{n-1}(t) y^{(n-1)} + \cdots + p_0(t) y = 0 \tag{12.7}$$

on the interval $a < t < b$. Then the Wronskian $W(t) = W(y_1, \ldots, y_n)(t)$ satisfies the first-order differential equation

$$W'(t) = -p_{n-1}(t) W(t).$$

Hence for fixed t_0 with $a < t_0 < b$, the Wronskian $W(t)$ can be explicitly computed by **Abel's formula***:*

$$W(t) = W(t_0) \exp\left(-\int_{t_0}^{t} p_{n-1}(s)\,ds\right). \tag{12.8}$$

If $W(t_0) \neq 0$, then $W(t) \neq 0$ for all t with $a < t < b$.

The proof of Theorem 12.3 uses the same ideas as that of Theorem 8.5 but is more complicated because $n \times n$ determinants are involved.

To define the Wronskian of n functions in *Mathematica*, we need to use a *list* of n functions instead of the individual functions. In this way the Wronskian becomes a function of one argument, the list, instead of a function of n arguments.

```
Wronskian[ls_List][t_]  :=
    Module[{tmp},tmp=Through[ls[t]];
       Simplify[Det[FoldList[D,tmp,
          Table[t,{Length[tmp]-1}]]]]]]
```

Example 12.1. *Use Mathematica to compute the Wronskian of* sin, cos *and* log.

Solution. We use

```
Wronskian[{Sin,Cos,Log}][t]
```

to get

```
 -2
t    - Log[t]
```

The discussion of linear dependence and independence given in Section 8.5 can also be generalized.

Definition. *Functions $y_1, \ldots, y_n$ defined on an interval $a < t < b$ are said to be* **linearly dependent** *if there exist constants $C_1, \ldots, C_n$, not all zero, such that*

$$C_1 y_1(t) + \cdots + C_n y_n(t) \equiv 0 \tag{12.9}$$

for $a < t < b$. If it is not possible to find such constants, then the functions are said to be **linearly independent***.*

The following lemma generalizes Lemma 8.6.

Lemma 12.4. *Functions $y_1, \ldots, y_n$ defined on an interval $a < t < b$ are linearly independent if and only if*

$$C_1 y_1(t) + C_2 y_2(t) + \cdots + C_n y_n(t) \equiv 0$$

implies that $C_1 = C_2 = \cdots = C_n = 0$.

There are also superposition and subtraction principles generalizing those for second-order equations given on page 243.

Superposition principle

If y_1 and y_2 are solutions to the **homogeneous equation**

$$y^{(n)} + p_{n-1}(t)y^{(n-1)} + \cdots + p_0(t)y = 0,$$

then so is $C_1 y_1(t) + C_2 y_2(t)$ for any choice of the constants C_1 and C_2.

Subtraction principle

If y_1 and y_2 are both solutions to the second-order linear **inhomogeneous equation**

$$y^{(n)} + p_{n-1}(t)y^{(n-1)} + \cdots + p_0(t)y = r(t),$$

then $y = y_1 - y_2$ is a solution to the homogeneous equation

$$y^{(n)} + p_{n-1}(t)y^{(n-1)} + \cdots + p_0(t)y = 0.$$

Exercises for Section 12.1

Reduce each of the following higher-order equations to the leading-coefficient-unity-form (12.2).

1. $\log(t)y^{(4)} + e^t y''' + 3y'' + y = 0$ **3.** $\dfrac{y'''}{t} + y'' + t\,y' + y = 1$

2. $t\,y^{(5)} + 4y^{(4)} + t\,y''' = t\,\sin(t)$ **4.** $t^8 y^{(8)} + 4t^4 y^{(4)} + 4y = 0$

For each of the following differential equations, determine intervals on which a solution is certain to exist.

5. $y^{(8)} + 2y^{(4)} + y = 0$ **7.** $t^5 y^{(5)} = 1$

6. $t^3 y''' + t^2 y'' + t\,y' + y = 0$ **8.** $t(t-1)y''' + y = t^2$

Use *Mathematica* to compute the Wronskian of each of the following sets of functions.

9. $\{e^t, t\,e^t, e^{3t}\}$ **10.** $\{\sin(t), t^3\sin(t), t^5\sin(t), t^7\sin(t)\}$

11. Show that the operator L defined by

$$L[y] = y^{(n)} + p_{n-1}(t)y^{(n-1)} + \cdots + p_0(t)y$$

is linear in the following sense:

$$L[C_1 y_1 + C_2 y_2] = C_1 L[y_1] + C_2 L[y_2],$$

where C_1 and C_2 are constants and y_1 and y_2 are differentiable functions.

12.2   *Constant-Coefficient Higher-Order Homogeneous Equations*

We now turn to the study of the homogeneous differential equation

$$a_n y^{(n)} + a_{n-1} y^{(n-1)} + \cdots + a_1 y' + a_0 y = 0, \tag{12.10}$$

where $a_n, \ldots, a_0$ are constants with $a_n \neq 0$. Given our experience with constant-coefficient second-order equations in Chapter 9, we expect that (12.10) will have solutions of the form e^{rt} for suitable r. Therefore, we make the following definition.

Definition. *The* **associated characteristic equation** *of the homogeneous differential equation* (12.10) *is*

$$a_n r^n + a_{n-1} r^{n-1} + \cdots + a_1 r + a_0 = 0. \tag{12.11}$$

In fact, it is very easy to check that e^{rt} is a solution to the *differential equation* (12.10) if and only if r is a solution to the *algebraic equation* (12.11). However, n^{th}-degree polynomials are much more complicated than 2nd-degree polynomials.

The **fundamental theorem of algebra** states that a polynomial of degree n has n roots, say $r_1, \ldots, r_n$, so we can (in principle) factor the left-hand side of (12.11) as

$$a_n(r - r_1) \cdots (r - r_n). \tag{12.12}$$

However, the actual determination of the roots $r_1, \ldots, r_n$ can be quite complicated, or even impossible, by the extraction of roots. There is indeed Cardano's[2] formula for the roots of

---

2

Girolamo Cardano (1501–1576). Italian mathematician and physician. Cardano devised a method for solving the general cubic equation by transforming it into a cubic without a second-degree term and then applying "Cardano's rule" to the solution of this reduced equation. The solution in specialized cases was known to Tartaglia and Scipione dal Ferro.

a cubic polynomial ($n = 3$) and Ferrari's[3] formula for the roots of a biquadratic polynomial ($n = 4$). However, these formulas are so involved that they are of limited practical use. Moreover, a famous result of Galois[4] states that no formula in terms of radicals exists for the roots of the general polynomial $a_n r^n + a_{n-1} r^{n-1} + \cdots + a_1 r + a_0$ when $n \geq 5$.

Since the formulas for the roots of a polynomial of degree $n \geq 3$ are either nonexistent or extremely complicated, there is little point in giving a general *algebraic* technique for solving the differential equation (12.10) when $n \geq 3$. On the other hand, there are many important special cases of higher-order polynomials for which we can find the roots.

Higher-Order Differential Equations Whose Associated Characteristic Equation Can Be Factored

The first technique to try when solving a higher-order homogeneous differential equation with constant-coefficients is to try to factor the associated characteristic equation. Knowledge of a root r_1 automatically produces a factor $r - r_1$, which allows us to reduce to an equation of one lower degree. This technique is quite effective in some cases.

If the roots $r_1, \ldots, r_n$ of the associated characteristic equation are real and distinct, then the functions $e^{r_1 t}, \ldots, e^{r_n t}$ constitute a fundamental set of solutions for (12.10). Hence the general solution to (12.10) is

$$y(t) = C_1 e^{r_1 t} + C_2 e^{r_2 t} + \cdots + C_n e^{r_n t},$$

where $C_1, \ldots, C_n$ are constants.

Example 12.2. *Find the general solution to $y''' - 2y'' - 5y' + 6y = 0$.*

Solution. The associated characteristic equation is $r^3 - 2r^2 - 5r + 6 = 0$, so we try to factor the polynomial $r^3 - 2r^2 - 5r + 6$. In fact, it is easily checked that 1 is a root. We can use long division to compute

$$\frac{r^3 - 2r^2 - 5r + 6}{r - 1} = r^2 - r - 6 = (r + 2)(r - 3).$$

---

[3]Ludovico Ferrari (1522–1565). Italian mathematician, student of Cardano. Ferrari's solution to the general fourth-degree equation was published by Cardano in his **Ars Magna**, which also contained Cardano's solution to the general third degree equation. Cardano and Ferrari became embroiled with Tartaglia in an acrimonious priority dispute.

[4]

 Evariste Galois (1811–1832). French mathematician. Galois used group theory to show that the general fifth-degree equation is not solvable algebraically. The highly original work of Galois eventually asserted a profound influence on the development of algebra.

Hence $r^3 - 2r^2 - 5r + 6 = (r-1)(r+2)(r-3)$, and so the general solution to $y''' - 2y'' - 5y + 6 = 0$ is

$$y = C_1 e^t + C_2 e^{-2t} + C_3 e^{3t},$$

where C_1, C_2, and C_3 are constants. $\blacksquare$

If the roots $r_1, \ldots, r_n$ of the associated characteristic equation are not distinct, then the functions $e^{r_1 t}, \ldots, e^{r_n t}$ no longer constitute a fundamental set of solutions because they are linearly dependent. In the case of a second-order equation with a repeated root r_1, we observed that both $e^{r_1 t}$ and $t\, e^{r_1 t}$ are solutions to the differential equation. In the case of an nth-order equation for which r_1 is a root of the associated characteristic equation repeated k times, it turns out that

$$e^{r_1 t}, t\, e^{r_1 t}, \ldots, t^{k-1} e^{r_1 t}$$

are linearly independent solutions.

Example 12.3. *Find the general solution to*

$$y^{(4)} - 12y''' + 54y'' - 108y' + 81y = 0. \tag{12.13}$$

Solution. The associated characteristic equation of (12.13) factors as

$$r^4 - 12r^3 + 54r^2 - 108r + 81 = (r-3)^4,$$

so that 3 is a root of multiplicity 4. Thus not only is e^{3t} a solution to (12.13), but also $t\, e^{3t}$, $t^2 e^{3t}$, and $t^3 e^{3t}$. These four functions are linearly independent, and so they form a fundamental set of solutions to (12.13). Hence the general solution to (12.13) is

$$y = C_1 e^{3t} + C_2 t\, e^{3t} + C_3 t^2 e^{3t} + C_4 t^3 e^{3t},$$

where $C_1, \ldots, C_4$ are constants. $\blacksquare$

A solution to a higher-order equation can contain both real and complex exponentials. It is usually true that the coefficients $a_n, \ldots, a_0$ in (12.10) are real; when this is the case, the roots occur in complex conjugate pairs; the resulting complex exponentials should always be converted to trigonometric functions using Euler's formula (9.42).

Example 12.4. *Find the general solution to*

$$y''' + y'' + y' + y = 0. \tag{12.14}$$

Solution. Since the associated characteristic equation of (12.14) factors as

$$r^3 + r^2 + r + 1 = (r^2 + 1)(r + 1) = (r + i)(r - i)(r + 1),$$

it has roots $-1, i$, and $-i$. In terms of complex exponentials, the general solution to (12.14) is

$$y = A_1 e^{-t} + A_2 e^{it} + A_3 e^{-it}, \tag{12.15}$$

where A_1, A_2, and A_3 are complex constants. We can use Euler's formula (9.42) to convert (12.15) to

$$y = C_1 e^{-t} + C_2 \cos(t) + C_3 \sin(t), \tag{12.16}$$

where C_1, C_2, and C_3 are real constants. It is not necessary to worry about the exact relation between the constants in (12.15) and (12.16). Thus (12.16) is the preferred form of the solution of (12.14). ∎

In the next example we solve an equation whose associated characteristic equation has double complex roots.

Example 12.5. *Find the general solution to*

$$y^{(4)} + 2y'' + y = 0. \tag{12.17}$$

Solution. This time the associated characteristic equation factors as

$$r^4 + 2r^2 + 1 = (r^2 + 1)^2 = (r - i)^2 (r + i)^2,$$

so that i and $-i$ are double roots. In terms of complex exponentials the general solution to (12.17) is

$$y = A_1 e^{it} + A_2 e^{-it} + A_3 t\, e^{it} + A_4 t\, e^{-it}, \tag{12.18}$$

where A_1, A_2, A_3, and A_4 are complex constants. Just as in Example 12.4 we use Euler's formula (9.42) to convert (12.15) to

$$y = C_1 \cos(t) + C_2 \sin(t) + C_3 t \cos(t) + C_4 t \sin(t), \tag{12.19}$$

where C_1, C_2, C_3, and C_4 are real constants. Then (12.19) is the preferred form of the solution to (12.17). ∎

Next, we solve a fourth-order initial value problem.

Example 12.6. *Solve the initial value problem*

$$\begin{cases} y^{(4)} + 4y = 0, \\ y(0) = 0,\ y'(0) = 1,\ y''(0) = 0,\ y'''(0) = 2. \end{cases} \tag{12.20}$$

Solution. The associated characteristic equation has a tricky factorization:

$$r^4 + 4 = (r^2 + 2r + 2)(r^2 - 2r + 2).$$

The roots of each quadratic factor are easily found; thus, the four roots of the polynomial $r^4 + 4$ are $\pm 1 \pm i$. Therefore, the general solution to $y^{(4)} + 4y = 0$ in terms of complex exponentials is

$$y = e^t (C_1 e^{it} + C_2 e^{-it}) + e^{-t} (C_3 e^{it} + C_4 e^{-it}). \tag{12.21}$$

We can use Euler's formula (9.42) to convert (12.21) to the form

$$y = e^t \big(A \cos(t) + B \sin(t) \big) + e^{-t} \big(C \cos(t) + D \sin(t) \big). \qquad (12.22)$$

(As usual, it is not necessary to find the explicit formulas relating A, B, C, D to C_1, C_2, C_3, C_4.) To determine A, B, C, D from the initial conditions, we need to differentiate (12.22) three times:

$$\left\{ \begin{aligned} y' &= e^t \big((A + B) \cos(t) + (-A + B) \sin(t) \big) \\ &\quad + e^{-t} \big((-C + D) \cos(t) + (-C - D) \sin(t) \big), \\ y'' &= e^t \big(2B \cos(t) - 2A \sin(t) \big) + e^{-t} \big(-2D \cos(t) + 2C \sin(t) \big), \qquad (12.23) \\ y''' &= e^t \big((-2A + 2B) \cos(t) + (-2A - 2B) \sin(t) \big) \\ &\quad + e^{-t} \big((2C + 2D) \cos(t) + (-2C + 2D) \sin(t) \big). \end{aligned} \right.$$

From (12.22), (12.23), and the initial conditions we obtain the following system of four linear equations:

$$\left\{ \begin{aligned} 0 &= A + C, \\ 1 &= A + B - C + D, \\ 0 &= 2B - 2D, \\ 2 &= -2A + 2B + 2C + 2D. \end{aligned} \right. \qquad (12.24)$$

The solution to (12.24) is

$$A = C = 0, \qquad B = D = \frac{1}{2},$$

and so the solution to the initial value problem (12.20) is

$$y = \frac{1}{2} \big(e^t + e^{-t} \big) \sin(t) = \cosh(t) \sin(t). \ \blacksquare$$

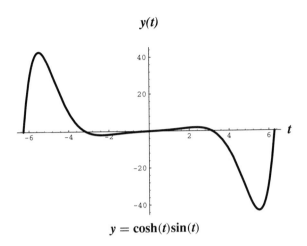

$$y = \cosh(t)\sin(t)$$

Finally, we show how to solve a simple third-order equation with complex coefficients.

Example 12.7. *Find the general solution of* $y''' + i\,y = 0$.

Solution. Since the associated characteristic equation factors as

$$r^3 + i = (r - i)(r^2 + i\,r - 1),$$

the general solution is

$$y(t) = C_1 e^{it} + C_2 \exp\left(\frac{(-i + \sqrt{2})\,t}{2}\right) + C_3 \exp\left(\frac{(-i - \sqrt{2})\,t}{2}\right),$$

where C_1, C_2, and C_3 are constants. ∎

Exercises for Section 12.2

Find the general solution to each of the following differential equations:

1. $y''' + y' = 0$

2. $y''' - y'' + y' - y = 0$

3. $y^{(4)} - 25y = 0$

4. $y^{(7)} + 2y^{(5)} - y''' - 2y' = 0$

5. $y^{(6)} - 6y^{(5)} + 15y^{(4)} - 20y''' + 15y'' - 6y' + y = 0$

6. $y^{(4)} - 10y''' + 35y'' - 50y' + 24y = 0$

7. $y^{(8)} - 2y^{(6)} - 3y^{(4)} + 4y'' + 4y = 0$

8. $y^{(8)} - 3y^{(7)} + 2y^{(6)} - 6y^{(5)} + y^{(4)} - 3y''' = 0$

9. $y^{(6)} - 12y^{(5)} + 55y^{(4)} - 120y''' + 135y'' - 108y' + 81y = 0$

10. $y^{(8)} - 2y^{(7)} + 7y^{(6)} - 12y^{(5)} + 17y^{(4)} - 22y''' + 17y'' - 12y' + 6y = 0$

Solve each of the following initial value problems and plot the solution.

11. $\begin{cases} y''' + y'' + y' + y = 0, \\ y(0) = 0, \, y'(0) = 1, \, y''(0) = 0 \end{cases}$

12. $\begin{cases} y^{(4)} - 5y'' + 4y = 0, \\ y(0) = 1, \, y'(0) = 1, \, y''(0) = 0, \, y'''(0) = 0 \end{cases}$

13. $\begin{cases} y^{(5)} + 4y''' + 4y' = 0, \\ y(0) = 1, \, y'(0) = -1, \, y''(0) = 1, \, y'''(0) = 1, \, y^{(4)}(0) = 1 \end{cases}$

14. $\begin{cases} y^{(7)} + 6y^{(5)} + 9y''' = 0, \\ y(0) = 1, \, y'(0) = -1, \, y''(0) = 0, \, y'''(0) = 1, \, y^{(4)}(0) = 1, \\ y^{(5)}(0) = 1, \, y^{(6)}(0) = 1 \end{cases}$

15. $\begin{cases} y^{(8)} - 2y^{(7)} + 10y^{(6)} - 18y^{(5)} + 36y^{(4)} - 54y''' + 54y'' - 54y' + 27y = 0, \\ y(0) = 0, \, y'(0) = -10, \, y''(0) = 10, \, y'''(0) = 1, \, y^{(4)}(0) = 0, \\ y^{(5)}(0) = 0, \, y^{(6)}(0) = 0, \, y^{(7)}(0) = 0 \end{cases}$

16. $\begin{cases} y^{(6)} - 2y^{(5)} - y^{(4)} + 4y''' - y'' - 2y' + y = 0, \\ y(0) = 10, \, y'(0) = 10, \, y''(0) = -10, \, y'''(0) = 0, \\ y^{(4)}(0) = 0, \, y^{(5)}(0) = 0 \end{cases}$

12.3   Variation of Parameters for Higher-Order Equations

In this section we extend to higher-order equations the method of variation of parameters developed in Section 9.4. Our goal is to find a particular solution to an inhomogeneous linear differential equation written in the leading-coefficient-unity-form

$$y^{(n)} + p_{n-1}(t)y^{(n-1)} + \cdots + p_0(t)y = r(t). \tag{12.25}$$

As in the second-order case, it is first necessary to find a fundamental set of solutions $\{y_1, \ldots, y_n\}$ for the homogeneous equation

$$y^{(n)} + p_{n-1}(t)y^{(n-1)} + \cdots + p_0(t)y = 0. \tag{12.26}$$

The general solution to (12.26) has the form

$$y_H = C_1 y_1 + C_2 y_2 + \cdots + C_n y_n, \tag{12.27}$$

where $C_1, \ldots, C_n$ are constants.

We shall need a generalization of Cramer's rule, which we explained on page 248. This is Cramer's rule for systems of n linear *algebraic* equations in n unknowns.

Theorem 12.5. (**Cramer's rule for systems of** n **linear equations in** n **unknowns**) *Suppose that*

$$\det(A) = \det \begin{pmatrix} a_{11} & a_{12} & \cdots & a_{1n} \\ a_{21} & a_{22} & \cdots & a_{2n} \\ \vdots & \vdots & & \vdots \\ a_{n1} & a_{n2} & \cdots & a_{nn} \end{pmatrix} \neq 0.$$

Then the solution to the system of linear equations

$$\begin{cases} a_{11}x_1 + \cdots + a_{1n}x_n = b_1, \\ a_{21}x_1 + \cdots + a_{2n}x_n = b_2, \\ \quad\vdots \\ a_{n1}x_1 + \cdots + a_{nn}x_n = b_n \end{cases}$$

is given by

$$x_1 = \frac{\det(A_1)}{\det(A)}, \quad x_2 = \frac{\det(A_2)}{\det(A)}, \ldots, \quad x_n = \frac{\det(A_n)}{\det(A)}, \tag{12.28}$$

where A_m is the matrix obtained from A by replacing the mth column with $(b_1, \ldots, b_n)^T$.

Proof. Theorem 12.5 can be proved using the same method that we used to prove Theorem 8.3. Or one may carefully examine the formulas used to compute the inverse of a matrix in equation (A.4) of Appendix A to deduce the above form of Cramer's rule. To do this in detail, we recall that the inverse matrix is defined in terms of $\mathrm{adj}(A) = (A_{ij})^T$ in equation (A.4), where A_{ij} is the ij^{th} cofactor defined in Section A.2 of Appendix A, and $(A_{ij})^T$ is the transpose of A_{ij}. The solution of the equation $A\mathbf{x} = \mathbf{b}$ is obtained by applying $\mathrm{adj}(A)$ to the column vector $(b_1, \dots, b_n)^T$; the first component of the result is

$$x_1 = \frac{A_{11}b_1 + \cdots + A_{n1}b_n}{\det A}. \tag{12.29}$$

But the numerator of (12.29) is precisely $\det(A_1)$, which one can see by expanding $\det(A_1)$ by the minors of its first column. Exactly the same reasoning allows us to compute the other components $x_2, \dots, x_n$ and to conclude equation (12.28) from equation (A.4) of Appendix A. ∎

Armed with Cramer's rule, we return to differential equations. As in the second-order linear case, a good guess for a particular solution to (12.25) is

$$y_{\mathrm{p}}(t) = u_1(t)y_1(t) + u_2(t)y_2(t) + \cdots + u_n(t)y_n(t), \tag{12.30}$$

where the constants $C_1, \dots, C_n$ in (12.27) have been replaced by possibly nonconstant functions $u_1(t), \dots, u_n(t)$. In order to determine the n functions $u_1(t), \dots, u_n(t)$ we need n conditions. One condition is the requirement that y_{p} satisfy (12.25). The other $n-1$ conditions generalize (9.72) and are chosen to eliminate higher derivatives of $u_1(t), \dots, u_n(t)$. Here are the n conditions:

$$\begin{cases} y_1 u_1' & + & y_2 u_2' & + \cdots + & y_n u_n' & = 0, \\ y_1' u_1' & + & y_2' u_2' & + \cdots + & y_n' u_n' & = 0, \\ y_1'' u_1' & + & y_2'' u_2' & + \cdots + & y_n'' u_n' & = 0, \\ & \vdots & & & & \vdots \\ y_1^{(n-2)} u_1' & + & y_2^{(n-2)} u_2' & + \cdots + & y_n^{(n-2)} u_n' & = 0, \\ y_1^{(n-1)} u_1' & + & y_2^{(n-1)} u_2' & + \cdots + & y_n^{(n-1)} u_n' & = r. \end{cases} \tag{12.31}$$

We consider (12.31) to be a system of n algebraic equations in the unknown functions $u_1', \dots, u_n'$. This system can be solved by Cramer's rule (Theorem 12.5). Note that the coefficients on the left-hand side of (12.31) form a matrix the determinant of which is precisely the Wronskian $W = W(y_1, y_2, \dots, y_n)$. Therefore, we find that

$$u_m' = \frac{W_m r}{W} \tag{12.32}$$

for $m = 1, \ldots, n$, where $W_m(t)$ is the determinant formed from $W(t)$ by replacing the mth column by the column vector $(0, \ldots, 0, 1)^T$, that is,

$$
W_m(t) = \det \begin{pmatrix}
y_1(t) & \cdots & 0 & \cdots & y_n(t) \\
y_1'(t) & \cdots & 0 & \cdots & y_n'(t) \\
\vdots & & \vdots & & \vdots \\
y_1^{(n-1)}(t) & \cdots & 1 & \cdots & y_n^{(n-1)}(t)
\end{pmatrix}.
$$

From (12.32) it follows that

$$
u_m(t) = \int \frac{W_m(s)r(s)}{W(s)} ds \bigg|_{s \to t} \tag{12.33}
$$

for $m = 1, \ldots, n$. We substitute (12.33) into (12.30) to obtain

$$
y_P(t) = \sum_{m=1}^{n} y_m(t) \int \frac{W_m(s)r(s)}{W(s)} ds \bigg|_{s \to t}. \tag{12.34}
$$

We can write (12.34) more compactly as

$$
y_P(t) = \int \sum_{m=1}^{n} \left(\frac{y_m(t)W_m(s)r(s)}{W(s)} \right) ds \bigg|_{s \to t}. \tag{12.35}
$$

Summarizing, we have

Theorem 12.6. (**Variation of parameters**) *Suppose that* $\{y_1, \ldots, y_n\}$ *is a fundamental set of solutions of the homogeneous differential equation*

$$
y^{(n)} + p_{n-1}(t)y^{(n-1)} + \cdots + p_0(t)y = 0.
$$

Then a particular solution y_P *to the inhomogeneous equation*

$$
y^{(n)} + p_{n-1}(t)y^{(n-1)} + \cdots + p_0(t)y = r(t)
$$

is given by (12.34) *or* (12.35).

Example 12.8. *Find the general solution to* $y''' + y' = \tan(t)$.

Solution. The general solution to the homogeneous equation $y''' + y' = 0$ is

$$
y_H = C_1 + C_2 \cos(t) + C_3 \sin(t),
$$

where C_1, C_2, and C_3 are constants. Thus $\{y_1, y_2, y_3\}$ constitutes a fundamental set of solutions of the homogeneous equation, where

$$y_1(t) = 1, \qquad y_2(t) = \cos(t), \qquad \text{and} \qquad y_3(t) = \sin(t).$$

It is easy to compute the Wronskian: $W(t) = W(y_1, y_2, y_3)(t) = 1$. Furthermore,

$$W_1(t) = \det \begin{pmatrix} 0 & \cos(t) & \sin(t) \\ 0 & -\sin(t) & \cos(t) \\ 1 & -\cos(t) & -\sin(t) \end{pmatrix} = 1;$$

similarly, $W_2(t) = -\cos(t)$ and $W_3(t) = -\sin(t)$. Therefore, a particular solution is given by

$$y_P(t) = \int \sum_{m=1}^{3} \left(\frac{y_m(t) W_m(s) \tan(s)}{W(s)} \right) ds \bigg|_{s \to t}$$

$$= \int \big(y_1(t) W_1(s) + y_2(t) W_2(s) + y_3(t) W_3(s) \big) \tan(s) ds \bigg|_{s \to t}$$

$$= \int \big(1 - \cos(t)\cos(s) - \sin(t)\sin(s) \big) \tan(s) ds \bigg|_{s \to t}$$

$$= \left(\int \frac{\sin(s) ds}{\cos(s)} - \cos(t) \int \sin(s) ds - \sin(t) \int \frac{\sin(s)^2 ds}{\cos(s)} \right) \bigg|_{s \to t}$$

$$= -\log\big(\cos(t)\big) + \cos(t)^2 - \sin(t) \int \frac{1 - \cos(t)^2 dt}{\cos(t)}$$

$$= -\log\big(\cos(t)\big) + 1 - \sin(t) \log\big(\tan(t) + \sec(t)\big). \tag{12.36}$$

When we compute the general solution of the inhomogeneous equation, we can subsume the 1 on the right-hand side of (12.36) into one of the constants of integration of the homogeneous equation. Therefore, the general solution of the inhomogeneous equation is given by

$$y = C_1 + C_2 \cos(t) + C_3 \sin(t) - \log\big(\cos(t)\big) - \sin(t) \log\big(\tan(t) + \sec(t)\big),$$

where C_1, C_2, and C_3 are constants. ∎

Exercises for Section 12.3

First use the method of variation of parameters to find a particular solution of each of the following differential equations. Then find the general solution.

1. $y''' - 2y'' - y' + 2y = e^t$

2. $y''' - y'' + y' - y = \sin(2t)$

3. $y^{(4)} + 5y'' + 4y = \sec(t)$

4. $y^{(4)} - y = t\sin(2t)$

5. $y''' - 4y'' + 3y' = te^t$

6. $y''' - 4y'' + 3y' = \dfrac{e^t}{1 + e^t}$

7. $y^{(6)} - 6y^{(5)} + 15y^{(4)} - 20y''' + 15y'' - 6y' + y = t$

8. $y^{(4)} - 10y''' + 35y'' - 50y' + 24y = \cosh(t)$

9. $y^{(4)} - 2y''' - 13y'' + 14y' + 24y = t\sinh(t)$

10. $y^{(6)} - 12y^{(5)} + 55y^{(4)} - 120y''' + 135y'' - 108y' + 81y = \sin(t) + \cos(t)$

11. $y^{(6)} - 3y^{(5)} + 4y^{(4)} - 6y''' + 5y'' - 3y' + 2y = t^2e^t$

12. $y^{(8)} - 2y^{(7)} + 3y^{(6)} - 4y^{(5)} + 3y^{(4)} - 2y''' + y'' = t\,e^t\sin(t)$

13. If $y_1 = t$, $y_2 = t\log(t)$, and $y_3 = t(\log(t))^2$ are solutions to the homogeneous equation associated with

$$t^3y''' + t\,y' - y = 24t\,\log(t), \qquad t > 0,$$

find the general solution.

14. Assume that f is a continuous function and consider

$$t^3y''' - 3t^2y'' + 6t\,y' - 6y = t^3 f(t), \qquad t > 0.$$

By trying $y = t^m$, find solutions to the homogeneous equation and then find the general solution in terms of f. Does this pattern continue to an nth-order equation?

12.4   *Higher-Order Differential Equations via* ODE

Simple higher-order linear differential equations with constant-coefficients can be solved with *Mathematica*'s built-in command **DSolve**, but more complicated ones need **ODE**. Symbolic solutions can be obtained using **ODE** with **Method->NthOrderLinear** or **Method->ApproximateNthOrderLinear**. We explain the option **Method->NthOrderLinear** in the present section, and in Section 12.5 we explain approximate method **Method->ApproximateNthOrderLinear**.

ODE's *option* Method->NthOrderLinear

ODE with the option **Method->NthOrderLinear** attempts to solve an nth-order linear differential equation with constant-coefficients by symbolically solving the associated characteristic equation.

Example 12.9. *Use* **ODE** *(or* **DSolve***) to find the general solution to the third-order equation* $y''' + 2y'' - 4y' - 8y = 0$.

Solution. Either of the commands

```
DSolve[y'''[t] + 2y''[t] - 4y'[t] - 8y[t] == 0,y[t],t]
```

or

```
ODE[y'''[t] + 2y''[t] - 4y'[t] - 8y[t] == 0,y[t],t,
Method->NthOrderLinear]
```

yields

$$\{\{y[t] \rightarrow \frac{C[1]}{E^{2\,t}} + \frac{t\;C[2]}{E^{2\,t}} + E^{2\,t}\;C[3]\}\}$$

so the general solution is

$$y = C_1 e^{-2t} + C_2 t\, e^{-2t} + C_3 e^{2t}.\ \blacksquare$$

Next, we solve an initial value problem.

Example 12.10. *Use* **ODE** *(or* **DSolve***) to solve the fourth-order initial value problem*

$$\begin{cases} y^{(4)} + y = 0, \\ y(0) = 1,\ y'(0) = 0,\ y''(0) = 3,\ y'''(0) = 0. \end{cases}$$

Plot the solution.

Solution. The command

```
DSolve[{y''''[t] + y[t] == 0,
y[0] == 1,y'[0] == 0,y''[0] == 3,y'''[0] == 0},y[t],t]
```

yields a complicated expression involving complex numbers. However,

```
ODE[{y''''[t] + y[t] == 0,
y[0] == 1,y'[0] == 0,y''[0] == 3,y'''[0] == 0},y[t],t,
Method->NthOrderLinear]
```

gives

```
                 t          t/Sqrt[2]        t
          Cos[-------]    E          Cos[-------]
              Sqrt[2]                    Sqrt[2]
{{y[t] -> ------------- + ---------------------- -
           t/Sqrt[2]               2
          2 E

            t          t/Sqrt[2]        t
     3 Sin[-------]  3 E          Sin[-------]
           Sqrt[2]                    Sqrt[2]
     ------------- + ----------------------}}
       t/Sqrt[2]              2
      2 E
```

Thus, the solution is

$$y = \frac{1}{2}\left(e^{t/\sqrt{2}}\left(\cos\left(\frac{t}{\sqrt{2}}\right) + 3\sin\left(\frac{t}{\sqrt{2}}\right)\right) + e^{-t/\sqrt{2}}\left(\cos\left(\frac{t}{\sqrt{2}}\right) - 3\sin\left(\frac{t}{\sqrt{2}}\right)\right)\right)$$

$$= \cosh\left(\frac{t}{\sqrt{2}}\right)\cos\left(\frac{t}{\sqrt{2}}\right) + 3\sinh\left(\frac{t}{\sqrt{2}}\right)\sin\left(\frac{t}{\sqrt{2}}\right). \ \blacksquare$$

The command

```
ODE[{y''''[t]  + y[t] == 0,
y[0] == 1,y'[0] == 0,y''[0] == 3,y'''[0] == 0},y[t],t,
Method->NthOrderLinear,PlotSolution->{{t,-5,5}}]
```

yields the plot of the solution. $\blacksquare$

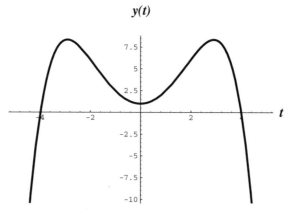

Solution to $y^{(4)} + y = 0$ **with the initial conditions**
$$y(0) = 1, \quad y'(0) = 0, \quad y''(0) = 3, \quad y'''(0) = 0$$

Exercises for Section 12.4

Use **ODE** with the option **Method->NthOrderLinear** to work all of the exercises in
Sections 12.2 and 12.3.

12.5   Seminumerical Solutions of Higher-Order Constant-Coefficient Equations

Numerical approximations to the roots of the associated characteristic equation of a constant-coefficient homogeneous differential equation can be used to find a numerical approximation
to the general solution to the equation.

Definition.   *A **seminumerical approximation** to the general solution of a homogeneous
linear constant-coefficient equation*

$$a_n \frac{d^n y}{dt^n} + a_{n-1} \frac{d^{n-1} y}{dt^{n-1}} + \cdots + a_1 \frac{dy}{dt} + a_0 y = 0 \tag{12.37}$$

is the function

$$y = C_1 e^{s_1 t} + \cdots + C_n e^{s_n t}, \tag{12.38}$$

where $s_1, \ldots, s_n$ are approximations to the roots of the associated characteristic equation of (12.37). *In the case of a root s_j of multiplicity $a_j \geq 2$, the term $C_j e^{s_j t}$ on the right-hand side of* (12.38) *is to be replaced by $C_j t^{a_j-1} e^{s_j t}$.*

We first show how to find seminumerical approximations using *Mathematica*'s command **NSolve**; then we show how to automate the procedure with **ODE**.

Example 12.11. *Find a seminumerical approximation to the general solution of*

$$y''' + y' + y = 0. \tag{12.39}$$

Solution. The associated characteristic equation of (12.39) cannot be easily factored. We can find approximations to the roots of $r^3 + r + 1$ by hand; alternatively, *Mathematica*'s command **NSolve** can be used:

```
NSolve[r^3 + r + 1 == 0,r]
```

```
{{r -> -0.682328}, {r -> 0.341164 - 1.16154 I},
 {r -> 0.341164 + 1.16154 I}}
```

Hence a seminumerical approximation to the general solution of (12.39) is

$$y(t) = C_1 e^{-0.682328t} + e^{0.341164t} \big(C_2 \cos(1.16154t) + C_3 \sin(1.16154t) \big). \ \blacksquare$$

Solutions to initial value problems can be approximated in the same way. We define the **seminumerical approximation to the solution of an initial value problem** to be the function obtained from a seminumerical approximation to the general solution that results when the initial conditions are used to determine the constants of integration.

Example 12.12. *Find a seminumerical approximation to the solution of the initial value problem*

$$\begin{cases} y^{(4)} + 2y''' - 3y'' + 4y' - 5y = 0, \\ y(0) = 1, \ y'(0) = 0, \ y''(0) = 0, \ y'''(0) = -1. \end{cases} \tag{12.40}$$

Solution. To find numerical approximations to the roots of the associated characteristic equation we use

```
NSolve[r^4 + 2r^3 - 3r^2 + 4r - 5 == 0,r]
```

obtaining

```
{{r -> -3.37192}, {r -> 0.130817 - 1.14823 I},
 {r -> 0.130817 + 1.14823 I}, {r -> 1.11029}}
```

Therefore, we define in *Mathematica* the function

```
y[t_]:= c1 E^(-3.37192t)   + c2 E^(1.11029t) +
       E^(0.130817t)(c3 Cos[1.14823t] + c4 Sin[1.14823t])
```

The constants of integration **c1, c2, c3**, and **c4** can be found with **NSolve**:

```
sol=NSolve[{y[0] == 1,y'[0] == 0,y''[0] == 0,y'''[0] == -1},
     {c1,c2,c3,c4}]
```

the answer is

```
{{c1 -> 0.0407675, c2 -> 0.343144, c3 -> 0.616089, c4 -> -0.282277}}
```

Mathematica determines the solution to the initial value problem with the command

```
y[t] /. sol
```

$$\left\{\frac{0.0407675}{E^{3.37192\ t}} + 0.343144\ E^{1.11029\ t} + \right.$$

$$\left. E^{0.130817\ t}\ (0.616089\ Cos[1.14823\ t] - 0.282277\ Sin[1.14823\ t])\right\}$$

Thus, in ordinary notation the solution to the initial value problem is given by

$$y(t) = (0.0407675)e^{-3.37192t} + (0.343144)e^{1.11029t}$$

$$+ e^{0.130817t}\big((0.616089)\cos(1.14823t) - (0.282277)\sin(1.14823t)\big).$$

To plot the solution to the initial value problem we use

```
Plot[y[t] /. sol, {t,-1,1}]
```

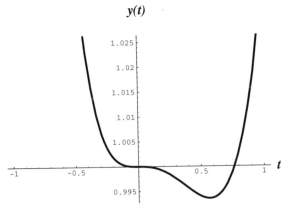

Solution to $y^{(4)} + 2y''' - 3y'' + 4y' - 5y = 0$ **with the**
initial conditions $y(0) = 1, y'(0) = 0, y''(0) = 0, y'''(0) = -1$

ODE's *option*
Method->ApproximateNthOrderLinear

In those cases where the associated characteristic equation of an nth order linear differential equation with constant-coefficients is either impossible or very complicated to solve, it is best to find the roots numerically. In this subsection we use **ODE** to automate the technique of finding a seminumerical approximation.

ODE with the option **Method->ApproximateNthOrderLinear** finds a seminumerical approximation to the solution of an nth-order linear differential equation with constant-coefficients in which the exponents are approximations to the roots of the associated characteristic equation.

Example 12.13. *Find a seminumerical approximation to the general solution of the third-order equation $y''' + y' + y = 0$. Compare the seminumerical approximation to the general solution itself.*

Solution. On the one hand,

```
ODE[y''' + y' + y == 0,y,t,
Method->ApproximateNthOrderLinear]
```

produces

$$
\{\{y \to \frac{C[1]}{E^{0.682\ t}} + 1.\ E^{0.341\ t}\ C[2]\ Cos[1.16\ t] +
$$

$$
E^{0.341\ t}\ C[3]\ Sin[1.16\ t]\}\}
$$

while

```
ODE[y''' + y' + y == 0,y,t,
Method->NthOrderLinear,PostSolution->{Simplify}]
```

yields

$$
E^{((-9 + Sqrt[93])^{1/3}/18^{1/3} - 2^{1/3}/(-27 + 3\ Sqrt[93])^{1/3})\ t}\ C[1] +
$$

$$
E^{(1/(12^{1/3}\ (-9 + Sqrt[93])^{1/3}) - (-9 + Sqrt[93])^{1/3}/(2\ 18^{1/3}))\ t}\ C[2]
$$

$$
Cos[\frac{(6 + 18^{1/3}\ (-9 + Sqrt[93])^{2/3})\ t}{2^{2/3}\ 5^{5/6}\ (-9 + Sqrt[93])^{1/3}}] +
$$

```
       2 2    3    (-9 + Sqrt[93])
         1/3                    1/3                  1/3           1/3
  (1/(12      (-9 + Sqrt[93])   )  -  (-9 + Sqrt[93])      /(2 18    )) t
  E                                                                        C[3]

              1/3                      2/3
        (6 + 18     (-9 + Sqrt[93])   ) t
   Sin[-----------------------------------]
         2/3  5/6                  1/3
        2 2    3    (-9 + Sqrt[93])
```

The seminumerical approximation is simpler than the exact solution. ∎

Note that **ODE** with the option **Method->NthOrderLinear** does the same thing as **Method->ApproximateNthOrderLinear** in the case that *Mathematica* is unable to find the roots of the associated characteristic equation exactly. This happens with many differential equations whose order is 5 or greater.

Example 12.14. *Find a seminumerical approximation to the general solution of the fifth order equation* $y^{(5)} + 2y' + y = 0$.

Solution. We use either of the commands

```
ODE[y''''' + 2y' + y == 0,y,t,
Method->NthOrderLinear,Form->Explicit]
```

or

```
ODE[y''''' + 2y' + y == 0,y,t,
Method->ApproximateNthOrderLinear,Form->Explicit]
```

to obtain

```
   C[1]           0.945 t                              C[2] Cos[0.88 t]
  --------  +  E           C[4] Cos[0.855 t] +       ------------------   +
  0.486 t                                                 0.702 t
  E                                                      E

        0.945 t                              C[3] Sin[0.88 t]
     E           C[5] Sin[0.855 t] +        ----------------
                                                 0.702 t
                                                E
```

Therefore, a seminumerical approximation to the general solution is

$$y = C_1 e^{-0.486t} + e^{-0.945t}(C_2 \cos(0.88t) + C_3 \sin(0.88t))$$

$$+ e^{0.945t}(C_4 \cos(0.855t) + C_5 \sin(0.855t)),$$

where C_1, C_2, C_3, C_4, and C_5 are constants. ∎

On the other hand, **Method->NthOrderLinear** and **Method->ApproximateNthOrderLinear** produce different output for equations whose order is 4 or less, because **Method->NthOrderLinear** can always find a symbolic solution.

Exercises for Section 12.5

Use **Method->NthOrderLinear** or **Method->ApproximateNthOrderLinear** to find seminumerical approximations to the general solution of each of the following differential equations:

1. $y^{(5)} + y''' + y'' - y' + y = 0$

2. $y^{(6)} + y' + y = 0$

3. $y^{(6)} + 2y''' + y'' - y' + y = 0$

4. $y^{(7)} + y''' + y = 0$

5. $y^{(7)} + 2y^{(4)} + 3y''' + 4y = 0$

6. $y^{(8)} + y''' + y = 0$

7. $y^{(8)} + y^{(4)} - y''' + y = 0$

8. $y^{(10)} + y^{(5)} - y''' + y = 0$

Use **Method->NthOrderLinear** or **Method->ApproximateNthOrderLinear** to solve and plot each of the following initial value problems:

9. $\begin{cases} y''' + 4y' = \sec(2t), \\ y(0) = y'(0) = 1, \ y''(0) = 0 \end{cases}$

10. $\begin{cases} 2y''' - 6y'' = t^2, \\ y(0) = y'(0) = 1, \ y''(0) = 0 \end{cases}$

11. $\begin{cases} y^{(4)} + 2y'' + y = t, \\ y(0) = y'(0) = y''(0) = y'''(0) = 0 \end{cases}$

12. $\begin{cases} y^{(5)} + y''' + y = 0, \\ y(0) = 1, \ y'(0) = y''(0) = y'''(0) = y^{(4)}(0) = 0 \end{cases}$

13. $\begin{cases} y^{(5)} + y''' + y'' + y = 0, \\ y(0) = y'(0) = 1, \ y''(0) = y'''(0) = y^{(4)}(0) = 0 \end{cases}$

14. $\begin{cases} y^{(4)} + 2y'' + y = \cosh(t), \\ y(0) = y'(0) = y''(0) = y'''(0) = 0 \end{cases}$

15. $\begin{cases} y^{(4)} - 2y''' - 13y'' + 14y' + 24y = \sinh(t), \\ y(0) = y'(0) = 1, \ y''(0) = y'''(0) = 0 \end{cases}$

16. $\begin{cases} y^{(5)} + y''' + y = 0, \\ y(0) = y'(0) = 1, \, y''(0) = y'''(0) = y^{(4)}(0) = 0 \end{cases}$

13

NUMERICAL SOLUTIONS OF DIFFERENTIAL EQUATIONS

Most differential equations do not have exact solutions. Even if a differential equation has an exact solution, the solution may be so complicated that it is useless for practical purposes.

On the other hand, we can try to find an approximation to the solution of a differential equation with an initial condition. For example, suppose we are given a first-order initial value problem

$$\begin{cases} y' = f(t, y), \\ y(a) = Y_0 \end{cases} \tag{13.1}$$

on an interval $a \leq t \leq b$. Although we may not be able to obtain a formula for the solution of (13.1), we can subdivide the interval as

$$a = t_0 < t_1 < \cdots < t_N = b$$

and try to assign approximate values $y(t_n)$ to t_n for $n = 1, \ldots, N$. Instead of a formula, we will have an approximation to the solution of (13.1) expressed as a table of the $y(t_n)$'s in terms of the t_n's. By graphing the table we can visualize the solution.

One of the main goals in this chapter is to explain the **Runge-Kutta method** for approximating the solution of a first-order initial value problem. This complicated algorithm needs some preparation, however. For that reason we begin by considering simpler methods of numerical approximations to solutions to differential equations. The simplest of all is the Euler method, which we describe in Section 13.1. The next simplest is the Heun

407

method, which is derived from the Euler method in Section 13.2. Then Section 13.3 is devoted to the Runge-Kutta method. These and other solution methods available in **ODE** are discussed in Section 13.4. We compare them briefly, and then in Section 13.5 we explain the details of how the Euler, Heun, and Runge-Kutta methods are implemented in **ODE**. In Section 13.6 we discuss *Mathematica*'s built-in numerical solver **NDSolve**. Adaptive step size is treated in Section 13.7. Finally, in Section 13.8 we present the theory and *Mathematica* implementation of the Numerov method for the numerical solution of second-order differential equations of the form $y'' = f(t)y + g(t)$.

Numerical solutions of systems of differential equations will be considered in Section 19.1.

13.1   The Euler Method

The Euler method was the first method used to find numerical solutions to differential equations. In spite of the fact that it is rarely used in practice, we need to study it because it serves as a model for more complicated methods such as the Runge-Kutta method that we shall study in Section 13.3.

The Euler method can be explained as follows. The objective is to construct an approximation to the solution of the initial value problem (13.1) for $a \leq t \leq b$. We divide the interval $a \leq t \leq b$ into equal subintervals:

$$a = t_0 < t_1 < \cdots < t_N = b,$$

where $h = t_{n+1} - t_n$ is called the **step size**. Let us first suppose that (13.1) has an exact solution $y(t)$, which we assume to be twice differentiable. The form of $y(t)$ will then be used to find a reasonable approximation to the solution. We need the finite Taylor expansion

$$y(t_1) = y(t_0) + (t_1 - t_0)y'(t_0) + \frac{(t_1 - t_0)^2}{2}y''(\xi_0), \qquad (13.2)$$

where $t_0 < \xi_0 < t_1$. Using (13.1), we can rewrite (13.2) as

$$y(t_1) = y(t_0) + h\,f(t_0, Y_0) + \frac{h^2}{2}y''(\xi_0). \qquad (13.3)$$

Now, if h is a positive number less than 1, the quantity h^2 is even smaller. Therefore, we have the approximate equality

$$y(t_1) \approx y(t_0) + h\,f(t_0, Y_0). \qquad (13.4)$$

The initial condition of (13.1) is $y(t_0) = Y_0$. Let us now *define* Y_1 by

$$Y_1 = Y_0 + h\,f(t_0, Y_0). \qquad (13.5)$$

Then Y_1 is an approximation to $y(t_1)$.

Suppose now that the exact solution of (13.1) cannot be easily or explicitly found. In fact, this is the case with most differential equations. (The function f in (13.1) must be quite simple to allow us to find a symbolic solution of (13.1); even then, ingenious methods are sometimes required to find the solution.) Although we do not know $y(t_1)$, we know its approximate value Y_1. Furthermore, if we put

$$Y_2 = Y_1 + h\,f(t_1, Y_1),$$

then we can expect Y_2 to be a reasonable approximation of $y(t_2)$. More generally, we define

$$Y_{n+1} = Y_n + h\,f(t_n, Y_n) \tag{13.6}$$

for $0 \le n \le N - 1$. The **Euler method**, or **tangent method**, consists in approximating the solution to (13.1) by means of (13.6), which we call the **Euler method formula**.

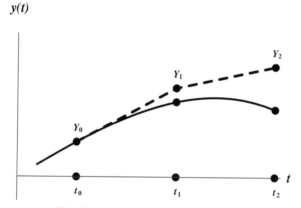

The Euler method approximation

The above picture shows the geometry of the Euler method. From (13.1) we know the point $\big(t_0, y(t_0)\big) = (t_0, Y_0)$ and the slope of the tangent line to the solution curve at $\big(t_0, y(t_0)\big)$, namely $f(t_0, Y_0)$. Thus, the tangent line is the graph of

$$t \longmapsto Y_0 + (t - t_0)f(t_0, Y_0).$$

We can obtain an approximation Y_1 to $y(t_1)$ by moving along this tangent line until t reaches t_1; then $Y_0 + (t_1 - t_0)f(t_0, Y_0) = Y_1$, say. Once Y_1 is determined, we can compute $f(t_1, Y_1)$, which is an approximation to $f\big(t_1, y(t_1)\big)$, which in turn approximates the slope of the tangent line to the actual solution at t_1. Continuing in this way, we obtain a broken line that approximates the solution curve.

To see how the Euler method works, we use it to approximate the solution to a simple first-order linear equation.

Example 13.1. *Use the Euler method to find an approximate solution to the initial value problem*

$$\begin{cases} y' = y + 1, \\ y(0) = 0, \end{cases} \tag{13.7}$$

for $0 \le t \le 1$. *Use step size* $h = 0.1$ *and compare the approximation with the exact solution.*

Solution. The Euler method formula (13.6) for the initial value problem (13.7) with step size $h = 0.1$ becomes

$$Y_{n+1} = Y_n + 0.1(Y_n + 1),$$

for $0 \le n \le N - 1$. The interval $0 \le t \le 1$ is divided into 10 equal pieces, so $N = 10$. We are given $Y_0 = 0$; then $Y_1 = Y_0 + 0.1(Y_0 + 1) = 0.1$, and

$$Y_2 = Y_1 + 0.1(Y_1 + 1) = 0.1 + 0.1(1.1) = 0.21.$$

Continuing in this way, we get the first three columns in the following table:

n	t_n	Y_n	$y_{\text{exact}}(t_n)$
0	0.0	0.0	0.0
1	0.1	0.1	0.10517
2	0.2	0.21	0.22140
3	0.3	0.331	0.34986
4	0.4	0.4641	0.49182
5	0.5	0.61050	0.64872
6	0.6	0.77156	0.82212
7	0.7	0.94871	1.01375
8	0.8	1.14359	1.22554
9	0.9	1.35795	1.45960
10	1.0	1.59370	1.71828

Since $y' = y + 1$ is a first-order linear differential equation, the exact solution of (13.7) is easily found to be

$$y_{\text{exact}}(t) = e^t - 1.$$

Then $y_{\text{exact}}(0.1) = e^{0.1-1} \approx 0.10517$, $y_{\text{exact}}(0.2) = e^{0.2-1} \approx 0.22140$, and so forth. Values for the exact solution are shown in the fourth column.

In the following plot we compare the exact solution of (13.7) with the approximate solution obtained by the Euler method.

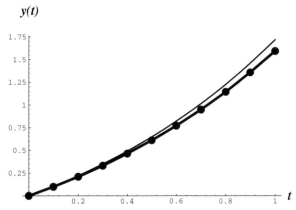

The exact solution of $y' = y + 1$, $y(0) = 0$ and
an approximate solution found by the Euler method

The difference $Y_{10} - y(t_{10})$ is approximately 0.12, an error of less than ten percent. However, we shall see that more sophisticated methods such as the Heun method and the Runge-Kutta method give much better results. ∎

Next, let us consider a differential equation for which it is impossible to find the solution, at least by elementary techniques.

Example 13.2. *Use the Euler method to find an approximate solution to the initial value problem*

$$\begin{cases} y' = 5 - t^2 y^3, \\ y(0) = 0, \end{cases} \tag{13.8}$$

for $0 \le t \le 1$. Use step sizes $h = 0.1$ and $h = 0.01$ and compare the results.

Solution. The Euler method formula (13.6) for the initial value problem (13.8) with step size h is

$$Y_{n+1} = Y_n + h\left(5 - t_n^2 Y_n^3\right). \tag{13.9}$$

Taking $h = 0.01$ and then $h = 0.001$ in (13.9) gives us the following table:

t	Y_n $(h = 0.1)$	Y_n $(h = 0.01)$
0.0	0.0	0.0
0.1	0.50000	0.49998
0.2	0.99987	0.99886
0.3	1.49588	1.48647
0.4	1.96575	1.92540
0.5	2.34422	2.24190
0.6	2.52216	2.36963
0.7	2.44457	2.32246
0.8	2.22875	2.18041
0.9	2.02021	2.01484
1.0	1.85237	1.86018

Here is a plot that shows the differences in the approximations obtained with the different step sizes:

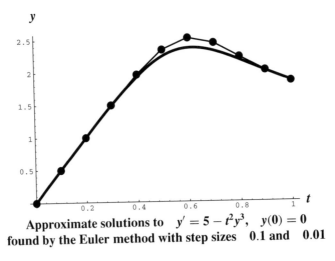

Approximate solutions to $\;y' = 5 - t^2 y^3, \;\; y(0) = 0$
found by the Euler method with step sizes $\;\;$ 0.1 and $\;\;$ 0.01

Finally, we give an alternate derivation of the Euler method formula (13.6). Suppose that we have an exact solution $y = z(t)$ of the initial value Problem (13.1). Then if we integrate $z'(t)$ from t_n to t_{n+1} we obtain

$$\int_{t_n}^{t_{n+1}} z'(t)dt = \int_{t_n}^{t_{n+1}} f\big(t, z(t)\big)dt,$$

which can be rewritten as

$$z(t_{n+1}) = z(t_n) + \int_{t_n}^{t_{n+1}} f\big(t, z(t)\big)dt. \qquad (13.10)$$

Consider the graph of the function $t \longmapsto f\big(t, z(t)\big)$. The integral on the right-hand side of (13.10) equals the area under this graph from t_n to t_{n+1}, as indicated by the following picture:

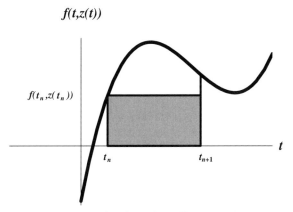

Integral derivation of the Euler method

We can approximate the actual area under the graph by the area of the shaded rectangle, whose area is

$$f\big(t_n, z(t_n)\big)(t_{n+1} - t_n).$$

From (13.10) we obtain the approximation

$$z(t_{n+1}) \approx z(t_n) + f\big(t_n, z(t_n)\big)(t_{n+1} - t_n) = z(t_n) + f\big(t_n, z(t_n)\big)h. \qquad (13.11)$$

In (13.11) we replace $z(t_n)$ by Y_n and $z(t_{n+1})$ by Y_{n+1}. In this way we obtain

$$Y_{n+1} = Y_n + h\,f(t_n, Y_n),$$

which is the same as Euler method formula (13.6).

Exercises for Section 13.1

In the following exercises, perform the indicated computations by hand, retaining only four significant digits at each step of the calculations.

1. Use the Euler method to compute an approximate solution to the initial value problem $y' = \sqrt{y+1}$, $y(0) = 0$ with step size $h = 0.1$. Compare with the exact solution $y(t) = t(4+t)/4$.

2. Consider the following initial value problem:

$$\begin{cases} y' = y - t\,y^2, \\ y(0) = 2. \end{cases}$$

a. Use the Euler method to compute an approximate solution at $t = 1.0$. Use the values $n = 2, 4, 8$, which correspond to step sizes of $h = 1/2, 1/4, 1/8$.

b. Determine the actual solution and compare the value of $y(1)$ with the results from part **a**.

Repeat Exercise 2 for the following initial value problems:

3. $\begin{cases} y' + 2y = t, \\ y(0) = 1 \end{cases}$ **5.** $\begin{cases} y' = t^3 e^{-2y}, \\ y(0) = 0 \end{cases}$

4. $\begin{cases} y' = t + 1, \\ y(0) = 1 \end{cases}$ **6.** $\begin{cases} y' = \dfrac{e^t}{y}, \\ y(0) = 1 \end{cases}$

7. Consider the following initial value problem:

$$\begin{cases} y' = y + t^2, \\ y(0) = 1. \end{cases}$$

a. Find the solution $y(t)$ and evaluate it for $t = 0.2, 0.4, \ldots, 1.0$.

b. Using the Euler method, with step size of $h = 0.2$, find approximate values for the solution at the t values in part **a**.

c. Repeat part **b** using $h = 0.1$.

d. Compare the results of part **b** with those of part **c** and the exact values. (The differences in the results for $h = 0.2$ and $h = 0.1$ tend to indicate whether a smaller step size must be used for the desired range of t values. The general rule of thumb is to use the smaller step size if the two solutions agree to the desired accuracy. If they do not agree, h should be reduced and the calculations repeated. This gives an indication but not a proof of the accuracy of the result.)

The initial value problems in the following exercises cannot be solved by symbolic methods. Use the Euler method with step size $h = 0.1$ to approximate the solution at $t = 1$ to three significant digits (see Exercise 7).

8. $\begin{cases} y' = \sin(y) + e^t, \\ y(0) = 0 \end{cases}$

10. $\begin{cases} y' = \sin(t) + \cos(y), \\ y(0) = 1 \end{cases}$

9. $\begin{cases} y' = y^{3/2} + t, \\ y(0) = 1 \end{cases}$

11. $\begin{cases} y' = e^{t^3}, \\ y(0) = 1 \end{cases}$

12. Consider the initial value problem:

$$\begin{cases} y' = t^2 y, \\ y(0) = 1. \end{cases}$$

    a. Find the exact solution at $t = 1.0$. Express this value to four decimal places.

    b. Use the Euler method with $h = 1/8$ to approximate the solution at $t = 1.0$. Compute the absolute error.

    c. Repeat part **b** with $h = 1/16, h = 1/32$. Create a table and a graph showing the absolute errors corresponding to the various step sizes. A theoretical analysis for the Euler method suggests a linear relationship between the absolute error and the step size. Do the numbers agree with the theory?

    d. Observe in part **c** that the error is roughly proportional to the step size. Use these data to estimate the constant of proportionality.

13. Apply the Euler method with successively smaller step sizes on the interval $0 \le t \le 1$ to verify empirically that the solution of the initial value problem

$$\begin{cases} y' = t^2 + y^2, \\ y(0) = 1 \end{cases}$$

has a vertical asymptote near $t = 0.97$.

13.2   *The Heun Method*

As a first attempt to obtain an improvement of the Euler method formula (13.6), let us replace $f(t_n, Y_n)$ with the *average* of $f(t_n, Y_n)$ and $f(t_{n+1}, Y_{n+1})$; this leads to the formula

$$Y_{n+1} = Y_n + \frac{h}{2}\bigl(f(t_n, Y_n) + f(t_{n+1}, Y_{n+1})\bigr). \tag{13.12}$$

Unfortunately, Y_{n+1} occurs on both sides of (13.12), so that we cannot obtain it without solving the equation for Y_{n+1}; this may be difficult. Fortunately, we already have an approximation for Y_{n+1}, namely the value

$$Y_n + h f(t_n, Y_n)$$

from the Euler method formula (13.6). Let us substitute $Y_n + h f(t_n, Y_n)$ for the Y_{n+1} occurring on the right-hand side of (13.12). The result is

$$Y_{n+1} = Y_n + \frac{h\Big(f(t_n, Y_n) + f\big(t_{n+1}, Y_n + h f(t_n, Y_n)\big)\Big)}{2}. \tag{13.13}$$

The **Heun**[1] **method**, or **improved Euler method**, consists in approximating the solution to

$$\begin{cases} y' = f(t, y), \\ y(a) = Y_0 \end{cases} \tag{13.14}$$

by means of (13.13), which we call the **Heun method formula**.

The Heun method is an example of a **predictor-corrector method**. The Euler method is used to "predict" a value for Y_{n+1}; this value is then used in (13.13) to obtain a better (or "more correct") approximation.

Example 13.3. *Use the Heun method to find an approximate solution to the initial value problem*

$$\begin{cases} y' = 5 - t^2 y^3, \\ y(0) = 0, \end{cases} \tag{13.15}$$

for $0 \le t \le 1$. Use step size $h = 0.1$. Compare the Heun method approximation with the Euler method approximation.

Solution. Let $f(t, Y) = 5 - t^2 Y^3$. Since the Euler method approximation to Y_{n+1} is $Y_n + h f(t_n, Y_n) = Y_n + h(5 - t_n^2 Y_n^3)$, we compute

$$f\big(t_{n+1}, Y_n + h f(t_n, Y_n)\big) = f\big(t_n + h, Y_n + h(5 - t_n^2 Y_n^3)\big)$$

$$= 5 - (t_n + h)^2 \big(Y_n + h(5 - t_n^2 Y_n^3)\big)^3.$$

---

[1]

Karl Heun (1859–1929). German mathematician, Professor at the Technische Hochschule Karlsruhe.

We also have $f(t_n, Y_n) = 5 - t_n^2 Y_n^3$. Thus the Heun method formula (13.13) becomes

$$Y_{n+1} = Y_n + \frac{h}{2}\left(10 - t_n^2 Y_n^3 - (t_n + h)^2\left(Y_n + h(5 - t_n^2 Y_n^3)\right)^3\right) \qquad (13.16)$$

We obtain $Y_1 = (h/2)(10 - h^5(125))$, and so forth. Taking $h = 0.1$ in (13.16) gives us the last column in the following table. The middle column is computed using the Euler method formula (13.6).

t	Y_n (Euler)	Y_n (Heun)
0.0	0.0	0.0
0.1	0.50000	0.49994
0.2	0.99987	0.99788
0.3	1.49588	1.48089
0.4	1.96575	1.90680
0.5	2.34422	2.20007
0.6	2.52216	2.30745
0.7	2.44457	2.26215
0.8	2.22875	2.14016
0.9	2.02021	1.99622
1.0	1.85236	1.85650

The data in this table can be visualized by means of the plot below.

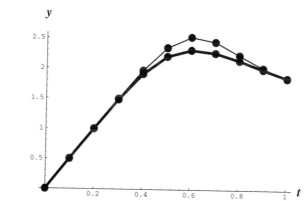

Comparison of approximate solutions to $y' = 5 - t^2 y^3$, $y(0) = 0$, **found by the Euler and Heun methods with step size** 0.1

Exercises for Section 13.2

In the following exercises, perform the indicated computations by hand, retaining only four significant digits at each step of the calculations.

1. Use the Heun method to compute an approximate solution to the initial value problem $y' = y + 1$, $y(0) = 0$ with step size $h = 0.1$. Compare with the exact solution $y(t) = e^t - 1$ and with the Euler method approximation from Example 13.1.

2. Consider the following initial value problem:

$$\begin{cases} y' = y - t\,y^2, \\ y(0) = 2. \end{cases}$$

   a. Use the Heun algorithm to compute an approximate solution at $t = 1.0$. Use the values $n = 2, 4, 8$, which correspond to step sizes of $h = 1/2, 1/4, 1/8$.

   b. Determine the actual solution and compare the value of $y(1)$ with the results from part **a**.

Repeat Exercise 2 for the following initial value problems:

3. $\begin{cases} y' + 2y = t, \\ y(0) = 1 \end{cases}$

5. $\begin{cases} y' = t^3 e^{-2y}, \\ y(0) = 0 \end{cases}$

4. $\begin{cases} y' = t + 1, \\ y(0) = 1 \end{cases}$

6. $\begin{cases} y' = \dfrac{e^t}{y}, \\ y(0) = 1 \end{cases}$

7. Consider the following initial value problem:

$$\begin{cases} y' = y + t^2, \\ y(0) = 1. \end{cases}$$

   a. Find the solution $y(t)$ and evaluate it for $t = 0.2, 0.4, \ldots, 1.0$.

   b. Using the Heun method with step size of $h = 0.2$, find approximate values for the solution at the t values in part **a**.

   c. Repeat part **b** using $h = 0.1$.

   d. Compare the results of part **b** with those of part **c** and the exact values. (The differences in the results for $h = 0.2$ and $h = 0.1$ tend to indicate whether a smaller step size must be used for the desired range of t values. The general rule of thumb is to use the smaller step size if the two solutions agree to the desired accuracy. If they do not agree, reduce h and repeat the calculations. This gives an indication but not a proof of the accuracy of the result.)

The initial value problems in the following exercises cannot be solved by exact methods. Use the Heun method with step size $h = 0.1$ to approximate the solution at $t = 1$ to three significant digits (see Exercise 7).

8. $\begin{cases} y' = \sin(y) + e^t, \\ y(0) = 0 \end{cases}$

10. $\begin{cases} y' = \sin(t) + \cos(y), \\ y(0) = 1 \end{cases}$

9. $\begin{cases} y' = y^{3/2} + t, \\ y(0) = 1 \end{cases}$

11. $\begin{cases} y' = e^{t^3}, \\ y(0) = 1 \end{cases}$

12. Consider the initial value problem:

$$\begin{cases} y' = t^2 y, \\ y(0) = 1. \end{cases}$$

   a. Find the exact solution at $t = 1.0$. Express this value to four decimal places.

   b. Use the Heun method with $h = 1/8$ to approximate the solution at $t = 1.0$. Compute the absolute error.

   c. Repeat part **b** with $h = 1/16, h = 1/32$. Create a table and a graph showing the absolute errors corresponding to the various step sizes. A theoretical analysis for the Heun method suggests a linear relationship between the absolute error and the square of the step size. Do the numbers agree with the theory?

   d. Observe in part **c** that the error is roughly proportional to the square of the step size. Use the data to estimate the constant of proportionality.

13. Apply the Heun method with successively smaller step sizes on the interval $0 \leq t \leq 1$ to verify empirically that the solution of the initial value problem

$$\begin{cases} y' = t^2 + y^2, \\ y(0) = 1 \end{cases}$$

has a vertical asymptote near $t = 0.97$.

13.3   *The Runge-Kutta Method*

The Runge[2]Kutta[3] method is an improvement of both the Euler method formula (13.6) and the Heun method formula (13.13) that involves a weighted average of four values of $f(t, y)$ taken at different points in the interval $t_n \le t \le t_{n+1}$. Explicitly, it is given by

$$Y_{n+1} = Y_n + \frac{h}{6}\left(a_{1,n} + 2a_{2,n} + 2a_{3,n} + a_{4,n}\right), \qquad (13.17)$$

where

$$a_{1,n} = f(t_n, Y_n),$$

$$a_{2,n} = f\left(t_n + \frac{h}{2}, Y_n + \frac{h}{2}a_{1,n}\right),$$

$$a_{3,n} = f\left(t_n + \frac{h}{2}, Y_n + \frac{h}{2}a_{2,n}\right),$$

$$a_{4,n} = f(t_n + h, Y_n + h\,a_{3,n}).$$

We call (13.17) the **Runge-Kutta method formula**; we omit its complicated derivation. Although the **Runge-Kutta method** (which consists of approximating the solution to a first-order initial value problem with the Runge-Kutta method formula) is significantly more complicated than the Euler and Heun methods, it is considerably more accurate. However, computers can easily handle the increased complexity.

Example 13.4.   *Use the Runge-Kutta method to find an approximate solution to the initial value problem*

$$\begin{cases} y' = 10(1 - t^2y^3 + t^4y^3), \\ y(0) = 0, \end{cases} \qquad (13.18)$$

for $0 \le t \le 1$. Use step size $h = 0.1$. Compare the Runge-Kutta method approximation with the Euler and Heun method approximations.

---

[2]

Carl David Tolmé Runge (1856–1927). Applied mathematics professor at Göttingen. Although his doctoral dissertation (1880) dealt with differential geometry, he worked mainly in applied mathematics. In particular, he studied the spectral lines of elements. Runge devised his numerical method about 1895.

[3]Martin Wilhelm Kutta (1867–1944). German applied mathematician who made important contributions to aerodynamics, in particular the Zhukovskii-Kutta aerofoil. Kutta extended Runge's method in 1901.

Solution. Let $f(t, y) = 10(1 - t^2y^3 + t^4y^3)$. We compute

$$a_{1,0} = f(0,0) = 10, \qquad\qquad a_{2,0} = f(0.05, 0.05 \times 10) = 9.99688,$$

$$a_{3,0} = f(0.05, 0.05 \times 9.99688) = 9.99688, \quad a_{4,0} = f(0.1, 0.1 \times 9.99688) = 9.90109.$$

Hence

$$Y_1 = 0.0166667(a_{1,0} + 2a_{2,0} + 2a_{3,0} + a_{4,0}) = 0.998144.$$

Continuing in this way we get the last column in the following table. Similar calculations using the Euler and Heun method formulas yield the second and third columns.

n	t_n	Y_n (Euler)	Y_n (Heun)	Y_n (Runge-Kutta)
0	0.0	0.0	0.0	0.0
1	0.1	1.0	0.99505	0.99814
2	0.2	1.99010	1.83994	1.90323
3	0.3	2.68744	1.99998	2.27305
4	0.4	2.09780	1.80606	2.08513
5	0.5	1.85703	1.64398	1.85736
6	0.6	1.65626	1.54332	1.70104
7	0.7	1.60945	1.50988	1.61893
8	0.8	1.56762	1.56258	1.61459
9	0.9	1.68005	1.75588	1.72850
10	1.0	1.95025	2.33931	2.19565

Visualization of the data in this table is provided by the plot below.

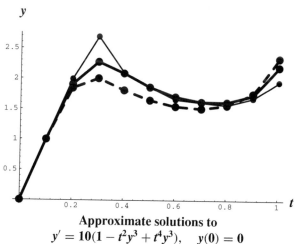

Approximate solutions to
$$y' = 10(1 - t^2y^3 + t^4y^3), \quad y(0) = 0$$
(Thin=Euler, Dashed=Heun, Thick=Runge-Kutta)

Exercises for Section 13.3

In the following exercises, perform the indicated computations by hand, retaining only four significant digits at each step of the calculations.

1. Use the Runge-Kutta method to compute an approximate solution to the initial value problem $y' = y + 1$, $y(0) = 0$ with step size $h = 0.1$. Compare with the exact solution $y(t) = e^t - 1$ and with the Euler method approximation from Example 13.1.

2. Consider the following initial value problem:

$$\begin{cases} y' = y - t\,y^2, \\ y(0) = 2. \end{cases}$$

   a. Use the Runge-Kutta algorithm to compute an approximate solution at $t = 1.0$. Use the values $n = 2, 4, 8$, which correspond to step sizes of $h = 1/2, 1/4, 1/8$.

   b. Determine the actual solution and compare the value of $y(1)$ with the results from part **a**.

Repeat Exercise 1 for the following initial value problems:

3. $\begin{cases} y' + 2y = t, \\ y(0) = 1 \end{cases}$

5. $\begin{cases} y' = t^3 e^{-2y}, \\ y(0) = 0 \end{cases}$

4. $\begin{cases} y' = t + 1, \\ y(0) = 1 \end{cases}$

6. $\begin{cases} y' = \dfrac{e^t}{y}, \\ y(0) = 1 \end{cases}$

7. Consider the following initial value problem:

$$\begin{cases} y' = y + t^2, \\ y(0) = 1. \end{cases}$$

   a. Find the solution $y(t)$ and evaluate it for $t = 0.2, 0.4, \ldots, 1.0$.

   b. Using the Runge-Kutta method with step size of $h = 0.2$, find approximate values for the solution at the t values in part **a**.

   c. Repeat part **b** using $h = 0.1$.

   d. Compare the results of part **b** with those of part **c** and the exact values. (The differences in the results for $h = 0.2$ and $h = 0.1$ tend to indicate whether a smaller step size must be used for the desired range of t values. The general rule of thumb is to use the smaller step size if the two solutions agree to the desired accuracy. If they do not agree, reduce h and repeat the calculations. This gives an indication but not a proof of the accuracy of the result.)

The initial value problems in the following exercises cannot be solved by exact methods. Use the Runge-Kutta method with step size $h = 0.1$ to approximate the solution at $t = 1$ to three significant digits (see Exercise 7).

8. $\begin{cases} y' = \sin(y) + e^t, \\ y(0) = 0 \end{cases}$

10. $\begin{cases} y' = \sin(t) + \cos(y), \\ y(0) = 1 \end{cases}$

9. $\begin{cases} y' = y^{3/2} + t, \\ y(0) = 1 \end{cases}$

11. $\begin{cases} y' = e^{t^3}, \\ y(0) = 1 \end{cases}$

12. Consider the initial value problem:

$$\begin{cases} y' = t^2 y, \\ y(0) = 1. \end{cases}$$

   a. Find the exact solution at $t = 1.0$. Express this value to four decimal places.

   b. Use the Runge-Kutta method with $h = 1/8$ to approximate the solution at $t = 1.0$. Compute the absolute error.

   c. Repeat part **b** with $h = 1/16$, $h = 1/32$. Create a table and a graph showing the absolute errors corresponding to the various step sizes. A theoretical analysis for the Runge-Kutta method suggests a linear relationship between the absolute error and the fourth power of the step size. Do the numbers agree with the theory?

   d. Observe in part **c** that the error is roughly proportional to the fourth power of the step size. Use the data to estimate the constant of proportionality.

13. Apply the Runge-Kutta method with successively smaller step sizes on the interval $0 \le t \le 1$ to verify empirically that the solution of the initial value problem

$$\begin{cases} y' = t^2 + y^2, \\ y(0) = 1 \end{cases}$$

has a vertical asymptote near $t = 0.97$.

14. This exercise describes the relation between the Runge-Kutta formula (13.17) and Simpson's rule.

    a. Suppose that $f(t)$ is a continuous function for $t_0 \leq t \leq t_0 + h$. Find a quadratic polynomial $y = a\,t^2 + b\,t + c$ that agrees with $f(t)$ at the points t_0, $t_0 + h/2$, and $t_0 + h$.

    b. Find the integral of this polynomial: $A = \displaystyle\int_{t_0}^{t_0+h} (a\,t^2 + b\,t + c)\,dt.$

    c. Show that

$$A = \frac{h(f(t_0) + 4\,f(t_0 + h/2) + f(t_0 + h))}{6},$$

consistent with (13.17).

13.4   *Solving Differential Equations Numerically with* ODE

In this section we explain how to use **ODE** to solve first-order differential equations numerically. Any of the following options can be used:

```
Method -> Euler               Method -> Heun
Method -> RungeKutta4         Method -> RungeKutta45
Method -> ImplicitRungeKutta  Method -> SecondOrderEuler
Method -> Milne               Method -> AdamsBashforth
Method -> SecondOrderEuler    Method -> BulirschStoer
Method -> NDSolve             Method -> Numerov
```

All of the above commands follow the same pattern

```
ODE[{diffeq,initcond},y,{t,a,b},Method,StepSize]
```

although not every option is supported by each method. The methods **NDSolve** and **Numerov** (discussed in Sections 13.6 and 13.8) are implemented somewhat differently from the others. The default step size is 0.1. Any numerical approximation found with **ODE** can be plotted using the option **PlotSolution**, as explained in Section 4.3. When plotting, it is frequently useful to suppress the numerical output by using the option **NumericalOutput->None**.

    All methods, with the exception of **Numerov**, can be used to find an approximate numerical solution of any differential equation (or system, see Section 19.1). The reason for considering more than one method is that for certain equations some methods give more accurate results than others.

Example 13.5.   *Use* **ODE** *with the option* **Method->Euler** *to find a numerical approximation to the solution of the initial value problem*

$$\begin{cases} y' + y = -y^3, \\ y(0) = 1 \end{cases}$$

over the interval $0 \le t \le 1$. *Use 5 steps. Also, plot the solution.*

Solution. We use

```
ODE[{y' + y == -y^3,y[0] == 1},y,{t,0,1},
Method->Euler,StepSize->0.2,
PlotSolution->{{t,0,1}}]
```

to obtain

```
{{0,1.},{0.2,0.6},{0.4,0.4368},{0.6,0.332772},{0.8,0.258848},{1.,0.203609}}
```

and the graph

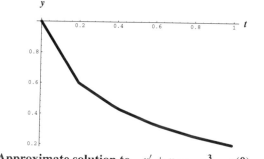

Approximate solution to $y' + y = -y^3$, $y(0) = 0$
found by the Euler method

In Example 13.5 any of the methods listed on page 424 can be substituted for **Euler**. Furthermore, the option

Method->AllNumerical

can be used to plot all of the approximations, one after another.

Example 13.6.   *Use* **ODE** *with the option* **Method->AllNumerical** *to find and plot numerical approximations to the solution of the initial value problem*

$$\begin{cases} y' - y = e^{5\sin(5y)}, \\ y(0) = 1 \end{cases}$$

over the interval $-1 \le t \le 1$. *Use step size 0.02.*

Solution. We use

```
ODE[{y' - y  == E^(5Sin[5y]),y[0] == 1},y,{t,-1,1},
Method->AllNumerical,
StepSize->0.02,NumericalOutput->None,
PlotSolution->{{t,-1,1}}]
```

to obtain the following graphs:

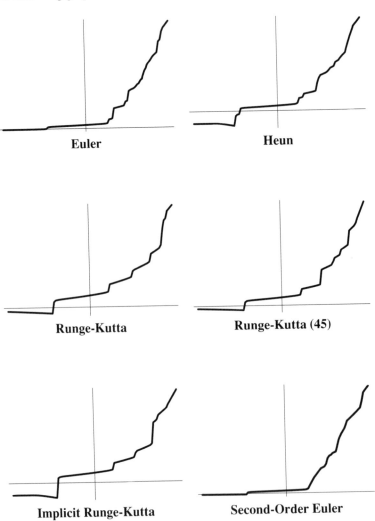

Euler   **Heun**

Runge-Kutta   **Runge-Kutta (45)**

Implicit Runge-Kutta   **Second-Order Euler**

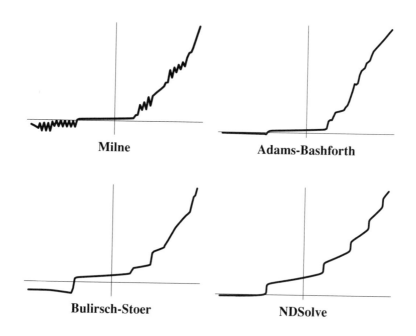

Milne   **Adams-Bashforth**

Bulirsch-Stoer   **NDSolve**

From these plots it is clear that some numerical methods are better than others. Inside of *Mathematica* the best method to use is **NDSolve**, as will be explained in Section 13.6. The Euler and Heun methods are too primitive to be of much practical use. The best all-around method is Runge-Kutta. The Milne and Adams[4]-Bashforth methods are examples of multistep predictor-corrector methods. The Milne method is included only for historical reasons; in practice it gives bad results. The Adams-Bashforth method provides the same accuracy as the Runge-Kutta method at twice the speed.

The reader is invited to experiment with the methods listed on page 424.

4

John Couch Adams (1819–1892). English astronomer and mathematician. In September 1845 Adams gave accurate information on the position of a new planet. The existence of Neptune was confirmed on September 23, 1846 by the Berlin Observatory. Adams became Regius Professor of Mathematics at St. Andrews in 1858 and director of the Cambridge Observatory in 1861.

13.5   ODE's *Implementation of Numerical Methods*

In this section[5] we give the details of how **ODE** finds approximate solutions using the Euler, Heun, and Runge-Kutta methods.

The Euler Method via ODE

We first describe a *Mathematica* miniprogram called **Euler**, which implements the Euler method, specifically formula (13.6). This routine is called by **ODE** with the option **Method->Euler**, but it can also be used independently of **ODE**.

We need two *Mathematica* functions, **Module** (described on page 88) and **NestList**, described below. We first define a function **emstep** of six variables as follows:

```
emstep[f_,{t_,y_},{tn_,yn_},h_]:=
    Module[{fn},
        fn = f /. {t->tn, y->yn};
        {tn + h,yn + h fn}]
```

The arguments of **emstep** are **f_, t_, y_, tn_, yn_** and **h_**; we separate some of them with braces in order to keep track of them. In the definition of **emstep** we use **Module** with one local variable **fn** and two *Mathematica* commands. The first command is

```
fn = f /. {t->tn, y->yn};
```

Here **fn** is the same as the expression **f**, but with **t** replaced by **tn** and **y** replaced by **yn**. The second command in the module is

```
{tn + h,yn + h fn}
```

It does nothing more than compute the ordered pair **{tn + h,yn + h fn}**.

Next, we define *Mathematica* function **Euler** which makes use of **emstep**:

```
Euler[f_,{t0_,y0_},h_,steps_][t_,y_]:=
    NestList[emstep[f,{t,y},#,h]&,{t0,y0},steps]
```

Note that **Euler** is a function of the seven variables

---

[5]This section contains technical *Mathematica* details on how **ODE** works. It is not needed to use **ODE**. However, the reader is encouraged to explore it because it contains useful information about programing techniques in *Mathematica*.

f, t0, y0, h, steps, t, y

The command **NestList** is useful for constructing a table of values of the result of repeated application of a function to an expression. Specifically

NestList[g, expr, n]

gives a list of the results of applying **g** to **expr** 0 through n times. For the command **Euler** we use

$$\textbf{steps} \quad for \quad \textbf{n}$$

$$\textbf{\{t0,y0\}} \quad for \quad \textbf{expr}$$

$$\textbf{emstep[f,\{t,y\}, \#,h]\&} \quad for \quad \textbf{g}$$

Here **emstep[f,t,y,#,h]&** is *Mathematica*'s abbreviation for the function that assigns the value **emstep[f,{t,y},{t0,y0},h]** to the ordered pair **{t0,y0}**. (See page 28.)

The following example demonstrates the use of the command **Euler**.

Example 13.7. *Use the command* **Euler** *to find a numerical approximation to the solution of the initial value problem*

$$\begin{cases} y' = \cos(15t\,y), \\ y(0) = 1 \end{cases}$$

over the interval $0 \le t \le 1$. *Use step size* $h = 0.1$.

Solution. We use

ex0 = Euler[Cos[15 t y],{0,1},0.1,10][t,y]

which results in

```
{{0,1}, {0.1, 1.1}, {0.2,1.09209}, {0.3,0.992993},
 {0.4,0.968843}, {0.5,1.05799}, {0.6,1.04991},
 {0.7,0.949935}, {0.8,0.864659},{0.9,0.806582},{1.0, 0.79593}}
```

The command

TableForm[ex0]

prints this information in a nice tabular form:

```
0.0    1.0
0.1    1.1
0.2    1.09209
0.3    0.992993
0.4    0.968843
0.5    1.05799
0.6    1.04991
0.7    0.949935
0.8    0.864659
0.9    0.806582
1.0    0.79593
```

The solution **ex0** can be plotted via the command **ListPlot**. This command is frequently used for plotting data as points. The option **PlotJoined->True**. draws line segments to connect the points. Here we use

ListPlot[ex0,PlotJoined->True];

to get

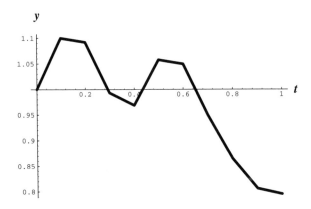

A better approximation is obtained if we take $h = 0.01$ and do 100 steps. Let us skip over printing out the numerical data and do the plot directly:

ListPlot[Euler[Cos[15 t y],{0,1},0.01,100][t,y],
PlotJoined->True];

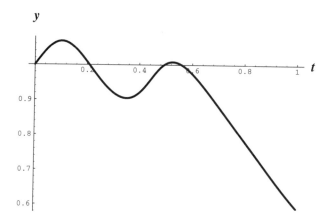

The Heun and Runge-Kutta Methods via ODE

There is a more complicated miniprogram for solving a differential equation numerically using the Heun method:

```
Heun[f_,{t0_, y0_},h_,steps_][t_,y_]:=
    NestList[iemstep[f,{t, y},#,h]&,{t0,y0},steps]
```

Here **iemstep** iterates for **Heun** in the same way that **emstep** iterates for **Euler**. The definition is as follows.

```
iemstep[f_,{t_, y_},{tn_,yn_},h_]:=
    Module[{k1, k2},
    k1 = f /. {t->tn, y->yn};
    k2 = f /. {t->tn + h, y->yn + h k1};
    {tn + h, yn +h (k1 + k2)/2}]
```

It should be clear that this is the *Mathematica* implementation of (13.13). Here is the corresponding program using the Runge-Kutta method:

```
RungeKutta4[f_,{t0_, y0_},h_,steps_][t_,y_]:=
    NestList[rkmstep[f,{t, y},#,h]&,{t0,y0},steps]
```

The analogue of **emstep** for **RungeKutta4** is given by

```
rkmstep[f_,{t_, y_},{tn_, yn_},h_]:=
    Module[{k1, k2, k3, k4},
        k1 = f /. {t->tn, y->yn};
        k2 = f /. {t->tn + h/2, y->yn + h k1/2};
        k3 = f /. {t->tn + h/2, y->yn + h k2/2};
        k4 = f /. {t->tn + h, y->yn + h k3};
        {tn + h, yn +h (k1 + 2 k2 + 2 k3 + k4)/6}]
```

This is the *Mathematica* version of (13.17).

Example 13.8. *Solve the initial value problem*

$$\begin{cases} y' = 1 - t + 4y, \\ y(0) = 1 \end{cases}$$

using the commands **Euler**, **Heun**, *and* **RungeKutta4** *and simultaneously plot the three approximations.*

Solution. We use

```
Show[
 {ListPlot[Euler[1 - t + 4y,{0,1},0.1,10][t,y],
 PlotJoined->True,DisplayFunction->Identity],
 ListPlot[Heun[1 - t + 4y,{0,1},0.1,10][t,y],
 PlotJoined->True,DisplayFunction->Identity],
 ListPlot[RungeKutta4[1 - t + 4y,{0,1},0.1,10][t,y],
 PlotJoined->True,DisplayFunction->Identity]},
 DisplayFunction->$DisplayFunction];
```

to get

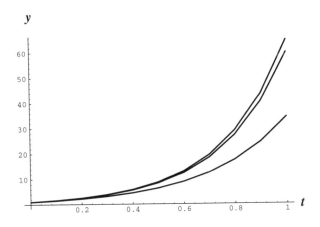

Exercises for Section 13.5

Use **ODE** with **Method->Euler** to plot the numerical solutions of the following differential equations. In each case use a step size $h = 0.01$ and do 100 steps. Try to find an exact solution for each problem using **ODE**. For those problems for which **ODE** yields an exact solution, compare the exact solution with the solution found by the Euler method.

1. $\begin{cases} y' = 1 - t\, y^{10}, \\ y(0) = 1 \end{cases}$

3. $\begin{cases} y' = t - y, \\ y(0) = 1 \end{cases}$

2. $\begin{cases} y' = \sin(t\, y), \\ y(0) = 1 \end{cases}$

4. $\begin{cases} y' = \dfrac{t}{y^5} - y, \\ y(0) = 1 \end{cases}$

Use *Mathematica* to plot the numerical solutions of the following differential equations using the Euler, Heun, and Runge-Kutta methods. In each problem use the indicated step size **h** and number of steps **steps** and combine the three graphs.

5. $y' = 1 - t\, y^{10}$, $y(0) = 1$,   **h=0.1, steps = 10**.

6. $y' = \sin(t\, y)$, $y(0) = 1$,   **h=0.2, steps = 10**.

7. $y' = t - \sin(y)$, $y(0) = 1$,   **h=0.2, steps = 10**.

8. $y' = t/y^5 - y$, $y(0) = 1$,   **h=0.1, steps = 15**.

For each of the initial value Problems 9–12, use *Mathematica* with the Euler, Heun, and Runge-Kutta methods to obtain a six-decimal approximation to $y(0.5)$. In each problem use the step size $h = 0.1$.

9. $\begin{cases} y' = e^{-y}, \\ y(0) = 0 \end{cases}$

11. $\begin{cases} y' = t\, y + \sqrt{y}, \\ y(0) = 1 \end{cases}$

10. $\begin{cases} y' = t + y^2, \\ y(0) = 0 \end{cases}$

12. $\begin{cases} y' = y - y^2, \\ y(0) = 0.5 \end{cases}$

13.6   *Using* `NDSolve`

`NDSolve` is *Mathematica*'s built-in numerical differential equations solver. Although it is very powerful, there is little documentation (at the present time) on how it works. It is known, however, that `NDSolve` uses the Adams predictor-corrector method to handle nonstiff problems, and the backwards differentiation formula (or Gear method) for stiff problems. Stiffness occurs in a problem where there are two or more very different scales of the independent variable on which the dependent variables are changing. Stiff problems arise in several fields of science, most notably in the theory of chemical kinetics, where, say, one part of a reaction occurs in a few milliseconds, while the remainder takes hours to complete. Many popular algorithms exhibit an extreme numerical instability that is not connected with any instability of the initial value problem.

`NDSolve` returns its solutions as `InterpolatingFunction` objects, which consist of internal *Mathematica* representations as piecewise cubic polynomials. Although it is possible to use *Mathematica*'s commands `InputForm` and `FullForm` to view an `InterpolatingFunction` object, it is not usually useful to do so.

`ODE` can call `NDSolve` using the option `Method->NDSolve`.

Example 13.9. *Use* `ODE` *with the option* `Method->NDSolve` *to find a numerical approximation to the solution of the initial value problem*

$$\begin{cases} y' = y + 1, \\ y(0) = 0, \end{cases} \tag{13.19}$$

for $0 \le t \le 1$. Use step size $h = 0.1$ and compare the approximation with the exact solution.

Solution. To get the exact solution, we define a *Mathematica* function **f** by

```
f[t_] = ODE[{y' == y + 1,y[0] == 0},y,{t,0,1},
        Method->FirstOrderLinear,Form->Explicit]
```

and to get the `NDSolve`-generated solution, we define another *Mathematica* function **g** by

```
g[t_] = ODE[{y' == y + 1,y[0] == 0},y,{t,0,1},
        Method->NDSolve,Form->Explicit]
```

(note that in both of these definitions we have used **=** and not **:=**). The following command generates a table of values comparing the exact solution with the approximate solution:

```
Table[{f[xx],g[xx]},{xx,0,1,0.1}]//TableForm
```

to get

```
0                0.
0.105171         0.105171
0.221403         0.221404
0.349859         0.34986
0.491825         0.491826
0.648721         0.648723
0.822119         0.822121
1.01375          1.01375
1.22554          1.22554
1.4596           1.45961
1.71828          1.71829
```

Thus, the solution generated by **NDSolve** is indeed very close to the exact solution. ▌

The output of **NDSolve** is best understood by plotting it.   We can consider an **InterpolatingFunction** object to be a pseudofunction; as such it can be plotted in much the same way that *Mathematica* plots an ordinary function. Furthermore, algebraic operations as well as differentiation can be performed on **InterpolatingFunction** objects, and the results of these operations can be plotted.

Example 13.10.   *Use* **ODE** *with the options* **Method->NDSolve** *and* **PlotSolution** *to plot the approximate solution to the initial value problem*

$$\begin{cases} y\,y' = t\sin(t), \\ y(0) = 1. \end{cases}$$

Solution. We use

```
ODE[{y y' == t Sin[t], y[0] == 1},y,{t,0,4.6},
Method->NDSolve,PlotSolution->{{t,0,4.6}}]
```

to get the plot

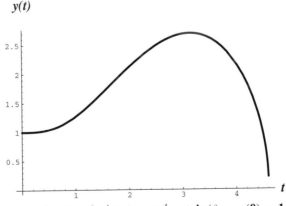

Approximate solution to $y\,y' = t\sin(t)$, $y(0) = 1$,
found by NDSolve

There are several options that control how **NDSolve** approximates the solution to an initial value problem. **MaxSteps** limits the number of steps for each calculation. With **MaxSteps->n**, the maximum number of steps taken is **n**. The default limitation is 500 steps. This limit can be effectively removed with **MaxSteps->Infinity**.

For example, to solve the initial value problem $y' = t\sin(t^2)$, $y(0) = 1$ on the interval $0 \le t \le 15$, we must increase the number of steps **NDSolve** will use by entering

```
ODE[{y' == y Sin[t^2],y[0] == 1},y,{t,0,15},
Method->NDSolve,MaxSteps->1000]
```

In general, *Mathematica* uses three terms to control the value of approximate numerical results. **WorkingPrecision** is simply the number of digits used in the arithmetic. The default **WorkingPrecision** is defined by **$MachinePrecision**, typically 16 on modern computers. **PrecisionGoal** defines the total number of correct significant digits, also related to the relative error in the calculation. **AccuracyGoal** defines the number of correct significant digits to the right of the decimal point, also related to the absolute error in the calculation. **NDSolve** uses the following defaults:

```
WorkingPrecision  -> 16
PrecisionGoal     -> Automatic
AccuracyGoal      -> Automatic
```

Here, **Automatic** means 10 digits less than **WorkingPrecision**, or 6 decimal digits on most computers. With **PrecisionGoal->Automatic** and **AccuracyGoal->Automatic**, convergence is determined by the first one of these to be satisfied. To insist on a particular convergence criterion, say **PrecisionGoal**, simply set **AccuracyGoal->Infinity**. To perform calculations at a higher level, say n, we normally begin with **WorkingPrecision->2n, PrecisionGoal->n+2** and **AccuracyGoal->n+2** and then experiment until we are satisfied with the result.

For example, we know that the exact solution to the initial value problem $y' = y$, $y(0) = 1$ is $y = e^t$. If we evaluate

```
f[t_] = ODE[{y' == y,y[0] == 1},y,{t,0,1},
        Method->NDSolve,Form->Explicit]
```

we can use

```
N[f[1],10]
```

to obtain a 10-digit approximation to the solution as

```
2.718289889
```

However, the correct 10-digit approximation is

```
2.718281828
```

To obtain a more precise solution, we can enter

```
g[t_] = ODE[{y' == y,y[0] == 1},y,{t,0,1},
          Method->NDSolve,WorkingPrecision->25,
          AccuracyGoal->12,PrecisionGoal->12];
```

Then when we enter **g[1]** we get the value of *e* with 25 digits of precision:

```
{{y -> 2.718281828477414818469359}}
```

Exercises for Section 13.6

In Exercises 1–10, plot the solutions to the initial value problems using **ODE** with the option **Method->NDSolve**. [Hint: It may be necessary to use *Mathematica*'s **?** to determine the name and syntax of certain functions.]

1. $\begin{cases} y' = \sin(t\,y^2)/(t-21), \\ y(0) = -1 \end{cases}$

2. $\begin{cases} y' = 1/t + 1/y, \\ y(1) = 1 \end{cases}$

3. $\begin{cases} y' = t\,y/(t^2-1), \\ y(0) = 1 \end{cases}$

4. $\begin{cases} y' = \left(t - \sqrt{1-y^2}\right)/y, \\ y(1) = 1 \end{cases}$

5. $\begin{cases} y' = \log\left(\sqrt{t^2 + y^2}\right), \\ y(0) = 1 \end{cases}$

6. $\begin{cases} y' = t^2 + y^2, \\ y(0) = 1 \end{cases}$

7. $\begin{cases} y' = \max(t,\,y), \\ y(0) = 0 \end{cases}$

8. $\begin{cases} y' = (y-t)^2, \\ y(0) = 0 \end{cases}$

9. $\begin{cases} y' = t^2/|y|, \\ y(1) = 1 \end{cases}$

10. $\begin{cases} y' = \text{sign}\,(t\,y)\sqrt{5\log(|y|)}, \\ y(1) = 2 \end{cases}$

11. Consider the initial value problem $y' = t^2 + y^2$, $y(0) = 1$. Use **ODE** and adjust the necessary **NDSolve** options to find the first positive vertical asymptote correct to six digits.

12. Consider the initial value problem $y' = 1/(t^2 + y^2)$, $y(0) = 1$. Use **ODE** with the option **Method->NDSolve** to estimate the set of $t > 0$ for which $y' < 10^{-3}$, sometimes referred to as the pseudo steady-state region.

NDSolve can also be used to solve higher-order differential equations numerically. Use **ODE** with the option **Method->NDSolve** to solve the following second-order initial value problems. Plot each solution on an interval containing the initial point.

13. $\begin{cases} y'' = t\,y, \\ y(0) = 1,\ y'(0) = 1 \end{cases}$

16. $\begin{cases} y'' + \sin(y) = 0, \\ y(0) = 1,\ y'(0) = 0 \end{cases}$

14. $\begin{cases} t^2 y'' + t\,y' + (t^2 - 9)y = 0, \\ y(1) = 1,\ y'(1) = 1 \end{cases}$

17. $\begin{cases} y'' - t\,y' + 7y = 0, \\ y(0) = 1,\ y'(0) = 1 \end{cases}$

15. $\begin{cases} t^2 y'' + t\,y' - (t^2 + 4)y = 0, \\ y(1) = 0,\ y'(1) = 1 \end{cases}$

18. $\begin{cases} t\,y'' + \left(\dfrac{1}{2} - t\right)y' - \dfrac{3y}{4} = 0, \\ y(1) = 1,\ y'(1) = 1 \end{cases}$

19. $\begin{cases} (1 - t^2)y'' - 2t\,y' + \left(\dfrac{8 + 8t^2 - 25}{1 - t^2}\right)y = 0, \\ y(0) = 1,\ y'(0) = 1 \end{cases}$

20. $\begin{cases} (1 - t^2)y'' + \left(\dfrac{1 - 3t^2}{t}\right)y' - y = 0, \\ y(1/2) = 1,\ y'(1/2) = 1 \end{cases}$

21. Consider the initial value problem:

$$\begin{cases} y'' + 0.1y' - y + y^5 = 0, \\ y(0) = -2.0,\ y'(0) = 0. \end{cases}$$

  a. Plot a numerical solution from $t = 0$ to $t = 50$.

  b. Experiment with **NDSolve**'s option **MaxSteps** to obtain an error free result. The option **PlotPoints** may also be needed to render an accurate plot.

  c. Use the **Parameter** option to produce a plot for each of the initial values $y(0) = -2.0, -1.5, \ldots, 2.0$.

  d. What can be inferred from this plot concerning the dependence of the long term behavior of the solutions on the initial values?

22. Consider the initial value problem:

$$\begin{cases} y'' + y + k\,y^3 = \cos(1.5t), \\ y(0) = 0,\ y'(0) = 0. \end{cases}$$

    a. Plot a numerical solution from $t = 0$ to $t = 100$ using $k = 0$.

    b. Experiment with **NDSolves**'s option **MaxSteps** to obtain an error free result. The option **PlotPoints** may also be needed to render an accurate plot.

    c. Use the **Table** command to produce a plot for each of the values $k = 0, -0.1, \ldots, -0.5$.

    d. What can be inferred from this plot concerning the values of k for which the solution is periodic?

    e. What can be inferred from this plot concerning the dependence of the long term behavior of the solutions on the parameter k?

13.7   *Adaptive Step Size and Error Control*

All modern numerical differential equation solvers exert some adaptive control over their own progress, making frequent adjustments to the step size. To see why the step size needs to be adjusted, let us consider a simple example using the Euler method.

Example 13.11. *Use the Euler method to find an approximate solution to*

$$\begin{cases} y' = -30y, \\ y(0) = 1 \end{cases} \tag{13.20}$$

over the interval $0 \le t \le 1$ using step size $h = 0.2$. Compare the approximate solution with the exact solution.

Solution. First, let us note that the exact solution of (13.20) is given by

$$y_{\text{exact}}(t) = e^{-30t};$$

consequently, $y_{\text{exact}}(t)$ is very nearly 0 for $t \ge 0.1$. However, the Euler method gives very different results. The Euler method formula for (13.20) is easy to find:

$$y_{n+1} = y_n + (0.2)(-30y_n) = -5y_n. \tag{13.21}$$

Using (13.21), we construct the following table:

n	t_n	Y_n	$y_{\text{exact}}(t_n)$
0	0.0	1	1
1	0.2	-5	0.002
2	0.4	25	0
3	0.6	-125	0
4	0.8	625	0
5	1.0	-3125	0

So the approximate solution obtained from the Euler method is very inaccurate. Instead of tending to 0, the Y_n's oscillate between positive and negative values whose absolute value is rapidly increasing. Here is the plot obtained with the command

```
ODE[{y' == -30y,y[0] == 1},y,{t,0,1},
Method->Euler,StepSize->0.2,
PlotSolution->{{t,0,1},PlotRange->All}]
```

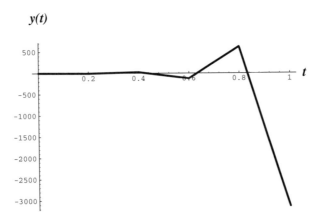

Euler approximation to $y' = -30y,\quad y(0) = 1$
plotted over the interval $0 < t < 1$ without adaptive step control

Of course, better results are obtained with smaller step size or with a more sophisticated technique such as Runge-Kutta. However, in this section we explore a different technique, that of **adaptive step size**. Adaptive step size control in **ODE** is activated using the option **VariableStepSize->True**. In addition to this option, several other options can affect the overall outcome. **StepSize** sets the initial step size, **MaxStepSize** sets the maximum value for the step size and **Tolerance** defines the maximum allowable relative error at each integration step.

Most of the methods in **ODE** use a technique called **step doubling**, when using the option **VariableStepSize->True**. Each step is taken twice, once as a full step, then as two half steps. Then the two estimates are compared to the tolerance defined by **Tolerance**. If the relative difference is greater than the tolerance, then the step size is halved and the process is begun again. If the relative difference is significantly smaller than the tolerance, then the step size is doubled for the next iteration.

Quite often **VariableStepSize->True** will produce an accurate solution, even for a stiff equation such as $y' = -30y$, but the price is a requirement for many small steps. These small steps are automatically created by *Mathematica*, so in general it will take longer for **ODE** to solve a problem with **VariableStepSize->True** than it would with **VariableStepSize->False**.

Example 13.12. *Use the Euler method to find an approximate solution to* (13.20) *over the interval* $0 \le t \le 1$ *using an adaptive step size control.*

Solution. We use

```
ODE[{y' == -30y,y[0] == 1},y,{t,0,1},Method->Euler,
VariableStepSize->True,StepSize->0.2,Tolerance->0.05,
PlotSolution->{{t,0,1},PlotRange->All}]
```

to obtain the plot

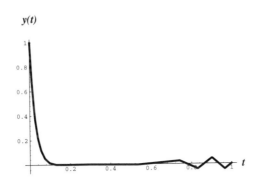

Euler approximation to $y' = -30y$, $y(0) = 1$
plotted over the interval $0 < t < 1$ **with adaptive step control**

Thus, the option **VariableStepSize->True** causes **ODE** to find a much more accurate solution. Even more accurate results would be obtained by setting **Tolerance** nearer to its default value of 0.00001. ▮

Adaptive step size control is more difficult to implement in multistep methods such as the **Milne** method and the **AdamsBashforth** method. These methods derive some of their simplicity by assuming a uniform step size over the interval of interpolation. Whenever this assumption changes, new points within the interval already covered must be computed. Two methods are typically used. Either interpolation is used to generate intermediate points, or the method is restarted at the appropriate new starting point using **RungeKutta4** or some other single step-method. **ODE** adopts the latter technique for all multistep methods.

For extrapolation techniques such as the **BulirschStoer** method, estimates of the solution can be generated by changing the order of the method, where the orders are defined by $\{2, 4, 6, 8, 12, 16, 24, 32, 48, 64, 96\}$. If the estimates computed for orders $n - 1$ and n lead to an acceptable tolerance, then the final value is returned. If the maximum order is reached, then the step size is halved and the process is started over.

Finally, implicit methods such as the **ImplicitRungeKutta** method iterate at each step until the tolerance is achieved. Because of this, these methods require another control parameter, called **ODEMaxSteps**, set to a default value of 500. Obviously, **Tolerance** and **ODEMaxSteps** affect each other, and some problems may require a larger value for **ODEMaxSteps**.

Exercises for Section 13.7

Use *Mathematica* to plot the numerical solutions of the following initial value problems. In each case first use **VariableStepSize->False** and **StepSize->0.1** and integrate over the interval $0 \leq t \leq 0.5$. Then use **VariableStepSize->True** and **Tolerance->0.001** and resolve the problem, again with **StepSize->0.1**. Solve each problem using **NDSolve** and compare it with the solutions found by the Euler, Runge-Kutta, Adams-Bashforth, Bulirsch-Stoer, and implicit Runge-Kutta methods. [Some of these problems take a while to complete.]

1. $\begin{cases} y' = -50y, \\ y(0) = 1/50 \end{cases}$

4. $\begin{cases} y' = \cos(99y) - \sin(98y), \\ y(0) = 1 \end{cases}$

2. $\begin{cases} y' = 10\sin(10y), \\ y(0) = 1 \end{cases}$

5. $\begin{cases} y' = \cos(99y) - \sin(98t), \\ y(0) = 1 \end{cases}$

3. $\begin{cases} y' = \sin(50y)\cos(50y), \\ y(0) = -1 \end{cases}$

6. $\begin{cases} y' = t\,\cos(100y) - t^5 y^2, \\ y(0) = 1 \end{cases}$

For each of the initial value Problems 7–12, use **ODE** with **VariableStepSize->True**, **Tolerance->0.0001**, and **StepSize->0.1**. Find the exact value at the indicated point and then obtain a four-decimal approximation at the indicated point using the Euler, Runge-Kutta, Adams-Bashforth, Bulirsch-Stoer, and implicit Runge-Kutta methods.

7. $\begin{cases} y' = y, \\ y(0) = 1/60, \quad y(1.0) \end{cases}$

10. $\begin{cases} y' = -t^2 + t + 1, \\ y(0) = 1, \quad y(2.0) \end{cases}$

8. $\begin{cases} y' = -y/2, \\ y(0) = 1, \quad y(2.0) \end{cases}$

11. $\begin{cases} y' = y - \cos(t) - \sin(t), \\ y(0) = 1, \quad y(\pi) \end{cases}$

9. $\begin{cases} y' = t, \\ y(0) = 1, \quad y(2.0) \end{cases}$

12. $\begin{cases} y' - y/t = t, \\ y(1) = 2, \quad y(5) \end{cases}$

13.8   The Numerov Method

In this section we derive a method for finding a numerical solution to a second-order linear differential equation with no first derivative present, written as

$$y'' = f(t)y + g(t).$$

This type of equation occurs frequently in scientific and engineering applications. The method we describe (found by B. Numerov in 1933, see [Har, page 142]) is popular with physicists because of its simplicity and speed.

To derive the Numerov method we need the following fact.

Lemma 13.1. *Suppose y is a solution of the second-order differential equation*

$$y'' = f(t)y + g(t). \tag{13.22}$$

Let $z(t) = \left(1 - \dfrac{h^2}{12} f(t)\right) y(t),$ *where h is some number; then*

$$z(t+h) + z(t-h) - \left(2 + \frac{h^2 f(t)}{1 - \dfrac{h^2}{12} f(t)}\right) z(t)$$

$$= \frac{h^2}{12}\big(g(t+h) + g(t-h) + 10 g(t)\big) + \text{higher-order terms}. \tag{13.23}$$

Proof. The Taylor expansion of $y(t \pm h)$ is

$$y(t \pm h) = y(t) \pm h\, y'(t) + \frac{h^2}{2} y''(t) \pm \frac{h^3}{6} y'''(t) + \frac{h^4}{24} y^{(4)}(t) + \cdots.$$

We add $y(t + h)$ and $y(t - h)$ to obtain

$$y(t + h) + y(t - h) = 2y(t) + h^2 y''(t) + \frac{h^4}{12} y^{(4)}(t) + \cdots. \tag{13.24}$$

Similarly,

$$y''(t + h) + y''(t - h) = 2y''(t) + h^2 y^{(4)}(t) + \cdots,$$

so that

$$h^2 y^{(4)}(t) = y''(t + h) + y''(t - h) - 2y''(t) + \cdots. \tag{13.25}$$

From (13.24) and (13.25) we obtain

$$y(t + h) + y(t - h) - 2y(t) = h^2 \left(y''(t) + \frac{1}{12} \left(y''(t + h) + y''(t - h) - 2y''(t) \right) \right) + \cdots$$

$$= \frac{h^2}{12} \left(y''(t + h) + y''(t - h) + 10y''(t) \right) + \cdots. \tag{13.26}$$

Now assume that (13.22) holds. From (13.26) we obtain

$$y(t + h) + y(t - h) - 2y(t) = \frac{h^2}{12} \big(f(t + h)y(t + h) + g(t + h) + f(t - h)y(t - h)$$

$$+ g(t - h) + 10f(t)y(t) + 10g(t) \big) + \cdots,$$

or

$$\left(1 - \frac{h^2}{12} f(t + h) \right) y(t + h) + \left(1 - \frac{h^2}{12} f(t - h) \right) y(t - h) - 2 \left(1 - \frac{h^2}{12} f(t) \right) y(t)$$

$$= \frac{h^2}{12} \big(g(t + h) + g(t - h) + 10g(t) + 12f(t)y(t) \big) + \cdots. \tag{13.27}$$

Then we can rewrite (13.27) as

$$z(t + h) + z(t - h) - 2z(t)$$

$$= \frac{h^2}{12} \left(g(t + h) + g(t - h) + 10g(t) + 12 \left(1 - \frac{h^2}{12} f(t) \right)^{-1} z(t) \right),$$

from which we get (13.23). ∎

The **Numerov method** finds a numerical approximation to the solution of the second-order linear initial value problem

$$\begin{cases} y'' = f(t)y + g(t), \\ y(a) = Y_0, \; y'(a) = Y_1 \end{cases} \tag{13.28}$$

on an interval $a \leq t \leq b$. To derive the iteration scheme for the Numerov method, we need a discrete version of Lemma 13.1. To this end, let h be a small positive number. We subdivide the interval $a < t < b$ as $a = t_0 < t_1 < \cdots < t_N = b$, where $t_{j+1} - t_j = h$ for $j = 1, \ldots, n$. Let

$$w_n = \frac{h^2 f(t_n)}{1 - \frac{h^2}{12} f(t_n)}. \tag{13.29}$$

Then z_0 and z_1 are given by

$$z_0 = \left(1 - \frac{h^2}{12} f(t_0)\right) Y_0 \quad \text{and} \quad z_1 = \left(1 - \frac{h^2}{12} f(t_1)\right) (Y_0 + h\, Y_1). \tag{13.30}$$

From (13.22) we get

$$z_{n+2} = (2 + w_{n+1}) z_{n+1} - z_n + \frac{h^2}{12} \big(g(t_{n+2}) + 10 g(t_{n+1}) + g(t_n)\big). \tag{13.31}$$

We can use (13.30) and (13.31) to determine z_n for $n = 2, 3, \ldots$. Finally, we determine Y_n for $n = 2, 3, \ldots$ by

$$Y_n = \left(1 - \frac{h^2}{12} f(t_n)\right)^{-1} z_n. \tag{13.32}$$

The *Mathematica* implementation of (13.29)–(13.32) is as follows:

```
Numerov[f_,g_,{t0_,Y0_,Y1_},h_,steps_][t_,y_]:=
    Module[{tt,ff,gg,z0,z1,tmp,z,zsol},
        ff[tt_,n_]:= f /. t -> tt + n h;
        gg[tt_,n_]:= g /. t -> tt + n h;
        z0 = Y0(1 - (h^2/12)ff[t0,0]);
        z1 = (1 - (h^2/12)ff[t0,1])(Y0 + h Y1);
        tmp = Table[{1,1/((1 - (h^2/12)ff[t0,k]))},
            {k,0,steps}];
        zsol = NestList[nmstep[ff,gg,{t,z},#,h]&,
            {t0,z0,z1},steps];
        Simplify[Map[Drop[#,{3}]&,zsol] tmp]]
```

```
nmstep[f_,g_,{t_,z_},{tn_,zn_,znp1_},h_]:=
    Module[{wnp1},
        wnp1 = h^2 f[tn,1]/(1 - (h^2/12)f[tn,1]);
        {tn + h, znp1, (2+wnp1)znp1 - zn +
            (h^2/12)(g[tn,2] + 10g[tn,1] + g[tn,0])}]
```

Let us see how the Numerov method works in practice.

Example 13.13. *Use the Numerov method to find a numerical approximation to the second-order initial value problem*

$$\begin{cases} y'' = -t^3 y + 1, \\ y(0) = 1, \, y'(0) = 0. \end{cases}$$

Solution. We use

```
ODE[{y'' == -t^3 y + 1,y[0] == 1,y'[0] == 0},y,{t,0,10},
Method->Numerov,StepSize->0.05,NumericalOutput->None,
PlotSolution->{{t,0,10}}]
```

or

```
ListPlot[Numerov[-t^3,1,{0,1,0},0.05,200][t,y],
PlotJoined->True]
```

to obtain the plot

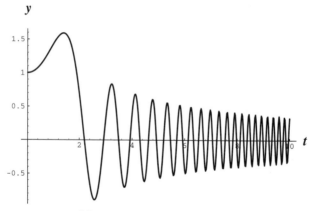

Numerov approximation to
$$y'' = -t^3 y + 1, \quad y(0) = 1, \quad y'(0) = 0$$

Although the above plot can also be obtained using **NDSolve**, for example using

```
ODE[{y'' == -t^3 y + 1,y[0] == 1,y'[0] == 0},y,{t,0,10},
Method->NDSolve,MaxSteps->2000,NumericalOutput->None,
PlotSolution->{{t,0,10},PlotPoints->400}]
```

Numerov is faster.

Exercises for Section 13.8

Use the Numerov method to find approximate solutions to the following initial value Problems 1–4. For each problem choose a suitable interval and step size and plot the solution.

1. $\begin{cases} y'' + t\,y = 0, \\ y(0) = 1, y'(0) = 1 \end{cases}$

3. $\begin{cases} y'' = \sin(t)y + t, \\ y(0) = 0, \quad y'(0) = 1 \end{cases}$

2. $\begin{cases} y'' = \dfrac{\cos(t)}{|\cos(t)|}y + t, \\ y(0) = 2, \quad y'(0) = 1 \end{cases}$

4. $\begin{cases} y'' = -\dfrac{y}{t+1} + \tan(5t), \\ y(0) = 2, \quad y'(0) = 1 \end{cases}$

For each of Problems 5–8 use **Numerov** to compute an approximate solution at $t = 1.0$. Use step sizes of $h = 1/2$, $1/4$, and $1/8$. Then determine the symbolic solution and compare the values of $y(1)$ from the symbolic and numerical solutions.

5. $\begin{cases} y'' + 2y = t, \\ y(0) = 1, \quad y'(0) = 0 \end{cases}$

7. $\begin{cases} y'' = y + t^3, \\ y(0) = 0, \quad y'(0) = 1 \end{cases}$

6. $\begin{cases} y'' = t + 1, \\ y(0) = 1, \quad y'(0) = 0 \end{cases}$

8. $\begin{cases} y'' = y + e^t \sin(t), \\ y(0) = 1, \quad y'(0) = 0 \end{cases}$

14

THE LAPLACE TRANSFORM

We have already observed that there is a strong analogy between linear differential equations and linear algebraic equations. In this chapter we study an operator called the **Laplace**[1] **transform** $\mathcal{L}$, which can be used to convert a linear differential equation with constant coefficients into an algebraic equation by a method developed by Heaviside[2]. The algebraic equation can then be solved and the **inverse Laplace transform** applied to obtain a solution to the original differential equation.

The reader may wonder why it is necessary to use the Laplace transform at all. Why not just solve the differential equation by the techniques developed in Chapters 9 and 12? There are two reasons why the Laplace transform is useful:

(i) The methods of Chapters 9 and 12 require three steps to solve a linear differential equation: (a) finding the solution of the homogeneous equation, (b) finding a particu-

---

[1]

Pierre Simon Marquis de Laplace (1749–1827). French mathematician and astronomer, known in his time as the Newton of France. His influential treatises were the five-volume **Mécanique Céleste** (which contains the second-order partial differential equation (1.2) now known as Laplace's equation) and **Théorie Analytique des Probabilités**. However, the theory and applications of the Laplace transform were not developed until much later and were due mainly to Heaviside.

[2]

Oliver Heaviside (1850–1925). English electrical engineer. Heaviside originally worked as a telegrapher until he became deaf. Although Heaviside introduced important results in operational calculus, they were first rejected on the grounds that they contained errors of substance and had inadequacies of proof. Eventually, Heaviside's operational calculus became one of the standard tools in analysis. The layer in the atmosphere that allows radio waves to follow the earth's curvature was predicted by Heaviside in 1902.

lar solution of the inhomogeneous equation, and finally, (c) substitution of the initial conditions.

In contrast, the Laplace transform handles not only a differential equation, but also its initial conditions. Hence the solution of a constant-coefficient linear differential equation (possibly inhomogeneous) can be obtained in fewer steps.

(ii) Laplace transforms permit the use of piecewise differentiable functions and impulse functions as forcing functions. These important mathematical concepts will be studied in Sections 14.2 and 14.7.

14.1   Definition and Properties of the Laplace Transform

The **Laplace transform** is an operator $\mathcal{L}$ that can be applied to certain functions $f(t)$ that are defined on the infinite interval $0 < t < \infty$. The explicit definition is the improper integral

$$\mathcal{L}\big(f(t)\big)(s) = \int_0^\infty f(t)e^{-st}\, dt = \lim_{T\to\infty} \int_0^T f(t)e^{-st}\, dt, \qquad (14.1)$$

provided the limit exists for sufficiently large s. Thus the Laplace transform is an operator that assigns one function to another. As preparation we first review properties of the improper integral.

Improper Integrals

Since an improper integral is more complicated than an ordinary integral, it is important to define the notion precisely. We assume that the reader is familiar with the notion of Riemann integration over finite intervals; this is the type of integration studied in most calculus courses. Continuous functions are Riemann integrable, as are piecewise continuous functions, defined later in this chapter.

Definition. *Suppose that g is a function defined for $t > t_0$ such that the integral*

$$\int_{t_0}^T g(t)\, dt$$

*exists in the sense of Riemann integration for every T with $t_0 < T < \infty$. The **improper integral***

$$\int_{t_0}^\infty g(t)\, dt$$

is defined to be the limit

$$\lim_{T \to \infty} \int_{t_0}^{T} g(t)\, dt, \tag{14.2}$$

provided the limit exists. When the limit in (14.2) exists, we say that the integral

$$\int_{t_0}^{\infty} g(t)\, dt$$

converges. *Otherwise, we say that the integral* **diverges**.

A few examples may be helpful.

Example 14.1. *For what values of the constant s does the integral*

$$\int_{0}^{\infty} e^{-st}\, dt \tag{14.3}$$

converge? When the integral converges, compute its value.

Solution. For $s \neq 0$ we compute

$$\int_{0}^{T} e^{-st}\, dt = \frac{1 - e^{-sT}}{s}. \tag{14.4}$$

If $s > 0$, the right-hand side of (14.4) has a limit when $T \longrightarrow \infty$, but if $s < 0$ the limit does not exist. If $s = 0$, then

$$\int_{0}^{T} e^{-st}\, dt = \int_{0}^{T} dt = T,$$

which also fails to have a limit as $T \longrightarrow \infty$. Therefore, we conclude that the integral (14.3) converges if and only if $s > 0$, in which case it has the value

$$\mathcal{L}(1)(s) = \frac{1}{s}. \quad \blacksquare \tag{14.5}$$

Example 14.2. *Let $b \neq 0$. For what values of s does the integral*

$$\mathcal{L}\big(\sin(b\,t)\big)(s) = \int_{0}^{\infty} e^{-st} \sin(b\,t)\, dt \tag{14.6}$$

converge? When the integral converges, compute its value.

Solution. In the case that $s = 0$, we have

$$\int_0^T \sin(b\,t)\,dt = \frac{1 - \cos(b\,T)}{b}. \tag{14.7}$$

The right-hand side of (14.7) has no limit when $T \longrightarrow \infty$. If $s \neq 0$, we integrate $e^{-st}\sin(b\,t)$ twice by parts to obtain

$$\int_0^T e^{-st}\sin(b\,t)\,dt = -e^{-st}\left(\frac{s\sin(b\,t) + b\cos(b\,t)}{s^2 + b^2}\right)\Big|_0^T. \tag{14.8}$$

Again, we see that for $s > 0$ the right-hand side of (14.8) has a limit when $T \longrightarrow \infty$, while for $s < 0$ the limit fails to exist. Moreover, when $s > 0$ we have

$$\int_0^{\infty} e^{-st}\sin(b\,t)\,dt = \lim_{T \to \infty} \int_0^T e^{-st}\sin(b\,t)\,dt \tag{14.9}$$

$$= \lim_{T \to \infty}\left(-e^{-st}\left(\frac{s\sin(b\,t) + b\cos(b\,t)}{s^2 + b^2}\right)\Big|_0^T\right)$$

$$= \lim_{T \to \infty}\left(-e^{-sT}\left(\frac{s\sin(b\,T) + b\cos(b\,T)}{s^2 + b^2}\right) + \frac{b}{s^2 + b^2}\right)$$

$$= \frac{b}{s^2 + b^2}.$$

Hence

$$\mathcal{L}\big(\sin(b\,t)\big)(s) = \frac{b}{s^2 + b^2}. \ \blacksquare$$

We could continue in this tedious fashion to compute Laplace transforms of many elementary functions. Instead, it is better to study the theory of the Laplace transform, so that we can deduce the values of more complicated Laplace transforms from simpler ones. Eventually, we shall construct a table of Laplace transforms. (This procedure is completely analogous to what the reader probably did when learning integration theory. After laboriously computing a few integrals from the definition, the reader learned techniques of integration from which the value of many integrals could be obtained from a table of integrals. See the table on page 471.)

First, we note a simple theorem that allows us to compute the Laplace transforms of linear combinations of functions whose Laplace transforms we already know.

Theorem 14.1. *Let f and g be functions whose Laplace transforms $\mathcal{L}(f)$ and $\mathcal{L}(g)$ exist, and let a and b be constants. Then the Laplace transform $\mathcal{L}(a\,f + b\,g)$ exists, and*

$$\mathcal{L}(a\,f + b\,g)(s) = a\mathcal{L}(f)(s) + b\mathcal{L}(g)(s). \tag{14.10}$$

Proof. (14.10) is a consequence of the facts that the integral of a sum is the sum of the integrals and that constants can be factored from integrals. ∎

Our next task is to give a general condition under which the Laplace transform exists. For that we need the following result.

Theorem 14.2. *Suppose that $f(t)$ and $g(t)$ are functions that are integrable on each finite interval of $t_0 < t < \infty$ and that*

$$|f(t)| \le g(t)$$

for $t \ge t_0$. Then the convergence of $\int_{t_0}^{\infty} g(t)\,dt$ implies the convergence of $\int_{t_0}^{\infty} f(t)\,dt$.

Theorem 14.2 is proved in courses on advanced calculus. It is made plausible as follows. The area under the graphs of $f(t)$ and $g(t)$ are represented by

$$\int_{t_0}^{\infty} f(t)\,dt \qquad \text{and} \qquad \int_{t_0}^{\infty} g(t)\,dt.$$

Since $f(t) < g(t)$ for all t, it follows that the area under the graph of $f(t)$ is less than the area under the graph of $g(t)$.

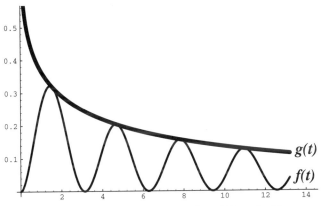

Comparison of the areas under $f(t)$ and $g(t)$

In order to make use of Theorem 14.2, we introduce the following notion.

Definition. *A function $g(t)$ defined for $t > t_0$ is said to have **exponential growth** if there exist constants r and M for which*

$$|g(t)| \le M\,e^{rt}.$$

*The number r is called a **growth exponent** of g.*

The next theorem gives the simplest and most useful criterion for the existence of a Laplace transform.

Theorem 14.3. *Suppose that $f(t)$ is a function that is integrable on each finite interval of $0 < t < \infty$ and that $f(t)$ has exponential growth with growth exponent r. Then the Laplace transform*

$$\mathcal{L}(f(t)) = \int_0^\infty e^{-st} f(t) dt \tag{14.11}$$

exists for $s > r$.

Proof. We must show that the improper integral on the right-hand side of (14.11) converges. We split the improper integral into two parts:

$$\int_0^\infty f(t) e^{-st} \, dt = \int_0^{t_0} f(t) e^{-st} \, dt + \int_{t_0}^\infty f(t) e^{-st} \, dt. \tag{14.12}$$

The first integral on the right-hand side of (14.12) exists because $f(t)$ is assumed to be integrable. Next, we estimate the second integral. Since $f(t)$ has exponential growth with growth exponent r, there exists a constant M such that $|f(t)| \leq M e^{rt}$ for $t > 0$. Therefore,

$$|e^{-st} f(t)| \leq M e^{-st} e^{rt} = M e^{(r-s)t}. \tag{14.13}$$

From Example 14.1 we know that when $r - s < 0$,

$$\int_0^\infty M e^{(r-s)t} dt$$

converges. Now Theorem 14.2 implies that

$$\int_{t_0}^\infty f(t) e^{-st} \, dt$$

converges, completing the proof of the theorem. ∎

Let us give some applications of Theorem 14.3.

Example 14.3. *Show that the Laplace transform*

$$\mathcal{L}(t^n)(s) = \int_0^\infty t^n e^{-st} dt \tag{14.14}$$

exists for all $s > 0$ and for all nonnegative integers n.

Solution. We must show that $f(t) = t^n$ has exponential growth rate r for every $r > 0$. Then we can apply Theorem 14.3. For $t \geq 0$ we have for any $r > 0$

$$e^{rt} = \sum_{k=0}^{\infty} \frac{(r\,t)^k}{k!} \geq \frac{(r\,t)^n}{n!}.$$

This inequality can be rewritten as

$$t^n \leq \frac{n!}{r^n} e^{rt}.$$

Hence t^n has exponential growth with growth exponent $r > 0$, so Theorem 14.3 implies that $\mathcal{L}(t^n)(s)$ exists for $s > 0$. ∎

Example 14.4.   *Show that the Laplace transform*

$$\mathcal{L}\big(t^n \sin(b\,t)\big)(s) = \int_0^{\infty} t^n e^{-st} \sin(b\,t)\,dt$$

exists for all $s > 0$ and for all nonnegative integers n.

Solution. We have $|\sin(b\,t)| \leq 1$, so by the solution of Example 14.3 we see that $t^n \sin(b\,t)$ has exponential growth with growth exponent r for every $r > 0$. Again Theorem 14.3 shows that $\mathcal{L}(t^n \sin(b\,t))(s)$ exists for $s > 0$. ∎

Here is a general way to construct new Laplace transforms from old; the proof is not hard.

Theorem 14.4.   *Suppose the Laplace transform of $f(t)$ exists. Then the Laplace transform of $e^{at}\,f(t)$ also exists, and*

$$\mathcal{L}\big(e^{at}\,f(t)\big)(s) = \mathcal{L}\big(f(t)\big)(s)\Big|_{s \to s-a}.$$

Exercises for Section 14.1

In Problems 14.1–14.1, determine the existence of the improper integral, and find its value if possible.

1. $\displaystyle\int_1^{\infty} \frac{dt}{t^3}$ 3. $\displaystyle\int_1^{\infty} \frac{e^t\,dt}{t^3}$

2. $\displaystyle\int_1^{\infty} \frac{dt}{t^5}$ 4. $\displaystyle\int_1^{\infty} \frac{\sin(t)\,dt}{t^3}$

5. Compute $\mathcal{L}\big(\cos(b\,t)\big)(s)$ from the definition.

6. For which values of the constants a and b does the improper integral

$$\int_{t_0}^{\infty} t^9 e^{-at} \sin(b\,t)\,dt$$

converge?

In Problems 7–10, determine which of the following functions is of exponential growth? Find the growth exponent if possible.

7. $f_1(t) = e^{t^2}$ **9.** $f_3(t) = e^{\sqrt{t}}$

8. $f_2(t) = t^3 e^{-t}$ **10.** $f_4(t) = \sin(t)\sinh(t)$

11. Prove Theorem 14.4.

14.2 *Piecewise Continuous Functions*

One of the most useful features of the Laplace transform $\mathcal{L}$ is that it can be applied to certain functions with discontinuities. In this section we define the notions of piecewise continuous and piecewise differentiable functions.

Definition. *Let f be a function defined on an interval $a < t < b$, and let $a < t_0 < b$. We say that f has a **left limit** as t approaches t_0 provided*

$$\lim_{\substack{t \to t_0 \\ t < t_0}} f(t)$$

*exists. Similarly, we say that f has a **right limit** as t approaches t_0 provided*

$$\lim_{\substack{t \to t_0 \\ t > t_0}} f(t)$$

exists. We use the following notation for the left and right limits:

$$f(t_0 - 0) = \lim_{\substack{t \to t_0 \\ t < t_0}} f(t) \qquad and \qquad f(t_0 + 0) = \lim_{\substack{t \to t_0 \\ t > t_0}} f(t).$$

In calculus a function is defined to be **continuous** at t_0 provided

$$f(t_0 - 0) = f(t_0 + 0) = f(t_0).$$

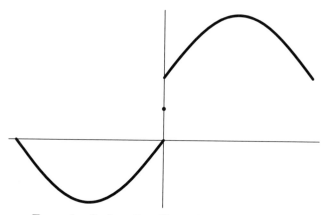

**Example of a function discontinuous at 0 whose
left and right limits exist at 0**

We require a generalization of the notion of continuous function.

Definition.   *A function $f(t)$ is* **piecewise continuous** *on an interval $a < t < b$ if there is a finite subdivision of points*

$$a = t_0 < t_1 < \cdots < t_p = b \qquad (14.15)$$

such that f is continuous on each subinterval (t_{i-1}, t_i) and the left and right limits exist at the end points of each subinterval that is interior to (a, b). These are denoted respectively by $f(t_i - 0)$ and $f(t_i + 0)$, for $i = 1, \ldots, p-1$. We also require that $f(t_0 + 0)$ and $f(t_p - 0)$ exist. A function defined on an infinite interval is defined to be piecewise continuous if it is piecewise continuous on each finite subinterval.

We remark that any continuous function is piecewise continuous, but the converse is not true.

Definition.   *A function $f(t)$ is* **piecewise differentiable** *on an interval $a < t < b$ provided that $f(t)$ is piecewise continuous and all of its derivatives $f'(t)$, $f''(t)$, $\ldots$ exist for $t \neq t_i$ and are piecewise continuous on $a < t < b$.*

The following fact is established in advanced calculus.

Lemma 14.5.   *A piecewise continuous function is integrable.*

Here is an important example of a piecewise continuous function that is not continuous.

Definition. *The* **integer part** *of a real number t is*

$$[t] = \text{greatest integer less than or equal to } t.$$

The corresponding Mathematica function is called **Floor**, *so we sometimes write*

$$\textbf{Floor}(t) = [t].$$

Floor(t)

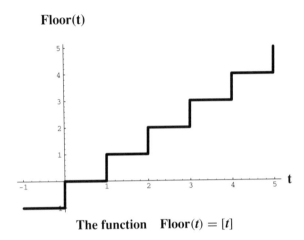

The function   Floor$(t) = [t]$

Example 14.5. *Show that the function* **Floor** *is piecewise continuous.*

Solution. For each integer n we have **Floor**$(t) = [t] = n$ for $n \le t < n+1$. The function **Floor** is continuous on the interval $n \le t < n+1$ because it is constant on that interval. It is easy to check that the left and right limits exist for each n; in fact,

$$\textbf{Floor}(n - 0) = \lim_{\substack{t \to n \\ t < n}} \textbf{Floor}(t) = n - 1 \quad \text{and} \quad \textbf{Floor}(n + 0) = \lim_{\substack{t \to n \\ t > n}} \textbf{Floor}(t) = n.$$

These relationships hold for all n, and so **Floor** is piecewise continuous. ∎

We can now give a result on the existence of Laplace transforms for piecewise continuous functions defined on $(0, \infty)$.

Theorem 14.6. *Suppose that* $f(t)$ *is piecewise continuous on any finite interval* $0 < t < T$ *and has exponential growth with growth exponent* r. *Then the Laplace transform*

$$\mathcal{L}(f)(s) = \int_0^\infty f(t) e^{-st}\, dt \tag{14.16}$$

exists for $s > r$.

Proof. Since a piecewise continuous function is integrable, the result is a consequence of Theorem 14.3. ∎

Exercises for Section 14.2

Which of the following functions are piecewise continuous?

1. $t - [t]$

3. $t \sin(1/t)$

2. $\sin(t)$

4. $e^{-1/t} \sin(1/t)$

The function **Ceiling** (corresponding to *Mathematica*'s `Ceiling`) is defined by

Ceiling(t) = the smallest integer greater than or equal to t.

5. Show that **Ceiling** is piecewise continuous.

6. Plot **Ceiling** for $-1 < t < 5$.

7. Compute **Ceiling**$(t) -$ **Floor**(t).

14.3   Using the Laplace Transform to Solve Initial Value Problems

In this section we use the Laplace transform to solve initial value problems for second-order differential equations of the form

$$\begin{cases} a\,y'' + b\,y' + c\,y = g(t), \\ y(t_0) = Y_0, \quad y'(t_0) = Y_1, \end{cases} \tag{14.17}$$

where a, b, c are constants and $g(t)$ is a function of exponential growth defined for $t > 0$. The key fact that makes Laplace transforms useful for solving differential equations is that there is a relation between the Laplace transforms of a function and its derivative.

Theorem 14.7.   *Suppose that f is continuous and that f' is piecewise continuous on every finite interval $0 < t < b$. If both f and f' have exponential growth with growth exponent r, then the Laplace transform $\mathcal{L}(f')(s)$ exists for $s > r$ and*

$$\mathcal{L}\big(f'(t)\big)(s) = s\mathcal{L}\big(f(t)\big)(s) - f(0).$$

Proof. Write $0 = x_0 < x_1 < \cdots < x_p = b$, where the x_i denote the points where f' fails to be continuous. The additivity of the Riemann integral implies that

$$\int_0^b e^{-st} f'(t)\,dt = \int_0^{x_1} e^{-st} f'(t)\,dt + \int_{x_1}^{x_2} e^{-st} f'(t)\,dt + \cdots + \int_{x_{p-1}}^b e^{-st} f'(t)\,dt.$$
(14.18)

We can integrate each term on the right-hand side of (14.18) by parts:

$$\int_{x_i}^{x_{i+1}} e^{-st} f'(t)\,dt = e^{-st} f(t)\Big|_{x_i}^{x_{i+1}} + s \int_{x_i}^{x_{i+1}} e^{-st} f(t)\,dt.$$
(14.19)

Cancellation occurs when we sum the integrated terms because of the assumption that $f(t)$ is continuous. Thus

$$e^{-st} f(t)\Big|_0^{x_1} + e^{-st} f(t)\Big|_{x_1}^{x_2} + \cdots + e^{-st} f(t)\Big|_{x_{p-1}}^{x_p} = e^{-sb} f(b) - f(0).$$
(14.20)

Also, it is clear that

$$s \int_0^{x_1} e^{-st} f(t)\,dt \quad + \quad s \int_{x_1}^{x_2} e^{-st} f(t)\,dt$$

$$+ \quad \cdots + s \int_{x_{p-1}}^{x_p} e^{-st} f(t)\,dt = s \int_0^b e^{-st} f(t)\,dt. \quad (14.21)$$

Now (14.18)–(14.21) imply that

$$\int_0^b e^{-st} f'(t)\,dt = e^{-sb} f(b) - f(0) + s \int_0^b e^{-st} f(t)\,dt.$$
(14.22)

Since f has exponential growth with growth exponent r, there exists a constant M such that

$$\left| e^{-sb} f(b) \right| \le M\, e^{(r-s)b}.$$

The hypothesis $r - s < 0$ implies that $M\, e^{(r-s)b} \longrightarrow 0$ as $b \longrightarrow \infty$. Therefore, when we take the limit as $b \longrightarrow \infty$ in (14.22) we get

$$\mathcal{L}\big(f'(t)\big)(s) = \int_0^\infty e^{-st} f'(t)\,dt = -f(0) + s \int_0^\infty e^{-st} f(t)\,dt \quad (14.23)$$

$$= s\mathcal{L}\big(f(t)\big)(s) - f(0). \quad \blacksquare$$

Here is a simple application of Theorem 14.7.

Example 14.6. *Let $b \neq 0$ and $s > 0$. Compute $\mathcal{L}\big(\cos(b\,t)\big)(s)$.*

Solution. We use Example 14.2 and Theorem 14.7 to compute

$$\mathcal{L}\big(\cos(b\,t)\big)(s) = \mathcal{L}\left(\frac{1}{b}\frac{d}{dt}\big(\sin(b\,t)\big)\right)(s)$$

$$= \frac{1}{b}\big(s\mathcal{L}\big(\sin(b\,t)\big)(s) - \sin(0)\big) = \frac{s}{s^2 + b^2}. \;\blacksquare$$

When we apply Theorem 14.7 twice, we get the formula for the Laplace transform of the second derivative.

Corollary 14.8. *Suppose that f and f' are continuous and that f'' is piecewise continuous on every finite interval $0 < t < b$. If f, f', and f'' have exponential growth with growth exponent r, then the Laplace transform $\mathcal{L}(f'')(s)$ exists for $s > r$ and*

$$\mathcal{L}\big(f''(t)\big)(s) = s^2\mathcal{L}\big(f(t)\big)(s) - s\,f(0) - f'(0).$$

We need the following important result, the proof of which is beyond the scope of this book.

Theorem 14.9. **(Lerch's[3] Theorem)** *Given a function $F(s)$, there is at most one continuous function $f(t)$ (defined for $t \geq 0$) for which the Laplace transform of $f(t)$ is $F(s)$, that is, $\mathcal{L}\big(f(t)\big) = F(s)$.*

Theorem 14.9 does *not* say that for a given $F(s)$ there is a function $f(t)$ for which $\mathcal{L}\big(f(t)\big) = F(s)$. Indeed, the function $F(s) = 1$ can never be the Laplace transform of any continuous function, as we shall see in Section 14.8. On the other hand, Theorem 14.9 *does* say that when there is a continuous function $f(t)$ for which $\mathcal{L}\big(f(t)\big) = F(s)$, then $f(t)$ is uniquely determined. Therefore, we can make the following definition.

Definition. *If it exists, the unique continuous function $f(t)$ defined for $t \geq 0$ for which $\mathcal{L}\big(f(t)\big) = F(s)$ is called the* **inverse Laplace transform** *of $F(s)$. We write*

$$f(t) = \mathcal{L}^{-1}\big(F(s)\big).$$

Armed with Theorem 14.9, we can now use the Laplace transform and the inverse Laplace transform to solve a differential equation. There are three steps to the process:

3

Mathias Lerch (1860–1922). Czech mathematician, who studied under Weierstrass. He became professor in Brno.

1. First, we use the Laplace transform to convert the differential equation to an algebraic equation.

2. Next, we solve the algebraic equation.

3. Finally, we apply the inverse Laplace transform to find the solution to the differential equation.

The following example illustrates the procedure.

Example 14.7. *Solve the differential equation*

$$y'' + y' - 12y = 0, \tag{14.24}$$

with the initial conditions

$$y(0) = 0, \qquad y'(0) = 1. \tag{14.25}$$

using the Laplace transform and inverse Laplace transform.

Solution. This simple problem is most easily solved by the methods of Section 9.1. Nevertheless, we will solve it by the Laplace transform method so that it will serve as a model for solving more complicated differential equations.

First, we apply the operator $\mathcal{L}$ to both sides of (14.24) and use the fact that $\mathcal{L}$ is linear:

$$0 = \mathcal{L}(y'') + \mathcal{L}(y') - 12\mathcal{L}(y). \tag{14.26}$$

Next, we apply Corollary 14.8 to (14.26) to obtain

$$0 = s^2 \mathcal{L}(y)(s) - s\, y(0) - y'(0) + s\mathcal{L}(y) - y(0) - 12\mathcal{L}(y). \tag{14.27}$$

When we use the initial conditions (14.25), equation (14.27) becomes

$$0 = (s^2 + s - 12)\mathcal{L}(y)(s) - 1,$$

or

$$\mathcal{L}(y)(s) = \frac{1}{s^2 + s - 12}. \tag{14.28}$$

Now we search for a function whose Laplace transform is the right-hand side of (14.28). This is most easily accomplished by the method of partial fractions. Since

$$s^2 + s - 12 = (s + 4)(s - 3),$$

we write

$$\frac{1}{s^2 + s - 12} = \frac{A}{s - 3} + \frac{B}{s + 4} = \frac{A(s + 4) + B(s - 3)}{s^2 + s - 12}. \tag{14.29}$$

From (14.29) it follows that

$$1 = A(s+4) + B(s-3) \tag{14.30}$$

for all s. In particular, if we set $s = 3$ in (14.30), we find that $A = 1/7$; similarly, setting $s = -4$ in (14.30) yields $B = -1/7$.

Now we can rewrite (14.28) as

$$\mathcal{L}(y)(s) = \frac{1/7}{s-3} + \frac{-1/7}{s+4} = \frac{1}{7}\left(\frac{1}{s-3} - \frac{1}{s+4}\right). \tag{14.31}$$

We know from Example 14.1 that

$$\mathcal{L}(e^{3t})(s) = \frac{1}{s-3} \quad \text{and} \quad \mathcal{L}(e^{-4t})(s) = \frac{1}{s+4};$$

Therefore, (14.31) becomes

$$\mathcal{L}(y)(s) = \frac{1}{7}\left(\mathcal{L}(e^{3t})(s) - \mathcal{L}(e^{-4t})(s)\right) = \mathcal{L}\left(\frac{1}{7}(e^{3t} - e^{-4t})\right)(s). \tag{14.32}$$

From (14.32) and Lerch's theorem (Theorem 14.9) we get the solution of the initial value problem ((14.24) and (14.25)):

$$y(t) = \frac{1}{7}(e^{3t} - e^{-4t}). \ \blacksquare$$

When we apply Theorem 14.7 n times, we get

Corollary 14.10. *Suppose that $f, f', \ldots, f^{(n-1)}$ are continuous and that $f^{(n)}$ is piecewise continuous on every finite interval $0 < t < b$. If $f, f', \ldots, f^{(n)}$ have exponential growth with growth exponent r, then the Laplace transform $\mathcal{L}(f^{(n)})(s)$ exists for $s > r$ and*

$$\mathcal{L}(f^{(n)}(t))(s) = s^n \mathcal{L}(f(t))(s) - s^{n-1} f(0) - \cdots - s\, f^{(n-2)}(0) - f^{(n-1)}(0).$$

Next, we give a useful theorem that has a form similar to that of Theorem 14.7. Instead of a formula for the Laplace transform of a derivative of a function, we derive a formula for the derivative of the Laplace transform of a function.

Theorem 14.11. *Suppose the Laplace transform of $f(t)$ exists. Then the Laplace transform of $t\, f(t)$ also exists, and*

$$\mathcal{L}(t\, f(t))(s) = -\frac{d}{ds}\left(\mathcal{L}(f(t))(s)\right).$$

Proof. Differentiation under the integral sign is permissible, so we compute

$$\frac{d}{ds}\mathcal{L}\big(f(t)\big)(s) = \frac{d}{ds}\int_0^\infty e^{-st}f(t)\,dt = \int_0^\infty \frac{d}{ds}\big(e^{-st}f(t)\big)dt$$

$$= \int_0^\infty -t\,e^{-st}f(t)\,dt = -\mathcal{L}\big(t\,f(t)\big)(s). \quad\blacksquare$$

More generally,

Corollary 14.12. *Suppose the Laplace transform of $f(t)$ exists. Then the Laplace transform of $t^n\,f(t)$ also exists, and*

$$\mathcal{L}\big(t^n f(t)\big)(s) = (-1)^n \frac{d^n}{ds^n}\big(\mathcal{L}\big(f(t)\big)(s)\big).$$

It would be quite tedious to find $\mathcal{L}\big(t\sin(b\,t)\big)(s)$ directly from the definition of Laplace transform; however, with Theorem 14.11 the computation is easy.

Example 14.8. *Let $b \neq 0$ and $s > 0$. Compute $\mathcal{L}\big(t\sin(b\,t)\big)(s)$*

Solution. We use Example 14.2 and Theorem 14.11 to compute

$$\mathcal{L}\big(t\sin(b\,t)\big)(s) = -\frac{d}{ds}\mathcal{L}\big(\sin(b\,t)\big)(s)$$

$$= -\frac{d}{ds}\left(\frac{b}{b^2+s^2}\right) = \frac{2b\,s}{(s^2+b^2)^2}. \quad\blacksquare$$

We conclude this section with the solution of a sixth-order initial value problem.

Example 14.9. *Solve the differential equation*

$$y^{(6)} + 6y^{(5)} + 15y^{(4)} + 20y''' + 15y'' + 6y' + y = 0, \tag{14.33}$$

with the initial conditions

$$y(0) = y'(0) = y''(0) = y'''(0) = y^{(4)}(0) = 0, \qquad y^{(5)}(0) = 1. \tag{14.34}$$

Solution. We apply the Laplace transform to (14.33), use Corollary 14.10 and substitute the initial values given by (14.34); the result is

$$s^6\mathcal{L}(y)+6s^5\mathcal{L}(y)+15s^4\mathcal{L}(y)+20s^3\mathcal{L}(y)+15s^2\mathcal{L}(y)+6s\mathcal{L}(y)+\mathcal{L}(y)-1 = 0. \tag{14.35}$$

Since $(s+1)^6 = s^6 + 6s^5 + 15s^4 + 20s^3 + 15s^2 + 6s + 1$, we can write (14.35) more simply as $(s+1)^6\mathcal{L}(y) - 1 = 0$, or

$$\mathcal{L}(y) = \frac{1}{(s+1)^6}. \tag{14.36}$$

On the other hand,

$$\frac{d^5}{ds^5}\left(\frac{1}{s+1}\right) = -\frac{120}{(s+1)^6},$$

so Corollary 14.10 implies that

$$\mathcal{L}(y)(s) = -\frac{1}{120}\frac{d^5}{ds^5}\left(\frac{1}{s+1}\right) = -\frac{1}{120}\frac{d^5}{ds^5}\left(\mathcal{L}(e^{-t})(s)\right) = \mathcal{L}\left(\frac{t^5e^{-t}}{120}\right)(s). \quad (14.37)$$

From Lerch's theorem (theorem 14.9) we conclude that

$$y = \frac{t^5e^{-t}}{120}. \quad \blacksquare$$

Example 14.10. *Solve the initial value problem*

$$\begin{cases} y'' - 4y' + 9y = t, \\ y(0) = y'(0) = 0. \end{cases} \quad (14.38)$$

Solution. We have $(s^2 - 4s + 9)\mathcal{L}(y)(s) = \dfrac{1}{s^2}$, so that

$$\mathcal{L}(y)(s) = \frac{1}{s^2(s^2 - 4s + 9)} = \frac{1}{9s^2} + \frac{4}{81s} + \frac{7 - 4s}{81(s^2 - 4s + 9)}. \quad (14.39)$$

We use Theorem 14.4 to calculate

$$\frac{7 - 4s}{81(s^2 - 4s + 9)} = \frac{-1 - 4(s - 2)}{81\left((s - 2)^2 + 5\right)} = \frac{-1 - 4s}{81(s^2 + 5)}\bigg|_{s \to s-2} \quad (14.40)$$

$$= \mathcal{L}\left(-\frac{4}{81}\cos\left(t\sqrt{5}\right) - \frac{1}{81\sqrt{5}}\sin\left(t\sqrt{5}\right)\right)(s)\bigg|_{s \to s-2}$$

$$= \mathcal{L}\left(e^{2t}\left(-\frac{4}{81}\cos\left(t\sqrt{5}\right) - \frac{1}{81\sqrt{5}}\sin\left(t\sqrt{5}\right)\right)\right).$$

From (14.39) and (14.40) it follows that

$$y(t) = \frac{t}{9} + \frac{4}{81} - e^{2t}\left(\frac{4}{81}\cos\left(t\sqrt{5}\right) + \frac{1}{81\sqrt{5}}\sin\left(t\sqrt{5}\right)\right). \quad \blacksquare$$

Exercises for Section 14.3

For each of the following initial value problems, find a formula for the Laplace transform $Y(s)$ and then find the solution $y(t)$ for which $Y(s) = \mathcal{L}(y)(s)$.

1.
$$\begin{cases} y'' + 2y' - 15y = 0, \\ y(0) = 2,\ y'(0) = 2 \end{cases}$$

6.
$$\begin{cases} y'' + 9y = \cos(3t), \\ y(0) = 1,\ y'(0) = 0 \end{cases}$$

2.
$$\begin{cases} y'' + 9y = 0, \\ y(0) = 9,\ y'(0) = 3 \end{cases}$$

7.
$$\begin{cases} y'' + 9y = \cos(3t) + \sin(3t), \\ y(0) = 16,\ y'(0) = 0 \end{cases}$$

3.
$$\begin{cases} y'' + 9y' = 0, \\ y(0) = 0,\ y'(0) = -2 \end{cases}$$

8.
$$\begin{cases} y^{(4)} - 16y = 0,\ y(0) = 7, \\ y'(0) = 20,\ y''(0) = -44, \\ y'''(0) = 58 \end{cases}$$

4.
$$\begin{cases} y'' - 3y' + 10y = \cos(t), \\ y(0) = 0,\ y'(0) = 0 \end{cases}$$

9.
$$\begin{cases} y'' - 3y' + 9y = e^{3t}, \\ y(0) = 1,\ y'(0) = 4 \end{cases}$$

5.
$$\begin{cases} y'' + c\,y = \cos(3t), \quad c \neq 9, \\ y(0) = 1,\ y'(0) = 0 \end{cases}$$

10.
$$\begin{cases} y'' - 3y' + 9y = t^4 e^{5t}, \\ y(0) = y'(0) = 0 \end{cases}$$

11. Prove Corollary 14.12.

14.4   The Gamma Function

To tackle the problem of actually computing $\mathcal{L}(t^p)(s)$ when p is not an integer, we need an important nonelementary function called the **gamma function**. It is a generalization of the well-known **factorial function** given by

$$n! = 1 \cdot 2 \cdot 3 \cdots (n - 1) \cdot n.$$

Here $n!$ is defined for each positive integer n and satisfies the functional equation

$$n! = n(n - 1)!. \tag{14.41}$$

How should $n!$ be generalized to the case when n is not a positive integer? Substitution of $n = 1$ into (14.41) tells us that $0! = 1$; then substitution of $n = 0$ into (14.41) tells us that $(-1)!$ should be infinite. More generally, the factorial of any negative integer should be infinite. To extend the definition of $n!$, we look for a formula for $n!$ which also makes sense for arbitrary positive real numbers

We look for a formula for $x!$ that makes sense for an arbitrary positive real number x and coincides with the usual factorial function on positive integers. In fact, there is a natural extension of the factorial function to arbitrary complex numbers called the **Gamma function** $\Gamma(z)$. Strictly speaking, it is a generalization of a modified form of the factorial function.

Definition. *Let z be a complex number. The* **gamma function** $\Gamma(z)$ *is defined by*

$$\Gamma(z) = \int_0^\infty t^{z-1} e^{-t} dt, \tag{14.42}$$

where the integral converges when $\mathfrak{Re}(z) > 0$.

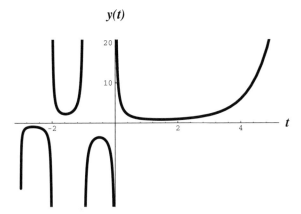

y(t)

The gamma function $y(t) = \Gamma(t)$ plotted for $-3 < t < 5$

The gamma function $\Gamma(z)$ is denoted in *Mathematica* by **Gamma[z]**.

Lemma 14.13. *For all complex numbers z with $\mathfrak{Re}(z) > 0$ we have*

$$\Gamma(z+1) = z\,\Gamma(z). \tag{14.43}$$

Proof. Integrating (14.42) by parts yields (14.43). ∎

Lemma 14.14. $\Gamma(1) = 1$, *and if n is an integer,*

$$\Gamma(n+1) = n!. \tag{14.44}$$

Proof. (14.44) follows by induction from (14.43). ∎

We can use (14.43) as a postulate to extend the definition of the factorial function $z!$ to any complex number z. Writing

$$\Gamma(z) = \frac{\Gamma(z+1)}{z} \tag{14.45}$$

allows us to extend the definition of $\Gamma(z)$ to values of z for which $\mathfrak{Re}(z) > -1$, where we take $\Gamma(0)$ to be infinite. Again we use (14.45) to extend the definition to values of z for which $\mathfrak{Re}(z) > -2$, with the convention that $\Gamma(-1)$ is infinite. Continuing in this way, we obtain a function that is defined and finite for all complex numbers with the exception of $0, -1, -2, \ldots$. From this construction, the resulting function also satisfies the functional equation (14.13) for all values of z. Finally, we define

$$z! = \Gamma(z + 1)$$

so that $0! = 1$ and the definition is consistent with the definition of the factorial function for positive integers.

Mathematica can be used to find numerical approximations to the gamma function. For example,

Gamma[1/3]//N

gives the response

2.67894

This tells us that $\Gamma(1/3) \approx 2.67894$. It is difficult to compute $\Gamma(z)$ exactly when z is not a positive integer, except in some special cases, such as:

Lemma 14.15. *We have*

$$\Gamma\left(\frac{1}{2}\right) = \left(-\frac{1}{2}\right)! = \sqrt{\pi}. \tag{14.46}$$

Proof. We use the change of variables $t = u^2$ in (14.42) to obtain

$$\Gamma\left(\frac{1}{2}\right) = \int_0^\infty t^{-1/2} e^{-t} dt = 2 \int_0^\infty e^{-u^2} du. \tag{14.47}$$

From (14.47) we get

$$\Gamma\left(\frac{1}{2}\right)^2 = 4\left(\int_0^\infty e^{-u^2} du\right)^2 = 4\left(\int_0^\infty e^{-u^2} du\right)\left(\int_0^\infty e^{-u^2} du\right) \tag{14.48}$$

$$= 4 \int_0^\infty \int_0^\infty e^{-u^2-v^2} du\, dv = \int_{-\infty}^\infty \int_{-\infty}^\infty e^{-u^2-v^2} du\, dv.$$

To compute the double integral on the right-hand side of (14.48), we change to polar coordinates by setting $u = r\cos(\theta)$ and $v = r\sin(\theta)$. We obtain

$$\int_{-\infty}^\infty \int_{-\infty}^\infty e^{-u^2-v^2} du\, dv = \int_0^{2\pi} \int_0^\infty r\, e^{-r^2} dr\, d\theta = 2\pi \int_0^\infty r\, e^{-r^2} dr \tag{14.49}$$

$$= 2\pi \left(\frac{-e^{-r^2}}{2}\right)\bigg|_0^\infty = \pi.$$

Now (14.46) follows from (14.48) and (14.49). ∎

Theorem 14.16. *For all real numbers $p > -1$ we have the formula*

$$\mathcal{L}(t^p)(s) = \frac{\Gamma(p+1)}{s^{p+1}} = \frac{p!}{s^{p+1}}. \tag{14.50}$$

Proof. Making the change of variables $u = st$ in the definition of the Laplace transform of t^p, we compute

$$\mathcal{L}(t^p)(s) = \int_0^\infty e^{-st} t^p \, dt = \frac{1}{s^{p+1}} \int_0^\infty e^{-u} u^p \, du = \frac{\Gamma(p+1)}{s^{p+1}}. \quad \blacksquare$$

Exercises for Section 14.4

1. Show that
$$\Gamma\left(\frac{3}{2}\right) = \frac{\sqrt{\pi}}{2} \quad \text{and} \quad \Gamma\left(\frac{5}{2}\right) = \frac{3\sqrt{\pi}}{4}.$$

2. Show that $\left(n + \dfrac{1}{2}\right)! = \dfrac{\sqrt{\pi}}{2^{2n}} \dfrac{(2n+1)!}{n!}$

3. Use *Mathematica* to find numerical approximations to $\Gamma(100)$, $\Gamma(0.01)$, and $\Gamma(i)$.

4. Use *Mathematica*'s command **Plot3D** to plot $\Gamma(u + iv)$ for $-1.5 \leq u \leq 1.5$ and $-2 \leq v \leq 2$.

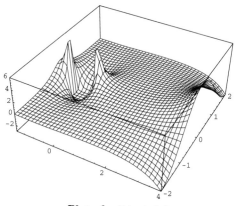

Plot of $\Gamma(u + iv)$

14.5   *Computation of Laplace Transforms*

In order to work with Laplace transforms, we need to find as many examples as possible. In this section we use elementary integration techniques to begin such a table of Laplace transforms.

We have already computed the Laplace transform of t^p in (14.50) and the Laplace transform of $\sin(bt)$ and $\cos(bt)$ in Example 14.2.

Example 14.11. *The hyperbolic functions* $\cosh(bt)$ *and* $\sinh(bt)$ *are defined by*

$$\cosh(bt) = \frac{e^{bt} + e^{-bt}}{2} \qquad \text{and} \qquad \sinh(bt) = \frac{e^{bt} - e^{-bt}}{2}.$$

Find their Laplace transforms.

Solution. We have

$$\mathcal{L}\big(\cosh(bt)\big) = \frac{1}{2}\big(\mathcal{L}(e^{bt}) + \mathcal{L}(e^{-bt})\big) = \frac{1}{2}\left(\frac{1}{s-b} + \frac{1}{s+b}\right) = \frac{s}{s^2 - b^2}.$$

Similarly,

$$\mathcal{L}\big(\sinh(bt)\big) = \frac{b}{s^2 - b^2}.$$

Of course, these formulas are valid only for $s > b$, that is, when the respective integrands are of exponential growth. ∎

In order to calculate Laplace transforms with *Mathematica*, we need the special package **LaplaceTransform.m**; it can be read into a *Mathematica* session with the command

Needs["Calculus'LaplaceTransform'"]

Now the commands **LaplaceTransform** and **InverseLaplaceTransform** are available.

Example 14.12. *Compute* $\mathcal{L}(t^2 \sin(t))$ *and* $\mathcal{L}^{-1}\left(\dfrac{s^3}{(s^2+1)^4}\right)$.

Solution. To the query

LaplaceTransform[t^2 Sin[t],t,s]

we get the response

```
     2
  8 s            2
------------ - ------------
     2  3         2  2
(1 + s )       (1 + s )
```

This tells us that

$$\mathcal{L}\big(t^2 \sin(t)\big)(s) = \frac{8s^2}{(1+s^2)^3} - \frac{2}{(1+s^2)^2}.$$

Similarly,

InverseLaplaceTransform[s^3/(s^2 + 1)^4,s,t]

yields

```
      2
 -(t   Cos[t])     t Sin[t]     t   Sin[t]
 --------------  +  ----------  +  ----------
      16               16             48
```

Hence

$$\mathcal{L}^{-1}\left(\frac{s^3}{(s^2+1)^4}\right) = \frac{-3t^2 \cos(t) + 3t \sin(t) + t^3 \sin(t)}{48}. \ \blacksquare$$

Mathematica also knows how to compute the Laplace transform of the derivative of a function. To the query

LaplaceTransform[y′[t],t,s]

we get the response

```
s LaplaceTransform[y[t], t, s] - y[0]
```

The following tables summarize useful information about Laplace transforms.

$f(t)$	$\mathcal{L}\big(f(t)\big)(s)$
1	$1/s$
e^{at}	$1/(s-a)$
t^p	$\dfrac{p!}{s^{p+1}} = \dfrac{\Gamma(p+1)}{s^{p+1}}$
$\sin(a\,t)$	$\dfrac{a}{s^2+a^2}$
$\cos(a\,t)$	$\dfrac{s}{s^2+a^2}$
$\sinh(a\,t)$	$\dfrac{a}{s^2-a^2}$
$\cosh(a\,t)$	$\dfrac{s}{s^2-a^2}$

$f(t)$	$\mathcal{L}\big(f(t)\big)(s)$
$e^{ct} f(t)$	$F(s-c)$, where $F(s) = \mathcal{L}\big(f(t)\big)(s)$
$-t f(t)$	$\dfrac{d}{ds}\mathcal{L}\big(f(t)\big)(s)$
$(-t)^n f(t)$	$\dfrac{d^n}{ds^n}\mathcal{L}\big(f(t)\big)(s)$
$f'(t)$	$s\mathcal{L}\big(f(t)\big)(s) - f(0)$
$f''(t)$	$s^2\mathcal{L}\big(f(t)\big)(s) - f(0)s - f'(0)$
$u_c(t) f(t-c)$	$e^{-cs} F(s)$, where $F(s) = \mathcal{L}\big(f(t)\big)(s)$
$\delta(t-c) f(t)$	$e^{-cs} f(c)$

$f(t)$	$\mathcal{L}\big(f(t)\big)(s)$
$\displaystyle\int_0^t f(\tau)g(t-\tau)\,d\tau$	$F(s)G(s),$ where $F(s) = \mathcal{L}\big(f(t)\big)(s),\, G(s) = \mathcal{L}\big(g(t)\big)(s)$
$f(t+T) = f(t)$	$\displaystyle\frac{1}{1-e^{-sT}}\int_0^T e^{-st}f(t)\,dt$
$f(at)$	$\displaystyle\frac{1}{a}\mathcal{L}\big(f(t)\big)\left(\frac{s}{a}\right)$
$\displaystyle\int_0^t f(\tau)\,d\tau$	$\displaystyle\frac{1}{s}\mathcal{L}\big(f(t)\big)(s)$
$\displaystyle\frac{f(t)}{t}$	$\displaystyle\int_s^\infty F(\tau)\,d\tau,$ where $F(s) = \mathcal{L}\big(f(t)\big)(s)$

(The unit step function $u_c(t)$ and the Dirac delta function $\delta(t)$ are defined later in this chapter.)

The list of Laplace transforms can be greatly extended by suitable use of **partial fractions**. This allows us to express a large class of rational functions

$$\frac{P(s)}{Q(s)}$$

as Laplace transforms of elementary functions, as follows.

From the fundamental theorem of algebra, we know that any polynomial $Q(s)$ of degree N can be expressed in terms of its roots $s_1, \ldots, s_M$ through the product

$$Q(s) = A \prod_{j=1}^{M}(s - s_j)^{m_j}$$

for some constant A. Here the roots $s_1, \ldots, s_M$ are distinct complex numbers and m_j is the **multiplicity** of the root s_j. Of course,

$$N = \sum_{j=1}^{M} m_j.$$

If $P(s)$ is another polynomial whose degree is less than N, the method of partial fractions asserts that the ratio can be expressed as

$$\frac{P(s)}{Q(s)} = \sum_{j=1}^{M}\sum_{m=1}^{m_j}\frac{A_j^m}{(s - s_j)^m}.$$

We recall that the Laplace transform of the function $\dfrac{t^m}{m!}e^{at}$ is $\dfrac{1}{(s-a)^{m+1}}$. Therefore, it is immediate that $P(s)/Q(s)$ is the Laplace transform of

$$\sum_{j=1}^{M}\sum_{m=1}^{m_j} A_j^m \frac{t^{m-1}e^{s_jt}}{(m-1)!}.$$

We illustrate with some examples.

Example 14.13. *Find the inverse Laplace transform of* $\dfrac{2s+3}{s^3-s}$.

Solution. The denominator has zeros at $s=0$, $s=1$, and $s=-1$. The method of partial fractions yields the decomposition

$$\frac{2s+3}{s^3-s} = \frac{-3}{s} + \frac{5/2}{s-1} + \frac{1/2}{s+1}.$$

Therefore, the required inverse Laplace transform is

$$f(t) = -3 + \frac{5}{2}e^t + \frac{1}{2}e^{-t}. \ \blacksquare$$

Exercises for Section 14.5

Find the following Laplace transforms.

1. $\mathcal{L}\big(t\cos(bt)\big)$

2. $\mathcal{L}\big(ae^{bt}\big)$

3. $\mathcal{L}\big(\sin(at)+\cos(at)\big)$

4. $\mathcal{L}\big((\sin(at))^2\big)$

5. $\mathcal{L}\big(e^t(t-\sin(t))\big)$

6. $\mathcal{L}\big((t+e^t)^3\big)$

7. $\mathcal{L}\big(t^2-e^{-9t}+5\big)$

8. $\mathcal{L}\big(e^{-t}\cosh(kt)\big)$

9. $\mathcal{L}\big(\sin(2t)\cos(2t)\big)$ [Hint: Consider $\sin(4t)$.]

10. $\mathcal{L}\big(\sin(t)\cos(2t)\big)$ [Hint: Consider $\sin(t_1\pm t_2)$.]

11. $\mathcal{L}\big((\cos(t))^2\big)$ [Hint: Consider $\cos(2t)$.]

12. $\mathcal{L}\big((\sin(t))^3\big)$ [Hint: $(\sin(t))^3 = \sin(t)(\sin(t))^2$.]

Solve the following initial value problems using Laplace transforms:

13. $\begin{cases} y'(t) + 2y(t) = 0, \\ y(0) = 1 \end{cases}$

17. $\begin{cases} y''(t) - y'(t) - 6y(t) = 0, \\ y(0) = 2, \ y'(0) = -1 \end{cases}$

14. $\begin{cases} y'(t) - y(t) = 2e^t + t\,e^t, \\ y(0) = 2 \end{cases}$

18. $\begin{cases} y''(t) + y(t) = \sin(2t), \\ y(0) = y'(0) = 0 \end{cases}$

15. $\begin{cases} y''(t) + y(t) = 2y'(t), \\ y(0) = 1, \ y'(0) = -1 \end{cases}$

19. $\begin{cases} y''(t) + y(t) = \exp(3t), \\ y(0) = y'(0) = 0 \end{cases}$

16. $\begin{cases} y''(t) + y(t) = \sinh(t), \\ y(0) = 1, \ y'(0) = 0 \end{cases}$

20. $\begin{cases} y''(t) + y(t) = t + \exp(t), \\ y(0) = y'(0) = 0 \end{cases}$

21. A function $f(t)$ is said to be **periodic** with period $T > 0$ provided that $f(t) = f(t + T)$ for all t. Suppose that $f(t)$ is piecewise continuous and of exponential growth. If $f(t)$ is periodic with period T, show that

$$\mathcal{L}\big(f(t)\big) = \frac{1}{1 - e^{-sT}} \int_0^T e^{-st} f(t)\, dt. \tag{14.51}$$

22. Show that if $a > 0$, then $\mathcal{L}\big(f(a\,t)\big) = \frac{1}{a} F\left(\frac{s}{a}\right)$.

23. If $F(s) = \mathcal{L}(f(t))$, show that $\mathcal{L}\left(\int_0^t f(\tau)\, d\tau\right) = \frac{1}{s} F(s)$.

24. If $F(s) = \mathcal{L}(f(t))$, show that $\mathcal{L}\left(\frac{f(t)}{t}\right) = \int_s^\infty F(\tau)\, d\tau$.

14.6   *Step Functions*

All of the inhomogeneous differential equations we have considered so far have been of the form

$$L[y] = f(t), \tag{14.52}$$

where L is a linear differential operator and the forcing function $f(t)$ is a differentiable function such as a trigonometric function or a polynomial. The Laplace transform can also be used to solve differential equations with discontinuous forcing functions. In this section we develop additional properties of the Laplace transform to deal with these more general forcing functions.

First, we define an important piecewise continuous function that serves as a building block for other piecewise continuous functions.

Definition.   *The* **unit step function** *with a jump at* 0 *is given by*

$$u(t) = \begin{cases} 0 & \text{if } t < 0, \\ 1 & \text{if } t \geq 0. \end{cases}$$

It is easy to construct from u a unit step function with a jump at some nonzero number a.

Lemma 14.17.   *The function* $t \longmapsto u(t-a)$ *is given by*

$$u(t-a) = \begin{cases} 0 & \text{if } t < a, \\ 1 & \text{if } t \geq a. \end{cases}$$

We sometimes write $u(t-a) = u_a(t)$ *and call* u_a *the* **unit step function** *with a jump at* a.

Proof. If $t < a$ then $t - a < 0$, and so $u(t-a) = 0$. Similarly, $t \geq a$ implies $t - a \geq 0$, so that $u(t-a) = 1$. ∎

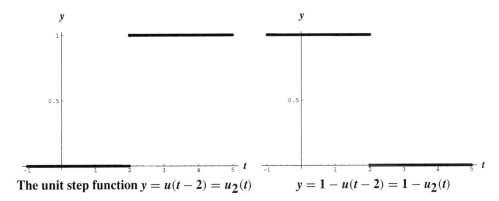

The unit step function $y = u(t-2) = u_2(t)$ $y = 1 - u(t-2) = 1 - u_2(t)$

Unit step functions can be used to build other step functions.

Example 14.14.   *Let* h *be the function defined by*

$$h(t) = \begin{cases} 0 & \text{if } t < 1, \\ 1 & \text{if } 1 \leq t < 2, \\ 0 & \text{if } t \geq 2. \end{cases}$$

Sketch the graph of h, *and write* h *in terms of unit step functions.*

Solution. It is clear that the plot of h is:

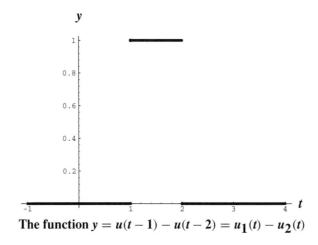

The function $y = u(t-1) - u(t-2) = u_1(t) - u_2(t)$

To write h in terms of step functions, we observe that the points of discontinuity of h are $t = 1$ and $t = 2$. This leads us to conjecture that h is a linear combination of $u_1(t)$ and $u_2(t)$:

$$h(t) = A u_1(t) + B u_2(t). \tag{14.53}$$

A and B can be found by substitution of well-chosen values of t into (14.53). For example,

$$1 = h\left(\frac{3}{2}\right) = A u_1\left(\frac{3}{2}\right) + B u_2\left(\frac{3}{2}\right) = A$$

and

$$0 = h(3) = A u_1(3) + B u_2(3) = A + B,$$

from which it follows that $A = -B = 1$. Thus we should take $h(t) = u_1(t) - u_2(t)$. In fact, it can be checked that

$$h(t) = u_1(t) - u_2(t)$$

for all t. ∎

Next, we compute the Laplace transform of the unit step function $u_a(t)$. Let $s > 0$; then

$$\mathcal{L}\big(u_a(t)\big)(s) = \int_0^\infty e^{-st} u_a(t)\, dt = \int_0^a e^{-st} u_a(t)\, dt + \int_a^\infty e^{-st} u_a(t)\, dt$$

$$= \int_a^\infty e^{-st}\, dt = \frac{e^{-as}}{s}. \tag{14.54}$$

There is an important generalization of (14.54). First, we need a definition.

Definition.   The **translate** *of a function f by a distance a is the function g defined by*

$$g(t) = f(t-a) \qquad \text{if } t < a \tag{14.55}$$

and zero otherwise. We can rewrite (14.55) in terms of step functions:

$$g(t) = u_a(t) f(t-a). \tag{14.56}$$

Theorem 14.18.   *If the Laplace transform $\mathcal{L}(f(t))(s)$ exists for $s > c \geq 0$ and if a is a positive constant, then the Laplace transform of the translate $u_a(t) f(t-a)$ exists for $s > c$ and*

$$\mathcal{L}\big(u_a(t) f(t-a)\big)(s) = e^{-as} \mathcal{L}\big(f(t)\big)(s). \tag{14.57}$$

Conversely, if $f(t) = \mathcal{L}^{-1}\big(F(s)\big)$, then

$$u_a(t) f(t-a) = \mathcal{L}^{-1}\big(e^{-cs} F(s)\big).$$

Proof.   We use the fact that $u_a(t) = 0$ for $t < a$ to compute

$$\mathcal{L}\big(u_a(t) f(t-a)\big)(s) = \int_0^\infty e^{-st} u_a(t) f(t-a)\, dt \tag{14.58}$$

$$= \int_0^a e^{-st} u_a(t) f(t-a)\, dt + \int_a^\infty e^{-st} u_a(t) f(t-a)\, dt$$

$$= \int_a^\infty e^{-st} f(t-a)\, dt.$$

We change variables in the integral on the right-hand side of (14.58) by putting $w = t - a$. Thus

$$\mathcal{L}\big(u_a(t) f(t-a)\big)(s) = \int_0^\infty e^{-s(w+a)} f(w)\, dw = e^{-sa} \int_0^\infty e^{-sw} f(w)\, dw$$

$$= e^{-sa} \mathcal{L}\big(f(t)\big)(s). \ \blacksquare$$

Example 14.15.   *Find the Laplace transform of the function $f(t) = u_\pi(t) \cos(t - \pi)$.*

Solution.   We use (14.57) to write

$$\mathcal{L}\big(u_\pi(t) \cos(t - \pi)\big)(s) = e^{-\pi s} \mathcal{L}(\cos(t))(s) = e^{-\pi s} \frac{s}{1 + s^2}. \ \blacksquare$$

Example 14.16.   *Find the inverse Laplace transform of $F(s) = e^{-4s}/(1 + s^2)$.*

Solution.   From the computations in (14.57), we see that the required function is $f(t) = u_4(t) \sin(t - 4)$. $\blacksquare$

Exercises for Section 14.6

Write the following functions in terms of unit step functions.

1. $f(t) = \begin{cases} 0 & \text{if } t < 0, \\ 1 & \text{if } t \geq 0 \end{cases}$

3. $f(t) = \begin{cases} 0 & \text{if } t < 0, \\ 2 & \text{if } 0 \leq t < 1, \\ -1 & \text{if } t \geq 1 \end{cases}$

2. $f(t) = \begin{cases} 0 & \text{if } t < 0, \\ t & \text{if } t \geq 0 \end{cases}$

4. $f(t) = \begin{cases} 0 & \text{if } t < 0, \\ \sin(t) & \text{if } 0 \leq t < \pi, \\ \cos(t) & \text{if } t \geq \pi \end{cases}$

5. $f(t) = \begin{cases} 1 & \text{if } 0 \leq t < 1, \\ e^t & \text{if } 1 \leq t < 2, \\ 2 & \text{if } t \geq 2 \end{cases}$

6. $f(t) = \begin{cases} 1 & \text{if } 0 \leq t < 1, \\ 4t - t^2 & \text{if } 1 \leq t < 2, \\ 1 & \text{if } t \geq 2 \end{cases}$

Graph the following functions:

7. $f(t) = e^t u_3(t)$

9. $f(t) = t - (t - 1)u_1(t)$

8. $g(t) = (t - 2)^2 u_2(t)$

10. $\begin{aligned} g(t) &= (t - 3)u_2(t) - (t - 2)u_3(t) \\ &\quad + 2u_5(t) \end{aligned}$

Find the Laplace transforms of the following functions:

11. $f(t) = \begin{cases} 3 & \text{if } 0 \leq t < 2, \\ 0 & \text{if } t \geq 2 \end{cases}$

14. $f(t) = u_\pi(t) \cos(t)$

12. $f(t) = u_\pi(t) \sin(t - \pi)$

15. $f(t) = u_{\pi/2}(t) \cos(t)$

13. $f(t) = \begin{cases} 4 & \text{if } 1 \leq t < 3, \\ 0 & \text{if } t < 1 \text{ or } \geq 3 \end{cases}$

16. $f(t) = \begin{cases} |\sin(t)| & \text{if } 0 \leq t < 2\pi, \\ 0 & \text{if } t < 0 \text{ or } \geq 2\pi \end{cases}$

Find the inverse Laplace transform of the following functions:

17. $\dfrac{e^{-s}}{s}$

20. $\dfrac{e^{-4s}}{s+4}$

18. $\dfrac{e^{-2s}}{s-3}$

21. $\dfrac{s+e^{-\pi s}}{a(s+1)^2}$

19. $\dfrac{e^{-3s}}{s^2+4}$

22. $\dfrac{e^{-s} - 2e^{-2s} + 2e^{-3s} - e^{-4s}}{s}$

14.7   Second-Order Equations with Piecewise Continuous Forcing Functions

In this section we solve the second-order initial value problem

$$\begin{cases} a\,y'' + b\,y' + c\,y = g(t), \\ y(t_0) = Y_0, \quad y'(t_0) = Y_1, \end{cases} \tag{14.59}$$

where the forcing function $g(t)$ is assumed to be only piecewise continuous. In particular, $g(t)$, may be a discontinuous function.

The integral of a piecewise continuous function is continuous. More generally:

Lemma 14.19. *Let $g(t)$ be a piecewise continuous function defined on the interval $a < t < b$. We subdivide the interval $a < t < b$ as*

$$a = x_0 < x_1 < \cdots < x_p = b,$$

so that g is continuous on each subinterval (x_{i-1}, x_i) and the left and right limits $g(x_i - 0)$ and $g(x_i + 0)$ exist for $i = 1, \ldots, p-1$, as well as $g(x_0 + 0)$ and $g(x_p - 0)$ Then there exists a unique solution of (14.59) on the interval $a < t < b$ in the following sense:

(i)   *both $y(t)$ and $y'(t)$ are continuous throughout $a < t < b$;*

(ii)   *$y''(t)$ is continuous on each subinterval (x_{i-1}, x_i);*

(iii)   *(14.59) holds on each subinterval (x_{i-1}, x_i).*

Proof. Clearly, (14.59) has a unique solution on the interval $a = x_0 < t < x_1$. Moreover, the left limits $y(x_1 - 0)$ and $y'(x_1 - 0)$ exist. We can use these as initial values for a solution $y(t)$ to (14.59) on the interval (x_1, x_2). Therefore,

$$y'(x_1 + 0) = y'(x_1 - 0) \qquad \text{and} \qquad y(x_1 + 0) = y(x_1 - 0).$$

Hence y and y' are continuous at x_1. By continuing in this way we construct y on the whole interval $a < t < b$ such that (i), (ii), and (iii) hold. ∎

The following example illustrates how to use the Laplace transform to solve an initial value problem with a discontinuous forcing function.

Example 14.17. *Use Laplace transforms to solve the differential equation*

$$y'' + y = g(t) = \begin{cases} 1 & \text{if } t < \pi, \\ 0 & \text{if } t \geq \pi, \end{cases} \qquad (14.60)$$

with the initial conditions $y(0) = 1$ *and* $y'(0) = 0$.

Solution. We take the Laplace transform of both sides of (14.60), obtaining

$$s^2 \mathcal{L}(y(t)) - s + \mathcal{L}(y(t)) = \mathcal{L}(g(t)) = \int_0^\pi e^{-st}\, dt = \frac{1 - e^{-\pi s}}{s}. \qquad (14.61)$$

Then we solve (14.61) algebraically for $\mathcal{L}(y(t))(s)$:

$$\mathcal{L}(y(t)) = \frac{1}{s^2 + 1}\left(\frac{1 - e^{-\pi s}}{s} + s\right) = \frac{1}{s} - \frac{e^{-\pi s}}{s(s^2 + 1)}$$

$$= \frac{1}{s} - \left(\frac{e^{-\pi s}}{s} - \frac{s\, e^{-\pi s}}{s^2 + 1}\right). \qquad (14.62)$$

From the table on page 471 we see that

$$\frac{1}{s} = \mathcal{L}(1) \qquad \text{and} \qquad \frac{e^{-\pi s}}{s} = \mathcal{L}(u_\pi(t)). \qquad (14.63)$$

Furthermore,

$$\frac{s\, e^{-\pi s}}{s^2 + 1} = e^{-\pi s}\mathcal{L}(\cos(t)) = \mathcal{L}(u_\pi(t)\cos(t - \pi)). \qquad (14.64)$$

From (14.62)–(14.64) we conclude that

$$\mathcal{L}(y(t)) = \mathcal{L}(1 - u_\pi(t) + u_\pi(t)\cos(t - \pi)).$$

By taking inverse Laplace transforms (that is, using Lerch's theorem (theorem 14.9)), we get

$$y(t) = 1 - u_\pi(t) + u_\pi(t)\cos(t - \pi) = 1 - u_\pi(t) - u_\pi(t)\cos(t)$$

$$= 1 - u_\pi(t)(1 + \cos(t)). \qquad (14.65)$$

We can rewrite (14.65) in the form

$$y(t) = \begin{cases} 1 & \text{if } t < \pi, \\ -\cos(t) & \text{if } \pi \le t < \infty. \end{cases} \blacksquare$$

Here is the plot of the solution to (14.60).

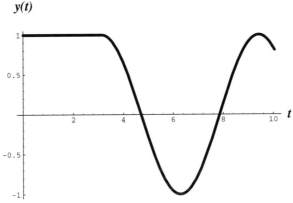

y(t)

The function $y = 1 - u_\pi(t)(1 + \cos(t))$

Exercises for Section 14.7

Solve and plot each of the following initial value problems.

1. $\begin{cases} y' = 1 - u_1(t), \\ y(0) = 0 \end{cases}$

3. $\begin{cases} y' + y = t - u_1(t), \\ y(0) = 1 \end{cases}$

2. $\begin{cases} y' = 1 - 2u_1(t) + u_2(t), \\ y(0) = 1 \end{cases}$

4. $\begin{cases} y'' + y = u_3(t), \\ y(0) = 0, \quad y'(0) = 1 \end{cases}$

5. $\begin{cases} y'' + y = t - t\, u_1(t), \\ y(0) = y'(0) = 0 \end{cases}$

8. $\begin{cases} y'' + 4y = u_{2\pi}(t)\sin(t), \\ y(0) = 1, \quad y'(0) = 0 \end{cases}$

6. $\begin{cases} y'' + y = u_\pi(t) - u_{2\pi}(t), \\ y(0) = 0, \quad y'(0) = 1 \end{cases}$

9. $\begin{cases} y'' + y = u_\pi(t) - u_{2\pi}(t), \\ y(0) = 0, \quad y'(0) = 1 \end{cases}$

7. $\begin{cases} y'' + 4y = t - u_{\pi/2}(t)(t - \pi/2), \\ y(0) = y'(0) = 0 \end{cases}$

10. $\begin{cases} y'' + 4y' + 3y = 1 - u_2(t) - u_4(t) + u_6(t), \\ y(0) = y'(0) = 0 \end{cases}$

14.8 *Impulse Functions*

In many applications it is necessary to deal with forcing functions of an impulsive nature. Such functions have a very large magnitude over a very short time interval. We are led to consider differential equations of the form

$$a\, y'' + b\, y' + c\, y = g_\varepsilon(t),$$

where $g_\varepsilon(t)$ is zero except in an interval $t_0 - \varepsilon < t < t_0 + \varepsilon$, where $g_\varepsilon(t)$ is very large.

Definition.   *The* **total impulse** $\mathbf{I}(g)$ *of a forcing function* $g(t)$ *is the integral*

$$\mathbf{I}(g) = \int_{-\infty}^{\infty} g(t)\, dt.$$

In the case of a forcing function g_ε that is zero outside an interval $t_0 - \varepsilon < t < t_0 + \varepsilon$, we have

$$\mathbf{I}(g_\varepsilon) = \int_{t_0-\varepsilon}^{t_0+\varepsilon} g(t)\, dt.$$

We impose the condition $\mathbf{I}(g_\varepsilon) = 1$ and let ε tend to 0. In the limit we get a mathematical object that resembles a function and is given by

$$g_0(t) = \begin{cases} 0 & \text{if } t \neq 0, \\ \infty & \text{if } t = 0, \end{cases} \qquad \text{and} \qquad \mathbf{I}(g_0) = \int_{-\infty}^{\infty} g_0(t)\, dt = 1. \qquad (14.66)$$

The integral of any ordinary function that vanishes except for one point must be zero. Therefore, g_0 cannot exist as a mathematical function in the ordinary sense of the term. Let us persevere and try to make sense of g_0.

We first take $t_0 = 0$ and define a function d_ε by

$$d_\varepsilon(t) = \begin{cases} \dfrac{1}{2\varepsilon} & \text{for } -\varepsilon < t < \varepsilon, \\ 0 & \text{otherwise.} \end{cases}$$

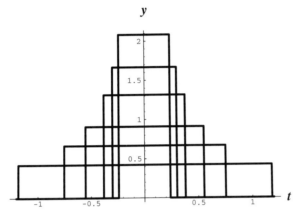

Approximations to the Dirac delta function

Clearly,

$$\mathbf{I}(d_\varepsilon) = \int_{-\infty}^{\infty} d_\varepsilon(t)\, dt = \int_{-\varepsilon}^{\varepsilon} \frac{1}{2\varepsilon} dt = 1.$$

Moreover,

$$\lim_{\varepsilon \to 0} d_\varepsilon(t) = 0 \text{ for } t \neq 0 \tag{14.67}$$

and

$$\lim_{\varepsilon \to 0} \mathbf{I}(d_\varepsilon) = 1. \tag{14.68}$$

We now use (14.67) and (14.68) to define an idealized function that imparts an impulse of enormous magnitude at $t = 0$ but is zero otherwise.

Definition.   *The* **Dirac**[4] **delta function** δ *is characterized by the properties*

$$\delta(t) = 0 \text{ for } t \neq 0 \tag{14.69}$$

---

[4]

Paul Adrien Maurice Dirac (1902–1982). British physicist, known for his work in quantum mechanics. He won a Nobel prize in 1933.

and

$$\mathbf{I}(\delta) = \int_{-\infty}^{\infty} \delta(t)\, dt = 1. \tag{14.70}$$

The Dirac delta function is an example of a "generalized function". From (14.69) and (14.70) it is clear that

$$\delta(t - t_0) = 0 \text{ for } t \neq t_0 \tag{14.71}$$

and

$$\int_{-\infty}^{\infty} \delta(t - t_0)\, dt = 1. \tag{14.72}$$

Since we have taken an excursion into the unfamiliar territory of generalized functions, we might as well go all the way and compute, at least formally, the Laplace transform of the Dirac delta function. First, of all, we must decide on a reasonable definition of $\mathcal{L}\big(\delta(t - t_0)\big)$. We cannot use (14.1) because $\delta(t - t_0)$ is not an ordinary function. Instead, we choose to define $\mathcal{L}\big(\delta(t - t_0)\big)$ as the limit

$$\mathcal{L}\big(\delta(t - t_0)\big) = \lim_{\varepsilon \to 0} \mathcal{L}\big(d_\varepsilon(t - t_0)\big). \tag{14.73}$$

In order to find the right-hand side of (14.73), we first prove

Lemma 14.20. *Let $g(t)$ be a continuous function defined for $-\infty < t < \infty$. Then for $-\infty \leq a < t_0 < b \leq \infty$ we have*

$$\int_a^b g(t)\delta(t - t_0)\, dt = g(t_0). \tag{14.74}$$

Proof. The mean value theorem for integrals from calculus states that

$$\int_{t_0-\varepsilon}^{t_0+\varepsilon} g(t)\, dt = 2\varepsilon\, g(t_1), \tag{14.75}$$

for some t_1, where $t_0 - \varepsilon < t_1 < t_0 + \varepsilon$. Choose ε so small that $a < t_0 - \varepsilon < t_0 + \varepsilon < b$; then (14.75) implies that

$$\int_a^b d_\varepsilon(t)g(t)\, dt = \frac{1}{2\varepsilon} \int_{t_0-\varepsilon}^{t_0+\varepsilon} g(t)\, dt = g(t_1). \tag{14.76}$$

Since $t_1 \longrightarrow t_0$ as $\varepsilon \longrightarrow 0$, when we take the limit as $\varepsilon \longrightarrow 0$ in (14.76), we get (14.74). ∎

Corollary 14.21. *The Laplace transform of the function $\delta(t - t_0)g(t)$ is given by*

$$\mathcal{L}\big(\delta(t - t_0)g(t)\big) = e^{-st_0} g(t_0). \tag{14.77}$$

In particular, the Laplace transform of the Dirac delta function is given by

$$\mathcal{L}\big(\delta(t - t_0)\big) = e^{-st_0} \qquad and \qquad \mathcal{L}\big(\delta(t)\big) = 1. \tag{14.78}$$

Proof. Taking $a = 0$, $b = \infty$, and $f(t) = e^{-st}g(t)$ in (14.74), we get (14.77). ∎

Example 14.18. *Solve the initial value problem*

$$\begin{cases} y'' + 4y = \delta(t - \pi/2), \\ y(0) = y'(0) = 0. \end{cases} \tag{14.79}$$

Solution. We take the Laplace transform of both sides of (14.79) and use (14.78) to obtain

$$(s^2 + 4)\mathcal{L}(y(s)) = e^{s\pi/2}.$$

Thus

$$\mathcal{L}(y(t))(s) \;=\; \frac{e^{s\pi/2}}{s^2 + 4} = \frac{e^{s\pi/2}}{2}\mathcal{L}\big(\sin(2t)\big)(s)$$

$$=\; \frac{1}{2}\mathcal{L}\big(u_{\pi/2}(t)\sin(2(t - \pi/2))\big)(s) = -\frac{1}{2}\mathcal{L}\big(u_{\pi/2}(t)\sin(2t)\big)(s).$$

Therefore

$$y(t) = -\frac{u_{\pi/2}(t)\sin(2t)}{2}. \qquad ∎$$

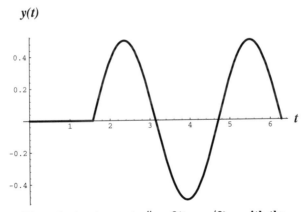

**The solution to $y + y'' = \delta(t - \pi/2)$ with the
initial conditions $y(0) = y'(0) = 0$**

Exercises for Section 14.8

Use the Laplace transform to solve the following initial value problems:

1.
$$\begin{cases} y'(t) + y(t) = \delta(t - 2), \\ y(0) = 2 \end{cases}$$

4.
$$\begin{cases} y''(t) + y(t) = \delta(t - \pi/2) + \delta(t - 3\pi/2), \\ y(0) = y'(0) = 0 \end{cases}$$

2.
$$\begin{cases} y'' + 2y' + 2y = t\,\delta(t - \pi), \\ y(0) = y'(0) = 0 \end{cases}$$

5.
$$\begin{cases} y''(t) - 2y'(t) = 1 + \delta(t - 2), \\ y(0) = 0, \ \ y'(0) = 1 \end{cases}$$

3.
$$\begin{cases} y'(t) - 3y(t) = t\,\delta(t - 2), \\ y(0) = 1 \end{cases}$$

6.
$$\begin{cases} y''(t) + 2y'(t) + y = \delta(t - 1), \\ y(0) = y'(0) = 0 \end{cases}$$

7.
$$\begin{cases} y''(t) + 4y'(t) + 13y = \delta(t - \pi) + \delta(t - 3\pi), \\ y(0) = 1, \ \ y'(0) = 0 \end{cases}$$

8.
$$\begin{cases} y''(t) - 7y'(t) + 6y = e^t + \delta(t - 2) + \delta(t - 4), \\ y(0) = y'(0) = 0 \end{cases}$$

9. Given $y'' + a\,y' + b\,y = f(t)$, where a and b are constants and f is piecewise continuous and of exponential growth, show that the effect of replacing $f(t)$ by $f(t) + c\,\delta(t)$ has the same effect as increasing the initial value of $y'(0)$ by the constant c. A similar result holds for higher-order equations.

10. With A, a, b, c constant and $c > 0$, consider the initial value problem

$$y'' + y = A\,\delta(t - c), \qquad y(0) = a, \qquad y'(0) = b.$$

Under what conditions, if any, is $y(t) = 0$ for $t \geq c$? Equivalently, is it possible to choose the strength and location of the impulse in such way as to cancel the oscillation completely?

14.9   *Convolution*

It is certainly *not* the case that the Laplace transform of a product is the product of the Laplace transforms. For example,

$$\left(\mathcal{L}\big(\cos(t)\big)\right)^2 = \frac{s^2}{(s^2+1)^2} \qquad \text{but} \qquad \mathcal{L}\big(\cos(t)^2\big) = \mathcal{L}\left(\frac{1+\cos(2t)}{2}\right) = \frac{2+s^2}{s(s^2+4)}.$$

There is, however, a nonstandard way to multiply functions called the convolution. The new multiplication is denoted by $*$; we shall show that $\mathcal{L}(f*g) = \mathcal{L}(f)\mathcal{L}(g)$.

Definition. *Let f and g be functions integrable on the interval $0 < t < \infty$. The* **convolution** *of f and g is the function $f*g$ defined by*

$$(f*g)(t) = \int_0^t f(t-u)g(u)du.$$

When it is convenient, we write $f(t) * g(t)$ for $(f*g)(t)$.

Example 14.19. *Compute the convolution $\cos(t)*\cos(t)$.*

Solution. We have

$$\cos(t)*\cos(t) = \int_0^t \cos(t-u)\cos(u)du = \int_0^t \cos(u)\big(\cos(t)\cos(u)+\sin(t)\sin(u)\big)du$$

$$= \cos(t)\int_0^t \big(\cos(u)\big)^2 du + \sin(t)\int_0^t \cos(u)\sin(u)du$$

$$= \cos(t)\left(\frac{t}{2}+\frac{\sin(2t)}{4}\right)+\sin(t)\left(\frac{1}{2}-\frac{\big(\cos(t)\big)^2}{2}\right)$$

$$= \frac{t\cos(t)+\sin(t)}{2}. \quad \blacksquare$$

Let us next prove that the Laplace transform of the convolution of two functions is the ordinary product of the Laplace transforms.

Theorem 14.22. *Suppose $\mathcal{L}\big(f(t)\big)$ and $\mathcal{L}\big(g(t)\big)$ exist for $s > a \geq 0$. Then $\mathcal{L}(f*g)(s)$ exists for $s > a$ and*

$$\mathcal{L}(f*g)(s) = \mathcal{L}(f)(s)\mathcal{L}(g)(s).$$

Proof. Using the change of variables $u = t - w$, have

$$\mathcal{L}(g)(s) = \int_0^\infty e^{su}g(u)du = \int_w^\infty e^{-s(t-w)}g(t-w)\,dt. \qquad (14.80)$$

We can extend the definition of g so that $g(t) = 0$ for $t < w$; then (14.80) becomes

$$\mathcal{L}(g)(s) = \int_w^\infty e^{-s(t-w)} g(t-w)\, dt = e^{sw} \int_0^\infty e^{-st} g(t-w)\, dt,$$

so that

$$e^{-sw} \mathcal{L}(g)(s) = \int_0^\infty e^{-st} g(t-w)\, dt. \tag{14.81}$$

We have

$$\mathcal{L}(f)(s)\mathcal{L}(g)(s) = \mathcal{L}(g)(s) \int_0^\infty e^{-sw} f(w)\, dw = \int_0^\infty e^{-sw} \mathcal{L}(g)(s) f(w)\, dw. \tag{14.82}$$

From (14.81) and (14.82) we get

$$\begin{aligned}
\mathcal{L}(f)(s)\mathcal{L}(g)(s) &= \int_0^\infty \left(\int_0^\infty e^{-st} g(t-w)\, dt \right) f(w)\, dw \\
&= \int_0^\infty e^{-st} \left(\int_0^t g(t-w) f(w)\, dw \right) dt \\
&= \int_0^\infty e^{-st} (f*g)(t)\, dt = \mathcal{L}(f*g)(s). \ \blacksquare
\end{aligned}$$

The convolution has some of the same properties as ordinary multiplication:

Lemma 14.23. *Let f, g, and h be integrable functions on the interval $0 < t < \infty$. Then*

(i) *Convolution is commutative: $f*g = g*f$;*

(ii) *Convolution is distributive: $f*(g_1 + g_2) = f*g_1 + f*g_2$;*

(iii) *Convolution is associative: $f*(g*h) = (f*g)*h$;*

(iv) $f*0 = 0*f = 0$.

The *Mathematica* definition of convolution is as follows:

```
Convolution[f_,g_][t_]:= Integrate[f[t - u]g[u],{u,0,t}]
```

Exercises for Section 14.9

Find the convolution $(f_1 * f_2)(t)$ in the following cases:

1. $f_1(t) = 1$, $f_2(t) = t$, $t > 0$

3. $f_3(t) = \sin(t)$, $f_2(t) = \cos(t)$, $t > 0$

2. $f_1(t) = t^2$, $f_2(t) = t$, $t > 0$

4. $f_1(t) = e^{at}$, $f_2(t) = e^{bt}$, $t > 0$

Use the convolution theorem to find the inverse Laplace transform of each of the following functions.

5. $\dfrac{1}{s(s+1)}$

8. $\dfrac{1}{(s+4)^2}$

6. $\dfrac{1}{s(s^2+4)}$

9. $\dfrac{s}{(s^2+9)^2}$

7. $\dfrac{1}{(s+1)(s-2)}$

10. $\dfrac{s}{(s^2+4s-5)^2}$

14.10   Laplace Transforms via ODE

Direct use of **LaplaceTransform** and **InverseLaplaceTransform** to solve a differential equation is somewhat complicated in *Mathematica*. Typically, it is a three-step procedure: computing a Laplace transform, solving an algebraic equation, and finally, applying the inverse Laplace transform. Instead, **ODE** with the option **Method->Laplace** automates the whole process. Here is an example that would be tedious to do by hand.

Example 14.20. *Use* ODE *to solve*

$$y'' + 4y = 4t^2 \cos(2t),$$

with the initial conditions $y(0) = 1$ *and* $y'(0) = 5$.

Solution. We use

```
ODE[{y'' + 4 y == 4 t^2 Cos[2t],y[0]  == 0,y'[0]  == 5},y,t,
Method->Laplace]
```

to get

$$\{\{y[t] \rightarrow \frac{t^2 \ Cos[2\ t]}{4} + \frac{5\ Sin[2\ t]}{2} - \frac{t\ Sin[2\ t]}{8} + \frac{t^3\ Sin[2\ t]}{3}\}\}$$

Hence the solution is

$$y = \frac{t^2 \cos(2t)}{4} + \frac{5\sin(2t)}{2} - \frac{t\sin(2t)}{8} + \frac{t^3\sin(2t)}{3}. \ \blacksquare$$

Mathematica's implementations of the unit step and Dirac delta functions

The *Mathematica* versions of the unit step function and Dirac delta function are contained in a special package called **DiracDelta.m**; this package can be read into a *Mathematica* session with the command

Needs["Calculus`DiracDelta`"]

Note that reading in either of the packages **LaplaceTransform.m** or **ODE.m** automatically reads in the package **DiracDelta.m**.

The following example shows how to use **ODE** to solve a differential equation with a forcing function containing a unit step function.

Example 14.21. *Use* **ODE** *to solve*

$$y'' + y = \cos(t)u_{\pi/2}(t),$$

with the initial conditions $y(0) = 1$ *and* $y'(0) = 5$. *Plot the solution.*

Solution. We use

ODE[{y'' + y == Cos[t] UnitStep[t - Pi/2],
y'[0] == 0,y[0] == 0},y,t,Method->Laplace]

to get

$$\{\{y[t] \rightarrow \frac{Cos[t]\ UnitStep[\frac{-Pi}{2} + t]}{2} - \frac{Pi\ Sin[t]\ UnitStep[\frac{-Pi}{2} + t]}{4} + $$

$$\frac{t\ Sin[t]\ UnitStep[\frac{-Pi}{2} + t]}{2}\}\}$$

Hence the solution is

$$y = \frac{\cos(t)u_{\pi/2}(t)}{2} - \frac{\pi \, \sin(t)u_{\pi/2}(t)}{4} + \frac{t \, \sin(t)u_{\pi/2}(t)}{2}.$$

To plot the solution we use

```
ODE[{y'' + y == Cos[t] UnitStep[t - Pi/2],
y'[0] == 0,y[0] == 0},y,t,
Method->Laplace,PlotSolution->{{t,0,2Pi}}]
```

to get

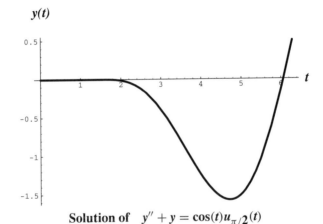

Solution of $y'' + y = \cos(t)u_{\pi/2}(t)$

with the initial conditions $y(0) = y'(0) = 0$

Example 14.22. *Solve the initial value problem*

$$\begin{cases} y'' + y = \cos(t)\delta(t - \pi), \\ y(0) = y'(0) = 0 \end{cases}$$

and plot the solution.

Solution. We use

```
ODE[{y'' + y == Cos[t] DiracDelta[t - Pi],y[0]==0,y'[0]==0},
y,t,Method->Laplace,PlotSolution->{{t,0,4Pi}}]
```

to get

```
{{y[t] -> Sin[t] UnitStep[-Pi + t]}}
```

and the plot

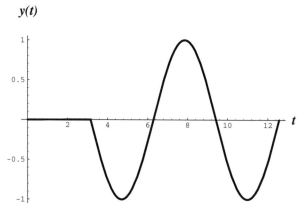

$y(t)$

The solution to $y + y'' = \cos(t)\delta(t - \pi)$ **with the initial conditions** $y(0) = y'(0) = 0$

Thus, the solution is $y = \sin(t)u(t - \pi) = \sin(t)u_\pi(t)$. ∎

Exercises for Section 14.10

Use *Mathematica* to find the Laplace transforms of the following functions of t:

1. $t^2 \sin(t)$

2. $t^2 y'(t)$

3. $\cos(m\,t)\sin(n\,t)$

4. $y'''(t) + 2y''(t) + y'(t) + 2y(t)$

5. $u_\pi(t) - u_{2\pi}(t)$

6. $t\cos(t)e^{2t}$

7. $t - u_{\pi/2}(t)(t - \pi)$

8. $\delta(t - 1) + 2\delta(t - 2) + 3\delta(t - 3)$

Use *Mathematica* to find the inverse Laplace transforms of the following functions of s:

9. $\dfrac{1}{s^2(s + 1)^3}$

10. $\dfrac{1}{s^3(s^2 + 4)^2}$

11. $s^2 \mathcal{L}(y(t))(s),\quad y(0) = y'(0) = 0$

12. $\dfrac{s}{(s^2 + 9)^6}$

13. $\dfrac{s}{(s^2 + 4s - 5)^2}$ **15.** $s + 5s^4$

14. $\dfrac{s}{(s^2 + 4s + 5)^2}$ **16.** $\dfrac{e^{-as}}{(s - b)}$

17. Produce a table of familiar Laplace transforms with the command

```
fns = {1,t,t^n,Exp[a t],Sin[a t],Cos[a t],Sinh[a t],
Cosh[a t],UnitStep[t],DiracDelta[t]};
TableForm[Transpose[{fns,LaplaceTransform[fns,t,s]}]]
```

18. Produce a table of familiar inverse Laplace transforms with the command

```
ifns = {1/s,1/(s + a),1/(s + a)^n,a/(s^2 + a^2),
s/(s^2 + a^2),E^-s,E^-s/s^n,s};
TableForm[Transpose[
{ifns,InverseLaplaceTransform[ifns,s,t]}]]//PowerExpand
```

Use *Mathematica* to solve and plot the following initial value problems:

19. $\begin{cases} y'' + 6y' + 9y = t^4 \sin(3t), \\ y(0) = 1,\ y'(0) = 4 \end{cases}$

22. $\begin{cases} y'' + y = u_2(t) + \delta(t - 2), \\ y(0) = y'(0) = 0 \end{cases}$

20. $\begin{cases} y'' + 9y = t^2 \sin(3t), \\ y(0) = 1,\ y'(0) = 4 \end{cases}$

23. $\begin{cases} y'' + y' + y = t\, u_1(t), \\ y(0) = y'(0) = 0 \end{cases}$

21. $\begin{cases} y^{(4)} + 2y'' + y = t^2 \cos(t), \\ y(0) = y'(0) = 0, \\ y''(0) = y'''(0) = 0 \end{cases}$

24. $\begin{cases} y'' + y = \delta(t - \pi) - \delta(t - 2\pi), \\ y(0) = y'(0) = 0 \end{cases}$

15

SYSTEMS OF LINEAR
DIFFERENTIAL
EQUATIONS

Up until now we have been concerned with single differential equations in one unknown function. In the present chapter we consider systems of n differential equations involving n unknown functions. Just as with single differential equations, we can speak of linear and nonlinear systems. This chapter, as well as Chapter 16, is devoted to linear systems; nonlinear systems will be discussed in Chapter 17.

Section 15.1 introduces the definitions and notation we need; we also explain why it is usually sufficient to limit ourselves to first-order systems instead of higher-order systems. Existence and uniqueness theorems for first-order systems are discussed in Section 15.2. To get started solving systems of linear differential equations, we show in Section 15.3 how to solve upper triangular linear systems, a particularly simple type.

The study of linear systems divides naturally into homogeneous and inhomogeneous systems; the situation is parallel to that of single higher-order linear equations. In particular, in Section 15.4 we define the notions of fundamental set of solutions, Wronskian, and general solution for homogeneous linear systems.

In Section 15.5 we turn to the most important type of homogeneous linear systems—those with constant-coefficients. Solution techniques for inhomogeneous linear systems are discussed in Sections 15.6 and 15.7. Finally, in Section 15.8 we show how to use Laplace transforms to solve linear systems.

Applications of first-order systems will be given in Chapters 18 and 19. Many concepts from linear algebra will be needed; for a review of this material see Appendix A.

15.1   Notation and Definitions for Systems

So far the differential equations we have discussed have involved only one unknown function. In many applications, however, there is more than one unknown function and more than one equation. A **system of first-order differential equations** is written in the form

$$
\begin{cases}
\dfrac{dx_1}{dt} = F_1(t, x_1, \ldots, x_n), \\[2mm]
\quad\vdots \qquad\qquad \vdots \\[2mm]
\dfrac{dx_n}{dt} = F_n(t, x_1, \ldots, x_n).
\end{cases}
\tag{15.1}
$$

The independent variable is t, and the unknown functions are $x_1(t), \ldots, x_n(t)$. The functions $F_1, \ldots, F_n$ are assumed to be given. A **solution** of (15.1) consists of a collection of differentiable functions $\{x_1(t), \ldots, x_n(t)\}$ that satisfy the system (15.1) for all values of t in some interval $a < t < b$.

We may also speak of an **initial value problem** corresponding to a system of differential equations. Such an initial value problem consists of a system of the form (15.1) together with n initial conditions

$$
x_1(t_0) = X_1, x_2(t_0) = X_2, \ldots, x_n(t_0) = X_n
\tag{15.2}
$$

for some t_0 satisfying $a < t_0 < b$.

It is useful to abbreviate (15.1) and (15.2) to

$$
\mathbf{x}'(t) = \mathbf{F}(t, \mathbf{x})
\tag{15.3}
$$

and

$$
\mathbf{x}(t_0) = \mathbf{X},
\tag{15.4}
$$

where

$$
\mathbf{x} = \begin{pmatrix} x_1 \\ \vdots \\ x_n \end{pmatrix}, \qquad
\mathbf{F} = \begin{pmatrix} F_1 \\ \vdots \\ F_n \end{pmatrix}, \qquad \text{and} \qquad
\mathbf{X} = \begin{pmatrix} X_1 \\ \vdots \\ X_n \end{pmatrix}.
$$

We can think of $\mathbf{X}$ as a point or vector in the space $\mathbb{R}^n$ of n-tuples of real numbers, and $t \longmapsto \mathbf{x}(t)$ as a vector-valued function or (parametrized) curve in $\mathbb{R}^n$. The vector notation in (15.3) and (15.4) contains no new mathematics; however, it is a very convenient abbreviation.

An important observation is that a higher-order single differential equation can be written in terms of first-order systems. The following example illustrates the procedure.

Example 15.1. *Show that the second-order equation*

$$y'' + p(t)y'(t) + q(t)y(t) = r(t) \qquad (15.5)$$

can be written as a system of two first-order differential equations.

Solution. By writing $x_1(t) = y(t)$ and $x_2(t) = y'(t)$, we convert (15.5) to the system

$$\begin{cases} \dfrac{dx_1}{dt} = x_2(t), \\[2mm] \dfrac{dx_2}{dt} = -p(t)x_2(t) - q(t)x_1(t) + r(t). \end{cases} \qquad (15.6)$$

Clearly, (15.6) is a special case of (15.1) (with $n = 2$) by taking

$$F_1(t, x_1, x_2) = x_2 \qquad \text{and} \qquad F_2(t, x_1, x_2) = -p(t)x_2 - q(t)x_1 + r(t). \ \blacksquare$$

More generally, an nth-order differential equation gives rise to n first-order differential equations.

Example 15.2. *Write $y''' + y'' + y' + y = 0$ as a system of three first-order differential equations.*

Solution. We put $x_1 = y$, $x_2 = y'$, and $x_3 = y''$. This yields the system

$$\begin{cases} x_1' = x_2, \\ x_2' = x_3, \\ x_3' = -x_1 - x_2 - x_3. \ \blacksquare \end{cases}$$

One could also consider more general systems of *higher-order* differential equations. However, any such system can be subsumed under the formalism of a first-order system by defining new functions as the derivatives of the original functions, just as in the above example of a single second-order equation. The following example illustrates this point.

Example 15.3. *Show that the second-order system*

$$\begin{cases} y_1''(t) = 3y_1(t) + y_1'(t) + 4y_2(t), \\ y_2''(t) = 3y_1(t) + 4y_1'(t) + 5y_2(t) \end{cases} \qquad (15.7)$$

can be written as a system of four first-order differential equations.

Solution. We simply define

$$x_1 = y_1, \qquad x_2 = y_1', \qquad x_3 = y_2, \qquad x_4 = y_2',$$

and compute the derivatives to obtain

$$
\begin{cases}
x_1' = x_2, \\[4pt]
x_2' = 3x_1 + x_2 + 4x_3, \\[4pt]
x_3' = x_4, \\[4pt]
x_4' = 3x_1 + 4x_2 + 5x_3.
\end{cases}
\tag{15.8}
$$

Then (15.8) is the first-order system equivalent to (15.7). ∎

Definition.   *The* **dimension** *of a system of first-order differential equations is the number n of equations.*

The system in Example 15.1 has dimension two, while the system in Example 15.2 has dimension three, and the system in Example 15.3 has dimension four.

A system of first-order differential equations (15.1) is classified as **linear** or **nonlinear** depending on the form of the functions $F_1, \ldots, F_n$. Thus a linear system has by definition the form

$$
\begin{cases}
\dfrac{dx_1}{dt} = F_1(t, x_1, \ldots, x_n) = p_{11}(t)x_1 + \cdots + p_{1n}(t)x_n + g_1(t), \\[6pt]
\;\;\vdots \qquad\qquad \vdots \qquad\qquad\qquad\qquad \vdots \\[6pt]
\dfrac{dx_n}{dt} = F_n(t, x_1, \ldots, x_n) = p_{n1}(t)x_1 + \cdots + p_{nn}(t)x_n + g_n(t).
\end{cases}
\tag{15.9}
$$

The notions of homogeneous and inhomogeneous for systems are completely analogous to those for single equations that we defined on page 240.

Definition.   *A linear system (15.9) is said to be* **homogeneous** *if all of the functions $g_i(t)$ are identically zero for $1 \le i \le n$; otherwise a linear system is said to be* **inhomogeneous**. *A* **constant-coefficient linear system** *is a linear system for which the functions $p_{ij}(t)$ are constant for $1 \le i, j \le n$.*

For instance, the systems in Examples 15.1–15.3 are all homogeneous, and the systems of Examples 15.2 and 15.3 have constant-coefficients.

It is also possible to write a linear system of first-order differential equations as a single higher-order equation, as we show in the next example.

Example 15.4.   *Write the three-dimensional first-order system*

$$
\begin{cases}
x_1' = x_2 - x_3, \\[4pt]
x_2' = x_3 - x_1, \\[4pt]
x_3' = x_1 + x_2 + x_3
\end{cases}
\tag{15.10}
$$

as a single third-order differential equation.

Solution. Let $y = x_1$; from (15.10) we compute

$$\begin{cases} y & = x_1, \\ y' & = x_2 - x_3, \\ y'' & = -2x_1 - x_2, \\ y''' & = x_1 - 2x_2 + x_3. \end{cases} \tag{15.11}$$

The fourth equation of (15.11) must be a consequence of the first three equations. In fact, we compute

$$y''' - y'' + y' - 3y = 0,$$

which is the required third-order equation. ∎

Summary of Techniques Introduced

Any single n^{th}-order linear differential equation can be written in terms of a system of n first-order equations, and conversely.

Exercises for Section 15.1

Write each of the following linear differential equations as a first-order system of differential equations.

1. $y'' + y' \sin(t) = 4e^t$

2. $y'' + 3t\, y' - 2y = 0$

3. $y''' - 3y'' + 2y' - t^2 y = e^t$

4. $\begin{cases} y_1'' + 2y_1' & = 3y_1 + 4y_2, \\ y_2'' - 3y_2' & = 4y_1 + 3y_2 \end{cases}$

For each of the following systems of equations, find a single differential equation satisfied by the functions $x_1(t)$ and $x_2(t)$.

5. $\begin{cases} x_1' & = 3x_1 - x_2, \\ x_2' & = 2x_1 + 5x_2 \end{cases}$

6. $\begin{cases} x_1' & = t\, x_1 - 3x_2 + 4e^t, \\ x_2' & = 3x_1 - t\, x_2 + 6e^t \end{cases}$

7. $\begin{cases} x_1''' & = x_2, \\ x_2''' & = -x_1 \end{cases}$

8. $\begin{cases} x_1' & = x_1 - x_2, \\ x_2' & = x_1 + x_2 \end{cases}$

9. $\begin{cases} x_1'' = 3x_1' + 2x_2' + x_1 - 4x_2 + 5, \\ x_2'' = 5x_1' - 3x_2' + 4x_1 + 5x_2 \end{cases}$ **10.** $\begin{cases} x_1' = 3x_1 - 2x_2, \\ x_2' = 7x_1 + 5x_2 \end{cases}$

15.2   Existence and Uniqueness Theorems for Systems

The theory of systems of first-order differential equations parallels that of a single first-order differential equation. Given a set of initial conditions, we look for a *local solution*, defined in some interval $a < t < b$ about the initial time point. The length of this interval depends on the initial conditions. Just as in Sections 5.1 and 8.2, we state the fundamental existence and uniqueness theorem in terms of an **initial value problem**.

Theorem 15.1.   (**Local existence and uniqueness**) *Suppose that the functions*

$$F_1(t, x_1, \ldots, x_n), \ldots, F_n(t, x_1, \ldots, x_n)$$

and their derivatives are continuous in a region containing the point $(t_0, X_1, \ldots, X_n)$ *in* $\mathbb{R}^{n+1}$. *Then there is a number* $h > 0$ *and unique functions* $\phi_1(t), \ldots, \phi_n(t)$ *defined for* $t_0 - h < t < t_0 + h$ *that satisfy the system of equations*

$$\begin{cases} \phi_1'(t) = F_1\big(t, \phi_1(t), \ldots, \phi_n(t)\big), \\ \quad\vdots \qquad\qquad \vdots \\ \phi_n'(t) = F_n\big(t, \phi_1(t), \ldots, \phi_n(t)\big), \end{cases} \tag{15.12}$$

together with the initial conditions

$$\phi_1(t_0) = X_1, \ldots, \phi_n(t_0) = X_n.$$

This is proved by writing the system of differential equations as a system of *integral equations* and applying the method of Picard iterations, which was discussed in Chapter 5 for a single equation. For a detailed demonstration one must check that the proof of Theorem 5.1 (page 122) can be generalized from a single equation to a system.

Theorem 15.1 provides a local solution, just as in the case $n = 1$ studied in Chapter 5. We can strengthen this statement to obtain a *global solution* in the case of a *linear* system of differential equations.

Theorem 15.2.   (**Global existence and uniqueness for linear systems**) *Suppose that for* $1 \leq i, j \leq n$ *the functions* $p_{ij}(t)$ *and* $g_i(t)$ *are continuous on the interval* $a < t < b$, *and* that $a < t_0 < b$. Then there exist unique functions $\phi_1(t), \ldots, \phi_n(t)$ defined for the whole interval $a < t < b$ that satisfy the system of linear equations

$$
\begin{cases}
\phi_1'(t) = \displaystyle\sum_{j=1}^{n} p_{1j}(t)x_j + g_1(t), \\
\vdots \qquad\qquad\qquad \vdots \\
\phi_n'(t) = \displaystyle\sum_{j=1}^{n} p_{nj}(t)x_j + g_n(t)
\end{cases}
\tag{15.13}
$$

together with the initial conditions

$$
\phi_1(t_0) = X_1, \ldots, \phi_n(t_0) = X_n.
$$

Some suggestions on the proof can be obtained by consulting the exercises.

The superposition and subtraction principles for single second-order linear differential equations (see page 243) are easily extended to first-order linear systems, as follows.

Theorem 15.3.   (Superposition principle for homogeneous linear systems) *Suppose that*

$$
\mathbf{x}(t) = \begin{pmatrix} x_1(t) \\ \vdots \\ x_n(t) \end{pmatrix} \qquad and \qquad \mathbf{y}(t) = \begin{pmatrix} y_1(t) \\ \vdots \\ y_n(t) \end{pmatrix}
$$

are both solutions of the same homogeneous linear system

$$
\begin{cases}
\dfrac{dx_1}{dt} = p_{11}(t)x_1 + \cdots + p_{1n}(t)x_n, \\
\vdots \qquad\qquad\qquad \vdots \\
\dfrac{dx_n}{dt} = p_{n1}(t)x_1 + \cdots + p_{nn}(t)x_n.
\end{cases}
\tag{15.14}
$$

Then for any constants a and b, the function

$$
a\,\mathbf{x}(t) + b\,\mathbf{y}(t) = \begin{pmatrix} a\,x_1(t) + b\,y_1(t) \\ \vdots \\ a\,x_n(t) + b\,y_n(t) \end{pmatrix}
$$

is also a solution to the homogeneous linear system (15.14).

Theorem 15.4.   (**Subtraction principle for linear systems**) *Suppose that*

$$\mathbf{x}(t) = \begin{pmatrix} x_1(t) \\ \vdots \\ x_n(t) \end{pmatrix} \qquad and \qquad \mathbf{y}(t) = \begin{pmatrix} y_1(t) \\ \vdots \\ y_n(t) \end{pmatrix}$$

are both solutions of the same inhomogeneous linear system

$$\begin{cases} \dfrac{dx_1}{dt} = p_{11}(t)x_1 + \cdots + p_{1n}(t)x_n + g_1(t), \\ \quad \vdots \qquad\qquad\qquad \vdots \\ \dfrac{dx_n}{dt} = p_{n1}(t)x_1 + \cdots + p_{nn}(t)x_n + g_n(t). \end{cases} \qquad (15.15)$$

Let $z_i(t) = x_i(t) - y_i(t)$ for $1 \le i \le n$. Then

$$\mathbf{z}(t) = \begin{pmatrix} z_1(t) \\ \vdots \\ z_n(t) \end{pmatrix}$$

is a solution to the homogeneous system

$$\begin{cases} \dfrac{dz_1}{dt} = p_{11}(t)z_1 + \cdots + p_{1n}(t)z_n, \\ \quad \vdots \qquad\qquad\qquad \vdots \\ \dfrac{dz_1}{dt} = p_{n1}(t)z_1 + \cdots + p_{nn}(t)z_n. \end{cases}$$

These principles can be used to find the general solution of certain simple systems of first-order differential equations. For their proofs see the exercises.

Example 15.5.   *Find the general solution to the linear system*

$$\begin{cases} \dfrac{dx_1}{dt} = x_2, \\ \dfrac{dx_2}{dt} = x_2 - 1. \end{cases} \qquad (15.16)$$

Solution.   It can be verified by inspection that a particular solution to (15.16) is given by $x_{1P}(t) = t$ and $x_{2P}(t) = 1$. This allows us to reduce the problem to that of finding the general solution to the homogeneous linear system

$$\begin{cases} \dfrac{dx_{1H}}{dt} = x_{2H}, \\ \dfrac{dx_{2H}}{dt} = x_{2H}. \end{cases} \qquad (15.17)$$

The second equation of (15.17) is a first-order linear equation whose general solution is

$$x_{2\text{H}}(t) = C_1 e^t, \tag{15.18}$$

for some constant C_1. Substituting (15.18) into the first equation of (15.17) and integrating, we obtain

$$x_{1\text{H}}(t) = C_1 e^t + C_2,$$

where C_2 is another constant of integration. This gives the required general solution to the homogeneous system. The desired general solution to the inhomogeneous system (15.16) can be obtained by adding the particular and homogeneous solutions; thus

$$\begin{cases} x_1(t) &= t + C_2 + C_1 e^t, \\ x_2(t) &= 1 + C_1 e^t. \end{cases} \blacksquare$$

Summary of Techniques Introduced

Any first-order system of differential equations can be solved, in principle, uniquely on a suitable interval, to obtain a local solution. In the case of a linear system we also have a global solution. A superposition principle holds for homogeneous linear systems, and a subtraction principle holds for general linear systems.

Exercises for Section 15.2

1. Prove the superposition principle for homogeneous linear systems.

2. Prove the subtraction principle for general linear systems.

The following exercises provide some guidance in the proof of the global existence theorem (Theorem 15.2) for linear systems.

3. Show that any solution to the first-order linear system

$$\begin{cases} x_1'(t) &= \displaystyle\sum_{j=1}^{n} p_{1j}(t) x_j(t) + g_1(t), \\ \vdots & \qquad\qquad \vdots \\ x_n'(t) &= \displaystyle\sum_{j=1}^{n} p_{nj}(t) x_j(t) + g_n(t), \end{cases} \tag{15.19}$$

with the initial conditions $x_1(t_0) = X_1, \ldots, x_n(t_0) = X_n$, is also a solution to the system of integral equations

$$
\begin{cases}
x_1(t) = X_1 + \displaystyle\int_{t_0}^{t} \left(\sum_{j=1}^{n} p_{1j}(s)x_j(s) + g_1(s) \right) ds, \\
\quad\vdots \qquad\qquad\qquad\qquad \vdots \\
x_n(t) = X_n + \displaystyle\int_{t_0}^{t} \left(\sum_{j=1}^{n} p_{nj}(s)x_j(s) + g_n(s) \right) ds.
\end{cases}
\tag{15.20}
$$

4. Show that any solution to the system of integral equations (15.20) is also a solution to the first-order linear system (15.19) with the initial conditions $x_1(t_0) = X_1, \ldots, x_n(t_0) = X_n$.

5. If $X_1, \ldots, X_n$ are given numbers, define $x_i^0(t) = X_i$, and for $i = 1, \ldots, n$ and for $m \geq 0$ define

$$
x_i^{m+1}(t) = X_i + \int_{t_0}^{t} \left(g_i(s) + \sum_{j=1}^{n} p_{ij}(s)x_j^m(s) \right) ds.
\tag{15.21}
$$

Show that

$$
\left| x_i^1(t) - x_i^0(t) \right| \leq P(t - t_0) \sum_{i=1}^{n} \left| x_i^0 \right| + \int_{t_0}^{t} G(s)\, ds
$$

and that for $m \geq 1$

$$
\left| x_i^{m+1}(t) - x_i^m(t) \right| \leq P \sum_{j=1}^{n} \int_{t_0}^{t} \left| x_j^m(s) - x_j^{m-1}(s) \right| ds,
$$

where P is an upper bound for the numbers $|p_{ij}(s)|$, for $1 \leq i, j \leq n$, $t_0 \leq s \leq t$, and

$$
G(t) = \sum_{i=1}^{n} |g_i(t)|.
$$

6. Use the results of Exercise 5 to show that for $m \geq 0$

$$
\left| x_i^{m+1}(t) - x_i^m(t) \right| \leq \frac{|P(t - t_0)|^{m+1}}{(m+1)!} \sum_{i=1}^{n} \left| x_i^0 \right| + P^m \int_{t_0}^{t} \frac{(t-s)^m}{m!} G(s)\, ds.
$$

7. Use the result of Exercise 6 to prove Theorem 15.2.

15.3 Solution of Upper Triangular Systems by Elimination

There is a special class of linear systems that can be solved directly by using the methods for a single first-order linear equation. These are the **upper triangular systems**, defined by the property that $p_{ij}(t) = 0$ whenever $i > j$. This means that the terms strictly below the main diagonal are all identically zero. Explicitly, an upper triangular system has the form

$$\begin{cases} x_1'(t) = p_{11}(t)x_1(t) + p_{12}(t)x_2(t) + \cdots + p_{1n}(t)x_n(t) + g_1(t), \\ x_2'(t) = \phantom{p_{11}(t)x_1(t) +} p_{22}(t)x_2(t) + \cdots + p_{2n}(t)x_n(t) + g_2(t), \\ \vdots \vdots \\ x_n'(t) = \phantom{p_{11}(t)x_1(t) + p_{12}(t)x_2(t) + \cdots +} p_{nn}(t)x_n(t) + g_n(t). \end{cases} \tag{15.22}$$

One could similarly define **lower triangular systems**, whose discussion is entirely parallel.

We use the method of **back substitution** to solve an upper triangular system of linear first-order differential equations. We first note that the equation for the function $x_n(t)$ does not involve the other functions and can be solved directly as a first-order linear equation by the methods of Section 3.2. Indeed, the last equation of (15.22) has the integrating factor

$$\exp\left(-\int p_{nn}(t)dt\right).$$

Once we have obtained the formula for $x_n(t)$, we can substitute it into the equation for $x_{n-1}(t)$, which reads

$$x_{n-1}'(t) = p_{n-1n-1}(t)x_{n-1}(t) + p_{n-1n}(t)x_n(t) + g_{n-1}(t). \tag{15.23}$$

The last two terms on the right-hand side of (15.23) are known, so that (15.23) can be solved using the integrating factor

$$\exp\left(-\int p_{n-1n-1}(t)dt\right).$$

By continuing this procedure we can obtain the remaining functions $x_{n-1}(t), \ldots, x_1(t)$.

Example 15.6. *Solve the upper triangular system*

$$\frac{dx_1}{dt} = 3x_1 + 4x_2 + 5, \tag{15.24}$$

$$\frac{dx_2}{dt} = 6x_2 + 1 \tag{15.25}$$

with the initial conditions $x_1(0) = 0$ and $x_2(0) = 0$ by the method of back substitution.

Solution. Equation (15.25) is a first-order linear equation with the integrating factor e^{-6t}, leading to

$$\frac{d}{dt}\left(x_2(t)e^{-6t}\right) = e^{-6t}.\tag{15.26}$$

When we integrate both sides of (15.26) and use the initial condition $x_2(0) = 0$, we obtain the explicit formula

$$x_2(t) = \frac{e^{6t} - 1}{6}.\tag{15.27}$$

We then substitute (15.27) into (15.24), producing

$$\frac{dx_1}{dt} = 3x_1 + \frac{4(e^{6t} - 1)}{6} + 5 = 3x_1 + \frac{2}{3}e^{6t} + \frac{13}{3}.\tag{15.28}$$

Multiplication by the integrating factor e^{-3t} converts (15.28) to

$$\frac{d}{dt}\left(x_1(t)e^{-3t}\right) = \frac{2}{3}e^{3t} + \frac{13}{3}e^{-3t}.\tag{15.29}$$

We integrate both sides of (15.29) and use the initial condition $x_1(0) = 0$, obtaining

$$x_1(t)e^{-3t} = \frac{2}{9}(e^{3t} - 1) + \frac{13}{9}(1 - e^{-3t})$$

and the explicit formula

$$x_1(t) = \frac{2}{9}(e^{6t} - e^{3t}) + \frac{13}{9}(e^{3t} - 1) = \frac{2}{9}e^{6t} + \frac{11}{9}e^{3t} - \frac{13}{9}. \ \blacksquare$$

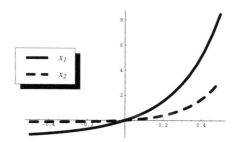

Solution of the system
$$x_1' = 3x_1 + 4x_2 + 5, \quad x_2' = 6x_2 + 1$$
with the initial conditions $x_1(0) = x_2(0) = 0$

Unfortunately, most systems of differential equations that arise in applications are *not* upper triangular. The methods of Section 15.3 are simple, and use of the techniques of this section will be made in Section 15.5.3. In order to deal with a general system of linear differential equations, we must develop new methods. This is most efficiently done by using the notations, concepts, and theorems of *linear algebra and matrix theory*, which the reader may have already studied in the latter stages of the calculus sequence. The necessary facts are developed without reference to previous course work in Appendix A. Rather than consider a finite set of ordered functions that satisfy a system of differential equations, we consider a single vector-valued function. This simultaneously provides notational simplicity and conceptual clarity, as the reader will quickly come to appreciate.

In the following sections we shall systematically develop the theory of first-order linear systems of differential equations using vectors and matrices.

Summary of Techniques Introduced

Any upper triangular system of linear differential equations can be solved by successively solving each of a series of first-order linear equations.

Exercises for Section 15.3

Find the general solution to each of the following upper/lower triangular systems.

1. $\begin{cases} x_1' = 4x_1 - 3x_2 + 7, \\ x_2' = 4x_2 + 5 \end{cases}$

4. $\begin{cases} x_1' = 5x_1 - 4x_2, \\ x_2' = x_2 + 1 \end{cases}$

2. $\begin{cases} x_1' = x_1 + 4x_2, \\ x_2' = 4x_2 - 6 \end{cases}$

5. $\begin{cases} x_1' = 3x_1 + 4, \\ x_2' = 5x_1 + 4x_2 + 6 \end{cases}$

3. $\begin{cases} x_1' = -x_1 + 2x_2 + x_3, \\ x_2' = x_2 + x_3, \\ x_3' = 2x_3 \end{cases}$

6. $\begin{cases} x_1' = x_1 + x_2 + x_3, \\ x_2' = x_2 + x_3, \\ x_3' = x_3 \end{cases}$

15.4   Homogeneous Linear Systems

In this section we apply linear algebra techniques to the theory of homogeneous first-order linear systems of differential equations. The development is entirely parallel to the theory of Section 8.3 for a single second-order differential equation.

15.4.1   *Fundamental Sets of Solutions of a Homogeneous System*

In this subsection we discuss the structure of the set of solutions of a first-order homogeneous system,

$$\begin{cases} \dfrac{dx_1}{dt} = p_{11}(t)x_1 + \cdots + p_{1n}(t)x_n, \\ \quad\vdots \qquad\qquad\qquad \vdots \\ \dfrac{dx_n}{dt} = p_{n1}(t)x_1 + \cdots + p_{nn}(t)x_n. \end{cases} \tag{15.30}$$

Notationally, the system (15.30) is easier to understand when we write it in the form

$$\mathbf{x}'(t) = \frac{d\mathbf{x}}{dt} = P(t)\mathbf{x}(t). \tag{15.31}$$

Then (15.30) is a special case of (15.3). Here $P(t)$ is a given $n \times n$ matrix-valued function defined on an interval $a < t < b$, and $\mathbf{x}(t)$ is the column vector-valued function to be found. Explicitly, the entries of $P(t)$ are the coefficients in (15.30), that is,

$$P(t) = \begin{pmatrix} p_{11}(t) & \cdots & p_{1n}(t) \\ \vdots & & \vdots \\ p_{n1}(t) & \cdots & p_{nn}(t) \end{pmatrix}.$$

The notion of fundamental set of solutions discussed in Sections 8.3 and 12.1 has a natural analogue for systems.

Definition.   *A set of column vector-valued functions*[1]

$$\mathbf{x}^1(t) = \begin{pmatrix} x_{11}(t) \\ \vdots \\ x_{n1}(t) \end{pmatrix}, \ldots, \mathbf{x}^n(t) = \begin{pmatrix} x_{1n}(t) \\ \vdots \\ x_{nn}(t) \end{pmatrix}$$

is called a **fundamental set of solutions** *of* (15.31), *provided that each column satisfies the system* (15.31) *and every other solution* $\mathbf{x}(t)$ *can be written in the form*

$$\mathbf{x}(t) = C_1\mathbf{x}^1(t) + \cdots + C_n\mathbf{x}^n(t) \tag{15.32}$$

for some choice of the complex numbers $C_1, \ldots, C_n$. *We call* (15.32) *the* **general solution** *of the homogeneous system* (15.31).

---

[1] We use superscripts as indices to distinguish one column vector from another. The superscripts are *neither* powers *nor* derivatives.

It is useful to combine fundamental solutions $\mathbf{x}^1(t), \ldots, \mathbf{x}^n(t)$ into a single matrix.

Definition.   *A **fundamental matrix** for (15.31) is the matrix $\Phi(t)$ consisting of a fundamental set of solutions, that is, the n column vectors $\mathbf{x}^1(t), \ldots, \mathbf{x}^n(t)$:*

$$\Phi(t) = \begin{pmatrix} x_{11}(t) & \cdots & x_{1n}(t) \\ \vdots & & \vdots \\ x_{n1}(t) & \cdots & x_{nn}(t) \end{pmatrix}.$$

Example 15.7. *Find a fundamental set of solutions and a fundamental matrix for the system*

$$\begin{cases} x_1' = x_2, \\ x_2' = 0. \end{cases} \tag{15.33}$$

Solution.   From the second equation of (15.33) we must have $x_2 = C_2$; then the first equation of (15.33) gives $x_1 = C_1 + t\,C_2$. Here C_1 and C_2 are arbitrary constants. In terms of column vectors we have the general solution

$$\mathbf{x}(t) = \begin{pmatrix} x_1 \\ x_2 \end{pmatrix} = C_1 \begin{pmatrix} 1 \\ 0 \end{pmatrix} + C_2 \begin{pmatrix} t \\ 1 \end{pmatrix}.$$

Thus, we may choose

$$\mathbf{x}^1(t) = \begin{pmatrix} 1 \\ 0 \end{pmatrix} \quad \text{and} \quad \mathbf{x}^2(t) = \begin{pmatrix} t \\ 1 \end{pmatrix}$$

as a fundamental system for (15.33). The fundamental matrix for (15.33) is

$$\Phi(t) = \begin{pmatrix} 1 & t \\ 0 & 1 \end{pmatrix}.$$

We also note that the determinant of the matrix $\Phi(t)$ is

$$\det\big(\Phi(t)\big) = \det \begin{pmatrix} 1 & t \\ 0 & 1 \end{pmatrix} = 1,$$

which shows directly that $\Phi(t)$ is nonsingular for all t. ∎

The *existence* of a fundamental set of solutions for a first-order linear homogeneous system (15.31) can be inferred from the existence and uniqueness theorem (Theorem 15.1).

Theorem 15.5.   *The system* $\mathbf{x}'(t) = \mathbf{P}(t)\mathbf{x}(t)$ *has a fundamental set of solutions* $\mathbf{x}^1(t), \ldots, \mathbf{x}^n(t)$ *on the interval* $a < t < b$ *such that for each* i

$$\mathbf{x}^i(t_0) = \begin{pmatrix} 0 \\ \vdots \\ 1 \\ \vdots \\ 0 \end{pmatrix},$$

where the entries are 1 in the ith *row and zero elsewhere. The formula that expresses any solution* $\mathbf{x}(t)$ *in terms of* $\mathbf{x}^1(t), \ldots, \mathbf{x}^n(t)$ *is*

$$\mathbf{x}(t) = C_1\mathbf{x}^1(t) + \cdots + C_n\mathbf{x}^n(t), \tag{15.34}$$

where $C_1, \ldots, C_n$ *are constants.*

Proof.   Theorem 15.1 guarantees the existence of $\mathbf{x}^1(t), \ldots, \mathbf{x}^n(t)$.   To show that $\mathbf{x}^1(t), \ldots, \mathbf{x}^n(t)$ form a fundamental set of solutions, let

$$\mathbf{x}(t) = \begin{pmatrix} x_1(t) \\ \vdots \\ x_n(t) \end{pmatrix}$$

be any other solution to (15.31). Choose constants $C_1, \ldots, C_n$ such that

$$C_1 = x_1(t_0), \ldots, C_n = x_n(t_0),$$

and form the vector function

$$\mathbf{y}(t) = C_1\mathbf{x}^1(t) + \cdots + C_n\mathbf{x}^n(t).$$

The superposition principle (Theorem 15.3) tells us that $\mathbf{y}(t)$ is a solution to (15.31). At the point t_0 we have

$$\mathbf{y}(t_0) = C_1\mathbf{x}^1(t_0) + \cdots + C_n\mathbf{x}^n(t_0) = \begin{pmatrix} C_1 \\ \vdots \\ C_n \end{pmatrix} = \mathbf{x}(t_0).$$

The uniqueness part of Theorem 15.1 tells us that $\mathbf{x}(t) = \mathbf{y}(t)$ for all t with $a < t < b$. Hence the general solution of $\mathbf{x}'(t) = \mathbf{P}(t)\mathbf{x}(t)$ is given by (15.34). ∎

15.4.2   *The Wronskian of a System of Solutions*

In practice, we need some computational mechanism for verifying that a *given set* of solutions to (15.31) indeed forms a fundamental set. For this purpose we generalize the concept of the Wronskian of Section 8.4 to systems of first-order differential equations.

Definition.   *Suppose that n column vector functions*

$$\mathbf{x}^1(t) = \begin{pmatrix} x_{11}(t) \\ \vdots \\ x_{n1}(t) \end{pmatrix}, \ldots, \mathbf{x}^n(t) = \begin{pmatrix} x_{1n}(t) \\ \vdots \\ x_{nn}(t) \end{pmatrix}$$

are given on the interval $a < t < b$. The **Wronskian** *of $\mathbf{x}^1(t), \ldots, \mathbf{x}^n(t)$ is defined as*

$$W(t) = W(\mathbf{x}^1, \ldots, \mathbf{x}^n)(t) = \det(\mathbf{x}^1(t), \ldots, \mathbf{x}^n(t))$$

$$= \det \begin{pmatrix} x_{11}(t) & \cdots & x_{1n}(t) \\ \vdots & & \vdots \\ x_{n1}(t) & \cdots & x_{nn}(t) \end{pmatrix}.$$

Clearly, the Wronskian is just the determinant of the fundamental matrix, that is,

$$W(t) = \det(\mathbf{\Phi}(t)).$$

It is essential to note that the number of column vector functions is equal to the dimension of the system. Otherwise, the Wronskian is not defined, since we can only take the determinant of a *square* matrix.

Theorem 15.6.   *Suppose that $\mathbf{x}^1(t), \ldots, \mathbf{x}^n(t)$ are n column vector solutions of the n-dimensional system*

$$\mathbf{x}'(t) = \mathbf{P}(t)\mathbf{x}(t) \tag{15.35}$$

on the interval $a < t < b$. The following are equivalent:

(i)   *the Wronskian $W(t)$ is nonzero for all t with $a < t < b$;*

(ii)   *the Wronskian $W(t_0) \neq 0$ for some t_0 with $a < t_0 < b$;*

(iii)   *the vectors $\mathbf{x}^1(t), \ldots, \mathbf{x}^n(t)$ are linearly independent for all t with $a < t < b$;*

(iv)   *the column vector functions $\mathbf{x}^1(t), \ldots, \mathbf{x}^n(t)$ form a fundamental set of solutions of (15.35) for $a < t < b$.*

Proof. The idea of the proof is to show that the Wronskian satisfies a first-order differential equation; the details of the proof are similar to those of Theorem 8.5. We do the case $n = 2$, where the details are transparent. (For the general case see the exercises.) In order to study the Wronskian $W(t)$, we determine a first-order differential equation that it satisfies. Since $n = 2$, we have two solutions to (15.35), which we write as

$$\mathbf{x}^1(t) = \begin{pmatrix} x_{11}(t) \\ x_{21}(t) \end{pmatrix} \quad \text{and} \quad \mathbf{x}^2(t) = \begin{pmatrix} x_{12}(t) \\ x_{22}(t) \end{pmatrix}.$$

The matrix $P(t)$ in (15.35) is a 2×2 matrix, which we write explicitly as

$$P(t) = \begin{pmatrix} p_{11}(t) & p_{12}(t) \\ p_{21}(t) & p_{22}(t) \end{pmatrix}.$$

Since both $\mathbf{x}^1$ and $\mathbf{x}^2$ are solutions of the system $\mathbf{x}'(t) = P(t)\mathbf{x}(t)$, we have

$$\left(\mathbf{x}^1, \mathbf{x}^2\right)' = \left((\mathbf{x}^1)', (\mathbf{x}^2)'\right) = \left(P(t)\mathbf{x}^1, P(t)\mathbf{x}^2\right) = P(t)\left(\mathbf{x}^1, \mathbf{x}^2\right). \tag{15.36}$$

Then (15.36) is written out in full as

$$\begin{pmatrix} x'_{11}(t) & x'_{12}(t) \\ x'_{21}(t) & x'_{22}(t) \end{pmatrix} = \begin{pmatrix} p_{11}(t) & p_{12}(t) \\ p_{21}(t) & p_{22}(t) \end{pmatrix}\begin{pmatrix} x_{11}(t) & x_{12}(t) \\ x_{21}(t) & x_{22}(t) \end{pmatrix}$$

$$= \begin{pmatrix} p_{11}(t)x_{11}(t) + p_{12}(t)x_{21}(t) & p_{11}(t)x_{12}(t) + p_{12}(t)x_{22}(t) \\ p_{21}(t)x_{11}(t) + p_{22}(t)x_{21}(t) & p_{21}(t)x_{12}(t) + p_{22}(t)x_{22}(t) \end{pmatrix}. \tag{15.37}$$

Since we are dealing with a two-dimensional system, the Wronskian $W(t)$ is given by the simple formula

$$W(t) = \det\begin{pmatrix} x_{11}(t) & x_{12}(t) \\ x_{21}(t) & x_{22}(t) \end{pmatrix} = x_{11}(t)x_{22}(t) - x_{12}(t)x_{21}(t). \tag{15.38}$$

We differentiate (15.38) to obtain

$$W' = \left(x_{11}x'_{22} + x'_{11}x_{22}\right) - \left(x_{12}x'_{21} + x'_{12}x_{21}\right). \tag{15.39}$$

From (15.37) and (15.38) it follows that

$$\begin{aligned} W' &= x_{11}\left(p_{21}x_{12} + p_{22}x_{22}\right) + x_{22}\left(p_{11}x_{11} + p_{12}x_{21}\right) \\ &\quad -x_{12}\left(p_{21}x_{11} + p_{22}x_{21}\right) - x_{21}\left(p_{11}x_{12} + p_{12}x_{22}\right). \end{aligned} \tag{15.40}$$

The terms in (15.40) involving $p_{12}x_{21}x_{22}$ cancel one another, as do the terms involving $p_{21}x_{11}x_{12}$. Thus, (15.40) reduces to

$$W' = (p_{11} + p_{22})(x_{11}x_{22} - x_{12}x_{21}) = (p_{11} + p_{22})W(t). \qquad (15.41)$$

The solution to the first-order linear equation (15.41) is immediately found by the methods of Section 3.2 as

$$W(t) = W(t_0) \exp\left(\int_{t_0}^{t} \left(p_{11}(s) + p_{22}(s) \right) ds \right). \qquad (15.42)$$

The exponential on the right-hand side of (15.42) is never zero. Therefore, $W(t_0) \neq 0$ implies that $W(t) \neq 0$ for all t, showing that (i) is equivalent to (ii).

The equivalence of (i) and (iii) follows from Theorem A.2 of Appendix A. Clearly, (iv) implies (iii). Finally, given a solution $\mathbf{x}(t)$ and t_0 with $a < t_0 < b$, we can solve the equation

$$\mathbf{x}(t_0) = C_1\mathbf{x}_1(t_0) + C_2\mathbf{x}_2(t_0) \qquad (15.43)$$

uniquely for the constants C_1 and C_2. By the uniqueness theorem (Theorem 15.1), the function $C_1\mathbf{x}_1(t) + C_2\mathbf{x}_2(t)$ is the only solution to $\mathbf{x}'(t) = \mathbf{P}(t)\mathbf{x}(t)$ satisfying the initial condition (15.43). Hence $\mathbf{x}(t) = C_1\mathbf{x}_1(t) + C_2\mathbf{x}_2(t)$ for all t, which proves the uniqueness. ∎

The key to the proof of Theorem 15.6 is equation (15.42), which can be considered to be Abel's formula for two-dimensional systems. In fact (see Exercises 17 and 18) we can prove **Abel's formula for n-dimensional first-order systems:**

Theorem 15.7.   *Suppose that $\mathbf{x}^1(t), \ldots, \mathbf{x}^n(t)$ are n column vector solutions of the n-dimensional system $\mathbf{x}'(t) = \mathbf{P}(t)\mathbf{x}(t)$ on the interval $a < t < b$. Fix t_0 with $a < t_0 < b$. Then the Wronskian $W(t) = W(\mathbf{x}^1, \ldots, \mathbf{x}^n)(t)$ can be explicitly computed by*

$$W(t) = W(t_0) \exp\left(\int_{t_0}^{t} \left(p_{11}(s) + \cdots + p_{nn}(s) \right) ds \right)$$

$$= W(t_0) \exp\left(\int_{t_0}^{t} \mathrm{tr}\left(\mathbf{P}(s) \right) ds \right), \qquad (15.44)$$

where $\mathrm{tr}\left(\mathbf{P}(s) \right)$ denotes the trace of the matrix $\mathbf{P}(s)$.

(See page 773 for the definition of trace.)

Summary of Techniques Introduced

Any homogeneous linear system of differential equations possesses a fundamental set of solutions. These can be displayed in matrix form as the fundamental matrix for the system.

The Wronskian can be defined and used to determine whether a set of solutions forms a fundamental set. The Wronskian satisfies a first-order linear differential equation, whose solution is represented by means of Abel's formula.

Exercises for Section 15.4

Find a fundamental set of solutions for each of the following systems.

1. $\begin{cases} x_1' = 0, \\ x_2' = x_1 \end{cases}$

5. $\begin{cases} x_1' = x_1 + x_2, \\ x_2' = 3x_2 \end{cases}$

2. $\begin{cases} x_1' = 3x_1, \\ x_2' = 4x_2 \end{cases}$

6. $\begin{cases} x_1' = 2x_1, \\ x_2' = x_1 + 3x_2 \end{cases}$

3. $\begin{cases} x_1' = 3x_1, \\ x_2' = 4x_1 \end{cases}$

7. $\begin{cases} x_1' = x_2, \\ x_2' = -x_1 \end{cases}$

4. $\begin{cases} x_1' = 3x_1, \\ x_2' = 4x_1, \\ x_2' = x_1 + x_2 \end{cases}$

8. $\begin{cases} x_1' = x_2, \\ x_2' = -x_1, \\ x_3' = 2x_3 \end{cases}$

Compute the Wronskian of each of the following sets of vector functions.

9. $\left\{ \begin{pmatrix} t \\ 1 \end{pmatrix}, \begin{pmatrix} 1 \\ 0 \end{pmatrix} \right\}$

12. $\left\{ \begin{pmatrix} e^t \\ e^t \\ e^t \end{pmatrix}, \begin{pmatrix} e^{2t} \\ 2e^{2t} \\ 4e^{2t} \end{pmatrix}, \begin{pmatrix} e^{3t} \\ 3e^{3t} \\ 9e^{3t} \end{pmatrix} \right\}$

10. $\left\{ \begin{pmatrix} e^t \\ e^t \end{pmatrix}, \begin{pmatrix} 4e^{3t} \\ 3e^{3t} \end{pmatrix} \right\}$

13. $\left\{ \begin{pmatrix} \cos(t) \\ \sin(t) \end{pmatrix}, \begin{pmatrix} -\sin(t) \\ \cos(t) \end{pmatrix} \right\}$

11. $\left\{ \begin{pmatrix} e^{2t} \\ e^{2t} \end{pmatrix}, \begin{pmatrix} t\,e^{2t} \\ t\,e^{2t} \end{pmatrix} \right\}$

14. $\left\{ \begin{pmatrix} e^t \cos(t) \\ e^t \sin(t) \end{pmatrix}, \begin{pmatrix} e^t \sin(t) \\ -e^t \cos(t) \end{pmatrix} \right\}$

15. $\left\{ \begin{pmatrix} e^t \\ 3e^t \end{pmatrix}, \begin{pmatrix} 4e^{5t} \\ 12e^{5t} \end{pmatrix} \right\}$

16. $\left\{ \begin{pmatrix} e^t \cos(t) \\ e^t \sin(t) \\ e^{2t} \end{pmatrix}, \begin{pmatrix} e^t \sin(t) \\ -e^t \cos(t) \\ e^{2t} \end{pmatrix}, \begin{pmatrix} e^t \\ e^t \\ e^t \end{pmatrix} \right\}$

17. Prove the following formula for differentiation of determinants of matrix-valued functions:

$$\frac{d}{dt} \det \begin{pmatrix} x_{11}(t) & \cdots & x_{1n}(t) \\ \vdots & \ddots & \vdots \\ x_{n1}(t) & \cdots & x_{nn}(t) \end{pmatrix} = \det \begin{pmatrix} x'_{11}(t) & \cdots & x'_{1n}(t) \\ \vdots & \ddots & \vdots \\ x_{n1}(t) & \cdots & x_{nn}(t) \end{pmatrix}$$

$$+ \cdots + \det \begin{pmatrix} x_{11}(t) & \cdots & x_{1n}(t) \\ \vdots & \ddots & \vdots \\ x'_{n1}(t) & \cdots & x'_{nn}(t) \end{pmatrix}.$$

18. Apply the result of Problem 17 to complete the details of the proof of Theorem 15.6 for general n.

15.5   Constant-Coefficient Homogeneous Systems

In this section we develop the structure of the solution to constant-coefficient systems, in parallel with the theory of Section 9.1 for a single second-order linear differential equation. The basic constant-coefficient homogeneous system is written

$$\mathbf{x}'(t) = A\mathbf{x}(t), \tag{15.45}$$

where A is an $n \times n$ matrix of complex numbers. In order to formulate a trial solution to (15.45), we recall that for a single second-order differential equation

$$a\,y'' + b\,y' + c\,y = 0,$$

we found the general solution in the form $y(t) = C\,e^{rt}$, where C is a constant. By analogy we look for the trial solution to (15.45) in the form

$$\mathbf{x}(t) = \boldsymbol{\xi} e^{rt},$$

where $\boldsymbol{\xi}$ is a constant vector in $\mathbb{R}^n$. For this choice we compute

$$\mathbf{x}'(t) = r\,\boldsymbol{\xi} e^{rt} \qquad \text{and} \qquad A\mathbf{x}(t) = e^{rt} A\boldsymbol{\xi}.$$

Then (15.45) holds if and only if the vector ξ is an eigenvector with eigenvalue r, that is,

$$A\xi = r\,\xi.$$

This leads us immediately to the following result, which applies in the case of diagonalizable matrices. (See Section A.4 of Appendix A for a discussion of eigenvalues and eigenvectors and page 777 for the definition of diagonalizable matrix.)

Theorem 15.8.   *Suppose that the $n \times n$ complex matrix A is diagonalizable, with linearly independent eigenvectors $\xi_1, \ldots, \xi_n$ and eigenvalues $r_1, \ldots, r_n$. Then a fundamental set of solutions of (15.45) is provided by the vector functions*

$$\mathbf{x}^1(t) = \xi_1 e^{r_1 t}, \ldots, \mathbf{x}^n(t) = \xi_n e^{r_n t}.$$

The general solution of (15.45) is

$$\mathbf{x}(t) = C_1 \xi_1 e^{r_1 t} + \cdots + C_n \xi_n e^{r_n t}, \tag{15.46}$$

where $C_1, \ldots, C_n$ are constants.

Proof.   We have just shown that $\mathbf{x}^1(t), \ldots, \mathbf{x}^n(t)$ are solutions of (15.45). Furthermore, the Wronskian of $\mathbf{x}^1(t), \ldots, \mathbf{x}^n(t)$ is easily computed to be

$$W(t) = e^{(r_1 + \cdots + r_n)t} \det\big((\xi_1, \ldots, \xi_n)\big),$$

which is nonzero, since the vectors $\xi_1, \ldots, \xi_n$ are linearly independent. Hence, $\mathbf{x}^1(t), \ldots, \mathbf{x}^n(t)$ form a fundamental set of solutions to (15.45). ∎

Note that the fundamental solutions given by Theorem 15.8 are in general complex-valued. We shall examine separately the several special cases of Theorem 15.8.

15.5.1   The Case of a Diagonalizable Matrix with Real Eigenvalues

When we specialize Theorem 15.8 to the case of a real matrix A with real eigenvalues, we obtain a fundamental set of real solutions as follows:

Theorem 15.9.   *Suppose that the real matrix A is diagonalizable, with real eigenvalues $r_1, \ldots, r_n$. Then the eigenvectors $\xi_1, \ldots, \xi_n$ are real, and a fundamental set of real-valued solutions to*

$$\mathbf{x}'(t) = A\mathbf{x}(t)$$

is provided by the real vector functions

$$\mathbf{x}^1(t) = \xi_1 e^{r_1 t}, \ldots, \mathbf{x}^n(t) = \xi_n e^{r_n t}.$$

The general solution is (15.46).

Proof. The eigenvectors can be selected real because they are obtained by solving linear equations with real coefficients. The rest of the theorem is a consequence of Theorem 15.8. ∎

It may be difficult to solve $\mathbf{x}' = A\mathbf{x}$ when A is an $n \times n$ matrix, because the eigenvalues of A are roots of the nth-order polynomial $\det(A - r\, I_{n \times n})$, which may be hard to factor. However, if A is a real 2×2 matrix with real distinct eigenvalues, the solution to $\mathbf{x}' = A\mathbf{x}$ can be found in a straightforward manner, as we now illustrate.

Example 15.8. *Find a fundamental set of solutions of the linear system*

$$\begin{cases} x_1' = x_1 + x_2, \\ x_2' = 4x_1 + x_2, \end{cases} \tag{15.47}$$

and also the general solution.

Solution. We first find the eigenvalues and eigenvectors of the matrix A of the system, which is

$$\begin{pmatrix} 1 & 1 \\ 4 & 1 \end{pmatrix}.$$

The eigenvalues of A are determined by solving

$$0 = \det\begin{pmatrix} 1 - r & 1 \\ 4 & 1 - r \end{pmatrix} = (1 - r)^2 - 4$$

$$= r^2 - 2r - 3 = (r + 1)(r - 3).$$

Thus, the eigenvalues of A are 3 and -1. Since the eigenvalues are distinct, we know that A is diagonalizable, and we can find the fundamental set using Theorem 15.9. An eigenvector $\begin{pmatrix} c_1 \\ c_2 \end{pmatrix}$ corresponding to the eigenvalue 3 must satisfy

$$\begin{pmatrix} 0 \\ 0 \end{pmatrix} = (A - 3I)\begin{pmatrix} c_1 \\ c_2 \end{pmatrix} = \begin{pmatrix} -2 & 1 \\ 4 & -2 \end{pmatrix}\begin{pmatrix} c_1 \\ c_2 \end{pmatrix},$$

where I denotes the 2×2 identity matrix. This leads to the system of redundant equations

$$\begin{cases} -2c_1 + c_2 = 0, \\ 4c_1 - 2c_2 = 0, \end{cases}$$

whose solution is $c_2 = 2c_1$. Hence any eigenvector corresponding to the eigenvalue 3 must be a multiple of $\begin{pmatrix} 1 \\ 2 \end{pmatrix}$. Similarly, an eigenvector $\begin{pmatrix} c_1 \\ c_2 \end{pmatrix}$ corresponding to the eigenvalue -1 must satisfy the system of redundant equations

$$\begin{cases} 2c_1 + c_2 = 0, \\ 4c_1 + 2c_2 = 0, \end{cases}$$

whose solution is $c_2 = -2c_1$. Thus any eigenvector corresponding to the eigenvalue -1 must be a multiple of $\begin{pmatrix} 1 \\ -2 \end{pmatrix}$. It follows that a fundamental set of solutions to (15.47) is provided by

$$\mathbf{x}_1(t) = \begin{pmatrix} 1 \\ 2 \end{pmatrix} e^{3t} \qquad \text{and} \qquad \mathbf{x}_2(t) = \begin{pmatrix} 1 \\ -2 \end{pmatrix} e^{-t}.$$

The general solution to (15.47) is

$$\mathbf{x}(t) = C_1 \begin{pmatrix} 1 \\ 2 \end{pmatrix} e^{3t} + C_2 \begin{pmatrix} 1 \\ -2 \end{pmatrix} e^{-t} = \begin{pmatrix} C_1 e^{3t} + C_2 e^{-t} \\ 2C_1 e^{3t} - 2C_2 e^{-t} \end{pmatrix},$$

where C_1 and C_2 are constants. ∎

15.5.2   The Case of a Real Diagonalizable Matrix with Complex Eigenvalues

If some of the eigenvalues of the matrix A are complex numbers, we can still find solutions to $\mathbf{x}' = A\mathbf{x}$ in the form $\mathbf{x}(t) = \boldsymbol{\xi} e^{rt}$, where r is an eigenvalue with eigenvector $\boldsymbol{\xi}$. In case the matrix A is diagonalizable, this procedure provides a fundamental set of complex-valued solutions.

If the matrix A is *real* (this is usually the case), it is almost always desirable to obtain real-valued solutions of the system. To do this, we write the complex eigenvalue r in the form

$$r = \lambda + i\,\mu$$

and the corresponding eigenvector $\boldsymbol{\xi}$ as

$$\boldsymbol{\xi} = \mathbf{a} + i\,\mathbf{b},$$

where λ and μ are real numbers and $\mathbf{a}$ and $\mathbf{b}$ are real vectors. Then we decompose the complex-valued solution $\mathbf{x}(t)$ to $\mathbf{x}'(t) = A\mathbf{x}(t)$ as follows:

$$\begin{aligned} \mathbf{x}(t) &= (\mathbf{a} + i\,\mathbf{b})e^{(\lambda+i\mu)t} = e^{\lambda t}(\mathbf{a} + i\,\mathbf{b})\big(\cos(\mu\,t) + i\sin(\mu\,t)\big) \\ &= e^{\lambda t}\big(\mathbf{a}\cos(\mu\,t) - \mathbf{b}\sin(\mu\,t)\big) + i\,e^{\lambda t}\big(\mathbf{a}\sin(\mu\,t) + \mathbf{b}\cos(\mu\,t)\big). \end{aligned}$$

Thus, we can decompose $\mathbf{x}(t)$ into its real and imaginary parts, that is, $\mathbf{x}(t) = \mathfrak{Re}\big(\mathbf{x}(t)\big) + i\,\mathfrak{Im}\big(\mathbf{x}(t)\big) = \mathbf{u}(t) + i\,\mathbf{v}(t)$, where

$$\mathbf{u}(t) = e^{\lambda t}\big(\mathbf{a}\cos(\mu\,t) - \mathbf{b}\sin(\mu\,t)\big) \qquad \text{and} \qquad \mathbf{v}(t) = e^{\lambda t}\big(\mathbf{a}\sin(\mu\,t) + \mathbf{b}\cos(\mu\,t)\big).$$

In fact, $\mathbf{u}(t)$ and $\mathbf{v}(t)$ are real linearly independent solutions to $\mathbf{x}' = A\mathbf{x}$ (see Exercise 25). One can show that the same real-valued solutions are obtained if we use instead the complex-conjugate eigenvalue $\lambda - i\,\mu$. We summarize the findings as follows.

Theorem 15.10.   *Suppose that the $n \times n$ matrix A is real and diagonalizable, so that the complex conjugate eigenvalues $r_1, r_2, \ldots, r_{2k}$ can be labeled in pairs:*

$$\begin{cases} r_{2\alpha-1} = \lambda_\alpha + i\,\mu_\alpha, \\ r_{2\alpha} = \lambda_\alpha - i\,\mu_\alpha, \end{cases}$$

for $1 \le \alpha \le k$, where each $\mu_\alpha \ne 0$. Let the real eigenvalues be labeled r_α for $2k+1 \le \alpha \le n$. Let ξ_α denote an eigenvector corresponding to r_α. For $1 \le \alpha \le k$ write

$$\xi_\alpha = \mathbf{a}_\alpha + i\,\mathbf{b}_\alpha,$$

where $\mathbf{a}_\alpha$ and $\mathbf{b}_\alpha$ are real vectors, and assume that ξ_α is real for $2k+1 \le \alpha \le n$. Then a fundamental set of complex-valued solutions is provided by

$$\mathbf{x}^\alpha(t) = \xi_\alpha\, e^{r_\alpha t}$$

for $1 \le \alpha \le n$. Furthermore, a fundamental set of real-valued solutions is given by

$$\left\{ \mathbf{u}^1(t), \mathbf{v}^1(t), \ldots, \mathbf{u}^k(t), \mathbf{v}^k(t), \mathbf{x}^{2k+1}(t), \ldots, \mathbf{x}^n(t) \right\},$$

where

$$\begin{cases} \mathbf{u}^\alpha(t) = e^{\lambda_\alpha t}\big(\mathbf{a}_\alpha \cos(\mu_\alpha t) - \mathbf{b}_\alpha \sin(\mu_\alpha t)\big), \\ \mathbf{v}^\alpha(t) = e^{\lambda_\alpha t}\big(\mathbf{a}_\alpha \sin(\mu_\alpha t) + \mathbf{b}_\alpha \cos(\mu_\alpha t)\big) \end{cases}$$

for $1 \le \alpha \le k$.

We illustrate Theorem 15.10 in the case of a two-dimensional system.

Example 15.9.   *Find a fundamental set of solutions of the system*

$$\begin{cases} x_1' = x_1 + 2x_2, \\ x_2' = -2x_1 + x_2, \end{cases} \tag{15.48}$$

and also the general solution.

Solution.   The eigenvalues of (15.48) are determined from the equation

$$0 = \det\begin{pmatrix} 1-r & 2 \\ -2 & 1-r \end{pmatrix} = (1-r)^2 + 4,$$

which is solved to yield the complex numbers $1 \pm 2i$. The complex eigenvector $\xi = \begin{pmatrix} c_1 \\ c_2 \end{pmatrix}$ corresponding to the eigenvalue $1+2i$ is found by solving the system of redundant equations

$$\begin{cases} -2i\,c_1 + 2c_2 = 0, \\ -2c_1 - 2i\,c_2 = 0. \end{cases}$$

Thus, $c_2 = i\,c_1$, so that any eigenvector corresponding to the eigenvalue $1 + 2i$ is a nonzero multiple of $\begin{pmatrix} 1 \\ i \end{pmatrix}$. We have generated the complex solution

$$\mathbf{x}^1(t) = \begin{pmatrix} 1 \\ i \end{pmatrix} e^{(1+2i)t} \tag{15.49}$$

of (15.48). There is no need to carry out similar computations using the eigenvalue $1 - 2i$; instead, we find the real and imaginary parts of $\mathbf{x}^1(t)$ by expanding (15.49) as

$$\mathbf{x}^1(t) = \begin{pmatrix} 1 \\ i \end{pmatrix} e^t \big(\cos(2t) + i\sin(2t) \big) = e^t \begin{pmatrix} \cos(2t) + i\sin(2t) \\ -\sin(2t) + i\cos(2t) \end{pmatrix}.$$

Therefore

$$\mathbf{u}(t) = \mathfrak{Re}\big(\mathbf{x}^1(t)\big) = e^t \begin{pmatrix} \cos(2t) \\ -\sin(2t) \end{pmatrix} \quad \text{and} \quad \mathbf{v}(t) = \mathfrak{Im}\big(\mathbf{x}^1(t)\big) = e^t \begin{pmatrix} \sin(2t) \\ \cos(2t) \end{pmatrix}.$$

Then $\mathbf{u}(t)$ and $\mathbf{v}(t)$ are linearly independent real solutions of (15.48). Furthermore, the general solution to (15.48) is

$$\mathbf{x}(t) = e^t \left(C_1 \begin{pmatrix} \cos(2t) \\ -\sin(2t) \end{pmatrix} + C_2 \begin{pmatrix} \sin(2t) \\ \cos(2t) \end{pmatrix} \right)$$

$$= \begin{pmatrix} e^t \big(C_1 \cos(2t) + C_2 \sin(2t) \big) \\ e^t \big(-C_1 \sin(2t) + C_2 \cos(2t) \big) \end{pmatrix},$$

where C_1 and C_2 are real constants. ∎

15.5.3   *The Case of Repeated Eigenvalues.*
Generalized Eigenvectors

Finally, we examine the important case when the matrix A has repeated eigenvalues. It may happen that A is not diagonalizable. Since the details in the general case are quite complicated, we first illustrate the procedure with a two-dimensional system.

Consider $\mathbf{x}' = A\mathbf{x}$, where A is a 2×2 matrix with a double eigenvalue r. If there are two linearly independent eigenvectors $\boldsymbol{\xi}_1$ and $\boldsymbol{\xi}_2$ corresponding to r, then the general solution to $\mathbf{x}' = A\mathbf{x}$ can be found by Theorem 15.9; the result is

$$\mathbf{x} = \big(C_1 \boldsymbol{\xi}_1 + C_2 \boldsymbol{\xi}_2 \big) e^{rt}.$$

However, it may happen that there is only one linearly independent eigenvector ξ corresponding to r. In that case, we can find one nonzero solution to $\mathbf{x}' = A\mathbf{x}$ by the method of Section 15.5.1, namely $\mathbf{x}^1(t) = \xi e^{rt}$. Our work in Section 9.1.2 with the repeated-root case of $a\,y'' + b\,y' + c\,y = 0$ suggests $t\,\mathbf{x}_1(t)$ as a candidate for a second solution to $\mathbf{x}' = A\mathbf{x}$. Since this does not quite work, we use instead

$$\mathbf{x}^2(t) = t\,e^{rt}\xi + e^{rt}\eta,$$

where η is a vector yet to be determined. We find that in order that $\mathbf{x}^2(t)$ be a solution of $\mathbf{x}' = A\mathbf{x}$ it is necessary that η satisfy

$$(A - r\,I)\eta = \xi;$$

In other words, η is a generalized eigenvector of A (see page 782).

Example 15.10. *Find a fundamental set of solutions for the system*

$$\begin{cases} x_1' = 2x_1 + x_2, \\ x_2' = 2x_2, \end{cases} \tag{15.50}$$

and also the general solution.

Solution. The coefficient matrix for the system (15.50) is

$$A = \begin{pmatrix} 2 & 1 \\ 0 & 2 \end{pmatrix},$$

whose eigenvalues are obtained by solving

$$0 = \det\begin{pmatrix} 2 - r & 1 \\ 0 & 2 - r \end{pmatrix} = (2 - r)^2.$$

Thus, $r = 2$ is a double eigenvalue of A. Let us find the eigenvectors corresponding to this eigenvalue. We must solve

$$\begin{pmatrix} c_2 \\ c_1 \end{pmatrix} = \begin{pmatrix} 0 & 1 \\ 0 & 0 \end{pmatrix}\begin{pmatrix} c_1 \\ c_2 \end{pmatrix} = \begin{pmatrix} 1 \\ 0 \end{pmatrix}.$$

Thus, c_1 is arbitrary and $c_2 = 0$; it follows that any eigenvector corresponding to the eigenvalue 2 must be a multiple of

$$\xi = \begin{pmatrix} 1 \\ 0 \end{pmatrix}.$$

There is no other linearly independent eigenvector, so instead we look for a generalized eigenvector $\eta = \begin{pmatrix} \eta_1 \\ \eta_2 \end{pmatrix}$. For this we must solve the vector equation

$$(A - 2I)\eta = \xi,$$

or

$$\begin{pmatrix} 0 & 1 \\ 0 & 0 \end{pmatrix} \begin{pmatrix} \eta_1 \\ \eta_2 \end{pmatrix} = \begin{pmatrix} 1 \\ 0 \end{pmatrix},$$

which reduces to the system of equations

$$\begin{cases} \eta_2 = 1, \\ \quad 0 = 0. \end{cases}$$

Hence $\eta = \begin{pmatrix} 0 \\ 1 \end{pmatrix}$. A fundamental set of solutions for (15.50) consists of

$$\mathbf{x}^1(t) = \begin{pmatrix} 1 \\ 0 \end{pmatrix} e^{2t} \qquad \text{and} \qquad \mathbf{x}^2(t) = \begin{pmatrix} 1 \\ 0 \end{pmatrix} t\, e^{2t} + \begin{pmatrix} 0 \\ 1 \end{pmatrix} e^{2t}.$$

The general solution of (15.50) is

$$\mathbf{x}(t) = C_1 \begin{pmatrix} 1 \\ 0 \end{pmatrix} e^{2t} + C_2 \left(\begin{pmatrix} 1 \\ 0 \end{pmatrix} t\, e^{2t} + \begin{pmatrix} 0 \\ 1 \end{pmatrix} e^{2t} \right) = \begin{pmatrix} C_1 e^{2t} + C_2 t\, e^{2t} \\ C_2 e^{2t} \end{pmatrix},$$

where C_1 and C_2 are constants. ∎

Next, we outline how to find a fundamental set of solutions for $\mathbf{x}' = A\mathbf{x}$ when A is an $n \times n$ nondiagonalizable matrix. This can be done most efficiently by appealing to the Jordan canonical form (see Section A.4 of Appendix A). Thus there is an invertible matrix U and an upper triangular matrix Λ for which

$$A U = U \Lambda.$$

Therefore, the system of differential equations $\mathbf{x}'(t) = A\mathbf{x}(t)$ can be rewritten as

$$\mathbf{x}'(t) = U \Lambda U^{-1} \mathbf{x}(t). \tag{15.51}$$

In terms of a new unknown vector $\mathbf{y}(t) = U^{-1}\mathbf{x}(t)$, we can rewrite (15.51) in the upper triangular form

$$\mathbf{y}'(t) = \Lambda \mathbf{y}(t).$$

This upper triangular system can be solved by back substitution, as discussed in Section 15.3. We first solve the last equation for y_n and then substitute in the previous equation to obtain

y_{n-1}, and so forth. To examine this in detail, we consider the case of a Jordan block (see page 781).

The system corresponding to an $m \times m$ Jordan block B with eigenvalue r is

$$\mathbf{y}'(t) = \begin{pmatrix} r & 1 & 0 & \cdots & 0 & 0 \\ 0 & r & 1 & \cdots & 0 & 0 \\ \vdots & \vdots & \vdots & \ddots & \vdots & \vdots \\ 0 & 0 & 0 & \cdots & r & 1 \\ 0 & 0 & 0 & \cdots & 0 & r \end{pmatrix} \mathbf{y}(t) = B\mathbf{y}(t). \tag{15.52}$$

Written out in components, the system of equations (15.52) becomes

$$\begin{cases} y_1' = r\,y_1 + y_2, \\ y_2' = r\,y_2 + y_3, \\ \qquad\vdots \\ y_{m-1}' = r\,y_{m-1} + y_m, \\ y_m' = r\,y_m. \end{cases} \tag{15.53}$$

The system (15.53) can be solved by the techniques described in Section 15.3 to obtain the general solution as

$$\begin{cases} y_m(t) = C_m e^{rt}, \\ y_{m-1}(t) = (C_m t + C_{m-1})e^{rt}, \\ y_{m-2}(t) = \left(C_m \dfrac{t^2}{2} + C_{m-1}t + C_{m-2}\right)e^{rt}, \\ \qquad\vdots \\ y_1(t) = \left(C_{m-1}\dfrac{t^{m-1}}{m-1!} + \cdots + C_2 t + C_1\right)e^{rt}, \end{cases} \tag{15.54}$$

where $C_1, \ldots, C_m$ are arbitrary constants. We can rewrite (15.54) in terms of vectors as

$$\mathbf{y}(t) = C_1 \begin{pmatrix} e^{rt} \\ 0 \\ \vdots \\ 0 \end{pmatrix} + C_2 \begin{pmatrix} t\,e^{rt} \\ e^{rt} \\ \vdots \\ 0 \end{pmatrix} + \cdots + C_m \begin{pmatrix} \dfrac{e^{rt}t^{m-1}}{(m-1)!} \\ \vdots \\ t\,e^{rt} \\ e^{rt} \end{pmatrix}.$$

In terms of the coordinate vectors

$$\mathbf{e}_1 = \begin{pmatrix} 1 \\ 0 \\ \vdots \\ 0 \end{pmatrix}, \ldots, \mathbf{e}_m = \begin{pmatrix} 0 \\ \vdots \\ 0 \\ 1 \end{pmatrix},$$

we have

$$\mathbf{y}(t) = C_1 e^{rt}\mathbf{e}_1 + C_2 e^{rt}(t\,\mathbf{e}_1 + \mathbf{e}_2) + \cdots + C_m e^{rt}\left(\frac{t^{m-1}}{(m-1)!}\mathbf{e}_1 + \cdots + \mathbf{e}_m\right). \quad (15.55)$$

But $\mathbf{e}_1, \ldots, \mathbf{e}_n$ are the eigenvector and generalized eigenvectors of this Jordan block B; in detail,

$$\begin{cases} (B - r\,I_{m\times m})\mathbf{e}_1 = 0, \\[2mm] (B - r\,I_{m\times m})\mathbf{e}_2 = \mathbf{e}_1, \\[2mm] \qquad\qquad\vdots \\[2mm] (B - r\,I_{m\times m})\mathbf{e}_m = \mathbf{e}_{m-1}. \end{cases}$$

When we transform back to the original representation, we have the eigenvector and generalized eigenvectors of the original matrix A, written

$$\begin{cases} \boldsymbol{\xi} = U\mathbf{e}_1, \\[2mm] \boldsymbol{\eta}_1 = U\mathbf{e}_2, \\[2mm] \qquad\vdots \\[2mm] \boldsymbol{\eta}_{m-1} = U\mathbf{e}_m. \end{cases} \qquad (15.56)$$

From (15.55) and (15.56) it follows that the original unknown function $\mathbf{x}(t)$ is

$$\mathbf{x}(t) = U\mathbf{y}(t) = C_1 \boldsymbol{\xi} e^{rt} + C_2 e^{rt}(t\,\boldsymbol{\xi} + \boldsymbol{\eta}_1) + \cdots + C_m e^{rt}\left(\frac{t^{m-1}}{(m-1)!}\boldsymbol{\xi} + \cdots + \boldsymbol{\eta}_{m-1}\right),$$

where $\boldsymbol{\xi}$ is the eigenvector and $\boldsymbol{\eta}_1, \ldots, \boldsymbol{\eta}_{m-1}$ are the generalized eigenvectors corresponding to this Jordan block; in detail

$$\begin{cases} (A - r\,I_{m\times m})\boldsymbol{\xi} = 0, \\[2mm] (A - r\,I_{m\times m})\boldsymbol{\eta}_1 = \boldsymbol{\xi}, \\[2mm] \qquad\qquad\vdots \\[2mm] (A - r\,I_{m\times m})\boldsymbol{\eta}_{m-1} = \boldsymbol{\eta}_{m-2}. \end{cases}$$

We summarize the above work as follows.

Theorem 15.11. *Suppose that the eigenvalue r corresponds to a Jordan block with eigenvector $\boldsymbol{\xi}$ and generalized eigenvectors $\boldsymbol{\eta}_1, \ldots, \boldsymbol{\eta}_{m-1}$. Then a corresponding fundamental set of solutions of the system $\mathbf{x}'(t) = A\mathbf{x}(t)$ of differential equations is provided by the vector solutions*

$$\begin{cases} \mathbf{x}_1(t) = \boldsymbol{\xi}\, e^{rt}, \\[2mm] \mathbf{x}_2(t) = (t\,\boldsymbol{\xi} + \boldsymbol{\eta}_1)e^{rt}, \\[2mm] \mathbf{x}_3(t) = \left(\dfrac{t^2}{2}\boldsymbol{\xi} + t\,\boldsymbol{\eta}_1 + \boldsymbol{\eta}_2\right)e^{rt}, \\[2mm] \qquad \vdots \\[2mm] \mathbf{x}_m(t) = \left(\dfrac{t^{m-1}}{(m-1)!}\boldsymbol{\xi} + \cdots + t\,\boldsymbol{\eta}_{m-2} + \boldsymbol{\eta}_{m-1}\right)e^{rt}. \end{cases}$$

Proof. We illustrate below the computations necessary to verify that $\mathbf{x}_2(t)$ is a solution. We have

$$\mathbf{x}_2(t) = (t\,\boldsymbol{\xi} + \boldsymbol{\eta}_1)e^{rt}.$$

The derivative of $\mathbf{x}_2(t)$ is computed as

$$\mathbf{x}_2'(t) = \boldsymbol{\xi}\, e^{rt} + (t\,\boldsymbol{\xi} + \boldsymbol{\eta}_1)r\, e^{rt} = (r\,t\,\boldsymbol{\xi} + r\,\boldsymbol{\eta}_1 + \boldsymbol{\xi})e^{rt}.$$

Application of the matrix A produces

$$A\mathbf{x}_2(t) = (t\, A\boldsymbol{\xi} + A\boldsymbol{\eta}_1)e^{rt}.$$

We have $\mathbf{x}_2'(t) = A\mathbf{x}_2(t)$ provided that

$$A\boldsymbol{\xi} = r\,\boldsymbol{\xi} \qquad \text{and} \qquad A\boldsymbol{\eta}_1 = r\,\boldsymbol{\eta}_1 + \boldsymbol{\xi}.$$

The first of these equations holds, since $\boldsymbol{\xi}$ is an eigenvector with eigenvalue r. The second equation also holds, since $\boldsymbol{\eta}_1$ is a *generalized eigenvector* corresponding to the eigenvalue r.

It is left as an exercise for the reader to verify directly that the higher vector functions $\mathbf{x}_3(t), \ldots, \mathbf{x}_m(t)$ are solutions of the system. $\blacksquare$

Example 15.11. *Find the general solution to the system*

$$\begin{cases} x_1' = 2x_1 + x_2, \\[1mm] x_2' = 2x_2 + x_3, \\[1mm] x_3' = 2x_3 + x_4, \\[1mm] x_4' = 2x_4. \end{cases} \tag{15.57}$$

Solution. The coefficient matrix

$$A = \begin{pmatrix} 2 & 1 & 0 & 0 \\ 0 & 2 & 1 & 0 \\ 0 & 0 & 2 & 1 \\ 0 & 0 & 0 & 2 \end{pmatrix}$$

has 2 as a quadruple eigenvalue. The eigenvector and generalized eigenvectors of A can be chosen to be

$$\xi = \begin{pmatrix} 1 \\ 0 \\ 0 \\ 0 \end{pmatrix}, \quad \eta_1 = \begin{pmatrix} 0 \\ 1 \\ 0 \\ 0 \end{pmatrix}, \quad \eta_2 = \begin{pmatrix} 0 \\ 0 \\ 1 \\ 0 \end{pmatrix}, \quad \text{and} \quad \eta_3 = \begin{pmatrix} 0 \\ 0 \\ 0 \\ 1 \end{pmatrix}.$$

Hence the general solution to (15.57) is

$$\begin{pmatrix} x_1 \\ x_2 \\ x_3 \\ x_4 \end{pmatrix} = e^{2t} \left(C_1 \xi + C_2 (t\,\xi + \eta_1) + C_3 \left(\frac{t^2}{2} \xi + t\,\eta_1 + \eta_2 \right) \right. $$
$$\left. + C_4 \left(\frac{t^3}{6} \xi + \frac{t^2}{2} \eta_1 + t\,\eta_2 + \eta_3 \right) \right)$$

$$= e^{2t} \begin{pmatrix} C_1 + t\,C_2 + \dfrac{t^2}{2} C_3 + \dfrac{t^3}{6} C_4 \\ C_2 + t\,C_3 + \dfrac{t^2}{2} C_4 \\ C_3 + t\,C_4 \\ C_4 \end{pmatrix}. \ \blacksquare$$

15.5.4   *The Fundamental Matrix by an Exponential Formula*

In this subsection we outline an alternative procedure for finding a fundamental matrix for a linear system of first-order differential equations with constant-coefficients. Fundamental matrices are useful for solving inhomogeneous systems, as we shall see in Section 15.7. The procedure does not involve explicit computation of the eigenvalues, eigenvectors, and generalized eigenvectors. Its practical success depends upon the ability to manipulate infinite series of matrices. For more information on the exponential of a matrix see

Section A.5 of Appendix A and Exercise 26, which allows one to compute the matrix exponential from a fundamental matrix.

Lemma 15.12. *A fundamental matrix for the constant-coefficient system* $\mathbf{x}' = A\mathbf{x}$ *may be obtained concisely as the matrix exponential*

$$\Phi(t) = e^{tA} = \sum_{k=0}^{\infty} \frac{t^k A^k}{k!}. \tag{15.58}$$

Proof. Equation (15.58) follows from the fact that

$$\Phi(0) = e^0 = I_{n \times n}$$

and the differentiation formula

$$\frac{d}{dt} e^{tA} = A\, e^{tA}.$$

The columns of $\Phi(t)$ are solutions of the equation

$$\frac{d\mathbf{x}}{dt} = A\mathbf{x},$$

and for $t = 0$ they are linearly independent. Therefore, they are linearly independent for all $t \neq 0$ as well by Theorem 15.6. ∎

Example 15.12. *Find the fundamental matrix for the system* $\mathbf{x}'(t) = A\mathbf{x}(t)$ *when*

$$A = \begin{pmatrix} 0 & -3 \\ 3 & 0 \end{pmatrix}.$$

Solution. The exponential of the matrix $t A$ is

$$e^{tA} = \begin{pmatrix} \cos(3t) & -\sin(3t) \\ \sin(3t) & \cos(3t) \end{pmatrix}.$$

This is the required fundamental matrix. ∎

The structure of the fundamental matrix is especially simple if A is presented in Jordan canonical form. Computation of e^{tA} reduces to consideration of the block matrices B_i. In the case of a 1×1 block $B_i = r_i I_{1 \times 1}$, it is immediate that $e^{tB_i} = e^{r_i t} I_{1 \times 1}$. In the case of the other Jordan blocks we must compute higher powers.

Example 15.13. *Find the fundamental matrix in the case that*

$$A = \begin{pmatrix} \lambda & 1 & 0 \\ 0 & \lambda & 1 \\ 0 & 0 & \lambda \end{pmatrix},$$

where λ *is real.*

Solution. We must compute the higher powers of the matrix A. These are found directly as follows:

$$A^2 = \begin{pmatrix} \lambda^2 & 2\lambda & 1 \\ 0 & \lambda^2 & 2\lambda \\ 0 & 0 & \lambda^2 \end{pmatrix} \qquad \text{and} \qquad A^3 = \begin{pmatrix} \lambda^3 & 3\lambda^2 & 3\lambda \\ 0 & \lambda^3 & 3\lambda^2 \\ 0 & 0 & \lambda^3 \end{pmatrix}.$$

In computing the higher powers we find that the diagonal elements of A^k are the powers λ^k and the off-diagonal elements are $k\,\lambda^{k-1}$ and $\left(k(k-1)/2\right)\lambda^{k-2}$. This leads to the result

$$e^{tA} = \begin{pmatrix} e^{\lambda t} & t\,e^{\lambda t} & (t^2/2)e^{\lambda t} \\ 0 & e^{\lambda t} & t\,e^{\lambda t} \\ 0 & 0 & e^{\lambda t} \end{pmatrix}. \quad \blacksquare$$

Summary of Techniques Introduced

In this section we solved the constant-coefficient linear system $\mathbf{x}'(t) = A\mathbf{x}(t)$. In the case that the matrix A is diagonalizable, a fundamental set of solutions is obtained in terms of the eigenvalues and eigenvectors through the formula $\mathbf{x}(t) = \boldsymbol{\xi} e^{\lambda t}$. In the case that the matrix A is not diagonalizable, additional linearly independent solutions can be obtained in terms of the eigenvalues and *generalized eigenvectors*. In every case the fundamental matrix may be expressed through the matrix exponential as $\Phi(t) = e^{tA}$. Its columns are linearly independent solutions of the linear system $\mathbf{x}'(t) = A\mathbf{x}(t)$.

In the case of a real matrix A, the real-valued solutions can be obtained as the real and imaginary parts of the complex solutions, which occur in complex conjugate pairs.

Exercises for Section 15.5

Find a fundamental set of solutions and the fundamental matrix for each of the following two-dimensional homogeneous systems of linear differential equations.

1. $\begin{cases} x_1' = 3x_1 + 3x_2, \\ x_2' = x_1 + 5x_2 \end{cases}$

4. $\begin{cases} x_1' = x_1 - x_2, \\ x_2' = x_1 + x_2 \end{cases}$

2. $\begin{cases} x_1' = -6x_1 + 4x_2, \\ x_2' = -4x_1 + 2x_2 \end{cases}$

5. $\begin{cases} x_1' = 3x_1 + 3x_2, \\ x_2' = x_1 \end{cases}$

3. $\begin{cases} x_1' = 4x_1 + 2x_2, \\ x_2' = -x_1 + x_2 \end{cases}$

6. $\begin{cases} x_1' = 4x_1 + 4x_2, \\ x_2' = -4x_1 - 4x_2 \end{cases}$

Find a fundamental set of solutions and the fundamental matrix for each of the following three-dimensional homogeneous systems of linear differential equations.

7.
$$\begin{cases} x_1' = x_1, \\ x_2' = 2x_1 + x_2 - 2x_3, \\ x_3' = 3x_1 + 2x_2 + x_3 \end{cases}$$

10.
$$\begin{cases} x_1' = x_1 + 2x_2 - 3x_3, \\ x_2' = x_1 + x_2 + 2x_3, \\ x_3' = x_1 - x_2 + 4x_3 \end{cases}$$

8.
$$\begin{cases} x_1' = 3x_1 + 2x_2 + 4x_3, \\ x_2' = 2x_1 + 2x_3, \\ x_3' = 4x_1 + 2x_2 + 3x_3 \end{cases}$$

11.
$$\begin{cases} x_1' = x_1 + x_2, \\ x_2' = x_2, \\ x_3' = 3x_3 \end{cases}$$

9.
$$\begin{cases} x_1' = -2x_1 + x_2 + x_3, \\ x_2' = x_1 - 2x_2 + x_3, \\ x_3' = x_1 + x_2 - 2x_3 \end{cases}$$

12.
$$\begin{cases} x_1' = x_1 + x_2 - x_3, \\ x_2' = -x_1 - x_2, \\ x_3' = x_3 \end{cases}$$

Solve the following two-dimensional initial value problems.

13.
$$\begin{cases} x_1' = 3x_1 + 3x_2, & x_1(0) = 3, \\ x_2' = x_1 + 5x_2, & x_2(0) = 5 \end{cases}$$

16.
$$\begin{cases} x_1' = x_1 - x_2, & x_1(0) = 4, \\ x_2' = x_1 + x_2, & x_2(0) = 5 \end{cases}$$

14.
$$\begin{cases} x_1' = -6x_1 + 4x_2, & x_1(0) = 1, \\ x_2' = -4x_1 + 2x_2, & x_2(0) = 3 \end{cases}$$

17.
$$\begin{cases} x_1' = 3x_1 + 3x_2, & x_1(0) = 0 \\ x_2' = x_1, & x_2(0) = 4 \end{cases}$$

15.
$$\begin{cases} x_1' = 4x_1 + 2x_2, & x_1(0) = 2, \\ x_2' = -x_1 + x_2, & x_2(0) = -1 \end{cases}$$

18.
$$\begin{cases} x_1' = 4x_1 + 4x_2, & x_1(0) = 3 \\ x_2' = -4x_1 - 4x_2, & x_2(0) = 5 \end{cases}$$

Solve the following three-dimensional initial value problems.

19.
$$\begin{cases} x_1' = x_1, \\ x_2' = 2x_1 + x_2 - 2x_3, \\ x_3' = 3x_1 + 2x_2 + x_3, \\ x_1(0) = 4, x_2(0) = 3, x_3(0) = 1 \end{cases}$$

20.
$$\begin{cases} x_1' = 3x_1 + 2x_2 + 4x_3, \\ x_2' = 2x_1 + 2x_3, \\ x_3' = 4x_1 + 2x_2 + 3x_3, \\ x_1(0) = 2, x_2(0) = 5, x_3(0) = 6 \end{cases}$$

21.
$$\begin{cases} x_1' = -2x_1 + x_2 + x_3, \\ x_2' = x_1 - 2x_2 + x_3, \\ x_3' = x_1 + x_2 - 2x_3, \\ x_1(0) = 3, x_2(0) = 5, x_3(0) = 0 \end{cases}$$

23.
$$\begin{cases} x_1' = x_1 + x_2, \\ x_2' = x_2, \\ x_3' = 3x_3, \\ x_1(0) = 3, x_2(0) = 2, x_3(0) = 1 \end{cases}$$

22.
$$\begin{cases} x_1' = x_1 + 2x_2 - 3x_3, \\ x_2' = x_1 + x_2 + 2x_3, \\ x_3' = x_1 - x_2 + 4x_3, \\ x_1(0) = 1, x_2(0) = 2, x_3(0) = 3 \end{cases}$$

24.
$$\begin{cases} x_1' = x_1 + x_2 - x_3, \\ x_2' = -x_1 - x_2, \\ x_3' = x_3, \\ x_1(0) = 8, x_2(0) = 4, x_3(0) = 2 \end{cases}$$

25. Suppose $\mathbf{x}^1$ and $\mathbf{x}^2$ are complex conjugate solutions to $\mathbf{x}' = A\mathbf{x}$, where A is a real matrix. Let $\mathbf{u}(t) = \mathfrak{Re}(\mathbf{x}^1(t))$ and $\mathbf{v}(t) = \mathfrak{Im}(\mathbf{x}^1(t))$. Show that $\mathbf{u}$ and $\mathbf{v}$ are real solutions of $\mathbf{x}' = A\mathbf{x}$.

26. In Lemma 15.12 it was shown that the matrix exponential e^{tA} is a fundamental matrix for the linear system $d\mathbf{x}/dt = A\mathbf{x}$. Suppose that $\mathbf{\Phi}(t)$ is *any* fundamental matrix for this system. Prove that the matrix exponential can be recovered from the formula

$$e^{tA} = \mathbf{\Phi}(t)\mathbf{\Phi}(0)^{-1}.$$

[Hint: Both sides of the proposed equation satisfy the same first-order linear system of equations; now appeal to the uniqueness theorem (Theorem 15.1) for first-order linear systems.]

27. Suppose that the matrix A is diagonalizable with eigenvalues $r_1, \ldots, r_n$ and eigenvectors $\boldsymbol{\xi}_1, \ldots, \boldsymbol{\xi}_n$.

    a. Show that a fundamental matrix is obtained in the form

$$\mathbf{\Phi}(t) = \begin{pmatrix} \boldsymbol{\xi}_1 e^{r_1 t} & \cdots & \boldsymbol{\xi}_n e^{r_n t} \end{pmatrix},$$

where the columns are the respective solutions $\boldsymbol{\xi}_j e^{r_j t}$.

    b. Under what condition on the eigenvalues and eigenvectors of a 2×2 matrix A can we be assured that the above fundamental matrices satisfy the commutativity property $\mathbf{\Phi}(t)\mathbf{\Phi}(s) = \mathbf{\Phi}(s)\mathbf{\Phi}(t)$ for all s and t?

c. Show by example that the commutativity property does not hold for the fundamental matrix

$$\begin{pmatrix} 2e^t & 3e^{-t} \\ e^t & 2e^{-t} \end{pmatrix}$$

belonging to the system

$$\frac{d\mathbf{x}}{dt} = \begin{pmatrix} 7 & -12 \\ 4 & -7 \end{pmatrix} \mathbf{x}.$$

d. Suppose that $\mathbf{\Phi}(t)$ is *any* fundamental matrix for the linear system $d\mathbf{x}/dt = A\mathbf{x}$. Show that $\mathbf{\Phi}(t) = e^{At}\mathbf{\Phi}(0)$. [Hint: Apply the uniqueness theorem (Theorem 15.1) to the columns of both sides of the asserted equation after checking that both sides satisfy the linear system with the same initial conditions.]

28. Let $\mathbf{\Phi}(t)$ be any fundamental matrix for the linear system $d\mathbf{x}/dt = A\mathbf{x}$. Show that the matrix exponential can be recovered through the formula

$$\mathbf{\Phi}(t)\mathbf{\Phi}(s)^{-1} = e^{(t-s)A}$$

for any real numbers s and t. [Hint: Use the result of Problem 27.]

15.6   *The Method of Undetermined Coefficients for Systems*

In Section 15.5 we showed how to solve the constant-coefficient linear system

$$\frac{d\mathbf{x}}{dt} = A\mathbf{x}.$$

In this section we consider the general inhomogeneous system

$$\frac{d\mathbf{x}}{dt} = P(t)\mathbf{x}(t) + \mathbf{g}(t). \tag{15.59}$$

As in the case of a single second-order equation, we can try to find a solution to (15.59) by the method of **undetermined coefficients** or by the method of **variation of parameters**. The method of undetermined coefficients, which we discuss in the present section, is applicable in the case when the matrix $P(t)$ is a constant matrix and the components of $\mathbf{g}(t)$ have a special form, namely polynomial, exponential, or sinusoidal functions or the sums and products of such functions.

The method of variation of parameters, discussed in Section 15.7, works for a general system in which $P(t)$ is a possibly nonconstant matrix-valued function and $\mathbf{g}(t)$ is a vector-valued function.

Let A be a constant $n \times n$ matrix and $\mathbf{v}$ a column vector in $\mathbb{R}^n$. We look for a particular solution to the system

$$\frac{d\mathbf{x}}{dt} = A\mathbf{x} + \mathbf{v}\,e^{\alpha t}, \tag{15.60}$$

where $\mathbf{v}$ is a constant vector and α may be any complex number. (This will include the case of *trigonometric right-hand side* by taking α to be a purely imaginary number.) A particular solution to (15.60) is sought in the form

$$\mathbf{x}_{\mathrm{P}}(t) = \mathbf{w}\,e^{\alpha t}, \tag{15.61}$$

where $\mathbf{w}$ is a constant vector and we use the *same* exponent α that appears on the right-hand side of (15.60). Substitution of (15.61) into (15.60) gives the condition

$$e^{\alpha t}\alpha\,\mathbf{w} = e^{\alpha t}\left(A\mathbf{w} + \mathbf{v}\right). \tag{15.62}$$

We cancel the exponential factors from both sides of (15.62) and rewrite it as the matrix equation

$$(\alpha\,I_{n\times n} - A)\mathbf{w} = \mathbf{v}. \tag{15.63}$$

If α is not an eigenvalue of the matrix A, then (15.63) has a unique solution vector $\mathbf{w}$. This is the

Nonresonant case: α is not an eigenvalue of the matrix A

A particular solution to (15.60) is obtained in the form (15.61), where $\mathbf{w}$ is the solution to the matrix equation (15.63).

Example 15.14. *Find a particular solution to the system*

$$\frac{d\mathbf{x}}{dt} = \begin{pmatrix} 2 & 3 \\ 3 & 4 \end{pmatrix}\mathbf{x} + \begin{pmatrix} 4 \\ 6 \end{pmatrix}e^{3t}. \tag{15.64}$$

Solution. The eigenvalues of the matrix

$$A = \begin{pmatrix} 2 & 3 \\ 3 & 4 \end{pmatrix}$$

are easily computed to be $\alpha = 3 \pm \sqrt{10}$ (see Example A.7 of Appendix A); hence 3 is not an eigenvalue of A. Writing $\mathbf{w} = \begin{pmatrix} w_1 \\ w_2 \end{pmatrix}$, we must solve the matrix equation

$$(3I_{2\times 2} - A)\mathbf{w} = \begin{pmatrix} 1 & -3 \\ -3 & -1 \end{pmatrix} \begin{pmatrix} w_1 \\ w_2 \end{pmatrix} = \begin{pmatrix} 4 \\ 6 \end{pmatrix}. \tag{15.65}$$

The solution to (15.65) found by Cramer's rule is

$$w_1 = -7/5 \quad \text{and} \quad w_2 = -9/5;$$

thus, a particular solution to (15.64) is

$$\mathbf{x}_{\mathrm{P}}(t) = \begin{pmatrix} -7/5 \\ -9/5 \end{pmatrix} e^{3t}. \ \blacksquare$$

We now pass to the various resonant cases. If α is an eigenvalue of the matrix A, then the matrix $\alpha I_{n\times n} - A$ is singular, and we cannot in general find a solution in the form (15.61). In order to find a suitable particular solution, we modify (15.61) by the addition of polynomial terms. This leads to the

Resonant case: α is an eigenvalue of the diagonalizable matrix A

In this case we look for a particular solution to (15.60) in the form

$$\mathbf{x}_{\mathrm{P}}(t) = (t\,\boldsymbol{\xi} + \mathbf{w})e^{\alpha t}. \tag{15.66}$$

By substituting (15.66) into the system (15.60), we obtain the equation

$$e^{\alpha t}\big(\boldsymbol{\xi} + \alpha(t\,\boldsymbol{\xi} + \mathbf{w})\big) = e^{\alpha t}\big(t\,A\boldsymbol{\xi} + A\mathbf{w} + \mathbf{v}\big),$$

and by canceling the common factor $e^{\alpha t}$, we find

$$\boldsymbol{\xi} + \alpha(t\,\boldsymbol{\xi} + \mathbf{w}) = t\,A\boldsymbol{\xi} + A\mathbf{w} + \mathbf{v}. \tag{15.67}$$

Equation (15.67) is satisfied for all t if and only if the coefficients of t agree and the constant terms agree. This entails the two vector equations

$$\begin{cases} \alpha\boldsymbol{\xi} = A\boldsymbol{\xi}, \\ \boldsymbol{\xi} + \alpha\mathbf{w} = A\mathbf{w} + \mathbf{v}. \end{cases} \tag{15.68}$$

The first equation of (15.68) states that either $\boldsymbol{\xi} = 0$ or $\boldsymbol{\xi}$ is an eigenvector corresponding to the eigenvalue α, while the second equation requires that the vector $\mathbf{w}$ be determined as a solution to the matrix equation

$$(\alpha I_{n\times n} - A)\mathbf{w} = \mathbf{v} - \boldsymbol{\xi}. \tag{15.69}$$

It is shown at the end of this section that (15.69) can always be solved if A is diagonalizable—for example if A has distinct eigenvalues or is a symmetric matrix. We emphasize that in the practical calculations to determine $\boldsymbol{\xi}$ and $\mathbf{w}$, we must use *both* equations of (15.68). In some cases we may in fact need to take $\boldsymbol{\xi} = 0$.

The following examples illustrate three different possibilities for the form of the solution:

(i) $\boldsymbol{\xi} = 0$ and $\mathbf{w} \neq 0$;

(ii) $\boldsymbol{\xi} \neq 0$ and $\mathbf{w} = 0$;

(iii) $\boldsymbol{\xi} \neq 0$ and $\mathbf{w} \neq 0$.

Example 15.15.   *Find a particular solution to the system*

$$\mathbf{x}'(t) = \begin{pmatrix} 7 & -2 \\ 2 & 2 \end{pmatrix} \mathbf{x} + \begin{pmatrix} 6 \\ 3 \end{pmatrix} e^{3t}. \tag{15.70}$$

Solution. The eigenvalues of the matrix

$$A = \begin{pmatrix} 7 & -2 \\ 2 & 2 \end{pmatrix}$$

are obtained by solving the associated characteristic equation

$$0 = \det \begin{pmatrix} 7 - \alpha & -2 \\ 2 & 2 - \alpha \end{pmatrix} = \alpha^2 - 9\alpha + 18 = (\alpha - 3)(\alpha - 6). \tag{15.71}$$

Thus, the eigenvalues of A are 3 and 6. The right-hand side of (15.70) contains the factor e^{3t}, which corresponds to the eigenvalue $\alpha = 3$. An eigenvector corresponding to 3 can be chosen to be $\begin{pmatrix} 1 \\ 2 \end{pmatrix}$; thus $\boldsymbol{\xi} = \begin{pmatrix} c \\ 2c \end{pmatrix}$ for some possibly zero constant c. The matrix equation (15.69) becomes

$$(3I - A)\mathbf{w} = \begin{pmatrix} -4 & 2 \\ -2 & 1 \end{pmatrix} \mathbf{w} = \mathbf{v} - \boldsymbol{\xi} = \begin{pmatrix} 6 \\ 3 \end{pmatrix} - \begin{pmatrix} c \\ 2c \end{pmatrix}. \tag{15.72}$$

Write $\mathbf{w} = \begin{pmatrix} w_1 \\ w_2 \end{pmatrix}$; the system of linear equations corresponding to (15.72) is

$$\begin{cases} -4w_1 + 2w_2 = 6 - c, \\ -2w_1 + w_2 = 3 - 2c. \end{cases} \tag{15.73}$$

Multiplying the second equation of (15.73) by 2 and subtracting from the first equation, we see that $c = 0$. Thus (15.73) reduces to the single equation

$$-2w_1 + w_2 = 3. \tag{15.74}$$

One solution to (15.74) is $w_1 = -2$ and $w_2 = -1$. Thus, we get the particular solution

$$\mathbf{x}_p(t) = \begin{pmatrix} -2 \\ -1 \end{pmatrix} e^{3t}$$

to (15.70). ∎

   The differential equation of Example 15.15 is "accidentally" in resonance. The vector $\mathbf{v}$ is in the range of $\alpha I - A$, so that we obtain the solution as in the nonresonant case when α is not an eigenvalue. The next example illustrates the other extreme: $\boldsymbol{\xi} \neq 0$ and $\mathbf{w} = 0$.

Example 15.16. *Find a particular solution to the system*

$$\mathbf{x}'(t) = \begin{pmatrix} 7 & -2 \\ 2 & 2 \end{pmatrix} \mathbf{x} + \begin{pmatrix} 2 \\ 4 \end{pmatrix} e^{3t}. \tag{15.75}$$

Solution. The eigenvalues of the matrix A were obtained as in Example 15.15 as 3 and 6. Also, we found that an eigenvector corresponding to 3 can be chosen to be $\begin{pmatrix} 1 \\ 2 \end{pmatrix}$; thus $\boldsymbol{\xi} = \begin{pmatrix} c \\ 2c \end{pmatrix}$ for some possibly zero constant c. The matrix equation (15.69) in the special case of (15.75) becomes

$$(3I - A)\mathbf{w} = \begin{pmatrix} -4 & 2 \\ -2 & 1 \end{pmatrix} \mathbf{w} = \mathbf{v} - \boldsymbol{\xi} = \begin{pmatrix} 2 \\ 4 \end{pmatrix} - \begin{pmatrix} c \\ 2c \end{pmatrix}. \tag{15.76}$$

The system of linear equations corresponding to (15.76) is

$$\begin{cases} -4w_1 + 2w_2 = 2 - c, \\ -2w_1 + w_2 = 4 - 2c, \end{cases} \tag{15.77}$$

where $\mathbf{w} = \begin{pmatrix} w_1 \\ w_2 \end{pmatrix}$. Multiplying the second equation of (15.77) by 2 and subtracting from the first equation, we see that $c = 2$. Thus (15.77) reduces to the single equation

$$-2w_1 + w_2 = 0. \tag{15.78}$$

One solution to (15.78) occurs when $w_1 = w_2 = 0$. Thus, we get the particular solution

$$\mathbf{x}_p(t) = \begin{pmatrix} 2t \\ 4t \end{pmatrix} e^{3t}$$

of (15.75). ∎

The third example illustrates the possibility $\boldsymbol{\xi} \neq 0$ and $\mathbf{w} \neq 0$.

Example 15.17. *Find a particular solution to the system*

$$\mathbf{x}'(t) = \begin{pmatrix} 7 & -2 \\ 2 & 2 \end{pmatrix} \mathbf{x} + \begin{pmatrix} -3 \\ 3 \end{pmatrix} e^{3t}. \tag{15.79}$$

Solution. The eigenvalues of the matrix A were obtained as in Example 15.15 as 3 and 6. Also, we found that an eigenvector corresponding to 3 can be chosen to be $\begin{pmatrix} 1 \\ 2 \end{pmatrix}$; thus $\boldsymbol{\xi} = \begin{pmatrix} c \\ 2c \end{pmatrix}$ for some possibly zero constant c. The matrix equation (15.69) in the special case of (15.79) becomes

$$(3I - A)\mathbf{w} = \begin{pmatrix} -4 & 2 \\ -2 & 1 \end{pmatrix} \mathbf{w} = \mathbf{v} - \boldsymbol{\xi} = \begin{pmatrix} -3 \\ 3 \end{pmatrix} - \begin{pmatrix} c \\ 2c \end{pmatrix}. \tag{15.80}$$

The system of linear equations corresponding to (15.80) is

$$\begin{cases} -4w_1 + 2w_2 = -3 - c, \\ -2w_1 + w_2 = 3 - 2c. \end{cases} \tag{15.81}$$

Multiplying the second equation of (15.81) by 2 and subtracting from the first equation, we see that $c = 3$. Thus (15.81) reduces to the single equation

$$-2w_1 + w_2 = -3. \tag{15.82}$$

One solution to (15.82) is $w_1 = 2$ and $w_2 = 1$. Thus, we get the particular solution

$$\mathbf{x}_p(t) = \begin{pmatrix} 3t + 2 \\ 6t + 1 \end{pmatrix} e^{3t}$$

of (15.79). ∎

Cases of higher resonance

If the matrix A is nondiagonalizable, then one must resort to additional polynomial terms in order to obtain a particular solution to the differential system. In this case, a particular solution to (15.60) is obtained for some k in the form

$$\mathbf{x}_P(t) = \left(t^k \boldsymbol{\xi} + t^{k-1} \mathbf{w}_1 + \cdots + \mathbf{w}_k\right) e^{\alpha t},$$

where $\boldsymbol{\xi}$ is an eigenvector corresponding to the eigenvalue α, and the generalized eigenvectors $\mathbf{w}_1, \ldots, \mathbf{w}_{k-1}$ are determined by solving the equations

$$\begin{cases}
(\alpha I_{n\times n} - A)\boldsymbol{\xi} & = 0, \\
(\alpha I_{n\times n} - A)\mathbf{w}_1 + k\boldsymbol{\xi} & = 0, \\
(\alpha I_{n\times n} - A)\mathbf{w}_2 + (k-1)\mathbf{w}_1 & = 0, \\
\qquad\qquad\qquad\vdots & \\
(\alpha I_{n\times n} - A)\mathbf{w}_k + \mathbf{w}_{k-1} & = \mathbf{v}.
\end{cases} \qquad (15.83)$$

The following example illustrates some of the various possibilities.

Example 15.18. *Find a particular solution of the system*

$$\begin{cases} x_1' = x_1 + 2x_2 + v_1 e^t, \\ x_2' = x_2 + v_2 e^t, \end{cases} \qquad (15.84)$$

where v_1 and v_2 are constants.

Solution. We remove the exponential terms in (15.84) by writing $x_1 = y_1 e^t$ and $x_2 = y_2 e^t$ and obtain the system

$$\begin{cases} y_1' = 2y_2 + v_1, \\ y_2' = v_2. \end{cases} \qquad (15.85)$$

The solution to the second equation of (15.85) is $y_2 = v_2 t$. This allows us to solve the first equation of (15.85) as $y_1 = v_2 t^2 + v_1 t$. Hence

$$x_1 = t^2 e^t v_2 + t e^t v_1 \qquad \text{and} \qquad x_2 = t e^t v_2.$$

Thus, we can write a particular solution to (15.84) as

$$\mathbf{x}_P(t) = t^2 e^t \begin{pmatrix} v_2 \\ 0 \end{pmatrix} + t e^t \begin{pmatrix} v_1 \\ v_2 \end{pmatrix}.$$

The general solution to (15.84) is

$$\mathbf{x}(t) = t^2 e^t \begin{pmatrix} v_2 \\ 0 \end{pmatrix} + t e^t \begin{pmatrix} v_1 \\ v_2 \end{pmatrix} + C_1 e^t \begin{pmatrix} 1 \\ 0 \end{pmatrix} + C_2 e^t \begin{pmatrix} 1 \\ 1/2 \end{pmatrix},$$

where C_1 and C_2 are constants. ∎

This example clearly illustrates that in the case that $v_2 \neq 0$ we must choose $k = 2$ in the proposed solution, whereas if $v_2 = 0$, we may obtain the particular solution by taking $k = 1$.

For inhomogeneous systems of three or more dimensions, it may be necessary to take higher values of k in order to find a particular solution. We illustrate with a three-dimensional example.

Example 15.19. *Find a particular solution to the system*

$$\begin{cases} x_1' = x_2 + 1, \\ x_2' = x_3, \\ x_3' = 2 \end{cases} \tag{15.86}$$

by the method of undetermined coefficients.

Solution. We rewrite (15.86) in matrix form:

$$\begin{pmatrix} x_1 \\ x_2 \\ x_3 \end{pmatrix}' = A \begin{pmatrix} x_1 \\ x_2 \\ x_3 \end{pmatrix} + \mathbf{v}, \tag{15.87}$$

where

$$A = \begin{pmatrix} 0 & 1 & 0 \\ 0 & 0 & 1 \\ 0 & 0 & 0 \end{pmatrix} \qquad \text{and} \qquad \mathbf{v} = \begin{pmatrix} 1 \\ 0 \\ 2 \end{pmatrix}.$$

The matrix A has all of its eigenvalues equal to zero, so we look for the particular solution to (15.60) in the form

$$\mathbf{x}_p(t) = t^3 \boldsymbol{\xi} + t^2 \mathbf{w}_1 + t \mathbf{w}_2 + \mathbf{w}_3, \tag{15.88}$$

where the vector $\boldsymbol{\xi}$ in (15.88) is an eigenvector of A; thus

$$\boldsymbol{\xi} = \begin{pmatrix} c \\ 0 \\ 0 \end{pmatrix}.$$

for some nonzero constant c. We have

$$\mathbf{v} = \mathbf{x}'_p - A\mathbf{x}_p = t^2(3\boldsymbol{\xi} - A\mathbf{w}_1) + t(2\mathbf{w}_1 - A\mathbf{w}_2) + (\mathbf{w}_2 - A\mathbf{w}_3);$$

hence the vectors $\mathbf{w}_1$, $\mathbf{w}_2$, and $\mathbf{w}_3$ in (15.88) must satisfy the equations

$$\begin{cases} A\mathbf{w}_1 = 3\boldsymbol{\xi}, \\ A\mathbf{w}_2 = 2\mathbf{w}_1, \\ A\mathbf{w}_3 = \mathbf{w}_2 - \mathbf{v}. \end{cases} \tag{15.89}$$

These are solved to yield

$$\boldsymbol{\xi} = \begin{pmatrix} 1/3 \\ 0 \\ 0 \end{pmatrix}, \qquad \mathbf{w}_1 = \begin{pmatrix} 0 \\ 1 \\ 0 \end{pmatrix}, \qquad \text{and} \qquad \mathbf{w}_2 = \begin{pmatrix} 1 \\ 0 \\ 2 \end{pmatrix}.$$

Hence we obtain the particular solution to (15.86) in the form

$$\mathbf{x}_p(t) = \begin{pmatrix} t^3/3 + t \\ t^2 \\ 2t \end{pmatrix}.$$

The general solution to (15.88) is

$$\mathbf{x}(t) = \begin{pmatrix} t^3/3 + t \\ t^2 \\ 2t \end{pmatrix} + C_1 \begin{pmatrix} 1 \\ 0 \\ 0 \end{pmatrix} + C_2 \begin{pmatrix} t \\ 1 \\ 0 \end{pmatrix} + C_2 \begin{pmatrix} t^2/2 \\ t \\ 1 \end{pmatrix},$$

where C_1, C_2, and C_3 are constants. ∎

We provide an additional example of the general situation for two-dimensional systems.

Example 15.20. *Find a particular solution of the system*

$$\frac{d\mathbf{x}}{dt} = \begin{pmatrix} 1 & 3 \\ 0 & 2 \end{pmatrix}\mathbf{x} + \begin{pmatrix} -2 \\ -2 \end{pmatrix}e^t \tag{15.90}$$

by the method of undetermined coefficients.

Solution. The matrix

$$A = \begin{pmatrix} 1 & 3 \\ 0 & 2 \end{pmatrix}$$

has eigenvalues $\alpha = 1$ and $\alpha = 2$, the first of which corresponds to the exponential on the right-hand side of (15.90). The eigenvector corresponding to the eigenvalue $\alpha = 1$ is of the form

$$\xi = \begin{pmatrix} c \\ 0 \end{pmatrix},$$

where c is a nonzero constant. The vector $\mathbf{v}$ is *not* in the range of $A - I_{2\times 2}$. With the choice $c = 4$ we have that $\mathbf{v} - \xi$ *is* in the range of $A - I_{2\times 2}$. We look for a particular solution to (15.90) in the form (15.66) with $\alpha = 1$. The vector $\mathbf{w}$ is determined as

$$\mathbf{w} = \begin{pmatrix} 6 \\ 2 \end{pmatrix},$$

and the particular solution is

$$\mathbf{x}_{\mathrm{P}}(t) = \begin{pmatrix} 4t + 6 \\ 2 \end{pmatrix} e^t. \ \blacksquare$$

In passing, we note that for *practical* purposes, the method in the first nonresonant case may be applied in greater generality than is apparent at first. A given matrix A has only a finite number of eigenvalues. If the right-hand sides of a system of differential equations contain the eigenvalue in the exponential $e^{\alpha t}$, then by slightly changing the value of α, we can *avoid* coincidence with an eigenvalue and fall into the case of nonresonance. If we are interested in the behavior of the system over a relatively short period of time, then such a small change of the exponent will not make an appreciable difference in the value of the solution. If we are interested in the behavior of the system over a very long period of time, then a slight change in the exponent can make an enormous difference in the solution and should be applied with great care.

The method of undetermined coefficients applies only when the right-hand side of the system is of a particular form. It is often necessary to treat right-hand sides that are of a more general form. Therefore, we turn in the next section to more systematic methods that simultaneously treat more general right-hand sides.

Proof of solvability

In this paragraph we give the proof of the following statement about matrix equations, which was used in the discussion of equations (15.68) for the resonant case of the method of undetermined coefficients:

If A is a diagonalizable matrix with eigenvalue α and $\mathbf{v} \in R^n$, then we can find vectors ξ, $\mathbf{w}$ that satisfy the pair of equations

$$(\alpha I - A)\xi = 0, \qquad\qquad (\alpha I - A)\mathbf{w} = \mathbf{v} - \xi.$$

To prove this, we first use the hypothesis of diagonalizability to write $A = U\Lambda U^{-1}$, where Λ is a diagonal matrix and U is an invertible matrix. The diagonal elements of Λ are the eigenvalues of A, and the columns of U are the eigenvectors of A. We define $\mathbf{v}' = U^{-1}\mathbf{v}$ and two new (unknown) vectors: $\xi' = U^{-1}\xi$ and $\mathbf{w}' = U^{-1}\mathbf{w}$. The given system of equations is now transformed into

$$(\alpha I - \Lambda)\xi' = 0, \qquad\qquad (\alpha I - \Lambda)\mathbf{w}' = \mathbf{v}' - \xi'.$$

The first equation requires that the components of the vector ξ' be zero except for those corresponding to the values of j for which $\Lambda_{jj} = \alpha$. The second equation is written in detail as $(\alpha - \Lambda_{jj})w_j' = v_j' - \xi_j'$. If $\Lambda_{jj} = \alpha$, then w_j is indeterminate, but we must have $\xi_j' = v_j'$. If $\Lambda_{jj} \neq \alpha$ then we can solve to obtain $w_j' = v_j/(\alpha - \Lambda_{jj})$. Therefore, a particular solution is obtained as

$$\xi_j' = v_j' \quad \text{if} \quad \Lambda_{jj} = \alpha, \qquad \xi_j' = 0 \quad \text{otherwise;}$$

$$w_j' = 0 \quad \text{if} \quad \Lambda_{jj} = \alpha, \qquad w_j' = \frac{v_j'}{(\alpha - \Lambda_{jj})} \quad \text{otherwise.}$$

The sought-after vectors ξ, $\mathbf{w}$ are obtained as $\xi = U\xi'$, $\mathbf{w} = U\mathbf{w}'$. $\blacksquare$

We note in passing that the above equations *cannot* be solved, in general, if the matrix A is not diagonalizable. For example, if we have

$$A = \begin{pmatrix} 0 & 1 & 0 \\ 0 & 0 & 1 \\ 0 & 0 & 0 \end{pmatrix}, \qquad \mathbf{v} = \begin{pmatrix} 0 \\ 0 \\ 1 \end{pmatrix}, \qquad \alpha = 0,$$

then any eigenvector of A must be of the form $\xi = (c, 0, 0)^T$ for some constant c, and the equations for the vector $\mathbf{w}$ take the form $w_2 = -c$, $w_3 = 0, 0 = 1$, which is impossible.

Exercises for Section 15.6

Find a particular solution to each of the following systems by the method of undetermined coefficients.

1. $\begin{cases} x_1' = x_1 + 4x_2 - e^t, \\ x_2' = x_1 + x_2 + 2e^t \end{cases}$

5. $\begin{cases} x_1' = -2x_1 + x_2 + t, \\ x_2' = -5x_1 + 2x_2 - \sin(2t) \end{cases}$

2. $\begin{cases} x_1' = -2x_1 + 3x_2 - e^t, \\ x_2' = -x_1 + 2x_2 + e^t \end{cases}$

6. $\begin{cases} x_1' = -x_1 + 4x_2 + 2e^{-t}\cos(t), \\ x_2' = -2x_1 + 3x_2 - e^{-t}\sin(t) \end{cases}$

3. $\begin{cases} x_1' = -x_1 + x_2 - e^t, \\ x_2' = -4x_1 + 3x_2 + e^t \end{cases}$

7. $\begin{cases} x_1' = -3x_1 + 2x_2 + 2e^{-t}, \\ x_2' = 2x_1 - 3x_2 + 4t \end{cases}$

4. $\begin{cases} x_1' = -4x_1 + 8x_2 + 1, \\ x_2' = -2x_1 + 4x_2 + 2t \end{cases}$

8. $\begin{cases} x_1' = 2x_1 - 5x_2 - \cos(t), \\ x_2' = x_1 - 2x_2 + \sin(t) \end{cases}$

Solve the following initial value problems.

9. $\begin{cases} x_1' = 2x_1 - x_2 + e^t, \\ x_2' = 3x_1 - 2x_2 - e^t, \\ x_1(0) = 1, \quad x_2(0) = 0 \end{cases}$

11. $\begin{cases} x_1' = 7x_1 - 2x_2 + 2e^{3t}, \\ x_2' = 2x_1 + 2x_2 + 4e^{3t}, \\ x_1(0) = 1, \quad x_2(0) = -1 \end{cases}$

10. $\begin{cases} x_1' = -x_2 + \cos(t), \\ x_2' = x_1 + \sin(t), \\ x_1(0) = 0, \quad x_2(0) = 0 \end{cases}$

12. $\begin{cases} x_1' = x_2 + e^{2t}\cos(t), \\ x_2' = -5x_1 + 4x_2, \\ x_1(0) = 1, \quad x_2(0) = 0 \end{cases}$

13. Show that the method of undetermined coefficients can be developed to solve a system of equations of the form $\mathbf{x}'(t) = A\mathbf{x} + \mathbf{v}\,t^m$. Assuming that $\alpha = 0$ is not an eigenvalue of the matrix A, show that a particular solution can be found in the form $\mathbf{x}(t) = \mathbf{w}_0 t^m + \mathbf{w}_1 t^{m-1} + \cdots + \mathbf{w}_m$ for suitably chosen vectors $\mathbf{w}_0, \ldots, \mathbf{w}_m$.

14. Show how to modify the particular solution proposed in Exercise 13 to find a particular solution of the system $\mathbf{x}'(t) = A\mathbf{x} + \mathbf{v}\,t^m$ in the case that $\alpha = 0$ is an eigenvalue of the matrix A.

15.7   The Method of Variation of Parameters for Systems

The problem is to find a particular solution to the inhomogeneous system

$$\frac{d\mathbf{x}}{dt} = P(t)\mathbf{x} + \mathbf{g}(t). \tag{15.91}$$

In case the right-hand side of (15.91) is of a complicated form (for example when the coefficients are time-dependent) it may be difficult or impossible to apply the method of undetermined coefficients to find a particular solution. In this subsection we develop the appropriate form of the method of variation of parameters for systems of differential equations.

We look for a particular solution to (15.91) in the form

$$\mathbf{x_p}(t) = \Phi(t)\mathbf{u}(t), \tag{15.92}$$

where the vector function $\mathbf{u}(t)$ is to be determined and $\Phi(t)$ is a fundamental matrix, that is, any nonsingular solution to

$$\Phi'(t) = P(t)\Phi(t). \tag{15.93}$$

Substituting (15.92) into the inhomogeneous equation (15.91), we obtain

$$\Phi'(t)\mathbf{u}(t) + \Phi(t)\mathbf{u}'(t) = P(t)\Phi(t)\mathbf{u}(t) + \mathbf{g}(t). \tag{15.94}$$

Because of (15.93) we can cancel the first term on each side of (15.94), leading to

$$\Phi(t)\mathbf{u}'(t) = \mathbf{g}(t). \tag{15.95}$$

We now apply the inverse matrix $\Phi(t)^{-1}$ to both sides of (15.95) and integrate the resulting equation to obtain the specific choice

$$\mathbf{u}(t) = \int_{t_0}^{t} \Phi(s)^{-1}\mathbf{g}(s)ds. \tag{15.96}$$

Then

$$\Phi(t) \int_{t_0}^{t} \Phi(s)^{-1}\mathbf{g}(s)ds$$

is a particular solution to (15.91) with the initial condition $\mathbf{x}(t_0) = 0$. To obtain the general solution of the inhomogeneous system (15.91), we add in the general solution to the homogeneous system $\mathbf{x}' = P(t)\mathbf{x}$ to obtain the formula

$$\mathbf{x}(t) = \Phi(t)\mathbf{x}(t_0) + \Phi(t) \int_{t_0}^{t} \Phi(s)^{-1}\mathbf{g}(s)ds. \tag{15.97}$$

Formula (15.97) becomes especially transparent in the case of constant-coefficient systems, for which we may use the fundamental matrix $\Phi(t) = e^{(t-t_0)A}$. In this case (15.97) reduces to

$$\mathbf{x}(t) = e^{(t-t_0)A}\mathbf{x}(t_0) + e^{tA}\int_{t_0}^t e^{-sA}\mathbf{g}(s)ds \qquad (15.98)$$

$$= e^{(t-t_0)A}\mathbf{x}(t_0) + \int_{t_0}^t e^{(t-s)A}\mathbf{g}(s)ds.$$

Equation (15.98) is reminiscent of the corresponding result for a single constant-coefficient first-order equation of the form $x' = ax + g(t)$, which was solved by the method of integrating factors in Section 3.2. In the present theory the matrix exponential e^{-tA} plays the role of the integrating factor e^{-at}, which was previously used in the case of a single first-order equation.

Example 15.21. *Find a particular solution to the system*

$$\mathbf{x}' = \begin{pmatrix} 0 & -1 \\ 1 & 0 \end{pmatrix}\mathbf{x} + \begin{pmatrix} \csc(t) \\ \sec(t) \end{pmatrix}. \qquad (15.99)$$

Solution. Let

$$A = \begin{pmatrix} 0 & -1 \\ 1 & 0 \end{pmatrix} \quad \text{and} \quad \mathbf{g}(t) = \begin{pmatrix} \csc(t) \\ \sec(t) \end{pmatrix};$$

then

$$e^{-sA} = \begin{pmatrix} \cos(s) & \sin(s) \\ -\sin(s) & \cos(s) \end{pmatrix}.$$

We compute

$$e^{-sA}\mathbf{g}(s) = \begin{pmatrix} \cos(s)\csc(s) + \sin(s)\sec(s) \\ -\sin(s)\csc(s) + \cos(s)\sec(s) \end{pmatrix} = \begin{pmatrix} \tan(s) + \cot(s) \\ 0 \end{pmatrix}.$$

We compute the definite integral (15.96) with $t_0 = \pi/4$, obtaining

$$\int_{\pi/4}^t e^{-sA}\mathbf{g}(s)ds = \begin{pmatrix} -\log(\cos(t)) + \log(\sin(t)) \\ 0 \end{pmatrix}.$$

A final application of the matrix e^{tA} yields the required particular solution

$$\mathbf{x}_p(t) = \begin{pmatrix} \cos(t)(\log(\sin(t)) - \log(\cos(t))) \\ \sin(t)(\log(\sin(t)) - \log(\cos(t))) \end{pmatrix} = \begin{pmatrix} \cos(t)\log(\tan(t)) \\ \sin(t)\log(\tan(t)) \end{pmatrix}. \quad \blacksquare$$

Summary of Techniques Introduced

A particular solution to an inhomogeneous linear system can always be obtained by the method of variation of parameters. This requires knowledge of a fundamental matrix and the calculation of certain integrals. In the case that the system has constant-coefficients and the right-hand side is a product of a fixed vector with an exponential function of time, one can often obtain a particular solution by a suitable version of the method of undetermined coefficients, which only requires the solution to a single system of linear equations.

Exercises for Section 15.7

Find a particular solution to each of the following systems by the method of variation of parameters.

1. $\begin{cases} x_1' = -2x_1 + 3x_2 + t, \\ x_2' = -x_1 + 2x_2 + e^t \end{cases}$

5. $\begin{cases} x_1' = 4x_1 - 2x_2 + t^{-3}, \\ x_2' = 8x_1 - 4x_2 - t^{-2} \end{cases}$

2. $\begin{cases} x_1' = -2x_1 + x_2 + \sin(t), \\ x_2' = -5x_1 + 2x_2 - \cos(t) \end{cases}$

6. $\begin{cases} x_1' = 4x_1 + 2x_2 + \cot(t), \\ x_2' = -10x_1 + 4x_2 \end{cases}$

3. $\begin{cases} x_1' = -2x_1 + 4x_2 - 2e^t, \\ x_2' = x_1 + x_2 + e^{-2t} \end{cases}$

7. $\begin{cases} x_1' = 2x_1 - 5x_2 + \csc(t), \\ x_2' = x_1 - 2x_2 + \sec(t) \end{cases}$

4. $\begin{cases} x_1' = x_2 + e^t, \\ x_2' = x_2 + 2e^t, \\ x_3' = x_1 - 3x_2 + 3x_3 + 3e^t \end{cases}$

8. $\begin{cases} x_1' = -2x_1 + x_2 - 2t^2, \\ x_2' = 2x_2 + x_3 - t^3 - 2, \\ x_3' = x_1 - 2x_2 - 3t^4 \end{cases}$

Solve the following initial value problems.

9. $\begin{cases} x_1' = -2x_1 + e^t, & x_1(0) = 0, \\ x_2' = x_1 - 2x_2 + 3t, & x_2(0) = 0 \end{cases}$

11. $\begin{cases} x_1' = 2x_1 + 3x_2 + t, & x_1(0) = 0, \\ x_2' = -3x_1 + 2x_2, & x_2(0) = -1 \end{cases}$

10. $\begin{cases} x_1' = x_1 + x_2 + e^{2t}, & x_1(0) = 2, \\ x_2' = x_1 + x_2 + e^{2t}, & x_2(0) = -1 \end{cases}$

12. $\begin{cases} x_1' = -4x_1 + x_2, & x_1(0) = 0, \\ x_2' = 4x_1 - x_2 + e^{-5t}, & x_2(0) = 1 \end{cases}$

13. The purpose of this exercise is to give an independent derivation of the variation of parameters formula for a single second-order equation, beginning with the variation of parameters formula for a system.

    a. Write the second-order equation $y'' + p(t)y' + q(t)y = g(t)$ as a system.

    b. Find a particular solution of this system by formula (15.98).

    c. Identify the first component of this solution as formula (9.83).

15.8   Solving Systems Using the Laplace Transform

The method of Laplace transforms described in Chapter 14 can also be used to treat initial value problems for systems of first-order linear differential equations with constant-coefficients. A typical such initial value problem is written

$$\begin{cases} \mathbf{y}'(t) = A\mathbf{y}(t) + \mathbf{g}(t), \\ \mathbf{y}(0) = \mathbf{Y}_0. \end{cases} \tag{15.100}$$

Here $\mathbf{Y}_0$ is a given column vector in the space $\mathbb{R}^n$, and A is an $n \times n$ matrix of constants. Furthermore, $t \longmapsto \mathbf{g}(t)$ is a given piecewise differentiable function that assigns to each t the vector $\mathbf{g}(t)$. The goal is to find the vector-valued function $t \longmapsto \mathbf{y}(t)$.

    Since the integral of a matrix of functions is defined to be the matrix of the integrals of the functions, the Laplace transforms of $\mathbf{y}(t)$ and $\mathbf{g}(t)$ are given by

$$\mathcal{L}\big(\mathbf{y}(t)\big)(s) = \int_0^\infty e^{-st}\mathbf{y}(t)\,dt \quad \text{and} \quad \mathcal{L}\big(\mathbf{g}(t)\big)(s) = \int_0^\infty e^{-st}\mathbf{g}(t)\,dt.$$

When we take the Laplace transform of both sides of $\mathbf{y}' = A\mathbf{y} + \mathbf{g}(t)$ and apply the initial condition $\mathbf{y}(t_0) = \mathbf{Y}_0$, we obtain

$$s\mathcal{L}\big(\mathbf{y}(t)\big)(s) - \mathbf{Y}_0 = A\mathcal{L}\big(\mathbf{y}(t)\big)(s) + \mathcal{L}\big(\mathbf{g}(t)\big)(s),$$

which can be rewritten as

$$(s\,I - A)\mathcal{L}\big(\mathbf{y}(t)\big)(s) = \mathbf{Y}_0 + \mathcal{L}\big(\mathbf{g}(t)\big)(s), \tag{15.101}$$

where I denotes the identity $n \times n$ matrix. If the matrix $s\,I - A$ is nonsingular, we may compute its inverse and apply the inverse to both sides of (15.101):

$$\mathcal{L}\big(\mathbf{y}(t)\big)(s) = (s\,I - A)^{-1}\mathbf{Y}_0 + (s - A)^{-1}\mathcal{L}\big(\mathbf{g}(t)\big)(s). \tag{15.102}$$

Then the solution to (15.100) can be found by applying the inverse Laplace transform to both sides of (15.102).

We illustrate with two examples. The first example is a two-dimensional homogeneous system.

Example 15.22. *Solve the two-dimensional initial value problem*

$$\mathbf{y}' = \begin{pmatrix} 0 & 2 \\ -2 & 0 \end{pmatrix} \mathbf{y}, \qquad \mathbf{y}(0) = \begin{pmatrix} 3 \\ 1 \end{pmatrix} \tag{15.103}$$

by means of Laplace transforms.

Solution. The inverse of the matrix

$$s I - A = \begin{pmatrix} s & -2 \\ 2 & s \end{pmatrix}$$

is given by

$$(s I - A)^{-1} = \frac{1}{s^2 + 4} \begin{pmatrix} s & 2 \\ -2 & s \end{pmatrix},$$

so that the Laplace transform of the solution to (15.103) is given by

$$\mathcal{L}(\mathbf{y}(t))(s) = (s I - A)^{-1} \mathbf{Y}_0 = \frac{1}{s^2 + 4} \begin{pmatrix} s & 2 \\ -2 & s \end{pmatrix} \begin{pmatrix} 3 \\ 1 \end{pmatrix}$$

$$= \frac{1}{s^2 + 4} \begin{pmatrix} 3s + 2 \\ -6 + s \end{pmatrix}. \tag{15.104}$$

The computation of the inverse Laplace transform of the right-hand side of (15.104) can be accomplished using the table on page 471. Thus

$$\mathbf{y}(t) = \begin{pmatrix} \mathcal{L}^{-1}\left(\dfrac{3s + 2}{s^2 + 4}\right) \\ \mathcal{L}^{-1}\left(\dfrac{-6 + s}{s^2 + 4}\right) \end{pmatrix} = \begin{pmatrix} 3\cos(2t) + \sin(2t) \\ -3\sin(2t) + \cos(2t) \end{pmatrix}. \ \blacksquare$$

The next example treats an inhomogeneous system.

Example 15.23. *Use Laplace transforms to solve the initial value problem*

$$\mathbf{y}' = \begin{pmatrix} 0 & 2 \\ -2 & 0 \end{pmatrix} \mathbf{y} + \begin{pmatrix} e^{3t} \\ 1 \end{pmatrix}, \qquad \mathbf{y}(0) = \begin{pmatrix} 0 \\ 0 \end{pmatrix}. \tag{15.105}$$

Solution. The Laplace transform of $\mathbf{g}(t) = \begin{pmatrix} e^{3t} \\ 1 \end{pmatrix}$ is

$$\mathcal{L}\big(g(t)\big) = \begin{pmatrix} \dfrac{1}{s-3} \\ \dfrac{1}{s} \end{pmatrix}.$$

We use the computation of $(s\,I - A)^{-1}$ from Example 15.22. Using (15.102) we find that the Laplace transform of the solution to (15.105) is

$$\mathcal{L}\big(\mathbf{y}(t)\big)(s) = \frac{1}{s^2+4}\begin{pmatrix} s & 2 \\ -2 & s \end{pmatrix}\begin{pmatrix} \dfrac{1}{s-3} \\ \dfrac{1}{s} \end{pmatrix} = \frac{1}{s^2+4}\begin{pmatrix} \dfrac{s}{s-3}+\dfrac{2}{s} \\ \dfrac{-2}{s-3}+1 \end{pmatrix}. \tag{15.106}$$

By expanding the right-hand side of (15.106) into partial fractions and using the tables of Section 14.5 to compute inverse Laplace transforms, we obtain the solution to (15.105) as

$$\mathbf{y}(t) = \begin{pmatrix} -\dfrac{1}{2}+\dfrac{e^t}{5}-\dfrac{7}{10}\cos(2t)+\dfrac{2}{5}\sin(2t) \\ -\dfrac{e^t}{5}+\dfrac{2}{5}\cos(2t)+\dfrac{7}{10}\sin(2t) \end{pmatrix}. \ \blacksquare$$

Exercises for Section 15.8

Use Laplace transforms to solve the following systems of differential equations.

1. $\begin{cases} x_1' = 12x_1 + 5x_2, & x_1(0) = 0, \\ x_2' = -6x_1 + x_2, & x_2(0) = 1 \end{cases}$

2. $\begin{cases} x_1' = 4x_1 - 2x_2, & x_1(0) = 2, \\ x_2' = 5x_1 + 2x_2, & x_2(0) = -2 \end{cases}$

3. $\begin{cases} x_1' = -2x_1 + x_2, & x_1(0) = 5, \\ x_2' = -9x_1 + 4x_2, & x_2(0) = -3 \end{cases}$

4. $\begin{cases} x_1' = -x_1 - x_2, & x_1(0) = 4, \\ x_2' = -x_2, & x_2(0) = 1, \\ x_3' = -2x_3, & x_3(0) = 1 \end{cases}$

5. $\begin{cases} x_1' = -6x_1 + 2x_2, & x_1(0) = 1, \\ x_2' = -7x_1 + 3x_2, & x_2(0) = 0 \end{cases}$

6. $\begin{cases} x_1' = 4x_1 - x_2, & x_1(0) = 1, \\ x_2' = 2x_1 + 5x_2, & x_2(0) = 0 \end{cases}$

7. $\begin{cases} x_1' = x_1 + x_2, & x_1(0) = 1, \\ x_2' = -4x_1 + x_2, & x_2(0) = 1 \end{cases}$

8. $\begin{cases} x_1' = 2x_2, & x_1(0) = 1, \\ x_2' = -2x_1, & x_2(0) = 1, \\ x_3' = -3x_4, & x_3(0) = 1, \\ x_4' = x_3, & x_4(0) = 0 \end{cases}$

9. $\begin{cases} x' + y' = -3x - 2y + e^{-2t}, & x(0) = 0 \\ 2x' + y' = -2x - y + 1, & y(0) = 0 \end{cases}$

10. $\begin{cases} x' = x - y - e^{-t}, & x(0) = 1, \\ y' = 2x + 3y + e^{-t}, & y(0) = 0 \end{cases}$

11. $\begin{cases} x' + y' = x, & x(0) = 1, \\ y' + z' = x, & y(0) = 1, \\ z' + x' = x, & z(0) = 1 \end{cases}$

12. $\begin{cases} x_1' = 3x_1, & x_1(0) = 1 \\ x_2' = x_1 + 3x_2, & x_2(0) = 1 \\ x_3' = 3x_3, & x_3(0) = 1 \\ x_4' = 2x_3 + 3x_4, & x_4(0) = 1 \end{cases}$

16

PHASE PORTRAITS OF LINEAR SYSTEMS

We have seen in previous chapters that the plot of a solution to a differential equation provides a very helpful way to understand the solution; a plot is often more useful and more interesting than a formula or a table of numbers.

Similarly, we can plot each component of the solution to a system of n differential equations on the same plot, provided that we are given n initial conditions. The situation is not as pleasant as that of a single equation, however. For example, if we want to plot 2 solutions to a system, then we must plot $2n$ functions. We must be careful to distinguish the graphs of components of one solution from the graphs of components of the other solution.

In the case that n is 2 or 3, a **phase portrait** is a more useful graphical representation of the solution to a system. Phase portraits of two-dimensional linear systems are described and classified in Section 16.1. In Section 16.2 we show how to use **ODE** to solve systems of differential equations. Methods of creating phase portraits with *Mathematica* are described in Section 16.3.

16.1 Phase Portraits of Two-Dimensional Linear Systems

In Chapter 15 we treated constant-coefficient linear systems by algebraic methods, using the eigenvalues and eigenvectors to find a fundamental matrix for the system. This method applies to systems of any dimension. In the case of two-dimensional systems, one can amplify the analysis by studying the **trajectories** of the system in the **phase plane**. Each trajectory is a parametrized curve (or vector-valued function) $t \longmapsto \mathbf{x}(t) = \big(x(t), y(t)\big)$. The information obtained in this chapter will be used to study nonlinear systems in Chapters 17 and 19.

A two-dimensional system with constant-coefficients is written in component form as

$$
\begin{cases}
\dfrac{dx}{dt} = a_{11}x + a_{12}y, \\
\dfrac{dy}{dt} = a_{21}x + a_{22}y,
\end{cases}
\tag{16.1}
$$

where (a_{ij}) is a 2×2 matrix of arbitrary real numbers. The single point $(0, 0)$ is always a solution of this system, the *trivial solution*. More general solutions are obtained by finding the eigenvectors of the coefficient matrix

$$
A = \begin{pmatrix} a_{11} & a_{12} \\ a_{21} & a_{22} \end{pmatrix}.
$$

If $\boldsymbol{\xi} = (c_1, c_2)$ is an eigenvector corresponding to a zero eigenvalue of A, then there is an entire straight line of solutions to (16.1) with slope c_2/c_1. Each solution is a point on the line. In this case we say that $(0, 0)$ is a **nonisolated singularity** of the system (16.1). These solutions are indicated graphically by a dashed line through the origin. Examples are furnished by a repelling center (page 556), an attracting center (page 557) and an improper center (page 562).

If an eigenvector $\boldsymbol{\xi} = (c_1, c_2)$ corresponds to a nonzero *real* eigenvalue of A, then the phase portrait of (16.1) contains trajectories that either tend to infinity (in case the eigenvalue is positive) or tend to $(0, 0)$ (in case the eigenvalue is negative). These solutions are indicated by curves through the origin together with arrows specifying the direction of the motion. These arrows are crucial to a detailed understanding of the phase portrait.

If the eigenvalues of A are complex, then the phase portrait of the solution exhibits rotation. The sign of the real part of a complex eigenvalue determines whether the corresponding trajectories tend toward $(0, 0)$ (negative real part) or tend to infinity (positive real part). The orientation (clockwise or counterclockwise) depends on the sign of the off-diagonal elements a_{12} and a_{21}.

With this preliminary information, we now describe the systematic classification of all possible cases of phase portraits for which the coefficient matrix A is not identically zero. The classification proceeds by considering all possibilities for the eigenvalues and eigenvectors of A. A summary of the classification is given on page 566; there are thirteen possibilities. In the next thirteen subsections we illustrate each possibility with a phase portrait. In some cases we also give a three-dimensional plot of **space trajectories**, which we define to be curves in $\mathbb{R}^3$ of the form $t \longmapsto \big(x(t), y(t), t\big)$, where $t \longmapsto \mathbf{x}(t) = \big(x(t), y(t)\big)$ is an ordinary trajectory.

Repelling proper node: two unequal positive real eigenvalues

In this case there are two linearly independent eigenvectors. All nonzero solutions of the system tend to infinity. The direction of approach is along an eigenvector corresponding to the *larger* eigenvalue. This type of phase portrait is called a **repelling proper node**. The phase portrait below gives selected solution curves for the system

$$\begin{cases} \dfrac{dx}{dt} = 2x + 3y, \\ \dfrac{dy}{dt} = x + 4y. \end{cases}$$

All eigenvectors of the eigenvalue $r = 5$ are multiples of $\boldsymbol{\xi} = (1, 1)$, whereas all eigenvectors of the eigenvalue $r = 1$ are multiples of $\boldsymbol{\xi} = (3, -1)$.

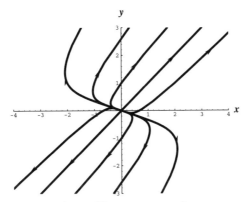

A repelling proper node

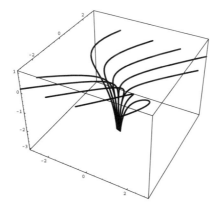

Space trajectories for a repelling proper node

Attracting proper node: two unequal negative real eigenvalues

In this case there are two linearly independent eigenvectors. All nonzero solutions of the system tend to $(0, 0)$. The direction of approach is along an eigenvector corresponding to the *larger eigenvalue*, in the sense of sign. This type of phase portrait is called an **attracting proper node**. The phase portrait below gives selected solution curves for the system

$$\begin{cases} \dfrac{dx}{dt} = -2x - 3y, \\ \dfrac{dy}{dt} = -x - 4y. \end{cases}$$

All eigenvectors of the eigenvalue $r = -5$ are multiples of $\boldsymbol{\xi} = (1, 1)$, whereas all eigenvectors of the eigenvalue $r = -1$ are multiples of $\boldsymbol{\xi} = (3, -1)$.

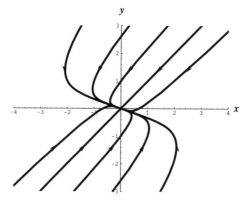

An attracting proper node

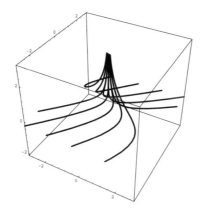

Space trajectories for an attracting proper node

Saddle: one positive and one negative real eigenvalue

In this case all nonzero solutions tend to infinity, with the exception of the solutions along an eigenvector with negative eigenvalue, which tend to $(0, 0)$. This type of phase portrait is called a **saddle**. The phase portrait below gives selected solution curves for the system

$$\begin{cases} \dfrac{dx}{dt} = 8y, \\ \dfrac{dy}{dt} = 2x. \end{cases}$$

All eigenvectors of the eigenvalue $r = 4$ are multiples of $\boldsymbol{\xi} = (2, 1)$, whereas all eigenvectors of the eigenvalue $r = -4$ are multiples of $\boldsymbol{\xi} = (-2, 1)$.

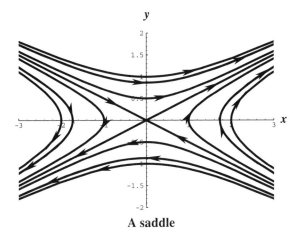

A saddle

Repelling center: one positive eigenvalue and one zero eigenvalue

In this case the constant solutions remain on a line whose direction is the same as that of an eigenvector corresponding to the zero eigenvalue. This line is represented in the phase portrait below as dashed. All other solutions tend to infinity along straight lines. The direction of approach is the same as that of an eigenvector corresponding to the positive eigenvalue. This type of phase portrait is called a **repelling center**. The phase portrait below gives selected solution curves for the system

$$\begin{cases} \dfrac{dx}{dt} = 4x - y, \\ \dfrac{dy}{dt} = -4x + y. \end{cases}$$

All eigenvectors of the eigenvalue $r = 0$ are multiples of $\boldsymbol{\xi} = (1, 4)$, whereas all eigenvectors of the eigenvalue $r = 5$ are multiples of $\boldsymbol{\xi} = (1, -1)$.

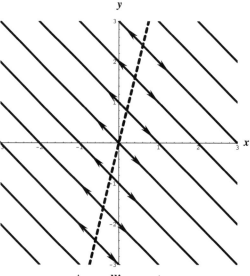

A repelling center

Attracting center: one negative eigenvalue and one zero eigenvalue

In this case the constant solutions remain on a line whose direction is the same as that of an eigenvector corresponding to the zero eigenvalue. This line is represented in the phase portrait below as dashed. All other solutions tend to the dashed line along straight lines. The direction of approach is the same as that of an eigenvector corresponding to the negative eigenvalue. This type of phase portrait is called an **attracting center**. The phase portrait below gives selected solution curves for the system

$$\begin{cases} \dfrac{dx}{dt} = -4x + y, \\ \dfrac{dy}{dt} = 4x - y. \end{cases}$$

All eigenvectors of the eigenvalue $r = 0$ are multiples of $\boldsymbol{\xi} = (1, 4)$, whereas all eigenvectors of the eigenvalue $r = -5$ are multiples of $\boldsymbol{\xi} = (1, -1)$.

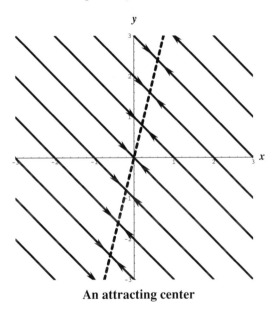

An attracting center

This concludes the classification of all possible cases of distinct real eigenvalues. We now pass to the cases of repeated real eigenvalues.

Repelling degenerate node: two equal positive real eigenvalues with two linearly independent eigenvectors

In this case A is a positive multiple of the identity matrix. All nonzero solutions tend to infinity along straight lines thorough the origin. This type of phase portrait is called a **repelling degenerate node**. The phase portrait below gives selected solution curves for the system

$$\begin{cases} \dfrac{dx}{dt} = 2x, \\ \dfrac{dy}{dt} = 2y. \end{cases}$$

The eigenvalue is $r = 2$, and all nonzero vectors are eigenvectors.

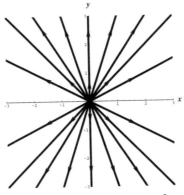

A repelling degenerate node

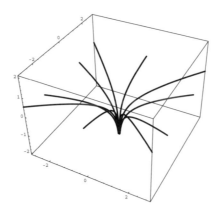

Space trajectories for a repelling degenerate node

Attracting degenerate node: two equal negative real eigenvalues with two linearly independent eigenvectors

In this case A is a negative multiple of the identity matrix. All nonzero solutions tend to $(0, 0)$ along straight lines through the origin. This type of phase portrait is called an **attracting degenerate node**. The phase portrait below gives selected solution curves for the system

$$\begin{cases} \dfrac{dx}{dt} = -2x, \\[2mm] \dfrac{dy}{dt} = -2y. \end{cases}$$

The eigenvalue is $r = -2$, and all nonzero vectors are eigenvectors.

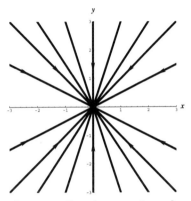

An attracting degenerate node

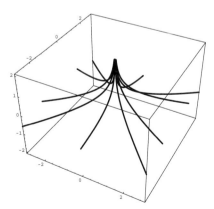

Space trajectories for an attracting degenerate node

Repelling improper node: two equal positive eigenvalues with only one linearly independent eigenvector

In this case all nonzero solutions tend to infinity. The direction of approach is parallel to an eigenvector, which divides the plane into two halves. The trajectories make an S-shaped display in the phase plane. From one half plane the trajectories move to infinity parallel to an eigenvector; from the other half plane the trajectories move to infinity in the opposite direction. This type of phase portrait is called a **repelling improper node**. The detailed orientation depends on the sign of the off-diagonal elements a_{12} and a_{21}. If $a_{12} > 0$, the upper half of the S curve will be oriented clockwise and the lower half oriented counterclockwise. The phase portrait below gives selected solution curves for the system

$$\begin{cases} \dfrac{dx}{dt} = -2x + y, \\ \dfrac{dy}{dt} = -9x + 4y. \end{cases}$$

The eigenvalue is $r = 1$, and all eigenvectors are multiples of $\xi = (1, 3)$.

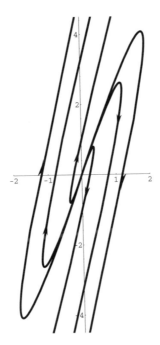

A repelling improper node

Attracting improper node: two equal negative eigenvalues with only one linearly independent eigenvector

In this case all nonzero solutions tend to $(0, 0)$. The direction of approach is along an eigenvector of the matrix. The trajectories again make an S-shaped display in the phase plane. The orientation of the S depends on the sign of the off-diagonal elements a_{12} and a_{21}. If $a_{12} > 0$, the upper half of the S curve will be oriented clockwise and the lower half will be oriented counterclockwise. This type of phase portrait is called an **attracting improper node**. The phase portrait below gives selected solution curves for the system

$$\begin{cases} \dfrac{dx}{dt} = 2x - y, \\ \dfrac{dy}{dt} = 9x - 4y. \end{cases}$$

The eigenvalue is $r = -1$, and all eigenvectors are multiples of $\boldsymbol{\xi} = (1, 3)$.

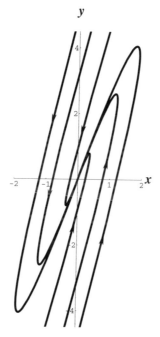

An attracting improper node

Improper center: two zero eigenvalues with only one eigenvector

In this case there is a line of solutions corresponding to an eigenvector with zero eigenvalue, indicated by a dashed line. All other solutions tend to infinity along straight lines parallel to this eigenvector. The orientation depends on the sign of the off-diagonal elements a_{12} and a_{21}. If $a_{12} > 0$, the solutions above the dashed line will move to the right and the solutions below the dashed line will move to the left.

This type of phase portrait is called an **improper center**. The phase portrait below gives selected solution curves for the system

$$\begin{cases} \dfrac{dx}{dt} = -3x + y, \\ \dfrac{dy}{dt} = -9x + 3y. \end{cases}$$

The eigenvalue is $r = 0$, and all eigenvectors are multiples of $\boldsymbol{\xi} = (1, 3)$.

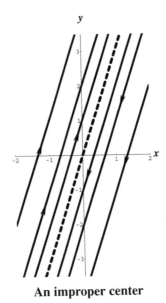

An improper center

This finishes the discussion of real eigenvalues. We conclude the classification by discussing the cases of complex eigenvalues.

Repelling spiral: complex conjugate eigenvalues with positive real part

In this case all nonzero solutions tend to infinity by spiraling around the origin. This type of phase portrait is called a **repelling spiral**. The orientation (clockwise or counterclockwise) depends on the sign of the off-diagonal elements a_{12} and a_{21}. If $a_{12} > 0$ the rotation is clockwise, whereas if $a_{12} < 0$ the rotation is counterclockwise. ($a_{12} = 0$ is impossible in this case, as shown in the exercises.) The phase portrait below gives selected solution curves for the system

$$\begin{cases} \dfrac{dx}{dt} = 2x + 3y, \\ \dfrac{dy}{dt} = -3x + 2y. \end{cases}$$

The eigenvalues are $r = 2 \pm 3i$ with the eigenvectors multiples of $\boldsymbol{\xi} = (\mp i, 1)$.

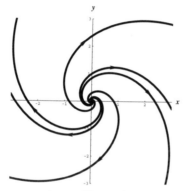

A repelling spiral

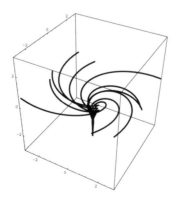

Space trajectories for a repelling spiral

Attracting spiral: complex conjugate eigenvalues with negative real part

In this case all nonzero solutions tend to $(0, 0)$ by spiraling around the origin. The orientation of the spiral depends on the sign of the off-diagonal elements a_{12} and a_{21}. If $a_{12} > 0$ the rotation is clockwise, whereas if $a_{12} < 0$ the rotation is counterclockwise. ($a_{12} = 0$ is impossible in this case, as shown in the exercises.) This type of phase portrait is called an **attracting spiral**. The phase portrait below gives selected solution curves for the system

$$\begin{cases} \dfrac{dx}{dt} = -2x + 3y, \\ \dfrac{dy}{dt} = -3x - 2y. \end{cases}$$

The eigenvalues are $r = -2 \pm 3i$ with eigenvectors multiples of $\boldsymbol{\xi} = (\mp i, 1)$.

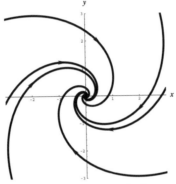

An attracting spiral

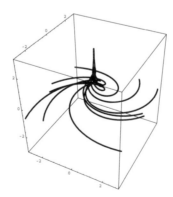

Space trajectories for an attracting spiral

Stable center: complex conjugate eigenvalues with zero real part

In this case all nonzero solutions are periodic functions that wind around the origin along ellipses (in particular, circles are included). This type of phase portrait is called a **stable center**. The orientation (clockwise or counterclockwise) depends on the signs of the off-diagonal elements a_{12} and a_{21}. If $a_{12} > 0$ the rotation is clockwise, whereas if $a_{12} < 0$ the rotation is counterclockwise. ($a_{12} = 0$ is impossible in this case, as shown in the exercises.) The phase portrait below gives selected solution curves for the system

$$\begin{cases} \dfrac{dx}{dt} = 4y, \\ \dfrac{dy}{dt} = -x. \end{cases}$$

The eigenvalues are $r = \pm 2i$ with eigenvectors multiples of $\boldsymbol{\xi} = (2, \pm i)$.

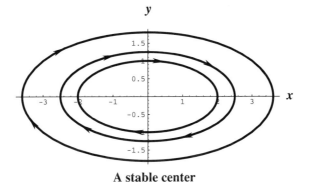

A stable center

This concludes the discussion of all possible cases of phase portraits of two-dimensional systems.

With hindsight the reader will note that many of these cases come in pairs. By reversing the direction of the arrows in the phase portraits we interchange repelling with attracting proper nodes, repelling with attracting centers, repelling with attracting degenerate nodes, repelling with attracting spirals, and repelling with attracting centers. The remaining single cases are those of saddles, stable centers, and improper centers, which are transformed into themselves by interchange of the arrows.

The following table summarizes the information obtained.

Eigenvalues	Conditions	Phase Portrait	Picture
$0 < \lambda_1 < \lambda_2$		Repelling proper node	page 553
$\lambda_1 < \lambda_2 < 0$		Attracting proper node	page 554
$\lambda_1 < 0 < \lambda_2$		Saddle	page 555
$0 = \lambda_1 < \lambda_2$		Repelling center	page 556
$\lambda_1 < \lambda_2 = 0$		Attracting center	page 557
$\lambda_1 = \lambda_2 > 0$	two eigenvectors	Repelling degenerate node	page 558
$\lambda_1 = \lambda_2 < 0$	two eigenvectors	Attracting degenerate node	page 559
$\lambda_1 = \lambda_2 > 0$	one eigenvector	Repelling improper node	page 560
$\lambda_1 = \lambda_2 < 0$	one eigenvector	Attracting improper node	page 561
$\lambda_1 = \lambda_2 = 0$	one eigenvector	Improper center	page 562
$\lambda_1, \lambda_2 = \alpha \pm i\beta$	$\alpha > 0, \beta \neq 0$	Repelling spiral	page 563
$\lambda_1, \lambda_2 = \alpha \pm i\beta$	$\alpha < 0, \beta \neq 0$	Attracting spiral	page 564
$\lambda_1, \lambda_2 = \pm i\beta$	$\beta \neq 0$	Stable center	page 565

There is one trivial case not included in the above table, namely $x' = y' = 0$, for which all points in the plane are solutions.

In the following example the system of differential equations contains a parameter c. Different values of c generate several different phase portraits.

Example 16.1. *Find the phase portraits of the two-dimensional systems*

$$\begin{cases} \dfrac{dx}{dt} = 2x + c\,y, \\ \dfrac{dy}{dt} = x + 2y, \end{cases}$$

for all possible values of the constant c.

Solution. We first find the eigenvalues of the corresponding matrix. These are obtained as the solutions of the quadratic equation $(2-r)^2 - c = 0$, thus, $r = 2 \pm \sqrt{c}$. The eigenvectors are multiples of $(\pm\sqrt{c}, 1)$. We consider separately five different cases:

(i) $c > 4$. The two eigenvalues are both real but have different signs; thus we have a saddle.

(ii) $c = 4$. One eigenvalue is positive and the other eigenvalue is zero, leading to a repelling center.

(iii) $0 < c < 4$. Both eigenvalues are real and positive, so that we have a repelling proper node.

(iv) $c = 0$. There is only one eigenvector, which has a positive eigenvalue, giving a repelling improper node. The orientation is determined from the lower right corner of the matrix: on the right half plane the trajectories move up with increasing time, whereas in the left half plane the trajectories move down with increasing time.

(v) $c < 0$. The eigenvalues are conjugate complex with a positive real part, leading to a repelling spiral. The sign of the off-diagonal element indicates that the orientation is counterclockwise. ∎

Summary of Techniques Introduced

In this section we have seen how to draw the phase portrait of a two-dimensional linear system from the knowledge of the eigenvalues and eigenvectors of the corresponding matrix. The positive eigenvalues cause repulsion from the origin, whereas the negative eigenvalues cause attraction to the origin. The eigenvectors determine the directions of motion to or from the origin. The off-diagonal elements of the matrix determine the clockwise or counterclockwise orientation of the phase portrait in the case of spirals and improper nodes.

Exercises for Section 16.1

In the following exercises draw the phase portraits of the indicated two-dimensional systems, taking care to label correctly the directions of the arrows and the orientation when appropriate.

1. $\begin{cases} x' = y, \\ y' = -29x - 4y \end{cases}$

3. $\begin{cases} x' = 2x - y, \\ y' = 3x - 2y \end{cases}$

2. $\begin{cases} x' = 5x - y, \\ y' = 3x + y \end{cases}$

4. $\begin{cases} x' = x - 4y, \\ y' = 4x - 7y \end{cases}$

5. $\begin{cases} x' = x - 5y, \\ y' = x - y \end{cases}$

8. $\begin{cases} x' = 3x - 4y, \\ y' = x - y \end{cases}$

6. $\begin{cases} x' = y, \\ y' = -2x + 2y \end{cases}$

9. $\begin{cases} x' = x + 2y, \\ y' = -5x - y \end{cases}$

7. $\begin{cases} x' = 3x - 2y, \\ y' = 4x - y \end{cases}$

10. $\begin{cases} x' = 13x + 4y, \\ y' = 4x + 7y \end{cases}$

The following problems deal with the Cartesian equations of the phase portraits in the x, y plane.

11. Suppose that we have a repelling proper node with $r_2 = 2r_1 > 0$. Show that the phase portrait consists of the eigenvector solutions together with a family of parabolas. Find the axis of symmetry of these parabolas.

12. Suppose that we have an attracting proper node with $r_1 = 2r_2 < 0$. Show that the phase portrait consists of the eigenvector solutions together with a family of parabolas. Find the axis of symmetry of these parabolas.

13. Suppose that we have a saddle with $r_1 + r_2 = 0$. Show that the phase portrait consists of the eigenvector solutions together with a family of hyperbolas that are asymptotic to the eigenvector solutions.

14. Suppose that we have a stable center. Show that the phase portrait consists of a family of ellipses centered at $(0, 0)$ and find the principal axes.

15. Suppose that we have a repelling or attracting spiral. Find a suitable system of polar coordinates (r, θ) in the x, y plane such that the phase portraits are represented by the equation $\theta = A \log(r) + B$ for suitable constants A and B.

16. Suppose that we have an improper center. Show that the phase portrait consists of a family of parallel lines and find the slope.

17. Suppose that we have a repelling or attracting center. Show that the phase portrait consists of a family of parallel lines and find the slope.

18. Suppose that we have an improper node, either repelling or attracting. Find the equation of the S-shaped phase portraits in the x, y-plane. [Hint: Rotate the coordinates so that the eigenvector is horizontal and write the system as a single differential equation for dy/dx, which can be solved in terms of logarithms.]

19. Suppose that the eigenvalues are complex, so that we have an attracting spiral, repelling spiral, or stable center. Show that the coefficients of the matrix A satisfy $a_{12}a_{21} < 0$; in particular, $a_{12} \neq 0$ and $a_{21} \neq 0$.

20. Suppose that the eigenvalues are equal, so that we have a degenerate node, improper node, or improper center. Show that the coefficients of the matrix A satisfy $a_{12}a_{21} \leq 0$.

16.2 *Using* **ODE** *to Solve Linear Systems*

ODE has four methods for solving a linear system of differential equations: **LinearSystem**, **ApproximateLinearSystem**, **DSolve**, and **Laplace**. Each of these methods can be used to solve inhomogeneous systems of the form studied in Sections 15.6–15.8, that is, systems of the form

$$\frac{d\mathbf{x}}{dt} = A\mathbf{x} + \mathbf{g}(t),$$

where A is a constant matrix and $\mathbf{g}(t)$ is a vector-valued function. Here is an example.

Example 16.2. *Use* **ODE** *with* **Method->LinearSystem**, **Method->DSolve** *and* **Method->Laplace** *to find the general solution to the system*

$$\begin{cases} x' = x + y, \\ y' = 4x + y + t. \end{cases} \tag{16.2}$$

Solution. A system of differential equations is represented in *Mathematica* as a list of equations, in this case

```
{x' == x + y, y' == 4x + y +t}
```

The dependent variables also form a list, namely **{x,y}**. To solve (16.2) we use

```
ODE[{x' == x  + y,y' == 4x + y + t},{x,y},t,
Method->LinearSystem]//TableForm
```

to obtain

$$x \to \frac{2}{9} - \frac{t}{3} + \left(\frac{1}{2E^t} + \frac{E^{3t}}{2}\right) C[1] + \left(\frac{-1}{4E^t} + \frac{E^{3t}}{4}\right) C[2]$$

$$y \to -\left(\frac{5}{9}\right) + \frac{t}{3} + \left(-E^{-t} + E^{3t}\right) C[1] + \left(\frac{1}{2E^t} + \frac{E^{3t}}{2}\right) C[2]$$

(**TableForm** is used to get a nice display.) Essentially the same output is also obtained with the command

```
ODE[{x' == x  + y,y' == 4x + y + t},{x,y},t,
Method->DSolve]//TableForm
```

A slightly different but equivalent output is obtained using **ODE** with the option **Method->Laplace**:

```
ODE[{x' == x  + y,y' == 4x + y + t},{x,y},t,
Method->Laplace]//TableForm
```

$$x \to \frac{2}{9} - \frac{1}{4E^t} + \frac{E^{3t}}{36} - \frac{t}{3} + \frac{t}{2E^t} + \frac{x[0]}{2} + \frac{E^{3t} x[0]}{2} - \frac{y[0]}{4E^t} + \frac{E^{3t} y[0]}{4}$$

$$y \to -\left(\frac{5}{9}\right) + \frac{1}{2E^t} + \frac{E^{3t}}{18} + \frac{t}{3} - \frac{t}{E^t} + E^{3t} x[0] + \frac{y[0]}{2E^t} + \frac{E^{3t} y[0]}{2}$$

Thus, the general solution to (16.2) is

$$
\begin{cases}
x(t) = \left(\dfrac{1}{2e^t} + \dfrac{e^{3t}}{2}\right)C_1 + \left(-\dfrac{1}{4e^t} + \dfrac{e^{3t}}{4}\right)C_2 + \dfrac{2}{9} - \dfrac{t}{3}, \\[2ex]
y(t) = \left(-e^{-t} + e^{3t}\right)C_1 + \left(\dfrac{1}{2e^t} + \dfrac{e^{3t}}{2}\right)C_2 - \dfrac{5}{9} + \dfrac{t}{3},
\end{cases}
$$

where C_1 and C_2 are constants. ∎

We can also use **ODE** to solve first-order systems with initial conditions.

Example 16.3.   *Use* **ODE** *to solve the system*

$$
\begin{cases}
x' = y, \\
y' = -4x
\end{cases}
\tag{16.3}
$$

with the initial conditions $x(0) = 1$ *and* $y(0) = 2$.

Solution.   Since the output of **Method->DSolve** gives output involving unwanted complex exponents, we use **Method->LinearSystem** instead. The command

```
ODE[{x' == y,y' == -4x,x[0] == 1,y[0] == 2},{x,y},t,
Method->LinearSystem]
```

yields

```
{{x -> Cos[2 t] + Sin[2 t], y -> 2 Cos[2 t] - 2 Sin[2 t]}}
```

Hence the solution to (16.3) is

$$\begin{cases} x(t) \; = \; \cos(2t) + \sin(2t), \\ y(t) \; = \; 2\cos(2t) - 2\sin(2t). \end{cases}$$ ∎

PlotSolution, PlotPhase, *and* PlotEvolvePhase

There are several methods to plot the solution to a system of differential equations; each displays graphical information about the solutions, but in quite different ways. The first way is to plot simultaneously the graphs of the individual functions x, y,... for which $(x, y, \ldots)$ is a solution to the system. This can be accomplished with **ODE**'s option **PlotSolution**.

Example 16.4.   *Plot the functions x and y of the system* (16.3) *on the same graph.*

Solution.   We use

```
ODE[{x' == y,y' == -4x,x[0] == 1,y[0] == 2},{x,y},t,
Method->LinearSystem,PlotSolution->{{t,0,2Pi},
PlotStyle->{{AbsoluteThickness[1.5]},
{AbsoluteThickness[1.5],AbsoluteDashing[{6,6}]}}}]
```

to obtain the plot

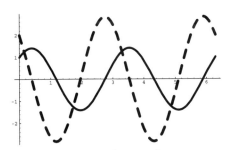

Solution to $x' = y,\;\; y' = -4x$
with the initial conditions $x(0) = 1,\;\; y(0) = 2$

(The option **PlotStyle** permits the specification of different styles for each of the curves in a simultaneous plot. The thickness of the curves can be controlled with **AbsoluteThickness**. Then **AbsoluteDashing[6,6]** creates a pattern of equally spaced dashes of moderate length.)

A more important way to display the solution to a system of differential equations graphically is to make a **phase portrait**, which consists of a plot of the curve $t \longmapsto \mathbf{x}(t) = \big(x_1(t), \ldots, x_n(t)\big)$, where $\mathbf{x}(t)$ is a solution to the system. Such a plot is possible when n equals 2 or 3. The classification of phase plots for two-dimensional systems $\mathbf{x}' = A\mathbf{x}$, where A is a constant matrix, was given in Section 16.1. The **ODE** plotting option to create a phase portrait is **PlotPhase**.

Example 16.5. *Make a phase portrait of the solution to the system* (16.3).

Solution. We use

```
ODE[{x' == y,y' == -4x,x[0] == 1,y[0] == 2},{x,y},t,
Method->LinearSystem,PlotPhase->{{t,0,2Pi}}]
```

to obtain the phase portrait

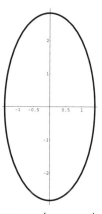

Solution to $x' = y, \quad y' = -4x$
with the initial conditions $x(0) = 1, \quad y(0) = 2$

PlotEvolvePhase is a variation of **PlotPhase**; whereas **PlotPhase** plots a plane curve $t \longmapsto \big(x(t), y(t)\big)$, **PlotEvolvePhase** plots the space curve $t \longmapsto \big(x(t), y(t), t\big)$. In many cases **PlotEvolvePhase** leads to increased understanding of the solution of a two-dimensional system.

Example 16.6. *Use* **ODE** *with the option* **PlotEvolvePhase** *to solve and plot the system*

$$\begin{cases} x' = \dfrac{x}{8} - y, \\ y' = x + \dfrac{y}{8}. \end{cases}$$

with the initial conditions

$$x(0) = 1, \quad y(0) = -1.$$

Solution. We use

```
ODE[{x' == x/8 - y,y' == x + y/8,x[0] == 1,y[0] == -1},
{x,y},t,Method->LinearSystem,
PlotEvolvePhase->{{t,-2Pi,2Pi}}]
```

to obtain the plot

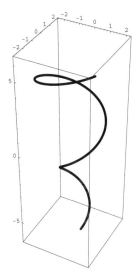

Solution to $x' = x/8 - y, \; y' = x + y/8$
with the initial conditions $x(0) = 1, \quad y(0) = -1$

Next, we tackle a three-dimensional system.

Example 16.7. *Use* **ODE** *to solve and plot the system*

$$\begin{cases} x' = x, \\ y' = 2x + y - 2z, \\ z' = 3x + 2y + z, \end{cases} \tag{16.4}$$

with the initial conditions

$$x(0) = 1, \quad y(0) = -1, \quad z(0) = 0.$$

Solution. To obtain the phase plot, we use

```
ODE[{x' ==   x,              x[0]  ==   1,
     y' == 2x +  y - 2z,y[0]  ==  -1,
     z' == 3x + 2y +  z,z[0]  ==   0},{x,y,z},t,
Method->LinearSystem,
PlotPhase->{{t,-Pi,Pi}}]
```

to get

$$\{\{x \to E^t, \; y \to \frac{-3\,E^t}{2} + \frac{E^t\,\mathrm{Cos}[2\,t]}{2} + E^t\,\mathrm{Sin}[2\,t],$$

$$z \to E^t - E^t\,\mathrm{Cos}[2\,t] + \frac{E^t\,\mathrm{Sin}[2\,t]}{2}\}\}$$

The solution is the space curve

$$t \longmapsto \left(e^t, \frac{e^t}{2}\big(-3 + \cos(2t) + 2\sin(t)\big), \frac{e^t}{2}\big(2 - 2\cos(2t) + \sin(2t)\big)\right).$$

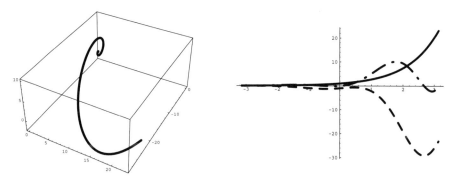

Phase and component plots of $x' = x, \quad y' = 2x + y - 2z, \quad z' = 3x + 2y + z$
with the initial conditions $x(0) = 1, \quad y(0) = -1, \quad z(0) = 0$

The command to produce the component plots is

```
ODE[{x'  ==   x,                  x[0]  ==   1,
      y'  == 2x  +  y  - 2z,y[0]  ==  -1,
      z'  == 3x + 2y  +   z,z[0]  ==   0},{x,y,z},t,
Method->LinearSystem,
PlotSolution->{{t,-Pi,Pi},PlotPoints->200,PlotRange->All,
  PlotStyle->{{AbsoluteThickness[1.5]},
  {AbsoluteThickness[1.5],AbsoluteDashing[{6,6}]},
  {AbsoluteThickness[1.5],RGBColor[1,0,0]}}}]
```

(Notice that **PlotStyle** is an option for **Plot** and **ParametricPlot**, but not for **ParametricPlot3D**. Consequently, **PlotStyle** is a valid plotting option for **PlotPhase** when it makes a phase portrait of a two-dimensional system, but not when it makes a phase portrait of a three-dimensional system.)

Seminumerical approximations to linear systems

A **seminumerical approximation to a constant-coefficient linear system** $x' = Ax$ is defined analogously to a seminumerical approximation to a constant-coefficient linear equation (see Section 12.5). The eigenvalues of the coefficient matrix A are obtained numerically and are then used to obtain the general solution of the system. Such a seminumerical approximation can be obtained using **ODE** with the option **Method->ApproximateLinearSystem**.

Example 16.8. *Use* ODE *with* **Method->ApproximateLinearSystem** *to solve and plot the system*

$$\begin{cases} x' = y, \\ y' = z, \\ z' = -x - y, \end{cases} \tag{16.5}$$

with the initial conditions $x(0) = 1, \quad y(0) = 2, \quad z(0) = 1/2.$

Solution. We use

```
ODE[{x'  ==   y,      x[0]  ==  1,
      y'  ==   z,      y[0]  ==  2,
      z'  == -x  - y,z[0]  ==  1/2},{x,y,z},t,
Method->ApproximateLinearSystem,
PostSolution->{Chop},PlotPhase->{{t,-4,4}}]
```

and

```
ODE[{x' == y,      x[0] == 1,
      y' == z,      y[0] == 2,
      z' == -x - y,z[0] == 1/2},{x,y,z},t,
Method->ApproximateLinearSystem,,PostSolution->{Chop}
PlotSolution->{{t,-4,4},
   PlotStyle->{{AbsoluteThickness[1.5]},
   {AbsoluteThickness[1.5],AbsoluteDashing[{6,6}]},
   {AbsoluteThickness[1.5],RGBColor[1,0,0]}}}]
```

to obtain

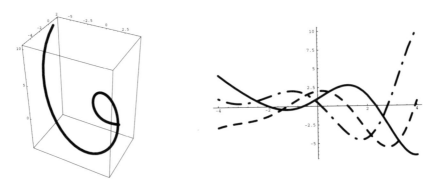

Phase and component plots of $x' = y, \quad y' = z, \quad z' = -x - y$
with the initial conditions $x(0) = 1, \quad y(0) = 2, \quad z(0) = 1/2$

Exercises for Section 16.2

Find the general solution to each of the following systems of differential equations using
ODE.

1. $\begin{cases} x' = x + 2y, \\ y' = 4x + 3y \end{cases}$

3. $\begin{cases} x' = 3x + 2y, \\ y' = -5x + y \end{cases}$

2. $\begin{cases} x' = y, \\ y' = -2x + 3y \end{cases}$

4. $\begin{cases} x' = 4x - 2y, \\ y' = -4x + 4y \end{cases}$

5. $\begin{cases} x' = 2x + y + 3e^{2t}, \\ y' = -4x + 2y + te^{2t} \end{cases}$ **7.** $\begin{cases} x' = -y - 5t, \\ y' = 3x - 4 \end{cases}$

6. $\begin{cases} x' = x + y + z - t, \\ y' = 2x + 2y + 2z - 4, \\ z' = 3x + 3y + 3z - \sin(t) \end{cases}$ **8.** $\begin{cases} x' + y' = x + y + z, \\ y' + z' = 2x + 2y + 2z, \\ z' + x' = 3x + 3y + 3z \end{cases}$

Solve and plot the following initial value problems using **ODE**.

9. $\begin{cases} x' = 2x + 4y, \quad x(0) = 2, \\ y' = -4x + 2y, \quad y(0) = -2 \end{cases}$

10. $\begin{cases} x' = -x + 3y, \quad x(0) = 0, \\ y' = 3x - y, \quad y(0) = 4 \end{cases}$

11. $\begin{cases} x' = 4x + 4y, \quad x(0) = 4, \\ y' = 3x - 4y, \quad y(0) = 1 \end{cases}$

12. $\begin{cases} x' = 5x/2 + 4y, \quad x(0) = 4, \\ y' = 3x - 4y, \quad y(0) = 1 \end{cases}$

13. $\begin{cases} x' = 3x + y - z, \quad x(0) = 1, \\ y' = x + 3y - z, \quad y(0) = 0, \\ z' = 3x + 3y - z, \quad z(0) = 0 \end{cases}$

14. $\begin{cases} x' = x - y - z, \quad x(0) = 1, \\ y' = \quad 2y + 3z, \quad y(0) = 0, \\ z' = \quad 3y + z, \quad z(0) = 0 \end{cases}$

15. $\begin{cases} x' = 7x - y + 6z, \quad x(0) = 1, \\ y' = -10x + 4y - 12z, \quad y(0) = 1, \\ z' = -2x + y - z, \quad z(0) = 1 \end{cases}$

16.3   Phase Portraits of Two-Dimensional Linear Systems via ODE

To create phase plots containing multiple trajectories, we use **ODE** with the plotting option **PlotPhase** in conjunction with the **Parameters** option. The following example illustrates the procedure.

Example 16.9. *Use* **ODE** *to solve the linear system*

$$\begin{cases} \dfrac{dx}{dt} = x, \\ \dfrac{dy}{dt} = 2y, \end{cases} \tag{16.6}$$

with the initial conditions $x(0) = a$ *and* $y(0) = b$. *Find the phase portrait for* $-2 \le a \le 2$ *at intervals of* $1/2$ *for the values* $b = \pm 1$.

Solution. We use the following command:

```
ODE[{x' == x,y' == 2y,x[0] == a,y[0] == b},{x,y},t,
Method->LinearSystem,Parameters->{{a,-2,2,0.5},{b,-1,1,2}},
PlotPhase->{{t,0,1.5},PlotRange->{{-6,6},{-7,7}}}]
```

to get

```
          t           2 t
{{x -> a E , y -> b E   }}
```

and the first plot below. The second plot is obtained by letting t vary between -2 and 1.5.

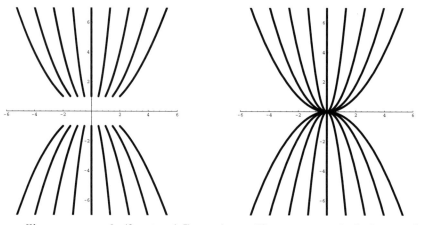

A repelling proper node $(0 \le t < 1.5)$ **A repelling proper node** $(-2 \le t < 1.5)$

The second phase plot is more visually appealing than the first. But to get it we must use negative as well as positive values of t. In fact, theoretically we should take all negative values of t. But in this case using the values of t between -2 and 0 is sufficient to bring the curves very close to the critical point $(0, 0)$. ∎

Exercises for Section 16.3

Draw the phase portraits of the following linear systems using **ODE**:

1. $\begin{cases} x'(t) = x, \\ y'(t) = 2y \end{cases}$

6. $\begin{cases} x'(t) = 2x + y, \\ y'(t) = -x + 2y \end{cases}$

2. $\begin{cases} x'(t) = x, \\ y'(t) = -2y \end{cases}$

7. $\begin{cases} x'(t) = x, \\ y'(t) = y \end{cases}$

3. $\begin{cases} x'(t) = y/2, \\ y'(t) = -2x \end{cases}$

8. $\begin{cases} x'(t) = -2x + 5y, \\ y'(t) = -x + y \end{cases}$

4. $\begin{cases} x'(t) = y, \\ y'(t) = -x \end{cases}$

9. $\begin{cases} x'(t) = x + y, \\ y'(t) = y \end{cases}$

5. $\begin{cases} x'(t) = \dfrac{3x + 5y}{2}, \\ y'(t) = \dfrac{-5x + 3y}{2} \end{cases}$

10. $\begin{cases} x'(t) = \dfrac{3x + 5y}{2}, \\ y'(t) = -\dfrac{5x + 3y}{2} \end{cases}$

17

STABILITY OF
NONLINEAR SYSTEMS

In this chapter we use the theory of linear systems to study the behavior of nonlinear systems of the form

$$\mathbf{x}'(t) = \boldsymbol{F}(t, \mathbf{x}), \tag{17.1}$$

where

$$\mathbf{x}(t) = \begin{pmatrix} x_1(t) \\ \vdots \\ x_n(t) \end{pmatrix} \quad \text{and} \quad \boldsymbol{F}(t, \mathbf{x}) = \begin{pmatrix} F_1(t, \mathbf{x}) \\ \vdots \\ F_n(t, \mathbf{x}) \end{pmatrix}.$$

In this generality we cannot hope to find an explicit formula for the general solution of (17.1). Instead, we study the qualitative aspects of (17.1) that can be ascertained without actually solving the equation. This has already been done in Section 5.6 for autonomous first-order equations.

We concentrate on the case of two simultaneous differential equations. Applications will be given in Chapters 18 and 19. Preparatory material concerning parametrically defined curves is given in Section 17.1. Differences between autonomous and nonautonomous systems are discussed in Section 17.2.

The basic approach given in this chapter to study nonlinear systems consists of two parts: first we determine certain solutions called *critical solutions*, just as we did in Section 5.6. Then we attempt to understand other solutions in terms of the critical solutions. The theory that we present gives a reasonable but rough idea of the noncritical solutions. Better approximations to noncritical solutions can be found by numerical methods, as we shall discover in Section 19.1.

Critical points for systems are defined and studied in Section 17.3. Section 17.4 is devoted to asymptotic stability. Linearization is described in Section 17.5. Lyapunov stability theory is described in Section 17.6.

17.1   Curves

Before proceeding further, let us say exactly what we mean by a curve. There are two versions.

Definition.   *Let $\alpha\colon (a, b) \longrightarrow \mathbb{R}^n$ be a function, where (a, b) is an open interval in $\mathbb{R}$. We allow the interval to be finite, infinite, or half-infinite. We write*

$$\alpha(t) = \big(a_1(t), \ldots, a_n(t)\big),$$

where each a_j is an ordinary real-valued function of a real variable. We say that α is **differentiable** *provided that a_j is differentiable for $j = 1, \ldots, n$. A* **parametrized curve** *in $\mathbb{R}^n$ is a differentiable function $\alpha\colon (a, b) \longrightarrow \mathbb{R}^n$.*

Parametrized curves are the ones that are easiest to deal with mathematically. The solution to a system of differential equations is naturally given as a curve.

On the other hand, many plane curves are given more often than not implicitly. For example, an ellipse centered at the origin with major axis of length a and minor axes of length b is given by the equation

$$\frac{x^2}{a^2} + \frac{y^2}{b^2} - 1 = 0.$$

This leads to the second version of the notion of curve.

Definition.   *An* **implicitly defined curve** *in $\mathbb{R}^2$ is the set of zeros of a differentiable function $F\colon \mathbb{R}^2 \longrightarrow \mathbb{R}$. Frequently, we refer to the set of zeros as "the implicitly defined curve $F(x, y) = 0$". Let S be a subset of $\mathbb{R}^2$. If $F\colon \mathbb{R}^2 \longrightarrow \mathbb{R}$ is a differentiable function whose set of zeros is S, we say that the equation $F(x, y) = 0$ is a* **nonparametric form** *of S. If $\alpha\colon (a, b) \longrightarrow \mathbb{R}^2$ is a curve whose trace is S (that is, $S = \{\alpha(t) \mid a < t < b\}$), we say that $t \longmapsto \alpha(t)$ is a* **parametrization**, *or* **parametric form**, *of S.*

Finding a parametrization of the implicitly defined curve $F(x, y) = 0$ requires finding a differentiable function $\alpha\colon (a, b) \longrightarrow \mathbb{R}^2$ such that $F\big(a_1(t), a_2(t)\big) \equiv 0$, where $\alpha(t) = \big(a_1(t), a_2(t)\big)$. On the other hand, finding an implicitly defined curve from a parametrized curve $\alpha = (a_1, a_2)$ in $\mathbb{R}^2$ requires finding a functional relation between a_1 and a_2.

Example 17.1.   *Show that the parametrized version of the folium of Descartes*

$$x^3 + y^3 - 3x\,y = 0 \tag{17.2}$$

is given by

$$\mathbf{folium}(t) = \left(\frac{3t}{1 + t^3}, \frac{3t^2}{1 + t^3} \right).$$

Solution. (See the picture on page 83.) The substitution $y = t\,x$ converts (17.2) to

$$0 = x^3 + t^3 x^3 - 3t\,x^2 = x^3(t^3 + 1) - 3x^2 t, \tag{17.3}$$

and from (17.3) we obtain

$$x = \frac{3t}{1 + t^3} \quad \text{and} \quad y = \frac{3t^2}{1 + t^3}.$$

Hence we obtain the parametrization (17.2). ∎

Exercises for Section 17.1

1. Find parametrizations for the ellipse $\dfrac{x^2}{a^2} + \dfrac{y^2}{b^2} - 1 = 0$ and the parabola $x^2 - 4a\,y = 0$.

2. A **cardioid** is the parametrized curve defined by

$$\mathbf{cardioid}[a](t) = \big(2a\,\cos(t)(1 + \cos(t)),\, 2a\,\sin(t)(1 + \cos(t))\big).$$

   Find the corresponding implicitly defined curve.

17.2 Autonomous Systems

Especially important among systems of differential equations are those that are time-independent in the following sense.

Definition. *An* **autonomous system of differential equations** *is one in which the independent variable t does not occur explicitly.*

Thus, a first-order autonomous system has the form

$$\begin{cases} \dfrac{dx_1}{dt} &= F_1(x_1, \ldots, x_n), \\[4pt] \quad\vdots & \quad\vdots \\[4pt] \dfrac{dx_n}{dt} &= F_n(x_1, \ldots, x_n). \end{cases} \tag{17.4}$$

An mth-order single equation of the form

$$y^{(m)} + F\left(y^{(m-1)}, \ldots, y\right) = 0 \tag{17.5}$$

is also autonomous. As we shall see, the time-independent nature of an autonomous system frequently allows us to determine qualitative properties of its solutions. Here is an important property of an autonomous system.

Lemma 17.1. *Suppose that* $t \longmapsto \phi(t)$ *is a solution to an autonomous system of the form* (17.4) *or* (17.5), *where*

$$\phi(t) = \left(\phi_1(t), \ldots, \phi_n(t)\right).$$

Then for any number t_0, *the function* $t \longmapsto \phi(t - t_0)$ *is also a solution.*

Proof. We give the details for (17.4); the proof for (17.5) is similar. We abbreviate (17.4) to

$$\mathbf{x}' = \boldsymbol{F}(\mathbf{x}), \tag{17.6}$$

where

$$\mathbf{x}(t) = \begin{pmatrix} x_1(t) \\ \vdots \\ x_n(t) \end{pmatrix} \quad \text{and} \quad \boldsymbol{F}(\mathbf{x}) = \begin{pmatrix} F_1(\mathbf{x}) \\ \vdots \\ F_n(\mathbf{x}) \end{pmatrix}.$$

Suppose that ϕ is a solution of (17.6), and define $\tilde{\phi}$ by $\tilde{\phi}(t) = \phi(t - t_0)$. The chain rule implies that

$$\tilde{\phi}'(t) = \phi'(t - t_0) = \boldsymbol{F}\left(\phi(t - t_0)\right) = \boldsymbol{F}\left(\tilde{\phi}(t)\right).$$

Thus, by definition, $\tilde{\phi}$ is a solution of (17.6). ∎

To explain geometrically the difference between autonomous and nonautonomous systems, let us consider an initial value problem of the form

$$\begin{cases} \mathbf{x}' = \boldsymbol{F}(t, \mathbf{x}), \\ \mathbf{x}(s) = \mathbf{p}, \end{cases} \tag{17.7}$$

where $\mathbf{p}$ is a point in $\mathbb{R}^n$. The solution of (17.6) is a curve passing through $\mathbf{p}$; in general, this curve will depend on s. However, for an *autonomous* system, the solution curves are formed from one another by means of a translation in the independent variable.

Example 17.2. *Solve the initial value problem*

$$\begin{cases} x' = x, & x(s) = 2, \\ y' = 2y, & y(s) = 1, \end{cases} \tag{17.8}$$

and show that all solutions lie on the parabola $4y = x^2$.

Solution. Since each equation of (17.8) can be solved separately, the solution of (17.8) is found to be the parametrically defined curve

$$\begin{cases} x(s) = 2e^{t-s}, \\ y(s) = e^{2(t-s)}. \end{cases} \tag{17.9}$$

It is easy to eliminate e^{t-s} from (17.9); thus $4y = x^2$. ∎

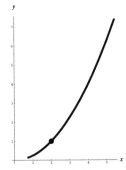

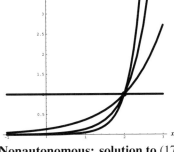

Autonomous: solution to (17.8)   **Nonautonomous: solution to** (17.10)

The solution curves of a nonautonomous system can be quite different from those of an autonomous system.

Example 17.3. *Solve the initial value problem*

$$\begin{cases} x' = x/t, & x(s) = 2, \\ y' = 2y & y(s) = 1. \end{cases} \tag{17.10}$$

Show that the solution curve depends on s.

Solution. The solution (as a parametric curve) of the linear system (17.10) is easily found to be

$$\begin{cases} x(t) = 2t/s, \\ y(t) = e^{2(t-s)}. \end{cases} \tag{17.11}$$

When we eliminate t from (17.11), we get $y = e^{s(x-2)}$. ∎

Exercises for Section 17.2

1. Show that if $t \longmapsto y(t)$ is a solution of the nth-order equation (17.5), then so is $t \longmapsto y(t - t_0)$.

2. Use **ODE** with **Method->DSolve** and **PlotPhase** to plot several solution curves of the nonautonomous system

$$\begin{cases} x' = t, & x(s) = 1, \\ y' = t^2, & y(s) = 1 \end{cases}$$

17.3   *Critical Points of Systems of Differential Equations*

An autonomous system of two differential equations can be written in the form

$$\begin{cases} \dfrac{dx}{dt} = F(x, y), \\[2mm] \dfrac{dy}{dt} = G(x, y). \end{cases} \tag{17.12}$$

Here the functions $F(x, y)$ and $G(x, y)$ are assumed to be smooth, so that local existence is satisfied via Theorem 15.1. If $F(x, y)$ and $G(x, y)$ are linear functions of x and y, we say that (17.12) is **linear**; otherwise (17.12) is said to be **nonlinear**.

If we solve an initial value problem corresponding to (17.12), we get a parametrized plane curve. Such a curve is often called a **trajectory** of the system. Alternatively, we can combine the two equations of (17.12) to obtain

$$\frac{dy}{dx} = \frac{G(x, y)}{F(x, y)}. \tag{17.13}$$

The solution of an initial value problem corresponding to (17.13) is an implicitly defined curve in $\mathbb{R}^2$. It may be difficult to solve (17.13) directly.

In Section 5.6 we found that constant or critical solutions of autonomous first-order equations were of particular significance. Let us generalize the notion of critical solution to autonomous systems.

Definition.   *A point (x_c, y_c) is called a* **critical point** *for the autonomous system* (17.12) *if it is simultaneously a zero of both $F(x, y)$ and $G(x, y)$, that is,*

$$F(x_c, y_c) = G(x_c, y_c) = 0.$$

If (x_c, y_c) is a critical point of (17.12), then we automatically have a solution of the system (17.12), namely,

$$\begin{cases} x(t) \equiv x_c, \\ y(t) \equiv y_c. \end{cases} \tag{17.14}$$

(The left-hand sides of (17.12) are zero because $x(t)$ and $y(t)$ are constants, while the right-hand sides of (17.12) are zero because x_c and y_c are critical points.) This leads to the following definition.

Definition. *Suppose that (x_c, y_c) is a critical point for the autonomous system* (17.12). *The constant solution*

$$\big(x(t), y(t)\big) \equiv (x_c, y_c)$$

is called a **critical solution,** *or* **equilibrium solution,** *of* (17.12).

Critical solutions can also be used to study other solutions of the system (17.12). The uniqueness theorem (Theorem 15.1) for a nonlinear system (17.12) states that there is a unique solution of the system passing through each point of the plane. In particular, the solution that passes through a critical point is the constant solution (17.14). Therefore, we see a second role of critical points:

Theorem 17.2. *A nonconstant solution of the system* (17.12) *cannot pass through a critical point.*

It must be understood that a nonconstant solution can *approach* a critical point as time increases without ever passing through the critical point at a finite time. For example, the system

$$\begin{cases} x'(t) = -x, \\ y'(t) = -2y \end{cases} \qquad (17.15)$$

has $(0, 0)$ as a critical point.

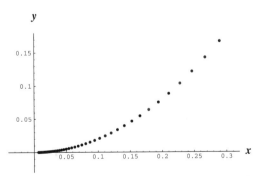

A nonconstant solution to the system $x' = -x \, , y' = -2y$
approaching the critical point $(0, 0)$

The general solution of (17.15) is easily seen to be

$$\begin{cases} x(t) = C_1 e^{-t}, \\ y(t) = C_2 e^{-2t}, \end{cases}$$

where C_1 and C_2 are constants. Assume that at least one of C_1 and C_2 is nonzero so that $t \longmapsto \big(x(t), y(t)\big)$ is nonconstant. Then $\big(x(t), y(t)\big)$ approaches $(0, 0)$ when $t \longrightarrow \infty$, but at any finite value of t we have $\big(x(t), y(t)\big) \neq (0, 0)$.

Example 17.4. *Find the critical points of the system*

$$\frac{dx}{dt} = x(1 - 4x + 2y),$$

$$\frac{dy}{dt} = y(1 - x - y).$$

Solution. We must find the simultaneous solutions of the equations

$$\begin{cases} x(1 - 4x + 2y) = 0, \\ y(1 - x - y) = 0. \end{cases} \tag{17.16}$$

From the first equation of (17.16) we must have either $x = 0$ or $4x - 2y = 1$. From the second equation we must have either $y = 0$ or $x + y = 1$. If $x = 0$, then the second equation dictates the choices $y = 0$ and $y = 1$. If $4x - 2y = 1$, then the second equation dictates either $y = 0$, $x = 1/4$, or $x + y = 1$. Thus the system (17.16) has four critical points:

$$(0, 0), \quad (0, 1), \quad (\frac{1}{4}, 0), \quad (\frac{1}{2}, \frac{1}{2}). \ \blacksquare$$

Linearized Approximation at a Critical Point

In order to study a system of differential equations in the neighborhood of a critical point, we must first obtain the associated linear system. To do this, we recall (see for example [Frank, page 64]) the appropriate form of Taylor's theorem for a function of two variables:

Theorem 17.3. **(First-order Taylor's theorem)** *Suppose that $f(x, y)$ is a differentiable function of x and y in some rectangle of the x, y plane that contains the point (a, b). Then $f(x, y)$ is well approximated by the linear function $f_1(x, y)$, where*

$$f_1(x, y) = f(a, b) + (x - a)\frac{\partial f}{\partial x}(a, b) + (y - b)\frac{\partial f}{\partial y}(a, b),$$

in the sense that

$$f(x, y) = f_1(x, y) + \varepsilon(x, y),$$

where the remainder term $\varepsilon(x, y)$ satisfies

$$\lim_{(x,y) \to (a,b)} \frac{\varepsilon(x, y)}{\sqrt{(x - a)^2 + (y - b)^2}} = 0.$$

We call the function $f_1(x, y)$ the **first-order Taylor approximation** *to the function $f(x, y)$.*

Geometrically, Theorem 17.3 allows us to approximate the surface $z = f(x, y)$ by the plane

$$z = f(a, b) + (x - a)\frac{\partial f}{\partial x}(a, b) + (y - b)\frac{\partial f}{\partial y}(a, b)$$

in a neighborhood of a point (a, b). The given function $f(x, y)$ may be very complicated, but the first-order Taylor approximation is always a linear function.

Example 17.5. *Find the first-order Taylor approximation of the function*

$$f(x, y) = x^2 \cos(y)$$

at the point $(1, \pi/3)$.

Solution. We have $f(1, \pi/3) = 1/2$; furthermore, the first partials of $f(x, y)$ at $(1, \pi/3)$ are given by

$$\frac{\partial f}{\partial x}(1, \pi/3) = 2x\cos(y)\bigg|_{\substack{x=1 \\ y=\pi/3}} = 1 \quad \text{and} \quad \frac{\partial f}{\partial y}(1, \pi/3) = -x^2\sin(y)\bigg|_{\substack{x=1 \\ y=\pi/3}} = -\frac{\sqrt{3}}{2}.$$

The first-order Taylor approximation is

$$F_1(x, y) = \frac{1}{2} + (x - 1) - \frac{\sqrt{3}}{2}\left(y - \frac{\pi}{3}\right). \quad \blacksquare$$

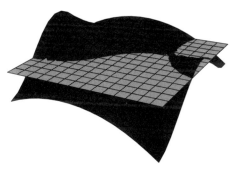

First-order Taylor approximation to
$z = x^2\cos(y)$ **at the point** $(1, \pi/3)$

The first-order Taylor approximation becomes especially simple in the case of a critical point of an autonomous system of nonlinear differential equations of the form (17.12). In that case we have $F(x_c, y_c) = G(x_c, y_c) = 0$, and the first-order Taylor approximations of the functions $F(x, y)$ and $G(x, y)$ reduce to

$$(x - x_c)\frac{\partial F}{\partial x}(x_c, y_c) + (y - y_c)\frac{\partial F}{\partial y}(x_c, y_c)$$

and

$$(x - x_c)\frac{\partial G}{\partial x}(x_c, y_c) + (y - y_c)\frac{\partial G}{\partial y}(x_c, y_c).$$

This leads us to consider a *linear system* of differential equations.

Definition. *The* **linearized system** *corresponding to the critical point* (x_c, y_c) *of the autonomous nonlinear system*

$$\begin{cases} \dfrac{dx}{dt} = F(x, y), \\[2mm] \dfrac{dy}{dt} = G(x, y) \end{cases}$$

is the constant-coefficient linear system

$$\begin{cases} \dfrac{dx_1}{dt} = a_{11}x_1 + a_{12}x_2, \\[2mm] \dfrac{dx_2}{dt} = a_{21}x_1 + a_{22}x_2, \end{cases} \tag{17.17}$$

where we make the identifications

$$x_1 = x - x_c, \qquad\qquad x_2 = y - y_c,$$

$$a_{11} = \frac{\partial F}{\partial x}(x_c, y_c), \qquad a_{12} = \frac{\partial F}{\partial y}(x_c, y_c),$$

$$a_{21} = \frac{\partial G}{\partial x}(x_c, y_c), \qquad a_{22} = \frac{\partial G}{\partial y}(x_c, y_c).$$

We emphasize that for each critical point there is a different linearized system; the solution of one linearized system may be radically different from another. When we "piece together" the different linearized systems, we may obtain an accurate picture of the nonlinear system.

Example 17.6. *For the system*

$$\begin{cases} \dfrac{dx}{dt} = y, \\[2mm] \dfrac{dy}{dt} = -\sin(x), \end{cases}$$

find the linearized systems at the critical points $(0, 0)$ *and* $(\pi, 0)$.

Solution. The Taylor expansion of $\sin(x)$ about 0 is

$$\sin(x) = x - \frac{x^3}{6} + \cdots = x + \varepsilon(x).$$

Thus, for the critical point $(0, 0)$ we have the first-order Taylor approximation

$$\begin{cases} F_1(x, y) = y, \\ G_1(x, y) = -x, \end{cases}$$

which yields the linearized system

$$\begin{cases} \dfrac{dx_1}{dt} = x_2, \\ \dfrac{dx_2}{dt} = -x_1. \end{cases}$$

The Taylor expansion of $\sin(x)$ about π is

$$\sin(x) = -(x - \pi) + \frac{(x - \pi)^3}{6} + \cdots = -(x - \pi) + \tilde{\varepsilon}(x - \pi).$$

Hence for the critical point $(\pi, 0)$ we have the first-order Taylor approximation

$$\begin{cases} F_1(x, y) = y, \\ G_1(x, y) = x - \pi, \end{cases}$$

which leads to the linearized system

$$\begin{cases} \dfrac{dx_1}{dt} = x_2, \\ \dfrac{dx_2}{dt} = x_1. \end{cases} \blacksquare$$

Example 17.6 is typical of the different possible behaviors. The linearized system for the critical point $(0, 0)$ has a stable center (as defined on page 565), while the linearized system for the critical point $(\pi, 0)$ has a saddle as a critical point (as defined on page 555).

Exercises for Section 17.3

For each of the following systems, find all critical points and the linearized system associated with each critical point.

1. $\begin{cases} x' = x + y^2, \\ y' = x + y \end{cases}$

6. $\begin{cases} x' = \sin(y), \\ y' = \cos(x) \end{cases}$

2. $\begin{cases} x' = 1 - x\,y, \\ y' = x - y^3 \end{cases}$

7. $\begin{cases} x' = (\sin(y))^2, \\ y' = (\cos(x))^2 \end{cases}$

3. $\begin{cases} x' = x - x^2 - x\,y, \\ y' = 3y - x\,y - 2y^2 \end{cases}$

8. $\begin{cases} x' = y, \\ y' = -\sin(x) - 4y \end{cases}$

4. $\begin{cases} x' = 1 - y, \\ y' = x^2 - y^2 \end{cases}$

9. $\begin{cases} x' = x(1 - x - y), \\ y' = y(3 - 2y - 2x) \end{cases}$

5. $\begin{cases} x' = y, \\ y' = -\sin(x) \end{cases}$

10. $\begin{cases} x' = x(3 - 2y), \\ y' = y(-4 + 2x) \end{cases}$

17.4   Stability and Asymptotic Stability of Nonlinear Systems

We next discuss the behavior of a two-dimensional autonomous system of the form (17.12) in the vicinity of an isolated critical point.

Definition.   *A critical point (x_c, y_c) of the system (17.12) is said to be* **isolated** *if there is some rectangle $R = \{(x, y) \mid a < x < b, c < y < d\}$ containing (x_c, y_c) for which there are no other critical points in this rectangle.*

Obviously, if a system of the form (17.12) has only finitely many critical points in each rectangle, they are all isolated. This is the only situation that we shall consider.

In Section 5.6 we defined the notions of stability and instability for a single first-order autonomous differential equation. Here is a similar definition for an autonomous system:

Definition.   *An isolated critical point (x_c, y_c) of the system (17.12) is said to be* **stable** *if for each $\varepsilon > 0$ there exists $\delta > 0$ such that*

$$\left(x(0) - x_c\right)^2 + \left(y(0) - y_c\right)^2 < \delta \quad \text{implies} \quad \left(x(t) - x_c\right)^2 + \left(y(t) - y_c\right)^2 < \varepsilon$$

for all $t > 0$. An isolated critical point (x_c, y_c) of the system (17.12) is said to be **unstable** *if it is not stable.*

Let us paraphrase this rather technical definition. In ordinary language, stability means that *any trajectory that starts close to the critical point will remain close to the critical point for all time.*

Example 17.7. *Show that* $(0, 0)$ *is an isolated stable critical point for the linear system*

$$\begin{cases} \dfrac{dx}{dt} = -y, \\[2mm] \dfrac{dy}{dt} = 2x. \end{cases} \tag{17.18}$$

Solution. That $(0, 0)$ is an isolated critical point follows from the fact that it is the only critical point. The system (17.18) can be solved by the methods of Chapter 15, but we can also solve it directly. If we differentiate the first equation of (17.18) and use the second equation, we get $x'' + 2x = 0$, the general solution of which is

$$x(t) = A \cos(\sqrt{2}\, t) + B \sin(\sqrt{2}\, t), \tag{17.19}$$

where A and B are constants of integration. From (17.19) and the first equation of (17.18) we find that

$$y(t) = \sqrt{2}\left(-A \sin(\sqrt{2}\, t) + B \cos(\sqrt{2}\, t) \right). \tag{17.20}$$

Then (17.19) and (17.20) imply that

$$2x^2 + y^2 = 2A^2 + B^2,$$

which is the nonparametric equation of an ellipse. Given $\varepsilon > 0$ we can take $\delta = \varepsilon/2$ to satisfy the definition of stability. ∎

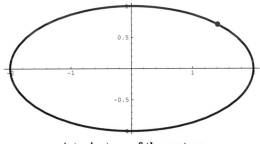

A trajectory of the system
$$x' = y, \ \ y' = -2x$$

Thus, the trajectory is an ellipse; we can imagine a nonconstant solution of (17.18) as a point moving around an ellipse.

Next, we give a simple example of an unstable critical point.

Example 17.8. *Show that the critical point* $(0, 0)$ *is an isolated unstable critical point for the linear system*

$$\begin{cases} \dfrac{dx}{dt} = y, \\[2mm] \dfrac{dy}{dt} = 2x. \end{cases} \tag{17.21}$$

Solution. The system (17.21) can also be solved by the methods of Chapter 15, or directly as in Example 17.7. We find that the solutions lie on hyperbolas of the form $2x^2 - y^2 = C$, where C is a constant. No matter how small we restrict the initial data $x(0)$ and $y(0)$, the solution $t \longmapsto \big(x(t), y(t)\big)$ tends to ∞ when $t \longrightarrow \infty$. In particular, it is impossible to satisfy the condition $x(t)^2 + y(t)^2 < \varepsilon$ for all t. Hence the critical point $(0, 0)$ is unstable. ∎

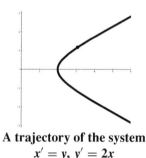

A trajectory of the system
$$x' = y, \ y' = 2x$$

A nonconstant solution of (17.21) can be considered to be a point moving along a hyperbola.

For many systems we may expect a stronger form of stability, corresponding to a loss of energy in a physical setting. For example, the motion of a mass-spring system with a frictional term may be expected to dissipate over the course of time. This leads us to the notion of asymptotic stability, expressed as follows:

Definition. *The isolated critical point of the system* (17.12)*,* (x_c, y_c)*, is said to be* **asymptotically stable** *provided that there exists* $\delta > 0$ *such that*

$$\big(x(0) - x_c\big)^2 + \big(y(0) - y_c\big)^2 < \delta \quad implies \quad \lim_{t \to \infty} \big(x(t), y(t)\big) = (x_c, y_c).$$

This definition can be paraphrased as stating that *any trajectory that starts close enough to the critical point is ultimately attracted to the critical point.*

Example 17.9. *Show that* $(0, 0)$ *is an isolated asymptotically stable critical point for the linear system*

$$\begin{cases} \dfrac{dx}{dt} = -x, \\[2mm] \dfrac{dy}{dt} = -2y. \end{cases} \tag{17.22}$$

Solution. Again $(0,0)$ is an isolated critical point because it is the only critical point. Each differential equation in the system (17.22) can be solved separately; we obtain

$$\begin{cases} x(t) = x(0)e^{-t}, \\ y(t) = y(0)e^{-2t}. \end{cases}$$

Clearly,

$$\lim_{t\to\infty} x(t) = \lim_{t\to\infty} y(t) = 0.$$

Therefore, the critical point $(0,0)$ is asymptotically stable. ∎

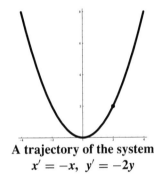

A trajectory of the system
$$x' = -x, \ \ y' = -2y$$

A nonconstant solution of (17.22) can be considered to be a point moving along a parabola toward $(0,0)$.

It is possible for a critical point to be stable without being asymptotically stable. For example, the system considered in Example 17.7 was shown to be stable at $(0,0)$. But the solutions lie on ellipses $2x^2 + y^2 = C > 0$ and hence do not satisfy the condition that

$$\lim_{t\to\infty} \big(x(t), y(t)\big) = (0,0),$$

which is required of asymptotic stability. The system of Example 17.7 is stable without being asymptotically stable.

17.5   Stability by Linearized Approximation

In this section we formulate conditions for stability and asymptotic stability that can be inferred from the linearized approximation at a critical point. In Section 17.6 we consider more general stability conditions obtained by the methods of Lyapunov.

Stability of Linear Systems with Constant Coefficients

Concrete conditions for stability and asymptotic stability are particularly easy to formulate in the case of two-dimensional homogeneous linear systems with constant-coefficients. Recall from Chapter 15 that these are systems of the form

$$\begin{cases} \dfrac{dx}{dt} = a_{11}x + a_{12}y, \\ \dfrac{dy}{dt} = a_{21}x + a_{22}y, \end{cases} \tag{17.23}$$

where the a_{ij}'s are constants. Clearly, $(0,0)$ is the unique critical point of (17.23). It will be an isolated critical point if and only if 0 is not an eigenvalue of the matrix A that defines the system. This, in turn, is equivalent to the statement that $\det(A) \neq 0$, or in detail,

$$a_{11}a_{22} - a_{12}a_{21} \neq 0.$$

The following theorem gives the stability criteria for linear systems.

Theorem 17.4.   *Suppose that the eigenvalues of the real matrix*

$$A = \begin{pmatrix} a_{11} & a_{12} \\ a_{21} & a_{22} \end{pmatrix}$$

have real parts less than zero, that is,

$$r = \alpha + i\beta \qquad with \qquad \alpha < 0.$$

Then the critical point $(0,0)$ of the two-dimensional system (17.23) is stable and asymptotically stable. If the eigenvalues are purely imaginary, that is, if

$$r = \pm i\beta \qquad with \qquad \beta \neq 0,$$

then the critical point $(0,0)$ is stable, but not asymptotically stable.

Proof.   We examine the cases separately. Let r_1 and r_2 be the eigenvalues of A. If $r_1 < r_2 < 0$, then the solution of (17.23) is written

$$\mathbf{x}(t) = C_1\boldsymbol{\xi}_1 e^{r_1 t} + C_2\boldsymbol{\xi}_2 e^{r_2 t},$$

where $\boldsymbol{\xi}_i$ is an eigenvector corresponding to r_i for $i = 1, 2$, and C_1 and C_2 are constants. Both $e^{r_1 t}$ and $e^{r_2 t}$ tend to zero when $t \longrightarrow \infty$, so that $\lim\limits_{t \to \infty} \mathbf{x}(t) = 0$. If $r_1 = r_2 = r < 0$, then the solution of (17.23) is written similarly as

$$\mathbf{x}(t) = e^{rt}\mathbf{c}_1 + t\, e^{rt}\mathbf{c}_2,$$

where $\mathbf{c}_1$ and $\mathbf{c}_2$ are constant vectors. We conclude similarly that $\lim_{t\to\infty} \mathbf{x}(t) = 0$. In each case we have both stability and asymptotic stability.

If the eigenvalues of A are purely imaginary, then $(0, 0)$ is a stable center, so that the solutions remain on ellipses of the form $a x^2 + 2b x y + c y^2 = C > 0$, where a, b, and c depend on a_{11}, a_{12}, a_{21}, and a_{22}, and C depends on $\mathbf{x}(0)$. (The proof is similar to that of Example 17.7.) Clearly, this satisfies the definition of stability: given $\varepsilon > 0$, we may choose C such that the ellipse lies within the circle of radius ε, so that for any starting point on this ellipse, the solution remains forever within the circle of radius ε. On the other hand, since the minimum distance of a point on the ellipse from $(0, 0)$ is nonzero, a solution can never approach $(0, 0)$. Hence $(0, 0)$ is not asymptotically stable. ∎

It is easy to determine the cases of stability and asymptotic stability for two-dimensional linear systems by using Theorem 17.4 combined with the classification of Section 16.1 given on page 566:

Corollary 17.5. *The critical point $(0, 0)$ of the two-dimensional system* (17.23) *is stable in the following cases:*

- $(0, 0)$ *is an* **attracting proper node**, *with $r_1 < r_2 < 0$;*

- $(0, 0)$ *is an* **attracting improper node**, *with $r_1 = r_2 < 0$;*

- $(0, 0)$ *is an* **attracting degenerate node**, *with $r_1 = r_2 < 0$;*

- $(0, 0)$ *is an* **attracting spiral**, *with $r = \alpha + i\beta$ where $\alpha < 0$ and $\beta \neq 0$;*

- $(0, 0)$ *is a* **stable center**, *with $r = i\beta$, where $\beta \neq 0$.*

In all of these cases, $(0, 0)$ is an asymptotically stable critical point of (17.23) *with the exception of a stable center.*

Proof. When the real parts of the eigenvalues of a linear system are strictly negative, we have both stability and asymptotic stability. In the case of a stable center, the real parts of the eigenvalues are zero, which leads to stability but not asymptotic stability, as noted on page 565. ∎

We now return to the general stability study for nonlinear systems.

Stability of Nonlinear Systems

In the case of a nonlinear system, it is natural to study stability and asymptotic stability in terms of a linearized system at each critical point. Recall that the linearized system $\mathbf{x}' = A\mathbf{x}$

associated with the critical point (x_c, y_c) of the two-dimensional system

$$\begin{cases} \dfrac{dx}{dt} = F(x, y), \\[2mm] \dfrac{dy}{dt} = G(x, y) \end{cases} \tag{17.24}$$

is defined by the matrix of partial derivatives

$$A = \begin{pmatrix} \dfrac{\partial F}{\partial x}(x_c, y_c) & \dfrac{\partial F}{\partial y}(x_c, y_c) \\[4mm] \dfrac{\partial G}{\partial x}(x_c, y_c) & \dfrac{\partial G}{\partial y}(x_c, y_c) \end{pmatrix}.$$

We have the following result.

Theorem 17.6. *Suppose that (x_c, y_c) is an isolated critical point of the nonlinear system (17.24) and that all eigenvalues of the matrix A of the linearized system at (x_c, y_c) have strictly negative real parts. Then the isolated critical point (x_c, y_c) is asymptotically stable for the nonlinear system.*

The proof of this theorem is deferred until Section 17.6, where we develop the general method of Lyapunov stability for n-dimensional systems. Specifically, it will follow from Theorem (17.9) and Theorem (17.10), which are proved there.

We illustrate the use of Theorem 17.6 with the system that governs a simple pendulum with damping, corresponding to the second-order equation

$$x'' + 2x' + \sin(x) = 0.$$

Example 17.10. *Show that the isolated critical point $(0, 0)$ is asymptotically stable for the system*

$$\begin{cases} \dfrac{dx}{dt} = y, \\[2mm] \dfrac{dy}{dt} = -2y - \sin(x). \end{cases} \tag{17.25}$$

Solution. We have $F(x, y) = y$ and $G(x, y) = -2y - \sin(x)$, so that the matrix of partial derivatives is computed as

$$A = \begin{pmatrix} 0 & 1 \\ -\cos(x) & -2 \end{pmatrix} \bigg|_{\substack{x=0 \\ y=0}} = \begin{pmatrix} 0 & 1 \\ -1 & -2 \end{pmatrix}.$$

The matrix A has -1 as a double eigenvalue. Therefore, Theorem 17.6 implies that $(0, 0)$ is an asymptotically stable critical point for the nonlinear system 17.25. ∎

Spiraling Behavior of Nonlinear Systems

We can also discuss the spiraling properties of the solution of a nonlinear system at a critical point for which the linearized system is an attracting spiral. According to the analysis of Section 16.1, spiraling for the linearized system occurs when the real parts of the eigenvalues of the linearized system are complex numbers with strictly negative real part: $r = \alpha \pm i\beta$ with $\alpha < 0$ and $\beta \neq 0$. The precise result for nonlinear systems is stated as follows.

Theorem 17.7. *Suppose that (x_c, y_c) is an isolated critical point of the nonlinear system (17.24) and that all of the eigenvalues of the matrix A of the corresponding linearized system are of the form $r = \alpha \pm i\beta$ with $\alpha < 0$ and $\beta \neq 0$. Let the solution of the nonlinear system be written in the form*

$$\begin{cases} x(t) = r(t)\cos(\theta(t)), \\ y(t) = r(t)\sin(\theta(t)) \end{cases}$$

for a suitable choice of the polar coordinates $r(t)$ and $\theta(t)$. Then there exists $\delta > 0$ such that whenever $(x(0) - x_c)^2 + (y(0) - y_c)^2 < \delta$, we have

$$\lim_{t\to\infty} r(t) = 0 \qquad and \qquad \lim_{t\to\infty} |\theta(t)| = \infty.$$

Proof. The statement concerning $r(t)$ is a reiteration of the asymptotic stability of the nonlinear system given in Theorem 17.6. The second part states that we have spiraling behavior. If

$$\lim_{t\to\infty} \theta(t) = +\infty,$$

then we have a counterclockwise spiral, while if

$$\lim_{t\to\infty} \theta(t) = -\infty,$$

then we have a clockwise spiral. To see this in detail, we consider the representative case

$$\begin{cases} x' = \alpha x + \beta y + \varepsilon_1(x, y), \\ y' = -\beta x + \alpha y + \varepsilon_2(x, y), \end{cases}$$

where $\alpha < 0$ and

$$\lim_{(x,y)\to(0,0)} \frac{\varepsilon_i(x, y)}{\sqrt{x^2 + y^2}} = 0$$

for $i = 1, 2$. The polar coordinates (r, θ) satisfy $r^2 = x^2 + y^2$ and $\theta = \arctan(y/x)$, so that

$$2r\,r' = 2x\,x' + 2y\,y' = 2x\big(\alpha x + \beta y + \varepsilon_1(x, y)\big) + 2y\big(-\beta x + \alpha y + \varepsilon_2(x, y)\big)$$

$$= 2\alpha(x^2 + y^2) + 2x\,\varepsilon_1(x, y) + 2y\,\varepsilon_2(x, y). \tag{17.26}$$

We summarize (17.26) as

$$r' = \alpha r + r \, \varepsilon(r, \theta),$$

where $\varepsilon(r, \theta) = (x \, \varepsilon_1 + y \, \varepsilon_2)/r^2$; then

$$\lim_{r \to 0} \varepsilon(r, \theta) = 0.$$

Similarly,

$$\theta' = \frac{x \, y' - x' \, y}{x^2 + y^2} = \frac{x\big(-\beta x + \alpha y + \varepsilon_2(x, y)\big) - y\big(\alpha x + \beta y + \varepsilon_1(x, y)\big)}{x^2 + y^2}$$

$$= \frac{-\beta \, (x^2 + y^2) + x \, \varepsilon_2(x, y) - y \, \varepsilon_1(x, y)}{x^2 + y^2},$$

which can be summarized in the form

$$\theta' = -\beta + \tilde{\varepsilon}(r, \theta), \tag{17.27}$$

where $\tilde{\varepsilon}(r, \theta) = (x \, \varepsilon_2 - y \, \varepsilon_1)/r^2$; then

$$\lim_{r \to 0} \tilde{\varepsilon}(r, \theta) = 0.$$

Integrating both sides of (17.27), we obtain

$$\theta(t) = \theta(0) - \beta t + \int_0^t \tilde{\varepsilon}\big(r(s), \theta(s)\big) \, ds. \tag{17.28}$$

We divide both sides of (17.28) by t and take the limit as $t \longrightarrow \infty$; we note that the integrand on the right-hand side of (17.28) tends to zero. Hence

$$\lim_{t \to \infty} \frac{1}{t} \int_0^t \tilde{\varepsilon}\big(r(s), \theta(s)\big) \, ds = 0,$$

and we have proved that

$$\lim_{t \to \infty} \frac{\theta(t)}{t} = -\beta,$$

as required. ∎

Example 17.11.   *For which values of the damping constant c can we expect asymptotic stability and spiraling behavior for the nonlinear system*

$$\begin{cases} \dfrac{dx}{dt} = y, \\[2mm] \dfrac{dy}{dt} = -\sin(x) - c \, y \end{cases} \tag{17.29}$$

with respect to the critical point $(0, 0)$?

Solution. We have $F(x, y) = y$ and $G(x, y) = -c\,y - \sin(x)$, so that the matrix of partial derivatives is computed as

$$A = \begin{pmatrix} 0 & 1 \\ -\cos(x) & -c \end{pmatrix} \Bigg|_{\substack{x=0 \\ y=0}} = \begin{pmatrix} 0 & 1 \\ -1 & -c \end{pmatrix}.$$

The eigenvalues of A are determined by solving the equation $0 = r(r + c) + 1 = 0$, whose solutions are

$$r = \frac{-c \pm \sqrt{c^2 - 4}}{2}.$$

If $c > 0$, then both eigenvalues have negative real part, so that we have asymptotic stability at $(0, 0)$. Similarly, if $c < 0$, we have instability. If $|c| < 2$, then the imaginary part of the eigenvalue r is nonzero, so that we can infer spiraling behavior from Theorem 17.7. ∎

It is also true in Example 17.11 that the spiraling is clockwise, as seen by noting that the coefficient $a_{12} = 1$ for the matrix of the linearized system. The system ((17.29)) is plotted below for $-0.2 \le c \le -0.05$ and then for $0 \le c \le 2$.

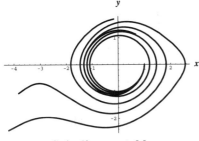
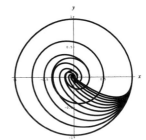

Spiraling unstable   **Spiraling asymptotically stable**

The proofs of Theorems 17.6, 17.7 and other theorems of this type can be obtained by a far-reaching idea introduced by Lyapunov[1] more than one hundred years ago. This method of *Lyapunov stability theory* will be explored in Section 17.6 and applied to both stability and asymptotic stability of nonlinear systems.

---

[1]

Aleksandr Mikhailovich Lyapunov (1857–1918). Russian mathematician. Lyapunov was a student of Chebyshev and a school friend of Markov.

Nonlinear Systems Compared with Linear Systems

The following table compares stability information of nonlinear systems at a critical point, corresponding to the thirteen phase portraits that were detailed in Chapter 16 on page 566. The term *indeterminate* means that no statement can be made in general; some nonlinear systems will destroy the stability of a linear system at a critical point, other nonlinear systems will preserve the stability of the critical point with the same linearized system.

Eigenvalues	Phase Portrait	Linearized Stability Type	Nonlinear Stability Type
$0 < r_1 < r_2$	Repelling proper node	Unstable	Unstable
$r_1 < r_2 < 0$	Attracting proper node	Asymptotically stable	Asymptotically stable
$r_1 < 0 < r_2$	Saddle	Unstable	Unstable
$0 = r_1 < r_2$	Repelling center	Unstable	Unstable
$r_1 < r_2 = 0$	Attracting center	Stable	Indeterminate
$r_1 = r_2 > 0$ two eigenvectors	Repelling degenerate node	Unstable	Unstable
$r_1 = r_2 < 0$ two eigenvectors	Attracting degenerate node	Asymptotically stable	Asymptotically stable
$r_1 = r_2 > 0$ one eigenvector	Repelling improper node	Unstable	Unstable
$r_1 = r_2 < 0$ one eigenvector	Attracting improper node	Asymptotically stable	Asymptotically stable
$r_1 = r_2 = 0$ one eigenvector	Improper center	Unstable	Indeterminate
$r_1, r_2 = \lambda \pm i\mu$ $\lambda > 0, \mu \neq 0$	Repelling spiral	Unstable	Unstable
$r_1, r_2 = \lambda \pm i\mu$ $\lambda < 0, \mu \neq 0$	Attracting improper node	Asymptotically stable	Asymptotically stable
$r_1, r_2 = \pm i\mu$ $\mu \neq 0$	Stable center	Stable	Indeterminate

It is observed from the above table that the asymptotic stability of a critical point is preserved in exactly four cases: an attracting proper node, an attracting degenerate node, an attracting improper node, and an attracting spiral. These are precisely the cases in which the eigenvalues are either strictly negative real numbers or complex numbers with a negative real part. In all cases we may state that *stability holds if and only if the eigenvalues of the linearized system have negative real parts*.

Exercises for Section 17.5

For each of the following nonlinear systems, verify that $(0, 0)$ is a critical point and determine whether or not it is unstable, stable, or asymptotically stable. Discuss spiraling behavior when applicable.

1. $\begin{cases} x' = x - y + xy, \\ y' = 3x - 2y - xy \end{cases}$

6. $\begin{cases} x' = x + 2x^2 - y^2, \\ y' = x - 2y + x^3 \end{cases}$

2. $\begin{cases} x' = x + x^2 + y^2, \\ y' = y - xy \end{cases}$

7. $\begin{cases} x' = y, \\ y' = -x + 4y(1 - x^2) \end{cases}$

3. $\begin{cases} x' = -2x - y - x(x^2 + y^2), \\ y' = x - y + y(x^2 + y^2) \end{cases}$

8. $\begin{cases} x' = 1 + y - e^{-x}, \\ y' = y - \sin(x) \end{cases}$

4. $\begin{cases} x' = y + x(1 - x^2 - y^2), \\ y' = -x + y(1 - x^2 - y^2) \end{cases}$

9. $\begin{cases} x' = (1 + x)\sin(y), \\ y' = 1 - x - \cos(y) \end{cases}$

5. $\begin{cases} x' = 2x + y + xy^3, \\ y' = x - 2y - xy \end{cases}$

10. $\begin{cases} x' = 1 - x + y - \cos(x), \\ y' = \sin(x - 3y) \end{cases}$

The following exercises illustrate the use of the word *indeterminate*.

11. Show that the system

$$\begin{cases} x' = -y - x(x^2 + y^2), \\ y' = x - y(x^2 + y^2) \end{cases}$$

has a stable critical point at $(0, 0)$. [Hint: Obtain a differential equation for $r^2 = x^2 + y^2$.]

12. Show that the system

$$\begin{cases} x' = -y + x(x^2 + y^2), \\ y' = x + y(x^2 + y^2) \end{cases}$$

has an unstable critical point at $(0, 0)$. [Hint: Obtain a differential equation for $r^2 = x^2 + y^2$.]

13. Conclude from Exercises 11 and 12 that a stable center has indeterminate stability type when nonlinear terms are added.

14. Show that the system

$$\begin{cases} x' = -x, \\ y' = y^2 \end{cases}$$

has an unstable critical point at $(0, 0)$.

15. Show that the system

$$\begin{cases} x' = -x, \\ y' = -y^3 \end{cases}$$

has a stable critical point at $(0, 0)$.

16. Conclude from Exercises 14 and 15 that an attracting center has indeterminate stability type when nonlinear terms are added.

17.6   *Lyapunov Stability Theory*

We consider an autonomous nonlinear system of differential equations

$$\begin{cases} \dfrac{dx_1}{dt} = F_1(x_1, \ldots, x_n), \\ \quad \vdots \qquad\qquad \vdots \\ \dfrac{dx_n}{dt} = F_n(x_1, \ldots, x_n) \end{cases} \tag{17.30}$$

with an isolated critical point $(x_1^c, \ldots, x_n^c)$. In detail, we have

$$F_1(x_1^c, , \ldots, x_n^c) = \cdots = F_n(x_1^c, \ldots, x_n^c) = 0.$$

In this generality there is no hope of our being able to solve (17.30) by any explicit method or by successive integrations. The method of Lyapunov allows us to abandon all such hopes and concentrate instead on solving certain *inequalities* to obtain a function $\Phi(x_1, \ldots, x_n)$ that can be thought of as playing the role of *energy* for the system of equations (17.30). In the case of (ordinary) stability we can expect that the function Φ remains bounded when computed along a solution trajectory $t \longmapsto \big(x_1(t), \ldots, x_n(t)\big)$. In the case of asymptotic stability we can expect that the function Φ tends to zero along a solution trajectory $t \longmapsto \big(x_1(t), \ldots, x_n(t)\big)$ when $t \longrightarrow \infty$. The precise notions are the following.

Definition.   *A function* $\Phi_1(x_1, \ldots, x_n)$ *is a* **Lyapunov function of the first type** *relative to the isolated critical point* $(x_1^c, \ldots, x_n^c)$ *of the nonlinear system* (17.30) *if it has the following properties:*

(i) $\Phi_1(x_1, \ldots, x_n)$ *is continuous with continuous first partial derivatives* $\partial\Phi_1/\partial x_i$ $(1 \leq i \leq n)$ *in a rectangle* R, *containing the critical point* $(x_1^c, \ldots, x_n^c)$.

(ii)   *At the critical point we have* $\Phi_1(x_1^c, \ldots, x_n^c) = 0$.

(iii)   *There is a positive strictly increasing continuous function* $\phi_1(r)$ *(called an* **under-estimator***) defined for* $0 < r < \delta$ *such that* $\Phi_1(x_1, \ldots, x_n) \geq \phi_1(r)$ *everywhere in the rectangle* R, *where*

$$r^2 = (x_1 - x_1^c)^2 + \cdots + (x_n - x_n^c)^2.$$

(iv)   *The first partial derivatives of* Φ_1 *satisfy the inequality*

$$F_1 \frac{\partial\Phi_1}{\partial x_1} + \cdots + F_n \frac{\partial\Phi_1}{\partial x_n} \leq 0 \tag{17.31}$$

everywhere in the rectangle R.

The crucial condition here is (iv), which states that the function Φ_1 is nonincreasing along trajectories of the nonlinear system (17.30). Indeed, the combination of partial derivatives displayed in (iv) is exactly the **total derivative**

$$\frac{d}{dt}\Big(\Phi_1\big(x_1(t), \ldots, x_n(t)\big)\Big)$$

of the composition of the function Φ_1 with the trajectory $t \longmapsto \big(x_1(t), \ldots, x_n(t)\big)$. Geometrically, (iv) states that the cosine of the angle between the gradient vector of Φ_1 and the vector field $(F_1, \ldots, F_n)$ that defines the nonlinear system is less than or equal to zero. From the geometric interpretation of gradient, this means that the function Φ_1 is nonincreasing in the direction of the vector field $(F_1, \ldots, F_n)$, to first-order of approximation.

We can use Lyapunov functions of the first type to prove the first theorem of Lyapunov.

Theorem 17.8.   (Lyapunov's first theorem) *Suppose that* $\Phi_1(x_1, \ldots, x_n)$ *is a Lyapunov function of the first type with respect to the isolated critical point* $(x_1^c, \ldots, x_n^c)$ *of the nonlinear system* (17.30). *Then the critical point is stable.*

The proof of Theorem (17.8) will be given at the end of the section. When the linearized approximation of a given nonlinear system has purely imaginary eigenvalues, it is not possible to determine the stability of the nonlinear system from the stability of the linearized approximation. Nevertheless, Theorem 17.8 can be used to determine stability. We illustrate with an example.

Example 17.12. *Show that the critical point* $(0, 0)$ *is stable for the nonlinear system*

$$\frac{dx}{dt} = y, \qquad \frac{dy}{dt} = -x - y^3. \tag{17.32}$$

Use the Lyapunov function $\Phi_1(x, y) = x^2 + y^2$.

Solution. The suggested Lyapunov function is clearly continuous and positive in any rectangle containing the critical point $(0, 0)$. The underestimator ϕ_1 can be chosen simply as $\phi_1(r) = r^2$. We must check the differential inequality (iv). For any solution of (17.32) we have

$$\frac{d}{dt}(x^2 + y^2) = 2x\frac{dx}{dt} + 2y\frac{dy}{dt} = 2x\,y + 2y(-x - y^3) = -2y^4 \le 0,$$

so that the differential inequality (17.31) is satisfied. We conclude that the critical point $(0, 0)$ of (17.32) is stable. ∎

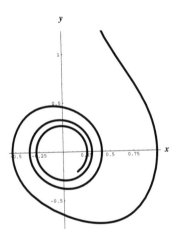

A trajectory of the system
$$x' = y, \quad y' = -x - y^3$$

We now pass to the discussion of *asymptotic stability* by Lyapunov's method.

Definition. *A function* $\Phi_2(x_1, \dots, x_n)$ *is a* **Lyapunov function of the second type** *relative to the isolated critical point* $(x_1^c, \dots, x_n^c)$ *of the nonlinear system* (17.30) *if it has the properties:*

(i') $\Phi_2(x_1, \dots, x_n)$ *is continuous with continuous first partial derivatives* $\partial \Phi_2 / \partial x_i$
     $(1 \le i \le n)$ *in a rectangle* R *containing the critical point* $(x_1^c, \dots, x_n^c)$.

(ii') *At the critical point we have* $\Phi_2(x_1^c, \ldots, x_n^c) = 0.$

(iii') *There is a positive strictly increasing continuous function $\phi_2(r)$ (called an* **under-estimator***) defined for $0 < r < \delta$ such that $\Phi_2(x_1, \ldots, x_n) \geq \phi_2(r)$ everywhere in the rectangle R, where*

$$r^2 = (x_1 - x_1^c)^2 + \cdots + (x_n - x_n^c)^2.$$

(iv') *The first partial derivatives of Φ_2 satisfy the inequality*

$$F_1 \frac{\partial \Phi_2}{\partial x_1} + \cdots + F_n \frac{\partial \Phi_2}{\partial x_n} < 0$$

everywhere in the rectangle R.

The reader will note that conditions (i')–(iii') are identical to the conditions (i)–(iii) in the definition of Lyapunov function of the first type. Condition (iv'), is a *strengthened* version of (iv); it requires that the time derivative of Φ_2 along the trajectory be *strictly negative* in the above sense. The following theorem shows the application of the Φ_2 function to asymptotic stability.

Theorem 17.9. (**Lyapunov's second theorem**) *Suppose that $\Phi_2(x_1, \ldots, x_n)$ is a Lyapunov function of the second type relative to the isolated critical point $(x_1^c, \ldots, x_n^c)$ of the nonlinear system* (17.30). *Then the critical point is asymptotically stable.*

The proof of Theorem 17.9 will be given at the end of this section.

Example 17.13. *Show that $(0, 0)$ is asymptotically stable for the nonlinear system*

$$\frac{dx}{dt} = y, \qquad \frac{dy}{dt} = -x - 2y - y^3. \tag{17.33}$$

Use the Lyapunov function $\Phi_2(x, y) = 6x^2 + 2xy + 4y^2.$

Solution.   The underestimator is found by noting that $2xy \geq -(x^2 + y^2)$; hence $\Phi_2(x, y) \geq 5x^2 + 3y^2 \geq 3(x^2 + y^2)$. Computing the derivatives, we have

$$\frac{d}{dt}(6x^2 + 2xy + 4y^2) = 12xy + 2y^2 + 2x(-x - 2y - y^3) + 8y(-x - 2y - y^3)$$

$$= -2x^2 - 14y^2 - 2xy^3 - 8y^4. \tag{17.34}$$

The fourth-degree term on the right-hand side of (17.34) is less than or equal to zero. In the unit square the term $2xy^3$ is less than or equal to $x^2 + y^2$. Putting these together we have

$$\frac{d}{dt}\Phi_2\big(x(t), y(t)\big) \leq -2x^2 - 14y^2 + x^2 + y^2 = -x^2 - 13y^2 \leq -(x^2 + y^2). \tag{17.35}$$

Dividing both sides of (17.35) by the positive function Φ_2, we obtain

$$\frac{1}{\Phi_2}\frac{d}{dt}\Phi_2\big(x(t),\,y(t)\big) \le \frac{-(x^2+y^2)}{\Phi_2}. \tag{17.36}$$

The fact that $2xy \le x^2 + y^2$ implies that the function Φ_2 also satisfies $\Phi_2(x,\,y) \le 7(x^2+y^2)$. Since $(x^2+y^2)/\Phi_2 \ge 1/7$, the negative term $-(x^2+y^2)/\Phi_2 \le -1/7$; thus (17.36) implies that

$$\frac{\dfrac{d}{dt}\big(\Phi_2(x(t),\,y(t))\big)}{\Phi_2\big(x(t),\,y(t)\big)} \le -\frac{1}{7}.$$

Therefore, Φ_2 is a Lyapunov function of the second type, from which we conclude the asymptotic stability of the critical point $(0,0)$ for (17.33). ∎

Below are plots of the nonlinear system (17.33) and its linearized approximation. From the graph we see that a trajectory of (17.33) remains close to the corresponding trajectory of the linearized approximation for small $t > 0$, diverges for moderate times, and converges for large times. Furthermore, the amount of divergence depends on the initial conditions.

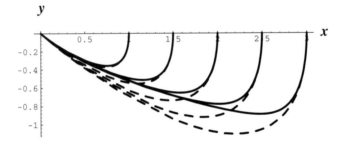

**Example Illustrating
Lyapunov's second theorem**

The reader will note that the computations in Example 17.13 are very cumbersome and depend crucially on a "good guess" for the Lyapunov function of the nonlinear system. It is possible in a wide variety of cases to obtain a suitable Lyapunov function of the second type directly from an analysis of the linearized system $d\mathbf{y}/dt = A\mathbf{y}$, which results from applying the first-order Taylor approximation to the functions $F_1, \ldots, F_n$ at the critical point. The method depends on the asymptotic stability of the linearized system and is described as follows.

Theorem 17.10. *Suppose that $(x_1^c, \ldots, x_n^c)$ is an isolated critical point for the nonlinear system* (17.30) *and that for the linearized system*

$$\frac{d\mathbf{y}}{dt} = A\mathbf{y},$$

all eigenvalues of the matrix A have strictly negative real part: $r = \alpha + i\beta$ with $\alpha < 0$. Then there is a Lyapunov function Φ_2 of the second type for the nonlinear system (17.30) that can be written as a homogeneous quadratic polynomial

$$\Phi_2(x_1, \ldots, x_n) = B(x_1 - x_1^c, \ldots, x_1 - x_n^c) \tag{17.37}$$

with

$$B(y_1, \ldots, y_n) = \sum_{i,j=1}^{n} b_{ij} y_i y_j, \tag{17.38}$$

where the symmetric matrix $B = (b_{ij})$ is given by

$$B = \int_0^\infty e^{t A^T} e^{t A} dt. \tag{17.39}$$

In particular, the critical point $(x_1^c, \ldots, x_n^c)$ is asymptotically stable.

Theorem 17.10 (whose proof will be given at the end of this section) allows us to avoid completely the tedious construction of a suitable Lyapunov function for the nonlinear system. All that is necessary is to check that all eigenvalues of the linearized system have strictly negative real part.

Example 17.14. *Use Theorem 17.10 to deduce the asymptotic stability of the critical point $(0, 0)$ of the nonlinear system*

$$\begin{cases} x' = y, \\ y' = -x - 2y - y^3. \end{cases} \tag{17.40}$$

Solution. The linearized system corresponding to (17.40) is defined by the matrix

$$A = \begin{pmatrix} 0 & 1 \\ -1 & -2 \end{pmatrix}.$$

Here A is the same matrix that we considered in Example 17.10. We found that -1 is a double eigenvalue for A. Therefore, we may infer the existence of a Lyapunov function of the second type and hence the asymptotic stability of the critical point $(0, 0)$. ∎

We can also use Theorem 17.10 to give a proof of part (i) of Theorem 5.5 in Chapter 5. In the case $n = 1$, the 1×1 matrix A that defines the linearized approximation near the critical point Y_i is simply the derivative $f'(Y_i)$. If this is negative, then the differential equation $y' = f(y)$ satisfies the hypotheses of Theorem 17.10, so that we may apply the conclusion to obtain the asymptotic stability of the critical point Y_i.

The above results allow us to obtain the stability and asymptotic stability by constructing an appropriate Lyapunov function. The following result allows us to infer instability by a parallel construction.

Theorem 17.11.   *Suppose that* $V(x_1, \ldots, x_n)$ *is defined and differentiable on a rectangle containing* $(0, 0, \ldots, 0)$ *such that* $V(0, 0, \ldots, 0) = 0$. *Suppose that there is a domain D for which* $(0, 0, \ldots, 0)$ *is a boundary point such that* $V(x_1, \ldots, x_n) > 0$ *in* D, $V(x_1, \ldots, x_n) = 0$ *on the boundary of* D, *and such that*

$$F_1 \frac{\partial V}{\partial x_1} + \cdots + F_n \frac{\partial V}{\partial x_n} > 0$$

in D. *Then* $(0, 0, \ldots, 0)$ *is an unstable critical point.*

The proof of Theorem 17.11 will be given at the end of the section.

Example 17.15.   *Show that* $(0, 0)$ *is unstable for the system*

$$\begin{cases} x' = y^3 + x^5, \\ y' = x^3 + y^5. \end{cases} \tag{17.41}$$

Solution. We use the function $V(x, y) = x^4 - y^4$. Here $D = \{(x, y) \mid |x| > |y|\}$, and

$$F_1 \frac{\partial V}{\partial x} + F_2 \frac{\partial V}{\partial y} = 4(x^8 - y^8) > 0$$

in D. Hence we can apply Theorem 17.11 to infer instability of $(0, 0)$ for (17.15). ∎

Proofs of the Lyapunov Theorems

Proof of Theorem 17.8

We have $\Phi_1(x_1, \ldots, x_n) \geq \phi_1(r)$, where $r^2 = (x_1 - x_1^c)^2 + \cdots + (x_n - x_n^c)^2$. Now

$$\frac{d}{dt} \Phi_1\big(x_1(t), \ldots, x_n(t)\big) = \frac{\partial \Phi_1}{\partial x_1} \frac{dx_1}{dt} + \cdots + \frac{\partial \Phi_1}{\partial x_n} \frac{dx_n}{dt}$$

$$= \frac{\partial \Phi_1}{\partial x_1} F_1 + \cdots + \frac{\partial \Phi_1}{\partial x_n} F_n \leq 0, \tag{17.42}$$

by hypothesis. Applying the fundamental theorem of calculus and using (17.42), we have

$$\Phi_1\big(x_1(t), \ldots, x_n(t)\big) = \Phi_1\big(x_1(0), \ldots, x_n(0)\big) + \int_0^t \frac{d}{ds} \Phi_1\big(x_1(s), \ldots, x_n(s)\big) ds$$

$$\leq \Phi_1\big(x_1(0), \ldots, , x_n(0)\big).$$

The function ϕ_1 is increasing and continuous and possesses a continuous inverse function ϕ^{-1}, from which we conclude that

$$r(t) \leq \phi_1^{-1}\left(\Phi_1\big(x_1(0), \ldots, x_n(0)\big)\right), \tag{17.43}$$

where

$$r(t) = \sqrt{\big(x_1(t) - x_1^c\big)^2 + \cdots + \big(x_n(t) - x_n^c\big)^2}.$$

Now, given $\varepsilon > 0$ we choose $\delta_1 > 0$ such that $\phi_1^{-1}(\rho) < \varepsilon$ whenever $0 < \rho < \delta_1$. With this value of δ_1, we choose a $\delta > 0$ such that $\Phi_1(x_1, \ldots, x_n) < \delta_1$ whenever $(x_1 - x_1^c)^2 + \cdots + (x - x_n)^2 < \delta^2$. This is possible by the assumed continuity of Φ_1 and the assumption $\Phi_1(x_1^c, \ldots, x_n^c) = 0$. It is now immediate from (17.43) that whenever the initial point $\big(x_1(0), \ldots, x_n(0)\big)$ satisfies

$$\big(x_1(0) - x_1^c\big)^2 + \cdots + \big(x_n(0) - x_n^c\big)^2 < \delta^2,$$

we have $r(t) < \varepsilon$ for all $t > 0$, which was to be proved. ∎

Proof of Theorem 17.9

From the hypothesis we have

$$\frac{d}{dt}\Phi_2\big(x_1(t), \ldots, x_n(t)\big) < 0,$$

so that $t \longrightarrow \Phi_2\big(x_1(t), \ldots, x_n(t)\big)$ is decreasing and bounded below by zero. Therefore, the limit

$$L = \lim_{t \to \infty} \Phi_2\big(x_1(t), \ldots, x_n(t)\big)$$

exists. If $L = 0$, we have $\phi_2(r(t)) \leq \Phi_2\big(x_1(t), \ldots, x_n(t)\big)$, so that

$$r(t) \leq \phi_2^{-1}\big(\Phi_2(x_1(t), \ldots, x_n(t))\big) \longrightarrow 0$$

when $t \longrightarrow \infty$, and the proof is complete. Otherwise, $L > 0$, and $\big(x_1(t), \ldots, x_n(t)\big)$ remains at a positive distance from the origin; thus

$$\frac{d}{dt}\Phi_2\big(x_1(t), \ldots, x_n(t)\big) \leq -A^2 < 0.$$

This implies that

$$\Phi_2\big(x_1(t), \ldots, x_n(t)\big) - \Phi_2\big(x_1(0), \ldots, x_n(0)\big) \leq -A^2 t \longrightarrow -\infty,$$

which contradicts $\Phi_2 \geq 0$. Hence in fact $L = 0$. ∎

Proof of Theorem 17.10

The integral (17.39) defining the matrix $B = (b_{ij})$ is a convergent improper integral because of the exponential decrease of the integrand. This is seen by writing the exponential of the matrix A in terms of the eigenvalues $r_j = \alpha_j + i\beta_j$ as

$$e^{tA} = \sum_{j=1}^{n} e^{tr_j} Z_j(t), \tag{17.44}$$

where the $Z_j(t)$ are polynomial matrices obtained from the Jordan decomposition of A. A corresponding formula holds for e^{tA^T}. Since the eigenvalues of A are assumed to have negative real parts, we must have $\alpha_j < 0$. In both cases all of the eigenvalues have negative real parts, and the convergence of the integrals depends on the convergence of the real-valued integrals that result from (17.44), namely

$$\int_0^\infty t^k e^{\alpha_j t} dt < \infty,$$

since $\alpha_j < 0$ for each j and $k = 1, 2, \ldots$.

The matrix $B = (b_{ij})$ defined by (17.39) is positive definite, since we can write

$$\langle B\mathbf{y}, \mathbf{y} \rangle = \int_0^\infty \langle e^{tA^T} e^{tA} \mathbf{y}, \mathbf{y} \rangle dt = \int_0^\infty |e^{tA}\mathbf{y}|^2 \, dt. \tag{17.45}$$

The integral (17.45) is that of a strictly positive function if $\mathbf{y} \neq \mathbf{0}$; hence the quadratic function $\langle B\mathbf{y}, \mathbf{y} \rangle$ is positive. As an underestimator of Φ_2 we use $b_1|\mathbf{y}|^2$, where $b_1 > 0$ is the smallest eigenvalue of the symmetric matrix B; thus, $\Phi_2(\mathbf{y}) \geq b_1|\mathbf{y}|^2$.

We now prove that the matrix B satisfies the matrix equation

$$A^T B + B A = -I_{n\times n}. \tag{17.46}$$

To do this we differentiate the product $e^{tA^T} e^{tA}$, obtaining

$$\frac{d}{dt}\left(e^{tA^T} e^{tA}\right) = \left(A^T e^{tA^T}\right)e^{tA} + e^{tA^T}\left(e^{tA} A\right). \tag{17.47}$$

We integrate (17.47) over $0 \leq t \leq \infty$ and use the fundamental theorem of calculus. Since

$$e^{tA^T} e^{tA}\Big|_{t=0} = I_{n\times n} \qquad \text{and} \qquad \lim_{t\to\infty} e^{tA^T} e^{tA} = 0,$$

we obtain

$$0 - I_{n\times n} = e^{tA^T} e^{tA}\Big|_{t=0}^{t=\infty} = \int_0^\infty \left(A^T e^{tA^T} e^{tA} + e^{tA^T} e^{tA} A\right) dt = A^T B + B A,$$

proving (17.46).

Now we write the nonlinear system in the form

$$\frac{d\mathbf{y}}{dt} = A\mathbf{y} + \varepsilon(\mathbf{y}),$$

where

$$\lim_{\mathbf{y} \to \mathbf{0}} \frac{\varepsilon(\mathbf{y})}{\|\mathbf{y}\|} = 0.$$

Applying the derivative to Φ_2, we have

$$\frac{d}{dt}\Phi_2(\mathbf{y}(t)) = \frac{d}{dt}\langle B\mathbf{y}(t), \mathbf{y}(t) \rangle$$

$$= \langle B(A\mathbf{y} + \varepsilon(\mathbf{y})), \mathbf{y} \rangle + \langle B\mathbf{y}(t), A\mathbf{y}(t) + \varepsilon(\mathbf{y}) \rangle$$

$$= -\langle \mathbf{y}, \mathbf{y} \rangle + |\mathbf{y}|\hat{\varepsilon}(\mathbf{y}),$$

where $\hat{\varepsilon}(\mathbf{y})$ is another function for which

$$\lim_{\mathbf{y} \to \mathbf{0}} \frac{\hat{\varepsilon}(\mathbf{y})}{\|\mathbf{y}\|} = 0.$$

This shows that Φ_2 satisfies condition (iv') for a Lyapunov function of the second type, and the theorem is proved. ∎

Proof of Theorem 17.11

We must show that there are points arbitrarily close to $(0, 0, \ldots 0)$ for which the solutions leave a region of fixed size. Let S_R be the ball of radius R centered at $(0, 0, \ldots, 0)$, chosen so that $dV/dt > 0$ in the intersection $D \cap S_R$. By hypothesis, the region D contains points that are arbitrarily close to $(0, 0, \ldots, 0)$. If the solution starting at such a point leaves $D \cap S_R$ through the boundary of D, then we must have $V = 0$ at a time $t_0 > 0$, which contradicts

$$V|_{t=0} > 0 \qquad \text{and} \qquad dV/dt > 0 \qquad \text{for } 0 \le t < t_0.$$

If on the other hand the solution remains in $D \cap S_R$ for all time, then we must have

$$\frac{dV}{dt}\big(x_1(t), \ldots, x_n(t)\big) \ge m > 0$$

everywhere. Hence $V\big(x_1(t), \ldots, x_n(t)\big)$ becomes arbitrarily large, so it must leave $D \cap S_R$, a contradiction. Therefore, the solution must leave through the boundary of S_R, which contradicts stability. ∎

Exercises for Section 17.6

1. Show that $V(x, y) = x^2 + y^2$ is a Lyapunov function for the system

$$\begin{cases} x' = -x^3 + 2x\, y^2, \\ y' = -2x^2 y - 5y^3. \end{cases}$$

2. Find a Lyapunov function of the form $V(x, y) = a\, x^2 + b\, y^2$ for the system

$$\begin{cases} x' = -x^3 + x\, y^2 + x^5, \\ y' = -2x^2 y - y^3 - y^5. \end{cases}$$

3. Find a Lyapunov function of the form $V(x, y) = a\, x^2 + b\, y^2$ for the system

$$\begin{cases} x' = -x^3 + x\, y^2 + 3x^7, \\ y' = -2x^2 y - 4y^3 - 3y^5. \end{cases}$$

4. Suppose that we have a linear system with an attracting spiral point corresponding to a 2×2 matrix (a_{ij}). Let

$$V(x_1, x_2) = a_{21}x_1^2 + (a_{22} - a_{11})x_1 x_2 - a_{12}x_2^2.$$

    a. Show that $V(x_1, x_2) \neq 0$ for $(x_1, x_2) \neq (0, 0)$ and that for any solution of the linear system $d\mathbf{x}/dt = A\mathbf{x}$ we have

$$\frac{d}{dt}V\big(x_1(t), x_2(t)\big) = (a_{11} + a_{22})V\big(x_1(t), x_2(t)\big).$$

    b. Conclude from part a. that all solutions of the linear system satisfy the limiting relation

$$\lim_{t \to \infty} \frac{1}{t} \log |\mathbf{x}(t)| = \frac{1}{2}(a_{11} + a_{22}) < 0.$$

5. Suppose that we have the nonlinear system $x' = F(x, y)$, $y' = G(x, y)$ with an isolated critical point at $(0, 0)$. Suppose that the corresponding linearized system defines an attracting spiral point with eigenvalues $r = \lambda \pm i\mu$ with $\lambda < 0$ and $\mu \neq 0$.

    a. Show that the solutions of the nonlinear system satisfy

$$\lim_{t \to \infty} \frac{1}{t} \log |\mathbf{x}(t)| = \lambda$$

provided that $x(0)^2 + y(0)^2 < \delta^2$ for a suitable $\delta > 0$. [Hint: Compute the time derivative of $\log V(x_1(t), x_2(t))$ from the previous exercise.]

b. Apply the result of part **a** to the damped pendulum system

$$\begin{cases} x' = y, \\ y' = -\dfrac{g}{l}\sin(x) - c\,y \end{cases}$$

with $0 < c < 2\sqrt{g/l}$.

Find a suitable Lyapunov function to investigate the stability of the following systems in the neighborhood of $(0, 0)$.

6. $\begin{cases} x' = -3y - 2x^3, \\ y' = 2x - 3y^3 \end{cases}$

7. $\begin{cases} x' = -x\,y^4, \\ y' = x^4 y \end{cases}$

8. $\begin{cases} x' = x + 2x\,y^2, \\ y' = -2y + 4x^2 y \end{cases}$

9. $\begin{cases} x' = -y - x/2 - x^3/4, \\ y' = x - y/2 - y^3/4 \end{cases}$

10. $\begin{cases} x' = y + x^3, \\ y' = -x + y^3 \end{cases}$

11. $\begin{cases} x' = y + x^2 y^2 - x^5/4, \\ y' = -2x - x^3 y - y^3/2 \end{cases}$

12. $\begin{cases} x' = x\,y^4 - 2x^3 - y, \\ y' = 2x^2 y^3 - y^7 + 2x \end{cases}$

13. $\begin{cases} x' = -2x - 3y, \\ y' = x - y \end{cases}$

14. $\begin{cases} x' = x\,y - x^3 + y, \\ y' = x^4 - x^2 y - x^3 \end{cases}$

15. $\begin{cases} x' = -2y - x(x - y)^2, \\ y' = 3x - (3/2)y(x - y)^2 \end{cases}$

16. $\begin{cases} x' = x + x^3, \\ y' = -y - y^3 \end{cases}$

17. $\begin{cases} x' = x^5 + y^3, \\ y' = x^3 - y^5 \end{cases}$

18. $\begin{cases} x' = x^3 + 2x\,y^2, \\ y' = x^2 y \end{cases}$

18

APPLICATIONS OF LINEAR SYSTEMS

This chapter gives applications of linear systems of differential equations. Applications of nonlinear systems of differential equations will be discussed in Chapter 19. Section 18.1 treats coupled systems of oscillators, and Section 18.2 treats systems of electric circuits. These two sections are system analogues of Sections 11.1 and 11.3. Markov chains are discussed in Section 18.3.

18.1   Coupled Systems of Oscillators

We encounter interesting systems of equations when we study a system of masses that are attached to one another through a set of springs. We may have fixed springs as well as frictional effects, as we had for a single mass studied in Chapter 11.

One of the simplest systems consists of a set of n particles on a line, interacting with one another by means of springs. Let $x_j(t)$ denote the displacement of the jth particle from its equilibrium position. The equations of motion of the particles can be written as

$$m_j \frac{d^2 x_j}{dt^2} + c_j \frac{dx_j}{dt} + k_j x_j = k_{j\,j+1}(x_{j+1} - x_j) + k_{j\,j-1}(x_{j-1} - x_j) \tag{18.1}$$

for $1 \leq j \leq n$, where we define $k_{10} = k_{nn+1} = x_0 = x_{n+1} = 0$. Physically, the jth particle has mass m_j. We call k_j the **spring constant** and c_j the **frictional constant** of the jth particle, just as we did for the single mass-spring systems studied in Chapter 11.

The motion of a system of particles such as (18.1) is much more complicated than that of a single particle because of **interaction** between the jth particle and its adjacent neighbors; we measure the interaction by means of **coupling constants** $k_{j\,j+1}$ and $k_{j\,j-1}$. As we shall see, a system of particles with nonzero coupling constants behaves differently from the n individual systems. We can rewrite (18.1) as a first-order system by introducing the velocities $y_j = dx_j/dt$, leading to the $2n$-dimensional first-order system

$$
\begin{cases}
x_1' = y_1, \\[4pt]
y_1' = \dfrac{1}{m_1}\bigl(-c_1 y_1 - (k_1 + k_{12})x_1 + k_{12}x_2\bigr), \\[4pt]
x_2' = y_2, \\[4pt]
y_2' = \dfrac{1}{m_2}\bigl(-c_2 y_2 + k_{21}x_1 - (k_2 + k_{21} + k_{23})x_2 + k_{23}x_3\bigr), \\[4pt]
\quad\vdots \\[4pt]
x_n' = y_n, \\[4pt]
y_n' = \dfrac{1}{m_n}\bigl(-c_n y_n - (k_n + k_{nn-1})x_n + k_{nn-1}x_{n-1}\bigr).
\end{cases}
\tag{18.2}
$$

We abbreviate (18.2) to $\dfrac{d\mathbf{x}}{dt} = A\mathbf{x}$, where the coefficient matrix A is

$$
\begin{pmatrix}
0 & 1 & 0 & 0 & 0 & \ldots & 0 & 0 \\[4pt]
-\dfrac{k_1 + k_{12}}{m_1} & -\dfrac{c_1}{m_1} & \dfrac{k_{12}}{m_1} & 0 & 0 & \ldots & 0 & 0 \\[4pt]
0 & 0 & 0 & 1 & 0 & \ldots & 0 & 0 \\[4pt]
\dfrac{k_{21}}{m_2} & 0 & -\dfrac{k_2 + k_{21} + k_{23}}{m_2} & -\dfrac{c_2}{m_2} & \dfrac{k_{23}}{m_2} & \ldots & 0 & 0 \\[4pt]
\vdots & \vdots & \vdots & \vdots & \vdots & \ldots & \vdots & \vdots \\[4pt]
0 & 0 & 0 & 0 & 0 & \ldots & 0 & 1 \\[4pt]
0 & 0 & 0 & 0 & 0 & \ldots & -\dfrac{k_n + k_{n\,n-1}}{m_n} & -\dfrac{c_n}{m_n}
\end{pmatrix}.
$$

We assume that the coefficient matrix A is diagonalizable; we enumerate the eigenvalues of A as $\{r_1, \ldots, r_n\}$. In general, each r_j will be complex, so we put

$$\lambda_j = \mathfrak{Re}(r_j) \qquad \text{and} \qquad \mu_j = \mathfrak{Im}(r_j)$$

for $j = 1, \ldots, n$. Let $\boldsymbol{\xi}_j$ be an eigenvector corresponding to r_j and write

$$\boldsymbol{\xi}_j = \mathbf{a}_j + i\,\mathbf{b}_j,$$

where $\mathbf{a}_j$ and $\mathbf{b}_j$ are real vectors. According to Theorem 15.10, a fundamental set of real-valued solutions for (18.2) is given by

$$\bigl\{\mathbf{u}^1(t), \mathbf{v}^1(t), \ldots, \mathbf{u}^n(t), \mathbf{v}^n(t)\bigr\},$$

where

$$
\begin{cases}
\mathbf{u}^j(t) = e^{\lambda_j t}\big(\mathbf{a}_j \cos(\mu_j t) - \mathbf{b}_j \sin(\mu_j t)\big), \\
\mathbf{v}^j(t) = e^{\lambda_j t}\big(\mathbf{a}_j \sin(\mu_j t) + \mathbf{b}_j \cos(\mu_j t)\big)
\end{cases}
\tag{18.3}
$$

for $1 \le j \le n$.

Definition. *The* **normal modes of oscillations** *of the system* (18.2) *are the solutions given by the vector-valued functions* (18.3). *The* j^{th} **angular frequency** *and* **quasiperiod** *are defined to be* μ_j *and* $T_j = 2\pi/\mu_j$ *for* $j = 1, \dots, n$.

In the case of no frictional forces ($c_1 = \cdots = c_n = 0$) it can be shown that all of the eigenvalues r_j are purely imaginary, and we have a system of harmonic oscillators with angular frequency μ_j. We now illustrate this in detail for the case $n = 2$.

Example 18.1. *Suppose that two masses interact in the absence of friction. Show that the eigenvalues are purely imaginary.*

Solution. In this case (18.2) reduces to

$$
\begin{cases}
x_1' = y_1, \\
y_1' = \dfrac{1}{m_1}\big(-(k_1 + k_{12})x_1 + k_{12}x_2\big), \\
x_2' = y_2, \\
y_2' = \dfrac{1}{m_2}\big(k_{21}x_1 - (k_2 + k_{21})x_2\big).
\end{cases}
\tag{18.4}
$$

Let B be the coefficient matrix of (18.4). The possible angular frequencies of (18.4) are determined by solving the associated characteristic equation of B, which is

$$
0 = \det(B - r\, I_{4\times4}) = \det
\begin{pmatrix}
-r & 1 & 0 & 0 \\
-\dfrac{k_1 + k_{12}}{m_1} & -r & \dfrac{k_{12}}{m_1} & 0 \\
0 & 0 & -r & 1 \\
\dfrac{k_{21}}{m_2} & 0 & -\dfrac{k_2 + k_{21}}{m_2} & -r
\end{pmatrix}
$$

$$
= r^4 + r^2\left(\frac{k_2 + k_{21}}{m_2} + \frac{k_1 + k_{12}}{m_1}\right) + \frac{(k_1 + k_{12})(k_2 + k_{21}) - k_{12}k_{21}}{m_1 m_2}.
$$

The quadratic formula yields the two roots

$$2r^2 = -\left(\frac{k_1 + k_{12}}{m_1} + \frac{k_2 + k_{21}}{m_2}\right)$$

$$\pm \sqrt{\left(\frac{k_1 + k_{12}}{m_1} - \frac{k_2 + k_{21}}{m_2}\right)^2 + \frac{4k_{12}k_{21}}{m_1 m_2}}. \qquad (18.5)$$

Since k_{12}, k_{21}, m_1, and m_2 are all nonnegative, the radical on the right-hand side of (18.5) is real; furthermore,

$$\left(\frac{k_1 + k_{12}}{m_1} - \frac{k_2 + k_{21}}{m_2}\right)^2 + \frac{4k_{12}k_{21}}{m_1 m_2}$$

$$\leq \left(\frac{k_1 + k_{12}}{m_1} - \frac{k_2 + k_{21}}{m_2}\right)^2 + \frac{4(k_1 + k_{12})(k_2 + k_{21})}{m_1 m_2}$$

$$= \left(\frac{k_1 + k_{12}}{m_1} + \frac{k_2 + k_{21}}{m_2}\right)^2. \qquad (18.6)$$

From (18.5) and (18.6) it follows that each root r has the property that r^2 is real and negative. Thus, each root is purely imaginary. ∎

If the coupling constants k_{12} and k_{21} both vanish, then (18.5) simplifies to

$$r^2 = -\frac{k_1}{m_1} \qquad \text{and} \qquad r^2 = -\frac{k_2}{m_2}$$

for the angular frequencies. Since $k_{12} = k_{21} = 0$ implies that there is no interaction between the masses, the motion of each mass is as described in Section 11.1. We next specialize Example 18.1 in a different way by assuming that there is simple but nonzero interaction between the masses.

Example 18.2.   *Solve for the angular frequencies and normal modes of Example* 18.1 *in the case of equal masses, equal spring constants, and equal coupling constants.*

Solution.   When $m_1 = m_2 = m$, $k_1 = k_2 = k$, and $k_{12} = k_{21}$, the system (18.4) becomes

$$\begin{cases} x_1' = y_1, \\[2mm] y_1' = -\dfrac{k + k_{12}}{m}x_1 + \dfrac{k_{12}}{m}x_2, \\[2mm] x_2' = y_2, \\[2mm] y_2' = \dfrac{k_{12}}{m}x_1 - \dfrac{k + k_{12}}{m}x_2. \end{cases} \qquad (18.7)$$

Furthermore, (18.5) simplifies to

$$r^2 = -\frac{k + k_{12}}{m} \pm \frac{k_{12}}{m},$$

which yields the angular frequencies in the form

$$\mu_1 = \sqrt{\frac{k}{m}} \quad \text{and} \quad \mu_2 = \sqrt{\frac{k + 2k_{12}}{m}}. \tag{18.8}$$

To find the normal modes of oscillation of (18.7), we must find the eigenvectors of the coefficient matrix M of the system (18.7). These are obtained by solving the matrix equation

$$(M - r\, I_{4 \times 4})\boldsymbol{\xi} = 0 \tag{18.9}$$

successively as r takes on the values $i\,\mu_1, -i\,\mu_1, i\,\mu_2, -i\,\mu_2$. The matrix M has only real entries; thus we need only do the computations for $i\,\mu_1$ and $i\,\mu_2$. Let $\boldsymbol{\xi} = (c_1, c_2, c_3, c_4)^T$; then (18.9) in the case that $r = i\mu_1$ is written in detail as

$$\begin{cases}
-i\,\mu_1 c_1 + c_2 = 0, \\[4pt]
-\dfrac{k + k_{12}}{m} c_1 - i\,\mu_1 c_2 + \dfrac{k_{12}}{m} c_3 = 0, \\[4pt]
-i\,\mu_1 c_3 + c_4 = 0, \\[4pt]
\dfrac{k_{12}}{m} c_1 - \dfrac{k_1 + k_{12}}{m} c_3 - i\,\mu_1 c_4 = 0.
\end{cases} \tag{18.10}$$

From the first and third equations of (18.10) we see that $c_2 = i\,\mu_1 c_1$ and $c_4 = i\,\mu_1 c_3$. When $c_2 = i\,\mu_1 c_1$ is substituted into the second equation of (18.10), we get

$$0 = -\frac{k + k_{12}}{m} c_1 - i\,\mu_1(i\,\mu_1 c_1) + \frac{k_{12}}{m} c_3 = -\frac{k + k_{12}}{m} c_1 + \mu_1^2 c_1 + \frac{k_{12}}{m} c_3. \tag{18.11}$$

Then (18.8) implies that $\mu_1^2 = k/m$, so that (18.11) becomes

$$-\frac{k_{12}}{m} c_1 + \frac{k_{12}}{m} c_3 = 0.$$

In other words, $c_1 = c_3$. Thus, we have shown that any eigenvector corresponding to $i\,\mu_1$ must be a multiple of

$$\boldsymbol{\xi}_1 = \begin{pmatrix} 1 \\ i\,\mu_1 \\ 1 \\ i\,\mu_1 \end{pmatrix}.$$

For the eigenvalue $i\,\mu_2$ given by (18.8), a calculation similar to that for μ_1 shows that

$$\frac{k_{12}}{m}c_1 + \frac{k_{12}}{m}c_3 = 0.$$

In other words, $c_1 = -c_3$. Thus, we have shown that any eigenvector corresponding to $i\,\mu_2$ must be a multiple of

$$\xi_2 = \begin{pmatrix} 1 \\ i\,\mu_2 \\ -1 \\ -i\,\mu_2 \end{pmatrix}.$$

In order to find the real-valued solutions of (18.7), it suffices to take the real and imaginary parts of the complex-valued solutions corresponding to $i\mu_1$ and $i\mu_2$. For the eigenvalue $i\,\mu_1$ we have

$$e^{i\mu_1 t}\begin{pmatrix} 1 \\ i\,\mu_1 \\ 1 \\ i\,\mu_1 \end{pmatrix} = \big(\cos(\mu_1 t) + i\sin(\mu_1 t)\big)\left(\begin{pmatrix} 1 \\ 0 \\ 1 \\ 0 \end{pmatrix} + i\begin{pmatrix} 0 \\ \mu_1 \\ 0 \\ \mu_1 \end{pmatrix}\right)$$

$$= \begin{pmatrix} \cos(\mu_1 t) \\ -\mu_1 \sin(\mu_1 t) \\ \cos(\mu_1 t) \\ -\mu_1 \sin(\mu_1 t) \end{pmatrix} + i\begin{pmatrix} \sin(\mu_1 t) \\ \mu_1 \cos(\mu_1 t) \\ \sin(\mu_1 t) \\ \mu_1 \cos(\mu_1 t) \end{pmatrix},$$

which yields the real-valued solutions

$$\mathbf{Z}_s(t) = \begin{pmatrix} \cos(\mu_1 t) \\ -\mu_1 \sin(\mu_1 t) \\ \cos(\mu_1 t) \\ -\mu_1 \sin(\mu_1 t) \end{pmatrix} \quad \text{and} \quad \mathbf{W}_s(t) = \begin{pmatrix} \sin(\mu_1 t) \\ \mu_1 \cos(\mu_1 t) \\ \sin(\mu_1 t) \\ \mu_1 \cos(\mu_1 t) \end{pmatrix}.$$

For the eigenvalue $i\,\mu_2$ we have

$$e^{i\mu_2 t}\begin{pmatrix} 1 \\ i\,\mu_2 \\ -1 \\ -i\,\mu_1 \end{pmatrix} = \left(\cos(\mu_2 t) + i\sin(\mu_2 t)\right)\left(\begin{pmatrix} 1 \\ 0 \\ -1 \\ 0 \end{pmatrix} + i\begin{pmatrix} 0 \\ \mu_2 \\ 0 \\ -\mu_2 \end{pmatrix}\right)$$

$$= \begin{pmatrix} \cos(\mu_2 t) \\ -\mu_2\sin(\mu_2 t) \\ -\cos(\mu_2 t) \\ \mu_2\sin(\mu_2 t) \end{pmatrix} + i\begin{pmatrix} \sin(\mu_2 t) \\ \mu_2\cos(\mu_2 t) \\ -\sin(\mu_2 t) \\ -\mu_2\cos(\mu_2 t) \end{pmatrix},$$

which yields the real-valued solutions

$$\mathbf{Z}_a(t) = \begin{pmatrix} \cos(\mu_2 t) \\ -\mu_2\sin(\mu_2 t) \\ -\cos(\mu_2 t) \\ \mu_2\sin(\mu_2 t) \end{pmatrix} \quad\text{and}\quad \mathbf{W}_a(t) = \begin{pmatrix} \sin(\mu_2 t) \\ \mu_2\cos(\mu_2 t) \\ -\sin(\mu_2 t) \\ -\mu_2\cos(\mu_2 t) \end{pmatrix}.$$

Note that the solutions $\mathbf{Z}_s(t)$ and $\mathbf{Z}_a(t)$ start from rest, that is, the second and fourth components are both zero at 0, whereas the solutions $\mathbf{W}_s(t)$ and $\mathbf{W}_a(t)$ start from the equilibrium position, that is, the first and third components are both zero at 0. The vector-valued functions $\mathbf{Z}_s(t)$ and $\mathbf{W}_s(t)$ are called the **symmetric modes of oscillation**, while $\mathbf{Z}_a(t)$ and $\mathbf{W}_a(t)$ are called the **antisymmetric modes of oscillation**.

Example 18.3. *Solve the system (18.7) in the case that $x_1(0) = x_2(0) = 1$ and $y_1(0) = y_2(0) = 0$.*

Solution. The general solution of (18.7) is written

$$\mathbf{Z}(t) = C_1\mathbf{Z}_s(t) + C_2\mathbf{W}_s(t) + C_3\mathbf{Z}_a(t) + C_4\mathbf{W}_a(t).$$

The initial conditions require that

$$\begin{pmatrix} 1 \\ 0 \\ 1 \\ 0 \end{pmatrix} = C_1\begin{pmatrix} 1 \\ 0 \\ 1 \\ 0 \end{pmatrix} + C_2\begin{pmatrix} 0 \\ \mu_1 \\ 0 \\ \mu_1 \end{pmatrix} + C_3\begin{pmatrix} 1 \\ 0 \\ -1 \\ 0 \end{pmatrix} + C_4\begin{pmatrix} 0 \\ \mu_2 \\ 0 \\ -\mu_2 \end{pmatrix},$$

for which the unique solution is $C_1 = 1$ and $C_2 = C_3 = C_4 = 0$. Thus, the solution is $\mathbf{Z}_a(t)$, or written in detail, $x_1(t) = \cos(\mu_1 t)$ and $x_2(t) = \cos(\mu_1 t)$. ∎

Example 18.4. *Consider a mass-spring system consisting of two particles satisfying $m_1 = m_2 = k_1 = k_2 = k_{12} = 1$, in which the particles start from rest. Assume that the equilibrium positions of the two masses are separated by a distance of 6 units. Find and plot the motion given each of the following sets of assumptions:*

(i) $x_1(0) = x_2(0) = 1$ *(symmetric mode case);*

(ii) $x_1(0) = 1$ *and* $x_2(0) = -1$ *(antisymmetric mode case);*

(iii) $x_1(0) = 2$ *and* $x_2(0) = 0$ *(mixed mode case).*

Solution. The initial value problem to be solved in case (i) is

$$\begin{cases} x_1' = y_1, & x_1(0) = 1, \\ y_1' = -2x_1 + x_2, & y_1(0) = 0, \\ x_2' = y_2, & x_2(0) = 1, \\ y_2' = x_1 - 2x_2, & y_2(0) = 0. \end{cases} \tag{18.12}$$

To solve (18.12) we use

```
symmode = ODE[{x1'  ==    y1,        x1[0] == 1,
               y1'  == -2x1 + x2, y1[0] == 0,
               x2'  ==    y2,       x2[0] == 1,
               y2'  ==    x1 - 2x2,y2[0] == 0},
              {x1[t],y1[t],x2[t],y2[t]},t,
              Method->LinearSystem]
```

and get the output

```
{{x1[t] -> Cos[t], x2[t] -> Cos[t], y1[t] -> -Sin[t], y2[t] -> -Sin[t]}}
```

For the plot of the motion of the two particles we need only x_1 and x_2. We assume the two particles to be positioned on the vertical axis. The equilibrium position of the first particle is positioned 3 units above the horizontal axis, while the equilibrium position of the second mass is positioned 3 units below the horizontal axis. Thus we use

```
Plot[Evaluate[{x1[t] + 3,x2[t] - 3} /. symmode],{t,0,4Pi}]
```

The result is

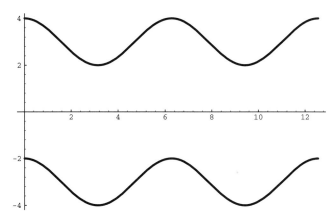

A mass-spring system vibrating in symmetric mode

Similarly, in case (ii) the initial value problem to be solved is

$$\begin{cases} x_1' = y_1, & x_1(0) = 1, \\ y_1' = -2x_1 + x_2, & y_1(0) = 0, \\ x_2' = y_2, & x_2(0) = -1, \\ y_2' = x_1 - 2x_2, & y_2(0) = 0. \end{cases} \qquad (18.13)$$

To solve (18.13) we use

```
antimode = ODE[{x1' == y1,      x1[0] ==  1,
               y1' == -2x1 + x2,y1[0] ==  0,
               x2' == y2,       x2[0] == -1,
               y2' == x1 - 2x2, y2[0] ==  0},
           {x1[t],y1[t],x2[t],y2[t]},t,
           Method->LinearSystem]
```

obtaining the output

```
{{x1[t] -> Cos[Sqrt[3] t], x2[t] -> -Cos[Sqrt[3] t],
 y1[t] -> -(Sqrt[3] Sin[Sqrt[3] t]), y2[t] -> Sqrt[3] Sin[Sqrt[3] t]}}
```

We want the same translations as in (i), so we use

```
Plot[Evaluate[{x1[t] + 3,x2[t] - 3} /. antimode],{t,0,4Pi}]
```

The result is

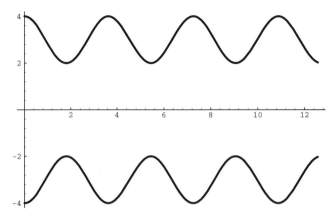

A mass-spring system vibrating in antisymmetric mode

Finally, in case (iii) the initial value problem to be solved is

$$\begin{cases} x_1' = y_1, & x_1(0) = 2, \\ y_1' = -2x_1 + x_2, & y_1(0) = 0, \\ x_2' = y_2, & x_2(0) = 0, \\ y_2' = x_1 - 2x_2, & y_2(0) = 0. \end{cases} \tag{18.14}$$

To solve (18.13) we use

```
mixedmode = ODE[{x1' == y1,          x1[0] == 2,
                 y1' == -2x1 +   x2,y1[0] == 0,
                 x2' == y2,          x2[0] == 0,
                 y2' ==    x1 - 2x2,y2[0] == 0},
            {x1[t],y1[t],x2[t],y2[t]},t,
            Method->LinearSystem]
```

obtaining the output

```
{{x1[t] -> Cos[t] + Cos[Sqrt[3] t], x2[t] -> Cos[t] - Cos[Sqrt[3] t],
  y1[t] -> -Sin[t] - Sqrt[3] Sin[Sqrt[3] t],
  y2[t] -> -Sin[t] + Sqrt[3] Sin[Sqrt[3] t]}} t]}}
```

We create the plot in the same way as we did in (i) and (ii) using

```
Plot[Evaluate[{x1[t] + 3,x2[t] - 3} /. mixedmode],{t,0,4Pi}]
```

The result is

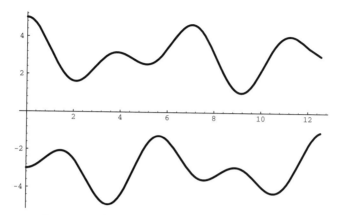

A mass-spring system vibrating in mixed mode

Exercises for Section 18.1

1. Solve the system (18.7) of coupled oscillators in the case that $m_1 = m_2 = 1$, $k_1 = k_2 = 4$, and $k_{12} = 2.5$ with the initial conditions $x_1(0) = 4$, $x_2(0) = 2$, and $y_1(0) = y_2(0) = 0$. Assume that the equilibrium positions of the two masses are separated by 8 units.

2. Solve the system (18.7) of coupled oscillators in the case that $m = 1$, $k = 4$, and $k_{12} = 6$ with the initial conditions $x_1(0) = 4$, $x_2(0) = 2$, $y_1(0) = 0$, and $y_2(0) = 0$. Assume that the equilibrium positions of the two masses are separated by 8 units.

3. Show that the solutions $\mathbf{Z}_a$ and $\mathbf{W}_a$ satisfy the identity $\mathbf{Z}_a(t - (\pi/2\mu_1)) = \mathbf{W}_a(t)$ for all t. (This is paraphrased as the statement that these two solutions are ninety degrees out of phase with one another.)

4. Show that the solutions $\mathbf{Z}_s$ and $\mathbf{W}_s$ satisfy the identity $\mathbf{Z}_s(t - (\pi/2\mu_1)) = \mathbf{W}_s(t)$ for all t. (This is paraphrased as the statement that these two solutions are ninety degrees out of phase with one another.)

5. Suppose that two equal masses interact in the absence of friction with equal spring constants and equal coupling constants. Let $w_1 = x_1 - x_2$ and $w_2 = x_1 + x_2$.

    a. Show that w_1 and w_2 satisfy the *decoupled* system

    $$\begin{cases} m\,w_1'' + (k + 2k_{12})w_1 = 0, \\ m\,w_2'' + k\,w_2 = 0, \end{cases} \qquad (18.15)$$

where m is the common value of the masses, k is the common value of the spring constants, and k_{12} is the common value of the coupling constants.

    b. Solve the system (18.15) to obtain w_1 and w_2; then find x_1 and x_2.

6. Suppose that the initial conditions are such that $x_1(0) = -x_2(0)$ and $x_1'(0) = -x_2'(0)$. Show that the resulting solution only contains terms with frequency $\mu_2 = \sqrt{(k + 2k_{12})/m}$.

7. Suppose that the initial conditions are such that $x_1(0) = x_2(0)$ and $x_1'(0) = x_2'(0)$. Show that the resulting solution only contains terms with frequency $\mu_1 = \sqrt{k/m}$.

8. Find the eigenvectors of the general matrix that defines the two *different* masses that interact with *different* spring constants and *different* coupling constants.

9. Suppose that two equal masses interact with the same spring constants and the same coupling constants in the presence of identical frictional terms $c\, dx_1/dt$ and $c\, dx_2/dt$. Find a fourth-degree polynomial equation to determine the quasiperiod and the relaxation times of this system.

10. Use **ODE** to solve the initial value problem of Exercise 9 in the case that $m = 1$, $k = 4$, $k_{12} = 6$, and $c = 0.5$, with the initial conditions $x_1(0) = 4$, $x_2(0) = 2$, $y_1(0) = 0$, and $y_2(0) = 0$. Assume that the equilibrium positions of the two masses are separated by 8 units.

18.2   *Electrical Circuits*

In Sections 7.3 and 11.3 we used first-order and second-order constant-coefficient differential equations to model the flow of current through a single circuit. Multiple circuits are naturally studied by means of systems of first-order differential equations with constant-coefficients. These are governed by the Kirchhoff voltage and current laws given on page 220. Voltage drops are described on page 369.

Example 18.5.   *Find the system of first-order linear differential equations that models the circuit shown below,*

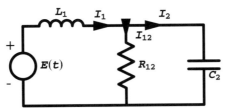

A system of electrical circuits

where $E(t)$ is an electromotive source, L_1 an inductor, R_{12} a resistor, C_2 a capacitor, I_1 and I_2 are the currents flowing through the left and right loops, respectively, and I_{12} is the current through the path common to both loops.

Solution. The circuit has two loops. We designate the elements that belong to only one loop by a single index and the elements that belong to two loops by a double index, and the same for currents and charges.

We need three relations to determine the currents I_1, I_2, and I_{12}. The first relation results when the Kirchhoff current law is applied to either of the two junctions; thus

$$I_1 = I_2 + I_{12}. \qquad (18.16)$$

We recall from Section 11.3 that the voltage drops across the inductor **L1** and the capacitor **C2** are given by $L_1 I_1'$ and Q_2/C_2, where L_1 is the inductance of **L1** and C_2 is the capacitance of **C2**. Ohm's law implies that the voltage drop across the resistor **R12** is $R_{12}I_{12}$, where R_{12} is the resistance of **R12**. The Kirchhoff voltage law applied to each of the two loops gives the other two relations, namely

$$\left\{ \begin{aligned} L_1\frac{dI_1}{dt} + R_{12}I_{12} &= E(t), \\ -R_{12}I_{12} + \frac{Q_2}{C_2} &= 0. \end{aligned} \right. \qquad (18.17)$$

We differentiate the second equation of (18.17), and use the fact that $Q_2' = I_2$. Then we solve (18.16) for I_{12} in terms of I_1 and I_2 and substitute this value into (18.17) to obtain a system of two linear differential equations in the primary loop currents I_1 and I_2:

$$\left\{ \begin{aligned} L_1\frac{dI_1}{dt} + R_{12}(I_1 - I_2) &= E(t), \\ -R_{12}\left(\frac{dI_1}{dt} - \frac{dI_2}{dt}\right) + \frac{I_2}{C_2} &= 0. \end{aligned} \right. \qquad (18.18)$$

We solve (18.18) for dI_1/dt and dI_2/dt, obtaining

$$
\begin{cases}
\dfrac{dI_1}{dt} = -\dfrac{R_{12}}{L_1}I_1 + \dfrac{R_{12}}{L_1}I_2 + \dfrac{1}{L_1}E(t), \\[3mm]
\dfrac{dI_2}{dt} = \dfrac{dI_1}{dt} - \dfrac{I_2}{R_{12}C_2} = -\dfrac{R_{12}}{L_1}I_1 + \left(\dfrac{R_{12}}{L_1} - \dfrac{1}{R_{12}C_2}\right)I_2 + \dfrac{1}{L_1}E(t).
\end{cases}
\tag{18.19}
$$

The coefficient matrix for the homogeneous system corresponding to (18.19) is

$$
A = \begin{pmatrix}
-\dfrac{R_{12}}{L_1} & \dfrac{R_{12}}{L_1} \\[3mm]
-\dfrac{R_{12}}{L_1} & \dfrac{R_{12}}{L_1} - \dfrac{1}{R_{12}C_2}
\end{pmatrix};
$$

let us also write

$$
B(t) = \begin{pmatrix}
\dfrac{E(t)}{L_1} \\[3mm]
\dfrac{E(t)}{L_1}
\end{pmatrix}.
$$

Then (18.19) can be abbreviated to

$$
\mathbf{I}' = A\,\mathbf{I} + B(t), \qquad \text{where} \qquad \mathbf{I} = \begin{pmatrix} I_1 \\ I_2 \end{pmatrix}. \ \blacksquare
$$

Example 18.6. *Solve* (18.19) *when* $E(t) = 1$ volt, $L_1 = 1$ henry, $R_{12} = 1$ ohm *and* $C_2 = 0.1$ farad.

Solution. In this case (18.19) becomes

$$
\begin{cases}
\dfrac{dI_1}{dt} = -I_1 + I_2 + 1, \\[3mm]
\dfrac{dI_2}{dt} = -I_1 - 9I_2 + 1.
\end{cases}
$$

This system can be solved with **ODE** using

```
ODE[{I1' == -I1 +  I2 + 1,I2' == -I1 - 9 I2 + 1,
I1[0] == 0, I2[0] == 0},{I1,I2},t,
Method->ApproximateLinearSystem,Form->Explicit,
PlotSolution->{{t,0,0.9}},PostSolution->{Chop}]
```

to obtain

$$\{1. \ + \ \frac{0.0164}{E^{8.87 \ t}} \ - \ \frac{1.02}{E^{1.13 \ t}}, \ \frac{-0.129}{E^{8.87 \ t}} \ + \ \frac{0.129}{E^{1.13 \ t}}\}$$

(*Mathematica*'s command **Chop** replaces very small numbers by 0.) Hence

$$I_1(t) \approx 1 + 0.0164e^{-8.87t} - 1.02e^{1.13t} \qquad \text{and} \qquad I_2(t) \approx -0.129e^{-8.87t} + 0.129e^{1.13t}. \ \blacksquare$$

Exercises for Section 18.2

1. Obtain the system of differential equations for the circuit

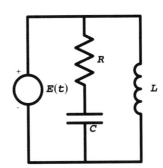

2. Obtain the system of differential equations for the circuit

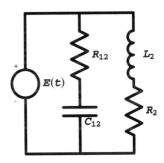

3. Consider the following circuit.

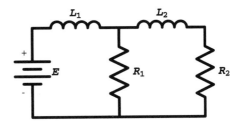

Assume **E** is a constant electromotive force of E volts, **R1** is a resistor of R_1 ohms, **R2** is a resistor of R_2 ohms, **L1** is an inductor of L_1 henrys, **L2** is an inductor of L_2 henrys, and the currents are initially zero.

    a. Write down the system of differential equations for the currents I_1 and I_2 in each loop.

    b. Solve the system explicitly in the case that $E = 100$ volts, $R_1 = 20$ ohms, $R_2 = 40$ ohms, $L_1 = 0.01$ henry, and $L_2 = 0.02$ henry.

4. Consider the following circuit.

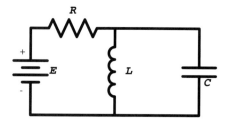

Assume that **E** is a constant electromotive force of E volts, **R** is a resistor of R ohms, **L** is an inductor of L henrys, **C** is an inductor of C farads, and the currents are initially zero.

    a. Write down the system of differential equations for the currents I_1 and I_2 in each loop.

    b. Solve the system explicitly in the case that $E = 100$ volts, $R = 20$ ohms, $L = 40$ henrys, and $C = 0.02$ farad.

5. Consider the following circuit.

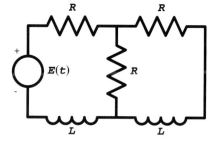

Assume **E** is an electromotive force with impressed voltage $E(t) = 110\sin(3t)$ volts, **R** is a resistor of 10 ohms, **L** is an inductor of 10 henrys, and the currents are initially zero. Find and plot the currents in each loop.

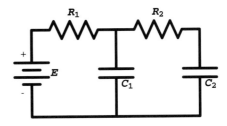

6. Consider the following circuit.

If **E** is a constant electromotive force of 100 volts, **R1** is a resistor of 20 ohms, **R2** is a resistor of 10 ohms, **C1** is a capacitor of 0.0001 farad, **C2** is a capacitor of 0.0002 farad, and the currents are initially zero, find the currents in each loop.

7. Consider the following circuit, called a two-stage low-pass filter.

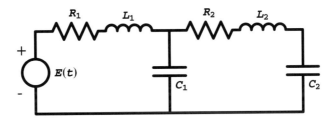

Suppose that R_1 and R_2 are resistors of 1000 ohms, L_1 and L_2 are inductors of 1.1 henrys, C_1 and C_2 are capacitors of 10^{-6} farad, and the currents are initially zero. If $E(t)$ is an electromotive force of $110\sin(377\,t)$ volts, use **ODE** with **Method->NDSolve** to compute the voltage across the C_2 capacitor by first finding the current in the second loop, $I_2(t)$, and then computing

$$V_{C_2}(t) = \frac{1}{C_2}\int_0^t I_2(s)\,ds.$$

8. Consider the following circuit, called a two-stage high-pass filter.

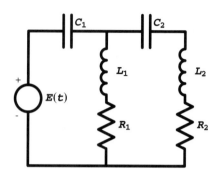

Suppose that R_1 and R_2 are resistors of 1000 ohms, L_1 and L_2 are inductors of 0.1 henry, C_1 and C_2 are capacitors of 10^{-6} farad, and the currents are initially zero. If $E(t)$ is an electromotive force of $110 \sin(377\,t)$ volts, use **ODE** with **Method->NDSolve** to compute the voltage across the $\widehat{L_2 R_2}$ circuit by first finding the current in the second loop, $I_2(t)$, and then computing

$$V_{\widehat{L_2 R_2}}(t) = L_2 \frac{dI_2(t)}{dt} + R_2 I_2(t).$$

18.3 *Markov Chains*

Systems of differential equations occur also in the theory of probability. In contrast with the other models in this chapter, probability models provide examples of systems of linear differential equations of arbitrary dimension, in contrast with the even-dimensional systems that arise from mechanics and electricity. Furthermore, the differential equations are deduced from the basic hypotheses of these models, rather than assumed at the outset.

A **Markov**[1] **chain** consists of a set of **states** labeled $[1, \ldots, n]$ together with **transition probabilities** $p_{ij}(t)$, signifying the probability that the system is in the state j at time instant t, given that the system is in the state i at time instant $t = 0$. From the interpretation of probability, we have

$$p_{ij}(t) \geq 0, \quad (1 \leq i, j \leq n), \qquad \sum_{j=1}^{n} p_{ij}(t) = 1, (1 \leq i \leq n), \tag{18.20}$$

---

[1]

Andrei Andreevich Markov (1856–1922). Russian probabilist, who laid the foundations of the theory of finite Markov chains. In particular, in 1906 he proved the existence of stationary initial distributions for finite-state chains.

and

$$\lim_{t \to 0} p_{ij}(t) = 0, \quad (i \neq j), \qquad \lim_{t \to 0} p_{ii}(t) = 1, \quad (1 \leq i \leq n). \tag{18.21}$$

The matrix $P(t) = \big(p_{ij}(t)\big)$ is called the **transition matrix** of the Markov chain.
The **Markov semigroup property** is the statement that

$$p_{ij}(t+s) = \sum_{k=1}^{n} p_{ik}(t) p_{kj}(s) \tag{18.22}$$

for $1 \leq i, j \leq n$ and $t, s > 0$. Intuitively, this means that the probability of going from state i to state j in time $t+s$ can be computed by summing over the possible intermediate states k that the system may assume at time t. More formally, (18.22) follows from the independence of the past and future, given the present state. (See the exercises for further information.)

The Markov semigroup property (18.22) can equivalently be written in terms of matrix multiplication, in the form

$$P(t+s) = P(t) P(s). \tag{18.23}$$

From these hypotheses it can be shown (see the exercises) that $P(t) = \big(p_{ij}(t)\big)$ is a differentiable matrix-valued function. The **infinitesimal transition matrix** $Q = (q_{ij})$ is defined as the matrix of derivatives

$$Q = \frac{dP}{dt}\bigg|_{t=0} = \lim_{t \to 0} \frac{P(t) - I_{n \times n}}{t}.$$

Lemma 18.1.   *The matrix Q has the properties*

(i) $q_{ij} \geq 0$ *for* $i \neq j$;

(ii) $\displaystyle\sum_{j=1}^{n} q_{ij} = 0$ *for* $1 \leq i \leq n$;

(iii) $q_{ii} \leq 0$ *for* $1 \leq i \leq n$.

Proof. $Q = (q_{ij})$ is given by $q_{ij} = p_{ij}'(0)$. To prove (i), we note that $p_{ij}(t) \geq 0$ for all i and j, in particular for $i \neq j$, with $p_{ij}(0) = 0$ for $i \neq j$. Thus for $i \neq j$, we have

$$q_{ij} = \lim_{t \to 0} \frac{p_{ij}(t)}{t} \geq 0.$$

To prove (ii), we differentiate

$$\sum_{j=1}^{n} p_{ij}(t) = 1$$

and set $t = 0$ to obtain

$$\sum_{j=1}^{n} q_{ij} = 0.$$

Finally, (iii) follows by noting from (ii) that

$$q_{ii} = -\sum_{j \neq i} q_{ij} \leq 0,$$

from (i). ∎

The parameters q_{ii} and q_{ij} have the following intuitive interpretations: $q_i = -q_{ii}$ is the **holding rate** of the state i, meaning that q_i^{-1} is the mean exit time of the system from the state i, in other words the average time that the system spends in state i before moving elsewhere. If q_i is large, then the system will move quickly out of state i, whereas if q_i is small, then the system will remain in state i for a long time, on the average. The numbers q_{ij} are interpreted as **jump rates**, meaning that $\pi_{ij} = q_{ij}/q_i$ is the probability that the system will move to state j when it leaves state i for the first time. Indeed, from parts (i)and (ii)of Lemma 18.1, the numbers π_{ij} are nonnegative and sum to 1; hence they may be regarded as a probability distribution on a finite set.

The relation with differential equations is described as follows.

Theorem 18.2 *The transition matrix $P(t)$ satisfies the two systems of differential equations*

$$\frac{dP}{dt} = P(t)\,Q = Q\,P(t).$$

Proof. We apply (18.23) twice to write

$$\frac{P(t+s) - P(t)}{s} = P(t)\left(\frac{P(s) - I}{s}\right) = \left(\frac{P(s) - I}{s}\right)P(t).$$

Now let $s \longrightarrow 0$ to obtain the equations as required . ∎

From our work in Section 15.5.4, we know that for any square matrix A, the matrix exponential e^{tA} is the unique solution of the linear system $\mathbf{x}' = A\mathbf{x}$ with initial matrix the identity. Applying this, we see the relation between the transition matrix $P(t)$ and the infinitesimal transition matrix Q is

$$P(t) = e^{tQ}.$$

In order to make effective use of the methods of Chapter 15, we need to know the eigenvalues and eigenvectors of the matrix Q. We have the following general fact.

Lemma 18.3.   *The matrix Q has 0 as an eigenvalue. Any nonzero multiple of $\boldsymbol{\xi} = (1, \ldots, 1)^T$ is an eigenvector corresponding to this eigenvalue. Any other eigenvalue r of Q satisfies $\mathfrak{Re}(r) < 0$.*

See the exercises for the proof of Lemma 18.3.

    We illustrate with an example.

Example 18.7. *A machine is in one of two states: $i = 1$ signifies that the machine is in working order, while $i = 2$ signifies that the machine is out of order. Assume the following holding rates and jump rates:*

$$q_{11} = -2, \quad q_{12} = 2, \quad q_{21} = 6, \quad q_{22} = -6.$$

Find the transition matrix $P(t)$ and $\lim_{t \to \infty} P(t)$.

Solution. We know from Section 15.5.4 that the transition matrix $P(t)$ can be computed in terms of a fundamental matrix $\Phi(t)$ by the formula

$$P(t) = e^{tQ} = \Phi(t)\Phi(0)^{-1}.$$

The fundamental matrix can be found in terms of the eigenvalues and eigenvectors of the matrix

$$Q = \begin{pmatrix} q_{11} & q_{12} \\ q_{21} & q_{22} \end{pmatrix} = \begin{pmatrix} -2 & 2 \\ 6 & -6 \end{pmatrix}.$$

The associated characteristic equation of Q is

$$0 = (2+r)(6+r) - 12 = r^2 + 8r,$$

whose roots are $r = 0$ and $r = -8$. The associated eigenvectors are multiples of $\xi_1 = (1, 1)^T$ and $\xi_2 = (1, -3)^T$, leading to the fundamental matrix

$$\Phi(t) = \begin{pmatrix} 1 & e^{-8t} \\ 1 & -3e^{-8t} \end{pmatrix}.$$

From this we have

$$\Phi(0)^{-1} = \begin{pmatrix} \frac{3}{4} & \frac{1}{4} \\ \frac{1}{4} & -\frac{1}{4} \end{pmatrix}.$$

Thus, the transition matrix is

$$P(t) = \frac{1}{4} \begin{pmatrix} 3 + e^{-8t} & 1 - e^{-8t} \\ 3(1 - e^{-8t}) & 1 + 3e^{-8t} \end{pmatrix}.$$

Taking the limit $t \longrightarrow \infty$, we obtain

$$\lim_{t \to \infty} P(t) = \begin{pmatrix} \frac{3}{4} & \frac{1}{4} \\ \frac{3}{4} & \frac{1}{4} \end{pmatrix}.$$

The form of this matrix shows the long-term probability of the states of the machine. Both terms of the first column are $3/4$, signifying that the long-term probability that the machine is in working order is $3/4$, no matter whether or not it was in working order initially. Similarly, both terms of the second column are $1/4$, signifying that the long-term probability that the machine is broken is $1/4$, whether or not it was in working order initially. ∎

For Markov chains with $n = 3$ states, it is always possible to find the eigenvalues of the matrix Q. Since $r_1 = 0$ is an eigenvalue with eigenvector $\boldsymbol{\xi} = (1, 1, 1)^T$, we can remove the common factor r from the (cubic) characteristic polynomial of Q to obtain a quadratic equation for the remaining eigenvalues r_2 and r_3. In detail,

$$0 = \det(r\, I_{3\times 3} - Q) = \det \begin{pmatrix} r - q_{11} & -q_{12} & -q_{13} \\ -q_{21} & r - q_{22} & -q_{23} \\ -q_{31} & -q_{32} & r - q_{33} \end{pmatrix}$$

$$= (r - q_{11})(r - q_{22})(r - q_{33}) - q_{12}q_{23}q_{31} - q_{13}q_{32}q_{21}$$

$$-q_{13}q_{31}(r - q_{22}) - q_{12}q_{21}(r - q_{33}) - q_{23}q_{32}(r - q_{11})$$

$$= r^3 - r^2(q_{11} + q_{22} + q_{33}) + r(q_{11}q_{22} + q_{22}q_{33} + q_{11}q_{33} - q_{13}q_{31} - q_{32}q_{23} - q_{12}q_{21}).$$

Specific solvable cases are contained in the exercises. The following example illustrates the main points.

Example 18.8. *A particle jumps among the vertices of an equilateral triangle according to the infinitesimal transition matrix*

$$Q = \begin{pmatrix} -2 & 1 & 1 \\ 1 & -2 & 1 \\ 1 & 1 & -2 \end{pmatrix}.$$

Find $P(t)$ and $\lim_{t\to\infty} P(t)$.

Solution. The characteristic equation of Q is

$$0 = \det(r\, I - Q) = (r + 2)^3 - 2 - 3(r + 2) = r^3 + 6r^2 + 9r = r(r + 3)^2,$$

whose roots are $r_1 = 0$, $r_2 = -3$, and $r_3 = -3$. Eigenvectors corresponding to $r_1 = 0$ are multiples of $\boldsymbol{\xi}_1 = (1, 1, 1)^T$. An eigenvector $\boldsymbol{\xi} = (c_1, c, c_3)$ corresponding to $r_1 = -3$ satisfies $c_1 + c_2 + c_3 = 0$, which is satisfied by taking $\boldsymbol{\xi}_2 = (1, -1, 0)$ and $\boldsymbol{\xi}_3 = (0, 1, -1)$. We form the matrix U whose columns are the eigenvectors $\boldsymbol{\xi}_1, \boldsymbol{\xi}_2, \boldsymbol{\xi}_3$:

$$U = \begin{pmatrix} 1 & 1 & 0 \\ 1 & -1 & 1 \\ 1 & 0 & -1 \end{pmatrix} \quad \text{with} \quad U^{-1} = \begin{pmatrix} \frac{1}{3} & \frac{1}{3} & \frac{1}{3} \\ \frac{2}{3} & -\frac{1}{3} & -\frac{1}{3} \\ \frac{1}{3} & \frac{1}{3} & -\frac{2}{3} \end{pmatrix}.$$

A fundamental matrix $\Phi(t)$ is obtained by listing the exponential terms in the respective columns of U. Then $P(t)$ is the matrix exponential e^{tQ}, which can be obtained in terms of the fundamental matrix $\Phi(t)$ as

$$
P(t) = \Phi(t)\Phi^{-1}(0) = \begin{pmatrix} 1 & e^{-3t} & 0 \\ 1 & -e^{-3t} & e^{-3t} \\ 1 & 0 & -e^{-3t} \end{pmatrix} \begin{pmatrix} \frac{1}{3} & \frac{1}{3} & \frac{1}{3} \\ \frac{2}{3} & -\frac{1}{3} & -\frac{1}{3} \\ \frac{1}{3} & \frac{1}{3} & -\frac{2}{3} \end{pmatrix}
$$

$$
= \begin{pmatrix} \frac{1}{3}+\frac{2}{3}e^{-3t} & \frac{1}{3}-\frac{1}{3}e^{-3t} & \frac{1}{3}-\frac{1}{3}e^{-3t} \\ \frac{1}{3}-\frac{1}{3}e^{-3t} & \frac{1}{3}+\frac{2}{3}e^{-3t} & \frac{1}{3}-\frac{1}{3}e^{-3t} \\ \frac{1}{3}-\frac{1}{3}e^{-3t} & \frac{1}{3}-\frac{1}{3}e^{-3t} & \frac{1}{3}+\frac{2}{3}e^{-3t} \end{pmatrix}.
$$

We see that $p_{ii}(t) = 1/3 + (2/3)e^{-3t}$, for $i = 1, 2, 3$ whereas $p_{ij}(t) = 1/3 - (1/3)e^{-3t}$, for $i \neq j$. In particular,

$$
\lim_{t \to \infty} p_{ij}(t) = \frac{1}{3}
$$

for all i and j. After a long time, the particle is equally likely to be at any one of the three vertices, for any initial state . ∎

The following theorem identifies the limiting value of the transition matrix in a wide variety of practical circumstances.

Theorem 18.4. *Suppose in addition that the jump rates satisfy $q_{ij} > 0$ for all $i \neq j$. Then all eigenvalues of the matrix Q satisfy $\mathfrak{Re}(r) < 0$ with the exception of the eigenvalue $r = 0$, which has a unique normalized left eigenvector $\pi = (\pi_1, \ldots, \pi_n)$ satisfying $\pi_i > 0$ for $1 \leq i \leq n$,*

$$
\sum_{i=1}^{n} \pi_i = 1, \quad \text{and} \quad \sum_{i=1}^{n} \pi_i q_{ij} = 0
$$

for $1 \leq j \leq n$. When $t \longrightarrow \infty$ we have

$$
\lim_{t \to \infty} p_{ij}(t) = \pi_j
$$

for $1 \leq i, j \leq n$.

Theorem 18.4 gives conditions under which the long-term probabilities are independent of the initial state. The limiting matrix has n rows, all of which are the *same* row vector $\pi = (\pi_i)$. For the proof of Theorem 18.4 see the exercises.

Markov chains can also be used to model a process of **random walk**. We think of a particle that can move from one state to another by jumping from time to time. Each state is characterized by its **holding rate** q_{ii} and its **jump rates** q_{ij}. We illustrate with a random walk on the points of a circular ring.

Example 18.9.   *A particle is situated at one of the n points of a regular n-sided polygon. It can move to the neighbor on the right (clockwise direction) with probability p or to the neighbor on the left (counterclockwise direction) with probability q, where $p > 0, q > 0$, and $p + q = 1$. Each state has the same holding rate $\lambda > 0$. Find the infinitesimal transition matrix of this Markov chain.*

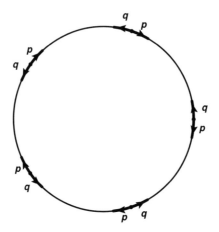

A Markov chain on a circle

Solution.   The Markov chain has n states, labeled $[1, \ldots, n]$, where we identify the first and last as neighbors. The infinitesimal transition matrix is written

$$
Q = \begin{pmatrix}
-\lambda & \lambda p & 0 & \ldots & 0 & \lambda q \\
\lambda q & -\lambda & \lambda p & \ldots & 0 & 0 \\
\vdots & \vdots & \vdots & \ddots & -\lambda & \lambda p \\
\lambda p & 0 & 0 & \ldots & \lambda q & -\lambda
\end{pmatrix} \cdot \blacksquare
\tag{18.24}
$$

The eigenvalues and eigenvectors for the random walk transition matrix (18.24) can be found by a special technique that exploits the symmetry of the problem. We think of the states of the system as the set of complex numbers $[1, z, z^2, \ldots, z^{n-1}]$, where $z = e^{2\pi i / n}$. These are equally spaced on the unit circle in the complex number plane. To move one step counterclockwise corresponds to multiplication by z, while moving one step clockwise corresponds to division by z. For any function f on this set, the Q matrix acts according to the formula

$$
Qf(z^j) = \lambda\big(p\, f(z^{j+1}) + q\, f(z^{j-1}) - f(z^j)\big).
$$

for $j = 1, \ldots, n$. It is immediately verified that any function of the form $f_k(z^j) = z^{jk}$ is an eigenvector of Q; indeed,

$$Q f_k(z^j) = \lambda\left(p\, z^{(j+1)k} + q\, z^{(j-1)k} - z^{jk}\right) = \lambda\, z^{jk}\left(p\, z^k + q\, z^{-k} - 1\right).$$

Therefore, the eigenvalues of Q are the numbers

$$r_k = \lambda(p\, z^k + q\, z^{-k} - 1)$$

for $k = 0, \ldots, n - 1$, and the eigenvectors are

$$\boldsymbol{\xi}_k = \left(z^k, z^{2k}, \ldots, z^{(n-1)k}\right),$$

where $z = e^{2\pi i/n}$.

Example 18.10.   *Find the eigenvalues and eigenvectors of the transition matrix Q of Example 18.9 in the case of $n = 3$ states and $\lambda = 1$.*

Solution. According to the above discussion, we have

$$z = e^{2\pi i/3} = \frac{-1 + i\sqrt{3}}{2}.$$

For $k = 0$ we have the eigenvalue $r_1 = 0$ with eigenvector $\boldsymbol{\xi} = (1, 1, 1)^T$. For $k = 1$ we have the eigenvalue

$$r_2 = p z + q/z - 1 = -3/2 + i\sqrt{3}(p - q)/2$$

with eigenvector $\boldsymbol{\xi} = (1, z, z^2)$. For $k = 2$ we have the eigenvalue

$$r_3 = p z^2 + q/z^2 - 1 = -3/2 - i\sqrt{3}(p - q)/2$$

with eigenvector $\boldsymbol{\xi} = (1, z^2, z)^T$. ∎

Exercises for Section 18.3

Exercises 1–7 involve explicit computations with Markov chains. The remaining exercises are more theoretical in nature.

1. The general two-state Markov chain is defined by two parameters $\lambda > 0$, $\mu > 0$ with $q_{11} = -\lambda$, $q_{12} = \lambda$, $q_{21} = \mu$, $q_{22} = -\mu$. Find $P(t)$ and $\lim_{t\to\infty} P(t)$.

2. The three-state Ehrenfest Markov chain is defined by the matrix

$$Q = \begin{pmatrix} -2 & 2 & 0 \\ 1 & -2 & 1 \\ 0 & 2 & -2 \end{pmatrix}.$$

Find the eigenvalues and eigenvectors of Q and use these to find $P(t) = e^{tQ}$ and $\lim_{t \to \infty} P(t)$.

3. Consider the three-state Markov chain with

$$Q = \begin{pmatrix} -\lambda_0 & \lambda_0 & 0 \\ \lambda_1 & -2\lambda_1 & \lambda_1 \\ 0 & \lambda_0 & -\lambda_0 \end{pmatrix}.$$

Find the eigenvalues and eigenvectors of Q and use these to find $P(t) = e^{tQ}$ and $\lim_{t \to \infty} P(t)$.

4. Consider the three-state Markov chain with

$$Q = \begin{pmatrix} -\lambda_0 & \lambda_0 & 0 \\ \lambda_1 & -(\lambda_1 + \mu_1) & \mu_1 \\ 0 & \lambda_0 & -\lambda_0 \end{pmatrix}.$$

Find the eigenvalues and eigenvectors of Q and use these to find $P(t) = e^{tQ}$ and $\lim_{t \to \infty} P(t)$.

5. Consider the three-state Markov chain of Example 18.10 in the case $p = 2/3$, $q = 1/3$. Find the eigenvalues and eigenvectors and use these to find $P(t)$ in terms of real-valued functions.

6. Find the eigenvalues and eigenvectors of the random walk model on $n = 4$ points.

7. Consider a random walk on n points with $p = q$. Show that the eigenvalues are obtained through the formula

$$r_k = \lambda \big(\cos(2\pi k/n) - 1 \big)$$

for $k = 0, 1, \dots, n-1$.

8. The purpose of this exercise is to show that the transition matrix $P(t)$ must be a continuous function.

    a. Write the Markov semigroup property in the form (18.23).

    b. Use (18.21) to conclude continuity from the right: $\lim P(t) = P(t_0)$ when t approaches t_0 from the right.

    c. Use (18.21) to show that there exists $\delta > 0$ such that $P(s)$ has an inverse $P(s)^{-1}$ for $0 < s < \delta$, which satisfies $\lim_{s \to 0} P(s)^{-1} = I$.

    d. Establish the equation $P(t_0 - s) = P(t_0)P(s)^{-1}$ for $0 < s < \delta$.

    e. Conclude that $\lim P(t) = P(t_0)$ when t approaches t_0 from the left.

9. The purpose of this exercise is to show that the transition matrix $P(t)$ is a differentiable function.

    a. Integrate the Markov semigroup property (18.23) to obtain the equation

    $$\int_t^{t+\delta} P(u)\,du = R\,P(t), \qquad R = \int_0^\delta P(u)\,du.$$

    b. Argue from the result of Exercise 8 that the matrix R has an inverse and that

    $$P(t) = R^{-1} \int_t^{t+\delta} P(u)\,du.$$

    c. Use the fundamental theorem of calculus to deduce the differentiability of the matrix-valued function $t \longrightarrow P(t)$.

10. This problem discusses the background of the Markov semigroup property for readers who are familiar with the notions of random variable and conditional probability, written **Prob**$(A|B)$. The random variable $X(t)$ represents the state of the systems a time t. The **temporally homogeneous Markov property** is the statement that

    $$\textbf{Prob}\big[X(t_k + s) = j | X(t_0) = i_0, \ldots, X(t_k) = i_k\big] = \textbf{Prob}\big[X(s) = j | X(0) = i_k\big],$$

    where $t_0 < \cdots < t_k$ is any finite set of time points and $i_0, \ldots, i_k, j$ are any members of the set $[1, \ldots, n]$. Suppose that the system of probabilities satisfies the temporally homogeneous Markov property, and let

    $$p_{ij}(t) = \textbf{Prob}\big[X(t) = j | X(0) = i\big].$$

    Show that the matrix $P(t) = \big(p_{ij}(t)\big)$ satisfies the Markov semigroup property (18.22) [Hint: Sum over the "intermediate states" that may be assumed by the random variable $X(t)$.]

11. In this problem we indicate how the temporally homogeneous Markov property may be derived from the property that *future and past are conditionally independent, given the present.*

   a. Suppose that the system of conditional probabilities has the property that for any three times $t - s_1 < t < t + s_2$ and any integers i_1, i and k we have

$$\mathbf{Prob}\left[X(t - s_1) = i_1, X(t + s_2) = k | X(t) = i\right]$$

$$= \mathbf{Prob}\left[X(t - s_1) = i_1 | X(t) = i\right]\mathbf{Prob}\left[X(t + s_2) = k | X(t) = i\right].$$

Show that this may be written in terms of **absolute probabilities** as

$$\mathbf{Prob}[X(t + s_2) = k, X(t) = i, X(t - s_1) = i_1]\mathbf{Prob}[X(t) = i]$$

$$= \mathbf{Prob}[X(t + s_2) = i_2, X(t) = k]\mathbf{Prob}[X(t) = k | X(t - s_1) = i_1].$$

   b. Divide both sides by the final factors on the left and right to obtain the conditional probability statement

$$\mathbf{Prob}\left[X(t + s_2) = i_2 | X(t) = k, X(t - s_1) = i_1\right] = \mathbf{Prob}\left[X(t + s_2) = i_2 | X(t) = k\right].$$

   c. Indicate how one would generalize the hypothesis of conditional independence in part **a** to obtain the general Markov property in the form

$$\mathbf{Prob}\left[X(t + s) = k | X(t) = i,\right.$$

$$\left. X(t - s_1) = i_1, X(t - s_2) = i_2, \ldots, X(t - s_p) = i_p\right]$$

$$= \mathbf{Prob}\left[X(t + s) = k | X(t) = i\right].$$

12. The purpose of this exercise is to show that the nonzero eigenvalues of the Q-matrix lie in the left half plane.

   a. Suppose that $\boldsymbol{\xi} = (c_i)$ is an eigenvector of Q with eigenvalue r. Show that

$$\sum_{j \neq i} q_{ij} c_j = (r + q_{ii}) c_i.$$

   b. Deduce from part **a** that

$$|(r + q_{ii}) c_i| \leq |q_{ii}| \max_{1 \leq j \leq n} |c_j|$$

for $1 \leq i \leq n$.

    c.   Choose i such that $|c_i| = \max_{1 \le j \le n} |c_j|$ to obtain $|r + q_{ii}| \le |q_{ii}|$ for that value of i.

    d.   Conclude that either $r = 0$ or the complex number r satisfies $\mathfrak{Re}(r) < 0$.

13.   In this exercise we show that if $q_{ij} > 0$ for all $i \ne j$, then the eigenvalue $r = 0$ has only the eigenvector $\boldsymbol{\xi} = (1, \ldots, 1)^T$. To prove this, let $\boldsymbol{\xi} = (c_1, \ldots, c_n)^T$ be an eigenvector, and write $c_j = \alpha_j + i\beta_j$.

    a.   Write the eigenvector condition $Q\boldsymbol{\xi} = \mathbf{0}$ in the form

$$\sum_{k \ne j} q_{jk}(c_k - c_j) = 0$$

for $1 \le j \le n$.

    b.   Take the real and imaginary parts of this equation to conclude that

$$\sum_{k \ne j} q_{jk}(\alpha_k - \alpha_j) = 0 \qquad \text{and} \qquad \sum_{k \ne j} q_{jk}(\beta_k - \beta_j) = 0$$

for $1 \le j \le n$.

    c.   Conclude that α_k and β_k are constants independent of k.

    d.   Conclude that the eigenvector $\boldsymbol{\xi}$ is a constant multiple of $(1, \ldots, 1)^T$.

14.   In this exercise we show that the eigenvalue $r = 0$ of the Q-matrix has no generalized eigenvectors.

    a.   Suppose that $\boldsymbol{\eta}$ is a vector that solves the equation $Q\boldsymbol{\eta} = c(1, \ldots, 1)^T$ for some complex constant c. Show that $P(t)\boldsymbol{\eta} = \boldsymbol{\eta} + t\, c(1, \ldots, 1)^T$. [Hint: Use the power series definition of the matrix exponential e^{tQ}.]

    b.   Conclude that $c = 0$. [Hint: Take $t \longrightarrow \infty$ in part **a** and use the inequalities $|p_{ij}(t)| \le 1$.]

    c.   Generalize to the case $Q^k \boldsymbol{\eta} = c(1, \ldots, 1)^T$ for $k > 1$.

15.   Use the results of Exercises 12, 13, and 14 to show that if $q_{ij} > 0$ for $i \ne j$, then the transition matrix $P(t)$ satisfies $\lim_{t \to \infty} p_{ij}(t) = \pi_j$ for $1 \le i, j \le n$, where π_j is the solution of the equations

$$\sum_{i=1}^{n} \pi_i q_{ij} = 0$$

for $1 \le j \le n$.

16. A Markov chain is said to be **irreducible** if for each pair i, j there exists a set of indices $i_1, \ldots, i_N$ such that $q_{ii_1} > 0, q_{i_1 i_2} > 0, \ldots, q_{i_N j} > 0$. The problem is to show that an irreducible Markov chain also satisfies Exercises 12, 13, 14, and hence also Theorem 18.4.

In each of the following exercises, find the transition matrix $P(t)$ and $\lim_{t \to \infty} P(t)$ for the given infinitesimal matrix Q. Use the *Mathematica* command **MatrixExp**.

17. $Q = \begin{pmatrix} -1 & 0 & 1 \\ 1 & -2 & 1 \\ 2 & 0 & -2 \end{pmatrix}$ **19.** $Q = \begin{pmatrix} -1 & \frac{1}{2} & \frac{1}{2} \\ 1 & -2 & 1 \\ 1 & 0 & -1 \end{pmatrix}$

18. $Q = \begin{pmatrix} -1 & 1 & 0 \\ 1 & -3 & 2 \\ 1 & 1 & -2 \end{pmatrix}$ **20.** $Q = \begin{pmatrix} -2 & 1 & 1 \\ 2 & -4 & 2 \\ 1 & 1 & -2 \end{pmatrix}$

19

APPLICATIONS OF
NONLINEAR SYSTEMS

Many systems of differential equations cannot be solved conveniently by analytical methods; therefore, it is important to obtain qualitative information about their solutions without actually solving the systems. The pioneering work in this area was done by Poincaré[1].

In order to solve and plot nonlinear systems, we discuss numerical methods for solving first-order systems in Section 19.1. Qualitative and numerical solutions of predator-prey problems are discussed in Section 19.2. We use the numerical programs of Section 19.1 to solve and plot the van der Pol equation in Section 19.3. The differential equation that models a simple pendulum is solved exactly in terms of elliptic functions and also numerically in Section 19.4. In Section 19.5 the fundamental theorem of plane curves is used with numerical solution techniques to construct interesting plane curves with specified curvature.

Jules Henri Poincaré (1854–1912), French mathematician, widely regarded as the leading mathematician of his time. He developed the concept of automorphic functions and invented much of of algebraic topology. He also worked in algebraic geometry, analytic functions of several complex variables and number theory. The qualitative theory of differential equations was created by Poincaré in several papers between 1880 and 1886.

19.1   Numerical Solutions of Systems of Differential Equations

In Chapter 13 we discussed numerical methods for solving initial value problems associated with first-order differential equations. Most such methods can be extended to systems of first-order differential equations more or less automatically. The trick is to use vectors instead of scalars. For example, suppose we are given a first-order initial value problem

$$\begin{cases} \mathbf{y}'(t) = \mathbf{f}(t, \mathbf{y}), \\ \mathbf{y}(t_0) = \mathbf{Y}_0, \end{cases} \tag{19.1}$$

where now $t \longmapsto \mathbf{y}(t)$ and $(t, \mathbf{y}) \longmapsto \mathbf{f}(t, \mathbf{y})$ are vector-valued functions and $\mathbf{Y}_0$ is a vector. It is true that we can write out these objects in terms of their components:

$$\begin{cases} \mathbf{y}(t) & = \big(y_1(t), \ldots, y_n(t) \big), \\ \mathbf{f}(t, \mathbf{y}) & = \Big(f_1\big(t, (y_1(t), \ldots, y_n(t))\big), \ldots, f_n\big(t, (y_1(t), \ldots, y_n(t))\big) \Big), \\ \mathbf{Y}_0 & = \big(y_1(t_0), \ldots, y_n(t_0) \big). \end{cases} \tag{19.2}$$

But usually (19.2) is more of a hindrance than a help for understanding the theory; the compact notation using $\mathbf{y}(t)$, $\mathbf{f}(t, \mathbf{y})$ and $\mathbf{Y}_0$ is better [2]. Of course, when specific systems are solved components must eventually be inserted, although it is best to keep the compact boldface notation as long as possible in the solution process. Since boldface letters are used to denote vectors and vector-valued functions, a great deal of the theory given in Chapter 13 can be generalized to systems simply by replacing appropriate nonboldface letters by boldface letters. Notice that t and t_0 are never boldface.

For example, the Euler method for systems can be explained as follows. Just as with its single-equation counterpart, the objective is to construct an approximation to the solution of the initial value problem (19.1) for $a \leq t \leq b$. We divide the interval $a \leq t \leq b$ into equal subintervals:

$$a = t_0 < t_1 < \cdots < t_N = b,$$

where $h = t_{k+1} - t_k$ is the **step size**. The generalization to systems of the **Euler method formula** (13.6) is just

$$\mathbf{Y}_{k+1} = \mathbf{Y}_k + h\,\mathbf{f}(t_k, \mathbf{Y}_k), \tag{19.3}$$

where each $\mathbf{Y}_k$ is a vector.

---

[2] It was not until the beginning of the twentieth century that vectors began to replace lists of components in the mathematical literature. The driving force was vector analysis as formulated by Gibbs (see [Crowe]). In *Mathematica* a vector is a list.

Example 19.1. *Find a numerical approximation to the solution of the initial value problem*

$$\begin{cases} x' = y\sin(t), & x(0) = 1, \\ y' = -x\cos(t), & y(0) = 0 \end{cases} \tag{19.4}$$

over the interval $0 \le t \le 10$ with step size $h = 1.0$. Use the Euler method.

Solution. Write $\mathbf{Y}_k = (Y_{k0}, Y_{k1})$; the Euler method formula (19.3) for the initial value problem (19.4) with step size $h = 1.0$ becomes

$$\mathbf{Y}_{k+1} = \mathbf{Y}_k + 1.0\big(Y_{k1}\sin(t_k), -Y_{k0}\sin(t_k)\big)$$

for $0 \le n \le N - 1$. The interval $0 \le t \le 10$ is divided into 10 equal pieces, so $N = 10$. We are given $\mathbf{Y}_0 = (1, 0)$; then

$$\mathbf{Y}_1 = \mathbf{Y}_0 + 1.0f\big(0, (1, 0)\big) = (1, 0) + (0, -1) = (1, -1),$$

and so forth. We obtain the following table:

$k = t_k$	$\mathbf{Y}_k$
0	$(1.0, 0)$
1	$(1.0 - 1.0)$
2	$(0.1586, -1.5403)$
3	$(-1.2420, -1.4743)$
4	$(-1.45012, -2.7040)$
5	$(0.5962, -3.6518)$
6	$(4.0981, -3.8210)$
7	$(5.1657, -7.7558)$
8	$(0.07025, -11.6502)$
9	$(-11.4560, -11.6400)$
10	$(-16.2531, -22.0780)$

The first plot below is the phase plot of the data given by this table. If we had used a step size of 0.01, we would have obtained the second plot.

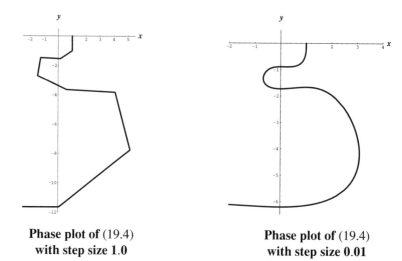

Phase plot of (19.4)   **Phase plot of** (19.4)
with step size 1.0   **with step size 0.01**

In the next example, we compare the phase portraits produced by the various numerical methods.

Example 19.2.   *Use* **ODE** *with the option* **Method->AllNumerical** *to solve numerically the initial value problem*

$$\begin{cases} x' = y, & x(0) = 1 \\ y' + x = -\sin(x^2 + t^2), & y(0) = 0 \end{cases}$$

over the interval $-\pi \le t \le 2\pi$ *using step size 0.02. Draw the phase portraits.*

Solution. We use

```
ODE[{x' == y,y' + x == -Sin[x^2 + t^2],x[0] == 1,y[0] == 0},
{x,y},{t,-Pi,2Pi},
Method->AllNumerical,ODETrace->False,
NumericalOutput->None,StepSize->0.2,
PlotPhase->{{t,-Pi,2Pi}}]
```

to obtain the following phase portraits:

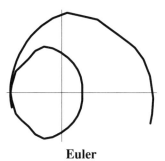

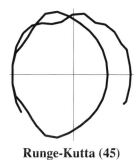

Euler   **Heun**

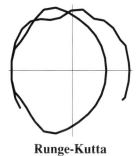

Runge-Kutta   **Runge-Kutta (45)**

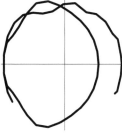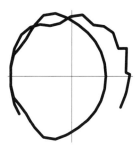

Implicit Runge-Kutta   **Second-order Euler**

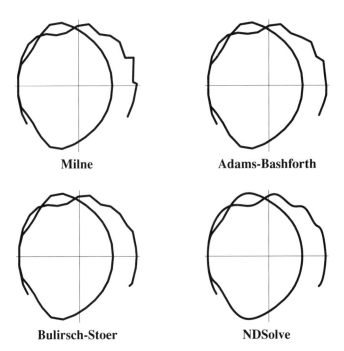

Milne   **Adams-Bashforth**

Bulirsch-Stoer   **NDSolve**

NDSolve is the best; the runner-up is **AdamsBashforth**.

Solving Higher-Order Equations Numerically Using ODE

NDSolve and all of **ODE**'s numerical solvers except **Numerov** can find a numerical solution to a differential equation of any order. Before a numerical solver begins its work, the command **Transformation->ConvertToSystem** is called to convert the differential equation to a system. The conversion can be done either automatically or more explicitly using **ConvertToSystem**. We illustrate the two procedures:

Example 19.3. *Use the Euler method with step size* 0.1 *to solve the second-order initial value problem*

$$\begin{cases} x'' + x'x = 1, \\ x(1) = 0, x'(1) = 0. \end{cases}$$

Find the phase plot of the solution over the interval $0 \leq t \leq 10$.

Solution. The simplest command to use is

```
ODE[{x'' + x' x == 1,x[1] == 0,x'[1] == 0},x,{t,0,10},
Method->Euler,PlotPhase->{{t,0,10}}];
```

A more complicated command that accomplishes the same thing is

```
ODE[ODE[{x'' + x' x == 1,x[1] == 0,x'[1] == 0},x,t,
Transformation->ConvertToSystem,TransformationVariable->w],
{w1,w2},{t,0,10},
Method->Euler,PlotPhase->{{t,0,10}}];
```

Using either command, we get the plot

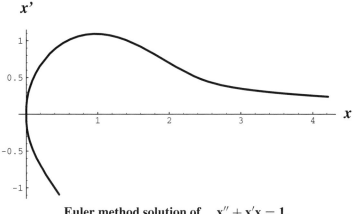

Euler method solution of $x'' + x'x = 1$,
$x(1) = 0$, $x'(1) = 0$ **plotted over** $0 \le t \le 10$

Exercises for Section 19.1

Use *Mathematica* to plot the numerical phase plane solutions of the following initial value problems over the indicated intervals. Use the Euler, Runge-Kutta, Adams-Bashforth, Bulirsch-Stoer, and implicit Runge-Kutta methods. In each case first use the options **VariableStepSize->False** and **StepSize->0.1** and integrate over the interval indicated in the problem. Then solve the problem again using **ODEMaxStepSize->0.1**, **VariableStepSize->True** and **Tolerance->0.001**, again with the option **StepSize->0.1**. If possible, find the exact solution.

1. $\begin{cases} x' = x - 2y, & x(0) = 0, \\ y' = 2x + y, & y(0) = 2, \\ 0 \le t \le 10 \end{cases}$

7. $\begin{cases} y'' + \sin(y) = 0, \\ y(0) = 1, \, y'(0) = 0, \\ 0 \le t \le 10 \end{cases}$

2. $\begin{cases} x' = -2x - 3y, & x(0) = 1, \\ y' = 2x + y, & y(0) = -1, \\ 0 \le t \le 10 \end{cases}$

8. $\begin{cases} y'' - t \sin(y') + y^2 = 0, \\ y(0) = 0, \quad y'(0) = 1, \\ 0 \le t \le 4 \end{cases}$

3. $\begin{cases} x' = 3\sin(x) - 4y, & x(0) = 1, \\ y' = 3x + 2y, & y(0) = 1, \\ 0 \le t \le 2 \end{cases}$

9. $\begin{cases} y'' + \sin(y') + y^2 = t, \\ y(0) = 1, \quad y'(0) = 1, \\ 0 \le t \le 10 \end{cases}$

4. $\begin{cases} x' = -2x - 5\cos(y), & x(0) = -2, \\ y' = 4x - 2y, & y(0) = 3, \\ 0 \le t \le 2 \end{cases}$

10. $\begin{cases} y'' - \sin(y') - \tan(y) = t^2, \\ y(0) = 1, \quad y'(0) = 1, \\ 0 \le t \le 5 \end{cases}$

5. $\begin{cases} x' = 2x - \tan(y), & x(0) = 1, \\ y' = \cos(x) + 2y, & y(0) = 1, \\ 0 \le t \le 1 \end{cases}$

11. $\begin{cases} y'' + y\cos(y') + y^2 = t, \\ y(0) = 0, \quad y'(0) = 0, \\ 0 \le t \le 10 \end{cases}$

6. $\begin{cases} x' = -x + y + y^3, & x(0) = 1, \\ y' = x^2, & y(0) = -1, \\ 0 \le t \le 2 \end{cases}$

12. $\begin{cases} y'' + y^2\sin(y') + y = \sqrt{t}, \\ y(0) = 0, \quad y'(0) = 0, \\ 0 \le t \le 20 \end{cases}$

19.2 *Predator-Prey Modeling*

In this section we use phase plane analysis to study the **predator-prey** problem from population dynamics. Let $\xi(t)$ and $\eta(t)$ denote the number of individuals at time t of two species, the first consisting of prey, the second consisting of predators. For example, $\xi(t)$ might be the number of deer and $\eta(t)$ might be the number of wolves in a closed forest. Just as we did in Section 6.1, we assume that $\xi(t)$ and $\eta(t)$ are well approximated by differentiable functions denoted by $x(t)$ and $y(t)$. Henceforth we refer to $x(t)$ as the number of prey and $y(t)$ as the number of predators. We make the following assumptions when we construct the model of the interaction of the two species:

- in the absence of the predator the prey grows at a rate proportional to its population size; thus, there is a constant $a > 0$ such that $x'(t) = a\,x(t)$ when $y(t) = 0$;

- in the absence of the prey the predator dies out at a rate proportional to its population size; thus, there is a constant $c > 0$ such that $y'(t) = -c\,y(t)$ when $x(t) = 0$;

- the number of encounters between the two species is proportional to the product of their population sizes. Each encounter increases the growth of the predator and inhibits the growth of the prey.

These assumptions are modeled by the nonlinear system

$$\begin{cases} x' = a\,x - b\,x\,y = x(a - b\,y), \\ y' = -c\,y + f\,x\,y = y(-c + f\,x), \end{cases} \tag{19.5}$$

called the **Lotka[3]-Volterra[4] equations**. In (19.5) the constants a, b, c and f are all positive; a and c measure the growth rate of the prey, while b and f measure the growth rate of the predators. In spite of their simplicity, the Lotka-Volterra equations can be used to model a wide class of problems.

Before working out a specific example, we derive some general properties of the Lotka-Volterra equations.

Example 19.4. *Find the implicit solutions of the Lotka-Volterra equations.*

Solution. Clearly, (19.5) implies that

$$\frac{dy}{dx} = \frac{y'}{x'} = \frac{y(-c + f\,x)}{x(a - b\,y)}; \tag{19.6}$$

this separable differential equation can be solved by rewriting (19.6) as

$$\frac{a - b\,y}{y}\,dy = \frac{-c + f\,x}{x}\,dx$$

---

[3] Alfred James Lotka (1880–1949). Lotka was born in Lvov, Ukraine (then a part of Austria). After immigrating to the United States he did important work in biomathematics.

[4]

 Vito Volterra (1860–1940). Professor in Turin and Rome, who is also known for his work on integral equations. As an Italian senator he took a stand against fascism, but eventually he was dismissed from his professorship in Rome. Volterra's ancestors came from a small town in Tuscany of the same name.

and integrating both sides. The result is

$$a \log(y) - b\, y = -c \log(x) + f\, x + \tilde{C}, \tag{19.7}$$

for some constant $\tilde{C}$. Exponentiation of (19.7) results in

$$y^a x^c e^{-b\, y - f\, x} = e^{\tilde{C}} = C, \tag{19.8}$$

where C is a positive constant. Any solution curve of (19.5) is described by (19.8). ▌

We remark that (19.6) can be solved using **ODE** as follows:

```
ODE[y' == (-c + f x)y/(x(a - b y)),y,x,Method->Separable]
```

with the result

```
   a       f x
  y       E    C[1]
{-----  ==  --------- ,  y == 0}
   b y         c
  E            x
```

Example 19.5. *Find and classify the critical points of the Lotka-Volterra equations* (19.5).

Solution. The critical points of (19.5) are clearly $(0, 0)$ and $(c/f, a/b)$. The linearized system for (19.5) at $(0, 0)$ is easily found by omitting the terms $-b\, x\, y$ and $f\, x\, y$:

$$\frac{d}{dt}\begin{pmatrix} x \\ y \end{pmatrix} = \begin{pmatrix} a & 0 \\ 0 & -c \end{pmatrix}\begin{pmatrix} x \\ y \end{pmatrix}. \tag{19.9}$$

The eigenvalues of (19.9) are a and $-c$, and the corresponding eigenvectors can be chosen to be

$$\begin{pmatrix} 1 \\ 0 \end{pmatrix} \quad \text{and} \quad \begin{pmatrix} 0 \\ -1 \end{pmatrix}.$$

Thus, $(0, 0)$ is a saddle point of (19.9) and hence unstable. Moreover, the general solution to (19.9) is

$$\begin{pmatrix} x \\ y \end{pmatrix} = C_1 \begin{pmatrix} 1 \\ 0 \end{pmatrix} e^{at} + C_2 \begin{pmatrix} 0 \\ -1 \end{pmatrix} e^{-ct},$$

where C_1 and C_2 are constants. Entrance to the saddle point is along the line $x = 0$; all other trajectories depart from a circle containing the critical point $(0, 0)$. The table on page 602 tells us that $(0, 0)$ is also a saddle point for the nonlinear system (19.5).

For the critical point $(c/f, a/b)$ we make the change of variables $x = u - c/f$ and $y = v - a/b$; then equations (19.5) become

$$\begin{cases} u' = \left(u - \dfrac{c}{f}\right)(-b\,v), \\[2mm] v' = \left(v - \dfrac{a}{b}\right)u\,f. \end{cases} \tag{19.10}$$

The linearized system to (19.10) is found by omitting the terms $-b\,u\,v$ and $f\,u\,v$:

$$\frac{d}{dt}\begin{pmatrix} u \\ v \end{pmatrix} = \begin{pmatrix} 0 & \dfrac{-b\,c}{f} \\[2mm] \dfrac{a\,f}{b} & 0 \end{pmatrix}\begin{pmatrix} u \\ v \end{pmatrix}. \tag{19.11}$$

The eigenvalues of (19.11) are $\pm i\,\sqrt{a\,c}$, so as a critical point of (19.11), $(c/f, a/b)$ is a stable center. In particular, the solution curves of the linear system at $(0,0)$ are closed curves. We cannot automatically conclude that $(c/f, a/b)$ is a stable center for (19.5). (See the table on page 602.) Nevertheless, this can be shown to be the case (see Exercise 2). ∎

Next, we put **ODE** to work to solve and plot a particular case of (19.5).

Example 19.6. *Find and classify the critical points of the system*

$$\begin{cases} x' = x\left(1 - \dfrac{y}{90}\right), \\[2mm] y' = -y\left(2 - \dfrac{x}{50}\right). \end{cases} \tag{19.12}$$

Then use **ODE** *with* **Method->LinearSystem** *to plot the linearized system at each of the critical points. Finally, use* **ODE** *with* **Method->NDSolve** *to plot the whole system.*

Solution. Clearly, (19.12) is a special case of (19.5) with $a = 1$, $b = 1/90$, $c = 2$, and $f = 1/50$. It follows from Example 19.6 that the critical points of (19.12) are $(0,0)$ and $(100, 90)$; these critical points can also be found by using the command

```
CriticalPoints[{x'[t] == x[t](1 - y[t]/90),
                y'[t] == -y[t](2 - x[t]/50)},{x[t],y[t]},t]
```

Since (19.12) is a special case of (19.5), we know that $(0,0)$ is an unstable saddle point both for (19.12) and the linearized system at $(0,0)$. To find and plot the linearized system at $(0,0)$, we use

```
ODE[{x'[t] == x[t](1 - y[t]/90),y'[t] == -y[t](2 - x[t]/50),
x[0] == a,y[0] == 50},{x,y},t,Method->LinearSystem,
Parameters->{{a,0.25,2,0.25}},LinearizeAt->{0,0},
PlotPhase->{{t,0,3},PlotRange->{{0,10},{0,10}}}]
```

obtaining the plot

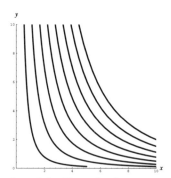

Phase plot of the linearized system at (0, 0)

This is clearly the first quadrant portion of a saddle; see page 555. Similarly, for the critical point (100, 90) we use

```
ODE[{x'[t] == x[t](1 - y[t]/90),y'[t] == -y[t](2 - x[t]/50),
x[0] == a,y[0] == 50},{x,y},t,
Method->LinearSystem,Parameters->{{a,10,50,10}},
LinearizeAt->{100,90},PlotPhase->{{t,0,2Pi}}]
```

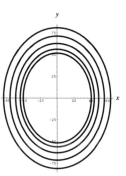

Phase plot of the linearized system at (100, 90)

To plot the whole system we use

```
ODE[{x'== x(1 - y/90),y' == -y(2 - x/50),
x[0] == a,y[0] == 50},{x,y},{t,0,8},Method->NDSolve,
Parameters->{{a,10,50,10}},PlotPhase->{{t,0,8}}];
```

obtaining the plot

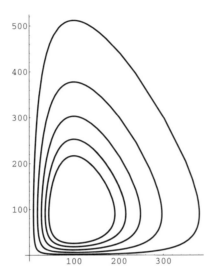

Predator-prey example

Exercises for Section 19.2

Use **ODE** to plot the solution to the following initial value problems.

1.
$$
\begin{cases}
x' = 2x(1 - 3y), & x(0) = 1, \\
y' = -4y(1 - x), & y(0) = b, \\
(0.5 \le b \le 4.5)
\end{cases}
$$

2.
$$
\begin{cases}
x' = x(1 - 1.2y), & x(0) = a, \\
y' = -y(1 - 0.9x), & y(0) = 3, \\
(1 \le a \le 7)
\end{cases}
$$

3. An extension of the Lotka-Volterra equations (19.5) includes a saturation effect, which is caused by a large number of prey. The more general equations are

$$
\begin{cases}
x' = ax - \dfrac{bxy}{1 + sx}, \\
y' = -cy + \dfrac{fxy}{1 + sx},
\end{cases}
\tag{19.13}
$$

where a, b, c, f, and s are constants. The first four parameters have the same meaning that they had in (19.5), while s measures the saturation. The parameters a, b, c, f are all positive, but there is no restriction on s.

a. Determine the critical solutions of (19.13) and their attraction types.

b. Solve and plot (19.13) in the case that $a = 1$, $b = 1/90$, $c = 2$, $f = 1/50$, and $s = 0.003$.

4. An improved version for the predator-prey model results when the the differential equation for the prey is modified so that it has the form of a logistic equation in the absence of predators. The new system is

$$\begin{cases} x' = x(a - g\,x - b\,y), \\ y' = y(-c + f\,x), \end{cases} \tag{19.14}$$

where a, b, c, f, and g are positive constants.

a. Determine the critical solutions of (19.14) and their attraction types.

b. Solve and plot (19.14) in the case that $a = 1$, $b = 1/90$, $c = 2$, $f = 1/50$, and $g = 1/130$.

5. Show that the solution curves of (19.5) are closed.

19.3   *The Van Der Pol Equation*

In the 1920s E. V. Appleton and B. van der Pol[5] conducted experimental research on the oscillation of the electric current and voltage of early radio sets. The circuits they considered were LRC loops, but with the resistor satisfying Ohm's law replaced by a vacuum tube. (Nowadays a semiconductor would be used instead of a vacuum tube.) The difference between a semiconductor and a resistor is this: a resistor dissipates energy at all levels, but a semiconductor pumps energy into a circuit at low levels and absorbs energy at high levels. Here is a typical van der Pol circuit.

Balthasar van der Pol (1889–1959). Dutch applied mathematician and engineer.

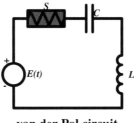

van der Pol circuit

Suppose that a power supply is attached to this circuit in order to energize it, and then the power supply is removed. Let us assume that the voltage drop across the semiconductor is given by the nonlinear function

$$V_S = I (I^2 - a)$$

of the current I, where a is a positive constant. The voltage drop across a resistor is given by a linear function, $V_R = I R$, and always has the same sign as I. In contrast, the nonlinear function V_S is negative when $|I| < \sqrt{a}$.

In addition to the voltage drop across the semiconductor we have the following voltage drops:

$$\begin{cases} \textbf{Voltage drop across an inductor } E_L = V_L = L \dfrac{dI}{dt}, \\[2mm] \textbf{Change of voltage across a capacitor } E_C = \dfrac{dV_C}{dt} = \dfrac{I}{C}. \end{cases}$$

The Kirchhoff voltage law dictates that

$$V_L + V_C + V_S = 0.$$

Thus

$$\frac{dI}{dt} = \frac{V_L}{L} = -\frac{V_S + V_C}{L} = \frac{-I^3 + a\,I - V_C}{L},$$

and we obtain the system

$$\begin{cases} I' = \dfrac{a}{L}I - \dfrac{V_C}{L} - \dfrac{I^3}{L}, \\[3mm] V_C' = \dfrac{I}{C}. \end{cases} \tag{19.15}$$

To convert (19.15) to a more manageable system, let

$$I = \alpha\, x, \qquad V_C = \beta\, y, \qquad \text{and} \qquad t = \gamma\, s.$$

Then

$$\frac{\alpha\, x}{C} = \frac{I}{C} = \frac{dV_C}{dt} = \frac{d(\beta\, y)}{ds}\frac{ds}{dt} = \frac{\beta}{\gamma}\frac{dy}{ds},$$

and

$$\frac{\alpha\, dx}{\gamma\, ds} = \frac{d(\alpha\, x)}{ds}\frac{ds}{dt} = \frac{dI}{dt} = \frac{a}{L}I - \frac{V_C}{L} - \frac{I^3}{L} = \frac{a\alpha}{L}x - \frac{\beta}{L}y - \frac{\alpha^3}{L}x^3.$$

Thus, if we choose α, β, and γ such that

$$\frac{\alpha}{C} = \frac{\beta}{\gamma} \qquad \text{and} \qquad \frac{\alpha^2\gamma}{L} = 1,$$

then (19.15) reduces to

$$\begin{cases} \dfrac{dx}{ds} = \mu\, x - y - x^3, \\[2mm] \dfrac{dy}{ds} = x, \end{cases}$$

where

$$\mu = \frac{a\,\gamma}{L}.$$

The nonlinear second-order differential equation

$$y'' + y = \mu(1 - y^2)y' \tag{19.16}$$

is called the **van der Pol equation**. We convert (19.16) to a first-order system:

$$\begin{cases} x_1' = x_2, \\ x_2' = -x_1 + \mu(1 - x_1^2)x_2. \end{cases}$$

Example 19.7. *Solve and plot the van der Pol system*

$$\begin{cases} x_1' = x_2, \\ x_2' = -x_1 + 1.5(1 - x_1^2)x_2. \end{cases}$$

with the initial conditions $x_2(0) = 0$ and $x_1(0) = a$ with a ranging from 0 to 3 at intervals of 0.5.

Solution. We use

```
ODE[{x1' == x2,x2' == -x1 + 1.5(1 - x1^2)x2,
x1[0] == a,x2[0] == 0},{x1,x2},{t,0,4Pi},
Method->NDSolve,NumericalOutput->No,
Parameters->{{a,0,3,0.5}},
PlotPhase->{{t,0,4Pi},PlotRange->All}]
```

to get the plot

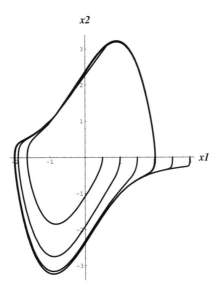

Solutions to the van der Pol equation
$$y'' + y = 1.5(1 - y^2)y'$$

19.4   *The Simple Pendulum*

A **simple pendulum** consists of a particle of mass m suspended by a rigid rod of negligible mass and length L from a fixed point **P**. The particle moves back and forth following a circular arc about the point **P**; it is presumed that all motion takes place in a plane. We wish to determine the equation of motion in terms of the **angle of displacement** θ, given an initial displacement and velocity.

The particle moves on a circle of radius L centered at **P**. Let $\mathbf{P} = (0, L)$; instead of describing this circle with the usual parametrization $t \longmapsto L\big(\sin(t), 1 - \cos(t)\big)$, we use the parametrization $t \longmapsto \mathbf{p}(t)$, where

$$\mathbf{p}(t) = L\big(\sin(\theta(t)), 1 - \cos(\theta(t))\big) = L\big(\sin(\theta), 1 - \cos(\theta)\big).$$

We now determine a second-order nonlinear differential equation that the angle of displacement θ must satisfy.

The velocity and acceleration of the curve $\mathbf{p}$ are given as follows. Let

$$\mathbf{T} = L\big(\cos(\theta), \sin(\theta)\big) \qquad \text{and} \qquad \mathbf{JT} = L\big(-\sin(\theta), \cos(\theta)\big).$$

Then

$$\begin{cases} \mathbf{p}'(t) &= L\big(\cos(\theta), \sin(\theta)\big)\theta' = L\mathbf{T}\theta' \\ \mathbf{p}''(t) &= L\big(-\sin(\theta), \cos(\theta)\big)\theta'^2 + L\big(\cos(\theta), \sin(\theta)\big)\theta'' = L\mathbf{T}\theta'' + L\mathbf{JT}\theta'^2. \end{cases}$$

Hence the tangential component of the acceleration is

$$\mathbf{p}''(t)_{\text{tangential}} = L\big(\cos(\theta), \sin(\theta)\big)\theta'' = L\,\mathbf{T}\,\theta''. \qquad (19.17)$$

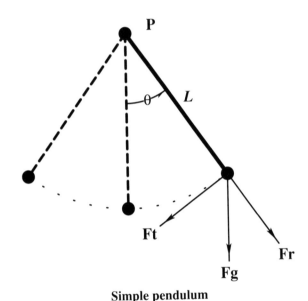

Simple pendulum

First, let us examine the force of gravity **Fg** pulling down on the particle. Since the magnitude of this force is $m\,g$, the force itself is given by

$$\mathbf{Fg} = (0, -m\,g).$$

The fact that $\sin(\theta)\mathbf{T} + \cos(\theta)\mathbf{JT} = (0, 1)$ implies that we can decompose **Fg** as

$$\mathbf{Fg} = -m\,g\sin(\theta)\mathbf{T} - m\,g\cos(\theta)\mathbf{JT}. \qquad (19.18)$$

Let **Ft** and **Fr** denote the tangential and radial parts of **Fg**. Then (19.18) implies that

$$\mathbf{Ft} = -m\,g\sin(\theta)\mathbf{T} \qquad \text{and} \qquad \mathbf{Fr} = -m\,g\cos(\theta)\mathbf{JT}. \qquad (19.19)$$

Newton's second law of motion (see page 5) states that

$$\mathbf{Ft} = m\,\mathbf{p}''(t) = F(t), \qquad (19.20)$$

where $F(t)$ is the resultant force acting on the particle at time t. We note that $F(t)$ is the sum of three forces:

$$F(t) = \mathbf{F}_g + \mathbf{F}_{\text{friction}} + \mathbf{F}_{\text{tension}}. \qquad (19.21)$$

The frictional force $\mathbf{F}_{\text{friction}}$ acts tangentially. We assume that the magnitude of $\mathbf{F}_{\text{friction}}$ is a constant multiple of the speed of the particle. Hence we can write

$$\mathbf{F}_{\text{friction}} = -c\,\mathbf{p}' = -c\,L\,\mathbf{T}\,\theta',$$

where c is a nonnegative constant.

Finally, the tension in the rod, denoted by $\mathbf{F}_{\text{tension}}$, acts along the rod. Thus we can write

$$\mathbf{F}_{\text{tension}} = h(t)\mathbf{JT},$$

for some function $h(t)$. From (19.17) and (19.21) we obtain

$$mL\,\mathbf{T}\,\theta'' = m\,\mathbf{p}''(t) = \big(-m\,g\,\sin(\theta) - c\,L\theta'\big)\mathbf{T} + \big(-m\,g\,\cos(\theta) + h(t)\big)\mathbf{JT}. \quad (19.22)$$

Equating the coefficients of $\mathbf{T}$ in (19.22) we get

$$m\,L\theta'' + c\,L\theta' + m\,g\,\sin(\theta) = 0. \qquad (19.23)$$

We call (19.23) the **pendulum equation**. The **undamped pendulum equation** is the special case $c = 0$.

Exact Solution of the Undamped Pendulum Equation in Terms of Elliptic Functions

Assume that a maximum value $\theta_{\max}$ of θ exists and satisfies $0 < \theta_{\max} < \pi$. Let t_0 be the first value of t for which this maximum occurs, that is, $\theta(t_0) = \theta_{\max}$. From calculus we know that $\theta'(t_0) = 0$. Thus, we have the initial value problem

$$\begin{cases} \dfrac{d^2\theta}{dt^2} + \dfrac{g}{L}\sin(\theta) = 0, \\[2mm] \theta(t_0) = \theta_{\max},\ \theta'(t_0) = 0. \end{cases} \qquad (19.24)$$

To solve it, we first multiply both sides of the differential equation by $d\theta/dt$, obtaining

$$\frac{d^2\theta}{dt^2}\frac{d\theta}{dt} + \frac{g}{L}\sin(\theta)\frac{d\theta}{dt} = 0. \qquad (19.25)$$

We can integrate both terms of (19.25); thus

$$\frac{1}{2}\left(\frac{d\theta}{dt}\right)^2 - \frac{g}{L}\cos(\theta) = C, \qquad (19.26)$$

where C is a constant. The fact that $\theta'(t_0) = 0$ can be used to evaluate the constant in (19.26):

$$-\frac{g}{L}\cos(\theta_{\max}) = C. \qquad (19.27)$$

From (19.26) and (19.27) we get

$$\frac{d\theta}{dt} = \pm\sqrt{\frac{2g}{L}\left(\cos(\theta) - \cos(\theta_{\max})\right)}. \qquad (19.28)$$

Since we expect θ to decrease, at least for small $t - t_0$, we choose the minus sign in (19.28). Let us rewrite (19.28) as

$$dt = -\sqrt{\frac{L}{2g}}\frac{d\theta}{\sqrt{\cos(\theta) - \cos(\theta_{\max})}}. \qquad (19.29)$$

In order to reduce the right-hand side of (19.29) to a standard form, let

$$\sin\left(\frac{\theta}{2}\right) = k\sin(\phi), \qquad \text{where} \qquad k = \sin\left(\frac{\theta_{\max}}{2}\right). \qquad (19.30)$$

We observe that

$$\cos(\theta) = 1 - 2k^2(\sin(\phi))^2, \qquad (19.31)$$

so that

$$\begin{aligned}
\cos(\theta) - \cos(\theta_{\max}) &= 1 - 2k^2(\sin(\phi))^2 - \left(1 - 2\sin(\theta_{\max}/2)^2\right) \\
&= 2\sin(\theta_{\max}/2)^2 - 2k^2(\sin(\phi))^2 \\
&= 2k^2\left(1 - (\sin(\phi))^2\right) = 2k^2(\cos(\phi))^2.
\end{aligned} \qquad (19.32)$$

Furthermore, differentiation of the equation $\cos(\theta) = 1 - 2k^2(\sin(\phi))^2$ yields

$$\sin(\theta)\,d\theta = 4k^2\sin(\phi)\cos(\phi)\,d\phi. \qquad (19.33)$$

Also,

$$\sin(\theta) = \sqrt{1 - \cos(\theta)^2} = 2k\sin(\phi)\sqrt{1 - k^2(\sin(\phi))^2}. \qquad (19.34)$$

We substitute (19.32), (19.33), and (19.34) into (19.29), obtaining

$$dt = -\sqrt{\frac{L}{g}}\frac{d\phi}{\sqrt{1 - k^2(\sin(\phi))^2}}. \qquad (19.35)$$

It follows from (19.30) that $\phi(t_0) = \pi/2$. Therefore, when we integrate (19.35) from t_0 to t we get

$$
t - t_0 = \sqrt{\frac{L}{g}} \int_{\phi}^{\pi/2} \frac{d\phi}{\sqrt{1 - k^2(\sin(\phi))^2}}
$$

$$
= \sqrt{\frac{L}{g}} \left(-\int_0^{\pi/2} \frac{d\phi}{\sqrt{1 - k^2(\sin(\phi))^2}} + \int_0^{\phi} \frac{d\phi}{\sqrt{1 - k^2(\sin(\phi))^2}} \right). \quad (19.36)
$$

The second integral on the right-hand side of (19.36) is called an **elliptic integral of the first kind**; it is denoted by $F(\phi, k^2)$.[6] Thus

$$
F(\phi, k^2) = \int_0^{\phi} \frac{d\phi}{\sqrt{1 - k^2(\sin(\phi))^2}}. \quad (19.37)
$$

Note that $F(\phi, 0) = \phi$. We assume that $-1 < k < 1$, so that the denominator on the right-hand side of (19.37) does not vanish. Although $\phi \longmapsto F(\phi, k^2)$ is an increasing function, it becomes increasingly wavy as k approaches ± 1.

The **period** $P(k)$ of a pendulum is defined to be the time required to make a complete oscillation between positions of maximum displacement. From (19.28) and (19.32) we get

$$
\frac{d\theta}{dt} = -2k\sqrt{\frac{g}{L}} \cos(\phi). \quad (19.38)
$$

To determine $P(k)$, we observe that the pendulum reverses direction at those values of θ for which $d\theta/dt = 0$; from (19.38), we see that these values correspond to $\phi = (n + 1/2)\pi$, where n is an integer. Clearly, $P(k)$ is 4 times the time required for ϕ to vary between 0 and $\pi/2$. Thus, from (19.36) it follows that the period is given by

$$
P(k) = 4\sqrt{\frac{L}{g}} \int_0^{\pi/2} \frac{d\phi}{\sqrt{1 - k^2(\sin(\phi))^2}}. \quad (19.39)
$$

The integral on the right-hand side of (19.39) is called a **complete elliptic integral of the first kind** and is denoted by $K(k^2)$. Then $K(0) = \pi/2$, so that $P(0) = 2\pi\sqrt{L/g}$.

---

[6]The notation varies from book to book; for example, the integral is sometimes denoted by $F(\phi, k)$ or $F(k, \phi)$. The notation we have chosen corresponds to *Mathematica*'s.

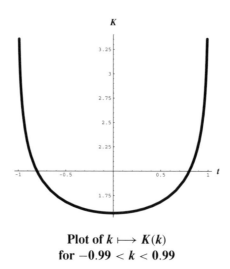

Plot of $k \longmapsto K(k)$
for $-0.99 < k < 0.99$

Since $\phi \longmapsto F(\phi, k^2)$ is an increasing function, it has an inverse, which is called the
Jacobi amplitude and is denoted by $am(u, k^2)$. Explicitly,

$$F\left(am(u, k^2), k\right) = u. \tag{19.40}$$

We then define

$$sn(u, k^2) = \sin\left(am(u, k^2)\right).$$

This function is one of twelve **Jacobi elliptic functions**. We can think of sn as an interpolation between the ordinary sine and the hyperbolic tangent, because

$$sn(u, 0) = \sin(u) \qquad \text{while} \qquad sn(u, 1) = \tanh(u).$$

(These formulas can be checked with *Mathematica*.) The following plots give some indication of the nature of $sn(t, k^2)$, $am(t, k^2)$, and $F(t, k^2)$.

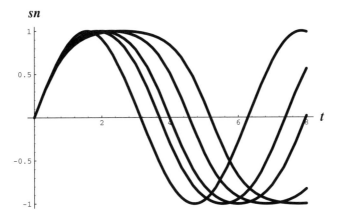

Plot of $t \longmapsto \text{sn}(t, k^2)$
for $k = 0, 0.7, 0.8, 0.9, 0.95$

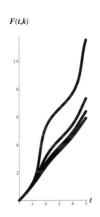

Plot of $t \longmapsto F(t, k^2)$
for $k = 0.69, 0.79, 0.89.0, 0.99$

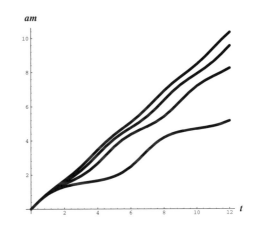

Plot of $t \longmapsto \text{am}(t, k^2)$
for $k = 0.69, 0.79, 0.89.0, 0.99$

From (19.36) and (19.39) we have

$$t - t_0 = \sqrt{\frac{L}{g}} \left(-\frac{P(k)}{4} + F(\phi, k^2) \right). \tag{19.41}$$

Then (19.41) and (19.40) imply that

$$\phi = \text{am}\left(\sqrt{\frac{g}{L}}(t - t_0) + \frac{P(k)}{4}, k^2\right),$$

so that

$$\sin(\theta/2) = k \sin(\phi) = k \sin\left(\text{am}\left(\sqrt{\frac{g}{L}}(t - t_0) + \frac{P(k)}{4}, k^2\right)\right)$$

$$= k \, \text{sn}\left(\sqrt{\frac{g}{L}}(t - t_0) + \frac{P(k)}{4}, k^2\right).$$

Therefore, the explicit solution of (19.24) is

$$\theta = 2 \arcsin\left(k \, \text{sn}\left(\sqrt{\frac{g}{L}}(t - t_0) + \frac{P(k)}{4}, k^2\right)\right).$$

The *Mathematica* versions of $F(\phi, k)$, $\text{am}(t, k)$, and $\text{sn}(t, k)$ are

EllipticF[phi,k],   JacobiAmplitude[phi,k]

and

JacobiSN[phi,k]

The Phase Portrait of the Undamped Pendulum

There are two methods available to plot the phase portrait of the undamped pendulum. The first method uses **Method->PlotPhase**:

```
ODE[{x' == y,y' == -2Sin[x],x[0] == a,y[0] == b},{x,y},
{t,-5,5},Method->NDSolve,
Parameters->{{a,-4,4,0.8},{b,-3,3,1}},
PlotPhase->{{t,-5,5},PlotPoints->100,
  AspectRatio->Automatic,PlotRange->{{-10,10},Automatic}}]
```

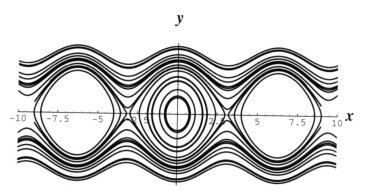

Phase portrait of an undamped pendulum

A simpler method is to use **ContourPlot** to plot the energy function of the undamped pendulum. To do this, we first define

```
penderg[x_,y_,w_]:=Integrate[
x''[t]x'[t] + w^2Sin[x[t]]x'[t],t] /. {x[t]->x,x'[t]->y}
```

Then

```
ContourPlot[penderg[x,y,1]//Evaluate,
{x,-10,10},{y,-3,3},ColorFunction->Hue,
AspectRatio->Automatic,Contours->Range[-10,10,0.25],
PlotPoints->100];
```

gives a nice color plot:

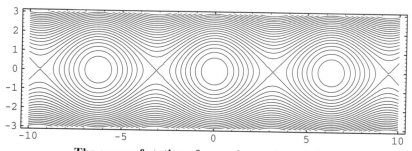

The energy function of an undamped pendulum

The energy function can also be visualized in 3 dimensions with the following plot:

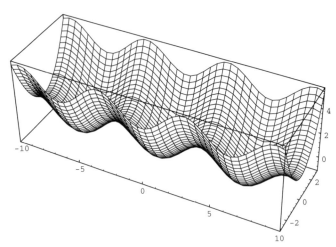

The energy function of an undamped pendulum

The command to produce this plot is

```
Plot3D[penderg[x,y,1]//Evaluate,
{x,-10,10},{y,-3,3},
ColorFunction->Hue,AspectRatio->Automatic,
PlotPoints->40,BoxRatios->{1,1/3,1/3}]
```

Notice that the syntax of the command to produce the **Plot3D** picture is quite similar to the command for the **ContourPlot** picture.

The Phase Portrait of the Damped Pendulum

The phase portrait of a damped pendulum can be drawn with

```
ODE[{x' == y,y' == -0.5y - 2Sin[x],x[0] == a,y[0] == b},
{x,y},{t,-5,5},Method->NDSolve,
Parameters->{{a,-4,4,1},{b,-3,3,2}},
PlotPhase->{{t,-5,5},PlotPoints->100,
AspectRatio->Automatic,PlotRange->{{-10,10},Automatic}}]
```

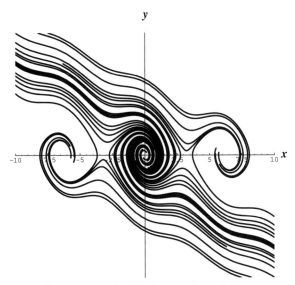

Phase portrait of a damped pendulum

Exercises for Section 19.4

1. A pendulum is displaced through an angle of $\pi/3$ radians. What is its period?

2. A *two-second pendulum* is one that makes a full swing in two seconds. Using the standard value $g = 32.174$ feet per second^2, find the length of the two-second pendulum.

3. If a two-second pendulum is displaced through an angle of $\pi/2$ radians, determine the time required for it to make one complete oscillation.

4. Repeat Exercise 3 if the initial displacement is $\pi/6$ radians.

5. Use **ODE** and plot the solution to a two-second pendulum with an initial displacement of $\pi/2$. Experiment with the plotting range to obtain an illustrative comparison with Exercise 3.

6. Use **ODE** to obtain the general solutions to the linear undamped pendulum and the nonlinear undamped pendulum, each with initial conditions $y(0) = Y_0$ and $y'(0) = 0$.

7. Use the results of the previous problem and compare the plots of the solutions to a two-second pendulum with an initial displacement of $\pi/3$. Experiment with the plotting range to obtain an illustrative comparison.

8. When we compare the exact solution to the undamped pendulum problem with the linear version obtained through a small angle approximation, initially we see that the solutions disagree more and more as t increases. Assuming a two-second pendulum with an initial displacement of $\pi/3$ radians, find the smallest time t for which the two solutions exhibit a maximum difference. That is, a time when their positions are exactly opposite one another.

19.5   *The Fundamental Theorem of Plane Curves*

The field of differential geometry is another source of systems of differential equations. We use the letter s to measure arc length along a plane curve $\alpha(s) = (x(s), y(s))$. The curvature $k(s)$ of α can be expressed in terms of the first and second derivatives of the functions $x(s)$ and $y(s)$. For example, a straight line has zero curvature, and the curvature of a circle is the reciprocal of the radius.

More generally, the **fundamental theorem of plane curves** states that given a piecewise continuous function k, there is a curve whose curvature is k. Furthermore, the curve is unique up to a rotation and translation of the plane. (See for example [Gray1], pages 110-111.) The system of differential equations to determine a curve from its curvature turns out to be

$$\begin{cases} x'(s) = \cos(\theta(s)), \\ y'(s) = \sin(\theta(s)), \\ \theta'(s) = k(s). \end{cases} \tag{19.42}$$

Example 19.8.   *Find the parametrization of a curve whose curvature is proportional to arc length.*

Solution.   We want to find a curve whose curvature is given by $k(s) = a\,s$, where a is a constant. Without loss of generality, we can assume that the curve starts at the origin and that $\theta(0) = 0$. Thus we need to solve the initial value problem

$$
\begin{cases}
x'(s) = \cos(\theta(s)), & x(0) = 0, \\[2mm]
y'(s) = \sin(\theta(s)), & y(0) = 0, \\[2mm]
\theta'(s) = a\,s, & \theta(0) = 0.
\end{cases}
\qquad (19.43)
$$

We can integrate the third equation of (19.43) immediately; thus

$$
\theta(s) = \frac{a\,s^2}{2}.
$$

Now we can also integrate the first two equations of (19.43):

$$
x(s) = \int_0^s \sin\!\left(\frac{a\,u^2}{2}\right) du \qquad \text{and} \qquad y(s) = \int_0^s \cos\!\left(\frac{a\,u^2}{2}\right) du. \qquad (19.44)
$$

It is impossible to express the integrals in (19.44) in terms of elementary functions. However, we can use the **Fresnel**[7] **integrals** that occur in mathematical physics; they are defined by

$$
\mathbf{S}(t) = \int_0^t \sin\!\left(\frac{\pi\,v^2}{2}\right) dv \qquad \text{and} \qquad \mathbf{C}(t) = \int_0^t \cos\!\left(\frac{\pi\,v^2}{2}\right) dv.
$$

Then clearly,

$$
x(s) = \frac{2}{\pi}\mathbf{C}\!\left(\frac{a\,s}{\sqrt{\pi}}\right) \qquad \text{and} \qquad y(s) = \frac{2}{\pi}\mathbf{S}\!\left(\frac{a\,s}{\sqrt{\pi}}\right).
$$

To solve the system (19.43) using *Mathematica*, we use

```
sol = ODE[{x' == Cos[theta[s]],x[0] == 0,
           y' == Sin[theta[s]],y[0] == 0,
           theta' == a s,       theta[0] == 0},{x,y,theta},s,
       Method->DSolve,DSolvePackage->True,Form->Explicit]
```

to obtain

```
                       Sqrt[a] s                        Sqrt[a] s
    2   Sqrt[Pi] FresnelC[---------]   Sqrt[Pi] FresnelS[---------]
   a s                    Sqrt[Pi]                        Sqrt[Pi]
 {---, -----------------------------, -----------------------------}
    2            Sqrt[a]                        Sqrt[a]
```

7

Augustin Jean Fresnel (1788–1827). French physicist. He was educated at the École Polytechnique and was active in the Corps des Ponts et Chaussées. He lost his engineering post temporarily during Napoleon's return from Elba in 1814 and around this time he did important work on optics where he was one of the founders of the wave theory of light.

Here **FresnelS** and **FresnelC** are *Mathematica*'s representations for the Fresnel integrals (see page 108). The curve that we obtain is called a **clothoid**; it is given by

```
clothoid[a_][s_] = Drop[sol,{1}]
```

$$\left\{\frac{\text{Sqrt[Pi] FresnelC}\left[\dfrac{\text{Sqrt[a] s}}{\text{Sqrt[Pi]}}\right]}{\text{Sqrt[a]}}, \frac{\text{Sqrt[Pi] FresnelS}\left[\dfrac{\text{Sqrt[a] s}}{\text{Sqrt[Pi]}}\right]}{\text{Sqrt[a]}}\right\}$$

Although *Mathematica* takes a very long time to produce it, a graph of the clothoid can be obtained with

```
ParametricPlot[Evaluate[clothoid[1][s]],
{s,-7,7},AspectRatio->Automatic]
```

yielding

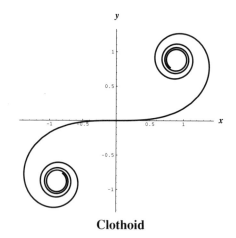

Clothoid

Since

$$\sqrt{\frac{\pi}{a}}\left(\mathbf{S}\left(\sqrt{\frac{a}{\pi}}s\right),\mathbf{C}\left(\sqrt{\frac{a}{\pi}}s\right)\right)=\left(\int_0^s \sin\left(\frac{a\,v^2}{2}\right)dv, \int_0^s \cos\left(\frac{a\,v^2}{2}\right)dv\right),$$

we define the clothoid more generally by

$$\mathbf{clothoid}[n,a](t)=a\left(\int_0^t \sin\left(\frac{u^{n+1}}{n+1}\right)du, \int_0^t \cos\left(\frac{u^{n+1}}{n+1}\right)du\right).$$

In the next example we show how to numerically find a curve with specified curvature.

Example 19.9. *Plot a curve α whose curvature is given by*

$$k(s) = s\sin(s).$$

Solution. To define the solution of (19.42) with $k(s) = s\sin(s)$ we use

```
ssin[s_] = ODE[{x' == Cos[theta[s]],x[0] == 0,
              y' == Sin[theta[s]],y[0] == 0,
              theta' == s Sin[s], theta[0] == 0},
          {x,y,theta},{s,-20,20},
          Method->NDSolve,Form->Explicit]
```

Then the curve α is plotted with

```
ParametricPlot[Evaluate[Drop[ssin[s],{1}]],{s,-20,20},
PlotPoints->80,AspectRatio->Automatic]
```

giving

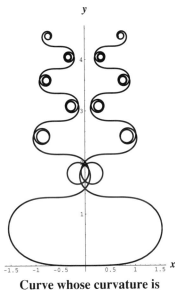

Curve whose curvature is
$$k(s) = s\sin(s)$$

Exercises for Section 19.5

1. Plot a curve α whose curvature is given by $k(s) = s^2$.

2. Plot a curve α whose curvature is given by $k(s) = s^2\sin(s)$.

20

POWER SERIES SOLUTIONS OF SECOND-ORDER EQUATIONS

In Chapter 9 we found the general solution of a second-order differential equation with *constant* coefficients:

$$a\,y'' + b\,y' + c\,y = g(t). \tag{20.1}$$

In the present chapter we describe the power series method for finding the solution of a second-order differential equation of the form

$$a(t)y'' + b(t)y' + c(t)y = g(t). \tag{20.2}$$

When the coefficients in (20.2) are *nonconstant*, (20.2) is much more complicated than (20.1), so much so that the method of trial solutions described in Section 9.1 does not work and must be modified in a nontrivial way.

Nevertheless, many differential equations arising in mathematical physics have the form (20.2), so it is imperative that we study solution techniques for (20.2). Some important special cases of (20.2) are:

- The **Cauchy-Euler equation**

$$a(t - t_0)^2 y'' + b(t - t_0)y' + c\,y = 0, \qquad \text{where } a, b, \text{ and } c \text{ are constants.} \tag{20.3}$$

679

- The **Bessel equation**

$$t^2 y'' + t\, y' + (t^2 - \lambda^2) y = 0, \qquad \text{where } \lambda \text{ is a constant.} \qquad (20.4)$$

- The **Legendre equation**

$$(1 - t^2) y'' - 2t\, y' + \lambda\, y = 0, \qquad \text{where } \lambda \text{ is a constant.} \qquad (20.5)$$

- The **Airy equation**

$$y'' - \lambda\, t\, y = 0, \qquad \text{where } \lambda \text{ is a constant.} \qquad (20.6)$$

- The **Hermite equation**

$$y'' - 2t\, y' + 2n\, y = 0, \qquad \text{where } n \text{ is an integer.} \qquad (20.7)$$

- The **Chebyshev equation**

$$(1 - t^2) y'' - t\, y' + n^2 y = 0, \qquad \text{where } n \text{ is an integer.} \qquad (20.8)$$

- The **Hypergeometric equation**

$$t(1 - t) y'' + \big(c - (a + b + 1)t\big) y' - a\, b\, y = 0, \qquad \text{where } a,\ b, \text{ and } c \text{ are constants.} \qquad (20.9)$$

In fact, (20.4)–(20.9) lead to new classes of functions called **Bessel functions, Legendre functions, Airy functions, Hermite functions, Chebyshev functions,**, and **hypergeometric functions**.

For the study of (20.2) it is important to distinguish two kinds of points.

Definition. *Suppose that the differential equation (20.2) has been reduced so that the functions $a(t)$, $b(t)$, and $c(t)$ have no common factors of the form $(t - t_0)^k$, where k is a positive integer.*

- *A point t_0 is called an **ordinary point** for (20.2) if $a(t_0) \neq 0$;*

- *A point t_0 is called a **singular point** for (20.2) if $a(t_0) = 0$.*

For example, the Airy and Hermite equations have only ordinary points. The Cauchy-Euler equation has t_0 as a singular point; all other points are ordinary. The singular and ordinary points of the Bessel, Legendre, Chebyshev, and hypergeometric equations are also easy to determine.

As the names imply, ordinary points are easier to deal with than singular points. The present chapter is devoted to solving (20.2) near an ordinary point using the **method of power series**. This technique is discussed in Section 20.3; we review power series in Section 20.1, and in Section 20.2 we discuss power series via *Mathematica*. The Airy and Legendre equations are discussed in Sections 20.4 and 20.5. The convergence of a series solution is established in Section 20.6.

Methods of finding singular solutions of (20.2) near a singular point are postponed to Chapter 21. Bessel and hypergeometric functions are also discussed in Chapter 21.

20.1   Review of Power Series

The simplest power series is a polynomial function

$$f(t) = a_0 + a_1 t + a_2 t^2 + \cdots + a_n t^n,$$

which is just a finite sum of terms, each of which is proportional to a power of t. From calculus we know that many elementary functions can be represented by the sum of infinitely many such terms, for example,

$$\begin{cases} e^t &= 1 + t + \dfrac{t^2}{2} + \cdots &= \displaystyle\sum_{n=0}^{\infty} \dfrac{t^n}{n!}, & -\infty < t < \infty, \\[2ex] \cos(t) &= 1 - \dfrac{t^2}{2} + \dfrac{t^4}{24} - \cdots &= \displaystyle\sum_{n=0}^{\infty} \dfrac{(-1)^n t^{2n}}{(2n)!}, & -\infty < t < \infty, \\[2ex] \sin(t) &= t - \dfrac{t^3}{6} + \dfrac{t^5}{120} - \cdots &= \displaystyle\sum_{n=0}^{\infty} \dfrac{(-1)^n t^{2n+1}}{(2n+1)!}, & -\infty < t < \infty, \\[2ex] \dfrac{1}{1-t} &= 1 + t + t^2 + \cdots &= \displaystyle\sum_{n=0}^{\infty} t^n, & -1 < t < 1. \end{cases} \qquad (20.10)$$

All of the series in (20.10) are expressed in terms of powers of t. For a more generally applicable theory, we need a general power series in the form

$$a_0 + a_1 (t - t_0) + a_2 (t - t_0)^2 + \cdots = \sum_{n=0}^{\infty} a_n (t - t_0)^n,$$

where each term is proportional to a power of $t - t_0$. This is useful, for example, in representing the natural logarithm as

$$\log(t) = \sum_{n=1}^{\infty} \frac{(-1)^{n-1} (t - 1)^n}{n}, \qquad (20.11)$$

for $0 < t \le 2$.

We summarize some important results about power series; for more details the reader should consult an advanced calculus book.

Definition.   *A power series* $\displaystyle\sum_{n=0}^{\infty} a_n (t - t_0)^n$ *is said to* **converge** *at a point t if the limit of the partial sums exists, that is,*

$$\lim_{m \to \infty} \sum_{n=0}^{m} a_n (t - t_0)^n$$

exists. A power series that does not converge at t is said to **diverge** *at t.*

Although we have given the definitions of convergence and divergence for power series of the form

$$\sum_{n=0}^{\infty} a_n (t - t_0)^n,$$

the definitions can be modified in obvious ways to apply to power series of the form

$$\sum_{n=M}^{\infty} a_n (t - t_0)^n,$$

where M is some positive integer. Clearly, any power series converges for $t = t_0$; in general a power series may converge for all t, or it may converge for some t and diverge for other values of t.

Definition.   *A power series* $\displaystyle\sum_{n=0}^{\infty} a_n (t - t_0)^n$ *is said to* **converge absolutely** *at a point t if the series*

$$\sum_{n=0}^{\infty} \left| a_n (t - t_0)^n \right|$$

converges.

Lemma 20.1.   *If a power series converges absolutely at t, then it also converges at t; the converse is not necessarily true.*

A useful test for the absolute convergence of a power series is the ratio test.

Lemma 20.2.   *Suppose that $a_n \ne 0$ for all n and that for a fixed value of t the limit*

$$\lim_{n \to \infty} \left| \frac{a_{n+1}(t - t_0)^{n+1}}{a_n (t - t_0)^n} \right| = |t - t_0| \lim_{n \to \infty} \left| \frac{a_{n+1}}{a_n} \right|$$

exists and equals L. Then the power series $\sum_{n=0}^{\infty} a_n (t - t_0)^n$ *converges at the value t if $L < 1$ and diverges if $L > 1$. If $L = 1$, the test is inconclusive.*

Lemma 20.3. *There exists a* **radius of convergence** $R \geq 0$ *such that the power series converges absolutely for $|t - t_0| < R$ and diverges for $|t - t_0| > R$. The set of t-values that satisfy $t_0 - R < t < t_0 + R$ is called the* **interval of convergence.** *For a power series that converges only for $t = t_0$ we define the radius of convergence R to be 0; for a power series that converges for all values of t we define the radius of convergence R to be ∞.*

In the first three examples of (20.10) the radius of convergence is $R = \infty$, while for the fourth example of (20.10) and for (20.11) the radius of convergence is $R = 1$. An example of a power series with radius of convergence 0 is

$$\sum_{n=0}^{\infty} n! t^n.$$

The behavior of the series when $|t - t_0| = R$ does not follow from general principles and must be determined on a case-by-case basis.

Frequently, Lemma 20.2 can be used to determine the radius of convergence.

Example 20.1. *Find the radius of convergence of the power series*

$$\sum_{n=0}^{\infty} (n^2 + 4n) t^n.$$

Solution. The ratio test (Lemma 20.2) tells us that convergence takes place if

$$1 > \lim_{n \to \infty} \left| \frac{a_{n+1} t^{n+1}}{a_n t^n} \right| = \lim_{n \to \infty} \frac{(n+1)^2 + 4(n+1)}{n^2 + 4n} |t|$$

$$= \lim_{n \to \infty} \left(\frac{(n+1)(n+5)}{n(n+4)} \right) |t| = |t|.$$

If $|t| > 1$, the terms of the series are unbounded and the series is divergent. Therefore, we see that the series converges if $|t| < 1$ and diverges if $|t| > 1$, which proves that the radius of convergence is $R = 1$. ∎

Convergent power series can be added and multiplied just like ordinary polynomials.

Lemma 20.4. *Suppose that* $\sum_{n=0}^{\infty} a_n (t - t_0)^n$ *and* $\sum_{n=0}^{\infty} b_n (t - t_0)^n$ *converge in some common interval. Then in the same interval we have*

$$\sum_{n=0}^{\infty} a_n (t - t_0)^n + \sum_{n=0}^{\infty} b_n (t - t_0)^n = \sum_{n=0}^{\infty} (a_n + b_n)(t - t_0)^n$$

and

$$\left(\sum_{n=0}^{\infty} a_n (t - t_0)^n\right)\left(\sum_{n=0}^{\infty} b_n (t - t_0)^n\right) = \sum_{n=0}^{\infty}\left(\sum_{k=0}^{n} a_k b_{n-k}\right)(t - t_0)^n.$$

Similarly, convergent power series can be differentiated.

Lemma 20.5. *Suppose that* $f(t) = \sum_{n=0}^{\infty} a_n (t - t_0)^n$ *converges in some open interval* $a < t < b$. *Then in* $a < t < b$ *we have*

$$f'(t) = \frac{d}{dt}\sum_{n=0}^{\infty} a_n (t - t_0)^n = \sum_{n=0}^{\infty} n\, a_n (t - t_0)^{n-1} = \sum_{n=1}^{\infty} n\, a_n (t - t_0)^{n-1},$$

$$f''(t) = \frac{d^2}{dt^2}\sum_{n=0}^{\infty} a_n (t - t_0)^n = \sum_{n=0}^{\infty} n(n-1)a_n (t - t_0)^{n-2} = \sum_{n=2}^{\infty} n(n-1)a_n (t - t_0)^{n-2},$$

and so forth.

These successive differentiated series yield the following basic formula linking the a_n's to $f(t)$ and its derivatives:

$$a_n = \frac{f^{(n)}(t_0)}{n!} \tag{20.12}$$

for $n = 0, 1, 2, \ldots$.

Power series can also be used to define new functions that are not expressible by elementary functions. For example, the Bessel function $J_0(t)$ of order 0 is defined by the everywhere convergent power series (that is, $R = \infty$):

$$J_0(t) = 1 - \frac{t^2}{4} + \frac{t^4}{64} - \cdots = \sum_{n=0}^{\infty}\frac{(-t^2)^n}{(n!)^2 2^{2n}}.$$

$J_0(t)$ cannot be expressed in terms of a finite number of elementary functions. We shall see that $J_0(t)$ is a solution of the Bessel equation (20.4) with $\lambda = 0$.

Let $f(t)$ be a continuous function that has derivatives of all orders for $|t - t_0| < R$, where $0 < R \le \infty$. Can $f(t)$ be represented by a power series? If we use (20.12) to define the a_n's, then it is natural to hope that

$$f(t) = \sum_{k=0}^{\infty}\frac{(t - t_0)^k}{k!} f^{(k)}(t_0). \tag{20.13}$$

More often than not (20.13) is true, but it can be false for pathological functions such as the function $g(t)$ defined by

$$g(t) = \begin{cases} \exp(-1/t^2) & \text{if } t \ne 0, \\ 0 & \text{if } t = 0. \end{cases}$$

We can investigate the validity of (20.13) for a specific value of t by using the **Taylor formula with remainder**

$$f(t) = \sum_{k=0}^{n} \frac{(t-t_0)^k}{k!} f^{(k)}(t_0) + R_n(t, t_0).$$

The Taylor remainder R_n can be expressed in terms of the next derivative at a nearby point ξ with $t_0 < \xi < t$ or equivalently as an integral by

$$R_n(t, t_0) = \frac{(t-t_0)^{n+1}}{(n+1)!} f^{(n+1)}(\xi) = \int_{t_0}^{t} \frac{(t-s)^n}{n!} f^{(n+1)}(s) \, ds.$$

If the Taylor remainder tends to zero when $n \longrightarrow \infty$, then we have the **Taylor series representation**

$$f(t) = \sum_{k=0}^{\infty} \frac{(t-t_0)^k}{k!} f^{(k)}(t_0). \tag{20.14}$$

This leads to the following definition.

Definition. *A function $f(t)$ that has a Taylor series representation about $t = t_0$ that converges to $f(t)$ for $|t - t_0| < R$ (where R is the radius of convergence) is said to be* **analytic** *at t_0.*

It is easy to prove that if $f(t)$ and $g(t)$ are analytic at $t = t_0$, then so are $f(t)g(t)$ and $a\,f(t) + b\,g(t)$ for any constants a and b. Furthermore, the composition of analytic functions is analytic. All of the series in (20.10) define analytic functions, and so they can be used as building blocks to construct more complicated analytic functions

Example 20.2. *Use (20.14) to find the Taylor series representation of the function*

$$f(t) = 3e^{2t}$$

about $t_0 = 1$.

Solution. The higher derivatives of $f(t)$ are obtained as

$$f^{(k)}(t) = 3(2^k)e^{2t};$$

thus, the Taylor series about $t_0 = 1$ is

$$f(t) = 3e^2 + 6e^2(t-1) + 6e^2(t-1)^2 + \cdots + 3e^2 \frac{2^n}{n!}(t-1)^n + \cdots$$

$$= 3e^2 \sum_{n=0}^{\infty} \frac{2^n}{n!}(t-1)^n. \ \blacksquare$$

We conclude this section by developing the **product rule for higher derivatives**, which is useful in dealing with power series solutions of ordinary differential equations. If A and B are two differentiable functions, the elementary product rule, usually attributed to Leibniz, states that the first derivative of the product $A\,B$ is given by

$$(A\,B)' = A\,B' + A'B.$$

If we apply this again and collect terms, we find

$$
\begin{aligned}
(A\,B)'' &= (A\,B'' + A'B') + (A'B' + A''B)\\
&= A\,B'' + 2A'B' + A''B.
\end{aligned}
$$

For the third derivative, we have

$$
\begin{aligned}
(A\,B)''' &= (AB''' + A'B'') + 2(A'B'' + A''B') + (A''B' + A'''B)\\
&= A\,B''' + 3A'B'' + 3A''B' + A'''B.
\end{aligned}
$$

We recognize the pattern of binomial coefficients, which are familiar from elementary algebra. This leads to the following:

Lemma 20.6. (**Product rule**). *The nth derivative of the product $A\,B$ is given by*

$$(A\,B)^{(n)} = \sum_{k=0}^{n} \binom{n}{k} A^{(k)} B^{(n-k)},$$

where $\binom{n}{k}$ is the binomial coefficient, given by the formula

$$\binom{n}{k} = \frac{n(n-1)\cdots(n-k+1)}{k!} = \frac{n!}{k!(n-k)!},$$

for $k = 0,\ldots,n$.

The first few binomial coefficients are given as follows:

$$\binom{n}{0} = 1, \quad \binom{n}{1} = n, \quad \binom{n}{2} = \frac{n(n-1)}{2}, \quad \binom{n}{3} = \frac{n(n-1)(n-2)}{6}.$$

In particular,

$$\binom{0}{0} = 1,$$

$$\binom{1}{0} = 1, \quad \binom{1}{1} = 1,$$

$$\binom{2}{0} = 1, \quad \binom{2}{1} = 2, \quad \binom{2}{2} = 1,$$

$$\binom{3}{0} = 1, \quad \binom{3}{1} = 3, \quad \binom{3}{2} = 3, \quad \binom{3}{3} = 1.$$

The proof of Lemma 20.6 can be carried out by mathematical induction, as indicated in Exercise 15. For example, we have the following applications of the formula:

$$(t\,A)^{(n)} = t\,A^{(n)} + n\,A^{(n-1)},$$

$$(t^2 A)^{(n)} = t^2 A^{(n)} + 2n\,t\,A^{(n-1)} + n(n-1)A^{(n-2)}.$$

Example 20.3. *Find the third derivative of* $t^2 e^t$.

Solution. $(t^2 e^t)^{(3)} = t^2 e^t + 3(2t)e^t + 6e^t = (t^2 + 6t + 6)e^t.$ ∎

Exercises for Section 20.1

Find the first few terms in the Taylor series representation of each of the following functions about the specified point t_0.

1. $f(t) = \sin(t), \quad t_0 = 0$

2. $f(t) = \sin(t), \quad t_0 = \pi/2$

3. $f(t) = e^{3t}, \quad t_0 = 1$

4. $f(t) = e^{t^2}, \quad t_0 = 0$

5. $f(t) = \dfrac{3}{2-t}, \quad t_0 = 0$

6. $f(t) = \dfrac{3+t}{3-t}, \quad t_0 = 0$

7. $f(t) = t\cos(3t), \quad t_0 = 0$

8. $f(t) = (t+2)e^{3t}, \quad t_0 = 0$

9. $f(t) = (t+2)e^{3t}, \quad t_0 = 1$

10. $f(t) = \cosh(t), \quad t_0 = 0$

Use the general product rule (Lemma 20.6) to compute the following higher derivatives:

11. $(t^3 e^{4t})''$

12. $\left(\sin(2t)\cosh(3t)\right)'''$

13. $\left(\cos(5t)e^{2t}\right)^{(4)}$

14. $\left(t^3 \log(t)\right)''$

15. Prove the general product rule (Lemma 20.6) by mathematical induction, assuming its truth for $n = N$ and concluding the truth for $n = N + 1$. [Hint: First, prove that the binomial coefficients satisfy an identity of the form

$$\binom{N+1}{k} = \binom{N}{k} + \binom{N}{k-1}.]$$

20.2   *Power Series via* **Mathematica**

The command **Series** can be used to find the power series expansion of a function of one or more variables. The syntax for one variable is as follows. The command **Series[f,{x,x0,n}]** generates a power series expansion for **f** about the point $x = x_0$ to order $(x - x_0)^n$.

Example 20.4. *Use Mathematica to find the terms of degree 5 or less of the power series expansion about $x = 0$ of $e^{\sin(x)}$. Plot the first five partial sums.*

Solution. We use

Series[E^Sin[x],{x,0,5}]

to get

```
          2    4     5
         x    x     x          6
1 + x + --- - --- - --- + O[x]
         2    8     15
```

The expression

```
      6
O[x]
```

stands for terms of degree 6 and higher. Hence

$$e^{\sin(x)} = 1 + x + \frac{x^2}{2} - \frac{x^4}{8} - \frac{x^5}{15} + \text{higher-order terms.} \quad \blacksquare$$

The command **Normal** gets rid of an **O[x]** term. This must be done before any plotting:

Normal[Series[E^Sin[x],{x,0,5}]]

yields

```
          2    4     5
         x    x     x
1 + x + --- - --- - ---
         2    8     15
```

To find the first five partial sums, we use

```
Table[Normal[Series[E^Sin[x],{x,0,k}]],{k,5}]
```

$$\{1 + x,\ 1 + x + \frac{x^2}{2},\ 1 + x + \frac{x^2}{2},\ 1 + x + \frac{x^2}{2} - \frac{x^4}{8},\ 1 + x + \frac{x^2}{2} - \frac{x^4}{8} - \frac{x^5}{15}\}$$

and to plot the first five partial sums we use

```
xxx = Plot[Evaluate[Table[Normal[Series[E^Sin[x],
    {x,0,k}]],{k,5}]],{x,-2Pi,2Pi}];
```

To plot $e^{\sin x}$ as a thick line, we use

```
yyy = Plot[E^Sin[x],{x,-2Pi,2Pi},
    PlotStyle->{{AbsoluteThickness[2]}}];
```

The two plots can be combined with

```
Show[{xxx,yyy}]
```

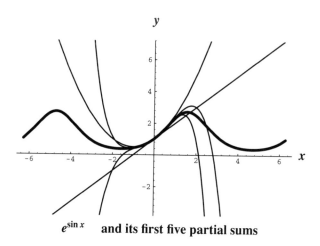

$e^{\sin x}$ **and its first five partial sums**

Series can be used to find the power series expansion about any point.

Example 20.5. *Use Mathematica to compute the first four terms in the power series expansion of* $\log(x)$ *about* $x = 1$.

Solution. We use

```
Series[Log[x],{x,1,4}]
```

to get

$$(-1 + x) - \frac{(-1 + x)^2}{2} + \frac{(-1 + x)^3}{3} - \frac{(-1 + x)^4}{4} + O[-1 + x]^5$$

Hence

$$\log(x) = (x - 1) - \frac{(x-1)^2}{2} + \frac{(x-1)^3}{3} - \frac{(x-1)^4}{4} + \text{higher-order terms.} \;\blacksquare$$

The usual algebraic operations work for series in much the same way that they work for polynomials.

Example 20.6. *Use Mathematica to find the sum, quotient, and composition of the power series expansions about* 0 *of* $\sin(x)$ *and* $\cos(x)$.

Solution. We enter the commands

```
f[x_] = Series[Sin[x],{x,0,6}]
```

and

```
g[x_] = Series[Cos[x],{x,0,6}]
```

Then the commands

```
f[y] + g[y]              f[y]/g[y]              g[f[x]]
```

yield

$$1 + y - \frac{y^2}{2} - \frac{y^3}{6} + \frac{y^4}{24} + \frac{y^5}{120} - \frac{y^6}{720} + O[y]^7$$

$$y + \frac{y^3}{3} + \frac{2 y^5}{15} + O[y]^7$$

and

$$1 - \frac{x^2}{2} + \frac{5 x^4}{24} - \frac{37 x^6}{720} + O[x]^7$$

The command **InverseSeries[s,z]** takes the series **s** generated by **Series** and gives a series for the inverse of the function using the variable **z**.

Example 20.7. *Compute the power series expansions about* 0 *of* tanh(x) *and its inverse.*

Solution. We enter the command

```
uu = Series[Tanh[x],{x,0,10}]
```

resulting in

$$x + \frac{x^3}{3} + \frac{2 x^5}{15} + \frac{17 x^7}{315} + \frac{62 x^9}{2835} + O[x]^{11}$$

Then

```
InverseSeries[uu,z]
```

yields

$$z + \frac{z^3}{3} + \frac{z^5}{5} + \frac{z^7}{7} + \frac{z^9}{9} + O[z]^{11}$$

Exercises for Section 20.2

1. Power series can be differentiated in *Mathematica* in the same way as ordinary expressions. Indefinite integration is also possible. Compute the derivative and indefinite integral of the first few terms in the power series expansion about 0 of $\sqrt{1-t}$.

2. Use *Mathematica* to do Exercises 1–10 of Section 20.1.

20.3   Power Series Solutions about an Ordinary Point

We now take up the problem of solving

$$a(t)y'' + b(t)y' + c(t)y = g(t) \tag{20.15}$$

in a neighborhood of an ordinary point $t = t_0$ using power series. There are two questions that must be considered:

- How do we formally determine[1] the coefficients in the power series solution?

- What is the radius of convergence of the power series solution?

We assume that the functions $a(t)$, $b(t)$, $c(t)$, and $g(t)$ are differentiable except at a finite number of singular points and that at all ordinary points these functions have power series expansions (see Section 20.1). When t_0 is an ordinary point for (20.15), there is a neighborhood of t_0 on which $a(t)$ is nonzero (since $a(t)$ is continuous). In that neighborhood we can divide by $a(t)$ and rewrite (20.15) in the leading-coefficient-unity-form

$$y'' + p(t)y' + q(t)y = r(t), \tag{20.16}$$

where $p(t) = b(t)/a(t)$, $q(t) = c(t)/a(t)$, and $r(t) = g(t)/a(t)$. To deal with (20.16), we need some formulas for the derivatives of a power series.

Lemma 20.7. *Suppose that*

$$w(t) = \sum_{k=0}^{\infty} \frac{W_k}{k!}(t - t_0)^k. \tag{20.17}$$

Then the following formulas hold for the derivatives of w:

$$w'(t) = \sum_{k=0}^{\infty} \frac{W_{k+1}}{k!}(t - t_0)^k, \tag{20.18}$$

and

$$w''(t) = \sum_{k=0}^{\infty} \frac{W_{k+2}}{k!}(t - t_0)^k. \tag{20.19}$$

More generally:

$$w^{(n)}(t) = \sum_{k=0}^{\infty} \frac{W_{k+n}}{k!}(t - t_0)^k. \tag{20.20}$$

Proof. We calculate

$$w'(t) = \sum_{k=1}^{\infty} \frac{k\,W_k}{k!}(t - t_0)^{k-1} = \sum_{k=1}^{\infty} \frac{W_k}{(k-1)!}(t - t_0)^{k-1} \tag{20.21}$$

---

[1]The method for determining the coefficients of a series solution given in this section and in Section 21.3 is nonstandard but effective. Its key ingredient is the evaluation of derivatives of a proposed solution at an ordinary point or a regular singular point to get a recurrence relation for the coefficients. It resembles a method used with Cauchy's problem in partial differential equations (see [Had]). The more traditional method for determining the coefficients consists in equating like powers of the independent variable to get a recurrence relation. That method is used in Section 21.9 to determine the coefficients for a series solution to the hypergeometric equation. Those readers who prefer the second method may want to read Section 21.9 and then derive the other series solutions of Chapters 20 and 21 as special cases.

We use the index shift $\ell = k - 1$ to convert the right-hand side of (20.21) to

$$\sum_{\ell=0}^{\infty} \frac{W_{\ell+1}}{\ell!}(t - t_0)^{\ell}. \tag{20.22}$$

Since ℓ is a dummy index, it can be replaced everywhere by the index k in (20.22). Thus we finally obtain

$$w'(t) = \sum_{k=0}^{\infty} \frac{W_{k+1}}{k!}(t - t_0)^k,$$

proving (20.18). Similar calculations yield (20.19), and (20.20) follows by mathematical induction . $\blacksquare$

Lemma 20.8. *Suppose that $w(t)$ is given by (20.17). Then*

$$W_n = w^{(n)}(t_0) = \frac{d^n w}{dt^n}\bigg|_{t=t_0}. \tag{20.23}$$

Proof. We write out the first few terms of (20.20):

$$w^{(n)}(t) = W_n + W_{n+1}(t - t_0) + \frac{W_{n+2}}{2}(t - t_0)^2 + \sum_{k=3}^{\infty} \frac{W_{n+k}}{k!}(t - t_0)^k. \tag{20.24}$$

Substitution of $t = t_0$ into (20.24) makes all the terms on the right-hand side of (20.24) vanish except the first; hence we get (20.23). $\blacksquare$

Now we are ready to determine the power series solution of (20.16). We assume that for some $R > 0$ the power series expansions about $t = t_0$ of the functions $p(t), q(t), r(t)$, namely

$$\begin{cases} p(t) = \sum_{k=0}^{\infty} \frac{p_k}{k!}(t - t_0)^k, \\[2mm] q(t) = \sum_{k=0}^{\infty} \frac{q_k}{k!}(t - t_0)^k, \\[2mm] r(t) = \sum_{k=0}^{\infty} \frac{r_k}{k!}(t - t_0)^k, \end{cases} \tag{20.25}$$

are valid in the interval $t_0 - R < t < t_0 + R$. The power series solution to (20.16) is sought in the form

$$y(t) = \sum_{k=0}^{\infty} \frac{Y_k}{k!}(t - t_0)^k. \tag{20.26}$$

We assume that Y_0 and Y_1 in (20.26) are known; for example, they might be given by initial conditions. The problem is to express the other Y_k's in terms of Y_0 and Y_1. For Y_2 we substitute (20.26) into (20.16) and then evaluate (20.16) at t_0. Using (20.23) we get

$$
0 = \left. \left(y''(t) + p(t)y'(t) + q(t)y(t) - r(t) \right) \right|_{t=t_0}
$$

$$
= Y_2 + p_0 Y_1 + q_0 Y_0 - r_0. \tag{20.27}
$$

Then (20.27) can be solved for Y_2. Next, we use the general product rule (Lemma 20.6) to compute

$$
0 = \left. \frac{d}{dt} \left(y''(t) + p(t)y'(t) + q(t)y(t) - r(t) \right) \right|_{t=t_0}
$$

$$
= Y_3 + (p_0 Y_2 + p_1 Y_1) + (q_0 Y_1 + q_1 Y_0) - r_1; \tag{20.28}
$$

clearly, (20.28) can be solved for Y_3. To find Y_4 we again use the general product rule (Lemma 20.6) to compute

$$
0 = \left. \frac{d^2}{dt^2} \left(y''(t) + p(t)y'(t) + q(t)y(t) - r(t) \right) \right|_{t=t_0}
$$

$$
= Y_4 + (p_0 Y_3 + 2p_1 Y_2 + p_2 Y_1) + (q_0 Y_2 + 2q_1 Y_1 + q_2 Y_0) - r_2; \tag{20.29}
$$

this equation can be solved for Y_4. Continuing in this way, we can compute all of the coefficients on the right-hand side of (20.26). In general, Y_{n+2} can be expressed in terms of known quantities by means of the equation

$$
Y_{n+2} + \sum_{k=0}^{n} \binom{n}{k} p_k Y_{n+1-k} + \sum_{k=0}^{n} \binom{n}{k} q_k Y_{n-k} = r_n, \tag{20.30}
$$

for $n = 0, 1, 2, \ldots$.

Equation (20.30) is called a **recurrence relation**. It allows us to determine the value of Y_{n+2} when we know the values of $Y_0, \ldots, Y_{n+1}$. The reader will note the structure of the indices therein; in the p_k-terms the indices add up to $n + 1$, whereas in the q_k-terms the indices add up to n.

From the theory of Chapter 8 (specifically Theorem 8.2 on page 246), we know that there exists a solution of the differential equation (20.16) for any given initial values $Y_0 = y(t_0)$ and $Y_1 = y'(t_0)$. Then Y_2 can be determined from (20.27); the result is then inserted into (20.28), which we can solve for Y_3. Clearly, the procedure can be continued indefinitely to obtain the higher coefficients.

The above method can be justified theoretically by showing that the numbers $Y_2, Y_3, \ldots$ that emerge from the solution process can be used to produce a convergent power series (20.26) according to the following theorem.

Theorem 20.9. *Suppose that the power series expansions of the functions $p(t), q(t), r(t)$ given by (20.25) are convergent in the interval $|t - t_0| < R$. Let the numbers Y_0 and Y_1 be prescribed arbitrarily. Then the equations (20.30) can be solved for Y_{n+2} for $n = 0, 1, \ldots$ to produce a power series (20.26) that is convergent in the interval $|t - t_0| < R$ and that satisfies the differential equation $y'' + p(t)y' + q(t)y = r(t)$.*

A proof of Theorem 20.9 is given in Section 20.6. However, this proof is not necessary in order to *solve* differential equations using power series.

Example 20.8. *Find the power series solution of the equation*

$$y'' + t\,y' - 2y = t$$

with $t_0 = 0, Y_0 = 0, Y_1 = 1$.

Solution. We have $p(t) = t$, $q(t) = -2$, and $r(t) = t$. Hence $p_1 = r_1 = 1$, $q_0 = -2$, and all other p_k's, q_k's, and r_k's vanish. The equations (20.30) become

$$
\begin{cases}
\quad\quad Y_2 - 2Y_0 = 0, & Y_3 + Y_1 - 2Y_1 = 1, \\
Y_4 + 2Y_2 - 2Y_2 = 0, & \quad\quad\quad Y_5 + Y_3 = 0, \\
\quad\quad Y_6 + 2Y_4 = 0, & \quad\quad\quad Y_7 + 3Y_5 = 0,
\end{cases}
\tag{20.31}
$$

and so forth. Solving (20.31) successively yields

$$Y_2 = 0,\; Y_3 = 2,\; Y_4 = 0,\; Y_5 = -Y_3 = -2,$$

and in general,

$$Y_{n+2} + (n - 2)Y_n = 0$$

for $n \geq 2$. Thus, $Y_n = 0$ for $n = 0, 2, 4, \ldots$, and we have the power series solution

$$y(t) = t + 2\frac{t^3}{3!} - 2\frac{t^5}{5!} + 6\frac{t^7}{7!} - 30\frac{t^9}{9!} + \cdots. \;\blacksquare$$

Returning to the general discussion, we now illustrate how the method of power series can be used to produce two linearly independent solutions of a linear homogeneous equation of second-order, written in the leading-coefficient-unity-form

$$y'' + p(t)y' + q(t)y = 0. \tag{20.32}$$

This is the special case of the above method when $r(t) \equiv 0$, but we want two solutions instead of one. Formula (20.30) becomes

$$Y_{n+2} = -\sum_{k=0}^{n} \binom{n}{k} p_k Y_{n+1-k} - \sum_{k=0}^{n} \binom{n}{k} q_k Y_{n-k}. \tag{20.33}$$

From (20.33) we see that the coefficients $Y_2, Y_3, \ldots$ are obtained recursively by

$$Y_2 = -p_0 Y_1 - q_0 Y_0,$$

$$
\begin{aligned}
Y_3 &= -(p_0 Y_2 + p_1 Y_1) - (q_0 Y_1 + q_1 Y_0) \\
&= -p_0(-p_0 Y_1 - q_0 Y_0) - p_1 Y_1 - q_0 Y_1 - q_1 Y_0, \\
&= Y_0(p_0 q_0 - q_1) + Y_1(p_0^2 - p_1 - q_0),
\end{aligned}
$$

and so forth. At each stage of the process the coefficient Y_n is obtained as a linear combination of Y_1 and Y_0, where the coefficients are expressed in terms of $p_0, q_0, p_1, q_1, \ldots$. The terms in the power series for a solution $y(t)$ to (20.32) can be separated into two classes: those containing Y_0 and those containing Y_1. Thus any solution $y(t)$ to (20.32) can be written in the form

$$y(t) = Y_0 \left(1 + \sum_{n=2}^{\infty} \tilde{A}_n(t - t_0)^n\right) + Y_1 \left((t - t_0) + \sum_{n=2}^{\infty} \tilde{B}_n(t - t_0)^n\right), \qquad (20.34)$$

where $Y_0 = y(t_0)$ and $Y_1 = y'(t_0)$. According to Theorem 8.2, there exist solutions $y_1(t)$ and $y_2(t)$ such that $y_1(t_0) = y_2'(t_0) = 1$ and $y_1'(t_0) = y_2(t_0) = 0$. From (20.34) it follows that

$$y_1(t) = 1 + \sum_{n=2}^{\infty} A_n(t - t_0)^n \qquad \text{and} \qquad y_2(t) = (t - t_0) + \sum_{n=2}^{\infty} B_n(t - t_0)^n,$$

for some constants A_n and B_n.

The procedure just explained is illustrated in terms of the following example of a homogeneous second-order equation with constant-coefficients.

Example 20.9. *Find two linearly independent power series solutions of the equation* $y'' + 4y = 0$ *around* $t_0 = 0$.

Solution. We have to solve the equations

$$Y_{n+2} + 4Y_n = 0$$

for $n = 0, 1 \ldots$; thus

$$Y_2 = -4Y_0, \quad Y_4 = (-4)^2 Y_0, \ldots \qquad \text{and} \qquad Y_3 = -4Y_1, \quad Y_5 = (-4)^2 Y_1, \ldots.$$

The general formulas are

$$Y_{2k} = (-4)^k Y_0 \qquad \text{and} \qquad Y_{2k+1} = (-4)^k Y_1$$

for $k = 0, 1 \ldots$. Thus the general solution to $y'' + 4y = 0$ expressed as a power series is given by

$$y(t) = Y_0 \sum_{k=0}^{\infty} (-4)^k \frac{t^{2k}}{(2k)!} + Y_1 \sum_{k=0}^{\infty} (-4)^k \frac{t^{2k+1}}{(2k+1)!}$$

$$= Y_0 \sum_{k=0}^{\infty} \frac{(-1)^k (2t)^{2k}}{(2k)!} + \frac{Y_1}{2} \sum_{k=0}^{\infty} \frac{(-1)^k (2t)^{2k+1}}{(2k+1)!}. \tag{20.35}$$

We recognize the familiar power series expansions of the sine and cosine functions on the right-hand side of (20.35); thus

$$y(t) = Y_0 \cos(2t) + Y_1 \frac{\sin(2t)}{2}. \blacksquare$$

Example 20.9 is somewhat exceptional. We *cannot* expect, in general, to be able to sum the power series solution in closed form in terms of elementary functions.

Exercises for Section 20.3

Solve the following initial value problems by the method of power series at the indicated points with the indicated initial values.

1. $\begin{cases} y'' - 4y = 0, \\ t_0 = 0, Y_0 = 1, Y_1 = 0 \end{cases}$

6. $\begin{cases} y'' + 9t^2 y = 0, \\ t_0 = 0, Y_0 = 1, Y_1 = 3 \end{cases}$

2. $\begin{cases} y'' - t\, y' - y = 0, \\ t_0 = 0, Y_0 = 2, Y_1 = 3 \end{cases}$

7. $\begin{cases} y'' + t\, y' + 4y = 0, \\ t_0 = 0, Y_0 = 3, Y_1 = 6 \end{cases}$

3. $\begin{cases} y'' - t\, y' - y = 0, \\ t_0 = 1, Y_0 = 1, Y_1 = 3 \end{cases}$

8. $\begin{cases} y'' - 3y' + 6y = 0, \\ t_0 = 0, Y_0 = 5, Y_1 = 3 \end{cases}$

4. $\begin{cases} y'' + 3t\, y' + 4y = 0, \\ t_0 = 0, Y_0 = 0, Y_1 = 1 \end{cases}$

9. $\begin{cases} y'' + 5t\, y' + 6t^2 y = 0, \\ t_0 = 0, Y_0 = 1, Y_1 = 0 \end{cases}$

5. $\begin{cases} y'' - 2t\, y' + 2y = 0, \\ t_0 = 0, Y_0 = 0, Y_1 = 1 \end{cases}$

10. $\begin{cases} y'' - 3y' + 6y = 0, \\ t_0 = 0, Y_0 = 5, Y_1 = 3 \end{cases}$

11. $\begin{cases} y'' + 5t\,y' + 6t^2 y = 0, \\ t_0 = 0,\, Y_0 = 0,\, Y_1 = 1 \end{cases}$

14. $\begin{cases} y'' - t\,y' = 0,\, t_0 = 1, \\ Y_0 = 1,\, Y_1 = 0 \end{cases}$

12. $\begin{cases} y'' - 2t\,y' + 5y = 0, \\ t_0 = 0,\, Y_0 = 1,\, Y_1 = 0 \end{cases}$

15. $\begin{cases} y'' + (t+1)y' + 4y = 0, \\ t_0 = 0,\, Y_0 = 0,\, Y_1 = 1 \end{cases}$

13. $\begin{cases} y'' - 2t\,y' + 5y = 0, \\ t_0 = 0,\, Y_0 = 0,\, Y_1 = 1 \end{cases}$

16. $\begin{cases} y'' + (t+1)y' + 4y = 0, \\ t_0 = -1,\, Y_0 = 0,\, Y_1 = 1 \end{cases}$

20.4   The Airy Equation

In this section we study in detail Airy[2] functions, which are obtained by finding the power series solution to an Airy equation (20.6).

Example 20.10. *Find the power series solutions about the ordinary point $t_0 = 0$ for a fundamental set of solutions of the Airy equation $y'' - \lambda t\, y = 0$, where λ is a constant.*

Solution. We determine the coefficients Y_n in the power series solution

$$y(t) = \sum_{k=0}^{\infty} \frac{Y_k}{k!} t^k,$$

with the initial conditions $y(0) = Y_0$ and $y'(0) = Y_1$. We have $p_n = q_n = 0$ for all $n \neq 1$, $p_1 = 0$ and $q_1 = -\lambda$. It follows from (20.33) that $Y_2 = 0$ and

$$Y_{n+2} - n\lambda Y_{n-1} = 0 \tag{20.36}$$

for all $n \geq 1$. Then repeated application of (20.36) implies that

$$\begin{cases} Y_{3k+2} = 0, \\ Y_{3k} = 1 \cdot 4 \cdot 7 \cdots (3k-2)\lambda^k Y_0, \\ Y_{3k+1} = 2 \cdot 5 \cdot 8 \cdots (3k-1)\lambda^k Y_1 \end{cases}$$

Sir George Biddel Airy (1801–1892). Astronomer Royal of England, was the first scientist to propose the notion of *diffraction rings*. He also developed the concept of the *transit circle*, which lasted for more than a century, and found the most accurate value for the *apex* of the universe.

for all k. Thus we find the following power series for the general solution of the Airy equation $y'' - \lambda t\, y = 0$:

$$y(t) = Y_0 \left(1 + \frac{(\lambda t)^3}{3!} + \frac{1 \cdot 4(\lambda t)^6}{6!} + \frac{1 \cdot 4 \cdot 7(\lambda t)^9}{9!} + \cdots \right)$$

$$+ Y_1 \left(1 + \frac{2(\lambda t)^4}{4!} + \frac{2 \cdot 5(\lambda t)^7}{7!} + \frac{2 \cdot 5 \cdot 8(\lambda t)^{10}}{10!} + \cdots \right).$$

In particular, if we put

$$y_1(\lambda, t) = \sum_{k=0}^{\infty} \frac{1 \cdot 4 \cdot 7 \cdots (3k-2)(\lambda t)^{3k}}{(3k)!} = 1 + \frac{(\lambda t)^3}{3!} + \frac{1 \cdot 4(\lambda t)^6}{6!} + \cdots \qquad (20.37)$$

and

$$y_2(\lambda, t) = \sum_{k=0}^{\infty} \frac{2 \cdot 5 \cdot 8 \cdots (3k-1)(\lambda t)^{3k+1}}{(3k+1)!} = t + \frac{2(\lambda t)^4}{4!} + \frac{2 \cdot 5(\lambda t)^7}{7!} + \cdots, \qquad (20.38)$$

then $y_1(\lambda, t)$ and $y_2(\lambda, t)$ constitute a fundamental set of solutions of $y'' - \lambda t\, y = 0$. ∎

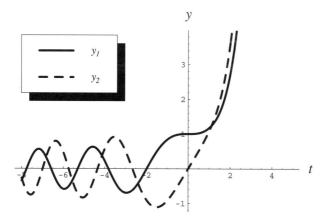

The solutions $y_1(1, t)$ **and** $y_2(1, t)$ **to the**
Airy equation $y'' - t\, y = 0$

From the graph it is clear that the functions $y_1(1, t)$ and $y_2(1, t)$ increase rapidly for $t > 0$, but they are highly oscillatory for $t < 0$. This is not surprising if we recall the behavior of the solution of the corresponding equation with constant-coefficients $y'' = A\, y$.

Two of the linear combinations of $y_1(1, t)$ and $y_2(1, t)$ have been singled out and are called Ai and Bi. Here are the exact formulas.

Definition. *The **Airy functions** Ai and Bi are defined as follows:*

$$Ai(t) = c_1 y_1(1, t) - c_2 y_2(1, t) \qquad and \qquad Bi(t) = \sqrt{3}\big(c_1 y_1(1, t) + c_2 y_2(1, t)\big),$$

where $y_1(\lambda, t)$ and $y_2(\lambda, t)$ are given by (20.37) and (20.38), and

$$c_1 = \frac{3^{-2/3}}{\Gamma(2/3)} \approx 0.355028 \qquad and \qquad c_2 = \frac{3^{-1/3}}{\Gamma(1/3)} \approx 0.258819.$$

Note that

$$c_1 = Ai(0) = \frac{Ai(0)}{\sqrt{3}} \qquad and \qquad c_2 = -Ai'(0) = \frac{Ai'(0)}{\sqrt{3}}.$$

Clearly, Ai and Bi constitute a fundamental set of solutions for the Airy equation $z'' - t z = 0$. Moreover, it is easy to prove

Lemma 20.10. *If $z(t)$ is a solution of $z'' - t z = 0$, then the function y defined by $y(t) = z(\lambda^{1/3} t)$ is a solution of $y'' - \lambda t y = 0$. Hence any solution to $y'' - \lambda t y = 0$ is expressible in terms of Airy functions.*

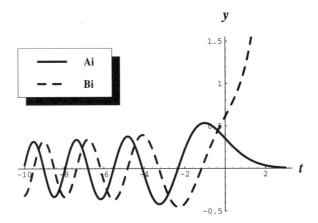

The Airy functions Ai(t) and Bi(t)

From the graph it is clear that $\lim_{t \to \infty} Ai(t) = 0$.

The Airy functions $Ai(t)$ and $Bi(t)$, which are solutions to the differential equation $z'' - t z = 0$, are somewhat analogous to the functions $\cos(t)$ and $\sin(t)$, which are solutions to the differential equation $z'' + z = 0$. The main difference, of course, is that $\cos(t)$ and $\sin(t)$ are oscillatory for all t, but the Airy functions are oscillatory only for negative t. Since $t \longmapsto \big(\cos(t), \sin(t)\big)$ parametrizes a circle, we can expect $t \longmapsto \big(Ai(t), Ai'(t)\big)$ and $t \longmapsto \big(Bi(t), Bi'(t)\big)$ to be spirals. Here are the plots:

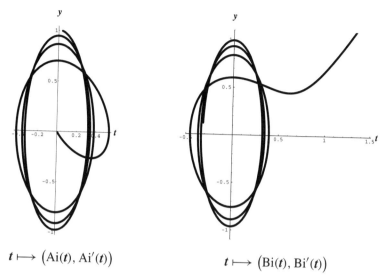

$$t \longmapsto \bigl(\text{Ai}(t),\ \text{Ai}'(t)\bigr) \qquad\qquad t \longmapsto \bigl(\text{Bi}(t),\ \text{Bi}'(t)\bigr)$$

The *Mathematica* versions of Ai and Bi are denoted **AiryAi** and **AiryBi**. Furthermore, the derivatives Ai$'$ and Bi$'$ exist in *Mathematica* as separate functions and are denoted by **AiryAiPrime** and **AiryBiPrime**.

Exercises for Section 20.4

Solve and plot the following Airy initial value problems over the stated ranges.

1. $\begin{cases} y'' - t\,y = 0, \\ y(0) = 3^{-2/3}(\Gamma(2/3))^{-1}, \\ y'(0) = -3^{-1/3}(\Gamma(1/3))^{-1}, \\ -15 < t < 5 \end{cases}$

3. $\begin{cases} y'' - 9t\,y = 0, \\ y(0) = 0,\ y'(0) = 1, \\ -1 < t < 4 \end{cases}$

2. $\begin{cases} y'' - t\,y = 0, \\ y(0) = -3^{-1/6}(\Gamma(2/3))^{-1}, \\ y'(0) = -3^{-1/3}(\Gamma(1/3))^{-1}, \\ -15 < t < 2 \end{cases}$

4. $\begin{cases} y'' + 36t\,y = 0, \\ y(0) = 0,\ y'(0) = 1, \\ -1 < t < 3 \end{cases}$

5. Use *Mathematica* to give a pseudoproof that the Wronskian of **AiryAi[t]** and **AiryBi[t]** is $1/\pi$. What about a mathematical proof?

6. Use *Mathematica* to show that the products **AiryAi[t]AiryAi[t]**, **AiryAi[t]AiryBi[t]** and **AiryBi[t]AiryBi[t]** are linearly independent solutions to $y''' - 4t\,y' - 2y = 0$.

20.5   The Legendre Equation

The method of power series solution can also be applied directly to certain equations that are not in leading-coefficient-unity-form. This is especially useful when the coefficients are *polynomial* functions. A general homogeneous equation of this type is written

$$a(t)y'' + b(t)y' + c(t)y = 0, \tag{20.39}$$

where $a(t_0) \neq 0$. If we were to transform this to leading-coefficient-unity-form, we would first have to find the power series expansions of the functions $b(t)/a(t)$ and $c(t)/a(t)$, which could be a quite lengthy process. Instead, we work directly with (20.39). Let

$$y(t) = \sum_{k=0}^{\infty} \frac{Y_k(t - t_0)^k}{k!}.$$

By differentiating (20.39) n times and using the general product rule (Lemma 20.6), we obtain

$$0 = \sum_{k=0}^{n} \binom{n}{k} a^{(k)}(t_0)Y_{n+2-k} + \sum_{k=0}^{n} \binom{n}{k} b^{(k)}(t_0)Y_{n+1-k}$$

$$+ \sum_{k=0}^{n} \binom{n}{k} c^{(k)}(t_0)Y_{n-k}. \tag{20.40}$$

Then (20.40) can be used to obtain $Y_2, Y_3, \ldots$ from Y_0 and Y_1.

We illustrate with the **Legendre**[3] equation (20.5).

3

Adrien Marie Legendre (1752–1833). French mathematician who made numerous contributions to number theory and the theory of elliptic functions. In 1782 Legendre determined the attractive force for certain solids of revolution by introducing an infinite series of polynomials that are now called Legendre polynomials. In his three-volume work **Traité des Fonctions Elliptiques** (1825, 1826, 1830), Legendre founded the theory of elliptic integrals.

Example 20.11. *Find the power series solution of the equation*

$$(1 - t^2)y'' - 2t\, y' + \lambda\, y = 0 \qquad (20.41)$$

about the ordinary point $t_0 = 0$, where λ is a constant, and find the radius of convergence.

Solution. We first note that $a(t) = (1 - t^2)$, $b(t) = -2t$, and $c(t) = \lambda$, so that

$$a^{(0)}(0) = 1, \qquad a^{(2)}(0) = b^{(1)}(0) = -2, \qquad c^{(0)}(0) = \lambda,$$

and all other coefficients $a^{(n)}(0)$, $b^{(n)}(0)$, $c^{(n)}(0)$ are zero. Substituting these values into (20.40), we obtain

$$Y_{n+2} - n(n-1)Y_n - 2nY_n + \lambda Y_n = 0,$$

which when simplified leads to

$$Y_{n+2} = \big(n(n+1) - \lambda\big)Y_n, \qquad (20.42)$$

for $n = 0, 1, 2, \ldots$. We can use (20.42) to express Y_n in terms of Y_0 and Y_1; thus

$$Y_2 = -\lambda Y_0, \qquad Y_3 = (2 - \lambda)Y_1, \qquad Y_4 = (6 - \lambda)Y_2 = (6 - \lambda)(-\lambda)Y_0,$$

and so forth. The general formulas are

$$Y_{2m} = Y_0 \prod_{\substack{k=0,2, \\ \ldots,2m}} \big(k(k+1) - \lambda\big) \qquad \text{and} \qquad Y_{2m+1} = Y_1 \prod_{\substack{k=1,3, \\ \ldots,2m+1}} \big(k(k+1) - \lambda\big) \qquad (20.43)$$

for $m = 1, 2, \ldots$. If we take $Y_0 = 1$ and $Y_1 = 0$ in (20.43), we are led to the solution

$$y_1(t) = \sum_{m=0}^{\infty} \frac{Y_{2m} t^{2m}}{(2m)!} \qquad (20.44)$$

of (20.41), where

$$Y_{2m} = \prod_{\substack{k=0,2, \\ \ldots,2m}} \big(k(k+1) - \lambda\big).$$

Similarly, the choice $Y_0 = 0$ and $Y_1 = 1$ leads to the solution

$$y_2(t) = \sum_{m=0}^{\infty} \frac{Y_{2m+1} t^{2m+1}}{(2m+1)!} \qquad (20.45)$$

of (20.41), where

$$Y_{2m+1} = \prod_{\substack{k=1,3, \\ \ldots,2m+1}} \big(k(k+1) - \lambda\big).$$

The general solution of (20.41) is then

$$y(t) = C_1 y_1(t) + C_2 y_2(t),$$

where C_1 and C_2 are constants.

To find the radius of convergence of

$$y(t) = \sum_{k=0}^{\infty} \frac{Y_k t^k}{k!}, \tag{20.46}$$

we note that

$$Y_{2m+1} = \big(2m(2m-1) - \lambda\big) Y_{2m-1}.$$

Thus, the ratio of the terms $Y_{2m+1} t^{2m+1}/(2m+1)!$ and $Y_{2m-1} t^{2m-1}/(2m-1)!$ of the power series (20.46) is

$$\frac{\dfrac{Y_{2m+1} t^{2m+1}}{(2m+1)!}}{\dfrac{Y_{2m-1} t^{2m-1}}{(2m-1)!}} = \frac{Y_{2m+1} t^{2m+1}}{2m(2m+1)Y_{2m-1} t^{2m-1}} = \frac{2m(2m-1) - \lambda}{2m(2m+1)} t^2.$$

A similar formula holds for the ratio of consecutive even terms. Hence

$$\lim_{m\to\infty} \left(\frac{\dfrac{Y_{2m+2} t^{2m+2}}{(2m+2)!}}{\dfrac{Y_{2m} t^{2m}}{(2m)!}} \right) = \lim_{m\to\infty} \left(\frac{\dfrac{Y_{2m+1} t^{2m+1}}{(2m+1)!}}{\dfrac{Y_{2m-1} t^{2m-1}}{(2m-1)!}} \right) = t^2.$$

Therefore, if $t^2 < 1$, we have convergence of the series (20.46). In other words, the radius of convergence of (20.46) is $R = 1$. ∎

The Legendre equation (20.41) has the further property that for certain values of the parameter λ, the power series solution reduces to a finite polynomial. This happens if λ is of the form $\lambda = n(n+1)$ for some integer n. In this case we have $Y_k = 0$ for $k = n+2, n+4, \cdots$ according to the recurrence relation. The corresponding solution is denoted by $P_n(t)$ and is called the nth **Legendre polynomial**, where we follow the customary normalization $P_n(1) = 1$. The first few Legendre polynomials are

$$P_0(t) = 1, \quad P_1(t) = t, \quad P_2(t) = \frac{3t^2 - 1}{2},$$

$$P_3(t) = \frac{5t^3 - 3t}{2}, \quad P_4(t) = \frac{35t^4 - 30t^2 + 3}{8}.$$

The general formula is

$$P_n(t) = \frac{1}{2^n} \sum_{k=0}^{[n/2]} \frac{(-1)^k (2n-2k)! \, t^{n-2k}}{k!(n-k)!(n-2k)!},$$

where $[n/2] = \mathbf{Floor}(n/2)$ denotes the greatest integer less than or equal to $n/2$.

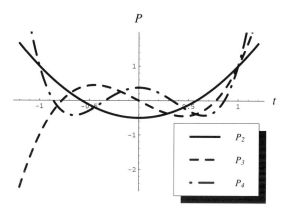

The Legendre polynomials $P_n(t)$, $n = 2, 3, 4$

A useful formula for computing Legendre polynomials is the **Rodrigues**[4] **formula**:

$$P_n(t) = \frac{1}{2^n n!} \frac{d^n}{dt^n} (t^2 - 1)^n. \tag{20.47}$$

Exercises for Section 20.5

1. Prove the orthogonality relation

$$\int_{-1}^{1} P_m(t) P_n(t) \, dt = 0, \qquad \text{for } m \neq n. \tag{20.48}$$

2. Use (20.48) to show that

$$\int_{-1}^{1} Q_n(t) P_n(t) \, dt = 0, \tag{20.49}$$

where $Q_n(t)$ is a polynomial of degree $n - 1$ or less.

---

[4]Olinde Rodrigues (1794–1851). French mathematician and banker.

3. Prove that if $Q_n(t)$ is a polynomial of degree n such that

$$\int_{-1}^{1} Q_n(t) P_k(t)\,dt = 0$$

for $k = 1, \ldots, n - 1$, then $Q_n(t) = c\,P_n(t)$ for some constant c.

4. To prove Rodrigues' formula (20.47), follow the following steps:

   a. Define $W_n(t) = \dfrac{d^n}{dt^n}(t^2 - 1)^n$ and show that $W_n(t)$ is a polynomial of degree n with leading coefficient $(2n)!/n!$.

   b. Use integration by parts n times to show that for any polynomial $Q_n(t)$ of degree $n - 1$ or less we have

$$\int_{-1}^{1} W_n(t) Q_n(t)\,dt = 0.$$

5. Use Exercise 4 to show that $P_n(t) = c\,W_n(t)$ for some constant c. Then compare the leading coefficients of $P_n(t)$ and $W_n(t)$ to conclude that $c = 1/(2^n n!)$.

6. The nth **Hermite[5] polynomial** is defined by the Rodrigues formula

$$H_n(t) = (-1)^n e^{t^2} \frac{d^n}{dt^n} e^{-t^2}.$$

Compute $H_n(t)$ for $0 \le n \le 6$.

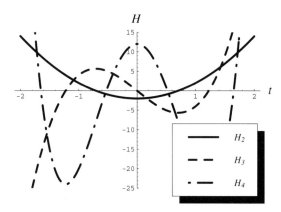

The Hermite polynomials $H_n(t)$, $n = 2, 3, 4$

7. Show that H_n satisfies the differential equation $y'' - 2t\, y' + 2n\, y = 0$.

8. Use *Mathematica* to give a pseudoproof of Exercise 7 as follows. Define functions **HH** and **HHeq** by

    ```
    HH[n_,t_]:= (-1)^n E^(t^2) D[E^(-t^2),{t,n}]
    ```

    and

    ```
    HHeq[n_,t_]:= Simplify[D[HH[n,t],{t,2}] -
                  2t D[HH[n,t],t] + 2n HH[n,t]]
    ```

    Then compute **Table[HHeq[n,t],n,0,6]**

9. It can be shown that

    $$H_n(t) = \sum_{k=0}^{[n/2]} \frac{(-1)^k n! (2t)^{n-2k}}{k!(n-2k)!}.$$

    Check this formula by hand for $0 \le n \le 6$.

10. Redo Exercise 9 in *Mathematica* as follows. Define a functions **HHH** by

    ```
    HHH[n_,t_]:= Sum[(-1)^k n! (2t)^(n-2k)/(k!(n-2k)!),
                 {k,0,Floor[n/2]}]
    ```

    Then compute **Table[HHH[n,t],n,0,6]**

11. The nth **Chebyshev**[6] **polynomial** is defined by

$$T_n(t) = \cos\bigl(n \arccos(t)\bigr).$$

Establish the recurrence relation

$$T_n(t) = 2t\, T_{n-1}(t) - T_{n-2}(t)$$

and compute $T_n(t)$ for $0 \le n \le 6$.

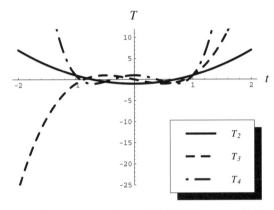

The Chebyshev polynomials $T_n(t)$, $n = 2, 3, 4$

12. Show that T_n satisfies the differential equation $(1 - t^2)y'' - t\, y' + n^2 y = 0$.

13. Use *Mathematica* to give a pseudoproof of Exercise 12 as follows. Define functions **TT** and **TTeq** by

```
TT[0,t_]:=  1
TT[1,t_]:=  t
TT[n_,t_]:= 2t TT[n - 1,t]  -  TT[n - 2,t]
```

and

6

Pafnuty Lvovich Chebyshev (1821–1894). Russian mathematician known for his work in approximation theory, differential geometry, orthogonal polynomials, probability theory.

```
TTeq[n_,t_]:= Simplify[(1 - t^2)D[TT[n,t],{t,2}] -
               t D[TT[n,t],t] + n^2 TT[n,t]]
```

Then compute `Table[TTeq[n,t],n,0,6]`

20.6   *Convergence of Series Solutions*

We now prove the convergence of the series in Theorem 20.9 by the **method of majorants**.[7] This consists in finding a related (majorant) equation that we can solve explicitly by the method of power series and whose coefficients are positive and greater in absolute value than the coefficients of the sought-after solution. We write out the details for the homogeneous equation $y'' + p(t)y' + q(t)y = 0$; for the inhomogeneous equation $y'' + p(t)y' + q(t)y = r(t)$ see Exercise 2.

For convenience we take $t_0 = 0$. The functions $p(t)$ and $q(t)$ are assumed to have convergent power series in the interval $|t| < R$. The related majorant equation is

$$z'' - \frac{P\,z'}{1 - t/r} - \frac{Q\,z}{(1 - t/r)^2} = 0, \qquad (20.50)$$

where P, Q, and r are suitable positive constants. This is the simplest second-order homogeneous equation that is both explicitly solvable and whose coefficients majorize the coefficients of the homogeneous equation $y'' + p(t)y' + q(t)y = 0$. The constants are chosen as follows: recall that the power series

$$\sum_{n=0}^{\infty} \frac{p_n t^n}{n!}$$

is convergent in the interval $|t| < R$. Fix r with $0 < r < R$. The terms $p_n r^n / n!$ remain bounded as $n \longrightarrow \infty$; thus

$$\left| \frac{p_n r^n}{n!} \right| \leq P, \qquad n = 0, 1, 2, \ldots$$

for some constant $P > 0$. Similarly, there is a number $Q > 0$ such that $|q_n r^n / n!| \leq Q$ for $n = 0, 1, 2, \ldots$. We define

$$\tilde{p}_n = \frac{P\,n!}{r^n} \qquad \text{and} \qquad \tilde{q}_n = (n + 1)\frac{Q\,n!}{r^n},$$

---

[7]This section contains theoretical material and can be omitted without loss of continuity.

which are strictly positive. The new coefficients $\tilde{p}_n$ and $\tilde{q}_n$ are related to the old coefficients by the inequalities

$$|p_n| \le \tilde{p}_n \quad \text{and} \quad |q_n| \le \tilde{q}_n.$$

We have the expansions

$$\frac{P}{1 - t/r} = \sum_{n=0}^{\infty} \frac{\tilde{p}_n t^n}{n!} \quad \text{and} \quad \frac{Q}{(1 - t/r)^2} = \sum_{n=0}^{\infty} \frac{\tilde{q}_n t^n}{n!}.$$

The general solution of (20.50) for $z(t)$ is written explicitly as

$$z(t) = A_1 \left(1 - \frac{t}{r}\right)^{-\alpha_1} + A_2 \left(1 - \frac{t}{r}\right)^{-\alpha_2}, \tag{20.51}$$

where α_1 and α_2 are roots of the quadratic equation

$$\alpha(\alpha + 1) - P \alpha r - Q r^2 = 0,$$

and A_1 and A_2 are chosen to fit the initial conditions. In fact, these roots are real and distinct, since by the quadratic formula we have

$$2\alpha = -(1 - P r) \pm \sqrt{(1 - P r)^2 + 4Q r^2}.$$

The solution (20.51) clearly has a convergent power series expansion

$$\sum_{n=0}^{\infty} \frac{z_n t^n}{n!},$$

valid for $|t/r| < 1$, whose coefficients can be found either by the binomial theorem or directly in terms of the corresponding system of recurrence relations:

$$z_2 - \tilde{p}_0 z_1 - \tilde{q}_0 z_0 = 0$$

$$z_{n+2} - (\tilde{p}_0 z_{n+1} + n\tilde{p}_1 z_n + \cdots + \tilde{p}_n z_1) - (\tilde{q}_0 z_n + n\tilde{q}_1 z_{n-1} + \cdots + \tilde{q}_n z_0) = 0.$$

In particular, we see that if $z_0 \ge 0$ and $z_1 \ge 0$, then $z_n \ge 0$ for all $n = 2, 3, \ldots$.

In order to obtain two linearly independent solutions of the given equation, we first take the initial conditions $Y_0 = 1$ and $Y_1 = 0$. Let $z(t)$ be the solution of the related majorant equation with $z_0 = 1$ and $z_1 = 0$. We now prove that $|Y_n| \le z_n$ for $n = 2, 3, \ldots$. For $n = 2$, we have

$$|Y_2| = |p_0 Y_1 + q_0 Y_0| \le P Y_1 + Q Y_0 = z_2.$$

Assuming that this has been proved for the values $n = 3, 4, \ldots, N + 1$, then

$$|p_0 Y_{N+1} + N p_1 Y_N + \cdots + p_N Y_1| \le \tilde{p}_0 z_{N+1} + N \tilde{p}_1 z_N + \cdots + \tilde{p}_N z_1$$

and

$$|q_0 Y_N + N\, q_1 Y_{N-1} + \cdots + q_N Y_0| \le \tilde{q}_0 z_N + N\, \tilde{q}_1 z_{N-1} + \cdots + \tilde{q}_N z_0.$$

But these upper bounds are exactly the two terms that figure in the equation for z_{N+2}; thus, we have $|Y_{N+2}| \le z_{N+2}$, and we have completed the proof by mathematical induction. Since the terms of the power series

$$\sum_{n=0}^{\infty} \frac{Y_n t^n}{n!}$$

are less than the terms of the convergent power series,

$$\sum_{n=0}^{\infty} \frac{z_n t^n}{n!},$$

we conclude convergence of the desired series. In order to obtain a second linearly independent solution, we let $Y_0 = 0$ and $Y_1 = 1$ and repeat the above procedure, which leads to the identical result $|Y_n| \le z_n$ and the existence of the second solution.

Exercises for Section 20.6

1. Show directly that the function $z(t) = (1 - t/r)^{-\alpha}$ is a solution of the equation

$$z'' - \frac{P\,z'}{(1 - t/r)} - \frac{Q\,z}{(1 - t/r)^2} = 0,$$

provided that α is a solution of the equation $\alpha(\alpha + 1) - P\,\alpha\,r - Q\,r^2 = 0$.

2. Show how to modify the above proof in order to obtain a power series solution of the equation $y'' + p(t)y' + q(t)y = r(t)$.

20.7   Series Solutions of Differential Equations Using ODE

A series solution to an initial value problem can be found with **ODE** using the option **Method->SeriesForm**.

Example 20.12.   Use **ODE** *with the option* **Method->SeriesForm** *to obtain and plot the power series through terms of degree* 20 *of the initial value problem*

$$\begin{cases} y'' + t^2 y = 0, \\ y(0) = 1,\ y'(0) = 0. \end{cases} \tag{20.52}$$

Solution. We use

```
ODE[{y'' + t^2 y == 0,y[0] == 1,y'[0] == 0},y,t,
Method->SeriesForm,Degree->20,
PlotSolution->{{t,-3.5,3.5}}]
```

to get

$$\{\{y \to 1 - \frac{t^4}{12} + \frac{t^8}{672} - \frac{t^{12}}{88704} + \frac{t^{16}}{21288960} - \frac{t^{20}}{8089804800}\}\}$$

The series solution to (20.52) is clear from this output. The plot is

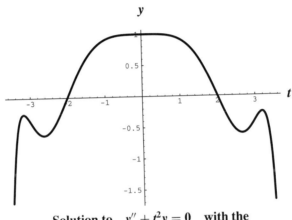

**Solution to $y'' + t^2 y = 0$ with the
initial conditions $y(0) = 1, \quad y'(0) = 0$**

Exercises for Section 20.7

1. Use *Mathematica* to do Exercises 1–16 on pages 697–698 of Section 20.3.

Use **ODE** to compute the exact solution and the series solution to each of the following equations and check that they are equivalent.

17. $y' = y$

18. $(1 + t^2)y' - y = 0$

19. $t^2 y'' + t\, y' - y = 0$

20. $t^2 y'' + y' + (t^2 - 1/4)y = 0$

21. $y'' - t\, y' + 4y = 0$

22. $(1 - t^2)y'' - 2t\, y' + 12y = 0$

21

FROBENIUS SOLUTIONS OF SECOND-ORDER EQUATIONS

In this chapter we continue our investigation of second-order linear equations of the form

$$a(t)y'' + b(t)y' + c(t)y = g(t). \qquad (21.1)$$

We assume that the coefficients $a(t)$, $b(t)$, and $c(t)$ are analytic at some point t_0, and we shall be concerned with the important case when $a(t_0) = 0$, that is when t_0 is a singular point of (21.1). We cannot apply the existence theorem, Theorem 8.1. Indeed, it is frequently difficult to determine the nature of solutions near such singular points. However, some singular points, which we call regular singular points, are such that the techniques of Chapter 20 can be modified in order to obtain solutions.

Regular singular points are introduced in Section 21.1. We solve the Cauchy-Euler equation in Section 21.2. In Section 21.3 we use the method of Frobenius to find one solution to a second-order linear equation in a neighborhood of a regular singular point, and in Section 21.5 we find a second linearly independent solution. Bessel functions are discussed in Section 21.4.

21.1   *Solutions about a Regular Singular Point*

In this section we establish the framework for studying the homogeneous differential equation

$$a(t)y'' + b(t)y' + c(t)y = 0 \tag{21.2}$$

in the neighborhood of a singular point t_0. We assume that (21.2) has been reduced so that the analytic functions $a(t)$, $b(t)$, and $c(t)$ have no common factors of the form $(t - t_0)^k$, where k is a positive integer. Recall from page 680 that the singular points of (21.2) are the points t_0 for which $a(t_0) = 0$.

Definition. *Let t_0 be a singular point for (21.2). We say that t_0 is a* **regular singular point** *provided that there exist functions $\widehat{p}(t)$ and $\widehat{q}(t)$, each having a convergent power series expansion in a neighborhood of the point t_0, such that*

$$\widehat{p}(t) = (t - t_0)\frac{b(t)}{a(t)} \qquad and \qquad \widehat{q}(t) = (t - t_0)^2\frac{c(t)}{a(t)}$$

around t_0.

In particular, $\widehat{p}(t_0)$ and $\widehat{q}(t_0)$ are well-defined. This means that both

$$\lim_{t \to t_0}(t - t_0)\frac{b(t)}{a(t)} \qquad and \qquad \lim_{t \to t_0}(t - t_0)^2\frac{c(t)}{a(t)}$$

exist and equal $\widehat{p}(t_0)$ and $\widehat{q}(t_0)$, respectively.

We shall consider only regular singular points. For example, 0 is a regular singular point of the Bessel equation (20.4). Note that the functions $\widehat{p}(t)$ and $\widehat{q}(t)$ are different from the functions $p(t)$ and $q(t)$ that we used for ordinary points in Chapter 20.

Lemma 21.1. *If t_0 is a regular singular point for (21.2), then (21.2) can be rewritten as*

$$(t - t_0)^2 y'' + (t - t_0)\widehat{p}(t)y' + \widehat{q}(t)y = 0. \tag{21.3}$$

Proof. We have

$$(t - t_0)b(t) = \widehat{p}(t)a(t) \qquad and \qquad (t - t_0)^2 c(t) = \widehat{q}(t)a(t). \tag{21.4}$$

Let us multiply (21.2) by $(t - t_0)^2$; this gives

$$(t - t_0)^2 a(t)y'' + (t - t_0)^2 b(t)y' + (t - t_0)^2 c(t)y = 0. \tag{21.5}$$

From (21.4) and (21.5) we get

$$(t - t_0)^2 a(t)y'' + (t - t_0)\widehat{p}(t)a(t)y' + \widehat{q}(t)a(t)y = 0. \tag{21.6}$$

Since $a(t)$ is nonzero in a neighborhood of t_0, we can divide both sides of (21.6) by $a(t)$ to get (21.3). ∎

Note that a differential equation of the type (21.3) *cannot* be subsumed under the ordinary point theory of Section 20.3, even in the case that $\widehat{p}(t)$ and $\widehat{q}(t)$ are constant functions. The theory must be generalized. In particular, we *cannot* expect to be able to solve an initial value problem corresponding to (21.3) for arbitrary choices of $y(t_0)$ and $y'(t_0)$. Because of the difficulty in defining the equation at $t = t_0$ and in defining the powers for negative argument, we require only that the solution exist for $0 < t - t_0 < R$ for some number R; this is sufficient for most applications.

21.2   The Cauchy-Euler Equation

In order to formulate a solution technique for (21.3) in the general case, we first consider in detail the important special case when $\widehat{p}(t)$ and $\widehat{q}(t)$ are constant. This is the (homogeneous second-order) **Cauchy[1]-Euler equation**

$$(t - t_0)^2 y'' + (t - t_0)p\, y' + q\, y = 0, \tag{21.7}$$

where p and q are constants. In order to solve (21.7), we first consider the special case $q = 0$.

Example 21.1. *Find the general solution to the Cauchy-Euler equation*

$$(t - t_0)^2 y'' + (t - t_0)p\, y' = 0, \tag{21.8}$$

where p is a constant not equal to 1.

Solution. Clearly, (21.8) is a first-order linear equation in y'; it can be rewritten in the form

$$(t - t_0)^p y'' + (t - t_0)^{p-1} p\, y' = 0. \tag{21.9}$$

Then (21.9) is an exact equation, which can be integrated as

$$(t - t_0)^p y' = A, \tag{21.10}$$

1

Augustin Louis Cauchy (1789–1857). One of the leading French mathematicians of the first half of the nineteenth century. He introduced rigor into calculus, including precise definitions of continuity and integration. His best known results include the Cauchy integral formula and Cauchy-Riemann equations in complex analysis, and the Cauchy-Kovalevskaya existence theorem for the solution of partial differential equations. Cauchy's scientific output was enormous, second only to that of Euler. In politics and religion, Cauchy was an archconservative.

where A is a constant; thus $y'(t) = A(t - t_0)^{-p}$. Hence the general solution to (21.8) is

$$y(t) = \frac{A(t - t_0)^{1-p}}{1 - p} + B,$$

where B is another constant. $\blacksquare$

Example 21.1 suggests that in the general case we should use a trial solution of (21.7) of the form

$$y(t) = A(t - t_0)^r,$$

for some constants A and r. When we substitute this trial solution into (21.7) and cancel the common factor, we find the so-called **indicial equation**

$$r(r - 1) + pr + q = 0. \tag{21.11}$$

This quadratic equation serves the same purpose in the study of Cauchy-Euler equations (and for more general equations, as we shall see in Section 21.3) that the associated characteristic equation serves in the study of constant-coefficient second-order equations (see page 266). The quadratic equation (21.11) has two solutions, r_1 and r_2, which we call **indicial roots**. The indicial roots may be real and distinct, real and repeated, or complex conjugate. Each indicial root of (21.11) gives rise to a solution of (21.9).

Case of real distinct roots $r_1 \neq r_2$

This is the simplest case. We find that $(t - t_0)^{r_1}$ and $(t - t_0)^{r_2}$ form a fundamental set of solutions of $(t - t_0)^2 y'' + (t - t_0)p\, y' + q\, y = 0$ and that the general solution is given by

$$y = A(t - t_0)^{r_1} + B(t - t_0)^{r_2},$$

where A and B are constants.

Example 21.2. *Find the general solution of the Cauchy-Euler equation*

$$t^2 y'' + 3t\, y' - 3y = 0. \tag{21.12}$$

Solution. Here $t_0 = 0$, and the indicial equation (21.11) becomes

$$0 = r(r - 1) + 3r - 3 = r^2 + 2r - 3 = (r - 1)(r + 3).$$

The roots are $r = 1$ and $r = -3$, leading to the general solution of (21.12) in the form

$$y(t) = A\, t + B\, t^{-3}, \tag{21.13}$$

where A and B are constants. $\blacksquare$

Example 21.3. *Solve the initial value problem*

$$\begin{cases} t^2 y'' + 3t\, y' - 3y = 0, \\ y(1) = 0,\ y'(1) = b \end{cases} \tag{21.14}$$

and plot the solutions corresponding to b ranging from −3 to 3 interspersed at intervals of 0.4.

Solution. From (21.13) we obtain

$$\begin{cases} 0 = y(1) \ = \ A + B, \\ b = y'(1) \ = \ A - 3B. \end{cases} \tag{21.15}$$

Thus

$$y(t) = \frac{b}{4}(t - t^{-3}).$$

The solutions can be plotted with

```
ODE[{t^2y'' + 3t y' - 3y == 0,y[1] == 0,y'[1] == b},y,t,
Method->NthOrderCauchyEuler,Parameters->{{b,-3,3,0.4}},
PlotSolution->{{t,-1,1},PlotRange->{Automatic,{-3,3}}}]
```

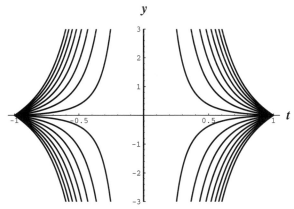

Solution to $t^2 y'' + 3t y' - 3y = 0$ with the
initial conditions $y(1) = 0$ and $y'(1) = b$

Case of real repeated roots $r_1 = r_2$

In the case that the indicial equation (21.11) has repeated roots, we cannot find the general solution of $(t - t_0)^2 y'' + (t - t_0) p\, y' + q\, y = 0$ solely in terms of powers; a logarithmic term must be included.

Example 21.4. *Find the general solution of the Cauchy-Euler equation*

$$t^2 y'' + t\, y' = 0. \tag{21.16}$$

Solution. We rewrite (21.16) as

$$(t\, y)' = t\, y'' + y' = 0. \tag{21.17}$$

Integration of (21.17) yields $t\, y' = C_1$. Integrating once more, we get

$$y = C_1 \log(t) + C_2,$$

where C_1 and C_2 are constants of integration. ∎

In general, we can always find one solution of the form $(t - t_0)^r$, where $r = r_1 = r_2$. The method of reduction of order (see Section 8.6) can be used to show that $(t - t_0)^r \log(t - t_0)$ is a second linearly independent solution. See Exercise 13.

Example 21.5. *Find the general solution of the Cauchy-Euler equation*

$$t^2 y'' - t\, y' + y = 0. \tag{21.18}$$

Solution. The indicial equation is $0 = r(r-1) - r + 1 = (r-1)^2$, so the general solution of (21.18) is

$$y = A\, t + B\, t \log(t),$$

where A and B are constants. ∎

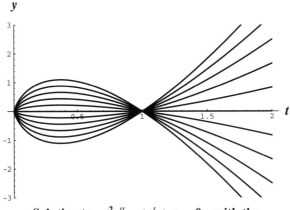

**Solution to $t^2 y'' - t y' + y = 0$ with the
initial conditions $y(1) = 0$ and $y'(1) = b$**

Case of complex conjugate roots $r_1, r_2 = \lambda \pm i\mu$

Assume that the coefficients p and q in the Cauchy-Euler equation $(t - t_0)^2 y'' + (t - t_0)p\, y' + q\, y = 0$ are real. In the case that the roots r_1 and r_2 of the indicial equation (21.11) are complex, they must be conjugate, so we can write

$$r_1 = \lambda + i\,\mu \qquad \text{and} \qquad r_2 = \lambda - i\,\mu,$$

where λ and μ are real. We can express the solution of $(t - t_0)^2 y'' + (t - t_0)p\, y' + q\, y = 0$ in terms of trigonometric functions by using the formula

$$t^{\lambda+i\mu} = t^\lambda t^{i\mu} = t^\lambda e^{i\mu \log(t)} = t^\lambda \big(\cos(\mu \log(t)) + i \sin(\mu \log(t)) \big). \tag{21.19}$$

Example 21.6. *Find the general solution of the Cauchy-Euler equation*

$$t^2 y'' - 3t\, y' + 13y = 0. \tag{21.20}$$

Solution. Here $t_0 = 0$ and the indicial equation (21.11) becomes

$$0 = r(r - 1) - 3r + 13 = r^2 - 4r + 13 = (r - 2)^2 + 9.$$

The roots are $r = 2 + 3i$ and $r = 2 - 3i$, leading to the general solution

$$y(t) = \tilde{A}\, t^{2+3i} + \tilde{B}\, t^{2-3i}. \tag{21.21}$$

where $\tilde{A}$ and $\tilde{B}$ are constants. To get rid of the complex exponents, we use (21.19) to convert (21.21) to

$$y(t) = A\, t^2 \cos(3 \log(t)) + B\, t^2 \sin(3 \log(t)).$$

As was the case when the associated characteristic equation had complex roots (see Section 9.1.3), the exact relation between the constants $\tilde{A}$, $\tilde{B}$, A, and B is immaterial. ∎

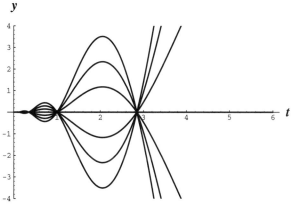

Solution to $t^2 y'' - 3t y' + 13y = 0$ with the
initial conditions $y(1) = 0$ and $y'(1) = b$

Cauchy-Euler equations via Mathematica

ODE's option **Method->NthOrderCauchyEuler** can be used to solve Cauchy-Euler equations. We give several examples. In each case the *Mathematica* output is so close to standard notation that it is not necessary to reformat the answers.

Example 21.7. *Find the general solution of $t^2 y'' + t\, y' - 4y = 0$ using* **ODE***.*

Solution. We use

```
ODE[t^2 y'' + t y' - 4y == 0,y,t,Method->NthOrderCauchyEuler]
```

obtaining

$$
\{\{y \to \frac{C[1]}{t^2} + t^2\, C[2]\}\}
$$

Next, we solve an initial value problem.

Example 21.8. *Use* **ODE** *to solve and plot the initial value problem*

$$
\begin{cases} t^2 y'' + t\, y' + y = 0, \\ y(1) = 1,\, y'(1) = 0. \end{cases}
$$

Solution. We use

```
ODE[{t^2 y'' + t y' + y == 0,y[1] == 1,y'[1] == 0},y,t,
Method->NthOrderCauchyEuler,PlotSolution->{{t,0.2,10}}]
```

obtaining

```
{{y -> Cos[Log[t]]}}
```

and the plot

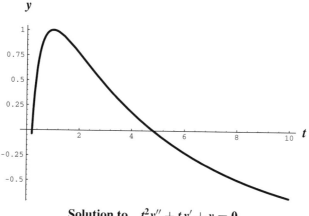

Solution to $t^2 y'' + t y' + y = 0$
with the initial conditions $y(1) = 1$ and $y'(1) = 0$

ODE can also solve higher-order Cauchy-Euler equations. By an nth-order Cauchy-Euler equation we mean an nth-order linear equation of the form

$$(t - t_0)^n a_n \frac{d^n y}{dt^n} + (t - t_0)^{n-1} a_{n-1} \frac{d^{n-1} y}{dt^{n-1}} + \cdots + (t - t_0) a_1 \frac{dy}{dt} + a_0\, y = g(t),$$

where $a_0, \ldots, a_n$ are constants.

Example 21.9. *Use* **ODE** *to find the general solution to*

$$t^3 y''' + 2t^2 y'' + 4t\, y' - 4y = 0.$$

Solution. We use

```
ODE[t^3 y''' + 2t^2 y'' + 4t y' - 4y == 0,y,t,
Method->NthOrderCauchyEuler]
```

to get

```
{{y -> t C[1] + C[2] Cos[2 Log[t]] + C[3] Sin[2 Log[t]]}}
```

Finally, we use **ODE** to solve an inhomogeneous Cauchy-Euler equation.

Example 21.10. *Use* **ODE** *to solve and plot the initial value problem*

$$\begin{cases} t^2 y'' + 3t\, y' + y = 80 \cos\big(10 \log(t)\big) \\ y(1) = 0,\ y'(1) = 2. \end{cases}$$

Solution. We use

```
ODE[{t^2 y'' + 3t y' + y == 80Cos[10Log[t]],
y[1] == 0,y'[1] == 2},y,t,Method->NthOrderCauchyEuler,
PlotSolution->{{t,0.2,10},PlotPoints->100}]
```

to get

$$\{\{y \rightarrow \frac{7920}{10201\ t} - \frac{7920\ \text{Cos}[10\ \text{Log}[t]]}{10201} + \frac{122\ \text{Log}[t]}{101\ t} + \frac{1600\ \text{Sin}[10\ \text{Log}[t]]}{10201}\}\}$$

and the plot

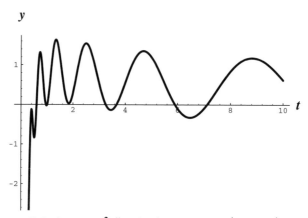

Solution to $t^2 y'' + 3t\, y' + y = 80\cos\bigl(10\log(t)\bigr)$
with the initial conditions $y(1) = 0$ **and** $y'(1) = 2$

Exercises for Section 21.2

Find the general solution of each of the following Cauchy-Euler equations.

1. $t^2 y'' - 7t\, y' + 15y = 0$

2. $t^2 y'' - 3t\, y' + 4y = 0$

3. $t^2 y'' + t\, y' - 4y = 0$

4. $t^2 y'' + t\, y' + 4y = 0$

5. $(t+1)^2 y'' + (t+1)y' - 9y = 0$

6. $(t-2)^2 y'' + 5(t-2)y' + 8y = 0$

Solve the following initial value problems.

7. $\begin{cases} t^2 y'' + t\, y' - 9y = 0, \\ y(1) = 1,\ y'(1) = -1 \end{cases}$ 10. $\begin{cases} t^2 y'' + 2t\, y' - 132y = 0, \\ y(1) = 0,\ y'(-1) = 2 \end{cases}$

8. $\begin{cases} t^2 y'' + ty' - 2.25y = 0, \\ y(2) = 0,\ y'(2) = 3 \end{cases}$ 11. $\begin{cases} t^3 y''' - t^2 y'' - 2t\, y' + 6y = 0, \\ y(1) = y'(1) = 0,\ y''(1) = 1 \end{cases}$

9. $\begin{cases} t^2 y'' - 9t\, y' + 25y = 0, \\ y(e) = 1,\ y'(e) = 0 \end{cases}$ 12. $\begin{cases} t^3 y''' - 3t^2 y'' - 12t\, y' + 60y = 0, \\ y(-2) = 1,\ y'(-2) = -1,\ y''(-2) = 2 \end{cases}$

13. Carry out the following steps to show that the Cauchy-Euler equation $t^2 y'' + t\, p\, y' + q\, y = 0$ can be transformed to an equation of constant-coefficients by the change of the independent variable. Let $z = \log(t)$.

   a. Show that

   $$\frac{dy}{dt} = \frac{1}{t}\frac{dy}{dz} \qquad \text{and} \qquad \frac{d^2 y}{dt^2} = \frac{1}{t^2}\left(\frac{d^2 y}{dz^2} - \frac{dy}{dz}\right).$$

   b. Show that the Cauchy-Euler equation becomes

   $$\frac{d^2 y}{dz^2} + (p-1)\frac{dy}{dz} + q\, y = 0.$$

14. The following problem develops a portion of the theory of undetermined coefficients for the inhomogeneous Cauchy-Euler equation. Suppose that p, q, and r are constants.

   a. (Nonresonant case) Show that a particular solution of the inhomogeneous Cauchy-Euler equation

   $$t^2 y'' + t\, p\, y' + q\, y = A\, t^r \tag{21.22}$$

   can be found in the form $y_p(t) = B\, t^r$ if and only if r *is not* a root of the indicial equation $r(r-1) + pr + q = 0$.

   b. (First resonant case) Suppose that r *is* a root of the indicial equation $r(r-1) + pr + q = 0$. Show that a particular solution of (21.22) can be found in the form $y_p(t) = B\, t^r \log(t)$ if and only if the roots of the indicial equation are distinct, that is, $2r - 1 + p \neq 0$.

   c. (Second resonant case) Suppose that r *is* a root of the indicial equation $r(r-1) + pr + q = 0$. Show that a particular solution of (21.22) can be found in the form $y_p(t) = B\, t^r \log(t)^2$ if and only if the roots of the indicial equation coincide, that is, $2r - 1 + p = 0$.

Find the general solution to the following inhomogeneous Cauchy-Euler equations.

15. $t^2 y'' - 5t\, y' + 5y = t^3$ **17.** $t^2 y'' - 9t\, y' + 25y = t^7$

16. $t^2 y'' - 15t\, y' + 63y = t^7$ **18.** $t^2 y'' + 11t\, y' + 25y = t^{-5}$

21.3 *Method of Frobenius: The First Solution*

In this section we show how to obtain a series solution of

$$(t - t_0)^2 y'' + (t - t_0)\widehat{p}(t)y' + \widehat{q}(t)y = 0 \tag{21.23}$$

with a regular singular point at $t = t_0$ by the **method of Frobenius.**[2]   (So $\widehat{p}(t)$ and $\widehat{q}(t)$ are analytic in a neighborhood of t_0.) A trial solution of (21.23) is sought in the form

$$y(t) = (t - t_0)^r \sum_{k=0}^{\infty} \frac{Z_k}{k!}(t - t_0)^k. \tag{21.24}$$

Notice that (21.24) is in general *not* a power series, because we are not requiring that r be an integer. In the case that $r(r - 1) + r\,\widehat{p}(t_0) + \widehat{q}(t_0) = 0$ has real roots, we show below that a solution can always be found in the form (21.24) corresponding to the *larger root*. This is called a **Frobenius solution** of the differential equation (21.23).

Let us write

$$y(t) = (t - t_0)^r z(t), \qquad \text{where} \qquad z(t) = \sum_{k=0}^{\infty} \frac{Z_k}{k!}(t - t_0)^k. \tag{21.25}$$

We can assume that $Z_0 \neq 0$; otherwise, we could replace r by some larger number in (21.24).

We already know how to solve (21.23) in the case of a Cauchy-Euler equation, where we can take $z(t) \equiv 1$. The Frobenius method can be considered to be a combination of the Cauchy-Euler solution method of Section 21.2 and the power series method of Section 20.3.

First, we derive a differential equation for z.

Lemma 21.2. *If $y = (t - t_0)^r z$ is a solution to (21.23), then z is a solution to*

$$(t - t_0)^2 z'' + (t - t_0)\big(2r + \widehat{p}(t)\big)z' + \big(r(r - 1) + r\,\widehat{p}(t) + \widehat{q}(t)\big)z = 0. \tag{21.26}$$

2

Georg Ferdinand Frobenius (1848–1917). German mathematician, most famous for his work in group theory. Frobenius was a student of Weieirstrass.

Proof. We have $y' = r(t - t_0)^{r-1}z + (t - t_0)^r z'$ and

$$y'' = r(r - 1)(t - t_0)^{r-2}z + 2r(t - t_0)^{r-1}z' + (t - t_0)^r z''.$$

When we substitute these two equations into (21.23), we get (21.26). $\blacksquare$

Next, we write the power series expansions of the coefficient functions $\widehat{p}(t)$ and $\widehat{q}(t)$ in the form

$$\widehat{p}(t) = \sum_{k=0}^{\infty} p_k \frac{(t - t_0)^k}{k!} \qquad \text{and} \qquad \widehat{q}(t) = \sum_{k=0}^{\infty} q_k \frac{(t - t_0)^k}{k!}.$$

The use of $p_k/k!$ and $q_k/k!$ instead of p_k and q_k for the coefficients in the power series expansions will simplify formulas later on.

Lemma 21.3. *If $y = (t - t_0)^r z$ is a solution to (21.23), then*

$$r(r - 1) + r\, p_0 + q_0 = 0. \tag{21.27}$$

Proof. When we substitute $t = t_0$ into (21.26) and use the fact that $Z_0 \neq 0$, we get

$$0 = r(r - 1) + r\, \widehat{p}(t_0) + \widehat{q}(t_0) = r(r - 1) + r\, p_0 + q_0. \quad \blacksquare$$

Since (21.27) generalizes (21.11), we make the following definition.

Definition. *We call (21.27) the* **indicial equation** *for the differential equation $(t - t_0)^2 y'' + (t - t_0)\widehat{p}(t)y' + \widehat{q}(t)y = 0$. The roots of (21.27) are called* **indicial roots**.

Just as in the case of a Cauchy-Euler equation, each indicial root gives rise to a solution of (21.23), but in certain cases the two solutions are linearly dependent.

We can obtain information about the Z_k's in the trial solution (21.25) by differentiating (21.26) n times and evaluating at $t = t_0$. Note that the higher derivatives of the function $(t - t_0)\widehat{p}(t)$ are obtained from the general product rule (Lemma 20.6) by computing

$$\left.\left((t - t_0)\widehat{p}(t)\right)^{(k)}\right|_{t=t_0} = \left.\left((t - t_0)\widehat{p}^{(k)}(t) + k\,\widehat{p}^{(k-1)}(t)\right)\right|_{t=t_0} = k\, p_{k-1}.$$

Let us find a relation between Z_1 and Z_0.

Lemma 21.4. *If $y = (t - t_0)^r z$ is a solution to (21.23), then*

$$(2r + p_0)Z_1 + (r\, p_1 + q_1)Z_0 = 0. \tag{21.28}$$

Proof. We take the first derivative of (21.26) and evaluate at t_0; the result is

$$0 = \left((t - t_0)^2 z'' + (t - t_0)(2r + \widehat{p}(t))z' + (r(r - 1) + r\,\widehat{p}(t) + \widehat{q}(t))z\right)'(t_0)$$

$$= \left((2r + \widehat{p}(t))z' + (r\,\widehat{p}'(t) + \widehat{q}'(t))z + (r(r - 1) + r\,\widehat{p}(t) + \widehat{q}(t))z'\right)(t_0)$$

$$= (2r + p_0)Z_1 + (r\, p_1 + q_1)Z_0 + (r(r - 1) + r\, p_0 + q_0)Z_1.$$

Then (21.28) follows from this equation and (21.27). ∎

With equation (21.28) we see the crucial difference between the Frobenius method and the power series method of Section 20.3. In the power series method, the first two coefficients Y_0 and Y_1 can be chosen independently, but in the Frobenius method Z_1 is determined by Z_0 (provided that $2r + p_0 \neq 0$, which is usually the case). See Lemma 21.7 for more details.

Next, we find the recursion formula for Z_n in terms of $Z_0, \ldots, Z_{n-1}$.

Lemma 21.5. *If $y = (t - t_0)^r z$ is a solution to (21.23), then*

$$n(n - 1 + 2r + p_0)Z_n + \sum_{k=1}^{n} \binom{n}{k}\left((n - k + r)p_k + q_k\right)Z_{n-k} = 0 \qquad (21.29)$$

for $n \geq 2$.

Proof. First, we compute

$$\left((t - t_0)^2 z''\right)^{(n)}(t_0)$$

$$= \left((t - t_0)^2 z^{(n+2)} + n\left((t - t_0)^2\right)' z^{(n+1)} + \frac{n(n - 1)}{2}\left((t - t_0)^2\right)'' z^{(n)}\right)(t_0)$$

$$= n(n - 1)z^{(n)}(t_0) = n(n - 1)Z_n. \qquad (21.30)$$

Similarly,

$$\left((t - t_0)(2r + \widehat{p})z'\right)^{(n)}(t_0) = \left((t - t_0)\left((2r + \widehat{p})z'\right)^{(n)} + n(t - t_0)'\left((2r + \widehat{p})z'\right)^{(n-1)}\right)(t_0)$$

$$= n\left((2r + \widehat{p})z'\right)^{(n-1)}(t_0) = n\left(\sum_{k=0}^{n-1}\binom{n-1}{k}(2r + \widehat{p})^{(k)}(z')^{(n-1-k)}\right)(t_0)$$

$$= n\left((2r + p_0)Z_n + \sum_{k=1}^{n-1}\binom{n-1}{k}p_k Z_{n-k}\right)$$

$$= \left(n(2r + p_0)Z_n + \sum_{k=1}^{n}\binom{n}{k}(n - k)p_k Z_{n-k}\right). \qquad (21.31)$$

Also, because of (21.27), we have

$$\left(\left(r(r - 1) + r\,\widehat{p} + \widehat{q}\right)z\right)^{(n)}(t_0)$$

$$= \left(r(r - 1) + r\,p_0 + q_0\right)z^{(n)}(t_0) + \sum_{k=1}^{n}\binom{n}{k}\left((r\,\widehat{p} + \widehat{q})^{(k)}(t_0)\right)z^{(n-k)}(t_0)$$

$$= \sum_{k=1}^{n} \binom{n}{k} \Big(r \, p_k + q_k \Big) Z_{n-k}. \tag{21.32}$$

We substitute (21.30)–(21.32) into the nth derivative of (21.26) evaluated at t_0; the result is (21.29). ∎

Equation (21.29) almost yields a formula for Z_n in terms of $Z_0, \dots, Z_{n-1}$. However, we must worry about whether the coefficient of Z_n vanishes or not. First, we consider the case of complex roots of the indicial equation.

Theorem 21.6. *If the roots r_1 and r_2 of the indicial equation (21.27) are complex, then each gives rise to a Frobenius solution of (21.3).*

Proof. Let r be r_1 or r_2. Since r is complex and p_0 is real, the quantity $n(n - 1 + 2r + p_0)$ is nonzero for all nonzero integers n. It follows that (21.29) can be solved for all n to express Z_n in terms of $Z_0, \dots, Z_{n-1}$. Thus, we get a solution of the form (21.25). ∎

When the roots of the indicial equation are real, the situation is more complicated.

Lemma 21.7. *Suppose that the roots r_1 and r_2 of the indicial equation (21.27) are real, and that $r_1 \geq r_2$. Then*

$$n - 1 + 2r_1 + p_0 > 0 \tag{21.33}$$

for all integers $n > 0$.

Proof. When we use the quadratic formula to solve (21.27), we must choose the plus sign; thus

$$2r_1 = 1 - p_0 + \sqrt{(1 - p_0)^2 - 4q_0} \geq 1 - p_0.$$

Hence (21.33) holds for all positive integers n. ∎

When $r_1 \geq r_2$, we say that "r_1 is the larger root" and "r_2 is the smaller root", even though it might be the case that $r_1 = r_2$.

Theorem 21.8. *Let t_0 be a regular singular point of the differential equation (21.23). Suppose that the roots r_1 and r_2 of the indicial equation (21.27) are real and that $r_1 \geq r_2$. Then the larger root r_1 gives rise to a Frobenius solution of (21.3). Furthermore, the series converges for all t with $0 \leq |t - t_0| < R$, where R is the distance from t_0 to the nearest other singular point (real or complex) of (21.23).*

Proof. From (21.33) we see that $n(n - 1 + 2r_1 + p_0)$ is nonzero for all positive integers n. It follows that (21.29) can be solved for all n to express Z_n in terms of $Z_0, \dots, Z_{n-1}$. Thus, we get a solution of the form (21.25). For the statement concerning the radius of convergence see [Ince, Chapter XVI]. ∎

The case of real roots of the indicial equation occurs much more frequently in applications than the case of complex roots. Unfortunately, the analysis of the real root case is considerably more complicated than that of the complex root case, because Theorem 21.8 guarantees only one solution. We must use another method, for example reduction of order, to determine the second solution. This will be done in Section 21.5. In the meantime, in Section 21.4 we examine in detail the Bessel equation (20.4) in order to see how Theorem 21.8 is applied in a concrete case.

Exercises for Section 21.3

Find the indicial equation and the Frobenius solution of each of the following equations, each of which has a regular singular point at $t = 0$.

1. $t^2 y'' + t\, y' - t^2 y = 0$

2. $t^2 y'' + t\, y' + t\, y = 0$

3. $t^2 y'' - 7t\, y' + (15 + t^2)y = 0$

4. $t^2 y'' - 3t\, y' + (4 + 3t^2)y = 0$

5. $t^2 y'' + t\, y' - 4y = 0$

6. $4t^2 y'' + (1 + t^2)y = 0$

7. $t^2 y'' + t\, y' + 3t^2 y = 0$

8. $2t^2 y'' + t\, y' + t^2 y = 0$

9. $3t^2 y'' + 2t\, y' + t^2 y = 0$

10. $t^2 y'' + (t + 3t^2)y = 0$

11. $t\, y'' + 2t\, y' + 6e^t y = 0$

12. $t^2 y'' - (t(2 + t))y' + (2 + t^2)y = 0$

13. $t(t - 1)y'' + 6t^2 y' + 3y = 0$

14. $t\, y'' + 4y' - t\, y = 0$

21.4   Bessel Functions I

In this section we study in detail Bessel[3] functions, which are solutions to the Bessel equation (20.4).

---

[3]

Friedrich Wilhelm Bessel (1784–1846). German astronomer and friend of Gauss. At the age of 26, Bessel was appointed director of the Königsberg Observatory, and he was granted the title of doctor by the University of Göttingen. In addition to his astronomical work, Bessel created a major tool for applied mathematics with his study of Bessel functions. Special cases of these important functions had been used by D. Bernoulli, Euler, Lagrange, and others, but the general treatment was given by Bessel in 1817 and 1824. The function $J_n(x)$ was used to study perturbations of planetary motion. Most of Bessel's mathematical works have applications in astronomy.

Example 21.11.   *Find the Frobenius solution of the general Bessel equation*

$$t^2 y'' + t\, y' + (t^2 - \lambda^2)y = 0, \tag{21.34}$$

where λ is a nonnegative integer.

Solution. In this case we have $\widehat{p}(t) = 1$ and $\widehat{q}(t) = t^2 - \lambda^2$, so that $p_0 = 1$, $q_0 = -\lambda^2$, $q_2 = 2$, and $p_k = q_k = 0$ for all other values of k. The indicial equation is

$$0 = r(r-1) + r - \lambda^2 = r^2 - \lambda^2,$$

whose larger root is $r = \lambda \geq 0$. Hence the series solution to (21.34) given by the method of Frobenius has the form

$$y(t) = t^\lambda Z(t), \qquad \text{where} \qquad Z(t) = \sum_{n=0}^{\infty} \frac{Z_n}{n!} t^n.$$

The recurrence relations (21.28) and (21.29) become

$$Z_1 = 0 \qquad \text{and} \qquad (n^2 + 2n\lambda)Z_n + n(n-1)Z_{n-2} = 0 \tag{21.35}$$

for $n \geq 2$. We solve (21.35) for Z_n by writing

$$Z_n = -\frac{n-1}{n+2\lambda}Z_{n-2};$$

thus,

$$Z_2 = -\frac{1}{2+2\lambda}Z_0, \qquad Z_4 = -\frac{3}{4+2\lambda}Z_2 = \frac{3 \times 1}{(4+2\lambda)(2+2\lambda)}Z_0,$$

$$Z_6 = -\frac{5}{6+2\lambda}Z_4 = -\frac{5 \times 3 \times 1}{(6+2\lambda)(4+2\lambda)(2+2\lambda)}Z_0.$$

In general, $Z_{2n+1} = 0$ for all n, and

$$Z_{2n} = -\frac{2n-1}{2n+2\lambda}Z_{2n-2} = \frac{(2n-1)(2n-3)}{2^2(\lambda+n)(\lambda+n-1)}Z_{2n-4} = \cdots$$

$$= \frac{(-1)^n (2n-1)(2n-3)\cdots 1}{2^n(\lambda+n)\cdots(\lambda+1)}Z_0. \tag{21.36}$$

Recall (see page 467) that $\lambda!$ is defined whether or not λ is an integer. Since

$$(2n-1)(2n-3)\cdots 1 = \frac{(2n)!}{2^n n!} \qquad \text{and} \qquad (\lambda+n)\cdots(\lambda+1) = \frac{(\lambda+n)!}{\lambda!},$$

we can rewrite (21.36) as

$$Z_{2n} = \frac{(-1)^n \left(\dfrac{(2n)!}{2^n n!} \right)}{\dfrac{2^n (\lambda + n)!}{\lambda!}} Z_0,$$

or

$$\frac{Z_{2n}}{(2n)!} = \frac{(-1)^n \lambda! Z_0}{2^{2n} n! (\lambda + n)!}. \tag{21.37}$$

From (21.37) we get

$$Z(t) = \sum_{n=0}^{\infty} \frac{Z_{2n} t^{2n}}{(2n)!} = \lambda! Z_0 \sum_{n=0}^{\infty} \frac{(-1)^n t^{2n}}{2^{2n} n! (\lambda + n)!}.$$

Hence we obtain the Frobenius solution

$$y(t) = t^\lambda Z(t) = t^\lambda \lambda! Z_0 \sum_{n=0}^{\infty} \frac{(-1)^n t^{2n}}{2^{2n} n! (\lambda + n)!}$$

$$= Z_0 t^\lambda \left(1 - \frac{t^2}{2(2 + 2\lambda)} + \frac{3t^4}{4!(4 + 2\lambda)(2 + 2\lambda)} \right.$$

$$\left. - \frac{5 \times 3 t^6}{6!(6 + 2\lambda)(4 + 2\lambda)(2 + 2\lambda)} + \cdots \right). \ \blacksquare$$

The standard practice is to take $Z_0 = 1/(\lambda! 2^\lambda)$, leading to the following definition.

Definition.   *The **Bessel function of the first kind of order** λ is the function $J_\lambda(t)$ given by*

$$J_\lambda(t) = \sum_{n=0}^{\infty} \frac{(-1)^n t^{2n+\lambda}}{2^{2n+\lambda} n! (\lambda + n)!}. \tag{21.38}$$

Notice that (21.38) makes sense for any complex number λ that is not a negative integer. It also makes sense when λ is a negative integer, provided that we interpret $1/(\lambda + n)!$ to be zero in each term in which $(\lambda + n)!$ is infinite. Thus the Bessel function $J_\lambda(t)$ of order λ, where λ is an arbitrary complex number, is given by (21.38).

Sometimes the general solution to the Bessel equation (21.34) can be given in terms of Bessel functions of the first kind.

Lemma 21.9. *If λ is not an integer, then $J_\lambda(t)$ and $J_{-\lambda}(t)$ are linearly independent. In this case the general solution to the Bessel equation (21.34) is*

$$y(t) = C_1 J_\lambda(t) + C_2 J_{-\lambda}(t),$$

where C_1 and C_2 are constants.

Proof. We compute

$$\frac{J_\lambda(t)}{J_{-\lambda}(t)} = \frac{\displaystyle\sum_{n=0}^\infty \frac{(-1)^n t^{2n+\lambda}}{2^{2n+\lambda} n!(\lambda+n)!}}{\displaystyle\sum_{n=0}^\infty \frac{(-1)^n t^{2n-\lambda}}{2^{2n-\lambda} n!(-\lambda+n)!}} = \frac{\frac{t^\lambda}{2^\lambda \lambda!}+\cdots}{\frac{t^{-\lambda}}{2^{-\lambda}(-\lambda!)}+\cdots} = \frac{\left(\frac{t}{2}\right)^{2\lambda}}{\frac{\lambda!}{(-\lambda!)}}+\cdots,$$

which is clearly a nonconstant function. Hence no linear combination of $J_\lambda(t)$ and $J_{-\lambda}(t)$ can vanish. ∎

In the case that λ is an integer m, the Bessel functions $J_m(t)$ and $J_{-m}(t)$ are linearly dependent. More precisely:

Lemma 21.10. *If m is an integer, then*

$$J_{-m}(t) = (-1)^m J_m(t).$$

Proof. Without loss of generality, we may assume that $m \geq 0$. Then

$$J_{-m}(t) = \sum_{n=0}^\infty \frac{(-1)^n t^{2n-m}}{2^{2n-m} n!(-m+n)!} = \sum_{n=m}^\infty \frac{(-1)^n t^{2n-m}}{2^{2n-m} n!(-m+n)!}, \qquad (21.39)$$

since the factors $1/(-m+n)!$ are zero for $n = 0, 1, \ldots, m-1$. We replace the dummy variable n in (21.39) by $n+m$. Then (21.39) becomes

$$J_{-m}(t) = \sum_{n=0}^\infty \frac{(-1)^{n+m} t^{2(n+m)-m}}{2^{2(n+m)-m}(n+m)!n!}(-1)^m = \sum_{n=0}^\infty \frac{(-1)^n t^{2n+m}}{2^{2n+m}(n+m)!n!}$$

$$= (-1)^m J_m(t). \ \blacksquare$$

Thus, when m is an integer, the method of Frobenius yields only one linearly independent solution to the Bessel equation $t^2 y'' + t y' + (t^2 - m^2)y = 0$, namely $J_m(t)$. In this case it is much more difficult to find a second solution that is linearly independent from $J_m(t)$. In Section 21.5 we take up this problem for a general Frobenius solution.

Exercises for Section 21.4

Establish the following identities:

1. $\dfrac{d}{dt} J_0(t) = -J_1(t)$

6. $J_{-1/2}(t) = \left(\dfrac{2}{\pi t}\right)^{1/2} \cos(t)$

2. $\dfrac{d}{dt} t J_1(t) = t J_0(t)$

7. $J_{\lambda-1}(t) + J_{\lambda+1}(t) = \dfrac{2\lambda}{t} J_\lambda(t)$

3. $\dfrac{d}{dt} t^\lambda J_\lambda(t) = t^\lambda J_{\lambda-1}(t)$

8. $J_{\lambda-1}(t) - J_{\lambda+1}(t) = 2\dfrac{d}{dt} J_\lambda(t)$

4. $\dfrac{d}{dt} t^{-\lambda} J_\lambda(t) = -t^{-\lambda} J_{\lambda+1}(t)$

9. $t\dfrac{d}{dt} J_\lambda(t) + \lambda J_\lambda(t) = t J_{\lambda-1}(t)$

5. $J_{1/2}(t) = \left(\dfrac{2}{\pi t}\right)^{1/2} \sin(t)$

10. $t\dfrac{d}{dt} J_\lambda(t) - \lambda J_\lambda(t) = -t J_{\lambda+1}(t)$

11. $\exp\left(\dfrac{x}{2}\left(t - \dfrac{1}{t}\right)\right) = J_0(x) + \displaystyle\sum_{n=1}^{\infty} J_n(x)\left(t^n + (-t)^{-n}\right)$

12. Let λ be any complex number. Check that $J_\lambda(t)$ defined by (21.38) satisfies the Bessel equation $t^2 y'' + t\, y' + (t^2 - \lambda^2)y = 0$.

Find the general solutions of the following Bessel equations:

13. $t^2 y'' + t\, y' + \left(t^2 - \dfrac{1}{9}\right) y = 0$

15. $t^2 y'' + t\, y' + \left(4t^2 - \dfrac{1}{25}\right) y = 0$

14. $t^2 y'' + t\, y' + \left(t^2 - 5\right) y = 0$

16. $t^2 y'' + t\, y' - \left(7t^2 + \dfrac{1}{49}\right) y = 0$

17. Show that if $y(t)$ is any solution of the Bessel equation $t^2 y'' + t\, y' + (t^2 - \lambda^2)y = 0$, then $w(z) = z^{-c} y(a\, z^b)$ is a solution of

$$z^2 w'' + (2c + 1)z\, w' + \left(a^2 b^2 z^{2b} + c^2 - \lambda^2 b^2\right)w = 0,$$

where a, b, and c are arbitrary complex numbers.

18. Use Problem 17 to show that any solution of the Airy equation $y'' + t\, y = 0$ can be written as

$$y(t) = t^{1/2} \left(C_1 J_{1/3}\left(\dfrac{2t^{3/2}}{3}\right) + C_2 J_{-1/3}\left(\dfrac{2t^{3/2}}{3}\right)\right).$$

21.5   Method of Frobenius: The Second Solution

In order to find a second, linearly independent solution of the differential equation (21.3) with a regular singular point, we apply the method of reduction of order introduced in Section 8.6. For convenience we assume that $t_0 = 0$. Let

$$y_1(t) = t^{r_1} \sum_{k=0}^{\infty} \frac{Y_k t^k}{k!}$$

be the Frobenius solution found in Section 21.3, with $Y_0 \neq 0$. To simplify the development below, we normalize by taking $Y_0 = 1$. Thus

$$y_1(t) = t^{r_1} \left(1 + \sum_{k=1}^{\infty} \frac{Y_k t^k}{k!} \right). \tag{21.40}$$

If $y_2(t)$ is a second solution (to be found), we define $w(t) = y_2(t)/y_1(t)$. Computing the derivatives by the product rule, we find that

$$y_2' = w \, y_1' + w' \, y_1 \qquad \text{and} \qquad y_2'' = w \, y_1'' + 2w' y_1' + w'' y_1. \tag{21.41}$$

From (21.41) and the fact that both y_1 and y_2 satisfy (21.3), we obtain

$$\begin{aligned}
0 &= t^2 \, y_2'' + t \, \widehat{p} \, y_2' + \widehat{q} \, y_2 \\
&= t^2 \big(w \, y_1'' + 2w' \, y_1' + w'' \, y_1 \big) + t \, \widehat{p} \big(w \, y_1' + w' \, y_1 \big) + \widehat{q} \, w \, y_1 \\
&= w'' t^2 y_1 + w' \big(2t^2 y_1' + t \, \widehat{p} \, y_1 \big).
\end{aligned} \tag{21.42}$$

We can consider (21.42) to be a separable first-order equation in w':

$$\frac{w''}{w'} = -\frac{2y_1'}{y_1} - \frac{\widehat{p}}{t}. \tag{21.43}$$

We integrate both sides of (21.43) and exponentiate to obtain

$$w' = \frac{1}{y_1^2} \exp\left(-\int \frac{\widehat{p}(t)dt}{t} \right), \tag{21.44}$$

where without loss of generality we have taken the arbitrary constant of integration to be 1. We have

$$\widehat{p}(t) = \sum_{k=0}^{\infty} \frac{p_k t^k}{k!},$$

so that

$$\int \frac{\widehat{p}(t)dt}{t} = C + p_0 \log(t) + \sum_{k=1}^{\infty} \frac{p_k t^k}{k\,k!}. \tag{21.45}$$

Without loss of generality the constant of integration C in (21.45) can be taken to be zero. Thus

$$-\int \frac{\widehat{p}(t)dt}{t} = -\left(p_0 \log(t) + \sum_{k=1}^{\infty} \frac{p_k t^k}{k\,k!} \right). \tag{21.46}$$

When we form the exponential of the series

$$\sum_{k=1}^{\infty} \frac{p_k t^k}{k\,k!},$$

we obtain another convergent power series, call it

$$1 + \sum_{k=1}^{\infty} a_k t^k.$$

Thus, (21.44) can be rewritten as

$$w'(t) = \frac{1}{y_1^2(t)t^{p_0}} \left(1 + \sum_{k=1}^{\infty} a_k t^k \right). \tag{21.47}$$

But the function $y_1(t)$ is a power series multiplied by the factor t^{r_1} as in (21.40). From (21.40) it follows that $y_1(t)^{-2}$ is of the form

$$\frac{1}{y_1^2(t)} = t^{-2r_1} \left(1 + \sum_{k=0}^{\infty} \tilde{a}_k t^k \right). \tag{21.48}$$

From (21.47) and (21.40) we get the expansion

$$w'(t) = t^{-2r_1 - p_0} \left(1 + \sum_{k=1}^{\infty} b_k t^k \right), \tag{21.49}$$

where the series on the right-hand side of (21.49) is convergent. Then (21.49) can be integrated and multiplied by y_1 to find y_2, the second linearly independent solution.

The precise form of (21.49) depends on further details of the coefficients, which we now explain in detail. Let

$$\Delta = r_1 - r_2 = \sqrt{(1 - p_0)^2 - 4q_0}.$$

Case I: $\Delta = 0$

We show that in this case y_2 has a logarithmic singularity. To see this, we note that $\Delta = 0$ implies $r_1 = r_2 = (1 - p_0)/2$, so that (21.49) becomes

$$w'(t) = \frac{1}{t}\left(1 + \sum_{k=1}^{\infty} b_k t^k\right). \qquad (21.50)$$

When we integrate both sides of (21.50), we get

$$w(t) = \log(t) + \sum_{k=1}^{\infty} c_k t^k, \qquad (21.51)$$

for a new power series

$$\sum_{k=1}^{\infty} c_k t^k. \qquad (21.52)$$

We have omitted the constant term in (21.52) since we are looking for a linearly independent solution $y_2(t)$, and inclusion of a constant would give a multiple of the first solution $y_1(t)$, which is already known. When we multiply (21.51) by the power series for $y_1(t)$, we obtain a new series,

$$\sum_{k=1}^{\infty} \frac{d_k t^k}{k!},$$

which also has no constant term. The second solution y_2 thus has the form

$$y_2(t) = \log(t)t^r \sum_{k=1}^{\infty} \frac{Y_k t^k}{k!} + t^r \sum_{k=1}^{\infty} \frac{d_k t^k}{k!}.$$

Case II: Δ *not equal to an integer*

We show in this case that $y_2(t)$ can be found in the form of a power series. Indeed, (21.49) becomes

$$w'(t) = \frac{1}{y_1^2(t)t^{p_0}}\left(1 + \sum_{k=1}^{\infty} a_k t^k\right) = t^{-(2r_1 + p_0)}\left(1 + \sum_{k=1}^{\infty} b_k t^k\right). \qquad (21.53)$$

From the quadratic formula we see that

$$r_1 = \frac{1 - p_0}{2} + \frac{\Delta}{2},$$

so that $2r_1 + p_0 = 1 + \Delta$ is not an integer. In particular, when we integrate both sides of (21.53), we never encounter the power -1. Hence the integral of w' contains only powers and no logarithmic term. When we integrate (21.53) we obtain a new power series

$$w(t) = \frac{t^{1-2r_1-p_0}}{1 - 2r_1 - p_0} \left(1 + \sum_{k=1}^{\infty} c_k t^k\right). \tag{21.54}$$

We multiply (21.54) by $y_1(t)$ and obtain the second solution in the form

$$y_2(t) = \frac{t^{1-r_1-p_0}}{1 - 2r_1 - p_0} \left(1 + \sum_{k=1}^{\infty} d_k t^k\right) = -\frac{t^{r_2}}{\Delta} \left(1 + \sum_{k=1}^{\infty} d_k t^k\right).$$

Case III: Δ equal to a nonzero integer

This case is indeterminate with respect to the appearance of a logarithmic term, as we now demonstrate. From (21.49) we have

$$w'(t) = t^{-1-\Delta} \left(1 + \sum_{k=1}^{\infty} b_k t^k\right). \tag{21.55}$$

Since Δ is an integer, we see that the right-hand side of (21.55) contains only integral powers of t. If the coefficient of t^{-1} is nonzero (and only then), there will be a logarithmic term appearing in the solution. This condition cannot be expressed simply in terms of the coefficients, in general. Both types of behavior are possible in this case, as we shall show.

The above discussion displays the *form* of the second solution in various cases but does not give a clear algorithm for computing the power series coefficients. In order to do this, we first note that in Case II the second solution is a power series of the same type as the Frobenius solution, with r_1 replaced by r_2. In this case the expansion coefficients can be obtained by the same method as in the original Frobenius method for the first solution. The same applies in Case III when we have a pure power series with no logarithmic term. In what follows, we illustrate the computation of the expansion coefficients in Case I, when we have the logarithmic term.

Since we have already illustrated the reduction of the general case to the case $r = 0$ in equation (21.26), we assume without loss of generality that $r = 0$ is the desired double root. In terms of the coefficients, we have $q_0 = 0$ and $p_0 = 1$. The second solution is written as

$$y_2(t) = \log(t) \sum_{k=0}^{\infty} \frac{Y_k t^k}{k!} + \sum_{k=1}^{\infty} \frac{d_k t^k}{k!}.$$

The differential equation $L[y_2] = 0$ reduces to

$$L\left(\sum_{k=1}^{\infty} \frac{d_k t^k}{k!}\right) = -L\left(\log(t) \sum_{k=0}^{\infty} \frac{Y_k t^k}{k!}\right). \tag{21.56}$$

The right-hand side of (21.56) consists of three terms, involving the function and its first and second derivatives. The function contributes

$$\widehat{q}(t) \log(t) \sum_{k=0}^{\infty} \frac{Y_k t^k}{k!}. \tag{21.57}$$

The first derivatives (multiplied by t) contribute the amount

$$t\,\widehat{p}(t) \log(t) \sum_{k=1}^{\infty} \frac{Y_k t^{k-1}}{(k-1)!} + \widehat{p}(t) \sum_{k=0}^{\infty} \frac{Y_k t^k}{k!}, \tag{21.58}$$

while the second derivatives (multiplied by t^2) contribute the amount

$$t^2 \log(t) \sum_{k=2}^{\infty} \frac{Y_k t^{k-2}}{(k-2)!} + 2 \sum_{k=1}^{\infty} \frac{Y_k t^{k-1}}{(k-1)!} - \sum_{k=0}^{\infty} \frac{Y_k t^k}{k!}. \tag{21.59}$$

When we add (21.57)–(21.59), the logarithmic terms cancel and the other terms in the second derivative simplify. The resulting equation is written

$$L\left(\sum_{k=1}^{\infty} \frac{d_k t^k}{k!} \right) = Y_0 - \sum_{k=1}^{\infty} (2k-1) \frac{Y_k t^k}{k!} - \widehat{p}(t) \sum_{k=0}^{\infty} \frac{Y_k t^k}{k!}. \tag{21.60}$$

The coefficients $d_1, d_2, \ldots$ in (21.56) are obtained by successive differentiation, just as for the Frobenius solution. For completeness we define $d_0 = 0$. In detail, we have

$$p_0 d_1 = -Y_1 - p_1 Y_0 - p_0 Y_1 = -2Y_1 - p_1 Y_0,$$

and for $n \geq 2$,

$$n^2 d_n + \sum_{k=2}^{n} \binom{n}{k}(n-k+1)\, p_{k-1} d_{n-k+1} + \sum_{k=1}^{n-1} \binom{n}{k} q_k d_{n-k} = -(2n-1)Y_n - \sum_{k=0}^{n} \binom{n}{k} p_k Y_{n-k}.$$

We illustrate with the Bessel equation of order zero.

Example 21.12. *Find a solution of the Bessel equation*

$$y'' + \frac{y'}{t} + y = 0$$

that is not a multiple of the Bessel function $J_0(t)$.

Solution. In this case we have

$$\widehat{p}(t) = 1, \qquad \widehat{q}(t) = t^2, \qquad Y_{2k-1} = 0, \qquad \text{and} \qquad Y_{2k} = \frac{(-1)^k}{2^{2k}} \binom{2k}{k}.$$

The recurrence relation for d_n is

$$n^2 d_n + n(n-1)d_{n-2} = -2n Y_n.$$

Taking into account that $Y_1 = 0$, we have $d_1 = 0$ and $d_{2k-1} = 0$ for all k. If $n = 2k$, we can cancel a factor of n and write

$$2k\, d_{2k} + (2k-1)\, d_{2k-2} = -2 Y_{2k} \tag{21.61}$$

for $k = 1, 2, \ldots$. From Example 21.11 with $m = 0$, the recurrence relation for Y_{2k} is

$$2k\, Y_{2k} + (2k-1)\, Y_{2k-2} = 0. \tag{21.62}$$

From (21.61) and (21.62) we have

$$\frac{d_{2k}}{Y_{2k}} - \frac{d_{2k-2}}{Y_{2k-2}} = -\frac{1}{k} \tag{21.63}$$

for $k = 1, 2, \ldots$. But $d_0 = 0$, so from (21.63) we get the telescoping sum

$$\frac{d_{2k}}{Y_{2k}} = \sum_{j=1}^{k} \left(\frac{d_{2j}}{Y_{2j}} - \frac{d_{2j-2}}{Y_{2j-2}} \right) = -\sum_{j=1}^{k} \frac{1}{j},$$

which is the partial sum of the *harmonic series*. Combining this with the above discussion of Case I, we have the second solution of the Bessel equation of order zero:

$$Y(t) = \log(t) \sum_{k=0}^{\infty} \frac{Y_k t^k}{k!} + \sum_{k=1}^{\infty} \frac{d_k t^k}{k!}$$

$$= \log(t) \sum_{k=0}^{\infty} \frac{(-1)^k t^{2k}}{2^{2k}(k!)^2} - \sum_{k=1}^{\infty} \frac{(-1)^k (1 + 1/2 + \cdots + 1/k) t^{2k}}{2^{2k}(k!)^2}.$$

Exercises for Section 21.5

1. Use the method of reduction of order to obtain a second linearly independent solution to each of the differential equations in Exercises 1–14 of Section 21.3.

2. Suppose that the roots of the indicial equation are $r = 0$ and $r = -1$. Show that the second linearly independent solution of (21.23) contains a logarithmic term if and only if the coefficients in the equation have the property that $q_1 - p_1 \neq 0$.

3. Apply the result of Exercise 2 to the equation $y'' + (2/t)y' + y = 0$. Find the power series solution for the first solution $y_1(t)$, and then show that the second solution $y_2(t)$ is also expressed in terms of a power series, with no logarithmic term.

4. Show that the power series solution for $y_1(t)$ in Exercise 2 can be expressed as an elementary trigonometric function. Use this, together with the method of reduction of order, to express $y_2(t)$ as an elementary trigonometric function.

5. Show how to modify the theory of the present section to find a particular solution of the inhomogeneous differential equation

$$t^2 y'' + t\, \widehat{p}(t) y' + \widehat{q}(t) y = t^r \widehat{r}(t),$$

where $\widehat{p}(t), \widehat{q}(t)$, and $\widehat{r}(t)$ have convergent power series about $t = 0$, and r is larger than any root of the corresponding indicial equation.

6. Suppose that we have an equation with a regular singular point at $t = 0$ written in the standard form $t^2 y'' + t\, \widehat{p}(t) y' + \widehat{q}(t) y = 0$. Suppose further that there are two solutions $y_1(t) = t^{r_1}$ and $y_2(t) = t^{r_2}$ with $r_1 \neq r_2$. Show that $\widehat{p}(t)$ and $\widehat{q}(t)$ are constants, that is, the equation is a Cauchy-Euler equation.

21.6   Bessel Functions II

At the end of Section 21.5 we used the method of reduction of order to find a solution to the Bessel equation of order 0. The calculations were complicated; in the present section we outline a simpler approach.

Definition. *First, suppose that v is not an integer. The* **Bessel function of the second kind of order** v *is the function $Y_v(t)$ given by*

$$Y_v(t) = \frac{\cos(v\pi) J_v(t) - J_{-v}(t)}{\sin(v\pi)}.$$

If m is an integer, we define

$$Y_m(t) = \lim_{v \to m} \frac{\cos(v\pi) J_v(t) - J_{-v}(t)}{\sin(v\pi)}. \tag{21.64}$$

In the case of an integer m, there is an alternate formula for $Y_m(t)$.

Lemma 21.11. *If m is an integer, then*

$$Y_m(t) = \frac{1}{\pi} \left(\frac{\partial J_v}{\partial v} - (-1)^v \frac{\partial J_{-v}}{\partial v} \right) \Bigg|_{v=m}. \tag{21.65}$$

Proof. Equation (21.65) is a consequence of applying l'Hôpital's rule to (21.64). ∎

Lemma 21.12. *$Y_\nu(t)$ is a solution of the Bessel equation*

$$t^2 y'' + t\, y' + (t^2 - \nu^2)y = 0 \tag{21.66}$$

for all ν.

Proof. If ν is not an integer, then $Y_\nu(t)$ is a solution of (21.66) because it is a constant linear combination of solutions. Thus

$$t^2 Y_\nu'' + t\, Y_\nu' + (t^2 - \nu^2)Y_\nu = 0. \tag{21.67}$$

When we take the limit of both sides of (21.67) as $\nu \longrightarrow m$ where m is an integer, we get

$$t^2 Y_m'' + t\, Y_m' + (t^2 - m^2)Y_m = 0,$$

showing that Y_m is also a solution of the Bessel equation of order m. ∎

The series expansion of an integral order Bessel function of the second kind is quite complicated. For a proof of the following theorem see [Watson, page 62].[4]

Theorem 21.13. *If m is an integer, then*

$$Y_m(t) = \frac{2}{\pi}\left(\log\left(\frac{t}{2}\right) + \gamma\right) J_m(t) - \frac{1}{\pi} \sum_{k=0}^{m-1} \frac{(m-k-1)!}{k!}\left(\frac{t}{2}\right)^{2k-m}$$

$$- \frac{1}{\pi} \sum_{k=0}^{\infty} \frac{(-1)^k (h_k + h_{m+k})}{k!(m+k)!}\left(\frac{t}{2}\right)^{2k+m}, \tag{21.68}$$

where $h_0 = 0$,

$$h_m = 1 + \frac{1}{2} + \cdots + \frac{1}{m}, \qquad \text{and} \qquad \gamma = \lim_{m\to\infty}\left(h_m - \log(m)\right) \approx 0.5772156649\ldots.$$

Here γ is called **Euler's constant**; it is denoted in *Mathematica* by **EulerGamma**.

We collect the information we have obtained about the general solution to the Bessel equation.

4

George Neville Watson (1886–1965). English mathematician. His clearly written book
A Treatise on the Theory of Bessel Functions [Watson] is the standard reference for
Bessel functions.

Theorem 21.14. *Let λ be a constant. The general solution to the Bessel equation*

$$t^2 y'' + t\, y' + (t^2 - \lambda^2) y = 0$$

is given by

$$y(t) = C_1 J_\lambda(t) + C_2 Y_\lambda(t),$$

where C_1 and C_2 are constants. If λ is not an integer, the general solution can also be written as

$$y(t) = C_3 J_\lambda(t) + C_4 J_{-\lambda}(t),$$

where C_3 and C_4 are constants.

Exercises for Section 21.6

1. The **modified Bessel equation** is

$$t^2 y'' + t\, y' - (t^2 + \lambda^2) y = 0, \qquad \text{where } \lambda \text{ is a constant.} \qquad (21.69)$$

   The **modified Bessel function of the first kind of order** λ is the function $I_\lambda(t)$ given by

$$I_\lambda(t) = \sum_{n=0}^{\infty} \frac{t^{2n+\lambda}}{2^{2n+\lambda} n! (\lambda + n)!}. \qquad (21.70)$$

   Show that $I_\lambda(t)$ is a solution of (21.69) and that $I_\lambda(t) = i^{-\lambda} J_\lambda(i\, t)$.

2. Suppose that ν is not an integer. The **modified Bessel function of the second kind of order** ν is the function $K_\nu(t)$ given by

$$K_\nu(t) = \left(\frac{\pi}{2} \right) \frac{I_{-\nu}(t) - I_\nu(t)}{\sin(\nu\, \pi)}.$$

   If m is an integer, we define

$$K_m(t) = \left(\frac{\pi}{2} \right) \lim_{\nu \to m} \frac{I_{-\nu}(t) - I_\nu(t)}{\sin(\nu\, \pi)}.$$

   Show that if m is an integer, then $I_{-m}(t) = I_m(t)$ and $K_{-m}(t) = K_m(t)$.

3. Prove the following result.

Theorem 21.15. *Let λ be a constant. The general solution to the modified Bessel equation*

$$t^2 y'' + t\, y' - (t^2 + \lambda^2) y = 0$$

is given by

$$y(t) = C_1 I_\lambda(t) + C_2 K_\lambda(t),$$

where C_1 and C_2 are constants. If λ is not an integer, the general solution can also be written as

$$y(t) = C_3 I_\lambda(t) + C_4 I_{-\lambda}(t),$$

where C_3 and C_4 are constants.

4. Show that $I_{1/2}(t) = \left(\dfrac{2}{\pi\, t}\right)^{1/2} \sinh(t)$ and $J_{-1/2}(t) = \left(\dfrac{2}{\pi\, t}\right)^{1/2} \cosh(t).$

Find the general solutions of the following differential equations:

5. $t^2 y'' + t\, y' + \left(t^2 - 25\right) y = 0$

6. $t^2 y'' + t\, y' + \left(t^2 + 25\right) y = 0$

7. $t^2 y'' + t\, y' - \left(t^2 + 25\right) y = 0$

8. $t^2 y'' + t\, y' - \left(t^2 - 25\right) y = 0$

9. $t^2 y'' + t\, y' + \left(4t^2 - 25\right) y = 0$

10. $t^2 y'' + t\, y' + \left(4t^2 + 25\right) y = 0$

11. $t^2 y'' + t\, y' - \left(4t^2 + 25\right) y = 0$

12. $t^2 y'' + t\, y' - \left(4t^2 - 25\right) y = 0$

21.7   Bessel Functions via *Mathematica*

The *Mathematica* notation for the Bessel function $J_m(x)$ is **BesselJ[m,x]**. The first few terms in the power series of $J_0(x)$ can be found with

Series[BesselJ[0,x],{x,0,6}]

```
      2     4      6
     x     x      x           8
 1 - -- +  -- -  ----  + O[x]
     4     64    2304
```

Here is the plot of $J_0(x)$, obtained with the command

```
Plot[BesselJ[0,x],{x,-20,20},PlotPoints->200];
```

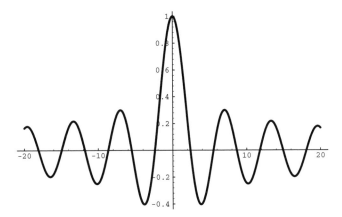

Similarly, for $J_1(x)$ we have

```
Series[BesselJ[1,x],{x,0,6}]
```

$$\frac{x}{2} - \frac{x^3}{16} + \frac{x^5}{384} + O[x]^7$$

and the plot

```
Plot[BesselJ[1,x],{x,-20,20},PlotPoints->200]
```

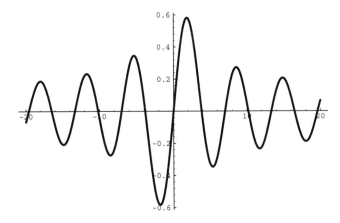

Finally, for $J_2(x)$ we have

```
Series[BesselJ[2,x],{x,0,6}]
```

$$\frac{x^2}{8} - \frac{x^4}{96} + \frac{x^6}{3072} + O[x]^8$$

```
Plot[BesselJ[2,x],{x,-20,20},PlotPoints->200]
```

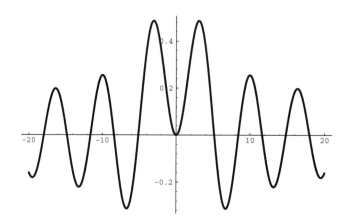

Solving a Bessel Equation with **ODE**

A Bessel equation can be solved with **ODE** using either of the options
Method->TableLookup or **Method->DSolve**.

Example 21.13. *Find the general solution of* $t^2 y'' + t y' + (3t^2 - 2)y = 0$.

Solution. The command

```
ODE[t^2 y'' + t y' + (3t^2 - 2)y == 0,y,t,
Method->TableLookup]
```

gives

```
{{y -> BesselJ[Sqrt[2], Sqrt[3] t] C[1]
      + BesselJ[-Sqrt[2], Sqrt[3] t] C[2]}}
```

while

```
ODE[t^2 y'' + t y' + (3t^2 - 2)y == 0,y,t,Method->DSolve]
```

gives

```
{{y -> BesselY[Sqrt[2], Sqrt[3] t] C[1]
     + BesselJ[Sqrt[2], Sqrt[3] t] C[2]}}
```

These two solutions are the same, since $\sqrt{2}$ is not an integer. Thus the general solution $t^2 y'' + ty' + (3t^2 - 2)y = 0$ is

$$y(t) = C_1 J_{\sqrt{2}}\left(t\sqrt{3}\right) + C_2 J_{-\sqrt{2}}\left(t\sqrt{3}\right),$$

where C_1 and C_2 are constants. ∎

Next, we do an initial value problem.

Example 21.14. *Solve and plot the solution to the initial value problem*

$$\begin{cases} t^2 y'' + t\, y' + (t^2 - 4)y = 0, \\ y(1) = 0, \, y'(1) = 1. \end{cases} \tag{21.71}$$

Solution. We use either of the commands

```
ODE[{t^2 y'' +t y' + (t^2 - 4)y == 0,y[1] == 0,y'[1] == 1},
y,t,Method->TableLookup,PostSolution->{N[#,3]&},
PlotSolution->{{t,0.001,20}}]
```

or

```
ODE[{t^2 y'' + t y' + (t^2 - 4)y == 0,y[1] == 0,y'[1] == 1},
y,t,Method->DSolve,PostSolution->{N[#,3]&},
PlotSolution->{{t,0.001,20}}]
```

to get the output

```
{{y -> 2.59 BesselJ[2., t] + 0.18 BesselY[2., t]}}
```

and the plot

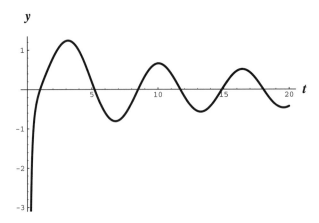

$$\textbf{Solution of}\quad t^2 y'' + t y' + (t^2 - 4)y = 0$$
$$\textbf{with the initial conditions}\quad y(1) = 0\quad \textbf{and}\quad y'(1) = 1.$$

Exercises for Section 21.7

Solve the following differential equations using **ODE**.

1. $t^2 y'' + t y' + (t^2 - 49)y = 0$ **2.** $t^2 y'' + t y' + (16t^2 - 2)y = 0$

21.8   An Aging Spring

Let us consider the problem of modeling a mass-spring system in which Hooke's "constant" is no longer assumed to be constant. For example, as the spring ages, the spring "constant" can be expected to decrease in value. One such model for an aging mass-spring system is

$$m u'' + c u' + k e^{-\eta t} u = F(t), \tag{21.72}$$

where m is the mass, $F(t)$ is a forcing function, and c, k, and η are constants with k and η positive and c nonnegative. Just as in (11.6) we interpret c as determining the amount of friction. The constant k measures the strength of the spring in a similar fashion to Hooke's constant in (11.6). The constant η measures the aging of the spring. As time goes on, $k\,e^{-\eta t}$ becomes less and less; hence we expect that the spring's ability to resist stretching becomes less and less.

Although (21.72) bears little superficial resemblance to the Bessel equation (20.4), it turns out that a transformation of variables will permit us to use Bessel functions to solve the homogeneous differential equation corresponding to (21.72), namely

$$m\,u'' + c\,u' + k\,e^{-\eta t} u = 0. \tag{21.73}$$

To this end, let $s = \alpha e^{\beta t}$, where α and β are constants that we shall choose shortly. We have

$$\frac{ds}{dt} = \beta\,s \quad \text{and} \quad \frac{d^2 s}{dt^2} = \beta^2 s;$$

then the chain rule implies that

$$\frac{du}{dt} = \frac{du}{ds}\frac{ds}{dt} = \beta\,s\frac{du}{ds}$$

and

$$\frac{d^2 u}{dt^2} = \frac{d}{ds}\left(\frac{du}{ds}\frac{ds}{dt}\right) = \frac{d^2 u}{ds^2}\left(\frac{ds}{dt}\right)^2 + \left(\frac{du}{ds}\right)\frac{d^2 s}{dt^2}$$

$$= \frac{d^2 u}{ds^2}\beta^2 s^2 + \frac{du}{ds}\beta^2 s.$$

Also,

$$e^{-\eta t} = (e^{\beta t})^{-\eta/\beta} = \left(\frac{s}{\alpha}\right)^{-\eta/\beta}.$$

Using these calculations, we transform (21.73) to

$$0 = m\left(\frac{d^2 u}{ds^2}\beta^2 s^2 + \frac{du}{ds}\beta^2 s\right) + c\,\beta\,s\frac{du}{ds} + k\left(\frac{s}{\alpha}\right)^{-\eta/\beta} u$$

$$= m\,\beta^2 s^2\frac{d^2 u}{ds^2} + (m\,\beta^2 + c\,\beta)s\frac{du}{ds} + k\left(\frac{s}{\alpha}\right)^{-\eta/\beta} u,$$

or

$$s^2\frac{d^2 u}{ds^2} + \left(1 + \frac{c}{m\,\beta}\right)s\frac{du}{ds} + \frac{k}{m\,\beta^2}\left(\frac{s}{\alpha}\right)^{-\eta/\beta} u = 0. \tag{21.74}$$

We choose α and β such that

$$-\frac{\eta}{\beta} = 2 \quad \text{and} \quad \frac{k}{m\,\beta^2\alpha^{-\eta/\beta}} = 1. \tag{21.75}$$

Then (21.74) becomes

$$s^2\frac{d^2 u}{ds^2} + \left(1 - \frac{2c}{m\,\eta}\right)s\frac{du}{ds} + s^2 u = 0. \tag{21.76}$$

We can solve (21.76) using

```
ODE[s^2 u'' + (1 - 2c/(m eta))s u' + s^2 u == 0,u,s,
Method->DSolve,PostSolution->{PowerExpand}]
```

obtaining the output

```
          c/(eta m)               c                        c
{{u -> s             (BesselY[-------, s] C[1] + BesselJ[-------, s] C[2])}}
                              eta m                     eta m
```

Thus, the general solution to (21.75) is

$$u = s^{c/(m\eta)} \left(C_1 J_{c/(m\eta)}(s) + C_2 Y_{c/(m\eta)}(s) \right), \tag{21.77}$$

where C_1 and C_2 are constants. It follows from (21.75) that $\alpha^2 = k/(m\beta^2)$. We can choose α to be positive; then

$$\alpha = \frac{1}{|\beta|}\sqrt{\frac{k}{m}} = \frac{2}{\eta}\sqrt{\frac{k}{m}}. \tag{21.78}$$

We have

$$s = \alpha\, e^{\beta t} = \frac{2}{\eta}\sqrt{\frac{k}{m}}\, e^{-\eta t/2} \quad \text{and} \quad s^{c/(m\eta)} = \left(\frac{2}{\eta}\sqrt{\frac{k}{m}}\right)^{c/(m\eta)} e^{-ct/(2m)}. \tag{21.79}$$

From (21.77)–(21.79) we find that the general solution to (21.73) is

$$
u(t) = \left(\frac{2}{\eta}\sqrt{\frac{k}{m}}\right)^{c/(m\eta)} e^{-ct/(2m)} \left(Q_1 J_{c/(m\eta)}\left(\frac{2}{\eta}\sqrt{\frac{k}{m}}\, e^{-\eta t/2}\right)\right.
$$

$$
\left. + \; Q_2 Y_{c/(m\eta)}\left(\frac{2}{\eta}\sqrt{\frac{k}{m}}\, e^{-\eta t/2}\right)\right), \tag{21.80}
$$

where Q_1 and Q_2 are constants.

There are two distinct motions that can occur for an aging spring: the motion can remain bounded, or it can become unbounded. Boundedness is characterized as follows:

Lemma 21.16.   *The following conditions are equivalent for the function $u(t)$ defined by* (21.80).

(i) *$u(t)$ is unbounded as $t \longrightarrow \infty$.*

(ii) *$c = 0$ and $Q_2 \neq 0$.*

For a proof of this lemma see Exercise 1.

A *Mathematica* version of (21.80) is given by

```
AgingSpring[Q1_,Q2_,m_,k_,c_,eta_][t_]:=
((2/eta)Sqrt[(k/m)])^(c/(m eta))Exp[-c t/(2m)](
Q1 BesselJ[c/(m eta),(2/eta)Sqrt[(k/m)]Exp[-eta t/2]] +
Q2 BesselY[c/(m eta),(2/eta)Sqrt[(k/m)]Exp[-eta t/2]])
```

A frictionless aging spring with bounded behavior

Lemma 21.16 implies that an aging spring for which $c = Q_2 = 0$ must have bounded motion. Moreover, in this case

$$\lim_{t \to \infty} u(t) = Q_1 J_0(0) = Q_1.$$

Thus, the end of the spring asymptotically approaches $Q_1 J_0(0) = Q_1$.

Example 21.15. *Plot the motion of an aging spring in the case that $Q_1 = m = k = 1$, $Q_2 = c = 0$, and $\eta = 0.15$.*

Solution. We use

```
Plot[Evaluate[AgingSpring[1,0,1,1,0,0.15][t]],{t,0,60},
PlotRange->{Automatic,{-0.45,1.1}},
Epilog->{AbsoluteDashing[{3,3}],Line[{{0,1},{60,1}}]}]
```

to get the plot

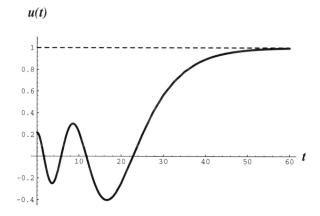

A frictionless aging spring with bounded motion

A frictionless aging spring with unbounded behavior

Lemma 21.16 implies that the motion of an aging spring is unbounded in the case that $c = 0$ and $Q_2 \neq 0$. What actually happens, of course, is that the spring stretches so much as to break.

Example 21.16. *Plot the motion of an aging spring in the case that $m = k = 1$, $Q_2 = -1$, $Q_1 = c = 0$, and $\eta = 0.15$.*

Solution. We use

```
Plot[Evaluate[AgingSpring[0,-1,1,1,0,0.03][t]],
{t,0,500},PlotPoints->100,PlotRange->All]
```

to get the plot

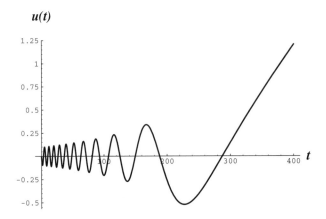

A frictionless aging spring with unbounded motion

Exercises for Section 21.8

1. Prove Lemma 21.16.

2. Plot the solution to the following initial value problem, which models an aging spring with friction:
$$\begin{cases} u'' + 0.08u' + e^{-0.15t}u = 0, \\ u(0) = 1, u'(0) = 0 \end{cases}$$

21.9   The Hypergeometric Equation

The second-order linear differential equation
$$t(1-t)y'' + \big(c - (a+b+1)t\big)y' - ab\,y = 0, \tag{21.81}$$

where a, b, and c are fixed parameters, is called the **hypergeometric equation**. This equation has regular singular points at $t_0 = 0$ and $t_0 = 1$. Theorem 21.8 implies that the

Frobenius solution for $t_0 = 0$ converges at least in the interval $|t| < 1$. Comparing (21.81) with (21.23), we see that $p_0 = c$ and $q_0 = 0$. Thus the indicial equation of (21.81) is

$$0 = r(r-1) + cr = r(r-1+c),$$

whose roots are 0 and $1-c$. We determine a Frobenius solution in the form

$$y(t) = \sum_{n=0}^{\infty} a_n t^{n+r}. \tag{21.82}$$

Substitution of (21.82) into (21.81) yields

$$0 = (t - t^2) \sum_{n=0}^{\infty} a_n (n+r)(n+r-1) t^{n+r-2}$$

$$+ \left(c - (a+b+1)t\right) \sum_{n=0}^{\infty} a_n (n+r) t^{n+r-1} - ab \sum_{n=0}^{\infty} a_n t^{n+r}. \tag{21.83}$$

We rewrite (21.83), collecting like powers:

$$0 = \sum_{n=0}^{\infty} a_n (n+r)(n+r+c-1) t^{n+r-1} - \sum_{n=0}^{\infty} a_n \big((n+r)(n+r+a+b)+ab\big) t^{n+r}. \tag{21.84}$$

Then an index shift converts (21.84) to

$$0 = a_0 r(r+c-1) t^{r-1} + \sum_{n=0}^{\infty} \Big(a_{n+1}(n+r+1)(n+r+c)$$

$$- a_n \big((n+r+a+b)(n+r)+ab\big)\Big) t^{n+r}. \tag{21.85}$$

Since (21.85) holds for all values of t, the coefficient of each power of t must vanish separately. This leads to the relations

$$a_0 r(r+c-1) = 0 \tag{21.86}$$

and

$$a_{n+1} = \frac{(n+r)(n+r+a+b)+ab}{(n+r+1)(n+r+c)} a_n. \tag{21.87}$$

From (21.87) we see that $a_0 = 0$ implies $a_n = 0$ for all n, and $y(t) = 0$ is the only solution. Therefore, to obtain a nontrivial solution, we must have $a_0 \neq 0$. Thus, (21.86) reduces to

$$r(r+c-1) = 0,$$

which is just the indicial equation of (21.81). The two roots of this indicial equation are $r_1 = 0$ and $r_2 = 1 - c$. We examine the case of each root separately. To avoid technical problems, we assume that c is not a negative integer.

The solution corresponding to $r_1 = 0$

We set $r = 0$ in (21.87) and rearrange the terms in the numerator, obtaining

$$a_{n+1} = \frac{(a+n)(b+n)}{(n+1)(c+n)} a_n. \tag{21.88}$$

Thus

$$a_1 = \frac{a\,b}{c} a_0,$$

$$a_2 = \frac{(a+1)(b+1)}{2(c+1)} a_1 = \frac{a(a+1)b(b+1)}{1 \times 2c(c+1)} a_0,$$

$$a_3 = \frac{(a+2)(b+2)}{3(c+2)} a_2 = \frac{a(a+1)(a+2)b(b+1)(b+2)}{1 \times 2 \times 3c(c+1)(c+2)} a_0,$$

and so forth. In general,

$$a_n = \frac{a(a+1)(a+2)\cdots(a+n-1)b(b+1)(b+2)\cdots(b+n-1)}{n!c(c+1)(c+2)\cdots(c+n-1)} a_0. \tag{21.89}$$

Thus, we obtain

$$y_1(t) = a_0 \left(1 + \sum_{n=1}^{\infty} \frac{a(a+1)(a+2)\cdots(a+n-1)b(b+1)(b+2)\cdots(b+n-1)t^n}{n!c(c+1)(c+2)\cdots(c+n-1)} \right)$$

as a solution to (21.81). (The assumption that c is not a negative integer or 0 prevents any factor in the denominator in (21.89) from vanishing.)

It is convenient to introduce the following notion:

Definition.   *The* **Pochhammer symbol** $(a)_n$ *is given by*

$$(a)_0 = 1 \qquad and \qquad (a)_n = a(a+1)(a+2)\cdots(a+n-1)$$

for $n = 1, 2, \ldots$.

Note that

$$(a)_n = \frac{(a+n-1)!}{(a-1)!} = \frac{\Gamma(a+n)}{\Gamma(a)}.$$

Using the Pochhammer symbol, we can rewrite $y_1(t)$ as

$$y_1(t) = a_0 F(a; b; c; t),$$

where $F(a; b; c; t)$ is given as follows.

Definition.   *The* **hypergeometric function** $F(a; b; c; t)$ *is the function defined by*

$$F(a; b; c; t) = \sum_{n=0}^{\infty} \frac{(a)_n (b)_n t^n}{n!(c)_n}.$$

Thus, $F(a; b; c; t)$ is a solution to (21.81) at the regular singular point $t_0 = 0$ corresponding to the root $r_1 = 0$ of the indicial equation. The name "hypergeometric" comes from the fact that $F(a; b; c; t)$ generalizes a geometric series; in fact, for any b and any $t < 1$ it is easy to check that

$$F(1; b; b; t) = \frac{1}{1 - t} = 1 + t + t^2 + t^3 + \cdots,$$

which is a geometric series. More generally,

$$F(k; b; b; t) = \frac{1}{(1 - t)^k}.$$

Hypergeometric functions were first discussed by Gauss in 1812. Many well-known functions can be written in terms of the hypergeometric function. See the exercises.

The solution corresponding to $r_2 = 1 - c$

We set $r = 1 - c$ in (21.87) and obtain

$$a_{n+1} = \frac{(n + 1 - c)(n + 1 - c + a + b) + a\,b}{(n + 1)(n + 2 - c)} a_n. \tag{21.90}$$

Let $c = 2 - \hat{c}$, $a = \hat{a} + c - 1$, and $b = \hat{b} + c - 1$; since

$$(n + 1 - c)(n + 1 - c + a + b) + a\,b = (n + 1 - c + a)(n + 1 - c + b) = (n + \hat{a})(n + \hat{b}),$$

we can rewrite (21.90) as

$$a_{n+1} = \frac{(n + \hat{a})(n + \hat{b})}{(n + 1)(n + \hat{c})} a_n. \tag{21.91}$$

Equation (21.91) has exactly the same form as (21.88), which determined the coefficients a_n for $r_1 = 0$. Thus corresponding to $r_2 = 1 - c$ we obtain the solution

$$
\begin{aligned}
y_2(t) &= t^{1-c} a_0 \left(1 + \sum_{n=1}^{\infty} \frac{\hat{a}(\hat{a}+1)(\hat{a}+2)\cdots(\hat{a}+n-1)\hat{b}(\hat{b}+1)(\hat{b}+2)\cdots(\hat{b}+n-1)t^n}{n!\hat{c}(\hat{c}+1)(\hat{c}+2)\cdots(\hat{c}+n-1)} \right) \\
&= t^{1-c} a_0 F(\hat{a}; \hat{b}; \hat{c}; t) \\
&= a_0 t^{1-c} F(1 + a - c; 1 + b - c; 2 - c; t).
\end{aligned}
\tag{21.92}
$$

(The assumption that c is not a negative integer prevents any denominator in (21.92) from vanishing. The assumption that $c \neq 1$ prevents y_1 from coinciding with y_2.)

We have completed most of the proof of the following theorem:

Theorem 21.17. *Suppose that c is not an integer. Then the general solution to (21.81) is*

$$y(t) = C_1 F(a; b; c; t) + C_2 t^{1-c} F(1 + a - c; 1 + b - c; 2 - c; t),$$

where C_1 and C_2 are constants.

Proof. It remains to show that $F(a; b; c; t)$ and $t^{1-c} F(1 + a - c; 1 + b - c; 2 - c; t)$ are linearly independent. But this is clear since $F(a; b; c; t)$ is a series involving only integral powers of t, while $t^{1-c} F(1 + a - c; 1 + b - c; 2 - c; t)$ is a series involving only nonintegral powers of t. ∎

Generalized hypergeometric functions

The hypergeometric function $F(a; b; c; t)$ has two numerator parameters, a and b, and one denominator parameter, c. It is natural to consider a function with p numerator parameters and q denominator parameters. This is the **generalized hypergeometric function** $_pF_q$ defined by

$$_pF_q(a_1, \ldots, a_p; b_1, \ldots, b_q; t) = \sum_{n=0}^{\infty} \frac{(a_1)_n \cdots (a_p)_n t^n}{(b_1)_n \cdots (b_q)_n n!}.$$

With this notation $F(a; b; c; t)$ becomes $_2F_1(a; b; c; t)$. Of particular importance is the **confluent hypergeometric function** $_1F_1$; it is given by

$$_1F_1(a; b; t) = \sum_{n=0}^{\infty} \frac{(a)_n t^n}{(c)_n n!}$$

and is well-defined provided that c is neither 0 nor a negative integer. Sometimes a confluent hypergeometric function is called a **Kummer**[5] function. Confluent hypergeometric functions satisfy the **confluent hypergeometric**, or **Kummer, equation**

$$t\,y'' + (c - t)y' - a\,y = 0. \tag{21.93}$$

Corresponding to (21.17) we have the following result:

---

[5] Ernst Eduard Kummer (1810–1893). Professor in Berlin remembered for his work in hypergeometric functions, number theory, and algebraic geometry.

Theorem 21.18.   *Suppose that c is not an integer. Then the general solution to (21.93) is*

$$y(t) = C_1 \, {}_1F_1(a; c; t) + C_2 t^{1-c} \, {}_1F_1(1 + a - c; 2 - c; t),$$

where C_1 and C_2 are constants.

For the proof of Theorem 21.18 see Exercise 15.

A Bessel function of nonintegral order λ is expressed in terms of a hypergeometric function ${}_0F_1$ by the formula

$$J_\lambda(t) = \frac{(t/2)^\lambda}{\lambda!} \, {}_0F_1\left(1; \lambda; -\frac{t^2}{4}\right).$$

(See [Rain2, page 105].)

Mathematica uses **Hypergeometric2F1[a,b,c,t]** for ${}_2F_1(a; b; c; t)$ and **Hypergeometric1F1[a,b,t]** for ${}_1F_1(a; b; t)$. More generally, the extended command **HypergeometricPFQ[numlist,denlist,z]** gives the generalized hypergeometric function ${}_pF_q$, where **numlist** is a list of the numerator parameters and **denlist** is a list of the denominator parameters.

Exercises for Section 21.9

1.   Show that $F(a; b; c; t) = \displaystyle\sum_{n=0}^{\infty} \frac{\Gamma(a+n)\Gamma(b+n)\Gamma(c)t^n}{\Gamma(a)\Gamma(b)\Gamma(c+n)n!}.$

2.   Show that if a or b is a negative integer, then $F(a; b; c; t)$ is a polynomial.

Establish the following identities both by hand and using *Mathematica*:

3. $\log(1 + t) = t\, F(1; 1; 2; -t)$
5. $\arctan(t) = t\, F\left(\dfrac{1}{2}; 1; \dfrac{3}{2}; \pm t^2\right)$

4. $\arcsin(t) = t\, F\left(\dfrac{1}{2}; \dfrac{1}{2}; \dfrac{3}{2}; t^2\right)$
6. $e^t = \displaystyle\lim_{t \to \infty} F\left(a; b; b; \dfrac{t}{a}\right)$

7. $\displaystyle\int_0^{\pi/2} \frac{d\phi}{\sqrt{1 - k(\sin(\phi))^2}} = F\left(\dfrac{1}{2}; \dfrac{1}{2}, 1, k\right)$

8. $\displaystyle\int_0^{\pi/2} \sqrt{1 - k(\sin(\phi))^2}\, d\phi = F\left(\dfrac{1}{2}; -\dfrac{1}{2}, 1, k\right)$

Find the general solution to each of the following differential equations.

9. $t(1-t)y'' + \left(\dfrac{1}{2} - 2t\right)y' - \dfrac{y}{4} = 0$ **11.** $2t(1-t)y'' + (3-4t)y' + 4y = 0$

10. $t(1-t)y'' + \left(\dfrac{1}{2} - t\right)y' + \dfrac{y}{4} = 0$ **12.** $t(1-t)y'' + (1-3t)y' - y = 0$

Use **ODE** with the option **Method->SecondOrderLinear** to solve each of the following initial value problems:

13. $\begin{cases} y'' = J_0(t), \\ y(0) = y'(0) = 0 \end{cases}$ **14.** $\begin{cases} y'' + y = J_0(t), \\ y(0) = y'(0) = 0 \end{cases}$

15. Prove Theorem 21.18 using the following steps.

    a. Assume that a Frobenius solution to (21.93) is of the form (21.82). Show that $a_0 r(r+c-1) = 0$ and

$$a_{n+1} = \frac{(n+r+a)}{(n+r+1)(n+r+c)} a_n. \tag{21.94}$$

    b. If $r = 0$, show that (21.94) implies that

$$a_n = \frac{(a)_n}{n!(c)_n} a_0. \tag{21.95}$$

Conclude that $_1F_1(a; c; t)$ is a solution of (21.93).

    c. If $r = 1 - c$, show that (21.94) implies that $_1F_1(a+1-c; 2-c; t)$ is a solution of (21.93).

    d. If c is not an integer, show that $_1F_1(a; c; t)$ and $_1F_1(a+1-c; 2-c; t)$ are linearly independent.

    e. Conclude that $y(t) = C_1 {}_1F_1(a; c; t) + C_2 t^{1-c} {}_1F_1(1+a-c; 2-c; t)$ is the general solution of (21.93) in the case that c is not an integer.

APPENDIX A: REVIEW OF LINEAR ALGEBRA AND MATRIX THEORY

This appendix is included as a review, or "short course", in linear algebra and matrix theory.

Vector and matrix notation is introduced Section A.1. Determinants and inverses of matrices are discussed in Section A.2. Section A.3 treats the relations between systems of linear equations and determinants. Section A.4 is devoted to eigenvalues and eigenvectors. In Section A.5 we provide information about the exponential of a matrix. We discuss abstract vector spaces in Section A.6. In Section A.7 we show how to use *Mathematica* to manipulate vectors and matrices. Solving equations with *Mathematica* is discussed in Section A.8. Finally, in Section A.9 we discuss computing eigenvalues and eigenvectors with *Mathematica*.

A.1 Vector and Matrix Notation

An n-dimensional **column vector** is written

$$\mathbf{x} = \begin{pmatrix} x_1 \\ x_2 \\ \vdots \\ x_n \end{pmatrix}, \tag{A.1}$$

where $x_1, \ldots, x_n$ are real or complex numbers. The set of all n-dimensional real column vectors is denoted by $\mathbb{R}^n$. We shall also find it useful to work with complex-valued column vectors, where the x_i are complex numbers. The set of all such complex n-dimensional vectors is denoted by $\mathbb{C}^n$. Unless mentioned otherwise, it will be assumed below that all scalars and vectors may be complex-valued. See Section 9.2.1 for a self-contained review of the principal properties of the complex number system that are used here.

The operations of addition and scalar multiplication of vectors are defined component by component:

$$
a\,\mathbf{x} + b\,\mathbf{y} = a \begin{pmatrix} x_1 \\ \vdots \\ x_n \end{pmatrix} + b \begin{pmatrix} y_1 \\ \vdots \\ y_n \end{pmatrix} = \begin{pmatrix} a\,x_1 + b\,y_1 \\ \vdots \\ a\,x_n + b\,y_n \end{pmatrix},
$$

where a and b are real or complex numbers. The zero vector consists of a single column of zeros:

$$
\mathbf{0} = \begin{pmatrix} 0 \\ \vdots \\ 0 \end{pmatrix}.
$$

The standard basis vectors are denoted by $\mathbf{e}_1, \ldots, \mathbf{e}_n$; explicitly,

$$
\mathbf{e}_1 = \begin{pmatrix} 1 \\ 0 \\ \vdots \\ 0 \end{pmatrix}, \ldots, \mathbf{e}_n = \begin{pmatrix} 0 \\ \vdots \\ 0 \\ 1 \end{pmatrix}.
$$

An $m \times n$ **matrix** is an array of real or complex numbers (a_{ij}), where $1 \leq i \leq m$ and $1 \leq j \leq n$; it is displayed as

$$
A = \begin{pmatrix} a_{11} & a_{12} & \cdots & a_{1n} \\ a_{21} & a_{22} & \cdots & a_{2n} \\ \vdots & \vdots & \ddots & \vdots \\ a_{m1} & a_{m2} & \cdots & a_{mn} \end{pmatrix}.
$$

This matrix A has m rows and n columns. We say that $m \times n$ matrices $A = (a_{ij})$ and $B = (b_{ij})$ are equal if and only if $a_{ij} = b_{ij}$ for $1 \leq i \leq m$ and $1 \leq j \leq n$. An $m \times 1$ matrix is a column vector of the form (A.1). A **square matrix** is a matrix with the same number of rows and columns, that is, $m = n$. In particular, the $n \times n$ **identity matrix** is defined by

$$
I_{n \times n} = \begin{pmatrix} 1 & 0 & \cdots & 0 \\ 0 & 1 & \cdots & 0 \\ \vdots & \vdots & \ddots & \vdots \\ 0 & 0 & \cdots & 1 \end{pmatrix}.
$$

More generally, a **diagonal matrix** is an $n \times n$ matrix of the form

$$
\begin{pmatrix}
d_1 & 0 & \cdots & 0 \\
0 & d_2 & \cdots & 0 \\
\vdots & \vdots & \ddots & \vdots \\
0 & 0 & \cdots & d_n
\end{pmatrix},
$$

and an **upper triangular matrix** is an $n \times n$ matrix $A = (a_{ij})$ for which $a_{ij} = 0$ for $i > j$. Similarly, a **lower triangular matrix** is a matrix for which $a_{ij} = 0$ for $i < j$.

Any $m \times n$ matrix A naturally defines a **linear transformation** L_A from $\mathbb{C}^n$ to $\mathbb{C}^m$ as follows. If

$$
\mathbf{x} = \begin{pmatrix} x_1 \\ x_2 \\ \vdots \\ x_n \end{pmatrix},
$$

then $A\mathbf{x}$ is the vector in $\mathbb{C}^m$ given by

$$
A\mathbf{x} = \begin{pmatrix} (A\mathbf{x})_1 \\ \vdots \\ (A\mathbf{x})_m \end{pmatrix} = \begin{pmatrix} \sum_{j=1}^{n} a_{1j} x_j \\ \vdots \\ \sum_{j=1}^{n} a_{mj} x_j \end{pmatrix}.
$$

We define $L_A \colon \mathbb{C}^n \longrightarrow \mathbb{C}^m$ by

$$
L_A(\mathbf{x}) = A\mathbf{x}.
$$

Example A.1. *Find the linear transformation from $\mathbb{C}^2$ to $\mathbb{C}^3$ determined by the matrix*

$$
A = \begin{pmatrix} 2 & 3 \\ 1 & 4 \\ 0 & 5 \end{pmatrix}.
$$

Solution. Let

$$
\mathbf{x} = \begin{pmatrix} x_1 \\ x_2 \end{pmatrix}.
$$

We compute

$$
L_A(\mathbf{x}) = A\mathbf{x} = A\begin{pmatrix} x_1 \\ x_2 \end{pmatrix} = \begin{pmatrix} 2 & 3 \\ 1 & 4 \\ 0 & 5 \end{pmatrix}\begin{pmatrix} x_1 \\ x_2 \end{pmatrix} = \begin{pmatrix} 2x_1 + 3x_2 \\ x_1 + 4x_2 \\ 5x_2 \end{pmatrix}. \ \blacksquare
$$

The term **linear transformation** refers to the property that for any two n-dimensional vectors $\mathbf{x}$ and $\mathbf{y}$ and any two real or complex numbers a and b we have

$$L_A(a\,\mathbf{x} + b\,\mathbf{y}) = A(a\,\mathbf{x} + b\,\mathbf{y}) = a\,A\mathbf{x} + b\,A\mathbf{y} = a\,L_A(\mathbf{x}) + b\,L_A(\mathbf{y}).$$

The number of components in the vector $\mathbf{x}$ must equal the number of *columns* of the matrix A. If the vector $\mathbf{x}$ has n components, then the vector $A\mathbf{x}$ has m components, one for each row of the matrix A.

The **transpose** of an $m \times n$ matrix A is the $n \times m$ matrix

$$A^T = \begin{pmatrix} a_{11} & a_{21} & \cdots & a_{m1} \\ a_{12} & a_{22} & \cdots & a_{m2} \\ \vdots & \vdots & \ddots & \vdots \\ a_{1n} & a_{2n} & \cdots & a_{mn} \end{pmatrix}.$$

In particular, the transpose of a row vector (that is, a matrix of the form $(x_1, \ldots, x_n)$) is a column vector:

$$(x_1, \ldots, x_n)^T = \begin{pmatrix} x_1 \\ \vdots \\ x_n \end{pmatrix}.$$

This is especially useful in displaying a column vector typographically as the transpose of a row vector, which we shall often do.

The **dot product** of two row vectors $\mathbf{x} = (x_1, \ldots, x_n)$ and $\mathbf{y} = (y_1, \ldots, y_n)$ is defined by

$$\mathbf{x} \cdot \mathbf{y} = \mathbf{x}\,\mathbf{y}^T = \sum_{i=1}^{n} x_i\,y_i.$$

Example A.2. *Find the dot product of the row vectors* $(2, 1, 3)$ *and* $(-1, -3, -4)$.

Solution. We compute

$$(2, 1, 3) \cdot (-1, -3, -4) = 2(-1) + 1(-3) + 3(-4) = -2 - 3 - 12 = -17. \ \blacksquare$$

The dot product is useful in expressing a characteristic property of the transpose operation:

$$(A\mathbf{x}) \cdot \mathbf{y} = \mathbf{x} \cdot (A^T\mathbf{y}).$$

Note that A^T naturally defines a linear transformation L_A^T from $\mathbb{C}^m$ to $\mathbb{C}^n$ by means of the equations

$$L_A^T(\mathbf{x}) = A^T\mathbf{x}.$$

Whereas L_A goes from $\mathbb{C}^n$ to $\mathbb{C}^m$, the transpose L_A^T goes in the opposite direction.

Addition of matrices and multiplication of matrices by scalars are defined component-wise. Thus, if A and B are $m \times n$ matrices whose ij^{th} elements are a_{ij} and b_{ij}, then the ij^{th} element of the matrix $a A + b B$ is by definition

$$a\, a_{ij} + b\, b_{ij}.$$

Here a and b are real or complex numbers. Note that $a A + b B$ is another $m \times n$ matrix.

The product of two matrices is defined in terms of the action of the corresponding linear transformations: If $A = (a_{ij})$ is an $m \times n$ matrix (defining a linear transformation from $\mathbb{C}^n$ to $\mathbb{C}^m$) and $B = (b_{ij})$ is an $n \times p$ matrix (defining a linear transformation from $\mathbb{C}^p$ to $\mathbb{C}^n$), then the product $C = A B = (c_{ij})$ is an $m \times p$ matrix that defines a linear transformation from $\mathbb{C}^p$ to $\mathbb{C}^m$. It is defined explicitly by

$$c_{ij} = \sum_{k=1}^{n} a_{ik} b_{kj},$$

for $1 \le i \le m$ and $1 \le j \le p$.

Matrix addition and multiplication satisfy the **associative** and **distributive** laws:

$$\begin{cases} (A+B)+C = A+(B+C), & A(BC) = (AB)C, \\ A(B+C) = AB+AC, & (B+C)A = BA+CA. \end{cases} \tag{A.2}$$

Matrix addition is **commutative**, that is, $A + B = B + A$, but matrix multiplication is generally *not* commutative. For example, if

$$A = \begin{pmatrix} a & b \\ c & d \end{pmatrix} \quad \text{and} \quad B = \begin{pmatrix} 0 & 1 \\ 1 & 0 \end{pmatrix},$$

then

$$A B = \begin{pmatrix} b & a \\ d & c \end{pmatrix} \quad \text{but} \quad B A = \begin{pmatrix} c & d \\ a & b \end{pmatrix}.$$

On the other hand, the identity matrix $I_{n \times n}$ commutes with *all* $n \times n$ matrices:

$$A I_{n \times n} = A = I_{n \times n} A.$$

A more interesting example of commutativity is that of two $n \times n$ diagonal matrices. If A and B are two such matrices, it is easy to check that $A B = B A$.

A.2 Determinants and Inverses

There is an important operation called the determinant and denoted by det that assigns to each $n \times n$ matrix a number. It has the following properties.

1. The determinant of a 1×1 matrix (a) is given by $\det(a) = a$.

2. The determinant of a 2×2 matrix is given by

$$\det\begin{pmatrix} a_{11} & a_{12} \\ a_{21} & a_{22} \end{pmatrix} = a_{11}a_{22} - a_{12}a_{21}.$$

3. The determinant of a 3×3 matrix is given by

$$\det\begin{pmatrix} a_{11} & a_{12} & a_{13} \\ a_{21} & a_{22} & a_{23} \\ a_{31} & a_{32} & a_{33} \end{pmatrix} = a_{11}a_{22}a_{33} + a_{12}a_{23}a_{31} + a_{21}a_{32}a_{13}$$
$$- a_{21}a_{12}a_{33} - a_{31}a_{13}a_{22} - a_{32}a_{23}a_{11}.$$

4. The determinant of a $n \times n$ diagonal matrix is the product of the diagonal elements:

$$\det\begin{pmatrix} d_1 & 0 & \cdots & 0 \\ 0 & d_2 & \cdots & 0 \\ \vdots & \vdots & \ddots & \vdots \\ 0 & 0 & \cdots & d_n \end{pmatrix} = d_1 d_2 \cdots d_n.$$

5. More generally, the determinant of an upper triangular or lower triangular matrix equals the product of the diagonal elements.

6. A matrix and its transpose have the same determinant: $\det(A^T) = \det(A)$.

7. If A and B are $n \times n$ matrices, then

$$\det(A\,B) = \det(A)\det(B).$$

In order to define the determinant of an $n \times n$ matrix for $n \geq 4$, we need the following notions.

Definition. *Let $A = (a_{ij})$ be an $n \times n$ matrix. The ij^{th} **minor** of A is the matrix M_{ij} obtained from A by omitting the ith row and the jth column. The ij^{th} **cofactor** is*

$$A_{ij} = (-1)^{i+j} \det(M_{ij}).$$

Of course, so far the notion of determinant is defined only for $n \times n$ matrices with $n < 4$. However, we can use cofactors of smaller matrices to define determinants of larger matrices.

Definition. *The determinant of an $n \times n$ matrix $A = (a_{ij})$ is defined inductively by*

$$\det(A) = \begin{cases} a_{11} & \text{if } n = 1, \\ a_{11}A_{11} + a_{12}A_{12} + \cdots + a_{1n}A_{1n} & \text{if } n > 1, \end{cases}$$

where the A_{1j}'s denote the cofactors associated with the entries of the first row of A.

It is not necessary to limit ourselves to the first row for the cofactor expansion. Here is another property of the determinant.

8. The determinant of an $n \times n$ matrix A can be expressed as a cofactor expansion using any row or column:

$$\det(A) = a_{i1}A_{i1} + a_{i2}A_{i2} + \cdots + a_{in}A_{in}$$

$$= a_{1j}A_{1j} + a_{2j}A_{2j} + \cdots + a_{nj}A_{nj}$$

for $1 \leq i, j \leq n$. Furthermore,

$$a_{i1}A_{j1} + a_{i2}A_{j2} + \cdots + a_{in}A_{jn} = \begin{cases} \det(A) & \text{if } i = j \\ 0 & \text{if } i \neq j. \end{cases} \tag{A.3}$$

Two further properties of the determinant are:

9. If A is a square matrix containing a row or column of zeros, then $\det(A) = 0$.

10. If A is a square matrix containing two identical rows or two identical columns, then $\det(A) = 0$.

Definition. *The **inverse** of an $n \times n$ matrix A is a matrix B that satisfies*

$$A B = B A = I_{n \times n}.$$

Not every $n \times n$ matrix has an inverse, but if the inverse exists it is unique. A necessary and sufficient condition that a matrix A have an inverse is that the determinant $\det(A)$ be nonzero. When that is the case, the inverse of A is denoted by A^{-1}.

There are two basic methods to compute the inverse of a matrix A with $\det(A) \neq 0$. To describe the first method, we first make the following definition.

Definition. *The **adjoint** of an $n \times n$ matrix A is a matrix $\mathrm{adj}(A)$ defined by*

$$\mathrm{adj}(A) = \begin{pmatrix} A_{11} & A_{12} & \cdots & A_{1n} \\ A_{21} & A_{22} & \cdots & A_{2n} \\ \vdots & \vdots & \ddots & \vdots \\ A_{n1} & A_{n2} & \cdots & A_{nn} \end{pmatrix}^T.$$

where A_{ij} denotes the ij^{th} cofactor of A.

Lemma A.1. *Let A be a matrix with nonzero determinant. Then the inverse of A is given by the formula*

$$A^{-1} = \frac{1}{\det(A)} \mathrm{adj}(A). \tag{A.4}$$

Proof. From (A.3) it follows that

$$A\big(\mathrm{adj}(A)\big) = \det(A)I_{n\times n}.$$

Hence we get (A.4). ∎

Unfortunately, using (A.4) to compute the inverse of a matrix can require a large number of additions and multiplications. Therefore, we describe a second procedure to compute the inverse of a matrix, called the method of **row reduction**; it requires far fewer additions and multiplications. The method of row reduction consists in performing a sequence of **elementary operations** on a matrix of the following three types:

   (i) Interchange of two rows of the matrix.

   (ii) Multiplication of a row by a nonzero constant.

   (iii) Addition of a multiple of one row to another row.

None of these operations changes the invertibility status of a matrix A. Each corresponds to multiplication by a square matrix of a special form. In detail,

   (i) Interchange of rows i and j corresponds to left multiplication by the matrix E_{ij} that has ones on the diagonal except for the ii^{th} and jj^{th} places and has zeros everywhere else except for the ij^{th} and ji^{th} places, where it has a one.

   (ii) Multiplication of row i by a constant $c \neq 0$ corresponds to left multiplication by the matrix M_i^c that has ones on the diagonal except for the ii^{th} place, where it has the constant c, and zeros everywhere off the diagonal.

   (iii) Addition to the ith row of the jth row multiplied by a constant c corresponds to left multiplication by the matrix E_{ij}^c that has ones everywhere on the diagonal and zero everywhere else, except the ij^{th} place where it has the constant c.

Of course, it is much simpler to use an elementary row operation than to multiply by an elementary matrix. But the two operations are completely equivalent.

Matrices of type (i), (ii) or (iii) are called **elementary matrices**. If A^{-1} exists, then there exist elementary matrices $E_1, \ldots, E_k$ such that

$$E_k E_{k-1} \cdots E_1 A = I_{n\times n}. \tag{A.5}$$

Multiplying both sides of (A.5) on the right by A^{-1}, we obtain

$$E_k E_{k-1} \cdots E_1 = A^{-1}. \tag{A.6}$$

This gives us a method for computing A^{-1}:

(i) Use elementary row operations to convert A to the identity matrix $I_{n \times n}$.

(ii) Each time an elementary row operation is applied to A, apply the same elementary row operation to $I_{n \times n}$.

(iii) After A has been completely row reduced by elementary matrices $E_k, E_{k-1}, \dots, E_1$, the matrix $I_{n \times n}$ will have been row reduced to A^{-1}.

For the sake of organization of our work, we **augment** A by placing it side by side with $I_{n \times n}$. The following example explains the procedure.

Example A.3. *Use the method of row reduction to find the inverse and determinant of the matrix*

$$A = \begin{pmatrix} 1 & 1 & 2 \\ -1 & 0 & 1 \\ 2 & -1 & 5 \end{pmatrix}.$$

Solution. First, we augment A by the identity matrix $I_{3 \times 3}$:

$$A_{\text{aug}} = \begin{pmatrix} 1 & 1 & 2 & 1 & 0 & 0 \\ -1 & 0 & 1 & 0 & 1 & 0 \\ 2 & -1 & 5 & 0 & 0 & 1 \end{pmatrix}.$$

Our goal is to perform those elementary row operations on A_{aug} that will convert it to a 3×6 matrix B whose left half is $I_{3 \times 3}$; then the right half of B will be A^{-1}. To this end we add the first row of A_{aug} to the second row and then add -2 times the first row of A_{aug} to the third row, giving us a new matrix

$$A_1 = \begin{pmatrix} 1 & 1 & 2 & 1 & 0 & 0 \\ 0 & 1 & 3 & 1 & 1 & 0 \\ 0 & -3 & 1 & -2 & 0 & 1 \end{pmatrix}.$$

We next add 3 times the second row of A_1 to the third row, resulting in the matrix

$$A_2 = \begin{pmatrix} 1 & 1 & 2 & 1 & 0 & 0 \\ 0 & 1 & 3 & 1 & 1 & 0 \\ 0 & 0 & 10 & 1 & 3 & 1 \end{pmatrix}.$$

These three operations preserve the determinant, hence $\det A = 10$. Next, we multiply the third row of A_2 by $1/10$ to obtain

$$A_3 = \begin{pmatrix} 1 & 1 & 2 & 1 & 0 & 0 \\ 0 & 1 & 3 & 1 & 1 & 0 \\ 0 & 0 & 1 & \dfrac{1}{10} & \dfrac{3}{10} & \dfrac{1}{10} \end{pmatrix}.$$

We multiply the third row of A_3 by -3 and add to the second row, then multiply the third row of A_3 by -2 and add to the first row. This produces the matrix

$$A_4 = \begin{pmatrix} 1 & 1 & 0 & 8/10 & -6/10 & -2/10 \\ 0 & 1 & 0 & 7/10 & 1/10 & -3/10 \\ 0 & 0 & 1 & 1/10 & 3/10 & 1/10 \end{pmatrix}.$$

We multiply the second row of A_4 by -1 and add to the first row, with the result

$$B = \begin{pmatrix} 1 & 0 & 0 & 1/10 & -7/10 & 1/10 \\ 0 & 1 & 0 & 7/10 & 1/10 & -3/10 \\ 0 & 0 & 1 & 1/10 & 3/10 & 1/10 \end{pmatrix}.$$

The left half of B will be the matrix A^{-1}; thus

$$A^{-1} = \begin{pmatrix} 1/10 & -7/10 & 1/10 \\ 7/10 & 1/10 & -3/10 \\ 1/10 & 3/10 & 1/10 \end{pmatrix}. \ \blacksquare$$

A.3 Systems of Linear Equations and Determinants

Definition. *Column vectors or row vectors* $\mathbf{x}_1, \ldots, \mathbf{x}_k$ *are said to be* **linearly dependent** *if there are constants* $c_1, \ldots, c_k$, *not all zero, for which*

$$c_1\mathbf{x}_1 + \cdots + c_k\mathbf{x}_k = \mathbf{0}.$$

Column vectors $\mathbf{x}_1, \ldots, \mathbf{x}_k$ *are said to be* **linearly independent** *if they are not linearly dependent. Otherwise stated,* $\mathbf{x}_1, \ldots, \mathbf{x}_k$ *are linearly independent provided*

$$c_1\mathbf{x}_1 + \cdots + c_k\mathbf{x}_k = \mathbf{0}$$

implies $c_1 = \cdots = c_n = 0$.

For example, the column vectors

$$\mathbf{x}_1 = \begin{pmatrix} 1 \\ 2 \\ -1 \end{pmatrix}, \quad \mathbf{x}_2 = \begin{pmatrix} 2 \\ 1 \\ 3 \end{pmatrix}, \quad \text{and} \quad \mathbf{x}_3 = \begin{pmatrix} -4 \\ 1 \\ -11 \end{pmatrix}$$

are linearly dependent, since they satisfy the equation $2\mathbf{x}_1 - 3\mathbf{x}_2 - \mathbf{x}_3 = \mathbf{0}$. On the other hand, the vectors

$$\mathbf{e}_1 = \begin{pmatrix} 1 \\ 0 \\ 0 \end{pmatrix}, \quad \mathbf{e}_2 = \begin{pmatrix} 0 \\ 1 \\ 0 \end{pmatrix}, \quad \text{and} \quad \mathbf{e}_3 = \begin{pmatrix} 0 \\ 0 \\ 1 \end{pmatrix}$$

are linearly independent, because they satisfy no linear equation of the form

$$c_1\mathbf{e}_1 + c_2\mathbf{e}_2 + c_3\mathbf{e}_3 = \mathbf{0}$$

except when $c_1 = c_2 = c_3 = 0$.

It is useful to record for future reference the basic dichotomy that pertains to a square $n \times n$ matrix $A = (a_{ij})$ and the associated system of linear equations.

Theorem A.2. *Any $n \times n$ matrix A falls into either* **Case I** *or* **Case II** *below.*
Case I: *All of the following are true:*

Ia: $\det(A) \neq 0$.

Ib: *The columns of A are linearly independent vectors.*

Ic: *The equation $A\mathbf{x} = \mathbf{0}$ has only one solution:* $\mathbf{x} = \mathbf{0}$.

Id: *The equation $A\mathbf{x} = \mathbf{y}$ has a solution for every* $\mathbf{y}$.

Ie: *The solution to the equation $A\mathbf{x} = \mathbf{y}$ is unique.*

If: *The matrix A can be row reduced to the $n \times n$ identity matrix $I_{n \times n}$.*

Ig: *The matrix A is invertible.*

Case II: *All of the following are true:*

IIa: $\det(A) = 0$.

IIb: *The columns of A are linearly dependent vectors.*

IIc: *The equation $A\mathbf{x} = 0$ has a solution $\mathbf{x} \neq \mathbf{0}$.*

IId: *The equation $A\mathbf{x} = \mathbf{y}$ fails to have a solution $\mathbf{x}$ for some $\mathbf{y} \neq \mathbf{0}$.*

IIe: *The equation $A\mathbf{x} = \mathbf{y}$ has more than one solution $\mathbf{x}$ for some $\mathbf{y} \neq \mathbf{0}$.*

IIf: *The matrix A can be row reduced to a matrix with a row of zeros.*

IIg: *The matrix is not invertible: there is no matrix B that satisfies either the equation $AB = I_{n \times n}$ or the equation $BA = I_{n \times n}$.*

We have two mutually exclusive lists of equivalent conditions that must be satisfied by a given matrix. Any one of the conditions implies all of the others, as is shown in courses in linear algebra.

Example A.4. *Show that the* 3×3 *matrix*

$$\begin{pmatrix} 3 & 0 & 1 \\ 2 & 1 & 5 \\ 1 & 2 & 6 \end{pmatrix}$$

satisfies the conditions of case I.

Solution. We compute the determinant, which is

$$\det \begin{pmatrix} 3 & 0 & 1 \\ 2 & 1 & 5 \\ 1 & 2 & 6 \end{pmatrix} = 3 \cdot 1 \cdot 6 + 2 \cdot 2 \cdot 1 - 1 \cdot 1 \cdot 1 - 5 \cdot 2 \cdot 3$$

$$= 18 + 4 - 1 - 30 = -9 \neq 0.$$

The determinant is nonzero, so we are in case I. ∎

In Case II, it is useful to have a simple criterion for the solvability of the equation $A\mathbf{x} = \mathbf{y}$. This is provided by the following simple criterion, sometimes called the **Fredholm alternative**:

Lemma A.3. *Suppose that the* $n \times n$ *matrix A has* $\det(A) = 0$. *Then the equation*

$$A\mathbf{x} = \mathbf{y}$$

has a solution if and only if the vector $\mathbf{y} = (y_i)$ *satisfies the equation*

$$\sum_{i=1}^{n} y_i z_i = 0 \tag{A.7}$$

for every solution $\mathbf{z} = (z_i)$ *of the equation* $A^T \mathbf{z} = \mathbf{0}$.

This proposition reduces the problem of solvability of the inhomogeneous equation $A\mathbf{x} = \mathbf{y}$ to the problem of solving the homogeneous **adjoint equation** $A^T \mathbf{z} = \mathbf{0}$. The set of solutions can be obtained by the process of row reduction, from which it can be tested whether or not the vector $\mathbf{y}$ satisfies (A.7).

Example A.5. *For which vectors* $\mathbf{y}$ *can we solve the system of equations*

$$\begin{pmatrix} -2 & 1 & 1 \\ 1 & -2 & 1 \\ 1 & 1 & -2 \end{pmatrix} \begin{pmatrix} x_1 \\ x_2 \\ x_3 \end{pmatrix} = \begin{pmatrix} y_1 \\ y_2 \\ y_3 \end{pmatrix}?$$

Solution. The general solution to the adjoint equation $A^T \mathbf{z} = 0$ is obtained by solving the system

$$\begin{cases} -2z_1 + z_2 + z_3 = 0, \\ z_1 - 2z_2 + z_3 = 0, \\ z_1 + z_2 - 2z_3 = 0, \end{cases}$$

leading to the equations

$$z_1 = z_2 = z_3,$$

whose general solution is $\mathbf{z} = c(1, 1, 1)^T$ for some constant c. Therefore, the equation

$$A\mathbf{x} = \mathbf{y}$$

has a solution if and only if the components of the vector $\mathbf{y}$ satisfy the equation $y_1 + y_2 + y_3 = 0$. ∎

Exercises for Section A.3

Solve the following systems of linear equations by the method of row reduction. Be careful to distinguish the cases of no solutions, one solution, and many solutions.

1. $$\begin{cases} x_1 + 3x_2 = 4, \\ x_1 - x_2 = 6 \end{cases}$$

2. $$\begin{cases} x_1 + 4x_2 = 5, \\ 2x_1 + 8x_2 = 11 \end{cases}$$

3. $$\begin{cases} x_1 + x_2 + x_3 = 4, \\ x_1 - 2x_2 + x_3 = 0, \\ 2x_1 - x_2 + 3x_3 = 5 \end{cases}$$

4. $$\begin{cases} x_1 + 2x_2 - x_3 = 0, \\ x_1 - x_2 + 2x_3 = 0, \\ 2x_1 + x_2 + x_3 = 0 \end{cases}$$

5. $$\begin{cases} -x_1 + x_2 + 2x_3 = 0, \\ x_1 - x_3 = 0, \\ 3x_1 + x_2 + 2x_3 = 0 \end{cases}$$

6. $$\begin{cases} x_1 - x_3 = 0, \\ 3x_1 + x_2 + x_3 = 0, \\ -x_1 + x_2 + 2x_3 = 0 \end{cases}$$

$$7. \quad \begin{cases} x_1 + x_2 + x_3 + x_4 = 4, \\ x_1 + 2x_2 + 3x_3 + 4x_4 = 3, \\ x_1 - x_2 + 3x_3 - 4x_4 = 0 \end{cases}$$

$$8. \quad \begin{cases} 5x_1 - 2x_2 + 9x_3 + 11x_4 = -29, \\ 7x_1 - x_2 + 4x_3 + 5x_4 = 18, \\ 6x_1 - 6x_2 + 2x_3 - 11x_4 = -28, \\ x_1 + x_2 + x_3 + x_4 = 0 \end{cases}$$

For each of the following matrices, determine whether it is of case I (invertible) or case II (noninvertible). Compute the inverse if the matrix is invertible.

9. $\begin{pmatrix} 1 & 4 \\ 2 & -3 \end{pmatrix}$

10. $\begin{pmatrix} 6 & 2 \\ 3 & -1 \end{pmatrix}$

11. $\begin{pmatrix} 1 & -1 & -1 \\ 3 & -1 & 2 \\ 2 & 2 & 3 \end{pmatrix}$

12. $\begin{pmatrix} 1 & 2 & 3 \\ 3 & 5 & 6 \\ 2 & 4 & 5 \end{pmatrix}$

13. $\begin{pmatrix} 2 & 3 & 1 \\ -1 & 2 & 1 \\ 4 & -1 & -1 \end{pmatrix}$

14. $\begin{pmatrix} 0 & 3 & 0 & 0 \\ -4 & 2 & 0 & 0 \\ 0 & -1 & 1 & 0 \\ 0 & 0 & 3 & 1 \end{pmatrix}$

15. Prove that the identity matrix commutes with all $n \times n$ matrices: $A I_{n \times n} = I_{n \times n} A = A$.

16. Prove that if B is an $n \times n$ matrix that commutes with *every* $n \times n$ matrix A, then $B = c I_{n \times n}$ for some constant c.

17. Prove that any two diagonal matrices commute: $\Lambda_1 \Lambda_2 = \Lambda_2 \Lambda_1$.

18. Prove that if B is an $n \times n$ matrix that commutes with *every* diagonal matrix, then B is a diagonal matrix.

19. Prove that the result of interchanging two rows of a matrix can be effected by the left multiplication of the matrix by a certain matrix.

20. Prove that the result of multiplying a row of a matrix by a fixed constant c can be effected by the left multiplication by a certain matrix.

21. Prove that the result of adding a multiple of one row to another row of a matrix can be effected by left multiplication by a certain matrix.

22. The **trace** of an $n \times n$ matrix

$$A = \begin{pmatrix} a_{11} & a_{12} & \cdots & a_{1n} \\ a_{21} & a_{22} & \cdots & a_{2n} \\ \vdots & \vdots & \ddots & \vdots \\ a_{n1} & a_{m2} & \cdots & a_{nn} \end{pmatrix}$$

is defined by $\mathrm{tr}\,(A) = a_{11} + \cdots + a_{nn}$. Show that $\mathrm{tr}\,(AB) = \mathrm{tr}\,(BA)$ for square matrices A and B.

A.4 *Eigenvalues and Eigenvectors*

Effective use of matrices in the study of systems of linear differential equations requires that we develop the notions of *eigenvalue* and *eigenvector*.[1]

Definition. *Let $A = (a_{ij})$ be an $n \times n$ matrix of real or complex numbers. We say that a complex number r is an **eigenvalue** of A provided that there is a nonzero vector $\mathbf{x}$ for which*

$$A\mathbf{x} = r\,\mathbf{x}. \tag{A.8}$$

Any vector $\mathbf{x} \neq \mathbf{0}$ for which (A.8) *holds is called an **eigenvector** of A.*

We can write out (A.8) more explicitly. Let $A = (a_{ij})$ and $\mathbf{x} = (x_1, \dots, x_n)^T$. Then

$$A\mathbf{x} - r\,\mathbf{x} = \begin{pmatrix} a_{11} & a_{12} & \cdots & a_{1n} \\ a_{21} & a_{22} & \cdots & a_{2n} \\ \vdots & \vdots & \ddots & \vdots \\ a_{n1} & a_{n2} & \cdots & a_{nn} \end{pmatrix} \begin{pmatrix} x_1 \\ x_2 \\ \vdots \\ x_n \end{pmatrix} - r \begin{pmatrix} x_1 \\ x_2 \\ \vdots \\ x_n \end{pmatrix}$$

$$= \begin{pmatrix} a_{11} - r & a_{12} & \cdots & a_{1n} \\ a_{21} & a_{22} - r & \cdots & a_{2n} \\ \vdots & \vdots & \ddots & \vdots \\ a_{n1} & a_{n2} & \cdots & a_{nn} - r \end{pmatrix} \begin{pmatrix} x_1 \\ x_2 \\ \vdots \\ x_n \end{pmatrix}$$

---

[1]

The linguistic derivation originates with the German word "Eigenwert," which literally translates as "proper value". The present terminology was introduced by Richard Courant (1888–1972), who brought many of the fine traditions of the Göttingen mathematical school to the United States in the 1930s.

$$= \begin{pmatrix} \displaystyle\sum_{j=1}^{n} a_{1j} x_j - r\, x_1 \\ \vdots \\ \displaystyle\sum_{j=1}^{n} a_{nj} x_j - r\, x_n \end{pmatrix}.$$

Thus, (A.8) is equivalent to the system of linear equations

$$\begin{cases} \displaystyle\sum_{j=1}^{n} a_{1j} x_j = r\, x_1, \\ \qquad\vdots \\ \displaystyle\sum_{j=1}^{n} a_{nj} x_j = r\, x_n. \end{cases} \tag{A.9}$$

Example A.6. *Suppose that* $\det(A) = 0$. *Show that* $r = 0$ *is an eigenvalue of matrix* A.

Solution. From Theorem A.2 we know that if $\det(A) = 0$ (Case I), then the vector equation $A\mathbf{x} = \mathbf{0}$ has a solution $\mathbf{x} \neq \mathbf{0}$. This vector is an eigenvector of the matrix A with eigenvalue zero. ∎

A given matrix A may have several different eigenvalues and eigenvectors. It is important to note that the *set* of eigenvalues is a well-defined collection of complex numbers determined by the matrix. But the eigenvectors corresponding to a given eigenvalue are *not* uniquely determined. For example, if $\mathbf{x}$ is an eigenvector, then so is the vector $c\,\mathbf{x}$ for any nonzero complex constant c. Some authors attempt to reduce this ambiguity by further requiring the normalization condition that the length of the eigenvector $\mathbf{x}$ be one. But this still leaves an ambiguity of at least ± 1 in the definition of the eigenvector $\mathbf{x}$. For this reason we shall not attempt to normalize the eigenvectors in our treatment.

We can compute the eigenvalues of an $n \times n$ matrix without knowing the eigenvectors, as follows.

Theorem A.4. *The complex number* r *is an eigenvalue of the* $n \times n$ *matrix* A *if and only if* r *is a solution to the* **characteristic equation**

$$\det(A - r\, I_{n \times n}) = 0. \tag{A.10}$$

Proof. If r is an eigenvalue of A, then by definition there is a nonzero vector $\mathbf{x}$ such that the following equation holds:

$$(A - r\, I_{n \times n})\,\mathbf{x} = \mathbf{0}. \tag{A.11}$$

By case II of Theorem A.2, equations (A.11) and (A.10) are equivalent. ∎

Example A.7.   *Find the eigenvalues of the matrix*

$$A = \begin{pmatrix} 2 & 3 \\ 3 & 4 \end{pmatrix}. \tag{A.12}$$

Solution. The eigenvalues of A are solutions of the equation

$$\det \begin{pmatrix} 2 - r & 3 \\ 3 & 4 - r \end{pmatrix} = 0. \tag{A.13}$$

But (A.13) is the quadratic equation

$$0 = (2 - r)(4 - r) - 9 = 0 = r^2 - 6r - 1 = 0. \tag{A.14}$$

The quadratic formula tells us that the solution to (A.14) is

$$r = 3 \pm \sqrt{10}.$$

These are the eigenvalues of A. ∎

In order to find the eigenvectors using the definition, we must solve the system of linear equations (A.9).

Example A.8.   *Find the eigenvectors of the matrix*

$$A = \begin{pmatrix} 2 & 3 \\ 3 & 4 \end{pmatrix}.$$

Solution. We have already found the eigenvalues of A in Example A.7. To find the eigenvectors, we consider each eigenvalue separately. For the eigenvalue $r_1 = 3 + \sqrt{10}$, we must solve the system

$$\begin{cases} (-1 - \sqrt{10})x_1 + 3x_2 = 0, \\ 3x_1 + (1 - \sqrt{10})x_2 = 0. \end{cases} \tag{A.15}$$

The two equations of (A.15) are redundant, that is, either is a consequence of the other. We can only express x_2 in terms of x_1, or vice versa. For example,

$$x_2 = \frac{1 + \sqrt{10}}{3}x_1. \tag{A.16}$$

If we choose $x_1 = 3$ in (A.16), then $x_2 = 1 + \sqrt{10}$. Thus

$$\begin{pmatrix} 3 \\ 1 + \sqrt{10} \end{pmatrix},$$

or any nonzero multiple of this vector, is an eigenvector corresponding to the eigenvalue $r_1 = 3 + \sqrt{10}$.

For the eigenvalue $r_2 = 3 - \sqrt{10}$ we must solve the system

$$\begin{cases} (-1 + \sqrt{10})x_1 + 3x_2 = 0, \\ 3x_1 + (1 + \sqrt{10})x_2 = 0. \end{cases} \tag{A.17}$$

From (A.17) it follows that an eigenvector corresponding to $r_2 = 3 - \sqrt{10}$ is any nonzero multiple of

$$\begin{pmatrix} 3 \\ 1 - \sqrt{10} \end{pmatrix}. \blacksquare$$

Definition. *We say that $n \times n$ matrices A and B are* **similar** *provided that there exists a nonsingular matrix S such that*

$$B = S^{-1} A S.$$

Lemma A.5. *Let A and B be similar matrices. Then A and B have the same characteristic polynomial.*

Proof. Suppose that $B = S^{-1} A S$. Then

$$\det(B - r \, I_{n \times n}) = \det(S^{-1} A S - r \, I_{n \times n}) = \det\big(S^{-1}(A - r \, I_{n \times n})S\big)$$

$$= \det(S)^{-1} \det(A - r \, I_{n \times n}) \det(S) = \det(A - r \, I_{n \times n}). \blacksquare$$

In Examples A.7 and A.8 we have been working with a matrix with the pleasant property of possessing two different eigenvalues whose associated eigenvectors are linearly independent. This desirable state of affairs cannot be expected to hold for *all* matrices.

Example A.9. *Find all of the eigenvalues and eigenvectors of the matrix*

$$A = \begin{pmatrix} 3 & 4 \\ 0 & 3 \end{pmatrix}.$$

Solution. The eigenvalues of A must satisfy the quadratic equation

$$0 = \det\begin{pmatrix} 3 - r & 4 \\ 0 & 3 - r \end{pmatrix} = (3 - r)(3 - r) - 4 \times 0 = (r - 3)^2,$$

which has the double root $r = 3$. The associated eigenvectors $\mathbf{x} = (x_1, x_2)^T$ must satisfy the system of equations

$$\begin{cases} (3 - r)x_1 + 4x_2 = 0, \\ 0 \times x_1 + (3 - r)x_2 = 0. \end{cases} \tag{A.18}$$

When we substitute $r = 3$ into (A.18), we see that it reduces to

$$4x_2 = 0.$$

Thus, $x_2 = 0$ and x_1 is arbitrary. Therefore, any eigenvector of A has the form

$$\mathbf{x} = C \begin{pmatrix} 1 \\ 0 \end{pmatrix},$$

where C is a constant. In particular, A has only one linearly independent eigenvector, not two. ∎

In light of Examples A.8 and A.9, one can ask which class of $n \times n$ matrices possess n linearly independent eigenvectors.

Definition. *An $n \times n$ matrix A is said to be* **diagonalizable** *if it possesses n linearly independent eigenvectors.*

The following conditions give some sufficient conditions for a matrix to be diagonalizable.

Theorem A.6. *An $n \times n$ matrix A is diagonalizable in any of the following cases:*

(i) *A has n distinct complex eigenvalues $r_1, \ldots, r_n$, that is, $r_i \neq r_j$ for $i \neq j$.*

(ii) *A is a* **real symmetric matrix***, that is, a_{ij} is real and*

$$a_{ij} = a_{ji} \tag{A.19}$$

for $1 \leq i, j \leq n$.

(iii) *A is a* **normal matrix***, that is,*

$$\sum_{k=1}^{n} a_{ik}\bar{a}_{jk} = \sum_{k=1}^{n} \bar{a}_{ki}a_{kj}. \tag{A.20}$$

Note that condition (A.19) can be written as

$$A = A^T,$$

and condition (A.20) can be written as

$$A\bar{A}^T = \bar{A}^T A.$$

For example,

$$\begin{pmatrix} -2 & 3 \\ 3 & -1 \end{pmatrix}$$

is a real symmetric matrix. Every real symmetric matrix is a normal matrix. The set of normal matrices also includes the following types of matrices:

Orthogonal matrices, that is, matrices for which $A^T = A^{-1}$;

Unitary matrices, that is, matrices for which $\overline{A}^T = A^{-1}$;

Antisymmetric matrices, that is, matrices for which $A^T = -A$.

Theorem A.6 does not describe all diagonalizable matrices. In particular, not all diagonalizable matrices have distinct eigenvalues. The following example illustrates a 4×4 matrix that is diagonalizable but has repeated eigenvalues and is not normal.

Example A.10.   *Show that the 4×4 matrix.*

$$A = \begin{pmatrix} 1 & 0 & 0 & 0 \\ -1 & 0 & 0 & 0 \\ 0 & 0 & 1 & 0 \\ 0 & 0 & -1 & 0 \end{pmatrix}$$

is diagonalizable.

Solution. The eigenvalues are determined from the fourth-order polynomial equation

$$r^4 - 2r^3 + r^2 = 0,$$

whose solutions are the double roots

$$r_1 = 0 \quad \text{and} \quad r_2 = 1.$$

Any eigenvector $\mathbf{x} = (x_1, x_2, x_3, x_4)^T$ of the eigenvalue $r_1 = 0$ must satisfy the conditions that $x_1 = x_3 = 0$. Thus, any eigenvector of r_2 is a linear combination of

$$\mathbf{e}_1 = \begin{pmatrix} 0 \\ 1 \\ 0 \\ 0 \end{pmatrix} \quad \text{and} \quad \mathbf{e}_2 = \begin{pmatrix} 0 \\ 0 \\ 0 \\ 1 \end{pmatrix}.$$

For the eigenvalue $r_2 = 1$ any eigenvector must satisfy the conditions $x_1 + x_2 = x_3 + x_4 = 0$. Thus, any eigenvector of r_2 is a linear combination of

$$\mathbf{e}_3 = \begin{pmatrix} 1 \\ -1 \\ 0 \\ 0 \end{pmatrix} \quad \text{and} \quad \mathbf{e}_4 = \begin{pmatrix} 0 \\ 0 \\ 1 \\ -1 \end{pmatrix}.$$

The four vectors $\mathbf{e}_1, \mathbf{e}_2\, \mathbf{e}_3, \mathbf{e}_4$ are linearly independent; hence the matrix A is diagonalizable. ∎

The term *diagonalizable* matrix is explained very simply as follows. For such a matrix A, let $\mathbf{x}_1, \ldots, \mathbf{x}_n$ be the eigenvectors of A, and assume that they are linearly independent column vectors. Let $r_1, \ldots, r_n$ be the corresponding eigenvalues, so that

$$A\mathbf{x}_i = r_i \mathbf{x}_i \tag{A.21}$$

for $1 \leq i \leq n$. Now define a matrix U to consist of the column vectors $\mathbf{x}_1, \ldots, \mathbf{x}_n$ and a matrix Λ to consist of zeros off of the main diagonal and the eigenvalues $r_1, \ldots, r_n$ listed along the main diagonal. In other words,

$$\Lambda = \begin{pmatrix} r_1 & 0 & \ldots & 0 \\ 0 & r_2 & \ldots & 0 \\ \vdots & \vdots & \ddots & \vdots \\ 0 & 0 & \ldots & r_n \end{pmatrix}.$$

Then (A.21) can be rewritten as the matrix equation

$$A U = U \Lambda, \tag{A.22}$$

which simply states that the columns of U are the eigenvectors of A and the diagonal elements of Λ are the corresponding eigenvalues. Now, since the columns of U are linearly independent, U is in case I of Theorem A.2 and therefore possesses an inverse matrix U^{-1}. Applying U^{-1} to both sides of (A.22), we obtain

$$U^{-1} A U = \Lambda = \begin{pmatrix} r_1 & 0 & \ldots & 0 \\ 0 & r_2 & \ldots & 0 \\ \vdots & \vdots & \ddots & \vdots \\ 0 & 0 & \ldots & r_n \end{pmatrix}.$$

We have *diagonalized* the matrix A by means of the matrix U, which explains the use of the term "diagonalizable". We summarize the above results as

Theorem A.7. *A matrix A is diagonalizable if and only if there is a matrix U such that $U^{-1} A U$ is a diagonal matrix. In this case, U can be chosen to consist of eigenvectors of A.*

For 2×2 matrices, the above concepts simplify, as follows.

Lemma A.8. *Suppose that the 2×2 matrix A is diagonalizable. Then either A has distinct eigenvalues $r_1 \neq r_2$ or A is a multiple of the identity matrix, that is,*

$$A = \begin{pmatrix} c & 0 \\ 0 & c \end{pmatrix}$$

for some constant c.

Proof. From Theorem A.7 we know that there exists a diagonal matrix Λ such that

$$A = U^{-1}\Lambda U.$$

If the eigenvalues of A coincide, then the matrix Λ is a multiple of the identity matrix, that is,

$$\Lambda = r\, I_{2\times 2}.$$

But the identity matrix commutes with any other 2×2 matrix, so that we must have the equation

$$A = U^{-1}\Lambda U = U^{-1}U\,\Lambda = I_{2\times 2}\Lambda = r\, I_{2\times 2}.$$

We have proved the assertion with the value $c = r$. ∎

When we pass to higher-dimensional matrices, Lemma A.8 must be modified. For example, the 3×3 matrix

$$A = \begin{pmatrix} 1 & 0 & 0 \\ 0 & 1 & 0 \\ 0 & 0 & 2 \end{pmatrix}$$

is diagonalizable, but neither does it have distinct eigenvalues nor is it a multiple of the identity matrix. However, the proof of Lemma A.8 can be extended to $n \times n$ matrices to show that if a diagonalizable matrix has all equal eigenvalues, then it is a multiple of the identity matrix.

Nondiagonalizable Matrices

Not every square matrix is diagonalizable. Nilpotent matrices form one of the simplest classes of nondiagonalizable matrices.

Definition. *An $n \times n$ square matrix N is **nilpotent** provided that there exists a nonnegative integer k such that N^k is the zero matrix. The unique integer k such that $N^k = 0$ but $N^{k-1} \neq 0$ is called the **order** of N.*

For example, the orders of the matrices

$$N_1 = \begin{pmatrix} 0 & 1 & 0 \\ 0 & 0 & 1 \\ 0 & 0 & 0 \end{pmatrix} \quad \text{and} \quad N_2 = \begin{pmatrix} 0 & 0 & 1 \\ 0 & 0 & 0 \\ 0 & 0 & 0 \end{pmatrix}$$

are 3 and 2, respectively.

Lemma A.9. *Let N be a nilpotent matrix. The only eigenvalue of N is 0.*

Proof. If $N\mathbf{x} = r\,\mathbf{x}$ with $\mathbf{x} \neq 0$, then

$$r^k \mathbf{x} = N^k \mathbf{x} = 0$$

for any integer k greater than or equal to the order of N. Hence $r = 0$. ∎

In case of a nondiagonalizable matrix, there is a substitute for a diagonal matrix of eigenvalues; it is known as the **Jordan[2] canonical form**.

Theorem A.10. *For any complex $n \times n$ matrix A, there exists an invertible matrix U such that the equation*

$$U^{-1} A U = \begin{pmatrix} B_1 & 0 & \dots & 0 \\ 0 & B_2 & \dots & 0 \\ \vdots & \vdots & \ddots & \vdots \\ 0 & 0 & \dots & B_k \end{pmatrix} \tag{A.23}$$

*holds, where the **Jordan blocks** $B_1, \dots, B_k$ are either of the form*

$$B_i = r_i \, I_{1 \times 1} \tag{A.24}$$

or of the form

$$B_i = \begin{pmatrix} r_i & 1 & 0 & \dots & 0 & 0 \\ 0 & r_i & 1 & \dots & 0 & 0 \\ \vdots & \vdots & \vdots & \ddots & \vdots & \vdots \\ 0 & 0 & 0 & \dots & r_i & 1 \\ 0 & 0 & 0 & \dots & 0 & r_i \end{pmatrix}. \tag{A.25}$$

$U^{-1} A U$ *is called the* **Jordan canonical form** *of A.*

In view of Theorem A.4 we make the following definition.

Definition. *The **multiplicity** of an eigenvalue r of a matrix A is the degree k of r as a root of the characteristic equation $\det(A - r\,I_{n \times n}) = 0$. If $k > 1$, we say that r is a* **multiple eigenvalue** *of A.*

An eigenvalue r of multiplicity k has at most k linearly independent eigenvectors associated with it. We say that r has a **full set of associated eigenvectors** provided that the number of linearly independent eigenvectors associated with it equals the multiplicity. Otherwise, r is called a **deficient eigenvalue**.

---

[2]

Camille Jordan (1838–1921). French mathematician known for his work in analysis and group theory.

Blocks of type (A.24) correspond to the eigenvalues of A that possess a full set of associated eigenvectors; blocks of type (A.25) correspond to the deficient eigenvalues. When the matrix is diagonalizable, the second type of block is not present.

Corollary A.11. *Let A be an $n \times n$ matrix. Then A can be expressed uniquely as a sum*

$$A = S + N, \tag{A.26}$$

where S is diagonalizable, N is nilpotent, and $S N = N S$.

Proof. Each Jordan block B_i in the Jordan canonical form of A can be decomposed as the sum of a diagonal matrix and a nilpotent matrix. When the blocks are reassembled using (A.23), we get (A.26).

To prove uniqueness, suppose $A = S_1 + N_1$, where S_1 is diagonalizable, N_1 is nilpotent, and $S_1 N_1 = N_1 S_1$. Then $N - N_1 = -S + S_1$, showing that $N - N_1$ is diagonal. But $(N - N_1)^\ell = 0$, where ℓ is an integer larger than the sum of the orders of N and N_1. Hence $N = N_1$ and $S = S_1$. ∎

In the case of the Jordan blocks corresponding to deficient eigenvalues, we may speak of **generalized eigenvectors**.

Definition. *Let A be an $n \times n$ square matrix. A **generalized eigenvector** corresponding to an eigenvalue r of A is a solution $\boldsymbol{\eta}$ of the equation*

$$(A - r\, I_{n \times n})^k \boldsymbol{\eta} = \mathbf{0}. \tag{A.27}$$

If $k = 1$, then (A.27) reduces to the definition of an ordinary eigenvector $\boldsymbol{\eta}_1$, which is just the solution to the equation $(A - r\, I)\boldsymbol{\eta}_1 = \mathbf{0}$. For $k \geq 2$ the generalized eigenvectors $\boldsymbol{\eta}_k$ may be obtained by solving the equations

$$(A - r\, I_{n \times n})\boldsymbol{\eta}_2 = \boldsymbol{\eta}_1, \qquad (A - r\, I_{n \times n})\boldsymbol{\eta}_3 = \boldsymbol{\eta}_2,$$

and so forth.

Example A.11. *Find the eigenvector and generalized eigenvectors of the matrix*

$$A = \begin{pmatrix} 3 & 1 & 0 \\ 0 & 3 & 1 \\ 0 & 0 & 3 \end{pmatrix}.$$

Solution. The characteristic equation of A is

$$0 = \det(A - r\, I_{3 \times 3}) = (3 - r)^3,$$

whose solution is $r = 3$. The corresponding eigenvector is any nonzero multiple of

$$\eta_1 = \begin{pmatrix} 1 \\ 0 \\ 0 \end{pmatrix}.$$

The generalized eigenvector η_2 is obtained by solving the equation

$$(A - 3I_{3\times3})\eta_2 = \begin{pmatrix} 1 \\ 0 \\ 0 \end{pmatrix}; \tag{A.28}$$

the solution to (A.28) is

$$\eta_2 = \begin{pmatrix} 0 \\ 1 \\ 0 \end{pmatrix}.$$

The generalized eigenvector η_3 is obtained by solving

$$(A - 3I_{3\times3})\eta_3 = \eta_2 = \begin{pmatrix} 0 \\ 1 \\ 0 \end{pmatrix},$$

whose solution is

$$\eta_3 = \begin{pmatrix} 0 \\ 0 \\ 1 \end{pmatrix}. \ \blacksquare$$

Example A.11 illustrates a general rule of operation. In order to determine the generalized eigenvectors of a matrix A, we first factor the characteristic polynomial of A. For each factor of the form $(r - r_1)^{k_1}$, we compute as many eigenvectors as possible corresponding to the eigenvalue r_1. If we find k_1 of these, then the corresponding Jordan blocks are 1-dimensional, of the form $B = r_1 I_{1\times1}$, and there is no need to search further; there are no generalized eigenvectors corresponding to this eigenvalue. However, if we find fewer than k_1 linearly independent eigenvectors with eigenvalue r_1, then we must begin to look for generalized eigenvectors by solving the equation $(A - r I_{n\times n})\eta_2 = \eta_1$, as outlined above.

Exercises for Section A.4

Find the eigenvalues and eigenvectors of each of the following 2×2 matrices.

1. $\begin{pmatrix} 3 & 3 \\ 1 & 5 \end{pmatrix}$

4. $\begin{pmatrix} 8 & 6 \\ 4 & 3 \end{pmatrix}$

2. $\begin{pmatrix} -6 & 4 \\ -4 & 10 \end{pmatrix}$

5. $\begin{pmatrix} 1 & -1 \\ 1 & 1 \end{pmatrix}$

3. $\begin{pmatrix} 4 & 2 \\ -1 & 1 \end{pmatrix}$

6. $\begin{pmatrix} 2 & 2 \\ -9 & 2 \end{pmatrix}$

Find the eigenvalues and eigenvectors of each of the following 3×3 matrices.

7. $\begin{pmatrix} 1 & 0 & 0 \\ 2 & 1 & -2 \\ 3 & 2 & 1 \end{pmatrix}$

9. $\begin{pmatrix} 2 & 2 & 1 \\ 3 & 4 & 2 \\ 1 & 2 & 1 \end{pmatrix}$

8. $\begin{pmatrix} 3 & 2 & 4 \\ 2 & 0 & 2 \\ 4 & 2 & 3 \end{pmatrix}$

10. $\begin{pmatrix} 1 & 2 & -3 \\ 1 & 1 & 2 \\ 1 & -1 & 4 \end{pmatrix}$

For each of the following matrices find the eigenvalues, eigenvectors and generalized eigenvectors:

11. $\begin{pmatrix} 3 & 3 \\ 0 & 3 \end{pmatrix}$

14. $\begin{pmatrix} 4 & 4 \\ -4 & -4 \end{pmatrix}$

12. $\begin{pmatrix} 1 & 1 & 0 \\ 0 & 1 & 0 \\ 0 & 0 & 3 \end{pmatrix}$

15. $\begin{pmatrix} 1 & 1 & -1 \\ -1 & -1 & 0 \\ 0 & 0 & 1 \end{pmatrix}$

13. $\begin{pmatrix} 1 & 1 & 1 \\ 1 & 0 & 1 \\ 0 & 1 & 0 \end{pmatrix}$

16. $\begin{pmatrix} 0 & 0 & 0 & 1 \\ 0 & 0 & 1 & 0 \\ 0 & 1 & 0 & 0 \\ 1 & 0 & 0 & 0 \end{pmatrix}$

A.5 The Exponential of a Matrix

The elementary function e^x has important generalizations in several advanced branches of mathematics, notably Riemannian geometry and Lie groups. In Section 9.2.1 we gave the basic theory of e^x in the case that x is a complex number. In this section we discuss the generalization of e^x to matrices.

The idea of the definition of e^A when A is a square matrix is simple, and essentially the same one used in Section 9.2.1: we use a power series expansion to define e^A.

Definition. *Let A be an $n \times n$ matrix. The* **exponential** *of A is*

$$e^A = \sum_{k=0}^{\infty} \frac{A^k}{k!} = I_{n \times n} + A + \frac{A^2}{2} + \cdots . \tag{A.29}$$

The proof that the power series on the right-hand side of (A.29) converges for any square matrix A is similar to the proof of convergence of e^x when x is a real number.

It is no longer the case that $e^{A+B} = e^A e^B$ for arbitrary matrices A and B, since A and B may not commute. However,

Lemma A.12. *Let A and B be $n \times n$ square matrices.*

(i) *e^A is always nonsingular with inverse e^{-A}.*

(ii) *If $A B = B A$, then $e^{A+B} = e^A e^B$.*

(iii) *If C is any nonsingular $n \times n$ matrix, then*

$$C^{-1} e^A C = e^{C^{-1}AC} .$$

Proof. The proof of (ii) is the same as that for complex exponents given in Lemma 9.5 on page 280. Then (i) follows from (ii). To prove (iii), we calculate as follows:

$$C^{-1} e^A C = C^{-1} \left(\sum_{k=0}^{\infty} \frac{A^k}{k!} \right) C = \sum_{k=0}^{\infty} C^{-1} \frac{A^k}{k!} C$$

$$= \sum_{k=0}^{\infty} \frac{(C^{-1} A C)^k}{k!} = e^{C^{-1}AC} . \blacksquare$$

It is easy to compute the exponential of a diagonal matrix.

Lemma A.13. *The exponential of a diagonal matrix*

$$D = \begin{pmatrix} d_1 & 0 & \cdots & 0 \\ 0 & d_2 & \cdots & 0 \\ \vdots & \vdots & \ddots & \vdots \\ 0 & 0 & \cdots & d_n \end{pmatrix} \quad \text{is given by} \quad e^D = \begin{pmatrix} e^{d_1} & 0 & \cdots & 0 \\ 0 & e^{d_2} & \cdots & 0 \\ \vdots & \vdots & \ddots & \vdots \\ 0 & 0 & \cdots & e^{d_n} \end{pmatrix} .$$

More generally, if A is a matrix such that $C^{-1} A C = D$ (where C is a nonsingular matrix), then

$$C^{-1} e^A C = e^D .$$

Proof. The formula for e^D follows directly from the definition, and the formula for e^A is a consequence of part (iii) of Lemma A.12. ∎

Even for a diagonalizable matrix A it is in general impractical to compute the exponential directly from the definition; too many powers of matrices need to be computed. Instead, the eigenvalues of A should be computed and Lemma A.13 used.

Example A.12. *Compute the exponential of the matrix* $\begin{pmatrix} 1 & 2 \\ 3 & 2 \end{pmatrix}$.

Solution. The eigenvalues of A are -1 and 4, and the corresponding eigenvectors are scalar multiples of $(-1, 1)^T$ and $(2, 3)^T$. Moreover,

$$\begin{pmatrix} -1 & 2 \\ 1 & 3 \end{pmatrix}^{-1} \begin{pmatrix} 1 & 2 \\ 3 & 2 \end{pmatrix} \begin{pmatrix} -1 & 2 \\ 1 & 3 \end{pmatrix} = \begin{pmatrix} -3/5 & 2/5 \\ 1/5 & 1/5 \end{pmatrix} \begin{pmatrix} 1 & 2 \\ 3 & 2 \end{pmatrix} \begin{pmatrix} -1 & 2 \\ 1 & 3 \end{pmatrix}$$

$$= \begin{pmatrix} -1 & 0 \\ 0 & 4 \end{pmatrix}.$$

Lemmas A.12 and A.13 imply that

$$\begin{pmatrix} -1 & 2 \\ 1 & 3 \end{pmatrix}^{-1} \exp\left(\begin{pmatrix} 1 & 2 \\ 3 & 2 \end{pmatrix} \right) \begin{pmatrix} -1 & 2 \\ 1 & 3 \end{pmatrix} = \begin{pmatrix} e^{-1} & 0 \\ 0 & e^4 \end{pmatrix}.$$

Hence

$$\exp\left(\begin{pmatrix} 1 & 2 \\ 3 & 2 \end{pmatrix} \right) = \begin{pmatrix} -1 & 2 \\ 1 & 3 \end{pmatrix} \begin{pmatrix} e^{-1} & 0 \\ 0 & e^4 \end{pmatrix} \begin{pmatrix} -3/5 & 2/5 \\ 1/5 & 1/5 \end{pmatrix}$$

$$= \begin{pmatrix} \dfrac{2e^5 + 3}{5e} & \dfrac{2e^5 - 2}{5e} \\[2mm] \dfrac{3e^5 - 3}{5e} & \dfrac{3e^5 + 2}{5e} \end{pmatrix}. \ ∎$$

Finally, we prove an important relation between the determinant of e^A and the trace of A.

Lemma A.14. *Let A be any matrix. Then* $\det(e^A) = e^{\operatorname{tr}(A)}$.

Proof. First, let B be a diagonalizable matrix, and let $\lambda_1, \ldots, \lambda_n$ be the eigenvalues of B. Then $e^{\lambda_1}, \ldots, e^{\lambda_n}$ are the eigenvalues of e^B, and so

$$\det(e^B) = e^{\lambda_1} \cdots e^{\lambda_n} = e^{\lambda_1 + \cdots + \lambda_n} = e^{\operatorname{tr}(B)}.$$

Furthermore, if N is a nilpotent matrix, then N is similar to an (upper triangular) matrix with zeros on the diagonal, and so

$$\det(e^N) = 1 = e^{\operatorname{tr}(N)}.$$

It follows from Corollary A.11 that for a general matrix A we can write $A = S + N$, where S and N are commuting matrices with S diagonalizable and N nilpotent.

$$\det(e^A) = \det(e^S)\det(e^N) = e^{\operatorname{tr}(S)} = e^{\operatorname{tr}(A)}. \quad \blacksquare$$

Exercises for Section A.5

Compute the exponentials of the following matrices using the definition of the exponential of a matrix.

1. $\begin{pmatrix} 1 & 0 \\ 0 & -1 \end{pmatrix}$

2. $\begin{pmatrix} 0 & t \\ -t & 0 \end{pmatrix}$

3. $\begin{pmatrix} 0 & 1 & t \\ 0 & 0 & 1 \\ 0 & 0 & 0 \end{pmatrix}$

4. $\begin{pmatrix} 0 & t \\ t & 0 \end{pmatrix}$

5. $\begin{pmatrix} 0 & t \\ 0 & 0 \end{pmatrix}$

6. $\begin{pmatrix} 0 & a & b & c \\ 0 & 0 & d & e \\ 0 & 0 & 0 & f \\ 0 & 0 & 0 & 0 \end{pmatrix}$

7. $\begin{pmatrix} 0 & i\pi/3 \\ 0 & -i\pi/3 \end{pmatrix}$

8. $\begin{pmatrix} a & b \\ a & b \end{pmatrix}$

A.6 *Abstract Vector Spaces*

In this section we give a summary of important facts about abstract vector spaces. These important mathematical entities enjoy many of the properties of $\mathbb{R}^n$ and $\mathbb{C}^n$, but because they are abstract they can be used to describe many other mathematical systems in which addition and scalar multiplication are defined. In particular, they are useful for describing the set of solutions of homogeneous linear differential equations and of systems of linear differential equations. In this section we use the symbol "$\in$" to mean "is a member of".

Definition.   *A* **real vector space** *consists of a set* **V** *and two operations, called* **addition** *and* **scalar multiplication***. By this we mean that to each pair of elements* **x** *and* **y** *in* **V** *we can associate a unique element* **x** + **y** *that is also in* **V***, and to each real number a we can associate a unique element a* **x** *in* **V***. We require that the following axioms be satisfied:*

A1. **x** + **y** = **y** + **x** *for all* **x**, **y** ∈ **V**.

A2. (**x** + **y**) + **z** = **x** + (**y** + **z**) *for all* **x**, **y**, **z** ∈ **V**.

A3. *There exists an element* **0** ∈ **V** *such that* **x** + **0** = **x** *for all* **x** ∈ **V**.

A4. *For each* **x** ∈ **V***, there exists an element* −**x** ∈ **V** *such that* **x** + (−**x**) = **0**.

A5. *a*(**x** + **y**) = *a* **x** + *a* **y** *for all real numbers a and all* **x**, **y** ∈ **V**.

A6. (*a* + *b*)**x** = *a* **x** + *b* **x** *for all real numbers a and b and all* **x** ∈ **V**.

A7. (*a b*)**x** = *a* (*b* **x**) *for all real numbers a and b and all* **x** ∈ **V**.

A8. 1 **x** = **x** *for all* **x** ∈ **V**.

A **complex vector space** *is defined in exactly the same way, except that complex numbers are used for scalar multiplication instead of real numbers. We use the term* **vector space** *to mean either a real or complex vector space.*

The elements of a vector space **V** are called **vectors,** and the real or complex numbers used for scalar multiplication are called **scalars**. It is straightforward to check that the set $\mathbb{R}^n$ of all $n \times 1$ real column vectors forms a real vector space, and the set $\mathbb{C}^n$ of all $n \times 1$ complex column vectors forms a complex vector space.

The notions of linear dependence and independence make sense for abstract vector spaces.

Definition.   *Let* **V** *be a vector space. Vectors* $\mathbf{x}_1, \ldots, \mathbf{x}_k \in \mathbf{V}$ *are said to be* **linearly dependent** *if there are constants* $c_1, \ldots, c_k$, *not all zero, for which*

$$c_1\mathbf{x}_1 + \cdots + c_k\mathbf{x}_k = \mathbf{0}.$$

If $\mathbf{x}_1, \ldots, \mathbf{x}_k$ *are not linearly dependent, they are said to be* **linearly independent** *Otherwise stated,* $\mathbf{x}_1, \ldots, \mathbf{x}_k$ *are linearly independent provided*

$$c_1\mathbf{x}_1 + \cdots + c_k\mathbf{x}_k = \mathbf{0}$$

implies $c_1 = \cdots = c_k = 0$.

Definition. *Let* $\mathbf{V}$ *be a vector space. Vectors* $\mathbf{x}_1, \ldots, \mathbf{x}_n \in \mathbf{V}$ *are said to* **span** V *provided that the set of all linear combinations* $c_1\mathbf{x}_1 + \cdots + c_n\mathbf{x}_n$ *exhausts* $\mathbf{V}$. *In other words,* $\mathbf{x}_1, \ldots, \mathbf{x}_n$ *span* $\mathbf{V}$ *if every element* $\mathbf{x}$ *of* $\mathbf{V}$ *can be expressed as a linear combination of* $\mathbf{x}_1, \ldots, \mathbf{x}_n$, *that is, there exist scalars* $c_1, \ldots, c_n$ *such that*

$$\mathbf{x} = c_1\mathbf{x}_1 + \cdots + c_n\mathbf{x}_n.$$

Definition. *The* **dimension** *of a vector space* $\mathbf{V}$ *is the minimum number of linearly independent vectors that span* $\mathbf{V}$. *If the dimension of* $\mathbf{V}$ *is* n, *we write* $\dim \mathbf{V} = n$. *We say that* $\mathbf{V}$ *is* **finite dimensional** *if its dimension is finite. Otherwise* $\mathbf{V}$ *is said to be* **infinite dimensional**.

Theorem A.15. *If* n *linearly independent vectors span a vector space* $\mathbf{V}$, *then* $\dim \mathbf{V} = n$.

Theorem A.15 is a consequence of the following two lemmas.

Lemma A.16. *A set of* m *homogeneous linear equations in* k *unknowns admits a nontrivial solution if* $m < k$.

Lemma A.17. *In an* n-*dimensional vector space* $\mathbf{V}$ *any set of* $p > n$ *vectors must be linearly dependent. In other words, the maximum number of linearly independent vectors in a finite-dimensional vector space equals the dimension of the vector space.*

For example, $\mathbb{R}^n$ has dimension n as a real vector space, and $\mathbb{C}^n$ has dimension n as a complex vector space. It is also possible to make $\mathbb{C}^n$ into a $2n$-dimensional real vector space. An example of an infinite-dimensional vector space is the set $C[a, b]$ of all continuous functions on the interval $a < t < b$.

Definition. *Let* $\mathbf{V}$ *be a finite-dimensional vector space. A* **basis** *for* $\mathbf{V}$ *consists of linearly independent vectors* $\mathbf{x}_1, \ldots, \mathbf{x}_n$ *that also span* $\mathbf{V}$.

Theorem A.15 can be rephrased as:

Lemma A.18. *Let* $\mathbf{V}$ *be an* n-*dimensional vector space, where* n *is finite. Then every basis of* $\mathbf{V}$ *has exactly* n *elements.*

Definition. *A* **subspace** *of a vector space* $\mathbf{V}$ *consists of a nonempty subset* $\mathbf{S}$ *of* $\mathbf{V}$ *that satisfies the conditions:*

S1. $\mathbf{x} + \mathbf{y} \in \mathbf{S}$ *whenever* $\mathbf{x} \in \mathbf{S}$ *and* $\mathbf{y} \in \mathbf{S}$.

S2. $a\,\mathbf{x} \in \mathbf{S}$ *whenever* $\mathbf{x} \in \mathbf{S}$ *and* a *is a scalar.*

If $\mathbf{V}$ is a vector space, then it is easy to check that $\mathbf{V}$ is a subspace of itself; also, the set $\{\mathbf{0}\}$ consisting of the zero vector alone is a subspace of $\mathbf{V}$. Furthermore, it is easy to prove the following lemma.

Lemma A.19. *Let* V_1 *and* V_2 *be subspaces of a vector space* V. *Then both the* **intersection** $V_1 \cap V_2$ *and* **sum** $V_1 + V_2$ *are subspaces of* V. *Here*

$$V_1 \cap V_2 = \{ \mathbf{x} \mid \mathbf{x} \in V_1 \text{ and } \mathbf{x} \in V_2 \}$$

and

$$V_1 + V_2 = \{ \mathbf{x}_1 + \mathbf{x}_2 \mid \mathbf{x}_1 \in V_1 \text{ and } \mathbf{x}_2 \in V_2 \}.$$

Exercises for Section A.6

1. Show that the real solutions to the homogeneous system of linear equations

$$\begin{cases} a_{11}x_1 + \cdots + a_{1n}x_n = 0, \\ a_{21}x_1 + \cdots + a_{2n}x_n = 0, \\ \quad\quad\quad \vdots \\ a_{n1}x_1 + \cdots + a_{nn}x_n = 0 \end{cases}$$

   form a real vector space and that the complex solutions form a complex vector space. Here solution means an n-dimensional column vector.

2. Show that the set of real solutions of a homogeneous n^{th}-order linear differential equation

$$a_n(t)y^{(n)} + a_{n-1}(t)y^{(n-1)} + \cdots + a_1(t)y' + a_0(t)y = 0$$

   form a real vector space and that the complex solutions form a complex vector space.

3. Show that the real solutions to the homogeneous system of linear differential equations

$$\begin{cases} \dfrac{dx_1}{dt} = p_{11}(t)x_1 + \cdots + p_{1n}(t)x_n, \\ \quad \vdots \quad\quad\quad\quad\quad \vdots \\ \dfrac{dx_n}{dt} = p_{n1}(t)x_1 + \cdots + p_{nn}(t)x_n \end{cases}$$

   form a real vector space and that the complex solutions form a complex vector space. Here solution means an n-dimensional vector.

4. Show that if $\mathbf{x}$ and $\mathbf{y}$ are linearly independent, then so are $\mathbf{x} + \mathbf{y}$ and $\mathbf{x} - \mathbf{y}$.

A.7   Vectors and Matrices with *Mathematica*

In *Mathematica* a vector is a list; a matrix is simply a list of lists of equal length, where the first list is the first row, the second list is the second row, and so forth. For example, if we define

```
xx = {7,-3}
aa = {{1,2},{3,4},{5,6}}
```

then **xx** is a vector in the plane $\mathbb{R}^2$ and **aa** is a 3 by 2 matrix. *Mathematica* has several commands that alter the output. To see the matrix **aa** as a rectangular array, we can specify

```
MatrixForm[aa]
```

```
1   2

3   4

5   6
```

Instead of **MatrixForm[aa]** it is frequently more convenient to write the equivalent command **aa//MatrixForm**. It is important to realize that **MatrixForm** does not create a matrix; it only specifies how the matrix is displayed.

If we wish to multiply the vector **xx** by the scalar 2, we simply enter

```
2 xx
```

```
{14, -6}
```

but if we want to multiply the vector **xx** by the matrix **aa**, we must use a period, or dot, (**.**) to indicate matrix multiplication. Thus

```
aa.xx
```

yields

```
{1, 9, 17}
```

Notice that *Mathematica* does not distinguish between row and column vectors, so that matrix multiplication **aa.xx** makes sense. However, **xx.aa** results in an error message.

The identity matrix $I_{n \times n}$ can be generated with the *Mathematica* command **IdentityMatrix[n]**. In order for this command to work, **n** must be given a specific integer value. For example, to generate $I_{4 \times 4}$ we use

```
IdentityMatrix[4]//MatrixForm
```

resulting in

```
1   0   0   0
0   1   0   0
0   0   1   0
0   0   0   1
```

Similarly, the command **DiagonalMatrix** creates a diagonal matrix. Thus

pp = DiagonalMatrix[{a,b,c,d}]

yields

```
{{a, 0, 0, 0}, {0, b, 0, 0}, {0, 0, c, 0}, {0, 0, 0, d}}
```

The transpose of a matrix is computed with **Transpose**. For example, the command

Transpose[aa]//MatrixForm

yields

```
1   3   5
2   4   6
```

Determinants and Inverses

The *Mathematica* command to compute the determinant of a square matrix is **Det**. With this useful command the user can use *Mathematica* to perform calculations that would be very tedious and time-consuming to do by hand.

Example A.13. *Use* **Det** *to compute the determinant of the matrix*

$$\begin{pmatrix} -5 & 1 & 7 & 6 \\ 5 & 0 & 1 & 18 \\ -2 & 11 & 7 & 4 \\ 1 & 4 & 8 & -9 \end{pmatrix}.$$

Solution. The command

Det[{{-5,1,7,6},{5,0,1,18},{-2,11,7,4},{1,4,8,9}}]

results in

```
-5923
```

It is also possible to compute the determinant of a matrix containing symbols.

Example A.14. *Use* **Det** *to compute the determinant of the matrix*

$$
\begin{pmatrix}
1 & 2 & 3 \\
a & b & c \\
-3 & -2 & -1
\end{pmatrix}.
$$

Solution. The command

Det[{{1,2,3},{a,b,c},{-3,-2,-1}}]

results in

```
-4 a + 8 b-4 c
```

Mathematica's command **Inverse** can be used to find the inverse of a matrix. This command also saves much time and energy.

Example A.15. *Use Mathematica to find the inverse and determinant of the matrix*

$$
A = \begin{pmatrix}
1 & 1 & 2 \\
-1 & 0 & 1 \\
2 & -1 & 5
\end{pmatrix}.
$$

Solution. First, we define the matrix A in *Mathematica* (using the symbol[3] **aaa** instead of A) with

aaa = {{1,1,2},{-1,0,1},{2,-1,5}}

and display it with

aaa//MatrixForm

```
1    1    2
-1   0    1
2    -1   5
```

The command

---

[3]Capital letters should be avoided when defining new symbols in *Mathematica*, since they might conflict with built-in symbols.

Det[aaa]

yields **10**, and the command

Inverse[aaa]//MatrixForm

gives the answer

$$
\begin{array}{ccc}
\dfrac{1}{10} & -\left(\dfrac{7}{10}\right) & \dfrac{1}{10} \\[2ex]
\dfrac{7}{10} & \dfrac{1}{10} & -\left(\dfrac{3}{10}\right) \\[2ex]
\dfrac{1}{10} & \dfrac{3}{10} & \dfrac{1}{10}
\end{array}
$$

Multiplication of Matrices

Mathematica defines two distinct ways to multiply matrices. To see the difference, let us define matrices **mmm** and **ppp** by

mmm = {{1,3,5},{2,a,-1},{b,0,4}}
ppp = {{2,-1,7},{-1,0,-1},{2,1,-1}}

We display them with

      MatrixForm[mmm]   and   **MatrixForm[ppp]**

yielding

$$
\begin{array}{ccc}
1 & 3 & 5 \\
2 & a & -1 \\
b & 0 & 4
\end{array}
\qquad \text{and} \qquad
\begin{array}{ccc}
2 & -1 & 7 \\
-1 & 0 & -1 \\
2 & 1 & -1
\end{array}
$$

To multiply each element of **mmm** by each element of **ppp** we use
MatrixForm[mmm ppp] (or **MatrixForm[mmm*ppp]**), obtaining

$$
\begin{array}{ccc}
2 & -3 & 35 \\
-2 & 0 & 1 \\
2\,b & 0 & -4
\end{array}
$$

Usually, however, we want the matrix product of **mmm** with **ppp**; for this we use **mmm.ppp**, as in the following:

`MatrixForm[mmm.ppp]`

```
9           4           -1
2 - a       -3          15 - a
8 + 2 b     4 - b       -4 + 7 b
```

Powers and Exponentials of Matrices

The command **MatrixPower** can be used to compute a matrix power. The commands **mat.mat** and **MatrixPower[mat,2]** are equivalent. To compute and display the seventh power of **ppp** we use

`MatrixPower[ppp,7]//MatrixForm`

with the result

```
23403    -939    39382
-7806    -602    -9332
10271    2465    5326
```

Notice that **ppp^7//MatrixForm** yields a quite different result:

```
128     -1      823543
-1       0      -1
128      1      -1
```

because **ppp^7** raises each element of **ppp** to the seventh power. We note that the commands **Inverse[ ]** and **MatrixPower[,-1]** are equivalent.

The exponential of a matrix can in principle be calculated using **MatrixExp**. For example

`MatrixForm[`
`    MatrixExp[{{1,1,0,0},{0,1,1,0},{0,0,1,1},{0,0,0,1}}]]`

yields

$$
\begin{array}{cccc}
E & E & \dfrac{E}{2} & \dfrac{E}{6} \\[2mm]
0 & E & E & \dfrac{E}{2}
\end{array}
$$

```
0    0    E    E


0    0    0    E
```

However, the matrices whose exponential can be computed with **MatrixExp** are quite limited, particularly those with symbolic entries. But see Exercise 2.

Exercises for Section A.7

1. Let

$$B = \begin{pmatrix} 0 & 0 & 0 & a \\ 0 & 0 & b & 0 \\ 0 & c & 0 & 0 \\ d & 0 & 0 & 0 \end{pmatrix}.$$

Use **MatrixPower** to compute B^k for $k = -1$, $k = 2$, and $k = 7$. In each case use **MatrixForm** to display the matrix. From this information guess what the matrix B^k is for an arbitrary integer k.

2. *Mathematica*'s command **Chop** replaces very small numbers by 0. (The default meaning of "very small" is less than 10^{-10}.) Compute the approximate value of the exponential of the matrix **ppp** using **Chop[N[MatrixExp[ppp]]]**.

3. The command **Reverse** when applied to a list creates a new list with the elements reversed. Consequently, an $n \times n$ matrix with 1's on the diagonal going from the upper right to the lower left can be generated for any integer **n** by the command **Reverse[IdentityMatrix[n]]**. This is the matrix

$$\begin{pmatrix} 0 & 0 & \dots & 0 & 1 \\ 0 & 0 & \dots & 1 & 0 \\ \vdots & \vdots & \ddots & \vdots & \vdots \\ 0 & 1 & \dots & 0 & 0 \\ 1 & 0 & \dots & 0 & 0 \end{pmatrix}.$$

For example,

Reverse[IdentityMatrix[5]]//MatrixForm

yields

```
0   0   0   0   1
0   0   0   1   0
0   0   1   0   0
0   1   0   0   0
1   0   0   0   0
```

Use **Table** to compute the determinant of **Reverse[IdentityMatrix[n]]** for $1 \leq n \leq 12$. What is the general formula?

A.8 *Solving Equations with* **Mathematica**

Using **Solve**

Linear and other simple algebraic equations can be solved with **Solve**. The basic command is

Solve[equationlist,unknownlist],

where **equationlist** consists of a list of equations such as

{2x + 3y == 4, 5x - y == 6}

and **unknownlist** consists of a list of unknowns such as **{x,y}**. The equations can be linear or simple polynomial equations. The output from **Solve** is a list of rules.

It is important to note that whenever *Mathematica* is asked to solve an equation, "**==**" must be used, whereas in ordinary mathematics one uses "**=**".

Example A.16. *Solve the system of equations*

$$\begin{cases} 2x + 3y = 4, \\ 5x - y = 6 \end{cases}$$

using **Solve**.

Solution. We use

Solve[{2x + 3y == 4, 5x - y == 6},{x,y}]

to get

$$\{\{x \,\to\, \frac{22}{17}, \; y \,\to\, \frac{8}{17}\}\}$$

So the solution is $x = \dfrac{22}{17}, \qquad y = \dfrac{8}{17}.$ ∎

When **equationlist** consists of a single equation, the braces can be omitted. A similar remark applies to **unknownlist**. The following example shows how to solve a simple quadratic equation.

Example A.17. *Use* **Solve** *to find the solutions to*

$$5x^2 - 2x + 3 = 0.$$

and simplify Mathematica's output.

Solution. We use

Solve[5x^2 - 2x + 3 == 0,x]//Simplify

to get

$$\{\{x \,\to\, \frac{1 - I \; Sqrt[14]}{5}\}, \; \{x \,\to\, \frac{1 + I \; Sqrt[14]}{5}\}\}$$

So the solutions are $x = \dfrac{1 \pm i\sqrt{14}}{5}.$ ∎

Inconsistent equations

Mathematica indicates that a set of equations has no solutions by returning the empty set **{}**.

Example A.18. *Use* **Solve** *to determine the solution set of the system of equations*

$$\begin{cases} x + y = 2, \\ x + y = 3. \end{cases}$$

Solution. To the command

Solve[{x + y == 2,x + y == 3},{x,y}]

Mathematica answers

{}

signifying that the solution set is empty. ∎

Redundant equations

When **Solve** is given a list of redundant equations, *Mathematica* chooses some of the variables and solves for the others in terms of the chosen variables.

Example A.19. *Solve the system*

$$\begin{cases} x + y = 1, \\ 2x + 2y = 2. \end{cases}$$

Solution. We use

Solve[{x + y == 1,2x + 2y == 2},{x,y}]

to get

```
{{x -> 1 - y}}
```

Thus, y is arbitrary and $x = 1 - y$. ∎

Using **NSolve**

Numerical solutions can be found using **NSolve** in place of **Solve**; the syntax of the two commands is the same except that optionally **NSolve** can specify the precision of the output. **NSolve** is the preferred command to use to solve an nth-order equation with $n \geq 3$, because for $n \geq 3$ it may happen that **Solve** either fails or gives a result so complicated that it is useless.

Example A.20. *Use* **NSolve** *to find the solutions to*

$$x^3 - 5x^2 - x + 15 = 0.$$

Solution. Although **Solve** can be used to solve this cubic equation:

Solve[x^3 - 5x^2 - x + 15 == 0,x]

the output is complicated and not particularly useful:

```
                         1/3
           5          28 2
{{x ->   - +   -------------------------  +
           3                        1/3
                3 (-110 + 6 I Sqrt[2103])

                              1/3
      (-110 + 6 I Sqrt[2103])
      -------------------------},
                 1/3
```

```
              3  2
                 1/3
       5     14 2    (1 + I Sqrt[3])
{x ->  - - ─────────────────────────── -
       3                          1/3
           3 (-110 + 6 I Sqrt[2103])

                                      1/3
   (1 - I Sqrt[3]) (-110 + 6 I Sqrt[2103])
   ──────────────────────────────────────},
                      1/3
                   6 2

                 1/3
       5     14 2    (1 - I Sqrt[3])
{x ->  - - ─────────────────────────── -
       3                          1/3
           3 (-110 + 6 I Sqrt[2103])

                                      1/3
   (1 + I Sqrt[3]) (-110 + 6 I Sqrt[2103])
   ──────────────────────────────────────}}
                      1/3
                   6 2
```

Instead, we use

NSolve[x^3 - 5x^2 - x + 15 == 0,x]

to get

```
{{x -> -1.58687}, {x -> 2.11268}, {x -> 4.47419}}
```

Thus, the approximate solutions are

$$x \approx -1.58687, \qquad x \approx 2.11268, \qquad \text{and} \qquad x \approx 4.47419. \quad \blacksquare$$

Using **LinearSolve**

Matrix equations can be solved with **LinearSolve**. In contrast to **Solve** and **NSolve**, the input to **LinearSolve** is not an equation; instead the command syntax is

LinearSolve[mat,vec]

where **mat** is a matrix and **vec** is a vector. The output is also a vector.

Example A.21. *Use* **LinearSolve** *to solve the matrix equation*

$$\begin{pmatrix} 5 & 7 & 3 \\ 2 & -1 & 6 \\ 0 & 5 & 5 \end{pmatrix} \begin{pmatrix} x_1 \\ x_2 \\ x_3 \end{pmatrix} = \begin{pmatrix} a \\ b \\ c \end{pmatrix}.$$

Solution. We first name the matrix **mm** with the command

```
mm = {{5,7,3},{2,-1,6},{0,5,5}}
```

As usual **MatrixForm[mm]** displays the matrix:

```
5    7    3
2   -1    6
0    5    5
```

Then the following command solves the matrix equation and simplifies the result.

```
LinearSolve[mm,{a,b,c}]//Simplify//MatrixForm
```

```
7 a + 4 b - 9 c
───────────────
      43
10 a - 25 b + 24 c
──────────────────
       215
-10 a + 25 b + 19 c
───────────────────
        215
```

Hence the solution is $\begin{pmatrix} x_1 \\ x_2 \\ x_3 \end{pmatrix} = \begin{pmatrix} \dfrac{7a + 4b - 9c}{43} \\ \dfrac{10a - 25b + 24c}{215} \\ \dfrac{-10a + 25b + 19c}{215} \end{pmatrix}$. ∎

Exercises for Section A.8

Find the solution sets to the following systems of linear equations using **Solve**.

1. $\begin{cases} x_1 + 2x_2 - 3x_3 - 4x_4 = 6, \\ x_1 + 3x_2 + x_3 - 2x_4 = 4, \\ 2x_1 + 5x_2 - 2x_3 - 5x_4 = 10 \end{cases}$

2. $\begin{cases} 2x_1 + x_2 + 5x_3 + x_4 = 5, \\ x_1 + x_2 - 3x_3 - 4x_4 = -1, \\ 3x_1 + 6x_2 - 2x_3 + x_4 = 8, \\ 2x_1 + 2x_2 + 2x_3 - 3x_4 = 2 \end{cases}$

3. $\begin{cases} x_1 + x_2 + 2x_3 + x_4 = 5, \\ 2x_1 + 3x_2 - x_3 - 2x_4 = 2, \\ 4x_1 + 5x_2 + 3x_3 = 7 \end{cases}$
  4. $\begin{cases} x_1 - x_2 = 1, \\ x_2 - x_3 = 2, \\ x_3 - x_4 = 3, \\ x_1 + x_2 + x_3 + x_4 = 4 \end{cases}$

5. Redo Exercises 1–4 using **LinearSolve**.

6. Use **NSolve** to find approximations to the solutions of the following equations:
   a. $x^3 + x^2 + x + 1 = 0$ **c.** $x^5 - a = 0$
   b. $x^3 + x + 1 = 0$ **d.** $x^2 - y^2 = 1, x + y = 2$

A.9   *Eigenvalues and Eigenvectors with Mathematica*

Mathematica's commands **Eigenvalues** and **Eigenvectors** can be used to find the eigenvalues and eigenvectors of a matrix.

Example A.22. *Find the eigenvalues and eigenvectors of the matrix*

$$A = \begin{pmatrix} 2 & 3 \\ 3 & 4 \end{pmatrix}$$

using Mathematica.

Solution. To find the eigenvalues we use

```
Eigenvalues[{{2,3},{3,4}}]//Simplify
```

obtaining

```
{3 - Sqrt[10], 3 + Sqrt[10]}
```

Similarly, the eigenvectors of A can be determined using *Mathematica*'s command **Eigenvectors**. Although this command gives a specific choice for the eigenvectors, we know from the general theory that any nonzero multiple of an eigenvector is an eigenvector. Using

```
Eigenvectors[{{2,3},{3,4}}]//Simplify
```

we obtain

```
    -1 - Sqrt[10]         -1 + Sqrt[10]
{{——————————— -,1},  {——————————— -,1}}
        3                     3
```

Finally, the command **Eigensystem** gives both the eigenvalues and eigenvectors of a matrix:

Eigensystem[{{2,3},{3,4}}]//Simplify

```
{{3 - Sqrt[10], 3 + Sqrt[10]},

    -1 - Sqrt[10]          -1 + Sqrt[10]
  {{———————————, 1},  {———————————, 1}}}
        3                     3
```

The output is a list consisting of two sublists; the first sublist is a list of the eigenvalues and the second sublist is a list of the eigenvectors. ▌

Next, we consider the case of a matrix with a defective eigenvalue. The command **Eigenvectors** still works, but some of the vectors in the list that it produces are zero vectors. We know that by definition an eigenvector is nonzero, so we must exclude zero vectors from the list.

Example A.23. *Use Mathematica to find the eigenvalues and eigenvectors of the matrix*

$$A = \begin{pmatrix} 3 & 4 \\ 0 & 3 \end{pmatrix}.$$

Solution. The eigenvalues can be found with the command

Eigenvalues[{{3,4},{0,3}}]

```
{3, 3}
```

Thus, 3 is an eigenvalue of multiplicity 2. But when we try

Eigenvectors[{{3,4},{0,3}}]

we get

```
{{1, 0}, {0, 0}}
```

Since A is a 2×2 matrix, *Mathematica* wants the list of eigenvectors to have 2 elements. We know better, so we exclude $(0, 0)$. The only eigenvector is **{1,0}**. ▌

Example A.24. *Use Mathematica to find the eigenvalues and eigenvectors of the matrix*

$$A = \begin{pmatrix} 1 & 0 & 0 & 0 \\ -1 & 0 & 0 & 0 \\ 0 & 0 & 1 & 0 \\ 0 & 0 & -1 & 0 \end{pmatrix}.$$

Solution. We first define the matrix A with the command

```
aa = {{ 1,  0,  0,  0},
      {-1,  0,  0,  0},
      { 0,  0,  1,  0},
      { 0,  0,-1,  0}};
```

Then we get the eigenvalues and eigenvectors with the commands
Eigenvalues[aa] and **MatrixForm[Transpose[Eigenvectors[aa]]]**
obtaining

```
                        0    0    0   -1

                        0    1    0    1
{0, 0, 1, 1}    and
                        0    0   -1    0

                        1    0    1    0
```

We use **Transpose** so that the eigenvectors will appear as column vectors. ∎

Finally, we give an example of numerical computation of eigenvalues and eigenvectors.

Example A.25. *Use Mathematica to compute the eigenvalues and eigenvectors of the matrix*

$$\begin{pmatrix} -5 & 1 & 7 & 6 \\ 5 & 0 & 1 & 18 \\ -2 & 11 & 7 & 4 \\ 1 & 4 & 8 & -9 \end{pmatrix}.$$

Solution. We define the matrix with

```
aa = {{-5,1,7,6},{5,0,1,18},{-2,11,7,4},{1,4,8,-9}}
```

To compute the eigenvalues we use

```
aa//Eigenvalues//N
```

yielding

{-5.53291 - 6.23785 I, -5.53291 + 6.23785 I, -11.8834, 15.9493}

The corresponding eigenvectors are computed with

aa//Eigenvectors//N

yielding

{{1.6721 + 2.66684 I, -3.26945 + 1.62396 I, 1.8591 - 1.92507 I, 1.},

  {1.6721 - 2.66684 I, -3.26945 - 1.62396 I, 1.8591 + 1.92507 I, 1.},

  {-1.0384, -1.10488, 0.32181, 1.}, {1.0911, 1.60723, 2.17865, 1.}}

Exercises for Section A.9

Use **Eigenvalues** and **Eigenvectors** to find the eigenvalues and eigenvectors of the following matrices. Also, compute the exponential of each matrix using **MatrixExp**.

1. $\begin{pmatrix} 3 & 2 \\ 4 & 1 \end{pmatrix}$

2. $\begin{pmatrix} 6 & -4 \\ 3 & -1 \end{pmatrix}$

3. $\begin{pmatrix} 3 & -1 \\ 1 & 1 \end{pmatrix}$

4. $\begin{pmatrix} 3 & -8 \\ 2 & 3 \end{pmatrix}$

5. $\begin{pmatrix} 1 & -1 \\ 2 & -3 \end{pmatrix}$

6. $\begin{pmatrix} 0 & 1 & 0 \\ 0 & 0 & 1 \\ 0 & 0 & 0 \end{pmatrix}$

7. $\begin{pmatrix} 1 & 1 & 1 \\ 0 & 2 & 1 \\ 0 & 0 & 1 \end{pmatrix}$

8. $\begin{pmatrix} 1 & 2 & 1 \\ 0 & 3 & 1 \\ 0 & 5 & -1 \end{pmatrix}$

9. $\begin{pmatrix} 4 & -5 & 1 \\ 1 & 0 & -1 \\ 0 & 1 & -1 \end{pmatrix}$

10. $\begin{pmatrix} -2 & 0 & 1 \\ 1 & 0 & -1 \\ 0 & 1 & -1 \end{pmatrix}$

11. $\begin{pmatrix} 2 & 0 & 0 & 0 \\ 0 & 2 & 0 & 0 \\ 0 & 0 & 3 & 0 \\ 0 & 0 & 0 & 4 \end{pmatrix}$

12. $\begin{pmatrix} 3 & 0 & 0 & 0 \\ 4 & 1 & 0 & 0 \\ 0 & 0 & 2 & 1 \\ 0 & 0 & 0 & 2 \end{pmatrix}$

APPENDIX B:
SYSTEMS OF UNITS

In this appendix we discuss some of the basic notions and other useful information connected with systems of units in physics.

The basic units of measurement in mechanics are those of **mass**, **length**, and **time**. In electricity we also encounter the unit of **charge**. These may be measured in any number of different systems: the standard metric system (c.g.s. or m.k.s.) or the British system are the most commonly used systems. All other quantities in physics are expressed in terms of these. For example we have the following equivalencies:

$$\text{velocity} = \frac{\text{distance}}{\text{time}}$$

$$\text{acceleration} = \frac{\text{velocity}}{\text{time}} = \frac{\text{distance}}{\text{time}^2}$$

$$\text{momentum} = \text{mass} \times \text{velocity} = \frac{\text{mass} \times \text{distance}}{\text{time}}$$

$$\text{force} = \text{weight} = \text{mass} \times \text{acceleration} = \frac{\text{mass} \times \text{distance}}{\text{time}^2}$$

$$\text{energy} = \text{work} = \text{force} \times \text{distance} = \text{mass} \times \text{velocity}^2 = \frac{\text{mass} \times \text{distance}^2}{\text{time}^2}.$$

In the c.g.s. system, length is measured in centimeters, mass is measured in grams, and time is measured in seconds.

In the m.k.s. system, length is measured in meters, mass is measured in kilograms, and time is measured in seconds.

In the British system, length is measured in feet, mass is measured in slugs, and time is measured in seconds.

The mass of a body is its intrinsic resistance to motion, while the **weight** of a body is

807

the force exerted by Earth's gravity at the surface of Earth. These are related by the formula

$$w = m\,g,$$

where g is the **acceleration of gravity**, given approximately in the various systems as

$$g = 32 \text{ feet/second}^2 = 9.8 \text{ meters/second}^2 = 980 \text{ centimeters/second}^2.$$

The following list of conversion factors is helpful in transcribing from one system of units to another:

Length:

1 meter	= 100 centimeters	= 1000 millimeters
1 kilometer	= 1000 meters	= 0.621371 miles
1 foot	= 30.48 centimeters	
1 inch	= 2.54 centimeters	
1 mile	= 5280 feet	= 1.60934 kilometers

Area:

1 centimeter^2	= 0.155 inches^2	
1 meter^2	= 10^4 centimeters^2	= 10.76 feet^2
1 inch^2	= 6.4516 centimeters^2	
1 foot^2	= 144 inches^2	= 0.0929 meters^2

Volume:

1 liter	= 1000 centimeters^3	= 0.001 meters^3	= 0.0353 feet^3
1 foot^3	= 1728 inches	= 0.0283 meters^3	= 28.32 liters
1 gallon^3	= 3.785 liters^3	= 231 inches^3	

Velocity:

$$1 \text{ centimeter/second} = 0.03281 \text{ feet/second}$$

$$1 \text{ foot/second} \qquad = 30.48 \text{ centimeters/second}$$

$$1 \text{ mile/minute} \qquad = 60 \text{ miles/hour} \qquad\qquad = 88 \text{ feet/second}$$

Acceleration:

$$1 \text{ centimeter/second}^2 = 0.03281 \text{ feet/second}^2 \qquad = 0.01 \text{ meters/second}^2$$

$$1 \text{ foot/second}^2 \qquad = 30.48 \text{ centimeters/second}^2 = 0.3048 \text{ meters/second}^2$$

Force:

$$1 \text{ dyne} \quad = 10^{-5} \text{ newtons} = 2.248 \times 10^{-6} \text{ pounds}$$

$$1 \text{ newton} = 10^5 \text{ dynes} \qquad = 0.2248 \text{ pounds}$$

Mass:

$$1 \text{ gram} \quad\quad = 10^{-3} \text{ kilograms} \quad = 6.85 \times 10^{-5} \text{ slugs}$$

$$1 \text{ kilogram} = 10^3 \text{ grams} \qquad\quad = 0.0685 \text{ slugs}$$

$$1 \text{ slug} \quad\quad = 1.459 \times 10^4 \text{ grams} = 14.59 \text{ kilograms}$$

Energy:

$$1 \text{ joule} \quad = 10^7 \text{ ergs} \qquad\qquad = 0.239 \text{ calories}$$

$$1 \text{ calorie} = 4.18 \text{ joule}$$

$$1 \text{ ev} \quad\quad = 10^{-6} \text{ Mev} \qquad\qquad = 1.60 \times 10^{-12} \text{ erg} = 1.07 \times 10^{-9} \text{ amu}$$

$$1 \text{ amu} \quad = 1.66 \times 10^{-24} \text{ grams} = 1.49 \times 10^{-3} \text{ erg} \quad = 931 \text{ Mev}$$

ANSWERS

Chapter 1

Section 1.1, page 3

1.1. ordinary, nonlinear, first-order

1.1. ordinary, nonlinear, second-order

5. ordinary, linear, first-order

7. partial, linear, second-order

Section 1.3, page 15

1. $\dfrac{2t^{5/2}}{5} + C$

3. $\log|1 - t| - \log|t| + C$

5. $\dfrac{8\sqrt{2}}{5} - \dfrac{2}{5} \approx 1.862$

7. $\log(2e - 1) - \log(e - 1) - \log(2)$

Chapter 2

Section 2.2, page 28

1. `y''[t] + t y'[t] + (1 + t^2)y[t] == t/(1 + t)`

3. The *Mathematica* defintion of the function is
   `f[t_]:= 1 + t + Sin[t] + E^t - Sqrt[5 - t^2]`
   Its exact and numerical values are computed with `f[2]` and `f[2]//N`.

5. The exact and numerical values of $(3 \cdot 5 \cdot 7 \cdot 9)^{-1}$ are computed with
   `1/(3 5 7 9)` and `1/(3 5 7 9)//N`

7. `Integrate[t^3 E^(5 t)Cos[t],t]`
   and

```
Integrate[(Sin[t])^7,{t,0,Pi}]
```

9. ```
   f = Function[t,t^3 + Tan[t]]
   f = (#^3 + Tan[#])&
   f[Pi/4]
   f[Pi/4]//N
   ```

Section 2.3, page 32

1. ```
   Plot[Sin[t^2]^2,{t,-2Pi,2Pi}]
   ```

3. ```
   ParametricPlot3D[{1 + Cos[t],Sin[t],2Sin[t/2]},{t,-2Pi,2Pi}]
   ```

Chapter 3

Section 3.1, page 38

1. $y = \dfrac{8t^3}{3} + 2t^6 + \dfrac{2t^9}{3} + \dfrac{t^{12}}{12} + C$

5. $y = 1 + 2t^2 + \dfrac{t^4}{4}$

3. $y = C - t\cos(t) + \sin(t)$

7. $y = \dfrac{-2e^3}{9} + e^{3t}\left(\dfrac{t}{3} - \dfrac{1}{9}\right)$

Section 3.2, page 44

1. $y(t) = \dfrac{t}{2} - \dfrac{1}{8} + \dfrac{3e^{3t}}{7} + C\,e^{-4t}$

13. $y(t) = \dfrac{t^4 e^{-t^2}}{4} + 4e^{-t^2}$

3. $y(t) = \dfrac{3t^2}{2e^{t^2}} + \dfrac{C}{e^{t^2}}$

15. $y(t) = \dfrac{t^3}{8} + \dfrac{7}{8t^5}$

5. $y(t) = \dfrac{t^3}{8} + \dfrac{C}{t^5}$

17. $y(t) = t^2 e^{e^t} + \dfrac{e^{e^t}}{e}$

7. $y(t) = t^2 e^{e^t} + C\,e^{e^t}$

19. $y(t) = \dfrac{\cos(t)}{t^2} + \dfrac{\sin(t)}{t} - \dfrac{\pi}{2t^2}$

9. $y(t) = \dfrac{C}{\sin(t)} - \dfrac{5e^{\cos(t)}}{\sin(t)}$

21. $t = \dfrac{e^y}{2} + \dfrac{C}{e^y}$

11. $y(t) = \dfrac{t}{2} - \dfrac{1}{8} + \dfrac{3e^{3t}}{7} + \dfrac{263e^{-4t}}{56}$

23. $e^y = t \pm \sqrt{t^2 - t_0^2}.$

25. Subtract the two differential equations and use the linearity of differentiation in the form $y_1' - y_2' = (y_1 - y_2)'$.
   ```

27. For part a, use the quotient rule for derivatives and the differential equations satisfied by the numerator and denominator. For part b, integrate the equation found in part a.

29. Divide both sides of the differential equation by $a(t)$ and apply steps 1–4 on page 44.

Section 3.3, page 52

1. $\dfrac{e^y}{y^3} = C\,t^5$

3. $y = -\sin\left(\dfrac{1}{2t^2} + C\right)$

5. $\exp\left(y + \dfrac{y^2}{2}\right) = (1+t)C$

7. $y\sin(y) = -t + \dfrac{t^2}{2} + t\log(t) + C$

9. $-y\cos(y) + \sin(y) = (t^2 - 2t + 2)e^t + C$

11. $y(t) = \tan\left(\dfrac{t^2}{2} + \dfrac{\pi}{4}\right)$

13. $y(t) = \sqrt[3]{3e^t + 64} - 3e$

15. $y(t) = \sqrt{2 - t^2}$

17. $y(t) = \pi$

19. $y(t) = \sqrt{1 + 2\arcsin(t)}$

Section 3.4, page 61

1. $\dfrac{t\,y(6 + t\,y)}{2} = C$

3. $t\,y + \dfrac{y^2}{2} = C$

5. $\begin{cases} \mu(y) = 1/y^2, \\ \log(y) + t/y = C \end{cases}$

7. $\begin{cases} \mu(t) = t, \\ \dfrac{t^4}{2} + t^3 y - \dfrac{t^2 y^2}{2} = C \end{cases}$

9. $\begin{cases} \mu(y) = e^{-y}, \\ \dfrac{y^2}{2} + t\,e^{t-y} = C \end{cases}$

11. $t^2 - t\,y + y^2 = 12$

13. $\begin{cases} \mu(t) = t^2 \sin(t), \\ y = \dfrac{\pi^2 \csc(t)}{4t^2} \end{cases}$

21. $\begin{cases} \mu(t\,y) = \dfrac{1}{(t\,y)^2}, \\ \dfrac{t}{e^{1/(t\,y)}} = C, \qquad y = 0 \end{cases}$

23. $\begin{cases} \mu(t\,y) = t\,y, \\ t^2 y^2 (y^2 - t^2) = C \end{cases}$

Section 3.5, page 65

1. $y = \dfrac{t}{2} + \dfrac{C}{t}$

3. $y = \pm t\sqrt{\dfrac{C\,t^8 - 1}{4}}$

5. $y^4 = t^4 \log(C\,t)$

7. $\dfrac{y^2}{(y^2 + 4t^2)^3} = C$

9. $y = \dfrac{t}{C\,t - 1},\ y = 0$

11. $\dfrac{y}{e^{8t/y}t(8 - y/t)^{17}} = \dfrac{t}{6^{16}3e^4}$

13. $y = \dfrac{1 - t^2}{2}$

15. $y = t\left(1 + \log(t)\right)$

17. $y = \dfrac{1 + \sqrt{1 + 4t - 4t^2}}{2}$

21. $\dfrac{\exp\left(\arctan\left(\dfrac{1/2 + y}{1/2 + t}\right)\right)}{\sqrt{1 + \left(\dfrac{1/2 + y}{1/2 + t}\right)^2}} = (1/2 + t)C$

23. $y - \log(1 + t + y) = C$

25. $y = -\dfrac{\log(t) + 1 + C}{t^2(\log(t) + C)}$

27. $y = Ct - \dfrac{1}{3t^2}$

Section 3.6, page 70

1. $y = \left(C\,e^{3t} + \dfrac{3}{10}\cos(t) + \dfrac{9}{10}\sin(t)\right)^{-1/3},\qquad y = 0$

3. $y = \pm\sqrt{\dfrac{C + e^t(-12 + 12t - 6t^2 + 2t^3)}{t^4}}$

5. $y = \dfrac{1}{t^{1/4}(1 - 4e + 4e^{\sqrt{t}})^{1/2}}$

7. $y = \left(-2 + 3e^{1-t^2}\right)^{-1/4}$

9. $y = \dfrac{\sqrt{t}}{\sqrt{2 - \cos(2) + \cos(t) - 2\sin(2) + t\sin(t)}}$

Chapter 4

Section 4.4, page 90

1. ```
ODE[y' - t y == t,y,t,
Method->FirstOrderLinear]
```
```

3. `ODE[t y' + y == t Sin[t^2],y,t,`
   `Method->FirstOrderLinear]`

5. `ODE[y' + y/(t Log[t]) == 1/t,y,t,`
   `Method->FirstOrderLinear]`

7. `ODE[y' - (3/t^2)y == 1/t^2,y,t,`
   `Method->FirstOrderLinear]`

9. `ODE[t y' + 2 y == Cos[t]^2,y,t,`
   `Method->FirstOrderLinear]`

11. `ODE[{t y' + y == t^4 + t^3,y[1] == 1/2},y,t,`
    `Method->FirstOrderLinear]`

13. `ODE[{y' + Cot[t]y == 2Csc[t],y[Pi/2] == 1},y,t,`
    `Method->FirstOrderLinear]`

15. `ODE[{y'Cot[t] + y == 4Sin[t],y[-Pi] == 0},y,t,`
    `Method->FirstOrderLinear]`

17. `ODE[{Cos[t] y' - Sin[t] y == t^3 E^(t^2),y[0] == 1},y,t,`
    `Method->FirstOrderLinear]`

19. `ODE[{0.05 t^2 y' + t y == t^2,y[0.1] == 0},y,t,`
    `Method->FirstOrderLinear]`

21. `ODE[{y' + y == t Sin[t],y[0] == 0},y,t,`
    `Method->FirstOrderLinear,`
    `PlotSolution->{{t,-2Pi,2Pi}}]`

23. `ODE[{y' - (2/t)y == t,y[1] == 1},y,t,`
    `Method->FirstOrderLinear,`
    `PlotSolution->{{t,0.01,1}}]`

25.
   1. `FirstOrderLinear[1&,4&,2#+3Exp[3#]&,C][t]`
   2. `FirstOrderLinear[1&,-2&,#Exp[3#]&,C][t]`
   3. `FirstOrderLinear[1&,2#&,3#Exp[-#^2]&,C][t]`
   4. `FirstOrderLinear[1&,1/#&,Exp[#^2]&,C][t]`
   5. `FirstOrderLinear[#&,5&,#^3&,C][t]`
   6. `FirstOrderLinear[(1+#^2)&,9&,0&,C][t]`
   7. `FirstOrderLinear[1&,-Exp[#]&,2#Exp[Exp[#]]&,C][t]`
   8. `FirstOrderLinear[1&,1&,Sin[#]&,C][t]`
   9. `FirstOrderLinear[1&,Cot[#]&,5Exp[Cos[#]]&,C][t]`
   10. `FirstOrderLinear[1&,2Cos[#]&,Sin[#]^2Cos[#]&,C][t]`

Section 4.5, page 94

1. `ODE[Sin[t] Sin[y[t]]^2 - Cos[t]^2 y'[t] == 0,y[t],t,`
   `Method->Separable]`

3. `ODE[1 + Sqrt[a^2 - t^2] y'[t] == 0,y[t],t,`
   `Method->Separable]`

5. `ODE[y[t]^2 + 2y[t] + 5 - y'[t] == 0,y[t],t,`
   `Method->Separable,Form->Equation]`

7. `ODE[y[t] + (1 + t^2)ArcTan[t]y'[t] == 0,y[t],t,`
   `Method->Separable,Form->Equation]`

Section 4.6, page 96

1. `ODE[t^3 + t y^4 + 2y^3y' == 0,y,t,`
   `Method->IntegratingFactor,Form->Equation]`

3. `ODE[2t y + (y^2 - 3t^2)y' == 0,y,t,`
   `Method->IntegratingFactor,Form->Equation]`

5. `ODE[y(1 + t y) +t(1 - t y)y' == 0,y,t,`
   `Method->IntegratingFactor,Form->Equation]`

7. `ODE[y^2 E^(t y^2) + 1 + (2t yE^(t y^2) - 3y^2)y' == 0,y,t,`
   `Method->FirstOrderExact]`

9. `ODE[2t Sin[t^2] + 3Cos[y] - 3t Sin[y]y' == 0,y,t,`
   `Method->FirstOrderExact]`

Section 4.7, page 99

1. `ODE[y' - y^2/(t^2 - t y) == 0,y,t,`
   `Method->FirstOrderHomogeneous]`

3. `ODE[y' == -(15t + 11y)/(9t + 5y),y,t,`
   `Method->FirstOrderHomogeneous,Form->Equation]`

5. `ODE[t y' - y == Sqrt[t^2 + y^2],y,t,`
   `Method->FirstOrderHomogeneous]`

7. `ODE[y' == (y - t - t Cos[y/t])/t,y,t,`
   `Method->FirstOrderHomogeneous,Form->LogEquation]`

9. `ODE[y' == (y - 2t Tanh[y/t])/t,y,t,`
   `Method->FirstOrderHomogeneous,Form->Equation]`

Section 4.8, page 101

1. `ODE[y' - y == t E^(5t)y^3,y,t,`
   `Method->Bernoulli,Form->Equation,`
   `PostSolution->{ExpandAll,PowerExpand}]`

3. `ODE[y' + y/(t - 3) == 5(t - 3)y^(3/2),y,t,`
   `Method->Bernoulli]`

5. `ODE[{t y' == t y^2 + (1 + t)y,y[1] == 1},y,t,`
   `Method->Bernoulli]`

7. `ODE[{y' + Tan[2t]y == y^2 Cos[2t]Sin[2t],y[0] == 1},y,t,`
   `Method->Bernoulli,Form->Equation]`

Section 4.9, page 106

1. ODE[y == t y' + (1 + y'^2)^(1/2),y,t,
   Method->Clairaut,Form->Equation]

3. ODE[y == t y' + Sin[y'],y,t,
   Method->Clairaut,Form->Equation]

5. ODE[y == t y' + E^y',y,t,
   Method->Clairaut,Form->Equation]

7. ODE[y == t y' + y'^4/4,y,t,
   Method->Clairaut,Form->Equation]

9. ODE[y == 2t y' + Log[y'],y,t,
   Method->Lagrange]

11. ODE[y == (3/2)t y' + Exp[y'],y,t,
    Method->Lagrange]

Section 4.10, page 111

1. ODE[{y' == Sin[t^2],y[0] == 0},y,t,
   Method->Separable,
   PlotSolution->{{t,-3Pi,3Pi},PlotPoints->100}]

3. ODE[{y' == t y + y^5,y[0] == 4},y,t,
   Method->Bernoulli,
   PlotSolution->{{t,-3Pi,3Pi},PlotPoints->100}]

Section 4.11, page 113

1. y1[t_]=ODE[{y' -t y == Sin[t],y[0] == 0},y,t,
       Method->FirstOrderLinear,Form->Explicit]
       Plot[y1[t],{t,-1,1}];

3. y3[t_]=ODE[{y' == Sin[E^t],y[0] == 0},y,t,
       Method->FirstOrderLinear,Form->Explicit]
       Plot[y3[t],{t,0,3Pi/2},PlotRange->All,PlotPoints->100];

5. y5[t_]=ODE[{y' + t y == Sin[t]/t,y[1] == 0},y,t,
       Method->FirstOrderLinear,Form->Explicit]
       Plot[y5[t],{t,-2Pi,2Pi},PlotRange->All];

7. y7[t_]=ODE[{y' == Exp[-t^2] Cos[t],y[0] == 1},y,t,
       Method->FirstOrderLinear,Form->Explicit]
       Plot[y7[t],{t,-Pi,Pi},PlotRange->All];

Section 4.12, page 116

1. $y(t) = \dfrac{2 - 3C\, e^t}{C\, e^t - 1}$

3. $y(t) = -\dfrac{e^{3t} - 6C}{e^{3t} + 3C}$

5. $y(t) = \tan(t) - \dfrac{\sec(t)}{C - \log\left(\dfrac{\cos(t)}{1 + \sin(t)}\right)}$

7. $y(t) = e^t + e^{-t}$

```
ODE[{y' E^-t + y^2 - 2y E^t == 1 - E^(2t),y[0] == 2},y,t,
Method->Riccati,KnownSolution->E^t]
```

9. $y(t) = \dfrac{e^t(1 - 2e^t)}{2e^t - 3}$

```
ODE[{y' == Exp[2t] + (1 + 2Exp[t])y + y^2,y[0] == 1},y,t,
Method->Riccati,KnownSolution->-Exp[t]]
```

Chapter 5

Section 5.1, page 126

1. The solution exists everywhere except the t-axis.

3. The solution exists everywhere except the circle $t^2 + y^2 = 4$.

5. We can expect a unique solution only if $\alpha \geq 1$.

7. A suitable code using **ODE** is

```
prob7[t_]=ODE[{y'== Sin[y]/Cosh[t],y[0] == Pi/2},y,{t,0,10},
        Method->NDSolve,MaxSteps->2000,Form->Explicit,
        PlotSolution->{{t,0,10},PlotRange->All}];
```

Section 5.2, page 130

1. Global existence does not hold.

3. Global existence holds.

5. Global existence does not hold.

7. Global existence holds.

9. Global existence holds.

11. $y(t) = (1 - 3t)^{-1/3}, \quad T = 1/3$

13. $y(t) = -2 + \dfrac{2}{2 - e^t}, \quad T = \log(2)$

Section 5.3, page 136

1. $y_0(t) = 1$, $y_1(t) = 1 + 2t$,
   $y_2(t) = 1 + 2t + 2t^2$,
   $y_3(t) = 1 + 2t + 2t^2 + 4t^3/3$

3. $y_0(t) = 2$, $y_1(t) = 2 + 7t$,
   $y_2(t) = 2 + 7t + 21t^2/2$,
   $y_3(t) = 2 + 7t + 21t^2/2 + 21t^3/2$

5. $y_0(t) = 1$, $y_1(t) = 1 - t/2$,
   $y_2(t) = 1 - t/2 + 3t^2/8 - t^3/8 + t^4/64$

7. $y_0(t) = 0$, $y_1(t) = t$,
   $y_2(t) = t + 2t^2 + 2t^3/3$,
   $y_3(t) = t + 2t^2 + 10t^3/3 + 8t^4/3 + 32t^5/15 + 8t^6/9 + 8t^7/63$

Section 5.5, page 144

1. `ODE[y' == (y + t)/(y -t),y,t,Method->None,`
   `PlotField->{{{t,-2,2},{y,-2,2}}}]`

3. `ODE[y' + y == t y^3,y,t,Method->None,`
   `PlotField->{{{t,-2,2},{y,-2,2}}}]`

5. `ODE[{y' == Sin[t]/t,y[0]==0},y,t,Method->None,`
   `PlotField->{{{t,-Pi,Pi},{y,-2,2}}}]`

7. `ODE[{y' == y + y^3,y[1]==1},y,t,`
   `PlotSolution->{{t,-2,2},`
   `PlotStyle->{{RGBColor[1,0,0],AbsoluteThickness[2]}}},`
   `PlotField->{{{t,-2,2},{y,-2,2}}},`
   `PlotRange->{{-2,2},{-2,2}}},`
   `SuperimposePlots->True]`

9. `ODE[{y' == Sin[t]/y,y[0]==1},y,t,`
   `PlotField->{{{t,-2Pi,2Pi},{y,-2Pi,2Pi}}},`
   `PlotSolution->{{t,-2Pi,2Pi},`
   `PlotStyle->{{RGBColor[1,0,0],AbsoluteThickness[2]}}},`
   `SuperimposePlots->True]`

11. `ODE[{y' == Cos[t]/Sin[y],y[0]==Pi/2},y,t,`
    `PlotField->{{{t,-2Pi,2Pi},{y,-2Pi,2Pi}}},`
    `PlotSolution->{{t,-2Pi,2Pi}},`
    `PlotStyle->{{RGBColor[1,0,0],AbsoluteThickness[2]}}},`
    `SuperimposePlots->True]`

13. `ODE[{y' == y - Cos[Pi/2 t],y[0]==0},y,t,`
    `Method->FirstOrderLinear,`
    `PlotField->{{{t,-2,2},{y,-2,2}}},`
    `PlotSolution->{{t,-2,2}},`
    `PlotStyle->{{RGBColor[1,0,0],AbsoluteThickness[2]}}},`
    `SuperimposePlots->True]`

Section 5.6, page 153

1. `ODE[y' == 2y - Exp[-y],y,t,Method->None,`
   `PlotField->{{{t,-2,2},{y,-2,2}}}]`
   `FindRoot[2y - Exp[-y] == 0,{y,0}]`

3. `ODE[y' == y^2 - y - Cos[y],y,t,Method->None,`
   `PlotField->{{{t,-2,2},{y,-2,2}}}]`
   `FindRoot[y^2 - y - Cos[y] == 0,{y,-2}]`
   `FindRoot[y^2 - y - Cos[y] == 0,{y,2}]`

5. $y = 0$ is unstable and $y = 1/3$ is stable.

7. The critical points are of the form $y = 2n\pi$ for integers n and they are all semistable.

9. $y = 0$ is semistable, $y = -4$ is unstable, and $y = 6$ is stable.

11. $y = 0$ is unstable.   15. $y = 0$ is semistable.

13. $y = 0$ is semistable.   17. $y = 0$ is semistable.

Chapter 6

Section 6.1, page 165

1. $k = \dfrac{0.693}{15600} = 0.000044 \text{ second}^{-1}$

3. **a.** $P(t) = 6e^{5t-10}$
   b. $P(t) = 3 \times 2^t$
   c. $P(t) = 3 \times 10^9 \, e^{0.0135(t-1993)}$

5. $0.012122/\text{years}$

7. $0.007499/\text{years}$

9. **a.** $14, 400$
   b. $345, 600$
   c. $126, 144, 000$
   d. 23.78 years

11. $P(t) = -3/2 + (10^{23} + 3/2)e^{2(t-1)}$

13. $k = 0.02$, $r = 0.05$

Section 6.2, page 169

1. $P(t) = 4t + \sin(t) + 3$

3. $P(t) = 7\exp\big(\sin(t)\big) - 4$

Section 6.3, page 176

1. **a.** 0 is an unstable critical point and 6 is a stable critical point.
   b. $P(t) = \dfrac{6P_0}{\big(6 - P_0\big)e^{-6t} + P_0}$ **c.** $t = \dfrac{\log(25)}{6} \approx 0.536$

3. **a.** $P(2) = 7, 720, 340$ kilograms   **b.** $t \approx 3.09$ years

5. **a.** 0 is a stable critical point and k/b is an unstable critical point.

7. **a.** 0 is an unstable critical point and k/b is a semistable critical point.

   b.

   $$-\frac{1}{k^2} \log\left(\frac{k - b\,P(t)}{P(t)}\right) + \frac{1}{k^2 - k\,b\,P(t)}$$

   $$= t - t_0 - \frac{1}{k^2} \log\left(\frac{k - b\,P_0}{P_0}\right) + \frac{1}{k^2 - k\,b\,P_0}$$

   c. $t = \dfrac{1}{k^2}\left(\dfrac{80}{9} + 2\log(9)\right) \approx \dfrac{13.2823}{k^2}$

9. $\mathbf{L}_d(25.4) = \dfrac{1}{3.06} \log\left(2 + \dfrac{2(0.02)(25.4)}{3.06 - 2(0.02)(25.4)}\right) = 0.089744$

11. If the initial amount of virus is below 0.33, then the virus will die out.

Section 6.4, page 183

1. $T = \displaystyle\int_0^{P(0)} \frac{dP}{(3 - P)(1 - P)} = \frac{1}{2} \log\left(\frac{3 - P(0)}{3(1 - P(0))}\right).$

3. $P = 1$ is an unstable critical point and $P = 5$ is a stable critical point. The solution $P(t)$ is given by
   $$\frac{P(t) - 5}{P(t) - 1} = e^{4t} \frac{P(0) - 5}{P(0) - 1}.$$

5. $P = 3$ is a semistable critical point. The solution is $P(t) = 3 + \dfrac{P(0) - 3}{1 + t(P(0) - 3)}$. If $P(0) < 3$
   then $P(t) = 0$ when $t = \dfrac{P(0)}{3(3 - P(0))}$. If $P(0) > 3$ then $P(t)$ is never zero and $P(t) \longrightarrow 3$ when
   $t \longrightarrow \infty.$

Section 6.6, page 199

1. $T(t) = 68 - 28e^{-3t}$

3. $T(t) = 68 + (3/4)e^{-4t}$

5. $\tau = 20.97$ minutes
   $T(15) = 51.86°F$

7. $T(2) = 80.902°F$
   $t_{85} = 2.1972$ hours

9. 13.9564 minutes

Chapter 7

Section 7.1, page 212

1. $v(t) = (F_0/m)t + V_0$, $y(t) = (F_0/2m)t^2 + V_0t + Y_0$

3.   **a.** 180.95 feet/second, 1388 feet

     b. 17.5 feet/second

     c. 270.4 seconds

5.   **a.** $v(t) = \dfrac{1}{\sqrt{k/(m\,g)}} \dfrac{V_0\sqrt{k/(m\,g)} - \tan(t\sqrt{kg/m})}{1 - V_0\sqrt{k/(m\,g)}\,\tan(t\sqrt{kg/m})}$

     b. $T = \sqrt{m/kg}\,\arctan(V_0\sqrt{k/(m\,g)})$

     c. $y_{max} = \dfrac{m}{2k}\log(k\,v^2 + m\,g)\Big|_0^{V_0} = \dfrac{m}{2k}\log(1 + k\,V_0^2/m\,g)$

7. $F(v) = -g - kv|v|/m$. The critical point is $v_{crit} = -\sqrt{mg/k}$, where $F'(v) = kv/m < 0$, hence a stable critical point.

9.   **a.** $y = \dfrac{m}{k}\left(v - \dfrac{m\,g}{k}\log(m\,g + k\,v)\right) - \dfrac{m}{k}\left(V_0 - \dfrac{m\,g}{k}\log(m\,g + k\,V_0)\right)$

     b. $y_{max} = -\dfrac{m\,V_0}{k} - \dfrac{m^2g}{k^2}\log\left(\dfrac{m\,g}{m\,g + k\,V_0}\right)$

     c. $y_{max} = 17.3273$ meters

     d. $v = -16.9217$ meters per second

Section 7.2, page 217

1.

Planet	Acceleration Due to Gravity (in kilometers/second^2)		Escape Velocity (in kilometers/second)	
Mercury	$g_{Mercury}$	$=$ 0.0037	$V_{Mercury}$	$=$ 4.24
Venus	g_{Venus}	$=$ 0.0088	V_{Venus}	$=$ 10.32
Earth	g_{Earth}	$=$ 0.0098	V_{Earth}	$=$ 11.18
Mars	g_{Mars}	$=$ 0.0037	V_{Mars}	$=$ 5.01
Jupiter	$g_{Jupiter}$	$=$ 0.0246	$V_{Jupiter}$	$=$ 59.31
Saturn	g_{Saturn}	$=$ 0.0104	V_{Saturn}	$=$ 35.41
Uranus	g_{Uranus}	$=$ 0.0089	V_{Uranus}	$=$ 21.33
Neptune	$g_{Neptune}$	$=$ 0.0111	$V_{Neptune}$	$=$ 23.45
Pluto	g_{Pluto}	$=$ 0.0006	V_{Pluto}	$=$ 1.17
Moon	g_{Moon}	$=$ 0.0016	V_{Moon}	$=$ 2.35

3. The position and velocity are related by $v^2/2 - g\,R^2/(R + y) = V_0^2/2 - g\,R$. Setting $v = 0$ and solving yields $y_{max} = R\,V_0^2/(2R\,g - V_0^2)$.

Section 7.3, page 224

1. $I(t) = \dfrac{E_0(1 - e^{-(R/L)t})}{R}$

3. $I(t) = \dfrac{E_0\left(R^2\cos(\omega\,t) + L\,R\,\omega\sin(\omega\,t) + L^2\omega^2 e^{-(R/L)t}\right)}{R(R^2 + \omega^2 L^2)}$

5.
   a. $I(t) = \dfrac{50(1 - e^{-15000t})}{3}$

   b. $\lim\limits_{t\to\infty} I(t) = \dfrac{50}{3}$

   c. $t_1 = 0.000046$ seconds

7. $I(t) = \dfrac{(15 + 44\pi + 60\pi^2)e^{-50t} + 22\sin(100\pi\,t) - 44\pi\cos(100\pi\,t)}{15 + 60\pi^2}$

Section 7.4, page 230

1. $\lim_{t\to 200} x(t) = 0$. The *Mathematica* commands are

    ```
    Clear[x]
    x[t_]=ODE[{x'[t] == 0.6 - 25 x[t]/(1000 - 5 t),x[0] == 5},x,t,
        Method->FirstOrderLinear,Form->Explicit]
    Limit[x[t],t->200]
    ```

3. **a.** $P(t)/K = \left(P(0)/K - 1\right)e^{-t/\tau} + 1$

    c. $P(t)/K_0 = \left(P(0)/K_0 - 1/(1 - \alpha\tau)\right)e^{-t/\tau} + e^{-\alpha t}/(1 - \alpha\tau)$

Section 7.5, page 234

1. The capture point is $(a, 2a/3)$.

3. Equation (7.63) can be used to define a function in *Mathematica* by

    ```
    linearpursuit[a_,k_][t_]:= (a k)/(k^2 - 1)
        + (k(a - t)(a - t)^k^(-1))/(2a^(1/k)(1 + k))
        - (a^(1/k)k(a - t)^(1 - 1/k))/(2(k - 1))
    ```

    The command

    ```
    Plot[Evaluate[linearpursuit[1,1.2][t]],{t,0,1}]
    ```

    yields the required plot.

Chapter 8

Section 8.1, page 239

1. $y'' + t\,\csc(t)y' + 3y\,\csc(t) = 1$ 7. $y(t) = C_1 \log(t) + C_2$

3. $y'' + t\,y' + t^2 y = t^4$

5. $y(t) = \dfrac{t^4}{4} + \dfrac{t^3}{3} - \sin(t) + C_1 t + C_2$ 9. $y(t) = C_1 \sin(3t) + C_2 \cos(3t)$

Section 8.2, page 244

1. In any interval that does not include $0, \pi, -\pi, 2\pi, -2\pi, \ldots$.

Section 8.3, page 246

1. $y_1(t) = 1, y_2(t) = e^{3t}$

5. $y(t) = -t + C_1 + C_2 e^{3t}$

3. $y_1(t) = \cos(4t), y_2(t) = \sin(4t)$

7. $y_1(t) = -2 + C_1 e^{3t} + C_2 e^{-3t}$

Section 8.4, page 253

1. $W(t) = -e^{2t}$

3. $W(t) = 2e^{2t}$

Section 8.5, page 258

1. Linearly independent

3. Linearly independent

Section 8.6, page 260

1. $y_2(t) = e^t$

5. $y_2(t) = \dfrac{\cos(t)}{\sqrt{t}}$

3. $y_2(t) = t^{-3}$

7. $y_2(t) = -2 - 2t - t^2$

Section 8.7, page 262

1. $y'' - \dfrac{2(t+1)y'}{t} + \dfrac{(t^2 + 2t + 2)y}{t^2} = 0$

3. $y'' + 2t' + 5y = 0$

5. $y'' + \dfrac{y'}{t} = 0$

Chapter 9

Section 9.1, page 275

1. $\{e^{-3t}, e^{-t}\}$

9. $\{e^{kt}, e^{-kt}\}$

3. $\{e^{3t}, t\,e^{3t}\}$

11. $y(t) = (5/2)e^t + (9/2)e^{3t}$

5. $\{e^{-2t}\cos(t), e^{-2t}\sin(t)\}$

7. $\{e^{-kt}, 1\}$

13. $y(t) = 4e^{3t} + 5t\,e^{3t}$

15. $y(t) = 3e^{-2t}\cos(t) + 4e^{-2t}\sin(t)$

19. $y(t) = \left(\dfrac{a}{2} + \dfrac{b}{2k}\right)e^{kt} + \left(\dfrac{a}{2} - \dfrac{b}{2k}\right)e^{-kt}$

17. $y(t) = -(b/k)e^{-kt} + a + b/k$

Section 9.2, page 282

1. $y(t) = e^{(3t/2)}\left(C_1\cos\left(\dfrac{\sqrt{3}\,t}{2}\right) + C_2\sin\left(\dfrac{\sqrt{3}\,t}{2}\right)\right)$

3. $y(t) = e^{2t}\left(C_1\cos(2t) + C_2\cos(2t)\right)$

5. $y(t) = C_1 e^{2it} + C_2 e^{3t}$

Section 9.3, page 292

1. $y_P(t) = \dfrac{e^{-5t}}{12}$

9. $y_P(t) = \dfrac{1}{4} + \dfrac{\cos(2t)}{68} - \dfrac{\sin(2t)}{17}$

3. $y_P(t) = \dfrac{t^2 e^{-2t}}{2}$

11. $y(t) = \dfrac{e^{-5t}}{12} + \dfrac{e^{-t}}{4} - \dfrac{e^{-2t}}{3}$

5. $y_P(t) = \dfrac{t\sin(2t)}{4}$

7. $y_P(t) = \left(\dfrac{19}{108} - \dfrac{5t}{18} + \dfrac{t^2}{6}\right)e^t$

13. $y(t) = \left(\dfrac{t^2}{2} + 4t + 2\right)e^{-2t}$

15. $y(t) = \left(\dfrac{t}{4} + \dfrac{5}{2}\right)\sin(2t) + 2\cos(2t)$

17. $y(t) = \dfrac{71e^{-t}}{4} - \dfrac{268e^{-2t}}{27} + \left(\dfrac{19}{108} - \dfrac{5t}{18} + \dfrac{t^2}{6}\right)e^t$

19. $y(t) = \dfrac{1}{4} + \dfrac{\cos(2t) - 4\sin(2t)}{68} + \dfrac{e^{-2t}\left(93\cosh(\sqrt{2}\,t) + 112\sqrt{2}\cosh(\sqrt{2}\,t)\right)}{68}$

21. $y_P(t) = A_0 + A_1 t + A_2 t^2 + A_3 t^3 + t(B_0 + B_1 t + B_2 t^2 + B_3 t^3 + B_4 t^4 + B_5 t^5)e^{-4t}$
    $+ C\cos(2t) + D\sin(2t)$

23. $y_P(t) = A\,e^{-t}\cos(2t) + B\,e^{-t}\sin(2t) + (C_0 + C_1 t + C_2 t^2)e^{2t}\sin(t)$
    $+ (D_0 + D_1 t + D_2 t^2)e^{2t}\cos(t)$

25. $y_P(t) = (A_0 t + A_1 t^2)e^{-t}\cos(2t) + (B_0 t + B_1 t^2)e^{-t}\sin(2t)$
    $+ (C_0 + C_1 t)e^{-2t}\cos(t) + (D_0 + D_1 t)e^{-2t}\sin(t)$

27. $y_P(t) = (A_0 + A_1 t)e^t \cos(2t) + (B_0 + B_1 t)e^t \sin(2t) + (C_0 + C_1 t + C_2 t^2)e^{5t}$

     $+ (D_0 t + D_1 t^2 + D_2 t^3 + D_3 t^4)e^{-5t}$

29. $y_P(t) = A t\, e^{-2t} + (B_0 + B_1 t)e^{-3t} + (C_0 t + C_1 t^2 + C_2 t^3)e^{-4t}$

     $+ (D_0 + D_1 t + D_2 t^2 + D_3 t^3)e^{-5t}$

Section 9.4, page 300

1. $y_P(t) = \dfrac{t^2 e^{2t}}{2} + \dfrac{t^3 e^{2t}}{6}$

3. $y_P(t) = \dfrac{1}{2e^t}$

5. $y_P(t) = \dfrac{e^t \sin(t)}{5} - \dfrac{2e^t \cos(t)}{5}$

7. $y_P(t) = t\, e^t (\log(t) - 1)$

9. $y_P(t) = \cos(t) \log\big(\cos(t)\big)$

     $- \cos(t) \log\big(1 + \cos(t)\big)$

11. $y_P(t) = \dfrac{t^2}{e^t}\left(\dfrac{\log(t)}{2} - \dfrac{3}{4}\right)$

13. $y_P(t) = \dfrac{t^{n+2} e^t}{n^2 + 3n + 2}$

15. $y(t) = 10 - 6e^t + 2e^{3t} + 8t + 3t^2$

17. $y(t) = 4e^{2t} + 3e^{3t} - 2e^{5t} + 3t\, e^{2t}$

19. $y(t) = \dfrac{1}{4e^{t/2}} + \dfrac{3e^{t/2}}{4}$

     $- \dfrac{t\, e^{t/2}}{4} + \dfrac{t^2 e^{t/2}}{8}$

21. $y(t) = e^{2t} - 2t\, e^{2t} - t^3 e^{2t} + t^4 e^{2t}$

23. $y(t) = \dfrac{3}{e^t} - \dfrac{2}{e^{2t}} + \dfrac{\cos(1)}{e^t}$

     $- \dfrac{\cos(1)}{e^{2t}} + \dfrac{\sin(1)}{e^{2t}} - \dfrac{\sin(e^t)}{e^{2t}}$

25. $y(t) = C_1 t + C_2 t^2 - t - t\, \log(t)$

27. $y(t) = C_1 t + C_2 t^2 + \dfrac{7}{4} + \dfrac{3}{2}\log(t)$

     $+ \dfrac{1}{2}\big(\log(t)\big)^2$

29. $y(t) = C_1 + C_2 t\, e^t - e^t$

31. $y(t) = t - t^2 + t^2 \log(t)$

33. $y(t) = -\dfrac{3}{4} + \dfrac{2}{3t} + \dfrac{t^2}{12} + \dfrac{\log(t)}{2}$

     $- \dfrac{\big(\log(t)\big)^2}{2}$

35. $y(t) = 1 - e^t + t\, e^t$

Chapter 10

Section 10.1, page 314

1. ```
ODE[3y'' + 4y' - 2y == 0,y,t,
Method->SecondOrderLinear]
```
```

3. `ODE[32y'' + 25y' - 4y == 0,y,t,`
   `Method->SecondOrderLinear]`

5. `ODE[y'' + 16y == E^t - 9t^2,y,t,`
   `Method->SecondOrderLinear]`

7. `ODE[I y'' + y' - 3y == Sin[I t],y,t,`
   `Method->SecondOrderLinear]`

9. `ODE[y'' - 4y' + 12y == 15E^t -4 Cos[t],y,t,`
   `Method->SecondOrderLinear]`

11. `ODE[{2y'' - 10y' - 3y == 0,y[0] == 5,y'[0] == 3},y,t,`
    `Method->SecondOrderLinear,PlotSolution->{{t,-1,1}}]`

13. `ODE[{y'' - 3y' + 10y == 0,y[0] == 1,y'[0] == -1},y,t,`
    `Method->SecondOrderLinear,PlotSolution->{{t,-3,3}}]`

15. `ODE[{8y'' + 5y == t^2,y[0] == 1,y'[0] == Pi/Sqrt[5/2]},y,t,`
    `Method->SecondOrderLinear,PlotSolution->{{t,-10,10}}]`

17. `ODE[{y'' - 4y == (1+t)Cos[10t],y[0] == 0,y'[0] == 0},y,t,`
    `Method->SecondOrderLinear,PlotSolution->{{t,-1,1}}]`

19. `ODE[{y'' - 2y == Sin[t] + Exp[t],y[0] == 1,y'[0] == 0},y,t,`
    `Method->SecondOrderLinear,PlotSolution->{{t,-1,1}}]`

21. `ODE[{y'' + y == Sin[2t],y[0] == 0,y'[0] == a},y,t,`
    `Method->SecondOrderLinear,Parameters->{{a,0,4,0.1}},`
    `StackPlotSolution->{{t,-2,2},PlotPoints->30},`
    `ViewPoint->{-1.582, 2.225, 2.000}]`

23. `ODE[{3y'' - 2y' + 15y == 0,y[0] == a,y'[0] == -1},y,t,`
    `Method->SecondOrderLinear,Parameters->{{a,0,4,0.2}},`
    `StackPlotSolution->{{t,-2Pi,2Pi},PlotPoints->50},`
    `ViewPoint->{0.109, -2.980, 1.600}]`

25. `ODE[{y'' + y == Cos[a t],y[0] == 1,y'[0] == 0},y,t,`
    `Method->SecondOrderLinear,Parameters->{{a,5,10,0.2}},`
    `StackPlotSolution->{{t,-2Pi,2Pi},PlotPoints->100},`
    `ViewPoint->{0.052, -2.647, 2.108}]`

27. `ODE[{a y'' + y' + y == 0,y[0] == 1,y'[0] == 0},y,t,`
    `Method->SecondOrderLinear,Parameters->{{a,0,5,0.2}},`
    `StackPlotSolution->{{t,-1,1},PlotPoints->100},`
    `ViewPoint->{0.479, -2.975, 1.540}]`

29. `ODE[{y'' - 4y' +4y == Sqrt[t],y[0] == 0,y'[0] == 0},y,t,`
    `Method->SecondOrderLinear,PlotSolution->{{t,0,2}}]`

31. `ODE[{y'' - y == E^(t^2),y[0] == 0,y'[0] == 0},y,t,`
    `Method->SecondOrderLinear,PlotSolution->{{t,0,3}}]`

Section 10.2, page 320

1. `HomogeneousSecondOrderLinear[1,0,1,C][t]`

3. `HomogeneousSecondOrderLinear[1,3,2,C][t]`

5. `HomogeneousSecondOrderLinear[1,-3,2,C][t]`

7. `HomogeneousSecondOrderLinear[1,0,I,C][t]`

9. `VariationOfParameters[{Sin,Cos},Sec][t]`

11. `VariationOfParameters[{Exp[-#]&,Exp[#]&},Cosh[#]&][t]`

13. `VariationOfParameters[{Exp[#]&,Exp[-#]&},2^#&][t]`

15. `VariationOfParameters[{Exp[#]Sin[#]&,Exp[#]Cos[#]&},Exp[#]Sec[#]&][t]`

Section 10.3, page 322

1. `ODE[12y[t]/t^2 - 6y'[t]/t + y''[t] == 0,y[t],t,`
   `Method->ReductionOfOrder,KnownSolution->t^3]`

3. `ODE[y''[t] - (1/t - 3/(16t^2))y[t] == 0,y[t],t,`
   `Method->ReductionOfOrder,KnownSolution->t^(1/4)Exp[2Sqrt[t]]]`

5. `ODE[t^2 y''[t] - 2y[t] == 0,y[t],t,`
   `Method->ReductionOfOrder,KnownSolution->t^2]`

7. `ODE[(t^2 - t) y''[t] + (3t-1)y'[t] + y[t] == 0,y[t],t,`
   `Method->ReductionOfOrder,`
   `KnownSolution->1/(1-t)]`

Section 10.4, page 324

1. `EquationFromSolutions[Sin[2#]&,Cos[2#]&][t,y]`

3. `EquationFromSolutions[#^2&,Log][t,y]`

5. `EquationFromSolutions[#E^#&,Log[#]&][t,y]`

7. `EquationFromSolutions[#^2+#&,Cos[#]&][t,y]`

Chapter 11

Section 11.1, page 346

1. $k = 9$ pounds/foot

3. $u(t) = 3\cos(2t) - 7\sin(2t)$

5. $u(t) = 1.2\cos(2t)$

7. $u(t) = -4\cos(t) - 4\sqrt{3}\sin(t)$

9. $u(t) = \dfrac{10\sqrt{2}}{7}\sin(7\sqrt{2}t) \approx 2.02\sin(9.9t)$

11. $u(t) = -10\cos(7\sqrt{2}t) + \dfrac{\sqrt{2}\sin(7\sqrt{2}t)}{7} \approx 10.002\cos(9.9t - 3.12)$

13. Let $\omega_0 = \sqrt{k/m}$. Then $u(t) = \dfrac{U_1\sin(\omega_0 t)}{\omega_0}$

15. $u(t) = \sqrt{\dfrac{kU_0^2 + mU_1^2}{k}}\cos\left(\omega_0 t - \arctan\left(\dfrac{U_1\omega_0}{U_0}\right)\right)$

17. We compute

$$\frac{d}{dt}E(t) = m\,u'(t)u''(t) + k\,u(t)u'(t) = u'(t)\Big(m\,u''(t) + k\,u(t)\Big) = 0.$$

Since its derivative is zero, the function $E(t)$ must be a constant.

19. Writing $x_1 = u(t) = R\cos(\omega_0 t - \alpha)$, we have

$$x_2 = u'(t) = -R\,\omega_0\sin(\omega_0 t - \alpha).$$

Thus

$$\frac{x_1^2}{R^2} + \frac{x_2^2}{R^2\omega_0^2} = \big(\cos(\omega_0 t - \alpha)\big)^2 + \big(\sin(\omega_0 t - \alpha)\big)^2 = 1,$$

which is the equation of an ellipse. From this equation it follows that the lengths of the axes of the ellipse are

$$2R = 2\sqrt{U_0^2 + \frac{U_1^2}{\omega_0^2}} \quad\text{and}\quad 2R\,\omega_0 = 2\sqrt{\omega_0^2 U_0^2 + U_1^2}.$$

21. $u(t) \approx e^{-0.875t}\big(3\cos(1.26t) - 1.89\sin(1.26t)\big)$

23. $u(t) = e^{-2t}\cosh\left(2\sqrt{\dfrac{2}{3}}\,t\right)$ and $\tau \approx 2.725$ seconds

25. Writing $x_1 = u(t) = C_1 e^{r_1 t} + C_2 e^{r_2 t}$, we have $x_2 = u'(t) = r_1 C_1 e^{r_1 t} + r_2 C_2 e^{r_2 t}$ and $|x_2 - r_1 x_1| = |(r_2 - r_1)C_2| e^{r_2 t}$, $|x_2 - r_2 x_1| = |(r_1 - r_2)C_1| e^{r_1 t}$. Thus $|x_2 - r_1 x_1|^{r_1} = |(r_2 - r_1)C_2|^{r_1} e^{r_1 r_2 t}$, $|x_2 - r_2 x_1|^{r_2} = |(r_1 - r_2)C_1|^{r_2} e^{r_1 r_2 t}$, which proves the assertion. The constant C is written in terms of the initial conditions as

$$C = \frac{|u'(0) - r_1 u(0)|^{r_1}}{|u'(0) - r_2 u(0)|^{r_2}}.$$

29. $E'(t) = m\,u(t)u''(t) + k\,u(t)u'(t) = u'(t)\Big(9m\,u''(t) + k\,u(t)\Big) = u'(t)(-c\,u'(t)) = -c\,u'(t)^2 \le 0.$
Since the derivative is less than or equal to zero, the function $E(t)$ must be a decreasing function of time.

31. If $u(t) = R e^{\lambda t} \cos(\mu t - \alpha)$, then $u' - \lambda u = -\mu R \sin(\mu t - \alpha)$. Therefore

$$r^2 = (\mu u)^2 + (u' - \lambda u)^2 = \mu^2 R^2 e^{-2\lambda t} \qquad \text{and} \qquad \log(r) = \log(R) + \lambda t.$$

But

$$\tan(\theta) = \frac{u' + \lambda u}{\mu u} = \tan(\mu t - \alpha),$$

so that $\theta = \mu t - \alpha - m\pi$ for some $m = 0, \pm 1, \pm 2, \ldots$. This proves that

$$\log(r) = \frac{\lambda t}{\mu} + C$$

for a constant C, from which the polar equation follows by taking the exponential of both sides.

Section 11.2, page 364

1. $u(t) = -\dfrac{163}{227} \cos\left(\dfrac{4t}{\sqrt{3}}\right) - \dfrac{163 \cos(9t)}{227}$

3.   **a.** $-2\cos(7t - \pi/2)\cos(t - \pi/2)$ **c.** $2\cos(8t)\cos(t - \pi/2)$

          b. $2\cos(7t)\cos(t)$ **d.** $2\cos(8t - \pi/2)\cos(t)$

5. For part **a** we write

$$
\begin{aligned}
u(t) &= A\cos(\omega t - \alpha)\cos(\nu t - \beta) \\[1em]
&= A\Big(\cos((\omega - \nu)t - (\alpha - \beta)) + \cos((\omega + \nu)t - (\alpha + \beta))\Big) \\[1em]
&= A\cos(\alpha - \beta)\cos\big((\omega - \nu)t\big) + A\sin(\alpha - \beta)\sin\big((\omega - \nu)t\big) \\
&\quad + A\cos(\alpha + \beta)\cos\big((\omega + \nu)t\big) + A\sin(\alpha + \beta)\big)\sin\big((\omega + \nu)t\big),
\end{aligned}
$$

which is of the required form. To do part **b**, we combine the result of part **a** with Exercise 4.

7. We modify the solution method of Exercise 6; specifically,

```
ODE[{m u'' + k u == F1 Sin[omega t],u[0] == U0,u'[0] == U1},u,t,
Method->SecondOrderLinear,Form->Explicit,
PostSolution->{(# /. k ->omega0^2 m)&,
PowerExpand,Expand,Collect[#,{U0,U1,F1}]&,Cancel}]
```

yields

```
                       U1 Sin[omega0 t]
U0 Cos[omega0 t] + ────────────────── +
                          omega0

   F1 (-(omega0 Sin[omega t]) + omega Sin[omega0 t])
  ──────────────────────────────────────────────────
                       2                2
          omega0 (m omega   - m omega0  )
```

Thus

$$u(t) = U_0 \cos(\omega_0 t) + + \left(U_1 - \frac{F_1 \omega}{m(\omega_0^2 - \omega^2)} \right) \frac{\sin(\omega_0 t)}{\omega_0} + \frac{F_1 \sin(\omega t)}{m(\omega_0^2 - \omega^2)}.$$

9. $\quad u(t) = \dfrac{F_0 \left(1 - \cos\left(t\sqrt{k/m} \right) \right)}{2k} + \dfrac{F_0 \left(\cos(2\omega t) - \cos\left(t\sqrt{k/m} \right) \right)}{4m\,\omega^2 - k} + 4\cos\left(t\sqrt{k/m} \right)$

11.   **a.** If $c = 0$, the amplification factor is $\dfrac{k}{\sqrt{m\,\omega^2 - k}}$, which is infinite when $\omega = \sqrt{k/m}$.

    b. If $0 < c^2 < 2mk$, then the amplification factor is $k/\sqrt{D}$, where $D = m^2(\omega^2 - \omega_0^2)^2 + c^2\omega^2$.
The maximum is obtained by finding the minimum value of D. This is a quadratic polynomial in ω^2
with an interior minimum that occurs when $2m^2\omega^2 + (c^2 - 2m^2\omega_0^2) = 0$.

    c. If $c^2 \geq 2mk$, then $m^2(\omega^2 - \omega_0^2)^2 + c^2\omega^2$ has no interior minimum. At the end point $\omega = 0$
we have $D = m^2\omega_0^4$, whereas $D \longrightarrow \infty$ for $\omega \longrightarrow \infty$. Hence the absolute minimum of D occurs
when $\omega = 0$.

13.   **a.** For $0 < t \leq \varepsilon$ we have $y(t) = \dfrac{F_0}{k}(1 - \cos(\omega_0 t))$, whereas for $t > \varepsilon$ we have

$$y(t) = y(\varepsilon) \cos(\omega_0(t - \varepsilon)) + \frac{y'(\varepsilon)}{\omega_0} \sin(\omega_0(t - \varepsilon)), \text{ where}$$

$$y(\varepsilon) = \frac{F_0}{k}(1 - \cos(\omega_0\varepsilon)) \quad \text{and} \quad y'(\varepsilon) = \frac{F_0\omega_0}{k} \sin(\omega_0\varepsilon)).$$

    b. When $\varepsilon \longrightarrow 0$, $F_0 \longrightarrow \infty$ with $F_0\varepsilon \longrightarrow I_0$, we have

$$y(\varepsilon) \longrightarrow 0 \quad \text{and} \quad y'(\varepsilon) \longrightarrow \frac{I_0\omega_0^2}{k} = \frac{I_0}{m},$$

by l'Hôpital's rule, for example. Therefore, $y(t)$ tends to $(I_0/m\,\omega_0)\sin(\omega_0 t)$.

Section 11.3, page 374

1.   We enter the following commands

```
Clear[Cp,L]
L /: Im[L] = 0;
Cp /: Im[Cp] = 0;
L /: Positive[L] = True;
Cp /: Positive[Cp] = True;
```

Then

```
ODE[{L Q'' + Q/Cp == 0,Q[0] == Q0,Q'[0] == Q1},Q,t,
Method->SecondOrderLinear,Form->Explicit,
PostSolution->{(# /. Cp ->1/(omega0^2 L)&),PowerExpand}]
```

yields

```
                     Q1 Sin[omega0 t]
Q0 Cos[omega0 t] + ──────────────────── .
                         omega0
```

Thus,

$$Q(t) = Q_0 \cos(\omega_0 t) + \frac{Q_1 \sin(\omega_0 t)}{\omega_0}.$$

3. We use

```
ODE[{L Q'' + Q/Cp == E1 Sin[omega t],Q[0] == Q0,Q'[0] == Q1},Q,t,
Method->SecondOrderLinear,Form->Explicit,
PostSolution->{(# /. Cp ->1/(omega0^2 L)&),
PowerExpand,Expand,Collect[#,{Q0,Q1,E1}]&,Cancel}]
```

to get

```
                   Q1 Sin[omega0 t]
Q0 Cos[omega0 t] + ─────────────────  +
                        omega0

  E1 (-(omega0 Sin[omega t]) + omega Sin[omega0 t])
  ─────────────────────────────────────────────────.
                      2              2
         omega0 (L omega   - L omega0 )
```

Thus,

$$Q(t) = Q_0 \cos(\omega_0 t) + \left(Q_1 - \frac{E_1 \omega}{L(\omega_0^2 - \omega^2)}\right)\frac{\sin(\omega_0 t)}{\omega_0} + \frac{E_1 \sin(\omega t)}{L(\omega_0^2 - \omega^2)}.$$

5.   **a.** $Q(t) = C_1 e^{-(750+50\sqrt{205})t} + C_2 e^{-(750-50\sqrt{205})t}$

         b. $Q(t) = C_1 e^{-(250+10\sqrt{615})t} + C_2 e^{-(250-10\sqrt{615})t}$

         c. $Q(t) = C_1 e^{-5t} \sin(5t) + C_2 e^{-5t} \cos(5t)$

7.   **a.** $I(t) = -25 e^{-4t} \sin(3t)$

         b. $I(t) = -(100/19)e^{-20t} + (50/9)e^{-10t} - (50/171)e^{-t}$

         c. $I(t) = -100\, t\, e^{-10t}$

9. $I(t) = A\cos(120\pi t) + B\sin(120\pi t)$ with

$$A = ((120\pi)^2 - 4 \times 10^6)/((120\pi)^2 - 4 \times 10^6 + 25 \times 10^6(120\pi)^2)$$

and

$$B = 5000 \times 110/((120\pi)^2 - 4 \times 10^6 + 25 \times 10^6(120\pi)^2)$$

11. We define the charge with

```
Q[t_]=ODE[{QQ''/2 + 20QQ' + QQ/10^-2 == 15,QQ[0] == 0,QQ'[0] == 0},
    QQ,t,Method->SecondOrderLinear,Form->Explicit]
```

Then evaluation of **Q[t]** and **Q'[t]//Expand** yields

```
       -20 t - 10 Sqrt[2] t        -20 t - 10 Sqrt[2] t
3    3 E                        3 E
── - ─────────────────────  +  ─────────────────────  -
20            40                   20 Sqrt[2]


    -20 t + 10 Sqrt[2] t       -20 t + 10 Sqrt[2] t
  3 E                        3 E
  ─────────────────────  -  ─────────────────────
           40                   20 Sqrt[2]
```

and

$$\frac{-3\ E^{-20\ t\ -\ 10\ Sqrt[2]\ t}}{2\ Sqrt[2]} + \frac{3\ E^{-20\ t\ +\ 10\ Sqrt[2]\ t}}{2\ Sqrt[2]}$$

Thus

$$Q(t) = \frac{3}{20} - \frac{3e^{-20t}\left(\cosh(10\sqrt{2}\,t) + \sqrt{2}\sinh(10\sqrt{2}\,t)\right)}{40}$$

and

$$I(t) = \frac{3\sqrt{2}\,e^{-20t}\,\sinh(10\sqrt{2}\,t)}{4}.$$

The charge and the current can be plotted with

```
Plot[Q[t]//Evaluate,{t,0,1}];
```

and

```
Plot[Q'[t]//Evaluate,{t,0,1}];
```

13. $Q(t) = \dfrac{15e^{-10t} - 11e^{-50t}}{200}$ and $I(t) = \dfrac{-3e^{-10t} + 11e^{-50t}}{4}$

15. $Q(t) = \dfrac{22\cos(100t) + 11\sin(100t)}{500} - \dfrac{e^{-100t}\left(44\cos(200t) + 33\sin(200t)\right)}{1000}$ and

$$I(t) = \frac{11\cos(100t) - 22\sin(100t)}{5} + \frac{e^{-100t}\left(-22\cos(200t) + 121\sin(200t)\right)}{10}$$

Chapter 12

Section 12.1, page 385

12.1. $y^{(4)} + \dfrac{e^t y''' + 3y'' + y}{\log(t)} = 0$

5. Any interval

7. Any interval not containing 0

12.1. $y''' + t\,y'' + t^2\,y' + t\,y = t$

9. $4e^{5t}$

Section 12.2, page 391

1. $y(t) = C_1 + C_2\cos(t) + C_3\sin(t)$

3. $y(t) = C_1 e^{-\sqrt{5}\,t} + C_2 e^{\sqrt{5}\,t} + C_3\cos(\sqrt{5}\,t) + C_4\sin(\sqrt{5}\,t)$

5. $y(t) = C_1 e^t + C_2 t\,e^t + C_3 t^2 e^t + C_4 t^3 e^t + C_5 t^4 e^t + C_6 t^5 e^t$

7. $y(t) = C_1 e^{-\sqrt{2}\,t} + C_2 t\,e^{-\sqrt{2}\,t} + C_3 e^{\sqrt{2}\,t} + C_4 t\,e^{\sqrt{2}\,t}$
    $+ C_5\cos(t) + C_6\sin(t) + C_7 t\cos(t) + C_8 t\sin(t)$

9. $y(t) = C_1 e^{3t} + C_2 t\,e^{3t} + C_3 t^2 e^{3t} + C_4 t^3 e^{3t} + C_5\cos(t) + C_6\sin(t)$

11. $y(t) = \sin(t)$

13. $y(t) = \dfrac{9 - (5 - t)\cos(\sqrt{2}\,t)}{4} - \dfrac{(5 - 3t)\sin(\sqrt{2}\,t)}{4\sqrt{2}}$

15. $y(t) = \dfrac{(2079 - 1026t)e^t}{128} + \dfrac{(1146t^2 + 4334t - 6237)\cos(\sqrt{3}\,t)}{384}$
    $+ \dfrac{(1389t^2 - 8622t - 11333)\sin(\sqrt{3}\,t)}{384\sqrt{3}}$

Section 12.3, page 397

1. $y_{\mathbf{p}}(t) = -\left(\dfrac{1}{4} + \dfrac{t}{2}\right)e^t$

5. $y_{\mathbf{p}}(t) = \dfrac{(-3 + 2t - 2t^2)e^t}{8}$

7. $y_{\mathbf{p}}(t) = 6 + t$

3. $y_{\mathbf{p}}(t) = \dfrac{\left(-1 + \cos(t)\right)\log\left(\cos(t)\right)}{3}$
    $+ \dfrac{t\sin(t)}{3} - \dfrac{\sin(2t)}{6}\log\left(\dfrac{1 + \sin(t)}{\cos(t)}\right)$

9. $y_{\mathbf{p}}(t) = -\dfrac{(181 + 30t + 450t^2)e^{-t}}{54000}$
    $+ \dfrac{(7 + 12t)e^t}{576}$

11. $y_p(t) = \dfrac{(-6t + 3t^2 - t^3)e^t}{12}$

13. $y(t) = C_1 t + C_2 t \log(t) + C_3 t\big(\log(t)\big)^2$
$\quad + \dfrac{8\,t^4\ (\log(t) - 1)}{9}.$

Section 12.5, page 405

1. `ODE[y''''' + y''' + y'' - y' + y == 0,y,t,`
    `Method->NthOrderLinear]`
    $$y(t) \approx C_1 e^{-1.14\,t} + e^{0.0863\,t}\big(C_2 \cos(1.26\,t) + C_3 \sin(1.26\,t)\big)$$
    $$+ e^{0.485\,t}\big(C_4 \cos(0.56\,t) + C_5 \sin(0.56\,t)\big)$$

3. `ODE[y'''''' + 2y''' + y'' - y' + y == 0,y,t,`
    `Method->NthOrderLinear]`
    $$y(t) \approx e^{-1.07\,t}\big(C_1 \cos(0.468\,t) + C_2 \sin(0.468\,t)\big)$$
    $$+ e^{0.385\,t}\big(C_3 \cos(0.534\,t) + C_4 \sin(0.534\,t)\big)$$
    $$+ e^{0.681\,t}\big(C_5 \cos(1.11\,t) + C_6 \sin(1.11\,t)\big)$$

5. `ODE[y''''''' + 2y'''' + 3y''' + 4y == 0,y,t,`
    `Method->NthOrderLinear]`
    $$y(t) \approx C_1 e^{-1.17\,t} + e^{-0.91\,t}\big(C_2 \cos(0.87\,t) + C_3 \sin(0.87\,t)\big)$$
    $$+ e^{0.613\,t}\big(C_4 \cos(0.862\,t) + C_5 \sin(0.862\,t)\big)$$
    $$+ e^{0.88\,t}\big(C_6 \cos(1.078\,t) + C_7 \sin(1.08\,t)\big)$$

7. `ODE[y'''''''' + y'''' - y''' + y == 0,y,t,`
    `Method->NthOrderLinear]`
    $$y(t) \approx e^{-0.902\,t}\big(C_1 \cos(0.647\,t) + C_2 \sin(0.647\,t)\big)$$
    $$+ e^{-0.469\,t}\big(C_3 \cos(0.715\,t) + C_4 \sin(0.715\,t)\big)$$
    $$+ e^{0.494\,t}\big(C_5 \cos(0.991\,t) + C_6 \sin(0.991\,t)\big)$$
    $$+ e^{0.877\,t}\big(C_7 \cos(0.367\,t) + C_8 \sin(0.367\,t)\big)$$

9. `ODE[{y''' + 4y' == Sec[2t],y[0] == 1,y'[0] == 1,y''[0] == 0},y,t,`
    `Method->NthOrderLinear,PlotSolution->{{t,0,5}}]`
    $$y(t) = 1 - \frac{t \cos(2t)}{4} + \frac{\sin(2t)}{2} + \frac{1}{8}\log\big((1 + \sin(2t))(1 + \tan(2t))\big)$$

11. `ODE[{y'''' + 2y'' + y == t,`
    `y[0] == 0,y'[0] == 0,y''[0] == 0,y'''[0] == 0},y,t,`
    `Method->NthOrderLinear,PlotSolution->{{t,-15,15}}]`
    $$y(t) = t + \frac{t \cos(t) - 3 \sin(t)}{2}$$

13. `ODE[{y''''' + y''' + y'' + y == 0,`
    `y[0] == 1,y'[0] == 1,y''[0] == 0,y'''[0] == 0,y''''[0] == 0},y,t,`
    `Method->NthOrderLinear,PlotSolution->{{t,0,5}}]`

$$y(t) = \cos(t) + \frac{2e^{t/2}\sin(\sqrt{3}\,t/2)}{\sqrt{3}}$$

15. `ODE[{y'''' - 2y''' - 13y'' + 14y' + 24y == Sinh[t],`
`y[0] == 1,y'[0] == 1,y''[0] == 0,y'''[0] == 0},y,t,`
`Method->NthOrderLinear,PlotSolution->{{t,-1,1}}]`

$$y(t) = -\frac{81e^{-3t}}{560} + \frac{103e^{-t}}{225} + \frac{e^t}{48} + \frac{37\,e^{2t}}{45} - \frac{82\,e^{4t}}{525} - \frac{t\,e^{-t}}{60}$$

Chapter 13

Section 13.1, page 413

n	t_n	Y_n	$y_{exact}(t_n)$
0	0.0	0.0	0.0
1	0.1	0.10	0.1025
2	0.2	0.2049	0.21
3	0.3	0.3146	0.3225
4	0.4	0.4293	0.44
5	0.5	0.5489	0.5625
6	0.6	0.6733	0.69
7	0.7	0.8027	0.8225
8	0.8	0.9369	0.96
9	0.9	1.076	1.1025
10	1.0	1.22	1.25

1.

3. $y(t) = \dfrac{5\,e^{2t}}{4} - \dfrac{t}{2} - \dfrac{1}{4}$

h	t_n	Y_n	$y_{exact}(t_n)$
1/2	1.0	0.25	0.4192
1/4	1.0	0.3281	0.4192
1/8	1.0	0.3751	0.4192

5. $y(t) = \log\left(\dfrac{\sqrt{2+t^4}}{\sqrt{2}}\right)$

h	t_n	Y_n	$y_{exact}(t_n)$
1/2	1.0	0.0625	0.2027
1/4	1.0	0.1333	0.2027
1/8	1.0	0.1691	0.2027

t_n	$Y_n(h = 0.2)$	$Y_n(h = 0.1)$	$y_{\text{exact}}(t_n)$
0.0	1.0	1.0	1.0
0.1		1.1	1.106
0.2	1.2	1.211	1.224
0.3		1.336	1.36
0.4	1.448	1.479	1.515
0.5		1.643	1.696
0.6	1.77	1.832	1.906
0.7		2.051	2.151
0.8	2.196	2.305	2.437
0.9		2.6	2.769
1.0	2.763	2.941	3.155

7. $y(t) = -2 + 3e^t - 2t - t^2$

9. 4.256

11. 2.263

13.

h	t_n	Y_n
1/10	1.0	7.19
1/20	1.0	12.32
1/30	1.0	17.8
1/40	1.0	23.93
1/50	1.0	30.92
1/60	1.0	39.03
1/70	1.0	48.59
1/80	1.0	59.99
1/90	1.0	73.8
1/100	1.0	90.76

Section 13.2, page 417

1.

n	t_n	Y_n	$y_{\text{exact}}(t_n)$
0	0.0	0.0	0.0
1	0.1	0.105	0.1052
2	0.2	0.221	0.2214
3	0.3	0.3492	0.3499
4	0.4	0.4909	0.4918
5	0.5	0.6474	0.6487
6	0.6	0.8204	0.8221
7	0.7	1.012	1.014
8	0.8	1.223	1.226
9	0.9	1.456	1.460
10	1.0	1.714	1.718

3. $y(t) = \dfrac{5e^{2t}}{4} - \dfrac{t}{2} - \dfrac{1}{4}$

h	t_n	Y_n	$y_{\text{exact}}(t_n)$
1/2	1.0	0.5625	0.4192
1/4	1.0	0.4407	0.4192
1/8	1.0	0.4235	0.4192

5. $y(t) = \log\left(\dfrac{\sqrt{2 + t^4}}{\sqrt{2}}\right)$

h	t_n	Y_n	$y_{\text{exact}}(t_n)$
1/2	1.0	0.2694	0.2027
1/4	1.0	0.2156	0.2027
1/8	1.0	0.2055	0.2027

7. $y(t) = -2 + 3e^t - 2t - t^2$

t_n	$Y_n(h = 0.2)$	$Y_n(h = 0.1)$	$y_{\text{exact}}(t_n)$
0.0	1.0	1.0	1.0
0.1		1.105	1.106
0.2	1.224	1.224	1.224
0.3		1.359	1.36
0.4	1.514	1.515	1.515
0.5		1.695	1.696
0.6	1.902	1.905	1.906
0.7		2.15	2.151
0.8	2.428	2.434	2.437
0.9		2.765	2.769
1.0	3.139	3.15	3.155

9. 5.286

11. 2.349

13.

h	t_n	Y_n
1/10	1.0	3.813×10^1
1/20	1.0	1.440×10^2
1/30	1.0	5.562×10^2
1/40	1.0	3.015×10^3
1/50	1.0	3.103×10^4
1/60	1.0	9.073×10^5
1/70	1.0	1.355×10^8
1/80	1.0	2.458×10^{11}
1/90	1.0	1.993×10^{16}
1/100	1.0	5.313×10^{23}

Section 13.3, page 422

1.

n	t_n	Y_n	$y_{\text{exact}}(t_n)$
0	0.0	0.0	0.0
1	0.1	0.1052	0.1052
2	0.2	0.2214	0.2214
3	0.3	0.3499	0.3499
4	0.4	0.4918	0.4918
5	0.5	0.6487	0.6487
6	0.6	0.8221	0.8221
7	0.7	1.014	1.014
8	0.8	1.226	1.226
9	0.9	1.46	1.460
10	1.0	1.718	1.718

3. $y(t) = \dfrac{5\,e^{2t}}{4} - \dfrac{t}{2} - \dfrac{1}{4}$

h	t_n	Y_n	$y_{\text{exact}}(t_n)$
1/2	1.0	0.4258	0.4192
1/4	1.0	0.4194	0.4192
1/8	1.0	0.4192	0.4192

5. $y(t) = \log\left(\dfrac{\sqrt{2+t^4}}{\sqrt{2}}\right)$

h	t_n	Y_n	$y_{\text{exact}}(t_n)$
1/2	1.0	0.2036	0.2027
1/4	1.0	0.2028	0.2027
1/8	1.0	0.2027	0.2027

7. $y(t) = -2 + 3e^t - 2t - t^2$

t_n	$Y_n(h=0.2)$	$Y_n(h=0.1)$	$y_{\text{exact}}(t_n)$
0.0	1.0	1.0	1.0
0.1		1.106	1.106
0.2	1.224	1.224	1.224
0.3		1.36	1.36
0.4	1.515	1.515	1.515
0.5		1.696	1.696
0.6	1.906	1.906	1.906
0.7		2.151	2.151
0.8	2.437	2.437	2.437
0.9		2.769	2.769
1.0	3.155	3.155	3.155

9. 5.372

11. 2.342

13.

h	t_n	Y_n
1/10	1.0	3.813×10^1
1/20	1.0	1.440×10^2
1/30	1.0	5.562×10^2
1/40	1.0	3.015×10^3
1/50	1.0	3.103×10^4
1/60	1.0	9.073×10^5
1/70	1.0	1.355×10^8
1/80	1.0	2.458×10^{11}
1/90	1.0	1.993×10^{16}
1/100	1.0	5.313×10^{23}

Section 13.5, page 433

The following command is to be used with each of the answers to Problems 1–4 in Section 13.5:

```
ODE[eq,y,{t,0,1},Method->Euler,StepSize->0.01,PlotSolution->{{t,0,1}}];
```

1. `eq = {y' == 1 - t y^10,y[0] == 1};`

3. `eq = {y' == t - Sin[y],y[0] == 1};`

The following set of commands is to be used with each of the answers to Problems 5–8 in Section 13.5:

```
$ODEPlotNumber = 0;
ODE[eq,y,range,Method->Euler,StepSize->0.1,
    PlotSolution->{range,PlotStyle->{GrayLevel[.6]}}];
ODE[eq,y,range,Method->Heun,StepSize->0.1,
    PlotSolution->{range,PlotStyle->{GrayLevel[.3]}}];
ODE[eq,y,range,Method->RungeKutta4,StepSize->0.1,
    PlotSolution->{range}];
Show[Graph[{1,2,3}]];
```

5. **eq = {y' == 1 - t y^10,y[0] == 1};**
    range = {t,0,5};

7. **eq = {y' == t - Sin[y],y[0] == 1};**
    range = {t,0,2};

The following set of commands is to be used with each of the answers to Problems 9–12 in Section 13.5:

```
ODE[eq,y,{t,0,0.5},Method->Euler,StepSize->0.1,
    ODEDigits->6,PostSolution->{Last[Last[#]]&}]
ODE[eq,y,{t,0,0.5},Method->Heun,StepSize->0.1,
    ODEDigits->6,PostSolution->{Last[Last[#]]&}]
ODE[eq,y,{t,0,0.5},Method->RungeKutta4,StepSize->0.1,
    ODEDigits->6,PostSolution->{Last[Last[#]]&}]
```

9. **eq = {y' == Exp[-y],y[0] == 0};**

```
Euler       = 0.419761
Heun        = 0.405281
RungeKutta  = 0.405465
```

11. **eq = {y' == t y + Sqrt[y],y[0] == 1};**

```
Euler       = 1.69024
Heun        = 1.75569
RungeKutta  = 1.75609
```

Section 13.6, page 437

1. **ODE[{y' == Sin[t y^2]/(t - 21),y[0] == -1},y,{t,0,20},**
    Method->NDSolve,PlotSolution->{{t,0,20}}];

3. **ODE[{y' == t y/(t^2 - 1),y[0] == 1},y,{t,0,2},**
    Method->NDSolve,PlotSolution->{{t,0,2}}];

5. **ODE[{y' == Log[Sqrt[t^2 + y^2]],y[0] == 1},y,{t,0,5},**
    Method->NDSolve,PlotSolution->{{t,0,5}}];

7. **ODE[{y' == Max[t,y],y[0] == 0},y,{t,0,1},**
    Method->NDSolve,PlotSolution->{{t,0,1}}];

9. **ODE[{y' == t^2/Abs[y],y[1] == 1},y,{t,1,5},**
    Method->NDSolve,PlotSolution->{{t,1,5}}];

11. **ODE[{y' == t^2 + y^2,y[0] == 1},y,{t,0,1},**
    Method->NDSolve,MaxSteps->1000];

13. `ODE[{y'' == t y,y[0] == 1,y'[0] == 1},y,{t,0,1},`
     `Method->NDSolve,PlotSolution->{{t,0,1}}];`

15. `ODE[{t^2 y'' + t y' - (t^2 + 4)y == 0,y[1] == 0,y'[1] == 1},`
     `y,{t,1,2},Method->NDSolve,`
     `PlotSolution->{{t,1,2}}];`

17. `ODE[{y'' - t y' + 7y == 0,y[0] == 1,y'[0] == 0},y,{t,0,5},`
     `Method->NDSolve,PlotSolution->{{t,0,5}}];`

19. `ODE[{(1 - t^2) y'' - 2t y' + (6 - 25/(1 - t^2))y == 0,y[0] == 1,`
     `y'[0] == 0},y,{t,0,0.5},Method->NDSolve,`
     `PlotSolution->{{t,0,0.5}}];`

21. `ODE[{y'' + 0.1 y' - y + y^5 == 0,y[0] == a,y'[0] == 0},`
     `y,{t,0,50},Method->NDSolve,MaxSteps->10000,`
     `Parameters->{{a,-2,2,0.5}},`
     `PlotSolution->{{t,0,50},PlotPoints->50}];`
     It appears that $\lim_{t \to \infty} y(t) = \pm 1$.

Section 13.7, page 442

The following command is to be used with each of the answers to Problems 1–6 in Section 13.7. It uses **Module** to combine several individual commands into a single command called **plt[vary]**, whose single argument, which must be **False** or **True**, deactivates or activates the **VariableStepSize** option, respectively. After defining **eq**, **range** and **interval**, first evaluate **plt[False]** and then **plt[True]**.

```
plt[vary_]:= Module[{},
$ODEPlotNumber = 0;
ODE[eq,y,interval,Method->FirstOrderLinear,StepSize->0.1,
    PlotSolution->{interval,PlotRange->range}];
ODE[eq,y,interval,VariableStepSize->vary,
    Tolerance->0.001,Method->Euler,StepSize->0.1,
    PlotSolution->{interval,PlotRange->range}];
ODE[eq,y,interval,VariableStepSize->vary,
    Tolerance->0.001,Method->RungeKutta4,StepSize->0.1,
    PlotSolution->{interval,PlotRange->range}];
ODE[eq,y,interval,VariableStepSize->vary,
    Tolerance->0.001,Method->AdamsBashforth,StepSize->0.1,
    PlotSolution->{interval,PlotRange->range}];
ODE[eq,y,interval,VariableStepSize->vary,
    Tolerance->0.001,Method->BulirschStoer,StepSize->0.1,
    PlotSolution->{interval,PlotRange->range}];
ODE[eq,y,interval,newoptions,
    Method->ImplicitRungeKutta,
    StepSize->0.1,Tolerance->0.001,
    PlotSolution->{interval,PlotRange->range}];
Show[GraphicsArray[{Graph[{1,2}],Graph[{3,4}],
    Graph[{5,6}]}]];]
```

Use *Mathematica* to plot the numerical solutions of the following initial value problems. In each case first use **VariableStepSize->False** and **StepSize->0.1** and integrate over the interval $0 \le t \le 1$.

Then resolve the problem using **VariableStepSize->True** and **Tolerance->0.001**, again with **StepSize->0.1**. Try to find an exact solution for each problem using **ODE**. For those problems for which **ODE** yields an exact solution, compare the exact solution with the solutions found by the Euler, Runge-Kutta, Adams-Bashforth, Bulirsch-Stoer, and implicit Runge-Kutta methods.

1. ```
   eq = {y' == -50y,y[0] == 1/50};
   range = {-1/50,1/50};
   interval = {t,0,1};
   ```
   $$y(t) = \frac{e^{-50t}}{50}$$

3. ```
   eq = {y' == -100y + t,y[0] == 1};
   range = {-1,1};
   interval = {t,0,1};
   ```
   $$y(t) = \frac{t}{100} - \frac{1}{10000} + \frac{10001e^{-100t}}{10000}$$

5. ```
   eq = {y' == -100(y - Cos[t]) - Sin[t],y[0] == 1};
   range = {-1,1};
   interval = {t,0,1};
   ```
   $$y(t) = \cos(t)$$

The following command is to be used with each of the answers to Problems 7–12 in Section 13.7. It uses **Module** to combine several individual commands into a single command called **fin[]**, which requires no argument. After defining **eq** and **interval**, evaluate **fin[]** to obtain the solutions.

```
fin[]:= Module[{},
Print[ODE[eq,y,interval,
    Method->FirstOrderLinear,Form->Equation]];
Print["Exact                    = ",N[ODE[eq,y,interval,
    Method->FirstOrderLinear,Form->Explicit] /.
    t -> Last[interval],4]];
Print["Euler                    = ",Last[Last[ODE[eq,y,interval,
    Method->Euler,StepSize->0.1,ODEDigits->4,
    VariableStepSize->True,Tolerance->0.0001]]]];
Print["Runge-Kutta              = ",Last[Last[ODE[eq,y,interval,
    Method->RungeKutta4,StepSize->0.1,ODEDigits->4,
    VariableStepSize->True,Tolerance->0.0001]]]];
Print["Adams-Bashforth          = ",Last[Last[ODE[eq,y,interval,
    Method->AdamsBashforth,StepSize->0.1,ODEDigits->4,
    VariableStepSize->True,Tolerance->0.0001]]]];
Print["Bulirsch-Stoer           = ",Last[Last[ODE[eq,y,interval,
    Method->BulirschStoer,StepSize->0.1,ODEDigits->4,
    VariableStepSize->True,Tolerance->0.0001]]]];
Print["Implicit Runge-Kutta = ",Last[Last[ODE[eq,y,interval,
    Method->ImplicitRungeKutta,StepSize->0.1,ODEDigits->4,
    VariableStepSize->True,Tolerance->0.0001]]]];]
```

For each of the initial value Problems 7–12, use **ODE** with **VariableStepSize->True**, **Tolerance->0.0001**, and **StepSize->0.1**. Obtain a four-decimal approximation to the indicated value using the Euler, Runge-Kutta, Adams-Bashforth, Bulirsch-Stoer, and implicit Runge-Kutta methods.

7. ```
   eq = {y' == y,y[0] == 1/60};
   ```

```
interval = {t,0,1};
```
$$y(t) = \frac{e^t}{60}$$

Exact	= 0.0453
Euler	= 0.04322
Runge-Kutta	= 0.04522
Adams-Bashforth	= 0.0453
Bulirsch-Stoer	= 0.0451
Implicit Runge-Kutta	= 0.0453

9.
```
eq = {y' == t,y[0] == 1};
interval = {t,0,2};
```
$$y(t) = 1 + \frac{t^2}{2}$$

Exact	= 3.
Euler	= 2.981
Runge-Kutta	= 3.
Adams-Bashforth	= 3.
Bulirsch-Stoer	= 3.
Implicit Runge-Kutta	= 3.

11.
```
eq = {y' == y - Cos[t] - Sin[t],y[0] == 1};
interval = {t,0,Pi};
```
$$y(t) = \cos(t)$$

Exact	= -1.
Euler	= -0.9033
Runge-Kutta	= -0.9979
Adams-Bashforth	= -1.
Bulirsch-Stoer	= -0.9988
Implicit Runge-Kutta	= -0.999

Section 13.8, page 447

1. ```
ODE[{y'' + t y == 0,y[0] == 1,y'[0] == 1},y,{t,0,8},
Method->Numerov,PlotSolution->{{t,0,8}}];
```

3. ```
ODE[{y'' == Sin[t] y + t,y[0] == 0,y'[0] == 1},y,{t,0,8},
Method->Numerov,PlotSolution->{{t,0,8}}];
```

The following set of commands is to be used with each of the answers to Problems 5-8 in Section 13.8:

```
f[t_] = ODE[eq,y,t,Method->SecondOrderLinear,Form->Explicit]
TableForm[Table[{1/n,Chop[N[f[1]]],Last[Last[ODE[eq,y,{t,0,1},
Method->Numerov,StepSize->1/n]]]},{n,2,8,2}]]
```

5. ```
eq = {y'' + 2y == t,y[0] == 1,y'[0] == 0}
```
$$y(t) = \frac{t}{2} + \cos(\sqrt{2}\,t) - \frac{\sin(\sqrt{2}\,t)}{2\sqrt{2}}$$

| $h$ | $Y_n(t = 1.0)$ | $y(1)$ |
|---|---|---|
| 1/2 | 0.64 | 0.307 |
| 1/4 | 0.476 | 0.307 |
| 1/8 | 0.392 | 0.307 |

7. ```
eq = {y'' == y + t^3,y[0] == 0,y'[0] == 1}
```
$$y(t) = \frac{7e^t}{2} - \frac{7}{2e^t} - 6t - t^3$$

h	$Y_n(t = 1.0)$	$y(1)$
1/2	1.18	1.23
1/4	1.21	1.23
1/8	1.22	1.23

Chapter 14

Section 14.1, page 455

1. $1/2$

3. Does not exist

5. $\dfrac{s}{s^2 + b^2}$

7. No

9. Yes

Section 14.2, page 459

1. Yes

3. Yes

Section 14.3, page 465

1. $y(t) = \dfrac{e^{-5t}}{2} + \dfrac{3e^{3t}}{2}$

3. $y(t) = -\dfrac{2}{9} + \dfrac{2e^{-9t}}{9}$

5. $y(t) = \dfrac{\cos(3t) - 10\cos(\sqrt{c}\,t) + c\cos(\sqrt{c}\,t)}{c - 9}$

7. $y(t) = 16\cos(3t) - \dfrac{t\cos(3t)}{6} + \dfrac{\sin(3t)}{18} + \dfrac{t\sin(3t)}{6}$

9. $y(t) = \dfrac{e^{3t}}{9} + \dfrac{8e^{3t/2}\cos\big((3/2)\sqrt{3}\,t\big)}{9} + \dfrac{14e^{3t/2}\sin\big((3/2)\sqrt{3}\,t\big)}{9\sqrt{3}}$

Section 14.4, page 469

1. $\Gamma\left(\dfrac{5}{2}\right) = \dfrac{3}{2}\Gamma\left(\dfrac{3}{2}\right) = \left(\dfrac{3}{2}\right)\left(\dfrac{1}{2}\right)\Gamma\left(\dfrac{1}{2}\right) = \dfrac{3\sqrt{\pi}}{4}$

3. `N[{Gamma[100],Gamma[0.01],Gamma[I]}]`

   a. 9.3326210^{155} b. 99.4326 c. $-0.15495 - 0.498016i$

Section 14.5, page 473

1. $\dfrac{2s^2}{(s^2+b^2)^2} - \dfrac{1}{s^2+b^2}$

3. $\dfrac{a}{s^2+a^2} + \dfrac{s}{s^2+a^2}$

5. $\dfrac{1}{(s-1)^2} - \dfrac{1}{(s-1)^2+1}$

7. $\dfrac{2}{s^3} + \dfrac{5}{s} - \dfrac{1}{s+9}$

9. $\dfrac{2}{s^2+16}$

11. $\dfrac{1}{2s} + \dfrac{s}{2(s^2+4)}$

13. $y(t) = e^{-2t}$

15. $y(t) = e^t - 2t\,e^t$

17. $y(t) = \dfrac{3e^{3t}}{5} + \dfrac{7e^{-2t}}{5}$

19. $y(t) = \dfrac{e^{3t} - \cos(t) - 3\sin(t)}{10}$

21. First, we write

$$\mathcal{L}\big(f(t)\big) = \int_0^T e^{-st} f(t)\,dt + \int_T^\infty e^{-st} f(t)\,dt.$$

Substitution of $t = u + T$ converts the last integral into

$$\int_T^\infty e^{-st} f(t)\,dt = e^{-sT}\int_0^\infty e^{-su} f(u)\,du = e^{-sT}\mathcal{L}\big(f(t)\big).$$

Hence,

$$\mathcal{L}\big(f(t)\big) = \int_0^T e^{-st} f(t)\,dt + e^{-sT}\mathcal{L}\big(f(t)\big).$$

When we solve this equation for $\mathcal{L}\big(f(t)\big)$ we get (14.51).

23. Let $G(t) = \displaystyle\int_0^t f(\tau)\,d\tau$. Then $G(0) = 0$ and $G'(t) = f(t)$. Using $\mathcal{L}\big(G'(t)\big) = s\mathcal{L}\big(G(t)\big) - 0$, we obtain $\mathcal{L}\big(G'(t)\big) = F(s) = s\mathcal{L}\big(G(t)\big)$. Now divide by s.

Section 14.6, page 478

14.6. $f(t) = u_0(t)$

14.6. $f(t) = 2u_0(t) - 3u_1(t)$

5. $f(t) = u_0(t) + (e^t - 1)u_1(t)$
$+ (2 - e^t)u_2(t)$

7. `Plot[E^t UnitStep[t - 3],`
`{t,-1,4}];`

9. `Plot[t - (t - 1)`
`UnitStep[t - 1],{t,-1,4},`
`PlotRange->All];`

11. $\dfrac{3(1 - e^{-2s})}{s}$

13. $4\left(\dfrac{e^{-s}}{s} - \dfrac{e^{-3s}}{s}\right)$

17. $u_1(t)$

19. $\dfrac{\sin(2t - 6)u_3(t)}{2}$

15. $-\dfrac{e^{-\pi s/2}}{(s^2 + 1)}$

21. $\dfrac{e^{\pi - t}(t - \pi)u_\pi(t)}{a} + \delta'(t)$

Section 14.7, page 481

1. $y(t) = t + (1 - t)u_1(t)$

3. $y(t) = e^{-t} - 1 + t + (e^{1-t} - 1)u_1(t)$

5. $y(t) = t - \sin(t) + \Big(\cos(1 - t) - \sin(1 - t) - t\Big)u_1(t)$

7. $y(t) = \dfrac{t}{4} - \dfrac{\sin(2t)}{8} + \left(\dfrac{\pi - 2t - 1}{8}\right)u_{\pi/2}(t)$

9. $y(t) = \sin(t) - 2(\sin(t/2))^2 u_{2\pi}(t) + 2(\cos(t/2))^2 u_\pi(t)$

Section 14.8, page 486

1. $y(t) = 2e^{-t} + e^{2-t}u_2(t)$

3. $y(t) = e^{3t} + 2e^{3t-6}u_2(t)$

5. $y(t) = \dfrac{-3 + 3e^{2t}}{4} - \dfrac{t}{2} + \dfrac{(e^{2t-4} - 1)u_2(t)}{2}$

7. $y(t) = e^{-2t}\cos(3t) + \dfrac{2e^{-2t}\sin(3t)}{3} - \dfrac{e^{6\pi - 2t}\sin(3t)u_{3\pi}(t)}{3}$

$\qquad - \dfrac{e^{2\pi - 2t}\sin(3t)u_\pi(t)}{3}$

Section 14.9, page 489

1. $\dfrac{t^2}{2}$

7. $\dfrac{e^{2t} - e^{-t}}{3}$

3. $\dfrac{t\sin(t)}{2}$

5. $1 - e^{-t}$

9. $\dfrac{t\sin(3t)}{6}$

Section 14.10, page 492

1. `LaplaceTransform[t^2Sin[t],t,s]//Simplify`

$$\frac{6s^2 - 2}{(s^2 + 1)^3}$$

3. `Integrate[Exp[-s t]Cos[m t]Sin[n t],{t,0,Infinity}]//Simplify`

$$\frac{\sqrt{(m + n)^2}}{2(s^2 + (m + n)^2)} - \frac{\sqrt{(m - n)^2}}{2(s^2 + (m - n)^2)}$$

5. `LaplaceTransform[UnitStep[t - Pi] - UnitStep[t - 2Pi],t,s]//Simplify`

$$\frac{e^{\pi s} - 1}{s e^{2\pi s}}$$

7. `LaplaceTransform[`
   `t - UnitStep[t - Pi/2](t - Pi),t,s]//Simplify`

$$\frac{\pi s + 2e^{\pi s/2} - 2}{2s^2 e^{\pi s/2}}$$

9. `InverseLaplaceTransform[1/(s^2(s + 1)^3),s,t]//Simplify`

$$\frac{6 - 6e^t + 4t + 2t\, e^t + t^2}{2e^t}$$

11. `InverseLaplaceTransform[s^2`
    `LaplaceTransform[y[t],t,s],s,t] /. {y[0]->0,y'[0]->0}`

$$y''(t)$$

13. `InverseLaplaceTransform[s/(s^2 + 4s - 5)^2,s,t]//Together`

$$\frac{-2 + 2e^{6t} - 15t + 3t\, e^{6t}}{108e^{5t}}$$

15. `InverseLaplaceTransform[s + 5s^4,s,t]`

$$\delta(t) + 5\delta^{(4)}(t)$$

17. `fns = {1,t,t^n,Exp[a t],Sin[a t],Cos[a t],Sinh[a t],`
    `Cosh[a t],UnitStep[t],DiracDelta[t]};`
    `TableForm[Transpose[{fns,LaplaceTransform[fns,t,s]}]]`

19. `ODE[{y'' + 6y' + 9y == t^4 Sin[3t],y[0] == 1,y'[0] == 4},y,t,`
    `Method->Laplace]`

$$y(t) = -\frac{2(119 + 849t)}{243} + \frac{10\cos(3t) - 24t\cos(3t) + 36t^3\cos(3t) - 27t^4\cos(3t)}{486}$$

$$+\frac{t\left(4\sin(3t) - 9t\sin(3t) + 6t^2\sin(3t)\right)}{81}$$

21. `ODE[{y'''' + 2y'' + y == t^2 Cos[t],`
    `y[0] == 0,y'[0] == 0,y''[0] == 0,y'''[0] == 0},y,t,`
    `Method->Laplace,Form->Explicit]`

$$y(t) = \frac{9t^2\cos(t) - t^4\cos(t) - 9t\sin(t) + 4t^3\sin(t)}{48}$$

23. `ODE[{y'' + y' + y == t UnitStep[t - 1],y[0] == 0,y'[0] == 0},y,t,`

```
Method->Laplace,Form->Explicit]
```

$$y(t) = \left(t - 1 + 2\sqrt{\frac{e}{3}}e^{-t/2}\sin\left(\sqrt{\frac{3}{2}}(t-1)\right)\right)u_1(t)$$

Chapter 15

Section 15.1, page 499

1. $\begin{cases} x_1' &= x_2, \\ x_2' &= -\sin(t)x_2 + 4e^t \end{cases}$

5. $y'' - 8y' + 17y = 0$

7. $y^{(6)} + y = 0$

3. $\begin{cases} x_1' &= x_2, \\ x_2' &= x_3, \\ x_3' &= 3x_3 - 2x_2 + t^2 x_1 + e^t \end{cases}$

9. $y^{(4)} - 25y'' + 9y' + 9y = -25$

Section 15.3, page 507

1. $\begin{cases} x_1(t) &= -\dfrac{43}{16} + C_1 t\, e^{4t} + C_2 t\, e^{4t}, \\ x_2(t) &= -\dfrac{5}{4} + C_2 e^{4t} \end{cases}$

3. $\begin{cases} x_1(t) &= C_1 e^{-t} + C_2\left(e^t - e^{-t}\right) + C_3\left(e^{2t} - e^t\right), \\ x_2(t) &= C_2 e^t + C_3\left(e^{2t} - e^t\right), \\ x_3(t) &= C_3 e^{2t} \end{cases}$

5. $\begin{cases} x_1(t) &= -\dfrac{4}{3} + C_1 e^{3t}, \\ x_2(t) &= \dfrac{1}{6} + C_1\left(e^{4t} - e^{3t}\right) + C_2 e^{4t} \end{cases}$

Section 15.4, page 514

1. $\left\{ \begin{pmatrix} 1 \\ t \end{pmatrix}, \begin{pmatrix} 0 \\ 1 \end{pmatrix} \right\}$

5. $\left\{ \begin{pmatrix} e^t \\ 0 \end{pmatrix}, \begin{pmatrix} e^{3t} \\ 2e^{3t} \end{pmatrix} \right\}$

3. $\left\{ \begin{pmatrix} 3e^{3t} \\ 4e^{3t} \end{pmatrix}, \begin{pmatrix} 0 \\ 1 \end{pmatrix} \right\}$

7. $\left\{ \begin{pmatrix} \cos(t) \\ -\sin(t) \end{pmatrix}, \begin{pmatrix} \sin(t) \\ \cos(t) \end{pmatrix} \right\}$

9. -1 13.   1

11.   0   15.   0

Section 15.5, page 528

In each of the following answers, a fundamental set of solutions consists of the columns of the fundamental matrix.

1.   A fundamental matrix is $\begin{pmatrix} e^{6t} & -3e^{2t} \\ e^{6t} & e^{2t} \end{pmatrix}$.

3.   A fundamental matrix is $\begin{pmatrix} -2e^{3t} & -e^{2t} \\ e^{3t} & e^{2t} \end{pmatrix}$.

5.   A fundamental matrix is

$$\begin{pmatrix} (3+\sqrt{21})\exp\left(\dfrac{3+\sqrt{21}}{2}t\right) & (3-\sqrt{21})\exp\left(\dfrac{3-\sqrt{21}}{2}t\right) \\ 2\exp\left(\dfrac{3+\sqrt{21}}{2}t\right) & 2\exp\left(\dfrac{3-\sqrt{21}}{2}t\right) \end{pmatrix}.$$

7.   A fundamental matrix is $\begin{pmatrix} 0 & 0 & 2e^{t} \\ e^{t}\cos(2t) & -e^{t}\sin(2t) & -3e^{t} \\ e^{t}\sin(2t) & e^{t}\cos(2t) & -2e^{t} \end{pmatrix}$.

9.   A fundamental matrix is $\begin{pmatrix} 1 & e^{-3t} & 0 \\ 1 & -e^{-3t} & e^{-3t} \\ 1 & 0 & -e^{-3t} \end{pmatrix}$.

11.   A fundamental matrix is $\begin{pmatrix} e^{t} & t\,e^{t} & 0 \\ 0 & e^{t} & 0 \\ 0 & 0 & e^{3t} \end{pmatrix}$.

13. $\begin{cases} x_1(t) = \dfrac{9e^{6t} - 3e^{2t}}{2}, \\ x_2(t) = \dfrac{9e^{6t} + e^{2t}}{2} \end{cases}$ 15. $\begin{cases} x_1(t) = 2e^{3t}, \\ x_2(t) = -e^{3t} \end{cases}$

17. $\begin{cases} x_1(t) = 2\sqrt{\dfrac{3}{7}}\sinh\left(\dfrac{(3+\sqrt{21})t}{2}\right), \\ x_2(t) = \left(1 + \sqrt{\dfrac{3}{7}}\right)\cosh\left(\dfrac{(3+\sqrt{21})t}{2}\right) \end{cases}$

15.5.4. $\begin{cases} x_1(t) = 4e^t, \\ x_2(t) = 3e^t \sin(2t) + 9e^t \cos(2t) - 6e^t, \\ x_3(t) = -3e^t \cos(2t) + 9e^t \sin(2t) + 4e^t \end{cases}$

21. $\begin{cases} x_1(t) = \dfrac{8 + e^{-3t}}{3}, \\[2mm] x_2(t) = \dfrac{8 + 7e^{-3t}}{3}, \\[2mm] x_3(t) = \dfrac{8 - 8e^{-3t}}{3} \end{cases}$

23. $\begin{cases} x_1(t) = 3e^t + 2t\,e^t, \\ x_2(t) = 2e^t, \\ x_3(t) = e^{3t} \end{cases}$

25. We have $\mathbf{u} = (1/2)\left(\mathbf{x}^1 + \mathbf{x}^2\right)$ and $\mathbf{v} = (1/2)\left(\mathbf{x}^1 - \mathbf{x}^2\right)$. By the superposition principle, $\mathbf{u}$ and $\mathbf{v}$ are real solutions of $\mathbf{x}' = A\mathbf{x}$.

Section 15.6, page 541

1. $\begin{cases} x_1 = -2e^t, \\ x_2 = e^t/4 \end{cases}$

7. $\begin{cases} x_1 = \dfrac{-48 + 40t}{25} + \dfrac{1 + 4t}{4\,e^t}, \\[2mm] x_2 = \dfrac{-52 + 60t}{25} + \dfrac{-1 + 4t}{4\,e^t} \end{cases}$

3. $\begin{cases} x_1 = 3t^2 e^t/2, \\ x_2 = 3t^2 e^t \end{cases}$

9. $\begin{cases} x_1 = (1 + 2t)e^t, \\ x_2 = 2t\,e^t \end{cases}$

5. $\begin{cases} x_1 = 1 - 2t + \dfrac{\sin(2t)}{3}, \\[2mm] x_2 = -5t + \dfrac{2\cos(2t) + 2\sin(2t)}{3} \end{cases}$

11. $\begin{cases} x_1 = e^{3t}(2t - 1) + 2e^{6t}, \\ x_2 = e^{3t}(4t - 2) + e^{6t} \end{cases}$

Section 15.7, page 543

1. $\begin{cases} x_1 = \dfrac{3t\,e^t}{2} - \dfrac{3e^t}{4} + 2t - 1, \\[2mm] x_2 = \dfrac{3t\,e^t}{2} - \dfrac{e^t}{4} + t \end{cases}$

3. $\begin{cases} x_1 = -e^{-2t}, \\ x_2 = e^t/2 \end{cases}$

5. $\begin{cases} x_1 = -2 - \dfrac{1}{2t^2} + \dfrac{2}{t} - 2\log(t), \\[2mm] x_2 = -4 + \dfrac{5}{t} - 4\log(t) \end{cases}$

7.
$$\begin{cases} x_1 &= -2t\cos(t) - 4t\sin(t) - 5\cos(t)\log\big(\cos(t)\big) \\ &\quad + \big(\cos(t) + 2\sin(t)\big)\log\big(\sin(t)\big), \\ x_2 &= -2t\sin(t) - 2\cos(t)\log\big(\cos(t)\big) - \sin(t))\log\big(\tan(t)\big) \end{cases}$$

9.
$$\begin{cases} x_1 &= \dfrac{-1 + e^{3t}}{3e^{2t}}, \\ x_2 &= \dfrac{23 - 12t - 27e^{2t} + 4e^{3t} + 54t\,e^{3t}}{36e^{2t}} \end{cases}$$

11.
$$\begin{cases} x_1 &= \dfrac{5 - 26t - e^{2t}\big(5\cos(3t) + 157\sin(3t)\big)}{169}, \\ x_2 &= -\dfrac{12 + 39t + e^{2t}\big(157\cos(3t) - 5\sin(3t)\big)}{169} \end{cases}$$

Section 15.8, page 548

1.
$$\begin{cases} x_1 &= -5e^{6t} + 5e^{7t}, \\ x_2 &= 6e^{6t} - 5e^{7t} \end{cases}$$

3.
$$\begin{cases} x_1 &= e^t(5 - 18t), \\ x_2 &= e^t(-3 + 54t) \end{cases}$$

5.
$$\begin{cases} x_1 &= -\dfrac{2}{5}e^{-4t} + \dfrac{2}{5}e^t, \\ x_2 &= -\dfrac{2}{5}e^{-4t} + \dfrac{7}{5}e^t \end{cases}$$

7.
$$\begin{cases} x_1 &= e^t\left(\cos(2t) + \dfrac{\sin(2t)}{2}\right), \\ x_2 &= e^t\big(\cos(2t) - 2\sin(2t)\big) \end{cases}$$

9.
$$\begin{cases} x &= 2 + e^{-2t} - 3e^{-t} - t\,e^{-t}, \\ y &= -3 - 2e^{-2t} + 5e^{-t} + 2t\,e^{-t} \end{cases}$$

11. $x = y = z = e^{t/2}$

Chapter 16

Section 16.1, page 567

1. Attracting spiral with clockwise rotation

3. Saddle point

5. Stable center with counterclockwise orientation

7. Repelling spiral with clockwise orientation

9. Stable center with clockwise orientation

Section 16.2, page 576

1.
$$\begin{cases} x(t) = \left(\dfrac{2e^{-t}}{3} + \dfrac{e^{5t}}{3}\right)x(0) + \left(-\dfrac{e^{-t}}{3} + \dfrac{e^{5t}}{3}\right)y(0), \\[3mm] y(t) = \left(-\dfrac{2e^{-t}}{3} + \dfrac{2e^{5t}}{3}\right)x(0) + \left(\dfrac{e^{-t}}{3} + \dfrac{2e^{5t}}{3}\right)y(0) \end{cases}$$

3.
$$\begin{cases} x(t) = \left(e^{2t}\cos(3t) + \dfrac{e^{2t}\sin(3t)}{3}\right)x(0) + \dfrac{2e^{2t}\sin(3t)}{3}y(0), \\[3mm] y(t) = \dfrac{-5e^{2t}\sin(3t)}{3}x(0) + \left(e^{2t}\cos(3t) - \dfrac{e^{2t}\sin(3t)}{3}\right)y(0) \end{cases}$$

5.
$$\begin{cases} x(t) = e^{2t}\cos(2t)x(0) + \dfrac{e^{2t}\sin(2t)}{2}y(0) + \dfrac{e^{2t}(2t + 11\sin(2t))}{8}, \\[3mm] y(t) = -2e^{2t}\sin(2t)x(0) + e^{2t}\cos(2t)y(0) + \dfrac{11e^{2t}(-1 + \cos(2t))}{4} \end{cases}$$

7.
$$\begin{cases} x(t) = x(0)\cos(\sqrt{3}\,t) - y(0)\dfrac{\sin(\sqrt{3}\,t)}{\sqrt{3}} - \dfrac{1 - \cos(\sqrt{3}\,t)}{3}, \\[3mm] y(t) = x(0)\sqrt{3}\sin(\sqrt{3}\,t) + y(0)\cos(\sqrt{3}\,t) - 5t + \dfrac{\sin(\sqrt{3}\,t)}{\sqrt{3}} \end{cases}$$

9.
$$\begin{cases} x(t) = 2e^{2t}\cos(4t) - 2e^{2t}\sin(4t), \\[2mm] y(t) = -2e^{2t}\cos(4t) - 2e^{2t}\sin(4t) \end{cases}$$

11.
$$\begin{cases} x(t) = 4\cosh(2\sqrt{7}\,t) + \dfrac{10\sinh(2\sqrt{7}\,t)}{\sqrt{7}}, \\[3mm] y(t) = \cosh(2\sqrt{7}\,t) + \dfrac{4\sinh(2\sqrt{7}\,t)}{\sqrt{7}} \end{cases}$$

13.
$$\begin{cases} x(t) = -e^{t} + 2e^{2t}, \\[1mm] y(t) = -e^{t} + e^{2t}, \\[1mm] z(t) = -3e^{t} + 3e^{2t} \end{cases}$$

15.
$$\begin{cases} x(t) = 3e^{2t} - 8e^{3t} + 6e^{5t}, \\[1mm] y(t) = -3e^{2t} + 16e^{3t} - 12e^{5t}, \\[1mm] z(t) = -3e^{2t} + 8e^{3t} - 4e^{5t} \end{cases}$$

Section 16.3, page 578

1. ```
   ODE[{x' == x,y' == 2y,x[0] == a,y[0] == b},{x,y},t,
   Method->LinearSystem,
   Parameters->{{a,-1,1,0.4},{b,-1,1,1}},
   PlotPhase->{{t,-2,2},
   PlotStyle->{{RGBColor[1,0,0],AbsoluteThickness[1]}},
   PlotRange->{{-3,3},{-4,4}}}]
   ```

3. ```
   ODE[{x' == y/2,y' == -2x,x[0] == a,y[0] == b},{x,y},t,
   Method->LinearSystem,
   Parameters->{{a,-1,1,1/2},{b,-1,1,1/2}},
   PlotPhase->{{t,0,Pi},
   PlotStyle->{{RGBColor[1,0,0],AbsoluteThickness[1]}},
   PlotRange->{{-1.6,1.6},{-2.5,2.5}}}]
   ```

5. ```
   ODE[{x' == 3x/2 + 5y/2,y' == -5x/2 + 3y/2,x[0] == a,
   y[0] == b},{x,y},t,Method->LinearSystem,
   Parameters->{{a,-1,1,1/2},{b,-1,1,1/2}},
   PlotPhase->{{t,-10,10},
   PlotStyle->{{RGBColor[1,0,0],AbsoluteThickness[1]}},
   PlotRange->{{-3.2,3.2},{-3.2,3.2}}}]
   ```

7. ```
   ODE[{x' == x,y' == y,x[0] == a,y[0] == b},{x,y},t,
   Method->LinearSystem,
   Parameters->{{a,-1,1,1/2},{b,-1,1,1/2}},
   PlotPhase->{{t,-10,10},
   PlotStyle->{{RGBColor[1,0,0],AbsoluteThickness[1]}},
   PlotRange->{{-2,2},{-2,2}}}]
   ```

9. ```
   ODE[{x' == x + y,y' == y,x[0] == a,y[0] == b},{x,y},t,
   Method->LinearSystem,
   Parameters->{{a,-1,1,1/2},{b,-1,1,1/2}},
   PlotPhase->{{t,-10,10},
   PlotStyle->{{RGBColor[1,0,0],AbsoluteThickness[1]}},
   PlotRange->{{-2,2},{-2,2}}}]
   ```

Chapter 17

Section 17.3, page 591

1.

Critical Point	Linearized System
$(0,0)$	$\begin{cases} x_1' = x_1, \\ x_2' = x_1 + x_2 \end{cases}$
$(-1,1)$	$\begin{cases} x_1' = x_1 + 2x_2, \\ x_2' = x_1 + x_2 \end{cases}$
   ```

Critical Point	Linearized System
$(0,0)$	$\begin{cases} x_1' = x_1, \\ x_2' = 3x_2 \end{cases}$
$(0, 3/2)$	$\begin{cases} x_1' = -(1/2), \\ x_2' = -(3/2)x_1 - 3x_2 \end{cases}$
$(1, 0)$	$\begin{cases} x_1' = -x_1 - x_2, \\ x_2' = 2x_2 \end{cases}$
$(-1, 2)$	$\begin{cases} x_1' = x_2, \\ x_2' = -2x_1 - 4x_2 \end{cases}$

3.

Critical Point	Linearized System
$(n\pi, 0)$ *n* even	$\begin{cases} x_1' = x_2, \\ x_2' = -x_1 \end{cases}$
$(n\pi, 0)$ *n* odd	$\begin{cases} x_1' = x_2, \\ x_2' = x_1 \end{cases}$

5.

7. The critical points are $((m + (1/2))\pi, n\pi)$, where m and n are integers. In every case the linearized system is $x_1' = 0, x_2' = 0$.

Critical Point	Linearized System
$(0,0)$	$\begin{cases} x_1' = x_2, \\ x_2' = 3x_2 \end{cases}$
$(0, 3/2)$	$\begin{cases} x_1' = -x_1/2, \\ x_2' = -3x_1 - 3x_2 \end{cases}$
$(1, 0)$	$\begin{cases} x_1' = -x_1 - x_2, \\ x_2' = x_2 \end{cases}$

9.

Section 17.5, page 603

1. $(0, 0)$ is a stable and asymptotically stable critical point for which the solutions spiral.

3. $(0, 0)$ is a stable and asymptotically stable critical point for whcih the solutions spiral.

5. $(0, 0)$ is an unstable critical point.

7. $(0, 0)$ is an unstable critical point.

9. $(0, 0)$ is a stable critical point.

11. The function $F = r^2$ satisfies the differential equation $F' = -2F^2$ with the initial condition $F(0) = r_0^2$, for which the solution is $F(t) = r_0^2/(1 + 2t\, r_0^2)$, so that $\lim\limits_{t \to \infty} F(t) = 0$, hence stability.

Section 17.6, page 614

1. $\frac{dV}{dt} = 2x(-x^3 + 2xy^2) + 2y(-2x^2y - 5y^3) = -2x^4 - 10y^4$,
   which is strictly negative, unless $(x, y) = (0, 0)$.

3. $\frac{dV}{dt} = -2ax^4(1 - 3x^2) + (2a - 2b)x^2y^2 - 8by^4 - 6by^6$. If in addition we have $0 < a \le b$, then the second term is non-positive. The first term is also non-positive, provided that we restrict x so that $|x| < 1/\sqrt{3}$. Therefore, V is a Lyapunov function in this region.

5. In the case of the damped pendulum, we have $a_{11} = 0, a_{12} = 1, a_{21} = -g/l, a_{22} = -c < 0$ and the eigenvalues are $2r = -c \pm \sqrt{c^2 - 4g/l}$, which satisfies the hypotheses of part **a**.

7. $V(x, y) = x^4 + y^4$ is a suitable Lyapunov function to prove stability.

9. $V(x, y) = x^2 + y^2$ is a suitable Lyapunov function to prove asymptotic stability.

11. $V(x, y) = 2x^2 + y^2$ is a suitable Lyapunov funciton to prove stability.

13. $V(x, y) = x^2 + 3y^2$ is a suitable Lyapunov function to prove asymptotic stability.

15. $V(x, y) = 3x^2 + 2y^2$ is a suitable Lyapunov function to prove stability.

17. $V(x, y) = x^4 - y^4$ proves instability.

Chapter 18

Section 18.1, page 627

1.

$$x_1(t) = 3\cos(2t) + \cos(3t), \qquad x_2(t) = 3\cos(2t) - \cos(3t),$$

$$y_1(t) = -6\sin(2t) - 3\sin(3t), \qquad y_2(t) = -6\sin(2t) = 3\sin(3t).$$

3. Use the identities $\cos(\alpha - \pi/2) = \sin(\alpha)$ and $\sin(\alpha - \pi/2) = -\cos(\alpha)$.

5. $t \longmapsto \big(x_1(t), x_2(t)\big)$ is a solution of the system

$$\begin{cases} m\, x_1'' + k\, x_1 = k_{12}(x_2 - x_1), \\ m\, x_2'' + k\, x_2 = -k_{12}(x_2 - x_1). \end{cases} \qquad (3.1)$$

When we add the two equations of (3.1) we get the second equation of (18.15), and when we subtract the two equations of (3.1) we get the first equation of (18.15).

The general solutions of each equation of (18.15) can be found separately. Thus

$$\begin{cases} y_1(t) = C_1\cos(\mu_2 t) + C_2\sin(\mu_2 t), \\ y_2(t) = C_3\cos(\mu_1 t) + C_4\sin(\mu_1 t), \end{cases} \tag{3.2}$$

where the C_j are constants and

$$\mu_1 = \sqrt{\frac{k}{m}} \quad \text{and} \quad \mu_1 = \sqrt{\frac{k+k_{12}}{m}}.$$

9. $r^2(r+c/m)^2 + 2r(r+c/m)(k+k_{12})/m + \big((k+k_{12})^2 - k_{12}^2\big)/m^2 = 0$

Section 18.2, page 631

1. $\begin{cases} R(I_1' - I_2') + \dfrac{I_1 - I_2}{C} = E'(t), \\ \qquad\qquad L\,I_2' = E(t) \end{cases}$

3. $\begin{cases} I_1 \approx 7.5 - 0.833e^{-4000t} - 6.67e^{-1000t}, \\ I_2 \approx 2.5 + 0.833e^{-4000t} - 3.33e^{-1000t}. \end{cases}$

5. $\begin{cases} I_1 = 0.00417e^{-3t} + 0.0075e^{t} - 0.0117\cos(3t) \\ I_2 = -0.00417e^{-3t} + 0.0075e^{t} - 0.00333\cos(3t). \end{cases}$

Section 18.3, page 641

1.

$$P(t) = \begin{pmatrix} \dfrac{\mu+\lambda e^{-(\lambda+\mu)t}}{\lambda+\mu} & \dfrac{\lambda-\lambda e^{-(\lambda+\mu)t}}{\lambda+\mu} \\[2ex] \dfrac{\mu-\mu e^{-(\lambda+\mu)t}}{\lambda+\mu} & \dfrac{\lambda+\mu e^{-(\lambda+\mu)t}}{\lambda+\mu} \end{pmatrix}, \qquad \lim_{t\to\infty} P(t) = \begin{pmatrix} \dfrac{\mu}{\lambda+\mu} & \dfrac{\lambda}{\lambda+\mu} \\[2ex] \dfrac{\mu}{\lambda+\mu} & \dfrac{\lambda}{\lambda+\mu} \end{pmatrix}$$

3.

$$p_{11}(t) = \frac{2\lambda_1 + (\lambda_0 + 2\lambda_1)e^{-\lambda_0 t} + \lambda_0 e^{-(\lambda_0+2\lambda_1)t}}{2(\lambda_0 + 2\lambda_1)}, \qquad p_{12}(t) = \frac{2\lambda_0 - 2\lambda_0 e^{-(\lambda_0+2\lambda_1)t}}{2(\lambda_0 + 2\lambda_1)},$$

$$p_{13}(t) = \frac{2\lambda_1 - (\lambda_0 + 2\lambda_1)e^{-\lambda_0 t} + \lambda_0 e^{-(\lambda_0+2\lambda_1)t}}{2(\lambda_0 + 2\lambda_1)}, \qquad p_{21}(t) = \frac{2\lambda_1 - 2\lambda_1 e^{-(\lambda_0+2\lambda_1)t}}{2(\lambda_0 + 2\lambda_1)},$$

$$p_{31}(t) = \frac{2\lambda_1 - (\lambda_0 + 2\lambda_1)e^{-\lambda_0 t} + \lambda_0 e^{-(\lambda_0+2\lambda_1)t}}{2(\lambda_0 + 2\lambda_1)}, \qquad p_{32}(t) = \frac{2\lambda_0 - 2\lambda_0 e^{-(\lambda_0+2\lambda_1)t}}{2(\lambda_0 + 2\lambda_1)},$$

$$p_{33}(t) = \frac{2\lambda_1 + (\lambda_0 + 2\lambda_1)e^{-\lambda_0 t} + \lambda_0 e^{-(\lambda_0+2\lambda_1)t}}{2(\lambda_0 + 2\lambda_1)}, \qquad p_{23}(t) = \frac{2\lambda_1 - 2\lambda_1 e^{-(\lambda_0+2\lambda_1)t}}{2(\lambda_0 + 2\lambda_1)},$$

$$p_{22}(t) = \frac{2\lambda_0 + 4\lambda_1 e^{-(\lambda_0+2\lambda_1)t}}{2(\lambda_0 + 2\lambda_1)}, \qquad \lim_{t\to\infty} P(t) = \frac{1}{\lambda_0 + 2\lambda_1}\begin{pmatrix} \lambda_1 & \lambda_0 & \lambda_1 \\ \lambda_1 & \lambda_0 & \lambda_1 \\ \lambda_1 & \lambda_0 & \lambda_1 \end{pmatrix}$$

5.

$$p_{11}(t) = \frac{1}{3} + \frac{2}{3}e^{-3t/2}\cos(t\sqrt{3}/6)$$

$$p_{12}(t) = \frac{1}{3} - \frac{1}{3}e^{-3t/2}\cos(t\sqrt{3}/6) + \frac{e^{-3t/2}}{\sqrt{3}}\sin(t\sqrt{3}/6)$$

$$p_{13}(t) = \frac{1}{3} - \frac{1}{3}e^{-3t/2}\cos(t\sqrt{3}/6) - \frac{e^{-3t/2}}{\sqrt{3}}\sin(t\sqrt{3}/6)$$

The others are obtained from these as follows:

$$p_{22}(t) = p_{33}(t) = p_{11}(t), \qquad p_{23}(t) = p_{31}(t) = p_{12}(t), \qquad p_{32}(t) = p_{13}(t) = p_{21}(t).$$

7. $r_k = \lambda\left(\cos(2\pi k/n) - 1\right)$ for $k = 0, 1, \dots, n-1$

17.

$$P(t) = \begin{pmatrix} 2/3 + (1/3)e^{-3t} & 0 & 1/3 - (1/3)e^{-3t} \\ 2/3 + (1/3)e^{-3t} - e^{-2t} & e^{-2t} & (1/3) - (1/3)e^{-3t} \\ 2/3 - (2/3)e^{-3t} & 0 & 1/3 + (2/3)e^{-3t} \end{pmatrix}$$

and

$$\lim_{t\to\infty} P(t) = \begin{pmatrix} 2/3 & 0 & 1/3 \\ 2/3 & 0 & 1/3 \\ 2/3 & 0 & 1/3 \end{pmatrix}$$

19.

$$P(t) = \begin{pmatrix} 1/2 + (1/2)e^{-2t} & 1/8 - (1/8)e^{-2t} + (t/4)e^{-2t} & 3/8 - (3/8)e^{-2t} - (t/4)e^{-2t} \\ 1/2 - (1/2)e^{-2t} & 1/8 + (7/8)e^{-2t} - (t/4)e^{-2t} & (3/8) - (3/8)e^{-2t} + (t/4)e^{-2t} \\ 1/2 - (1/2)e^{-2t} & 1/8 - (1/8)e^{-2t} - (t/4)e^{-2t} & 3/8 + (5/8)e^{-2t} + (t/4)e^{-2t} \end{pmatrix}$$

and

$$\lim_{t\to\infty} P(t) = \begin{pmatrix} 1/2 & 1/8 & 3/8 \\ 1/2 & 1/8 & 3/8 \\ 1/2 & 1/8 & 3/8 \end{pmatrix}$$

Chapter 19

Section 19.1, page 653

The following sets of commands are to be used with each of the answers to the problems in Section 19.1:

Part I:

```
ODE[sys,{x,y},interval,Method->Euler,PlotPhase->{interval}];
ODE[sys,{x,y},interval,Method->RungeKutta4,PlotPhase->{interval}];
ODE[sys,{x,y},interval,Method->AdamsBashforth,PlotPhase->{interval}];
ODE[sys,{x,y},interval,Method->BulirschStoer,PlotPhase->{interval}];
ODE[sys,{x,y},interval,Method->ImplicitRungeKutta,PlotPhase->{interval}];
```

Part II:

```
ODE[sys,{x,y},interval,Method->Euler,
PlotPhase->{interval},VariableStepSize->True,Tolerance->0.001];
ODE[sys,{x,y},interval,Method->RungeKutta4,
PlotPhase->{interval},VariableStepSize->True,Tolerance->0.001];
ODE[sys,{x,y},interval,Method->AdamsBashforth,
PlotPhase->{interval},VariableStepSize->True,Tolerance->0.001];
ODE[sys,{x,y},interval,Method->BulirschStoer,
PlotPhase->{interval},VariableStepSize->True,Tolerance->0.001];
ODE[sys,{x,y},interval,Method->ImplicitRungeKutta,
PlotPhase->{interval},VariableStepSize->True,Tolerance->0.001];
```

1.
```
sys = {x' == x - 2y,y' == 2x + y,x[0] == 0,y[0] == 2};
interval = {t,0,10};
```

3.
```
sys = {x' == 3Sin[x] - 4y,y' == 3x + 2y,x[0] == 1,y[0] == 1};
interval = {t,0,2};
```

5.
```
sys = {x' == 2x - Tan[y],y' == Cos[x] + 2y,x[0] == 1,y[0] == 1};
interval = {t,0,1};
```

7.
```
eq = {y'' + Sin[y] == 0,y[0] == 1,y'[0] == 0};
sys = ODE[eq,y,t,Transformation->ConvertToSystem,
TransformationVariable->w];
interval = {t,0,10};
```

9.
```
eq = {y'' + Sin[y'] + y^2 == t,y[0] == 1,y'[0] == 1};
sys = ODE[eq,y,t,Transformation->ConvertToSystem,
TransformationVariable->w];
interval = {t,0,10};
```

11.
```
eq = {y'' + y Cos[y'] + y^2 == t,y[0] == 0,y'[0] == 0};
sys = ODE[eq,y,t,Transformation->ConvertToSystem,
TransformationVariable->w];
interval = {t,0,10};
```

Chapter 20

Section 20.1, page 687

1. $f(t) = t - \dfrac{t^3}{3!} + \dfrac{t^5}{5!} - \cdots$

3. $f(t) = e^3 + 3e^3(t-1) + \dfrac{9e^3}{2}(t-1)^2 + \cdots$

5. $f(t) = \dfrac{3}{2} + \dfrac{3}{4}t + \dfrac{3}{8}t^2 + \dfrac{3}{16}t^3 + \cdots$

7. $f(t) = t - \dfrac{9}{2}t^3 + \dfrac{27}{8}t^5 - \cdots$

9. $f(t) = 3e^3 + 10e^3(t-1) + \dfrac{33e^3}{2}(t-1)^2 + \cdots$

11. $(16t^3 + 24t^2 + 6t)e^{4t}$

13. $41\cos(5t)e^{2t} + 840\sin(5t)e^{2t}$

Section 20.3, page 697

1. $y_{n+2} = 4y_n$,

   $y(t) = 1 + 2t^2 + \dfrac{2}{3}t^4 + \dfrac{4}{45}t^6 + \cdots$

3. $y_{n+2} = y_{n+1} + (n+1)y_n$,

   $y(t) = 1 + 3(t-1) + 2(t-1)^2 + \dfrac{5}{3}(t-1)^3 + \dfrac{11}{12}(t-1)^4 + \cdots$

5. $y_{n+2} = 2(n-1)y_n$, $y(t) = t$

7. $y_{n+2} = -(n+4)y_n$,

   $y(t) = 3 + 6t - 6t^2 - 5t^3 + \cdots$

9. $y_{n+2} = -5ny_n - 6n(n-1)y_{n-2}$,

   $y(t) = 1 - \dfrac{t^4}{2} + \dfrac{t^6}{3} - \dfrac{t^8}{8} + \cdots$

11. $y_{n+2} = -5ny_n - 6n(n-1)y_{n-2}$,

    $y(t) = t - \dfrac{5}{6}t^3 + \dfrac{13}{40}t^5 - \dfrac{25}{84}t^7 + \cdots$

13. $y_{n+2} = (2n-5)y_n$,

    $y(t) = t - \dfrac{t^3}{2} - \dfrac{t^5}{40} - \dfrac{t^7}{336} - \cdots$

15. $y_{n+2} = -y_{n+1} - (n+4)y_n$,

    $y(t) = t - \dfrac{t^2}{2} - \dfrac{2}{3}t^3 + \dfrac{7}{6}t^4 + \cdots$

Section 20.4, page 701

1. $y(t) = \mathbf{Ai}(t)$

3. $y(t) = -0.929\mathbf{Ai}(2.08t) + 0.536\mathbf{Bi}(2.08t)$

5. ```
N[Limit[Det[{{AiryAi[t],AiryBi[t]},{AiryAiPrime[t],
AiryBiPrime[t]}}],t->0]] == N[1/Pi]
```

# Section 20.6, page 709

17. The commands

```
ODE[y' == y,y,t,Method->SeriesForm,Form->Explicit]
```

```
Collect[Series[
ODE[y' == y,y,t,Method->Separable,Form->Explicit],
{t,0,5}],C[1]]
```

both yield

$$\left(1 + t + \frac{t^2}{2} + \frac{t^3}{6} + \frac{t^4}{24} + \frac{t^5}{120}\right) C[1]$$

19. The commands

```
Normal[Series[
ODE[t^2 y'' + t y' - y == 0,y,t,Method->CauchyEuler,Form->Explicit],
{t,0,5}]]
```

```
ODE[t^2 y'' + t y' - y == 0,y,t,Method->SeriesForm,
Form->Explicit]
```

are both equivalent to

$$\frac{C[1]}{t} + t\, C[2]$$

21. The commands

```
ODE[y'' - t y' + 4 y == 0,y,t,Method->SeriesForm,
Form->Explicit]
```

```
Collect[Series[
ODE[y'' - t y' + 4 y == 0,y,t,Method->TableLookup,Form->Explicit],
{t,0,5}],{C[1],C[2]}]
```

are both equivalent to

$$\left(1 - 2\, t^2 + \frac{t^4}{3}\right) C[1] + \left(t - \frac{t^3}{2} + \frac{t^5}{40}\right) C[2]$$
```

Chapter 21

Section 21.2, page 724

1. $y(t) = C_1 t^5 + C_2 t^3$

3. $y(t) = C_1 t^2 + C_2 t^{-2}$

5. $y(t) = C_1(t+1)^3 + C_2(t+1)^{-3}$

7. $y(t) = \dfrac{2t^{-3} + t^3}{3}$

9. $y(t) = \dfrac{t^5\left(6 - 5\log(t)\right)}{e^5}$

11. $y(t) = \dfrac{t^{-1} - 4t^2 + 3t^3}{12}$

15. $y(t) = C_1 t + C_2 t^5 - \dfrac{t^3}{4}$

17. $y(t) = C_1 t^5 + C_2 t^5 \log(t) + \dfrac{t^7}{4}$

Section 21.3, page 730

1. $r^2 = 0$,

$$y(t) = y_0 \left(1 + \frac{t^2}{4} + \frac{t^4}{64} + \frac{t^6}{2304} + \cdots \right)$$

3. $r^2 - 8r + 15 = 0$,

$$y(t) = y_0 t^5 \left(1 - \frac{t^2}{8} + \frac{t^4}{192} - \frac{t^6}{9216} + \cdots \right)$$

5. $r^2 - 4 = 0$,

$$y(t) = y_0 t^2$$

7. $r^2 = 0$,

$$y(t) = y_0 \left(1 - \frac{3}{4}t^2 + \frac{9}{64}t^4 - \frac{3}{256}t^6 + \cdots \right)$$

9. $3r^2 - r = 0$,

$$y(t) = y_0 t^{1/3} \left(1 - \frac{t^2}{14} + \frac{t^4}{364} - \cdots \right)$$

Section 21.4, page 733

13. $y = C_1 J_{1/3}(t) + C_2 J_{-1/3}(t)$

15. $y = C_1 J_{1/5}(2t) + C_2 J_{-1/5}(2t)$

Section 21.6, page 741

5. $y(t) = C_1 J_5(t) + C_2 Y_5(t)$

9. $y(t) = C_1 J_{\sqrt{3}}(\sqrt{2}\,t) + C_2 J_{-\sqrt{3}}(\sqrt{2}\,t)$

7. $y(t) = C_1 I_5(t) + C_2 K_5(t)$

11. $y(t) = C_1 I_{\sqrt{3}}(\sqrt{2}\,t) + C_2 I_{-\sqrt{3}}(\sqrt{2}\,t)$

Section 21.7, page 748

1. The command

```
ODE[t^2y'' + t y' + (t^2 -49)y,y,t,Method->TableLookup]
```

yields

```
{{y -> BesselJ[7, t] C[1] + BesselY[7, t] C[2]}}
```

Section 21.9, page 757

1. Since $(a)_n = \dfrac{\Gamma(a+n)}{\Gamma(a)}$, we have

$$F(a; b; c; t) = \sum_{n=0}^{\infty} \frac{(a)_n (b)_n t^n}{n!(c)_n} = \sum_{n=0}^{\infty} \frac{\Gamma(a+n)\Gamma(b+n)\Gamma(c)t^n}{\Gamma(a)\Gamma(b)\Gamma(c+n)n!}.$$

3. ```
t Hypergeometric2F1[1,1,2,-t]
```
produces
```
Log[1 + t]
```
```

5. `t Hypergeometric2F1[1/2,1,3/2,-t^2]//PowerExpand`
   produces
   `ArcTan[t]`

7. `(Pi/2)Hypergeometric2F1[1/2,1/2,1,k]`
   and

   `Integrate[1/Sqrt[1 - k Sin[phi]^2],phi]`

   both produce
   `EllipticK[k]`

9. $y(t) = C_1 \dfrac{1}{\sqrt{1-t}} + C_2 \dfrac{\arcsin(\sqrt{t})}{\sqrt{1-t}}$

11. $y(t) = C_1\left(1 - \dfrac{4t}{3}\right) + C_2\left(\dfrac{\sqrt{1-t}}{\sqrt{t}} - 4\sqrt{t}\sqrt{1-t}\right)$

13. `ODE[{y'' == BesselJ[0,t],y[0] == 0,y'[0] == 0},y,t,`
    `Method->SecondOrderLinear]`

Appendix A

Section A.3, page 771

1. $x_1 = 11/2, x_2 = -1/2$

3. $x_1 = 5/3, x_2 = 4/3, x_3 = 1$

5. $x_1 = 0, x_2 = 0, x_3 = 0$

7. $x_1 = 4 + 11s/6, x_2 = 1 - 8s/3$
   $x_3 = -1 - s/6, x_4 = s$
   $x_1 = 0, x_2 = 0, x_3 = 0$
   where s is any real number

9. $\begin{pmatrix} 3/11 & 4/11 \\ 2/11 & -1/11 \end{pmatrix}$

11. $\begin{pmatrix} 7/10 & -1/10 & 3/10 \\ 1/2 & -1/2 & 1/2 \\ -4/5 & 2/5 & -1/5 \end{pmatrix}$

13. noninvertible

Section A.4, page 783

	Eigenvalues	Eigenvectors
1.	$r_1 = 2$	$\mathbf{x}_1 = (-3, 1)$
	$r_2 = 6$	$\mathbf{x}_2 = (1, 1)$

	Eigenvalues	Eigenvectors
5.	$r_1 = 1 - i$	$\mathbf{x}_1 = (-i, 1)$
	$r_2 = 1 + i$	$\mathbf{x}_2 = (i, 1)$

	Eigenvalues	Eigenvectors
3.	$r_1 = 2$	$\mathbf{x}_1 = (-1, 1)$
	$r_2 = 3$	$\mathbf{x}_2 = (-2, 1)$

	Eigenvalues	Eigenvectors
7.	$r_1 = 1$	$\mathbf{x}_1 = (2, -3, 2)$
	$r_2 = 1 - 2i$	$\mathbf{x}_2 = (0, -i, 1)$
	$r_3 = 1 + 2i$	$\mathbf{x}_3 = (0, i, 1)$

	Eigenvalues	Eigenvectors
9.	$r_1 = 0$	$\mathbf{x}_1 = (0, -1, 2)$
	$r_2 = (7 - \sqrt{37})/2$	$\mathbf{x}_2 \approx (-0.85, 0.15, 1)$
	$r_3 = (7 + \sqrt{37})/2$	$\mathbf{x}_3 \approx (1.18, 2.18, 1)$

	Eigenvalue	Eigenvector	Generalized Eigenvector
11.	$r_1 = 3$	$\mathbf{x}_1 = (1, 0)$	$\boldsymbol{\eta}_{12} = (0, 1/3)$

	Eigenvalues	Eigenvectors	Generalized Eigenvectors
13.	$r_1 = 1$	$\mathbf{x}_1 = (0, -1, 0)$	
	$r_2 = 0$	$\mathbf{x}_1 = (1, 0, 0)$	
	$r_3 = 2$	$\mathbf{x}_3 = (3, 2, 1)$	

	Eigenvalues	Eigenvectors	Generalized Eigenvectors
15.	$r_1 = 0$	$\mathbf{x}_1 = (-1, 1, 0)$	$(0, -1, 0)$
	$r_2 = 1$	$\mathbf{x}_2 = (-2, 1, 1)$	

Section A.5, page 787

1. $\begin{pmatrix} e & 0 \\ 0 & e^{-1} \end{pmatrix}$

5. $\begin{pmatrix} 1 & t \\ 0 & 1 \end{pmatrix}$

3. $\begin{pmatrix} 1 & 1 & \frac{1}{2} + t \\ 0 & 1 & 1 \\ 0 & 0 & 1 \end{pmatrix}$

7. $\begin{pmatrix} 1 & \dfrac{1 + i\sqrt{3}}{2} \\ 0 & \dfrac{1 - i\sqrt{3}}{2} \end{pmatrix}$

Section A.7, page 791

1.

$$B^{2k} = \begin{pmatrix} a^{2k}d^{2k} & 0 & 0 & 0 \\ 0 & b^{2k}b^{2k} & 0 & 0 \\ 0 & 0 & b^{2k}b^{2k} & 0 \\ 0 & 0 & 0 & a^{2k}d^{2k} \end{pmatrix}$$

and

$$B^{2k+1} = \begin{pmatrix} 0 & 0 & 0 & a^{2k+1}d^{2k} \\ 0 & 0 & b^{2k+1}c^{2k} & 0 \\ 0 & c^{2k+1}b^{2k} & 0 & 0 \\ d^{2k+1}a^{2k} & 0 & 0 & 0 \end{pmatrix}$$

3. $\det(I\,R_{n\times n}) = (-1)^{n(n+1)/2}$

Section A.8, page 797

1. $x_1 = 10 + 11s,\, x_2 = -2 - 4s,\, x_3 = s,\, x_4 = 0$, where s is any real number

3. No solution

Section A.9, page 805

1.

Eigenvalues	Eigenvectors
$r_1 = -1$	$\mathbf{x}_1 = (-1, 2)$
$r_2 = 5$	$\mathbf{x}_2 = (1, 1)$

$$\exp(A) = \begin{pmatrix} \dfrac{2e^6 - 1}{3e} & \dfrac{e^6 - 1}{3e} \\ \dfrac{2e^6 - 2}{3e} & \dfrac{e^6 + 2}{3e} \end{pmatrix}$$

3.

Eigenvalues	Eigenvectors
$r_1 = 2$	$\mathbf{x}_1 = (1, 1)$
$r_2 = 2$	$\mathbf{x}_2 = (1, 1)$

$$\exp(A) = \begin{pmatrix} 2e^2 & -e^2 \\ e^2 & 0 \end{pmatrix}$$

5.

Eigenvalues	Eigenvectors
$r_1 = 2 - i$	$\mathbf{x}_1 = (-1/2 - i/2, 1)$
$r_2 = 2 + i$	$\mathbf{x}_2 = (-1/2 + i/2, 1)$

$$\exp(A) = \begin{pmatrix} e^2\big(\cos(1) - \sin(1)\big) & -e^2\sin(1) \\ 2e^2\sin(1) & e^2\big(\cos(1) + \sin(1)\big) \end{pmatrix}$$

7.

Eigenvalues	Eigenvectors
$r_1 = 1$	$\mathbf{x}_1 = (0, -1, 1)$
$r_2 = 1$	$\mathbf{x}_2 = (1, 0, 0)$
$r_3 = 2$	$\mathbf{x}_3 = (1, 1, 0)$

$$\exp(A) = \begin{pmatrix} e & -e + e^2 & -e + e^2 \\ 0 & e^2 & -e + e^2 \\ 0 & 0 & e \end{pmatrix}$$

9.

Eigenvalues	Eigenvectors
$r_1 = 0$	$\mathbf{x}_1 = (1, 1, 1)$
$r_2 = 1$	$\mathbf{x}_2 = (3, 2, 1)$
$r_3 = 2$	$\mathbf{x}_3 = (7, 3, 1)$

$$\exp(A) = \begin{pmatrix} \dfrac{7e^2 - 6e + 1}{2} & -7e^2 + 9e - 2 & \dfrac{7e^2 - 12e + 5}{2} \\[2ex] \dfrac{3e^2 - 4e + 1}{2} & -3e^2 + 6e - 2 & \dfrac{3e^2 - 8e + 5}{2} \\[2ex] \dfrac{e^2 - 2e + 1}{2} & -e^2 + 3e - 2 & \dfrac{e^2 - 4e + 5}{2} \end{pmatrix}$$

11.

Eigenvalues	Eigenvectors
$r_1 = 2$	$\mathbf{x}_1 = (1, 0, 0, 0)$
$r_2 = 2$	$\mathbf{x}_2 = (0, 1, 0, 0)$
$r_3 = 3$	$\mathbf{x}_3 = (0, 0, 1, 0)$
$r_4 = 4$	$\mathbf{x}_4 = (0, 0, 0, 1)$

$$\exp(A) = \begin{pmatrix} e^2 & 0 & 0 & 0 \\ 0 & e^2 & 0 & 0 \\ 0 & 0 & e^3 & 0 \\ 0 & 0 & 0 & e^4 \end{pmatrix}$$

PHOTO ACKNOWLEDGMENTS

BIBLIOGRAPHY

[Bahd]   T. Bahder, **Mathematica for Scientists and Engineers**,[1] Addison-Wesley, Reading, MA, 1995.

[Bau]   G. Bauman, **Mathematica in Theoretical Physics**,[1] Springer-Verlag, TELOS, Berlin-New York, 1996.

[BeCh]   J. K. Beatty and A. Chaikin, **The New Solar System**, Third Edition, Cambridge University Press, Cambridge, 1988.

[BiRo]   G. Birkhoff and G.-C. Rota, **Ordinary Differential Equations**, Fourth Edition, John Wiley and Sons, New York, 1989.

[Blach] N. Blachman, **Mathematica: A Practical Approach**, Prentice-Hall, Englewood Cliffs, NJ, 1992.

[BDH]   P. Blanchard, R. L. Devaney, and G. R. Hall, **Differential Equations**, Preliminary Edition, PWS, Boston, 1996.

[BC1]   R. L. Borrelli and C. Coleman, **Differential Equations: A Modeling Approach**, Prentice-Hall, Englewood Cliffs, NJ, 1987.

[BC2]   R. L. Borrelli and C. Coleman, **Differential Equations: A Modeling Perspective**, John Wiley and Sons, New York, 1992.

[BCB]   R. L. Borrelli, C. Coleman, and W. E. Boyce, **Differential Equations Laboratory Workbook**, John Wiley and Sons, New York, 1992.

[BdP]   W. E. Boyce and R. C. DiPrima, **Differential Equations and Boundary Value Problems**, Fourth Edition, John Wiley and Sons, New York, 1986.

[Braun] M. Braun, **Differential Equations and Their Applications**, Fourth Edition, Springer-Verlag, Berlin-New York, 1993.

[Cajori] F. Cajori, **A History of Mathematics**, Third Edition, Chelsea Publishing Co., New York, 1980.

---

[1]*Mathematica* programs for this book are available from **mathsource.wri.com**

[Codd]   E. A. Coddington, **An Introduction to Ordinary Differential Equations**, Prentice-Hall, Englewood Cliffs, NJ, 1961. Reprint by Dover Publications, New York, 1994.

[CHLO1] K. R. Coombes, B. R. Hunt, R. L. Lipsman, J. E. Osborne, and G. J. Stuck, **Differential Equations with Maple**, John Wiley and Sons, New York, 1996.

[CHLO2] K. R. Coombes, B. R. Hunt, R. L. Lipsman, J. E. Osborne, and G. J. Stuck, **Differential Equations with Mathematica**, John Wiley and Sons, New York, 1995.

[Crowe] M. J. Crowe, **A History of Vector Analysis**, Dover Publications, New York, 1994.

[Davis1] H. T. Davis, **Introduction to Nonlinear Differential and Integral Equations**, U.S. Atomic Energy Commission, Washington, DC, 1960. Reprint by Dover Publications, New York, 1962.

[Davis2] P. W. Davis, **Differential Equations for Mathematics, Science and Engineering**, Prentice-Hall, Englewood Cliffs, NJ, 1992.

[DeGr]   W. R. Derrick and S. I. Grossman, **A First Course in Differential Equations**, West Publishing Co., St. Paul, MN, 1987.

[EdPe1] C. H. Edwards and D. E. Penney, **Elementary Differential Equations with Boundary Value Problems**, Third Edition, Prentice-Hall, Englewood Cliffs, NJ, 1993.

[EdPe2] C. H. Edwards and D. E. Penney, **Differential Equations with Boundary Value Problems, Computing and Modeling**, Third Edition, Prentice-Hall, Englewood Cliffs, NJ, 1996.

[Frank] P. Franklin, **Methods of Advanced Calculus**, McGraw-Hill, New York, 1944.

[Frau]   J. C. Frauenthal, **Introduction to Population Modeling**, Birkhäuser Verlag, Basel-Boston, 1980.

[Gray1] A. Gray, **Modern Differential Geometry of Curves and Surfaces**,[2] CRC Press, Boca Raton, FL, 1993.

[Gray2] J. W. Gray, **Mastering Mathematica**,[1] Academic Press, New York, 1994.

[GrGl1] T. W. Gray and J. Glynn, **Exploring Mathematics with Mathematica**,[1] Addison-Wesley, Reading, MA, 1991.

[GrGl2] T. W. Gray and J. Glynn, **The Beginner's Guide to Mathematica**, Addison-Wesley, Reading, MA, 1992.

---

[2]*Mathematica* programs for this book are available from **bianchi.umd.edu**

[Had]   J. Hadamard, **Lectures on Cauchy's Problem in Linear Partial Differen-
           tial Equations**, Yale University Press, New York, 1923. Reprint by Dover
           Publications, New York, 1952.

[Har]   D. R. Hartree, **Numerical Analysis**, Second Edition, Clarendon Press, Oxford,
           1958.

[HPS]   P. G. Hoel, S. C. Port, and C. J. Stone, **Introduction to Stochastic Pro-
           cesses**, Waveland Press, Prospect Heights, IL, 1987.

[Ince]   E. L. Ince, **Integration of Ordinary Differential Equations**, Seventh Edi-
           tion, Oliver and Boyd, New York, 1967.

[Kamke]   E. Kamke, **Differentialgleichungen, Lösungenmethoden und Lös-
           ungen**, Tenth Edition, B. G. Teubner, Stuttgart, 1983.

[Kaplan]   W. Kaplan, **Ordinary Differential Equations**, Addison-Wesley, Reading,
           MA, 1958.

[KaBi]   Th. v. Kármán and M. A. Biot, **Mathematical Methods in Engineering**,
           McGraw-Hill, New York, 1940.

[KeWi]   J. B. Keiper and D. Withoff, **Numerical Computation in Mathematica**,
           Course Notes from the 1992 *Mathematica* Conference, Wolfram Research,
           Champaign, IL, 1992.

[Kocak]   H. Koçak, **Differential and Difference Equations Through Computer
           Experiments**, Second Edition, Springer-Verlag, New York, 1989.

[LaLe]   J. LaSalle and S. Lefschetz, **Stability by Liapunov's Direct Method**, Aca-
           demic Press, New York, 1961.

[Lawden]   D. F. Lawden, **Elliptic Functions and Applications**, Springer-Verlag, New
           York, 1989.

[Ma]   R. Maeder, **Programming in Mathematica**,[1] Second Edition, Addison-
           Wesley, The Advanced Book Program, Redwood City, CA, 1991.

[MaTh]   J. B. Marion and S. T. Thorton, **Classical Dynamics of Particles and
           Systems**, Fourth Edition, Saunders College Publishing, Fort Worth TX,
           1995.

[McEJo]   C. McEvedy and R. Jones, **Atlas of World Population History**, Penguin
           Books, Middlesex England, 1976.

[NaSa]   R. K. Nagle and E. B. Saff, **Fundamentals of Differential Equations**, Third
           Edition, Addison-Wesley, Reading, MA, 1993.

[Newton]   I. Newton, **Principia**, Third Edition, London 1727. Motte's translation revised
           by Cajori, University of California Press, Berkeley, CA, 1934.

[PeRe]   R. Pearl and L. J. Reed, "On the rate of growth of the population of the United States since 1790 and its mathematical representation", *Proceedings of the National Academy of Sciences* **6** (1920) 275–288.

[Pinsky] M. A. Pinsky, **Partial Differential Equations and Boundary Value Problems with Applications**, Second Edition, McGraw-Hill, New York, 1991.

[PlZa]   A. D. Ployanin and V. F. Zaitsev, **Handbook of Exact Solutions for Ordinary Differential Equations**, CRC Press, Boca Raton, FL, 1995.

[PFTV]   W. H. Press, B. P. Flannery, S. A. Teukolsky, and W. T. Vetterling, **Numerical Recipes in C**, Cambridge University Press, Cambridge, 1988.

[Rain1]   E. D. Rainville, **Intermediate Differential Equations**, Second Edition, Macmillan, New York, 1964.

[Rain2]   E. D. Rainville, **Special Functions**, Second Edition, Macmillan, New York, 1960. Reprint by Chelsea Publishing Co., New York, 1971.

[Sea]   J. B. Seaborn, **Hypergeometric Functions and Their Applications** Springer-Verlag, New York, 1995.

[ShTi]   W. T. Shaw and T. Tigg, **Applied Mathematica, Getting Started, Getting it Done**,[1] Addison-Wesley, The Advanced Book Program, Redwood City, CA, 1993.

[Sim]   G. F. Simmons, **Differential Equations, with Applications and Historical Notes**, Second Edition, McGraw-Hill, New York, 1991.

[SkKe]   R. K. Skeel and J. B. Keiper, **Elementary Numerical Computing with Mathematica**, McGraw-Hill, New York, 1993.

[SmBl]   C. Smith with N. Blachman, **The Mathematica Graphics Guide Book**,[1] Addison-Wesley, Reading, MA, 1994.

[StGa]   W. Stramp and V. Ganzha, **Differentialgleichungen mit Mathematica**, Friedr. Vieweg & Sohn, Braunschweig, 1995.

[TePo]   M. Tenenbaum and H. Pollard, **Ordinary Differential Equations**, Harper & Row, New York, 1963. Reprint by Dover Publications, New York, 1985.

[Thom]   W. T. Thomson, **Introduction to Space Dynamics**, John Wiley and Sons, New York, 1961. Reprint by Dover Publications, New York, 1986.

[Verh]   F. Verhulst, **Nonlinear Differential Equations and Dynamical Systems**, McGraw-Hill, New York, 1972.

[Vess]   E. Vessiot, "Sur l'intégration des equations differentielles lineares", *Annales Scientifiques de l'Ecole Normale Supérieure*, 3e série **9** (1892), 197–280.

[Walt]   P. Waltman, **A Second Course in Elementary Differential Equations**, Academic Press, New York, 1961.

[Watson] G. N. Watson, **A Treatise on the Theory of Bessel Functions**, Cambridge University Press, Cambridge, 1986.

[Wick] T. Wickham-Jones, **Graphics with Mathematica**,[1] Springer-Verlag, TELOS, Berlin-New York, 1994.

[Wm1] S. Wolfram, **Mathematica**, Second Edition, Addison-Wesley, Redwood City, CA, 1990.

[Wm2] S. Wolfram, The **Mathematica** Book, Third Edition, Wolfram Media, Champaign, Il, 1997.

[WyBa] C. R. Wylie and L. C. Barrett, **Advanced Engineering Mathematics**, Sixth Edition, McGraw-Hill, New York, 1995.

[ZiCu] D. G. Zill and M. R. Cullen, **Differential Equations with Boundary-Value Problems**, Third Edition, PWS-Kent, Boston, 1993.

[Zwill] D. Zwillinger, **Handbook of Differential Equations**, Second Edition, Academic Press, San Diego, CA, 1992.

GENERAL INDEX

NAME INDEX

Abel, Niels Henrik (1802–1829), 251
Adams, John Couch (1819–1892), 427
Airy, George Biddel (1801–1892), 698
Ampère, André Marie (1775–1836), 219

Bernoulli, Jakob (1654–1705), 6, 67
Bessel, Friedrich Wilhelm (1784–1846), 730
Biot, Jean Baptiste (1774–1862), 5

Cajori, Florian (1859–1930), 869
Cardano, Girolamo (1501–1576), 386
Cauchy, Augustin Louis (1789–1857), v, viii, 717
Chebyshev, Pafnuty Lvovich (1821–1894), 708
Clairaut, Alexis Claude (1713–1765), 102
Coulomb, Charles Augustin de (1736–1806), 370
Courant, Richard (1888–1972), 773
Cramer, Gabriel (1704–1752), 248, 393

D'Alembert, Jean le Rond (1717–1783), 258
Descartes, René du Perron (1596–1650), 83
Dirac, Paul Adrien Maurice (1902–1982), 483

Euler, Leonhard (1707–1783), 280, 408, 679, 717

Faraday, Michael (1791–1867), 220
Ferrari, Ludovico (1522–1565), 387
Fresnel, Augustin Jean (1788–1827), 675
Frobenius, Georg Ferdinand (1848–1917), 726

Galois, Evariste (1811–1832), 387
Gompertz, Benjamin (1779–1865), 176

Hadamard, Jacques (1865–1963), 871
Halley, Edmond (1656–1742), 214
Heaviside, Oliver (1850–1925), 449
Henry, Joseph (1797–1878), 219
Hermite, Charles (1822–1901), 706
Heun, Karl (1859–1929), 416
Hooke, Robert (1635–1703), 5, 210, 327

Ince, Edward Lindsay (1859–1930), 729, 871

Jordan, Camille (1838–1921), 522, 781

Kármán, Theodore v (1881–1963), 871
Kamke, Eric (1890–1961), 871
Kirchhoff, Gustav Robert (1824–1887), 220
Kummer, Ernst Eduard (1810–1893), 756
Kutta, Martin Wilhelm (1867–1944), 420

Lagrange, Joseph Louis (1736–1813), 106
Laplace, Pierre Simon, Marquis de (1749–1827), 449
Legendre, Adrien Marie (1752–1833), 702
Leibniz, Gottfried Wilhelm (1646–1716), 4, 686
Lerch, Mathias (1860–1922), 461
Lotka, Alfred James (1880–1949), 655
Lyapunov, Aleksandr Mikhailovich (1857–1918), 601

Malthus, Thomas Robert (1766–1834), 157

MINIPROGRAM AND MATHEMATICA INDEX

887

INTRODUCTION TO ORDINARY DIFFERENTIAL EQUATIONS WITH *MATHEMATICA*®

Since this field is fast-moving, we expect updates and changes to occur that might necessitate sending you the most current pertinent information by paper, electronic media, or both, regarding *Introduction to Ordinary Differential Equations with Mathematica*®. Therefore, in order to not miss out on receiving your important update information, please fill out this card and return it to us promptly. Thank you.

Name: _____

Title: _____

Company: _____

Address: _____

City: _____ State: _____ Zip: _____

Country: _____ Phone: _____

E-mail: _____

Areas of Interest/Technical Expertise: _____

Comments on this Publication: _____

_____

_____

❏ Please check this box to indicate that we may use your comments in our promotion and advertising for this publication.

Purchased from: _____
Date of Purchase: _____

❏ Please add me to your mailing list to receive updated information on *Introduction to Ordinary Differential Equations with Mathematica*® and other TELOS publications.

❏ I have a   ☐ IBM compatible   ☐ Macintosh   ☐ UNIX   ☐ other

Designate specific model _____

THE ELECTRONIC LIBRARY OF SCIENCE

PLEASE TAPE HERE

FOLD HERE

NO POSTAGE
NECESSARY
IF MAILED
IN THE
UNITED STATES

BUSINESS REPLY MAIL

FIRST CLASS MAIL PERMIT NO. 1314 SANTA CLARA, CA

POSTAGE WILL BE PAID BY ADDRESSEE

*THE
ELECTRONIC
LIBRARY
OF
SCIENCE*

3600 PRUNERIDGE AVE STE 200
SANTA CLARA CA 95051-9835

DATE DUE

GAYLORD No. 2333   PRINTED IN U.S.A.